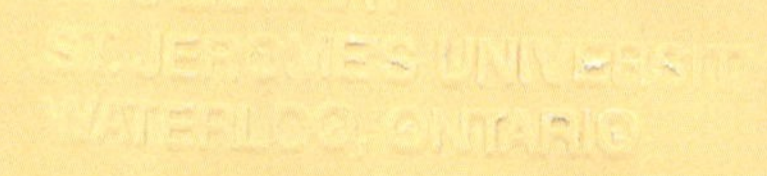

The Marriage and Family Experience

INTIMATE RELATIONSHIPS IN A CHANGING SOCIETY

Ninth Edition

Bryan Strong
Formerly of University of California, Santa Cruz

Christine DeVault

Theodore F. Cohen
Ohio Wesleyan University

Australia • Canada • Mexico • Singapore • Spain
United Kingdom • United States

THOMSON
WADSWORTH

Sociology Editor: Robert Jucha
Assistant Editor: Stephanie Monzon
Editorial Assistant: Melissa Walter
Technology Project Manager: Dee Dee Zobian
Marketing Manager: Matthew Wright
Marketing Assistant: Tara Pierson
Advertising Project Manager: Linda Yip
Project Manager, Editorial Production: Cheri Palmer
Print/Media Buyer: Doreen Suruki
Permissions Editor: Joohee Lee
Production Service: Andrea Fincke, Thompson Steele, Inc.
Text Designer: Kaelin Chappell
Photo Researcher: Sarah Evertson
Copy Editor: Stephanie Magean
Illustrator: Thompson Steele, Inc.
Cover Designer: Laurie Anderson
Cover Image: "Collective Conditioning" 1994. Machine-pieced and machine-quilted, 97 × 74 inch quilt by Rebecca Rohrkaste
Compositor: Thompson Steele, Inc.
Printer: R.R. Donnelley/Willard

Printed in the United States of America
1 2 3 4 5 6 7 08 07 06 05 04

For more information about our products, contact us at:
Thomson Learning Academic Resource Center
1-800-423-0563

For permission to use material from this text or product, submit a request online at
http://www.thomsonrights.com.
Any additional questions about permissions can be submitted by email to **thomsonrights@Thomson.com.**

Library of Congress Control Number: 2004102002
Student Edition: ISBN 0-534-60930-9
Instructor's Edition: ISBN 0-534-60932-5
0-534-60929-5

Thomson Wadsworth
10 Davis Drive
Belmont, CA 94002-3098
USA

Asia
Thomson Learning
5 Shenton Way #01-01
UIC Building
Singapore 068808

Australia/New Zealand
Thomson Learning
102 Dodds Street
Southbank, Victoria 3006
Australia

Canada
Nelson
1120 Birchmount Road
Toronto, Ontario M1K 5G4
Canada

Europe/Middle East/Africa
Thomson Learning
High Holborn House
50/51 Bedford Row
London WC1R 4LR
United Kingdom

Latin America
Thomson Learning
Seneca, 53
Colonia Polanco
11560 Mexico D.F.
Mexico

Spain/Portugal
Paraninfo
Calle Magallanes, 25
28015 Madrid, Spain

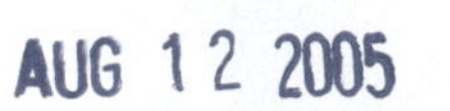

In the preface to the last edition, I noted that my wife and soul mate, Susan, was burdened daily by more than she should have been and received less assistance, credit, and recognition than she deserved. As I continue to try to get by without her, I realize how true but understated that was. As her surviving "lesser half," I dedicate this edition to her.

—Ted Cohen

BRIEF CONTENTS

CONTENTS

CHAPTER 2 Studying Marriage and the Family

PAGE 26

CHAPTER 3 Dynamics and Diversity of Families

PAGE 62

UNIT II ■ INTIMATE RELATIONSHIPS

CHAPTER 4 Contemporary Gender Roles

CHAPTER 5 *Friendship, Love, and Commitment*

PAGE 130

CHAPTER 6 *Communication, Power, and Conflict*

PAGE 164

CHAPTER 7

Singlehood, Pairing, and Cohabitation

CHAPTER 8 Understanding Sexuality

PAGE 232

CHAPTER 9 Family Processes, Family Life Cycles

PAGE 278

CHAPTER 10

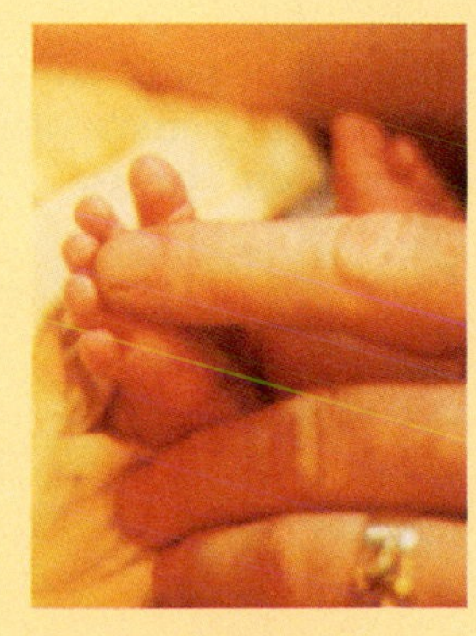

PAGE 316

Should We or Shouldn't We? Choosing Whether to Have Children

CHAPTER 11

PAGE 342

Experiencing Parenthood: Roles and Relationships of Parents and Children

CHAPTER 12 *Marriage, Work, and Economics*

UNIT IV ▪ FAMILY CHALLENGES AND STRENGTHS

CHAPTER 13 *Family Violence and Sexual Abuse*

PAGE 420

CHAPTER 14 *Coming Apart: Separation and Divorce*

CHAPTER 15 New Beginnings: Single-Parent Families, Remarriages, and Blended Families

PAGE 500

CHAPTER 16 Marriage and Family Strengths and Needs

PAGE 528

Preface

This, the Ninth Edition of *The Marriage and Family Experience: Intimate Relationships in a Changing Society,* is the second one on which I have had the pleasure of working. As originally written by Bryan Strong many years ago and developed, revised, and updated through subsequent editions by Bryan Strong, Christine DeVault, and later Barbara Sayad, this textbook has a long history on which to build. Through its many editions, *The Marriage and Family Experience* has acquired a wide and dedicated following from a number of different types of institutions and within a range of academic and applied disciplines. This edition is now the third to appear since Bryan Strong's untimely death. Despite his absence, the book retains his long-standing theme: Our families are what most count. In whatever form(s) we experience them, they shape who we are and become, they provide us with our most intimate and loving relationships, and they need to be cherished, honored, and supported.

This same perspective guides my personal and professional life. My wife, Susan Jablin Cohen, passed away between the last edition and this one. Her fourteen-month struggle with brain cancer, during which I was fortunate to be her full-time caregiver, only reaffirmed my conviction of the importance of marriage and family relationships. Although my own family continues to struggle with the devastating loss, the love and devotion we shared with her enriches us everyday.

Professionally, as a sociologist who teaches and writes about gender, masculinity, work and family, and new family forms (especially role-reversed households and shift-alternating dual earners), I have again brought my research interests and teaching emphases into this text. Without abandoning or significantly altering either the book's wonderful balance between an academic and more functional approach or its interdisciplinary appeal, I have drawn heavily from research in sociology, family studies and related fields. I have always taught my own family classes by incorporating anthropology, history, psychology, journalism, literature, economics, and gender studies into the sociological approach I take. And I have tried to ensure that this book continues to do the same.

New to This Edition

Among the major changes to this edition is the addition of a new chapter on Parenthood in the United States. Although longtime users may wonder whether this signals a return to the older format, with a chapter on pregnancy followed by one on parenthood, it does not. Instead, I have separated material on parenthood into a chapter on decisions whether or not to parent and a chapter on the experience of parenthood. The chapter on processes, "Family Processes, Family Life Cycles," has been moved up in this edition. Since it is the chapter that covers early marriage and adjustments to married life, it flows better in the midst of earlier discussions of such topics as relationships and sexuality.

New or Expanded Topics

Chapter 1, *The Meaning of Marriage and the Family.* The large emphasis on popular culture, especially television, has been removed so that the chapter can serve as more of an introduction to marriage and family studies. Up-to-date references and examples on the legal battles over same-sex marriages include discussion of the Massachusetts decision, the Defense of Marriage Act, and the ongoing controversy over whether and how to recognize same sex unions. A new *Perspective,* "The Rights and Benefits of Marriage," has been added to illustrate the legal and economic protections that unmarried couples lack. The discussion of "cultural constructions of family life" has been strengthened and now includes a distinction between liberal, conservative, and centrist positions. Also, a table detailing some of the areas of changing family patterns has been included.

Chapter 2, *Studying Marriage and the Family.* The chapter now more strongly focuses on the question "How do we know about families?" Beginning with a hypothetical situation involving a long-distance relationship and the contradictory common sense "wisdom" of *absence makes the heart grow fonder* and *out of sight, out of mind,* the chapter explores how family scientists approach studying family issues. The discussion of different theoretical perspectives now includes a discussion of ecological theory as well as expanded and strengthened discussions of developmental theory and systems theory. Ultimately, each of these perspectives is applied to the study of long-distance relationships. There are also new discussions of concepts, variables, deductive and inductive approaches to research and theory. The material on research techniques includes more detailed discussions of research ethics, secondary analysis, quantitative research, and clinical research. Ultimately, the chapter returns to long-distance relationships and summarizes what most research findings suggest about the outcomes of such relationships.

Also, a discussion of men's studies and the research that has been done on men in families has been incorporated into the discussion of the feminist perspective. There is a new boxed feature, a *Perspective* that summarizes some of the material on popular culture that was formerly in Chapter 1, which emphasizes the Advice/Information Genre —the one media form that most "competes" with research-based knowledge.

Chapter 3, *Dynamics and Diversity of Families.* Chapter 3 supplements its historical overview with a boxed feature, *Historical Perspectives of Marriage and Family Life,* that examines some of the ongoing differences in interpretations of family history. The discussion of social classes uses a more common model of class, in which both the upper class*es* and middle class*es* are pluralized and differentiated (for example, upper-upper and lower-upper; upper-middle and lower-middle). The material on the poor includes a brief discussion of how poverty is defined and what it does to families. The discussion of racial and ethnic diversity now includes a section on Amish families.

Chapter 4, *Contemporary Gender Roles.* This chapter has been reorganized to emphasize the importance of gender in family experience and the interconnections between gender and family. It no longer bounces back and forth from traditional male to traditional female to contemporary male to contemporary female roles. Instead, traditional and contemporary male experiences are joined and followed by a section on traditional and contemporary female family experiences. Other additions include: a brief section, Gender and Sexual Orientation, to show how the two are separate, although related, issues (for example, one cannot "read" a person's sexual preference from their gendered behavior); a new boxed feature summarizing research on "post-gender couples," to illustrate the range of possible role relationships; and an expanded discussion of the economic "gaps" separating women and men (in wages, mobility, and occupational status).

Chapter 5, *Friendship, Love, and Commitment.* The major addition here is in the discussion of unrequited love. There is also now a section on "stalking" as an extreme instance of jealousy and unrequited love.

Chapter 6, *Communication, Power, and Conflict.* There is new material on sexual communication, on cohabitation and later marital communication problems, and on the demand-withdraw pattern of marital communication,

Chapter 7, *Singlehood, Pairing, and Cohabitation.* The discussion of issues related to dating and romantic relationships is more expansive than in the prior edition; it includes more on how relationships develop (including a brief discussion of "pick-up lines") and relative costs and benefits of dating relationships. There is also a new version of an older boxed feature on Internet Personals and Computer Dating. The comparison of cohabitation and marriage has been expanded and updated and there is a new subsection on differences between cohabitants and marrieds in sexual relationships and infidelity. Additionally, the discussion of consequences of cohabitation on marriage has been brought in line with existing data and literature.

Chapter 8, *Understanding Sexuality.* There are two new sections: one addresses the social control of sexuality (the constraints on choices regarding *who* we can choose as partners, *what* we choose to do, and *when* we begin sexual relations), the other discusses nonconsensual sex (for example, sexual coercion, sexual abuse, and date and marital rape).

Chapter 9, *Family Processes, Family Life Cycles.* As noted above, this chapter has been moved forward from the last edition to better capitalize on the discussions of engagements, weddings, and early marriage. It stills includes the broad coverage of topics spanning the family life cycle.

Chapter 10, *Should We or Shouldn't We? Choosing Whether to Have Children* is one of the new chapters to this edition. It takes some material from the prior editions, but focuses more narrowly on the issues surrounding whether or not to have children and the patterns of childbearing in the United States.

Chapter 11, *Experiencing Parenthood: Roles and Relationships of Parents and Children.* This chapter begins with the transition into parenthood and proceeds to address the roles of mothers and fathers as experienced across race, ethnicity, class, and sexual orientation. It also includes material on children's and parents' needs, and considers issues related to outside caregiving.

Chapter 12, *Marriage, Work, and Economics.* The chapter includes new material on how children perceive and are affected by parents' jobs. Also, building on the last edition, there is more discussion of work-family conflict, the "time bind," and how family friendly workplace policies affect people with and without children.

Chapter 13, *Family Violence and Sexual Abuse.* There is a new discussion of social class and family violence that compares more common profiles of abuse with research on "upscale violence."

Chapter 14, *Coming Apart: Separation and Divorce.* There is more material on the different ways to look at the effects of divorce; the competing interpretations of why divorce has the consequences it has; and policy-related suggestions about how to reduce divorce rates. Data from Wallerstein's and Hetherington's studies of the long-term impact of divorce on families are compared. There is also updated discussion of covenant marriage.

Chapter 15, *New Beginnings: Single-Parent Families, Remarriages, and Blended Families.* This material has been updated with more recent statistical information.

Chapter 16, *Marriage and Family Strengths. and Needs.* The new material includes a wider discussion of family policies.

PEDAGOGY

Chapter Previews. Self-quiz chapter openers let students assess their existing knowledge of what will be discussed in the chapter. We have found these quizzes engage the student, drawing them into the material and stimulating greater interaction with the course.

Chapter Outlines. Each chapter contains an outline at the beginning of the chapter to allow students to organize their learning.

Exploring Diversity. These boxes let students see family circumstances from the vantage point of other cultures, other eras, or within different lifestyles in the contemporary United States.

Making Gender Matter Less, in Chapter 4, examines U.S. couples whose "post gender" lifestyles are efforts to live outside of more typical constraints.

Upscale Violence in Chapter 13 focuses on research into wife abuse among the upper class.

Understanding Yourself. These boxed inserts use research topics, findings, and instruments to stimulate students to examine their own family experiences or expectations. They help students see the personal meaning of otherwise abstract material.

Internet Personals, Computer Dating and Homogamy, in Chapter 6, is an updated look at the use of the Internet to locate potential mates.

Perspectives. These boxes focus on high interest topics. The Costs of Motherhood in Chapter 11 examines some of the economic impact motherhood has on women.

Are Family-Friendly Workplaces Unfriendly to Nonparents? in Chapter 12 examines how the move toward flexible and supportive "family friendly" workplaces may discriminate against those employees without children.

The material formerly found in margins—quotes, *Did You Know?* and *Reflections*—has been incorporated into the text but there are fewer of them. I have selected a quote to open each chapter, drawing upon the writing of poets, scholars, philosophers, and artists of all types. The *Did You Know?* items provide statistics and quick data relating to the concepts in the chapter. *Reflections* bring students closer to the material by encouraging them to consider their own ideas and beliefs.

Each chapter also has a *Chapter Summary,* list of *Key Terms,* and *Suggested Readings,* all of which are designed to maximize students' learning outcomes. The Chapter Summary reviews the main ideas of the chapter, making review easier and more effective. The Key Terms are boldfaced within the chapter and listed at the end, along with the page number where the term was introduced. Both Chapter Summaries and Key Terms assist students in test preparation. The *Suggested Readings* list includes material for further research or personal interest. We have also added *InfoTrac® College Edition Search terms* to each chapter so that this unique online database may be easily used by the students for extra-credit assignments, activities, and research.

Glossary. There is a comprehensive glossary of key terms included at the back of the textbook.

Appendixes. There are appendixes on sexual structure and the sexual response cycle, fetal development, and managing money.

Web Site Resource Center. Material that had been included in the Resource Center in prior editions is still available online at the Wadsworth Web site. This includes a self-help directory, and practical information on financial, health, sexual, and consumer matters that affect families. There are also study guides and topically organized lists of Web sites.

SUPPLEMENTS AND RESOURCES

The Marriage and Family Experience, Ninth Edition, is accompanied by a wide array of supplements prepared for both the instructor and student. Some new resources have been created specifically to accompany the Ninth Edition, and all of the continuing supplements have been thoroughly revised and updated.

Supplements for the Instructor

Instructor's Edition of *The Marriage and Family Experience,* Ninth Edition

The Instructor's Edition contains a visual preface, a walk-through of the text that provides an overview of its key features, themes, and supplements.

Instructor's Resource Manual with Test Bank (with Multimedia Manager CD-ROM)

This manual will help the instructor to organize the course and to captivate students' attention. The manual includes a chapter focus statement, key learning objectives, lecture outlines, in-class discussion questions, class activities, student handouts, extensive lists of reading and online resources, and suggested Internet sites and activities. The Test Bank portion includes approximately 40–50 multiple choice, 20 true/false, 10 short answer, and 5–10 essay questions, all with answers and page references, for each chapter of the text.

An all NEW *Multimedia Manager Instructor Resource CD-ROM* is now packaged with the IRM/TB. This new instructor resource includes book-specific PowerPoint® Lecture Slides, graphics from the book itself, the IRM/TB word docs, CNN Video Clips, and links to many of Wadsworth's important Sociology resources. All of your media teaching resources in one place!

ExamView® Computerized Testing

Create, deliver, and customize tests and study guides (both print and online) in minutes with this easy-to-use assessment and tutorial system. *ExamView* offers both a Quick Test Wizard and an Online Test Wizard that guide you step-by-step through the process of creating tests. The test appears on screen exactly as it will print or display online. Using *ExamView's* complete word processing capabilities, you can enter an unlimited number of new questions or edit existing questions.

Wadsworth's Marriage and Family 2005 Transparency Acetates

A selection of four-color acetates consisting of tables and figures from Wadsworth's marriage and family texts is available to help prepare lecture presentations. Free to qualified adopters.

CNN Today: Marriage and Family Video Series, Volumes I-VII

Illustrate how the principles that students learn in the classroom apply to the stories they see on television with the CNN Today Marriage and Family Video Series, an exclusive series jointly created by Wadsworth and CNN. Each video consists of approximately 45 minutes of footage originally broadcast on CNN and selected specifically to illustrate concepts relevant to the marriage and the family course. The videos are broken into short two- to five-minute segments, perfect for classroom use as lecture launchers or to illustrate key concepts. An annotated table of contents accompanies each video with descriptions of the segments. Special adoption conditions apply.

Wadsworth Sociology Video Library

This large selection of thought-provoking films, including some from the Films for the Humanities collection, is available to adopters based on adoption size.

Supplements for the Student

Study Guide

For each chapter of the text, this student study tool contains a chapter focus statement, key learning objectives, key terms, chapter outlines, assignments, internet activities and Web sites, and practice tests containing 20 multiple choice and 15 true/false with answers and page references and 5 short answer questions with page references.

Marriage and Family: Using Microcase ExplorIt, Third Edition

Written by Kevin Demmitt of Clayton College, this software-based workbook is an exciting way to get students to view marriage and the family from the sociological perspective. With this workbook and accompanying ExplorIt software and data sets, your students will use national and cross-national surveys to examine and actively learn marriage and family topics. This inexpensive workbook will add an exciting dimension to your marriage and family course.

Families and Society, First Edition

Written by Scott L. Coltrane, University of California, Riverside, this reader is designed to promote a sociological understanding of families, while at the same time demonstrating the diversity and complexity of contemporary family life. The different parts or sections of the reader are designed to "map" onto most sociology of the family textbooks and course syllabi. The articles emphasize a social constructionist and a sociological view of families. The reader is thus designed to dispel the myth that families are separate from society. Virtually every selection illustrates the myriad links that exist between families and their various social, cultural, economic, and political contexts. The first reading in each section provides a classic theoretical overview on the topic. These classic pieces provide students with an historical understanding of the development of the field and offer insight into some of the enduring sociological facts about families.

Readings in the Marriage and Family Experience, Third Edition

Written by Bryan Strong, Christine DeVault, and Barbara W. Sayad, this FREE reader contains articles, essays, excerpts from books, journals, magazines, and newspapers on the topic of marriage and family. They have all been carefully selected for their diverse and stimulating content. Critical thinking questions are included for each reading in order to encourage classroom discussion and student participation.

The Marriages and Families Activities Workbook

What are your risks of divorce? Do you have healthy dating practices? What is your cultural and ancestral heritage and how does it affect your family relationships? The answers to these and many more questions are found in this workbook of nearly a hundred interactive self-assessment quizzes designed for students studying marriage and the family. These self-awareness instruments, all based on known social science research studies, can be used as in-class activities or homework assignments to help students learn more about themselves and their family experience.

Marriage and Family Case Studies CD-ROM

This unique student CD-ROM includes a series of ten interactive case study videos that provide a dramatic enactment illustrating key topics and concepts from the text. Students watch each video and answer critical thinking questions, applying marriage and family theories to the case study videos. Students then compare their analysis with that of a marriage and family expert. Also on the CD-ROM is a direct link to InfoTrac College Edition where students can search for related articles from sociology periodicals and a link to Wadsworth's Virtual Society where Knox's companion Web site offers a wide range of book-specific study tools.

Online Resources

Wadsworth's Virtual Society: The Wadsworth Sociology Resource Center

http://www.wadsworth.com/sociology

Here you will find a wealth of resources for students and instructors, including the Marriage and Family Resource Center, a fun, interactive site filled with additional data, exercises, and enriching resources. For example: GSS Activities—students can compare their attitudes with GSS data in this interactive exercise; Marriage and Family Activities—fun self-quizzes for students to learn more about themselves; Marriage and Family Links—online resources; and Resources and Organizations—a library of useful resources.

Also from Wadsworth's Virtual Society, students will find the companion Web site for *Choice's in Relationships: An Introduction to Marriage and the Family,* Eighth Edition.

http://sociology.wadsworth.com/strong/marriage9e

Access useful learning resources for each chapter of the book. Some of these resources include:

- Tutorial Practice Quizzes that can be scored and e-mailed to the instructor
- Internet Exercises and Web Links
- Video Exercises
- InfoTrac College Edition Exercises
- Flashcards of the text's glossary
- Crossword Puzzles
- Essay Questions
- Learning Objectives
- And much more!

WebTutor™ Toolbox for WebCT or Blackboard

Preloaded with content and available free via pincode when packaged with this text, WebTutor toolbox pairs all the content of this text's rich book companion Web site with all the sophisticated course management functionality of a WebCT or Blackboard product. You can assign materials (including online quizzes) and have the results flow automatically to your grade book. Toolbox is ready to use as soon as you log on—or you can customize its preloaded content by uploading images and other resources, adding Web links, or creating your own practice materials. Students have access to only student resources on the Web site. Instructors can enter a pincode for access to password-protected instructor resources.

InfoTrac College Edition

Give your students anytime, anywhere access to reliable resources with InfoTrac College Edition, the online library. This fully searchable database offers 20 years' worth of full-text articles from almost 5000 diverse sources, such as academic journals, newsletters, and up-to-the-minute periodicals including *Time, Newsweek, Science, Forbes,* and *USA Today.* The incredible depth and breadth of material—available 24 hours a day from any computer with Internet access—makes conducting research so easy that your students will want to use it to enhance their work in every course! Through InfoTrac College Edition's InfoWrite, students now also have instant access to critical thinking and paper writing tools. Both adopters and their students receive unlimited access for four months.

Opposing Viewpoints Resource Center (OVRC)

Newly available from Wadsworth, this online center allows you to expose your students to all sides of today's most compelling issues! The Opposing Viewpoints Resource Center draws on Greenhaven Press's acclaimed social issues series, as well as core reference content from other Gale and Macmillan reference USA sources. The result is a dynamic online library of current event topics—the facts as well as the arguments of each topic's proponents and detractors. Special sections focus on critical thinking (walks student through how to critically evaluate point-counterpoint arguments) and researching and writing papers.

ACKNOWLEDGMENTS

Many people deserve to be recognized for the roles they played in the revision and production of this edition. I thank the following reviewers, whose comments and reactions were encouraging and whose suggestions were helpful: Augustine Ayree, Fitchburg State College; Brenda Bauch, Jefferson College; Henry Borne, Holy Cross College; Betsy Cullum-Swan, Michigan State University; Lynne Ann DeSpelder, Cabrillo College; Patricia Gibbs, Foothill College; Theodore Greenstein, North Carolina State University; Ron J. Hammond, Utah Valley State College; Jan Hare, University of Wisconsin–Stout; Carolyn Henry, Oklahoma State University; Howard Housen, Broward Community College; Melsome Nelson-Richards, SUNY Oswego.

This book remains the product of many hands. Bryan Strong and Christine DeVault created a wonderful book from which to branch. Their strong convictions about the meaning and importance of families, their well-conceived organization, and their reader-friendly prose make it obvious why this book has appealed to so many for so long. I am proud to, once again, follow on their heels.

A number of people at Wadsworth Publishing deserve thanks. Eve Howard, Editor-in-Chief for the Social Sciences, and Bob Jucha, my editor, showed considerable and continued, enthusiastic support for this book. They were patient and supportive at a very difficult time. Given the family tragedy of Sue's death, they delayed publication of this edition to allow me to grieve and mourn without the added burden of revision and writing. Bob, especially, through his thoughtful recommendations, his regular monitoring, and his friendly but persistent reminding made it possible to bring this edition to completion. I owe him an incredible debt of thanks. Melissa Walter, Editorial Assistant, performed tremendously in helping assemble the manuscript for production, Stephanie Monzon, the Assistant Editor, has put together the strong ancillary package that accompanies the Ninth Edition, Dee Dee Zobian, the Technology Project Manager, is responsible for the state of the art technology and Web site supporting the text, while Matthew Wright, Senior Marketing Manager, and Linda Yip, Advertisement Manager, have skillfully directed the introduction of the Ninth Edition to adopters and prospective adopters. I want to extend my thanks to Cheri Palmer, the Senior Production Project Manager at Wadsworth, who oversaw the complex production process with great skill. Thanks, too, to Joohee Lee in the permissions department at Wadsworth. Andrea Fincke and Thompson Steele were tremendously helpful and highly competent in the copyediting and production phases. The text looks and reads better just because of their involvement. My appreciation also goes to Sarah Evertson, photo editor and researcher, for finding such good examples of what were occasionally vaguely requested subjects.

Once again, I wish to express appreciation to my colleagues, friends, and students at Ohio Wesleyan University for the assistance, support, or sympathy they offered through the difficult times during Sue's illness, following her death, and through my attempt to juggle writing and teaching with mourning. The university generously provided me a semester leave without pay to add to my regular sabbatical, giving me a full year to be with Sue. They have also been consistently supportive of my efforts to complete this revision. My students again "cut me some slack," given how busy I became and how distracted or disorganized I occasionally seemed. My departmental colleagues, Jan Smith, Mary Howard, Akbar Mahdi, Jim Peoples, John Durst, and Pam Laucher make my academic life more comfortable and supportive than any other I know of. I appreciate their friendship, continued interest, and unwavering support. Other friends, especially Joan McLean and Cindy Johnson, Kurt and Danielle Clarke, Mario Sanchez, and Debra Peoples, have been constant sources of support for me and my family. By helping us heal they had a hand in making this edition possible. I want to express special thanks to Julie K. Pfister who has become a most special friend. In reaching out from her own grief, sharing support and empathy, and enabling me to look ahead to the future with hope in my heart, she has done more than I will ever be able to thank her for.

I want again to express my appreciation to my family. My parents, Kalman and Eleanor Cohen, and sisters, Laura Cohen and Lisa Merrill, continue to give me a supportive family to which to turn. My children, Dan and Allison, are my greatest joys and I am their biggest fan. As they move through adolescence and young adulthood, I marvel at the wonderful people they have become.

ABOUT THE AUTHOR

Theodore Cohen is a professor in the sociology/anthropology department at Ohio Wesleyan University. He earned his M.A. and Ph.D. in sociology from Boston University. His research and teaching specializations center around gender and family life, with special attention to men's family lives and to emergent family lifestyles. He is the editor of *Men and Masculinity: A Text Reader* (Wadsworth, 2000).

Preview

To gain a sense of what you already know about the material covered in this chapter, answer "True" or "False" to the statements below.

1. The majority of American families are traditional nuclear families in which the husband works and the wife stays at home caring for the children. True or false?
2. Families are easy to define and count. True or false?
3. No U.S. state prohibits interracial marriage. True or False?
4. All cultures traditionally divide at least some work into male and female work. True or false?
5. In the United States, all states now recognize same sex "civil unions."
6. There is widespread agreement about the nature and causes of change in American family patterns. True or false?
7. The majority of cultures throughout the world prefers monogamy, the practice of having only one husband or wife. True or false?
8. Married men tend to live longer than single men. True or false?
9. Most people who divorce eventually marry again. True or false?
10. Nuclear families, single-parent families, and stepfamilies are equally valid family forms. True or false?

"Home—the place, where, when you have to go there, they have to let you in."

ROBERT FROST

CHAPTER 1

The Meaning of Marriage and the Family

Outline

Courses in marriage and the family are unlike most other courses we take. At the very beginning of the term, before any books have been read or class material has been introduced, we believe we already "know" a lot about families. After all, each of us acquires much first-hand experience of family living before being formally instructed about what families are or what they do. But how much do we really know? If pressed, how would we describe American family life? Are our families "healthy" and stable? Are marriages enduring? Is marriage important for the well-being of adults and children? Are today's fathers and mothers sharing responsibility for raising their children? Do they share housework? How many cheat on each other? What happens when people divorce? Are stepfamilies different from biological families? How common are abuse and violence in families? Are they increasing? Such questions will be considered throughout this book, but for now they encourage us to think about what we know about families and where that knowledge comes from.

In this chapter, we examine how marriage and family are defined by individuals and society, paying particular attention to the discrepancies between the realities of family life as uncovered by social scientists and the impressions we have formed elsewhere. We then look at the functions that marriages and families fulfill and examine extended families and kinship. We close by identifying some major trends in American family life.

Personal Experience and Wishful Thinking

As we begin to study family patterns and issues, we need to consider how our attitudes and beliefs about families may affect and distort our efforts. In contemplating the wider issues about families that are the substance of this book, we may consider our own households and family experiences. How we respond to the questions in the opening paragraph above is quite likely influenced by what we ourselves have experienced in families. For some of us, those experiences have been largely loving and the relationships have remained stable. For others, family life has been characterized by conflict and bitterness, separations and reconfigurations. Most people have experienced both sides of family life, the love and the conflict, whether their families remained intact or not.

PREVIEW ANSWERS

1 False
2 False
3 True
4 True
5 False
6 False
7 False
8 True
9 True
10 True

We may be tempted to draw conclusions about families from our own personal experiences as they are related to particular families. Thinking that experience translates to expertise, we may find ourselves tempted to generalize from what we experience to what others must also encounter in family life. The dangers in that are clear; although the knowledge we have about our own families is vividly real, it is also highly subjective. We "see" things, in part, as we want to see them. The meanings we attach to our experiences are affected by the emotions we feel within the relationships that comprise our families. The other members of our own family are likely to have different perceptions and attach somewhat different meanings to those same relationships—which means that even the understanding we have of our own families is very likely a distorted one.

Furthermore, no other family is exactly like your family. We don't all live where you live or how you live, and we don't possess the exact same financial resources, draw from the same cultural backgrounds, and build on the same sets of experiences that make your family unique. As well as we might think we know our own families, they are poor sources of more general knowledge about the wider marital or family issues that are the focus of this book.

What Is Family? What Is Marriage?

In order to accurately understand marriage and family, it is important to define these terms. Before reading any further, think about what the words *marriage* and *family* mean to you. As you attempt to more systematically define these words, you will quickly discover the complexity involved in what seems to be such a simple and straightforward undertaking.

DEFINING "FAMILY"

As contemporary Americans, we live in a society composed of married couples, two-parent families, stepfamilies, single-parent families, multigenerational families, cohabiting adults, child-free families, families headed by gay men and by lesbians, and so on. With such variety, how can we define family? What are our criteria for identifying some (or all) of these groups as families?

For official counts of the numbers and characteristics of American families, the U.S. Census Bureau defines a **family** as "a group of two or more persons related by birth, marriage, or adoption and residing together in a household" (Statistical Abstract of the United States, 2001). A distinction is made between a family and a **household**, a grouping that consists of all people who share a living space, "a house, an apartment or other group of rooms, or a single room that constitutes 'separate living quarters'" (U.S. Census Bureau, 2001). Single people who live alone, roommates, lodgers, live-in domestic service employees, are all counted among members of households, as are family groups. Thus, the U.S. Census reports on characteristics of the nation's households *and* families (see Figure 1.1). Of the 109,297,000 households in the United States in 2002, 68 percent were *family* households (U.S. Census Bureau, Population Division, 2003).

When we asked our students who they included as family members, their lists included (alphabetically) the following:

aunt	grandfather	pet
best friend	grandmother	priest
boyfriend	great-grandparent	rabbi
cousin	half-sibling	second cousin
daughter	lover	sibling
father	minister	son
father-in-law	mother	stepfather
foster child	mother-in-law	stepmother
foster parent	neighbor	stepsibling
girlfriend	nephew	teacher
godchild	niece	uncle

Most of those who were designated as family members are individuals related by descent, marriage, remarriage, or adoption, but some are **affiliated kin**—unrelated individuals who feel and are treated as if they were relatives. (In a couple of instances, the family dog, cat, or bunny was included as a family member.)

FIGURE 1.1 ■ Household Composition: 2002

Married couples with children*	Married couples without children*	Other family households	Persons living alone	Other non-family households
24.1	28.7	16	25.5	5.7

*Own children under 18.

SOURCE: U.S. Census Bureau, *Current Population Reports*, pp. 20–547.

REFLECTIONS

Think about all the people you consider your family. What criteria—biological, legal, affectional—did you use? Did you exclude any biological or legal family? If so, who and why?

Being biologically related or related through marriage is not always sufficient to be counted as a family member or kin. One researcher (Furstenberg, 1987) found that 19 percent of the children with biological siblings living with them did not identify their brothers or sisters as family members. Sometimes an absent or divorced parent was not counted as a relative. Stepparents or stepchildren were the most likely not to be viewed as family members (Furstenberg, 1987; Ihinger-Tallman and Pasley, 1987). Emotional closeness may be more important than biology or law in defining family.

There are also ethnic differences as to what constitutes family. Among Latinos, for example, *compadres* (godparents) are considered family members. Among some Japanese Americans the *ie* (pronounced "ee-eh") is the traditional family. The *ie* consists of living members of the extended family (such as grandparents, aunts, uncles, and cousins) as well as deceased and yet-to-be-born family members (Kikumura and Kitano, 1988). Among many traditional Native-American tribes, the **clan,** a group of related families, is regarded as the fundamental family unit (Yellowbird and Snipp, 1994).

A major reason we have such difficulty in defining *family* is that we tend to think that the "real" family is the nuclear family, a household structure that we further perceive as the traditional family. The **nuclear family** is the family type consisting of mother, father, and children. But the nuclear family is merely an idea or model we have about families. The term itself is little more than 50 years old, coined by the anthropologist Robert Murdock in 1949 (Levin, 1993). What most Americans consider to be the **traditional family** is the middle-class, nuclear family in which women's primary roles are wife and mother, and men's primary roles are husband and breadwinner. The traditional family is the nuclear family wrapped in nostalgia and inequality. As we shall see in Chapter 3, the traditional family exists more in the imagination than it ever did in reality.

Because we believe that the nuclear or traditional family is the real family, we compare all other family forms against these models. To include the diverse

The strength and vitality of kin ties was a major theme in the popular movie My Big Fat Greek Wedding.

Photofest

forms, the definition of family needs to be expanded beyond the boundaries of the "official" census definition. A more contemporary and inclusive definition describes family as "two or more persons related by birth, marriage, adoption, or choice. Families are further defined by socioemotional ties and enduring responsibilities, particularly in terms of one or more members' dependence on others for support and nurturance" (Allen, Fine, and Demo, 2000). Such a definition more accurately reflects the diversity of contemporary American family experience.

DEFINING MARRIAGE

A **marriage** is a legally recognized union between a man and a woman in which they are united sexually, cooperate economically, and may give birth to, adopt, or rear children. The union is assumed to be permanent (although in reality it may be dissolved by separation or divorce). As simple as such a definition may make marriage seem, it differs among cultures and has changed considerably in our society.

Among non-Western cultures, who may marry whom and at what age varies greatly from our society. In some areas of India, Africa, and Asia, for example, children as young as 6 years may marry other children (and sometimes adults), although they may not live together until they are older. In many cultures, marriages are arranged by families who choose their children's partners. In many such societies, the "choice" partner is a first cousin. And in one region of China, marriages are sometimes arranged between unmarried young men and women who are dead. (See Exploring Diversity in this section, which discusses spirit marriage in Chinese society.)

Many Americans believe that marriage is divinely instituted; others, that it is a civil institution involving the state. The belief in the divine institution of marriage is common to many religions, such as Christianity, Judaism, and Islam, as well as many tribal religions throughout the world. But the Christian church only slowly became involved in weddings. In the early Middle Ages, for example, the priest's blessing was not important. As the church increased its power, however, it extended control over marriage. Traditionally, marriages had been arranged between families (the father "gave away" his daughter in exchange for goods or services); by the tenth century, marriages were valid only if they were performed by priests. By the thirteenth century, the ceremony was required to take place in a church (Gies and Gies, 1987). As states competed with organized religion for power, governments began to regulate marriage. In the United States today, for example, in order for marriages to be legal—whether they are performed by ministers, priests, rabbis, or imams—they must be validated through government-issued marriage licenses. This is a right for which many gay men and lesbians are fighting.

DID YOU KNOW?

In 2000, 56 percent of the adult population in the United States (age 18 and older) were married (U.S. Census Bureau, 2001, pp. 20–537, Table A1).

WHO MAY MARRY?

Who may marry whom has changed over the last 150 years in the United States. Laws once prohibited enslaved African Americans from marrying because they were regarded as property. Marriages between members of different races were illegal in more than half the states until 1966, when the U.S. Supreme Court declared such prohibitions unconstitutional. Each state enacts its own laws regulating marriage. In some states, first cousins may marry; other states prohibit such marriages as incestuous.

Certainly, the greatest current controversy regarding legal marriage is over the continuing question of same-sex marriage. As we revise the present edition of this book, we remain in the midst of potentially revolutionary change. Before we look at current developments, let's first look at the recent past.

Beginning in the 1990s, countries such as Germany, France, the Netherlands, Sweden, and Norway enacted legislation extending marital rights or marriage-like protections to gay couples. Some stopped short of allowing gay or lesbian couples to legally marry, but in the Netherlands and Belgium the right to marry extends to same-sex couples. In 2003, the province of Ontario, Canada's most populous, legalized same-sex marriage.

THE CROSS-CULTURAL PERSPECTIVE: THE SPIRIT MARRIAGE IN CHINESE SOCIETY

From *Time* magazine comes this report ("A Day in the Life of China," 1989):

> A beautiful day for a wedding—crisp, clear and, for China in mid-summer, relatively cool. The latest typhoon's high winds have swept away the air pollution, and under a brilliant blue sky the guests are chatting in the hollow of a terraced field beside a single spindly tree—symbolic decoration in a country whose scant arable land continues to disappear. Arranged neatly alongside the makeshift altar, the gifts intended for the bride's parents include a new refrigerator, a 24-in. color television set and a jet black Yamaha motorcycle. The presents are ogled, but atop the TV a photograph of Margaret Thatcher creates the greatest buzz, a reaction the bride, and perhaps the groom too, would undoubtedly have enjoyed. Were they still alive . . . I am transfixed by the marriage of the two coffins in front of me. The groom died in an automobile accident five days earlier at the age of 23. The body of his bride, dead of cancer for five months, cost $3 to exhume. They had never met.

After overcoming our surprise at the very notion of *spirit marriage,* we wonder why anyone would arrange a marriage between two dead persons. The case of spirit marriage, which was widely practiced in traditional Chinese society, underscores a fundamental point about the family. Anthropologists and historians tell us that families are organized differently in different cultures. Different cultural understandings about kinship, marriage, and residence give the family a decidedly different look across cultures.

Consider, for example, the case of kinship. Who counts as a relative? The answer to this question may seem self-evident to us, but we grew up learning who a relative is from within our own culture. Everyone everywhere learns this way, and cultural understandings of who counts as a relative vary. For example, among the traditional Iroquois of the northeastern United States, the important relatives were those who could be traced through the mother. This was a matrilineal society, and children took their clan membership, their names, and their chiefly titles (if any) from their mothers' families. Sons and daughters inherited from their mothers. A mother's relatives counted more than a father's. Thus, an aunt on your mother's side would be considered a very different kind of relative than one on your father's side.

Consider also the way in which marriage shapes the family. Who marries whom, and why? Who decides? These questions seem like nonquestions to us, because in our society, the individual generally makes these decisions. Yet cultural practices surrounding marriage vary tremendously, contributing further to the distinctive look and character of the family across cultures. In Bedouin society in Egypt, marriages are arranged by the elder males of the bride's and groom's families. The engaged couple themselves may never meet prior to marriage. Love as a "natural" basis for marriage turns out to be another of our cultural assumptions. In Bedouin society, parents may very well decide to arrange a marriage between a bride and groom who are first cousins on the male side—and definitely for reasons other than love! (A marriage between paternal cross-cousins was considered to be a particularly advantageous match.) Thus, who marries whom, why they marry, and who decides are cultural issues that greatly influence the shape of the family across cultures.

Finally, there is the question of residence. Where will the married couple live? And with whom? The answer to these questions—for us, a matter of individual choice (as well as economics)—gives the family its many different faces across cultures. In traditional

In the United States, the situation has been in flux for more than a decade. In the 1990s, U.S. courts rendered decisions that seemed to pave the way toward American legalization of same-sex marriage. The two most notable cases were in Hawaii and Vermont. In 1993, the Hawaii Supreme Court ruled that denying gay men and lesbians the right to marry was unconstitutional, in that it violated the equal protection clause of the state constitution. This decision led many to anticipate the eventual legalization of same-sex marriage. It also caused opponents of gay marriage to take action. A number of state legislatures along with the federal government passed laws that declared marriage to be the union of one man and one woman, which prevented the forced acceptance of gay and lesbian marriages should the Hawaiian decision stand up to an appeal.

In 1996, Congress passed the Defense of Marriage Act, and President Clinton signed it into law. This act denied federal recognition to same-sex couples and gave states the right to legally ignore gay or lesbian marriages should they gain legal

Chinese society, it was expected that upon marriage, the bride would live with her husband in his family home, which typically included his father and mother, his father's father (if alive), and his father's brothers and their wives and children. This was the family in traditional Chinese society. It was built upon the male descent line, as anthropologists call it. Unlike Iroquois families, Chinese families were patrilineal, the most important kinship ties being those traced on the father's side of the family. The father's relatives counted more.

In traditional Chinese society, male elders employed matchmakers to arrange marriages for their sons and daughters on the basis of economic and social criteria, not love. Marriages were arranged for sons in order that they themselves might have sons to continue the male descent line. (Marriages were arranged for daughters so that they would, as wives, produce sons for their husbands' descent lines.) Even the male ancestors—as deceased male elders—were thought to be very concerned about the continuation of their descent lines. It was a young man's duty—an almost religious obligation to his male ancestors and father—to continue the descent line unbroken into the future: *a man must have sons*.

Life did not always make it easy for men to continue their male descent lines. Many men in traditional Chinese society were too poor to marry, and thus poverty annihilated their descent lines. Death, too, could interfere with human plans for descent lines. Sons might die before they could marry and produce sons of their own. To remedy this situation, the practice of spirit marriage developed, guaranteeing that the descent line would continue even in the face of death. A family whose young son died would wait until the son's ghost (or spirit) reached proper marriageable age. Then, as in marriages arranged among the living, they engaged the services of a matchmaker to find an appropriate spouse—but in this case, a deceased bride for their dead son to marry.

For the groom's family, this spirit marriage would settle their son's restless spirit, for he would now have a wife. His parents would then adopt a son for him, one that they themselves would raise. This son would one day marry and have a son of his own—fate willing—and thus continue the line of his deceased father. The male descent line would remain intact.

For the bride's family, marrying off their deceased daughter also brought distinct advantages—or rather diverted distinct disadvantages and even disaster. The death of an unmarried daughter not only brought sorrow to her family; her unmarried ghost threatened the fertility of her brothers' wives, and hence their descent lines. The ghost of an unmarried daughter was troublesome to her family and brought misfortune, including bad harvests. But if she could be married off to a husband in a spirit marriage, she could then take her proper place as a married woman at his family home, diverting disaster at her own.

Thus, the practice and meaning of spirit marriage in traditional Chinese society was shaped by cultural notions of kinship (Who counts as a relative?), marriage (Who marries whom, why, and who decides?), and residence (Where does a couple live in marriage, and with whom?). The Chinese answers to all of these questions help to explain a marriage practice that to us may seem unfathomable. Of course, families change with time as new laws, technology, and other cultural influences reshape kinship, marriage, and residence practices. Yet with all the profound changes in Chinese society in recent decades, the continuing practice of spirit marriage speaks to the continuing belief in the male descent line as the foundation of Chinese families (Stockard,1989).

recognition in Hawaii or any other state. But the earlier Hawaiian decision did not stand. In a November 1998 ballot, 69 percent of Hawaiian voters voted to amend the state constitution, giving lawmakers the power to block same-sex marriage and limit legal marriage to heterosexual couples. Similar laws were passed in more than half of the 50 states by November 1998.

As 1999 drew to a close, the state of Vermont took a major step toward what some believed would be the eventual legal recognition of gay marriage. There, three same-sex couples filed lawsuits, challenging a 1975 state ruling prohibiting same-sex couples from marrying. On December 20, 1999, the Vermont Supreme Court ruled that the state legislature either had to grant marriage rights to same-sex couples or assure them a legal equivalent to marriage, providing them the same range of state benefits enjoyed by married heterosexuals.

On April 26, 2000, Vermont Governor Howard Dean signed into law legislation recognizing same sex "civil unions." Although they are not marriages,

Perspective

THE RIGHTS AND BENEFITS OF MARRIAGE

Heterosexuals rarely stop to think about the privileges that sexual orientation offers or withholds. One such privilege is the right to marry. Obviously, not all heterosexuals will marry nor do all desire to do so. Those couples who do marry receive many more rights and protections than those who don't marry. In other words, it is clear that couples suffer when they live together outside marriage. For heterosexual cohabitants, this is a matter of choice; they do so because they prefer the more informal arrangement. They could achieve the following by entering marriage but despite numerous marriage benefits they choose to remain unmarried. For many same-sex couples, the historical *inability* to marry when they want to has cost them the following protections. It is the lack of these rights and protections that state courts in the United States (Hawaii, Vermont, and Massachusetts, for example) have found unconstitutional.

- Accidental death benefit for the surviving spouse of a government employee
- Appointment as guardian of a minor
- Award of child custody in divorce proceedings
- Beneficial owner status of corporate securities
- Bill of Rights benefits for victims and witnesses
- Burial of service member's dependents
- Consent to postmortem examination
- Continuation of rights under existing homestead leases
- Control, division, acquisition, and disposition of community property
- Criminal injuries compensation
- Death benefit for surviving spouse for government employee
- Disclosure of vital statistics records
- Division of property after dissolution of marriage
- Eligibility for housing opportunity allowance program of the Housing, Finance and Development Corporation
- Exemption from claims of Department of Human Services for social services payments, financial assistance, or burial payments
- Exemption from conveyance tax
- Exemption from regulation of condominium sales to owner-occupants
- Funeral leave for government employees
- Income tax deductions, credits, rates exemption, and estimates
- Inheritance of land patents
- Insurance licenses, coverage, eligibility, and benefits organization of mutual benefits society
- Legal status with partner's children
- Making partner medical decisions
- Nonresident tuition deferential waiver
- Notice of guardian *ad litem* proceedings
- Notice of probate proceedings
- Payment of wages to a relative of deceased employee
- Payment of worker's compensation benefits after death
- Permission to make arrangements for burial or cremation
- Proof of business partnership
- Public assistance from the Department of Human Services
- Qualification at a facility for the elderly
- Right of survivorship to custodial trust
- Right to be notified of parole or escape of inmate
- Right to change names
- Right to enter into pre-marital agreement
- Right to file action for nonsupport
- Right to inherit property
- Right to sue for tort and death by wrongful act
- Right to support after divorce
- Right to support from spouse
- Rights and proceedings for involuntary hospitalization and treatment
- Rights to notice, protection, benefits, and inheritance under the uniform probate code
- Sole interest in property
- Spousal privilege and confidential marriage communications
- Spousal immigration benefits
- Status of children
- Support payments in divorce action
- Tax relief for natural disaster losses
- Vacation allowance on termination of public employment by death
- Veterans' preference to spouse in public employment
- In vitro fertilization coverage
- Waiver of fees for certified copies and searches of vital statistics

Of course, in addition to the above, there are also potential personal and emotional benefits related to the right to marry. Knowing that the wider society recognizes, accepts, or respects a relationship may cause one to feel greater self-validation and comfort within the relationship. On the other hand, knowing that people do not respect, accept, or recognize a commitment may cause additional emotional suffering and personal anguish.

SOURCE: "What Is Marriage, Anyway?" www.pflag.org/education/marriage.html.

"civil unions" are officially entered, offer the same rights and protections as marriages, and must be officially dissolved when they fail. As of January 2002, more than 4,000 such civil unions had been recorded in Vermont, involving residents of almost every state, the nation's capital, and several other countries, including Canada (Vermont Office of Legislative Council, 2002).

In other states the issue remains up in the air. During 2001, Connecticut, Rhode Island, Washington, Hawaii, and California considered legislation that would legalize same-sex marriages or domestic partnerships. Additionally, couples in Massachusetts, New Jersey, and Indiana filed suit for either the right to marry or recognition of their Vermont-enacted civil unions (www.datalounge.com, 2002).

In October 2001, California passed Chapter 893, a law granting gay or lesbian domestic partners many benefits (including tax benefits, step-parent adoption, sick leave, and permission to make medical decisions) otherwise restricted to married couples. Although far less sweeping in scope than Vermont's civil union legislation, Chapter 893 provides same-sex couples more benefits than found anywhere in the United States other than Vermont (Vermont Office of Legislative Council, 2002). In June 2002, Connecticut passed even more limited legislation giving gay and lesbian couples certain partnership rights and responsibilities.

On June 26, 2003, in the case of *Lawrence and Garner vs. Texas* the U.S. Supreme Court ruled 6–3 that existing laws against sodomy, in Texas and 12 other states, were illegal invasions of privacy. The ruling, which struck down the 13 remaining state sodomy statutes, stemmed from a 1998 arrest of two Houston men, John Lawrence and Tyron Garner, who were having sex when police entered their home on a false emergency call. The men were arrested, jailed overnight, and fined $200 under the Texas sodomy statute. Texas was one of four states whose sodomy statute pertained only to same-sex relations. The remaining nine statutes pertained to heterosexuals and homosexuals. All 13 were nullified with the Court decision. Although the ruling was about private, consensual sex, not about same-sex marriage, many perceived it as a potential step further down the path toward gay marriage.

Meanwhile, organized opposition to same-sex marriage still exists. In states such as Nebraska, Georgia, Ohio, and Missouri, court decisions opposed to gay marriage or legislation unfavorable to domestic partnerships have been enacted within the past two years. Lobbying by groups opposed to gay marriage continues to be seen, such as Massachusetts' Citizens for Marriage, or Maine's Christian Civic League. In fact, after the Supreme Court's ruling in *Lawrence and Garner vs. Texas*, many religious figures and politicians weighed in with comments explicitly and intensely opposed to further movement toward same-sex marriage. This opposition includes President George W. Bush, who in a July 30, 2003, press conference, declared that "marriage is between a man and a woman" (Fireman, 2003).

Same-sex marriage.

As the current edition of this book was being completed, the Massachusetts Supreme Court ruled the state's ban of same-sex marriage unconstitutional. The November 18, 2003, ruling gave the state legislature six months to remedy the situation. Although the remedy may come in the form of civil union protections that are the full equivalent of marriage, the court's decision specified the right to marry (i.e., not the right to enter something closely equivalent to marriage). Although the Massachusetts Legislature and governor remain opposed to same sex marriage, on February 4, 2004, the state supreme court ruled 4–3 that a "civil union" solution was unacceptable, in that it would constitute "an unconstitutional, inferior, and discriminatory status for same-sex couples."

Writing in the *Boston Globe,* journalist Raphael Lewis quoted the court: "For no rational reason the marriage laws of the Commonwealth discriminate against a defined class; no amount of tinkering with language will eradicate that stain. . . . The [civil unions] bill would have the effect of maintaining and fostering a stigma of exclusion that the Constitution prohibits" (Lewis, *Boston Globe,* February 12, 2004).

By now, as you read these words, the outcome of the Massachusetts ruling may have come about in the form of fully legal gay marriage, recognized in the United States for the first time.

It is difficult to predict what level of opposition or support will accompany the court's decision. It is also difficult to predict what may happen elsewhere in the United States. On February 12, 2004, for example, city officials in San Francisco *married* a lesbian couple, defying the state laws recognizing only heterosexual marriage. Supported by Mayor Gavin Newsom and other top city officials, the marriage of Phyllis Ryan, 79, and Del Martin, 83, was described by mayoral spokesperson Peter Ragone as the first, and they intend to issue marriage licenses to other gay or lesbian couples who apply (Leff, Associated Press, 2/12/2004). Some states may eventually recognize civil unions performed in Vermont or same-sex marriages performed in Massachusetts (should the court decision stand). Other state legislatures might decide to follow the legal logic and ruling of the Vermont Supreme Court and create their own domestic partner legislation. Still other states may enact and enforce legislation modeled on the Defense of Marriage Act, and pass constitutional amendments limiting marriage to heterosexual couples. This was the course taken by Ohio, which enacted some of the most restrictive defense of marriage legislation in the country. The bill, signed by Governor Bob Taft on February 5, 2004, defined and prohibited gay marriage as "against the strong public policy of the state." It further denies benefits to state employees' unmarried partners, whether they be heterosexuals, gay men, or lesbians. Without reciprocal recognition (i.e., other states acknowledging and supporting same-sex marriages performed in Massachusetts) more civil suits are certain to follow.

FORMS OF MARRIAGE

In Western cultures such as the United States, the only legal form of marriage is **monogamy,** the practice of having only one spouse—a husband or wife—at one time. Monogamy is the only form of marriage recognized in *all* other cultures, but it is not the *preference* of most. Among world cultures, only 24 percent of the known cultures perceive monogamy as the ideal form of marriage (Murdock, 1967). The preferred marital arrangement worldwide is **polygamy,** the practice of having more than one wife or husband. One study of 850 non-Western societies found that 84 percent of the cultures studied (representing, nevertheless, a minority of the world's population) practiced or accepted **polygyny,** the practice of having two or more wives (Gould and Gould, 1989). **Polyandry,** the practice of having two or more husbands, is quite rare: where it does occur, it coexists with polygyny. Even within polygynous societies, however, plural marriages are in the minority, primarily for simple economic reasons: They are a sign of status that relatively few people can afford. Although problems of jealousy may arise in plural marriages—the Fula in Africa, for example, call the second wife "the jealous one"—there are usually built-in control mechanisms to ease the problem. The wives may be related, especially as sisters; if they are unrelated, they usually have separate dwellings. Women in these societies often prefer that there be several wives; plural wives are a sign of status and, more important, ease the workload of individual wives. This last fact is apparent even in our culture; it is not uncommon to hear overworked American homemakers exclaim, "*I* want a wife."

In part, because of our culture's traditional roots in Christianity, polygamy is illegal in the United States. Beginning with a U.S. Supreme Court decision in 1879, polygamy was prohibited because it was considered a potential threat to the public order (Tracy, 2002). As a result, polygamy was looked upon as being strange or exotic. However, it may not really seem so strange if we look at actual American marital practices. Considering the high divorce rate in this country, monogamy may no longer be the best way of describing our marriage forms. For many, our marriage system might more accurately be called **serial monogamy** or **modified polygamy,** the practice where one person may have several spouses over his or her lifetime. In our nation's past, enslaved Africans tried unsuccessfully to continue their traditional polygamous practices when they first arrived in America; these attempts, however, were rigorously suppressed by their masters (Guttman, 1976). Mormons practiced polygamy until the late nineteenth century, when they officially abandoned the practice as a condition of Utah's becoming a state. Even today, though, an estimated 60,000 fundamentalist Mormon men, women, and children, as well as some members of the Nation of Islam, continue to practice polygamy despite its legal prohibition. In all such unions, only the first wife has legal status as a wife, however.

Functions of Marriages and Families

Whether it is the mother/father/child nuclear family, a married couple with no children, a single-parent family, a stepfamily, a dual-worker family, or a cohabiting family, the family generally performs four important functions: (1) it provides a source of intimate relationships; (2) it acts as a unit of economic cooperation and consumption; (3) it may produce and socialize children; and (4) it assigns social roles and status to individuals. Although these are the basic functions that families are "supposed" to fulfill, families do not necessarily have to fulfill them all (as in families without children), nor do they always fulfill them well (as in abusive families).

INTIMATE RELATIONSHIPS

Intimacy is a primary human need. Human companionship strongly influences rates of cancer, tuberculosis, suicide, accidents, and mental illness. Studies consistently show that married couples and adults living with others are generally healthier and have a lower mortality rate than divorced, separated, and never-married individuals (Ross, Mirowsky, and Goldsteen, 1991). This holds true for both whites and African Americans (Broman, 1988).

Family Ties

Marriage and the family usually furnish emotional security and support. This has probably been true from earliest times. Thousands of years ago, in the Judeo-Christian Bible, the book of Ecclesiastes (4:9–12) emphasized the importance of companionship:

> Two are better than one, because they have a good reward for their toil. For if they fall, one will lift up his fellow; but woe to him who is alone when he falls and has not another to lift him up. Again if two lie together, they are warm; but how can one be warm alone? And though a man might prevail against one, two will withstand him. A three-fold cord is not quickly broken.

In our families we generally find our strongest bonds. These bonds can be forged from love, attachment, loyalty, obligation or guilt. The need for intimate relationships, whether they are satisfactory or not, may hold unhappy marriages together indefinitely. Loneliness may be a terrible specter. Among

A major function of marriages and families is to provide us with intimacy and social support, thus protecting us from loneliness and isolation.

the newly divorced, it may be one of the worst aspects of the marital breakup.

Since the nineteenth century, marriage and the family have become even more important as sources of companionship and intimacy. They have become "havens in a heartless world" (Lasch, 1978). As society has become more industrialized, bureaucratic, and impersonal, it is within the family that we increasingly seek and expect to find intimacy and companionship. In the larger world around us, we are generally seen in terms of our roles. A professor may see us primarily as students; a used-car salesperson relates to us as potential buyers; a politician views us as voters. Only among our intimates are we seen on a personal level, as Maria or Will. Before marriage, our friends are our intimates. After marriage, our partners are expected to be the ones with whom we are most intimate. With our partners we disclose ourselves most completely, share our hopes, rear our children, grow old.

Pets and Intimacy

The need for intimacy is so powerful that many rely upon pets as additional or even substitute sources for satisfaction of those needs. Animals have been important human companions since prehistoric times (Siegel, 1993). They have been important emotional figures in our lives, especially if our other relationships are not fulfilling. Unmarried adults, for example, are more attached to their pets than are married men and women (Stallones et al., 1990). This does not mean, however, that you reject Fido or Fluffy when you become romantically involved or get married. What happens is that your pet becomes less important—he or she becomes more an "animal" and less "someone" to whom you are emotionally attached. As an object of attachment, your pet is replaced by your partner or children. You do not forget your pet, even in marriage or parenthood; your dog, cat, gerbil, parakeet, or turtle simply becomes less important.

Studies of the role of pets in human relationships suggest that the most prized aspects of pets, especially dogs and cats, are their attentiveness to their owners, their welcoming and greeting behaviors, and their role as confidants—qualities valued in our intimate relationships with humans as well. Pets give children an opportunity to nurture, and they provide a best friend, someone to love. Although there is interesting research on the ways in which we attach familial qualities to pets, the remainder of this text will consider human families.

ECONOMIC COOPERATION

The family is also a unit of economic cooperation that traditionally divides its labor along gender lines—that is, between males and females (Fox and

Pets are often considered to be members of the family. They often provide their owners with comfort and a sense of intimacy.

© Christine Mendes/Buena Vista Photography

Murry, 2000; Ferree, 1991; Thompson and Walker, 1989; Voydanoff, 1987). Although the division of labor by gender is characteristic of virtually all cultures, the work that males and females perform (apart from childbearing and breastfeeding) varies from culture to culture. Among the Namibikwara in Africa, for example, the fathers take care of the babies and clean them when they soil themselves; the chief's concubines, secondary wives in polygamous societies, prefer hunting over domestic activities. In American society, from the last century until recently, men were expected to work away from home, whereas women were to remain at home caring for the children and house. There is no reason, however, why these roles cannot be reversed. Such tasks are assigned by culture, not biology. Only a man's ability to impregnate and a woman's ability to give birth and produce milk are biologically determined. And some cultures practice *couvade,* ritualized childbirth in which a male gives birth to the child's spirit while his partner gives physical birth.

We commonly think of the family as a consuming unit, but it also continues to be an important producing unit. The husband does not get paid for building a shelf or attending to the children; the wife is not paid for fixing the leaky faucet or cooking. Although children contribute to the household economy by helping around the house, they generally are not paid for such things as cooking, cleaning their rooms, or watching their younger brothers or sisters (Coggle and Tasker, 1982; Gecas and Seff, 1991). Yet they are all engaged in productive labor.

Over the past decade, economists began to reexamine the family as a productive unit (Ferree, 1991). If men and women were compensated monetarily for the work done in their households, the total would be equal to the entire amount paid out in wages by every corporation in the United States. Household work and assets along with the production of goods (including food) all contribute to the family's productive activities.

As a service unit, the family is dominated by women. Because women's work at home is unpaid, the productive contributions of homemakers have been overlooked (Ciancanelli and Berch, 1987; Walker, 1991). Yet women's household work is equal to about 44 percent of the gross domestic product (GDP), and the value of such work is double the reported earnings of women. If women were paid wages for their labor as mothers and homemakers according to the wage scale for chauffeurs, physicians, babysitters, cooks, therapists, and so on, many women would make more for their work in the home than most men do for their jobs outside the home. One economic estimate of a typical housewife's work placed the yearly value at over $60,000 (Crittenden, 2001). Because family power is partly a function of who earns the money, paying women for their household work might have a significant impact on husband-wife relations.

REPRODUCTION AND SOCIALIZATION

The family makes society possible by producing (or adopting) and rearing children to replace the older members of society as they die off. Traditionally, reproduction has been a unique function of the married family. But single-parent and cohabiting families also perform reproductive and socialization functions. Technological change has also affected reproduction. Developments in contraception, artificial insemination, and in vitro fertilization have separated reproduction from sexual intercourse. Depending upon their contraceptive choices, couples can engage in sexual intercourse with relatively high confidence that they will not become parents. Innovations in reproductive technology permit many infertile couples to give birth. Such techniques have also made it possible for lesbian couples to become parents.

The family traditionally has been responsible for **socialization**—the shaping of individual behavior to conform to social or cultural norms. Children are helpless and dependent for years following birth. They must learn how to walk and talk, how to take care of themselves, how to act, how to love, how to touch and be touched. Teaching children how to fit into their particular culture is one of the family's most important tasks.

This socialization function, however, often includes caregivers outside of the family. The involvement of nonfamily in the socialization of children need not indicate a lack of parental commitment to their children or a lack of concern for the quality of care received by their children. Still, nonparental sources of child rearing may be one of the most significant societal changes in our lifetimes. Since the rise of compulsory education in the nineteenth century, the state has become responsible for a large

© David Young Wolff/PhotoEdit

Much childhood socialization occurs in nonfamily settings such as preschools or day-care centers.

part of the socialization of children older than age five. Increasing numbers of dual-earner households and employed single mothers have resulted in placing many infants, toddlers, and small children under the care of nonfamily members, thus broadening the role of others (such as neighbors, friends, or paid caregivers) and reducing the family's role in child rearing.

ASSIGNMENT OF SOCIAL ROLES AND STATUS

We fulfill various social roles as family members, and these roles provide us with much of our identity. During our lifetimes, most of us will belong to two families: the family of orientation and the family of procreation. The **family of orientation** (sometimes called the **family of origin**) is the family in which we grow up, the family that orients us to the world. The family of orientation may change over time if the marital status of our parents changes. Originally, it may be an intact nuclear family or a single-parent family; later it may become a stepfamily.

The common term for the family we form through marriage and childbearing is **family of procreation.** Because many families have stepchildren, adopted children, or no children, we can use a more recent term—**family of cohabitation**—to refer to the family we form through living or cohabiting with another person, whether we are married or unmarried. Most Americans will form families of cohabitation sometime in their lives. Much of our identity is formed in the crucibles of our families of orientation, procreation, and cohabitation.

In our families of orientation, we are given the roles of son or daughter, brother or sister, stepson or stepdaughter. We internalize these roles until they become a part of our being. In each of these roles, we are expected to act in certain ways. For example, children obey their parents, siblings help one another. Sometimes our feelings fit the expectations of our roles; other times they do not.

Our family roles as offspring and siblings are most important when we are living in our families of orientation. After we leave home, these roles gradually diminish in everyday significance, although they continue throughout our lives. In relation to our parents, we never cease being children; in relation to our siblings, we never cease being brothers and sisters. The roles change as we grow older. They begin the moment we are born and end only when we die.

As we leave our families of orientation, we usually are also leaving adolescence and entering adulthood. Being an adult in our society is defined in part by entering new family roles—those of husband, wife, or partner, or father or mother. These roles formed in our family of procreation take priority over the roles we had in our family of orientation. When we marry, we transfer our primary loyalties from our parents and siblings to our partners. Later, if we have children, we form additional bonds with them. When we assume the role of spouse or bonded partner, we assume an entirely new social identity linked with responsibility, work, and parenting. In earlier times such roles were considered to be lifelong in duration. Because of divorce or separation, however, these roles today may last for considerably less time.

The status or place we are given in society is acquired in large part through our families. Our families place us in a certain socioeconomic class, such as blue collar (working class), middle class, or upper class. We learn the ways of our class through identifying with our families. Different classes see the world through different eyes. These differences affect their perceptions of the role of women, how they value education, and how they rear children (Rubin, 1976, 1994).

Our families also give us our ethnic identities as African American, Latino, Jewish, Irish American, Asian American, Italian American, and so forth. Families also provide us with a religious tradition as Protestant, Catholic, Jewish, Greek Orthodox, Islamic, Hindu, or Buddhist—as well as agnostic, atheist, or "New Age." These identities help form our cultural values and expectations.

WHY LIVE IN FAMILIES?

As we look at the different functions of the family we can see that, theoretically, most of them can be fulfilled outside the family. In terms of reproduction, for example, artificial insemination permits a woman to be impregnated by a sperm donor, and embryonic transplants allow one woman to carry another's embryo. Children can be raised communally, cared for by foster families or child-care workers, or sent to boarding schools. Most of our domestic needs can be satisfied by eating frozen or prepared foods or going to restaurants, sending our clothes to the laundry, and hiring help to clean the bathroom and wash the mountains of dishes accumulating (or growing new life-forms) in the kitchen. Friends can provide us with emotional intimacy, therapists can listen to our problems, and sexual partners can be found outside of marriage. With the limitations and stresses of family life, why bother living in families?

Sociologist William Goode (1982) suggests that there are several advantages to living in families.

- Families offer continuity as a result of emotional attachments, rights, and obligations. Once we choose a partner or have children, we do not have to search continually for new partners or family members who can better perform a family task or function such as cooking, painting the kitchen, providing companionship, or bringing home a paycheck. We expect our family members—whether partner, child, parent, or sibling—to participate in family tasks over their lifetimes. If at one time we need to give more emotional support or attention to a partner or child than we receive, we expect the other person to reciprocate at another time. Or if we ourselves are down, we expect our family to help. We further expect that we can enjoy the fruits of our labors together. We count on our family to be there for us in multiple ways. We rarely have the same extensive expectations of friends.
- Families offer close proximity. We do not need to travel across town or cross-country for conversation or help. With families, we do not even need to go out of the house; a husband or wife, parent or child, or brother or sister is often right at hand (or underfoot, in the case of children). This close proximity facilitates cooperation and communication.
- Families offer us an abiding familiarity with others. Few people know us as well as our family members, for they have seen us in the most intimate circumstances throughout much of our lives. They have seen us at our best and our worst, when we are kind or selfish, understanding or intolerant. This familiarity and close contact teach us to make adjustments in living with others. As we do so, we also expand our own knowledge of ourselves and others.
- Families provide us with many economic benefits. They offer us economies of scale. Various activities, such as laundry, cooking, shopping, and cleaning, can be done almost as easily for several people as for one. It is almost as easy to prepare a meal for three people as it is for one, and the average cost per person in both time and money is usually less. As an economic unit, a family can cooperate to achieve what a single individual could not. It is easier for a working couple to purchase a house than a single individual, for example, because the couple can pool their resources.

Because most domestic tasks do not take great skill (a corporate lawyer can mop the floor as easily as anyone else), most family members can learn to do them. As a result, members do not need to go outside the family to hire experts. In fact, for many family tasks—from embracing a partner to bandaging a child's small cut or playing peekaboo with a baby—there are no experts to compete with family members.

These are only some of the theoretical advantages families offer to their members. Of course, not all families perform all these tasks or perform them well. But families, based on mutual ties of feeling and obligation, offer us greater potential for fulfilling our needs than do organizations based on profit (such as corporations) or compulsion (such as governments).

Extended Families and Kinship

Society "created" the family to undertake the task of making us human. According to some anthropologists, the nuclear family of man, woman, and child is universal, either in its basic form or as the building block for other family forms (Murdock, 1967). Other anthropologists disagree that the father is necessary, arguing that the basic family unit is the mother and child dyad, or pair (Collier, Rosaldo, and Yanagisako, 1982). The use of artificial insemination and new reproductive technologies, as well as the rise of female-headed single-parent families, are cited in support of the mother-child model.

EXTENDED FAMILIES

The **extended family,** as already described, consists not only of the cohabiting couple and their children but also of other relatives, especially in-laws, grandparents, aunts and uncles, and cousins. In the majority of non-European countries, the extended family is often regarded as the basic family unit.

For many Americans, especially those with strong ethnic identification and those in certain groups (discussed in Chapter 3), the extended family takes on great importance. Sometimes, however, we fail to recognize the existence of extended families because we uncritically accept the nuclear family model as our definition of family. We may even be blind to the reality of our own family structure. When someone asks us to name the members of our families, if we are unmarried, most of us will probably name our parents, brothers, and sisters. If we are married, we will probably name our husbands or wives and children. Only if questioned further may we include our grandparents, aunts or uncles, cousins, or even friends or neighbors who are "like family." We may not name all our blood relatives, but we will probably name the ones with whom we feel emotionally close, as we saw earlier in the chapter. Slight increases in the prevalence of

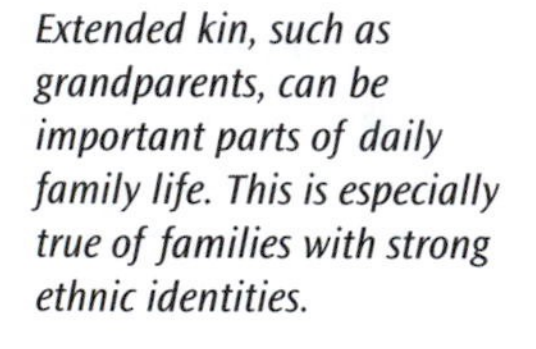

Extended kin, such as grandparents, can be important parts of daily family life. This is especially true of families with strong ethnic identities.

extended families in the United States is accounted for by the family structures of immigrants as well as by economic necessity (Glick, Bean, and Van Hook, 1997).

KINSHIP SYSTEMS

The **kinship system** is the social organization of the family. It is based on the reciprocal rights and obligations of the different family members, such as those between parents and children, grandparents and grandchildren, and mothers-in-law and sons-in-law.

Conjugal and Consanguineous Relationships

Family relationships are generally created in two ways: through marriage and through birth. Family relationships created through marriage are known as **conjugal relationships.** (The word *conjugal* is derived from the Latin *conjungere,* meaning "to join together.") In-laws, such as mothers-in-law, fathers-in-law, sons-in-law, and daughters-in-law, are created by law—that is, through marriage. **Consanguineous relationships** are created through biological (blood) ties—that is, through birth. (The word *consanguineous* is derived from the Latin *com-*, "joint," and *sanguineous,* "of blood.") Parents, children, grandparents, and grandchildren, for example, have consanguineous relationships. Aunts and uncles may be either consanguineous or conjugal.

Our families of orientation, procreation, and cohabitation provide us with some of the most important roles we will assume in life. These nuclear family roles (such as parent, child, husband, wife, and sibling) combine with extended family roles (such as grandparent, aunt, uncle, cousin, and in-law) to form the kinship system.

REFLECTIONS

Thinking about your own experiences, what different kinship roles—nephew or niece, aunt or uncle, in-law, grandson or granddaughter—do you play as a member of your extended family? What rights and obligations does each of these roles entail in your family? Can you identify any "affiliated kin"? How did they become "like family"?

Kin Rights and Obligations

In some societies, mostly non-Western or nonindustrialized cultures, kinship obligations may be very extensive. In cultures that emphasize kin groups, close emotional ties between a husband and wife are viewed as a threat to the extended family. A remarkable form of marriage that illustrates the precedence of the kin group over the married couple is the institution of spirit marriage, which continues today in Canton, China. According to anthropologist Janice Stockard (1989), a **spirit marriage** is arranged by two families whose son and daughter died unmarried. After the dead couple is "married," the two families adopt an orphaned boy and raise him as the deceased couple's son to provide family continuity. (See the Exploring Diversity box in this chapter.) In another Cantonese marriage form, women do not live with their husbands until at least three years after marriage, as their primary obligation remains with their own extended families. Among the Nayar of India, men have a number of clearly defined obligations toward the children of their sisters and female cousins, although they have few obligations toward their own children (Gough, 1968).

In American society, the basic kinship system consists of parents and children, but it may include other relatives as well, especially grandparents. Each person in this system has certain rights and obligations as a result of his or her position in the family structure. Furthermore, a person may occupy several positions at the same time. For example, an 18-year-old woman may simultaneously be a daughter, a sister, a cousin, an aunt, and a granddaughter. Each role entails different rights and obligations. As a daughter, the young woman may have to defer to certain decisions of her parents; as a sister, to share her bedroom; as a cousin, to attend a wedding; and as a granddaughter, to visit her grandparents during the holidays.

In our own culture, the nuclear family has many norms regulating behavior, such as parental support of children and sexual fidelity between spouses, but the rights and obligations of relatives outside the basic kinship system are less strong and less clearly articulated. Because there are neither culturally binding nor legally enforceable norms regarding the extended family, some researchers suggest that such kinship ties have become more or less voluntary. We are free to define our kinship relations much as we wish. Like friendship, these relations may be allowed to wane (Goetting, 1990).

Despite the increasingly voluntary nature of kin relations, our kin create a rich social network for us. Studies suggest that most people have a large number of kin living in their areas (Mancini and Blieszner, 1989). Adult children and their parents often live close to each other; make regular visits; and help each other with child care, housework, maintenance, repairs, loans, and gifts. The relations between siblings also are often strong throughout the life cycle (Lee, Mancini, and Maxwell, 1990).

We generally assume kinship to be lifelong. In the past, if a marriage was disrupted by death, in-laws generally continued to be thought of as kin. But today, divorce is as much a part of the American family system as marriage. Although shunning the former spouse may no longer be appropriate (or polite), no new guidelines on how to behave have been developed. The ex-kin role is a role-less role; that is, it is a role with no clearly defined rules.

Cultural Constructions of Contemporary Family Life

As we begin to study marriage and family, we need to recognize that family issues inspire much controversy and debate. This often makes it difficult to know what to think about much of the public discussion of family issues. Presently, there is considerable disagreement over both the status of contemporary families and the direction family life is heading. In fact, many of the so-called "culture wars" over such "hot-button" issues as the status of women, abortion, the effects of divorce, nonmarital births, and gay rights may really be conflicts over differing conceptions of family (Benokraitis, 2000; Glenn, 2000; Hunter, 1991).

For instance, those who believe that families of male providers, female homemakers, and their dependent children living together, 'til death do they part, are what families *should be*, cannot be encouraged by the continued high rates of divorce and cohabitation, or by the declining rates of marriage or full-time motherhood. Those on the "other" side, who claim that there are basic inequities within the traditional family, especially regarding the status of women, will not mourn the diminishing numbers of breadwinner-housewife families. Similarly, the question of gay marriage will divide those who believe marriage *must* be a relationship between a man and a woman from those who believe that we *must* recognize and support all kinds of families. Given the lack of societal consensus, it is easy to become confused or be misled about what American families are really like.

As we will see more extensively in Chapter 3, there is undeniable evidence that family life has changed, repeatedly and dramatically, throughout our history. In fact, Steven Mintz and Susan Kellogg suggest that "change . . . not stability . . . has been the norm" in American family life from Colonial days to the present (Mintz and Kellogg, 1988). But not everyone sees change through the same lens. To some, contemporary family life is weaker because of cultural and social changes and is now, to some extent, endangered (Wilson, 2002; Popenoe, 1993). More optimistic interpretations of changing family patterns celebrate the increased domestic diversity of numerous family types and the rich range of choices that are now available to Americans (Coontz, 1997; Stacey, 1993). Like the proverbial glass, some see the family as "half empty" others see it as "half full." What makes the "half full, half empty" metaphor even more apt is that even when looking at the same phenomenon or the same trend, some perceive family life as troubled while others see families as different or changing.

Ultimately, the ways we view families depend on what we conceive of *as* families. Such disagreements reflect both different definitions of family and different value orientations about particular kinds of families. Often the product of personal experience, these value positions reflect what we want families to be like and, thus, what we come to believe about the kinds of issues that will be raised throughout this book.

In the wider, societal discourse about families, we can see opposing ideological positions on the well-being of families (Glenn, 2000). The two extremes, conservative and liberal, are relatively familiar and differ in predictable ways. Conservatives are often pessimistic about the state of today's families. To **conservatives,** cultural values have shifted away from individual self-sacrifice toward self-fulfillment. This shift in values is seen as an important factor in some major changes in family life that occurred in the last three or four decades of the twentieth century (especially higher divorce rates, more cohabitation, and more births outside of marriage). Furthermore, as a result of such changes, today's families are seen as weaker and less able to meet the needs of children, adults, or the wider society (Glenn, 2000). Conservatives therefore recommend social policies to reverse or reduce the extent of such changes (recommendations to repeal no-fault divorce and the introduction of covenant marriage are two examples we will later examine).

Compared with conservatives, liberals are more optimistic about the status and future of family life in the United States. **Liberals** tend to believe that the changes in family patterns are just that—changes, not signs of familial decline (Benokraitis, 2000). The liberal position also sees these changing family patterns as products of wider social and economic changes rather than a shift in cultural values (Benokraitis, 2000; Glenn, 2000). Such changes in family experience create a wider range of contemporary household and family types and require greater tolerance of such diversity. Placing great emphasis upon economic issues, liberal family policies are often tied to the economic well-being of families. Additional examples would include supportive policies for the increasing numbers of employed mothers and two-earner households.

According to Norvall Glenn, unlike the situation of partially filled glasses, there is a third position in the discourse about families. **Centrists** share aspects of both conservative and liberal positions. Like conservatives, they believe that some familial changes have had negative consequences. Like liberals they identify wider social changes (e.g., economic or demographic) as major determinants of the changes in family life, but they assert greater emphasis than liberals do on the importance of cultural values. They note that too many people are too absorbed in their careers or too quick to surrender in the face of marital difficulties (Benokraitis, 2000; Glenn, 2000).

The assumptions within and the differences between these positions are more important than they might first appear to be. The perceptions one has of what accounts for the current status of family life or the directions in which it is heading influence what one believes families *need*. In so doing, they influence social policies regarding family life. As Nijole Benokraitis states, "Conservatives, centrists, liberals, and feminists who lobby for a variety of family-related 'remedies' affect our family lives on a daily basis" (Benokraitis, 2000, p. 19).

Contemporary Patterns of Marriage and Family Life

Before we begin examining marriages and families in detail, let's look at some of the changes that have occurred over the past four decades and sparked much debate.

- *Cohabitation:* In its technical sense, **cohabitation** refers to individuals sharing living arrangements in an intimate relationship, whether these individuals are married or unmarried. In common usage, cohabitation refers to relationships in which unmarried individuals share living quarters and are sexually involved. (*Cohabitation* and *living together* are often used interchangeably.) A cohabiting relationship may be similar to marriage in many of its functions and roles, but it does not have equivalent legal sanctions or rights. Cohabitation has increased dramatically over the past 40 years. In addition to the almost 5 million heterosexual couples, there are an additional 600,000 same-sex couples living together outside of marriage.
- *Marriage:* A combination of factors including the women's movement, shifting demographics, family policy, and changing values, particularly as they relate to sexuality, have altered the meaning of marriage and the role it plays in people's lives. Still, the vast majority of young women and men *will marry* at least once in their lifetimes.
- *Separation and Divorce:* Separation occurs when two married people no longer live together. It may or may not lead to divorce. Many more people separate than divorce. Divorce is the legal dissolution of a marriage. Over the last 40 years, divorce has changed the face of marriage and the family in America. At present, among adults 18

and over, there are nearly 20 million who are divorced. The divorce rate is two to three times what it was for our parents and grandparents. Slightly less than half of all those who currently marry will divorce within seven years.

- *The Normalization of Divorce:* Divorce has become so widespread that many scholars are beginning to view it as one variation of the normal life course of American marriages (Raschke, 1987). The high divorce rate does not indicate that Americans devalue marriage, however. Paradoxically, Americans may divorce because they value marriage so highly. If a marriage does not meet their standards, they divorce to marry again. They hope that their second marriages will fulfill the expectations that their first marriages failed to meet (Furstenberg and Spanier, 1987).
- *Remarriages, Stepfamilies, and Single-Parent Families:* Contemporary divorce patterns are largely responsible for three related versions of American marriages and families: single-parent families, remarriages, and stepfamilies. Because of their widespread incidence, these variations are becoming part of our normal marriage and family patterns. The majority of young Americans will probably have some experience of single-parent families, remarriage, or stepfamilies either as children or adults.
- *Remarriage:* Half of all recent marriages are remarriages for at least one partner (Coleman, Ganong, and Fine, 2000). Most individuals who divorce tend to remarry. Rates differ between men and women and across different ethnic groups. Those who remarry are usually older, have more experience in both life and work, and have different expectations than those who marry for the first time. Remarriages also may create stepfamilies. When remarriages include children, a person may become not only a husband or a wife but also a stepfather or stepmother. An estimated one-third of children will reside in a stepfamily household before reaching adulthood (Coleman et al., 2000). Ironically, despite the hopes and experience of those who remarry, their divorce rate is about the same as for those who marry for the first time.
- *Family:* During the twentieth century, Americans have tended to identify the nuclear family as the family. But the older (and wider) definition of

TABLE 1.1 ■ How Families Have Changed, 1970–2000

	1970	1980	1990	2000
Marriages	2,159,000	2,390,000	2,443,000	2,329,000
Marriage rate	10.6	10.6	9.8	8.5
Divorces	708,000	1,189,000	1,182,000	1,135,000
Divorce rate	3.5	5.2	4.7	4.2
Married Couples	44,728,000	49,112,000	52,317,000	55,311,000
Married Couples with Children	25,541,000	24,961,000	24,537,000	25,248,000
Percentage of all married couples with children	57	51	47	46
Unmarried Couple Households	523,000	1,589,000	2,856,000	4,486,000
Unmarried Couples with Children	196,000	431,000	891,000	1,563,000
Children Living with Two Parents	59,681,000	47,543,000	46,820,000	49,688,000
Children Living with One Parent	8,426,000	12,349,000	15,842,000	19,227,000
Births	3,731,000	3,612,000	4,158,000	4,063,000
Births to Unmarried Women	399,000	666,000	1,165,000	1,308,000
As percentage of all births	11	18	28	33

SOURCES: U.S. Census Bureau, *Statistical Abstract of the United States*, Washington, DC: U.S. Government Printing Office, 2000, Table 77; Jason Fields, *Children's Living Arrangements and Characteristics: March 2002; Births, Marriages, Divorces and Deaths: Provisional Data for 2001*, National Vital Statistics Reports, Vol. 50, No. 14; Jason Fields and Lynn M. Casper, *America's Families and Living Arrangements: March 2000.*

family also includes other kin, especially in-laws, grandparents, aunts and uncles, and cousins. To distinguish it from the the nuclear family, this wider family is known as the extended family.

Until recently, American researchers have focused on the traditional nuclear family, composed of a married couple and their biological children. The emphasis on the nuclear family as the American family has ignored a simple reality: The traditional nuclear family is no longer the dominant family form. In fact, today, only half of the contemporary American families fit this model. Instead, we have single-parent families, stepfamilies, cohabitating families, families without children, gay- and lesbian-headed families, and so on. The majority of Americans are members of one of these other family forms, not the traditional nuclear family.

The changes that have occurred over the past four decades have been considerable. In Chapter 3 we will look more closely at the extent of change in family experiences. Table 1.1 provides a sampling of many of the more significant characteristics of families and how they have changed between 1970 and 2000.

Hopefully, as you begin studying marriage and the family, you will see that such study is both abstract and personal. It is abstract insofar as you learn about the general structure, processes, and meanings associated with marriage and the family. But the study of marriage and the family is also personal. In the chapters that follow, you should learn things that help you to better understand your own families. In some ways, it is *your* present, *your* past, and *your* future that you are studying. It is the family from which you came, the family in which you are now living, and the family that you will create. What R. D. Laing (1972) wrote some years ago may ring true today: "The first family to interest me was my own. I still know less about it than I know about many other families. This is typical. Children are the last to be told what was really going on."

As you continue your study of marriage and the family, much of what was unknown in your own family may become known. You may discover new understanding, strength, complexity, and love. You will also be better able to appreciate the forces that shaped your experiences and how those experiences compare to those of others.

SUMMARY

- Marriage is a legally recognized union between a man and a woman in which they are united sexually; cooperate economically; and may give birth to, adopt, or rear children. The union is assumed to be permanent (although in reality it may be dissolved by separation or divorce). Marriage differs among cultures and has changed historically in our own society. Who may marry whom and at what age varies considerably from one society to another. In Western cultures, the preferred form of marriage is *monogamy*, in which there are only two spouses, the husband and wife. *Polygyny*, the practice of having more than one wife, is commonplace throughout many cultures in the world.
- Legal marriage provides a number of rights and protections to spouses that couples who live together lack.
- The current legal definitions of marriage are in the midst of change in both the United States and many other countries. The greatest change consists of the question of same-sex marriage.
- Defining the term *family* is complex because of the variety of family forms in contemporary America. Most definitions of family include individuals who are related by descent, marriage, remarriage, or adoption; some also include affiliative kin. *Family* may be defined as one or more adults related by blood, marriage, or affiliation who cooperate economically, who may share a common dwelling, and who may rear children. There are also ethnic differences as to what constitutes family. Among Latinos, for example, *compadres* (godparents) are considered family members. Among some Japanese Americans, the *ie* is the traditional family. Among many Native-American tribes, the clan is regarded as the fundamental family unit.
- Four important family functions are (1) the provision of intimacy, (2) the formation of a

cooperative economic unit, (3) reproduction and socialization, and (4) the assignment of social roles and status, which are acquired both in our *families of orientation* (in which we grow up) and our *families of cohabitation* (which we form by marrying or living together).

- Advantages to living in families include (1) continuity of emotional attachments, (2) close proximity, (3) familiarity with family members, and (4) economic benefits.
- The *extended family* consists of grandparents, aunts, uncles, cousins, and in-laws. It may be formed *conjugally* (through marriage), creating in-laws or stepkin, or *consanguineously* (by birth), through blood relationships. Extended families are especially important for African-American, Latino, and Asian-American families.
- The *kinship system* is the social organization of the family. In the *nuclear family*, it generally consists of parents and children, but it may also include members of the extended family, especially grandparents, aunts, uncles, and cousins. Kin can be *affiliated*, as when a nonrelated person is considered "as kin," or a relative may fulfill a different kin role, such as a grandmother's taking the role of a child's mother.
- Unmarried *cohabitation* is a relationship that occurs when a couple lives together and is sexually involved.
- *Divorce* is the legal dissolution of marriage following separation. High divorce rates have made single-parent families and stepfamilies important family forms in the contemporary United States. Divorce, remarriage, single-parent families, stepfamilies, and extended families are normal aspects of the contemporary American marriage system.

KEY TERMS

affiliated kin 5
centrists 21
clan 6
cohabitation 21
conjugal relationship 19
consanguineous relationship 19
conservatives 20
extended family 18
family 5
family of cohabitation 16
family of orientation 16
family of procreation 16
household 5
kinship system 19
liberals 21
marriage 7
modified polygamy 12
monogamy 12
nuclear family 6
polyandry 12
polygamy 12
pologyny 12
serial monogamy 12
socialization 15
spirit marriage 19
traditional family 6

SUGGESTED READINGS

Coontz, Stephanie. *The Way We Never Were: American Families and the Nostalgia Trap.* New York: Basic Books, 1992. A historian's view that contrasts the pop images conveyed through the media with the realities of American families since the 1950s.

Huston, Perdita. *Families as We Are: Conversations from Around the World.* New York: The Feminist Press at the City University of New York, 2001. Using the words of "ordinary people" in 11 countries, Huston depicts the forces that are affecting and changing families.

Marciano, Teresa, and Marvin Sussman. *Wider Families: New Traditional Family Forms.* New York: Haworth Press, 1991. A collection of scholarly essays that challenges the traditional concept of family. It includes essays on close relationships, gay and lesbian relationships, communes, and extended family networks.

Mintz, Steven, and S. Kellogg. *Domestic Revolutions: A Social History of American Family Life.* New York: Free Press, 1998.

Skolnik, Arlene. *Embattled Paradise: The American Family in an Age of Uncertainty.* New York: Basic Books, 1992. A thoughtful discussion of contemporary family diversity, pluralistic values, and the political debate about the family.

RESOURCES ON THE INTERNET

Companion Web Site for This Book

http://sociology.wadsworth.com/strong/marriage9e
Gain an even better grasp on this chapter by going to the companion Web site to take one of the Tutorial Quizzes, use the Flash Cards to master key terms, or check out the many other study aids you'll find there. Visit the Marriage and Family Resource Center on the site. You'll also find special features such as GSS Data and Census 2000 information that will put data and resources at your fingertips to help you with that special project or to do some research on your own.

InfoTrac College Edition: Search Word Summary

marriage	intimacy
family	socialization
kinship	cohabitation

To learn more about these central topics in the study of the family, you can conduct an electronic search using InfoTrac College Edition. To aid in your search and to gain useful tips, see the Student Guide to InfoTrac College Edition that you can access through the companion Web site for this book.

Preview

To gain a sense of what you already know about the material covered in this chapter, answer "True" or "False" to the statements below.

1. To answer questions about families we need to rely most on our "common sense." True or false?
2. The statement "Everyone should get married" is an example of an objective statement. True or false?
3. Many researchers believe that both love and conflict are normal features of families. True or false?
4. Stereotypes about families, ethnic groups, and gays and lesbians are easy to change. True or false?
5. We tend to exaggerate how much other people's families are like our own. True or false?
6. Family researchers formulate generalizations derived from carefully collected data. True or false?
7. Every method of collecting data on families is, in some ways, limited. True or false?
8. A belief that one's own ethnic group, nation, or culture is innately superior to another is an example of an ethnocentric fallacy. True or false?
9. According to some scholars, in marital relationships we tend to weigh the costs against the benefits of the relationship. True or false?
10. It is impossible to observe family behavior. True or false?

"Discovery consists of seeing what everybody has seen and thinking what nobody has thought."

ALBERT SZENT-GYORGYI

CHAPTER 2

Studying Marriage and the Family

Outline

The subjects covered in this book come up often and unexpectedly in everyday experience. The following situation, although hypothetical, is not an uncommon or unrealistic one. Imagine yourself sitting around having coffee with a couple of close friends. One confides that she is really worried about her relationship with her boyfriend of two years, now that they are separated by close to 600 miles while at different colleges. You feel for your friend, sensing the seriousness of her anxiety and the depth of her fears. You think hard about her predicament, smile, and offer the following: "I think you're worrying too much. After all, '*absence makes the heart grow fonder*.' Everyone knows that. Your relationship will probably get stronger and deeper through this separation. Try to relax." At that, a second, more cynical friend, chimes in: "I hate to be the one to have to tell you this, but you know what they say, '*Out of sight, out of mind*.' It's probably just a matter of time before this relationship is history. In fact, if I were you, I'd start looking for someone new. Now." Obviously both reactions can't be true. Moreover, how can "everyone know" one thing while "they" say the opposite? Surely, there must be a way to resolve such a contradiction.

In this chapter we'll examine how family researchers attempt to explore issues such as the one posed above. In that sense the chapter differs from all of the others. Instead of presenting material about different aspects of the marriage and family experience, it explains and illustrates *how we come to know* the information about relationships and families found in the rest of the book. It will enable you to understand and appreciate how much our knowledge and understanding of families can be enriched by the theories and research procedures we introduce. In learning how information is obtained and interpreted, we set the stage for the in-depth exploration of family issues in the chapters that follow.

PREVIEW ANSWERS

1 False
2 False
3 True
4 False
5 True
6 True
7 True
8 True
9 True
10 False

How Do We Know?

As sociologist Earl Babbie suggests, social research is one way we can come to know about things (Babbie, 2002). However, most of what we "know" about the social world we have "learned" elsewhere through other less systematic means (Babbie, 2002; Neuman. 2000). In the previous chapter we noted the dangers inherent in generalizing from our personal experiences. We all do this. In fact, if you or someone you know had an unfavorable experience with a long-distance relationship, you probably favor the "out of sight, out of mind" response more than the optimistic one credited to you.

The opening scenario illustrates the difficulty involved in relying on what are often called "common sense" based explanations or predictions (Neuman, 2000). Our common sense understanding of family life may be derived from "tradition" (what everyone knows because it has always been that way or been thought to be that way), from "authority figures" whose expertise we trust and whose knowledge we accept, or from various media sources.

The mass media are so pervasive that they become invisible, almost like the air we breathe. Yet they affect us. Popular culture, in all its forms, is a key source of both information and misinforma-

Dr. Phil McGraw is one of a number of television and radio talk show hosts who focuses on family issues.

TV sitcoms, such as the popular "Everybody Loves Raymond," influence our beliefs and attitudes about marriage and family. What messages and expectations do these programs convey?

tion about families. Television, pop music, the Internet, magazines, newspapers, and movies, together, help shape our attitudes and beliefs about the world in which we live. On average, each of us spends more than 3,400 hours a year using one of these media (U.S. Census Bureau, 2001, Table 1125). Popular culture conveys images, ideas, beliefs, values, myths, and stereotypes about every aspect of life and society, including the family.

DID YOU KNOW?

As of 2002, 98 percent of U.S. households had television sets. Preschool-aged children watched 24 hours of television a week; teens watched between 21 and 22 hours a week. The group with the greatest number of hours of viewing per week was those 55 and older. Men of that age group averaged 39 hours and 39 minutes a week, while females averaged 44 hours and 11 minutes (*Time Almanac*, 2003).

Television has a particularly powerful effect on our values and beliefs (see the Perspective box on families in the media). Because so much of the day-to-day stuff of family life (e.g., arguing, dividing chores, engaging in sexual behavior) takes place in private, behind closed doors, we do not have access to what really goes on. But we are privy to those behaviors on television and in movies and magazines. Thus, those depictions can influence what we *assume* happens in real families. If you have seen a movie or television show or read magazine articles in which couples in long-distance relationships thrived despite distance, those sources will likely influence you.

Cumulatively, these forms of common-sense knowledge (experience, tradition, authority, and media) are typically poor sources of accurate and reliable knowledge about social and family life. Often, what we consider and accept as common sense, is fraught with the kinds of contradictions depicted above (or, for example, "birds of a feather flock together," versus "opposites attract"). Even in the absence of contradiction, many common-sense beliefs are simply untrue. Thus, if we "really want to know" about how families work or what people in different kinds of family situations or relationships experience, we would be better informed by seeking and acquiring more trustworthy information.

Perspective

FAMILIES IN THE MEDIA

Popular culture, in all its forms is a key source of information and misinformation about families. Often, critics point to the pervasive influence of television and its distortions of reality, familial and otherwise. Television situation comedies unrealistically depict married life, understate the unique issues faced by various ethnic families, inaccurately depict single-parent family life, inaccurately portray the relative sexual activity levels of marrieds and singles, and portray conflict as something easily resolved within 20 minutes, especially with humor.

The combined portrayal of family life that results from soap opera families and daytime talk shows is both unrealistic and highly negative. Those who have scrutinized daytime soap operas note the extremely high rates of conflict, betrayal, infidelity, and divorce that afflict soap opera families (Pingree and Thompson, 1990; Benokraitis and Feagin, 1995). Characters go through multiple marriages, often carrying "deep, dark secrets" that they keep from their spouses. Soaps often stereotype women as starry-eyed romantics or scheming manipulators of men. Particularly unrealistic is the way soap operas portray sex, leading viewers to envision exaggerated estimates of how much sex does and should occur within relationships (Lindsey, 1997). Daytime talk shows, from Jerry Springer and Jenny Jones to Montel and Maury contribute to the idea that American family life is deeply dysfunctional, that parents are anything from "irresponsible fools" to "in-your-face monsters" (Hewlett and West, 1998), that spouses and partners routinely cheat on each other and often strike each other, and that teenagers are recklessly out of control.

As you will see throughout this book, although families are not without their share of even serious problems, daily family life is as poorly represented by daytime television as by prime time programming.

There is one form of media that deserves special attention here—what we call the advice/information genre. This form transmits information and conveys values and norms—cultural rules and standards—about marriage and family, often disguised as information and intended as entertainment. A veritable industry exists to support the advice/information genre. It produces self-help and child-rearing books, advice columns, radio and television shows, and numerous articles in magazines and newspapers.

In newspapers this genre has been represented by such popular newspaper "advice columnists" as Abigail Van Buren (real name Pauline Esther Friedman, whose column "Dear Abby" is now written by her daughter), Dan Savage (whose sex-advice column "Savage Love" is syndicated in 70 newspapers), Peg Winship (daughter of column creator, Beth Winship of "Ask Beth"), and the late Ann Landers (Abby's twin sister, Esther Pauline Friedman). Newer, Web-based columnists such as Alison Blackman Dunham and her late twin sister Jessica Blackman Freedman, the self-proclaimed "Advice Sisters," (or "Ann and Abby for the new millennium") carried this genre to the Internet. Self-help and pop psychology books written by "experts" frequent the best-seller lists, ranging from *The Power Principle* to *The Pleasure Prescription* to *The Rules.*

Radio therapists, such as Dr. Joy Browne and Dr. Laura Schlessinger, are sought out daily by callers seeking advice or information about relationships, family crises, and so on. Browne is a psychologist whose tenure on radio extends over 20 years, making her radio's longest running psychologist. She is the author of seven books, including *The Nine Fantasies that Will Ruin Your Life* and *Dating for Dummies.* Dr. Browne's radio show reaches an estimated 9 million listeners on 300 stations nationwide, and her weekly advice column is syndicated by the *New York Times.*

Dr. Laura Schlessinger is also a best-selling author. Her books include, *Ten Stupid Things Women Do to Mess Up Their Lives,* and *How Could You Do That: The Abdication of Character, Courage, and Conscience.* She publishes a monthly magazine, *The Dr. Laura Perspective,* and reaches an estimated 18 million listeners on radio stations across North America. She became controversial for her opposition to homosexuality and gay rights and was the target of boycotts that helped cancel her short-lived television program. She is also an articulate opponent of pre- and extra-marital sex, cohabitation, and divorce. However, she is also an outspoken advocate on behalf of adult self-sacrifice for their children. Her radio tag line, "I am my kid's mom," reflects her staunch advocacy for active, hands-on, even at-home, parenting. The fact that so many of her callers begin their conversations by saying, "Hi, Dr. Laura, I am my kid's mom" or "I am my kid's dad" suggests that she has struck a cultural nerve, especially with those who believe in the importance of intensive parenting (Hays, 1997).

On television, Dr. Philip McGraw's *Dr. Phil* has become a ratings success. McGraw, a psychologist of some 25 years, was featured frequently on *The Oprah Winfrey Show* before getting his own talk show in 2002. His shows cover a range of personal and family issues. In a recent two-week period, for example, episodes included, "Parenting 101: Biggest Nightmares," "Mom vs. Mom," "Wedding Nightmares," "A Family Divided," and "Are They Still Fighting?" . In addition to his highly successful television program, he is the author of a number of books, including *Self Matters: Creating Your Life from the Inside Out; Life Strategies: Doing What Works, Doing What Matters; Relationship Rescue: Seven Steps for Reconnecting With Your Partner;* and, his most recent, *The Ultimate Weight Solution.* He also has a Web site from which visitors can obtain a variety of suggestions for how to deal with the kinds of relationship and personal issues featured on his show.

Evaluating the Advice/Information Genre

The various radio or television talk shows, columns, articles, and advice books have several things in common. First, their primary purpose is to sell books or periodicals or to raise program ratings. This is in marked contrast to scholarly research, where the primary purpose is the pursuit of knowledge. Even the inclusion in magazines of survey questionnaires asking readers about their relationships or behaviors is ultimately designed to promote sales. We fill out the questionnaires for fun, much as we would crossword puzzles. Then we buy the subsequent issue or watch a later program to see how we compare with others.

Second, the media must entertain while disseminating information about marriage, family, and relationships. Because the genre seeks to entertain, the information and advice must be simplified. Complex explanations and analyses must be avoided because they would interfere with the entertainment purpose. Furthermore, the genre relies on high-interest or shocking material to attract readers or viewers. Consequently, we are more likely to read or view stories about finding the perfect mate or protecting our children from strangers than stories about new research methods or the process of gender stereotyping.

Third, the advice/information genre focuses on how-to-do-it information or morality. The how-to-do-it material advises us on how to improve our relationships, sex lives, child-rearing abilities, and so on. Advice and normative judgments (evaluations based on norms) are often mixed together. Advice columnists often give advice on issues of sexual morality: "Is it all right to have sex without commitment?" ("Yes, if you love him/her," "No, casual sex is empty," and so on.) Advice columnists act as moral arbiters, much as do ministers, priests, rabbis, and other religious leaders.

Fourth, the genre uses the trappings of social science without its substance. Writers and columnists interview social scientists and therapists to give an aura of scientific authority to their material. They rely especially heavily on therapists with clinical rather than academic backgrounds. Because clinicians tend to deal with people with problems, they often see relationships as problematical.

To reinforce their authority, the media also incorporate statistics, which are key features of social science research. But Susan Faludi (1991) offers this word of caution:

> The statistics that the popular culture chooses to promote most heavily are the very statistics we should view with the most caution. They may well be in wide circulation not because they are true but because they support widely held media preconceptions.

With the media awash in advice and information about relationships, marriage, and family, how can we evaluate what is presented to us? Here are some guidelines:

- *Be skeptical.* Remember: much of what you read or see is meant to entertain you. Are the sources scholarly or popular? Do they rely on self-described "experts" or "victims"? How representative are the people interviewed? If the story seems superficial, it probably is.
- *Search for biases, stereotypes, and lack of objectivity.* Information is often distorted by points of view. What conflicting information may have been omitted? How are women and members of ethnic groups portrayed? Does the media's idea of family include diverse family forms? Are nontraditional families stigmatized?
- *Look for moralizing.* Many times what passes as fact is really disguised moral judgment. What are the underlying values of the article or program?
- *Go to the original source or sources.* The media simplifies. Find out for yourself what the studies really said. How valid were their methodologies? What were their strengths and limitations?
- *Seek additional information.* The whole story is probably not told. In looking for additional information, consider information in scholarly books and journals, reference books, or college textbooks.

Throughout this book you will be exposed to a variety of information or data about families. This information may or may not reflect your experiences, but its value is this: It will enable you to learn about how other people experience family life. This knowledge of what other families experience and the results of different kinds of responses to family situations enables us to have a more informed understanding of families in general and of yourself as an individual. Finally, such information is important and necessary for a variety of professionals and practitioners, especially those who provide social services, medical care, or legal assistance, as they deal with family-related issues.

Thinking Critically about Marriage and the Family

Before we examine the specific theories and research techniques used by family researchers, it is important to emphasize that the attitudes of the researcher (or you, as you read research) are very important. In order to obtain valid research information we need to keep in mind the rules of critical thinking. The term *critical thinking* is another way of saying "clear and unbiased thinking."

We all have our own perspectives, values, and beliefs regarding marriage, family, and relationships. These can create blinders that keep us from accurately understanding the research information. We need instead to develop a sense of **objectivity** in our approach to information— to suspend the beliefs, biases, or prejudices we have about a subject until we really understand what is being said (Kitson et al., 1996). We can then take that information and relate it to the information and attitudes we already have. Out of this process a new and enlarged perspective may emerge.

One area in which we may need to be alert to maintaining an objective approach is that of family lifestyle. The values we have about what makes a successful family can cause us to decide ahead of time that certain family lifestyles are "abnormal" because they differ from our own experience or preference. We may refer to single-parent families as "broken" or say that adoptive parents are "not the real parents."

A clue that can sometimes help us "hear" ourselves and detect whether we are making value judgments or objective statements is as follows: A **value judgment** usually includes words that mean "should" and imply that our way is the correct way. An example is, "Everyone should get married." An **objective statement** presents information based on scientifically measured findings—for example, concluding that "about 90 percent of Americans marry."

Opinions, biases, and stereotypes are ways of thinking that lack objectivity. **Opinions** are based on our own experiences or ways of thinking. **Biases** are strong opinions that may create barriers to hearing anything that is contrary to our opinion. **Stereotypes** are sets of simplistic, rigidly held, and overgeneralized beliefs about the personal characteristics of a group of people. They form the "glasses" with which we "see" people and groups. Stereotypes are fairly resistant to change. Furthermore, stereotypes are often negative. Common stereotypes related to marriages and families include the following:

- Nuclear families are best.
- Stepfamilies are unhappy.
- Lesbians and gay men cannot be good parents.
- Latino families are poor.
- Women are instinctively nurturing.
- People who divorce are selfish.

We all have opinions and biases; most of us, to varying degrees, think stereotypically. But the commitment to objectivity requires us to become aware of these opinions, biases, and stereotypes and to put them aside in the pursuit of knowledge.

Fallacies are errors in reasoning. These mistakes come as the result of errors in our basic presuppositions. The *gambler's fallacy*, for example, is based upon the belief that following a stretch of bad luck at cards or dice, the next hand or roll has to be better. Or, having been "hot" one ought to quit because one's luck has or will soon "run out." However, every roll of two dice or hand of cards dealt is independent of whatever came before. Statistically, there is no truth to the gambler's fallacy.

Two common types of fallacies that especially affect our understanding of families are egocentric fallacies and ethnocentric fallacies. The **egocentric fallacy** is the mistaken belief that everyone has the same experiences and values that we have and therefore should think as we do. The **ethnocentric fallacy** is the belief that one's own ethnic group, nation, or culture is innately superior to others. In the next chapter, when we consider the differences and strengths of families from different ethnic and economic backgrounds, we will need to keep both of these fallacies from distorting our understanding.

As we have mentioned, from the day of your birth you have been forming impressions about human relationships and developing ways of behaving based on these impressions. Hence, you might feel a sense of "been there, done that" as you read about an aspect of personal development or family life. Initially, you may find it difficult to become interested in obtaining further background and insights. However, we would like to challenge you to remember that you are in a continuing research mode throughout life. You are constantly forming new insights and remodeling and reconfiguring your thinking and behaviors. Your study of the information in this book will provide you the opportunity to reconsider your present attitudes and past experiences and relate them to the experiences of others. As you do this you will be able to use the logic and problem-solving skills of critical thinking so that you can effectively apply that which is relevant to your life.

© Laurie DeVault Photography

We learn about families in many ways, including our own experience, the media, and the efforts of family researchers. The most common research method is the survey—a form that involves questionnaires or interviews.

Theories and Research Methods

Family researchers come from a variety of academic disciplines—from sociology, psychology, and social work to communication and family studies (sometimes known as "family and consumer sciences"). Although these disciplines may differ in terms of the specific questions they ask or the objectives of their research, they are unified in their pursuit of accurate and reliable information about families through the use of social scientific theories and research techniques. Scholarly research about the family brings together information and formulates generalizations about certain areas of experience. These generalizations help us to predict what happens when certain conditions or actions occur.

Family science researchers use the **scientific method**—well-established procedures used to collect information about family experiences. With scientifically accepted techniques, they analyze this information in a way that allows other people to know the source of the information and to be confident of the accuracy of the findings. Much of the research family scientists do is shared in specialized journals (for example, *Journal of Marriage and the Family, Journal of Family Issues, Family Relations, Journal of Sex Research,* and *Family and Consumer Sciences Research Journal*) or in book form. By communicating their results through such channels, other researchers can build on, refine, or further test research findings. Much of the information contained in this book originally appeared in scholarly journals.

Theories of Marriage and Families

One of the most important differences between the knowledge about marriage and family derived from family research and that which we acquire elsewhere is that family research is influenced or guided by **theories**—sets of general principles or concepts used to explain a phenomenon and to make predictions that may be tested and verified experimentally. While researchers collect and use a variety of kinds of data on marriages and families, these data alone do not automatically convey the meaning or importance of the information gathered. Concepts and theories supply the "story line" for the information we collect.

Concepts are abstract ideas that we use to represent the reality in which we are interested. We use concepts to focus our research and organize our data. Many examples of concepts—for example,—*nuclear families, monogamy, socialization*—were introduced in the previous chapter Family research involves the processes of **conceptualization,** the specification and definition of concepts used by the researcher, and **operationalization,** the identification and/or development of research strategies to observe or measure our concepts. For example, to study the relationship between social class and child-rearing strategies, we need to define and specify how we are going to identify and measure a person's social class position and child-rearing strategies.

In deductive research, concepts are turned into **variables,** concepts that can vary in some meaningful way. *Marital status* is an example of a variable used by family researchers. One may be married, divorced, widowed, or never married. As researchers explore the causes and/or consequences of one's marital status, they may formulate **hypotheses,** or predictions, about the relationships between marital status and other variables. One might hypothesize that one's race or social class influences whether one is married or not. In such an example, race is an **independent variable** and marital status the **dependent variable,** in that race is thought to influence the likelihood of getting or staying married. Marital status, on the other hand, may be a causal or independent variable in a hypothesized relationship between being married and life expectancy. Finally, marital status might be hypothesized as an **intervening variable,** affected by the independent variable, race, and in turn affecting the dependent variable, life expectancy. In that instance, the hypothesis suggests that race differences in marital status account for race difference in life expectancy (see Figure 2.1). Rarely do researchers construct theories with only two or three variables. In fact, they may hypothesize multiple inependent and intervening variables and seek to identify those having the greatest effect on the dependent variable (Neuman, 2000).

Inductive research is not hypothesis testing research. Instead, it begins with a topical interest and perhaps some vague concepts. As researchers gather their data, they refine their concepts, seek to identify recurring patterns out of which they can make generalizations, and, perhaps, end by building a theory (or asserting some hypotheses) based on the data collected. Theory that emerges in this inductive fashion is often referred to as **grounded theory,** in that it is *grounded* or "rooted in observations of specific, concrete details" (Neuman, 2000).

Theoretical Perspectives on Families

On a more abstract level of theory, we can identify major theoretical frameworks that guide much of the research about families. These frameworks (sometimes called **paradigms**) are sets of concepts and assumptions about how families work and how they fit into society. Theoretical frameworks guide the kinds of questions we raise, predictions we make, and where we look to find answers or construct explanations (Babbie, 1992).

We discuss several theories in this section: (1) ecological, (2) symbolic interaction, (3) social exchange, (4) developmental, (5) structural functional, (6) conflict, and (7) family systems. We also look at the influence of the feminist perspective on family studies. These theories are currently among the most influential ones used by sociologists and psychologists who study families. As you examine

FIGURE 2.1 ■ Marital Status as a Dependent, Independent, and Intervening Variable

Race ⟶ Marital status	Marital status as a dependent variable affected by race
Marital status ⟶ Life expectancy	Marital status as an independent variable, affecting life expectancy
Race ⟶ Marital status ⟶ Life expectancy	Marital status as an intervening variable, affected by race and, in turn, affecting life expectancy

them, notice how the choice of a theoretical perspective influences the way data are interpreted. Furthermore, as you read this book, ask yourself how different theoretical perspectives would lead to different conclusions about the same material. (For deeper exploration of family theories, see White and Klein, 2002; Winton, 1995; or Sussman, Steinmetz, and Peterson, 1999).

FAMILY ECOLOGY THEORY

The emphasis of **family ecology theory** is on how families are influenced by and in turn influence the wider environment. The theory developed was introduced in the late nineteenth century by plant and human ecologists. The German biologist Ernst Haeckel first used the term *ecology* (from the German word *oekologie*, which translates to "place of residence") and placed conceptual emphasis on **environmental influences.** This focus was soon picked up by Ellen Swallows Richards, the founder and first president of the American Home Economics Association (now known as the American Association of Family and Consumer Sciences). An MIT trained chemist, Richards believed that scientists needed to focus upon home and family, "for upon the welfare of the home depends the welfare of the commonwealth" (quoted in White and Klein, 2002).

The core concepts in ecological theory include **environment** and **adaptation.** Initially used to refer to the adaptation of plant and animal species to their physical environments, these concepts were later extended to humans and their physical, social, cultural, and economic environments (White and Klein, 2002). As applied to family issues, a key question is posed by the family ecology perspective: How is family life affected by the environments in which families live?

We use the plural *environments* to reflect the multiple environments that families encounter. In Urie Brofenbrenner's ecologically based theory of human development, the environment to which individuals adapt as they develop consists of four levels: (1) microsystem, (2) mesosystem, (3) exosystem, and (4) macrosystem. Cumulatively, these levels make up the environments in which we live. The **microsystem** contains the most immediate influences with whom individuals have frequent contact. For example, in adolescence our microsystem could include our families, peers, schools, and neighborhoods. In each of these, one has roles and relationships that exert influence over how one develops. The **mesosystem** consists of the interconnections between microsystems—for example, the ways school experiences and home experiences influence each other. The **exosystem** consists of settings in which the individual does not actively participate but which nonetheless affect his or her development. Parental work experiences—everything from their salaries to their schedules to their continued employment—will influence adolescent development. Finally, the **macrosystem** operates at the broadest level, encompassing the laws, customs, attitudes, and belief systems of the wider society, all of which influence individual development and experience (Rice and Dolgin, 2002).

Similarly, in constructing an ecological framework to better understand marriage relationships, Huston illustrated how marital and intimate unions are "embedded in a social context" (Huston, 2000). The social context includes the *macroenvironment*—the wider society, culture, and physical environment in which a couple lives—and their particular *ecological niche*—the behavior settings in which they function on a daily basis (e.g., a poor, urban neighborhood as opposed to a small town or suburb). Also included in the social context is the marriage relationship itself, especially as it is affected by a larger network of relationships. The final key element in Huston's ecological approach contains the physical, psychological, and social attributes of each spouse, including their attitudes and beliefs about their relationship and each other. As illustrated in Figure 2.2, each of these environments influences and is influenced by the others (Huston, 2000). One cannot fully understand marriage without exploring the interconnections between these three elements.

In a study of work-family stresses and problem drinking, Grzywacz and Marks (2000) applied an ecological approach. Emphasizing the "interaction between the individual and persons, objects, and symbols of prominent life domains (e.g., family and work)," problem drinking is seen as a consequence of "negative person-environment interactions," including, especially, high levels of work or family stress or issues arising from the mesosytem of work and family. Ecological factors, then, operate "above and beyond" individual factors in accounting for problem drinking. Negative "spillover" from work

FIGURE 2.2 ■ A Three-Level Model for Viewing Marriage

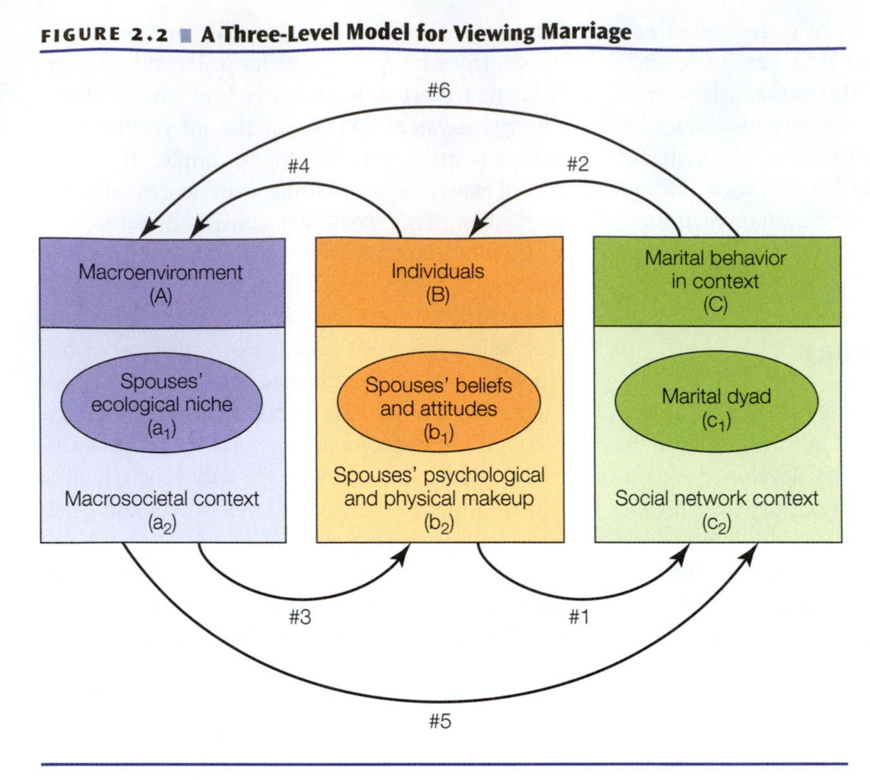

The various contexts and environments in which families live influence each other.

SOURCE: Huston, 2000.

to home included such things as job-induced irritability at home, job-induced fatigue inhibiting home involvement, and job worries that lead to distraction at home. All of these were factors that elevated the likelihood of problem drinking. Furthermore, positive person-environment interactions, such as positive work-family spillover, were associated with reduced likelihood of problem drinking (Grzywacz and Marks, 2000).

As each of these examples illustrates, the key concern of an ecological approach is how family experience is affected by the broader social environment. In many ways, much of what we examine in subsequent chapters has at least this level of ecological focus. We cannot understand what happens within families without considering the wider cultural, social, and economic environments within which family life takes place.

There have been a variety of criticisms of ecological theory (White and Klein, 2002). Here we note two. It is often unclear which level of analysis is most appropriate—individual, group, or population—to account for the behavior we attempt to explain. Additionally, there is a lack of specificity as to the process through which families are affected and what specifically is responsible for the outcomes we seek to explain. Also, some criticize the fact that the perspective seems to apply more easily to development and growth rather than decline or degeneration. Yet families are prone to decline and degeneration as much as they are to development and growth.

SYMBOLIC INTERACTION

Symbolic interaction is a theory that looks at how people interact with each other. Symbolic interactionists, like the rest of us, are concerned with relationships. When we feel that our partner really does (or does not) understand us, that we communicate well (or not well), that we live (or do not live) in harmony with each other , that our relationship can (or cannot) withstand the difficulties created by long-distance , we are expressing feelings that are at the heart of symbolic interaction research. Symbolic interactionists study the interactions that make up a relationship.

An **interaction** is a reciprocal act, the everyday words and actions that take place between people. For an interaction to occur, there must be at least two people who both act and respond to each other. When you ask your sister to pass the potatoes and she does it, an interaction takes place. Even if she intentionally ignores you or tells you to "Get the damn potatoes yourself," an interaction occurs (even if it is not a positive one). Such interactions are conducted through symbols, words, or gestures that stand for something else.

When we interact with people, we do more than simply react to them. We interpret or define their symbols. If your sister did not respond to your request for the potatoes, what did her nonresponse mean or symbolize? Hostility? Rudeness? Indifference? A hearing problem? We interpret the meaning and act accordingly. If we interpret the nonresponse as not hearing, we may repeat the request. If we believe it symbolizes hostility, rudeness, or indifference, we may become angry.

Family as the Unity of Interacting Personalities

In the 1920s, Ernest Burgess (1926) defined the family as a "unity of interacting personalities." This definition has been central to symbolic interaction theory and in the development of marriage and family studies. Marriages and families consist of individuals who interact with one another over a period of time. Such interactions and relationships define the nature of a family: a loving family, a dysfunctional family, a conflict-ridden family, an emotionally distant family, a high-achieving family, and so on.

In marital and family relationships, our interactions are partly structured by **social roles**—established patterns of behavior that exist independently of a person, such as the role of wife or husband existing independently of any particular husband or wife.) Each member in a marriage or family has one or more roles—such as husband, wife, mother, father, child, or sibling. These social roles help give us cues as to how we are supposed to act. They help create a "marriage," "family," or other intimate relationship. When we marry, for example, these roles help us "become" wives and husbands; when we have children, they help us "become" mothers and fathers.

Symbolic interactionists study how the sense of self is maintained in the process of acquiring these roles. We are, after all, more than simply the roles we fulfill. There is a core self that is independent of our being a husband or wife, father or mother, son or daughter. Symbolic interactionists ask how we fulfill our roles and continue to be ourselves and, at the same time, how our roles contribute to our sense of self. Our identities as human beings emerge from the interplay between our unique selves and our social roles.

Only in the most rudimentary sense are families created by society. According to symbolic interactionists, families are "created" by their members. Each family has its own unique personality and dynamics created by its members' interactions. To classify families by structure, such as nuclear family, stepfamily, and single-parent family, misses the point of families. Structures are significant only insofar as they affect family dynamics. It is what goes on inside families that is important.

By emphasizing the construction and communication of shared meanings, symbolic interactionists remind us that family behavior, like other social interaction, requires interpretation and sense making. This is illustrated nicely in the widely acclaimed book, *The Second Shift*, by sociologist Arlie Hochschild. Hochschild interviewed 50 dual-earner couples to see how they divided up housework and child care. She noted that only 20 percent of her sample couples shared housework responsibilities equally. In 70 percent of her sample couples, men did between one-third and one-half of the housework, and in the remaining 10 percent of sample households, men did less than one-third of the household tasks.

It is from such data that Hochschild derived the notion of a "leisure gap" that separates men's from women's experiences (with men enjoying more leisure time than women). But Hochschild went further and deeper. She examined what happened in households where what couples did (their actual behavior) conflicted with what each partner believed they should do (their "gender ideologies"). She described the strategic use of **family myths,** views of reality that couples construct and apply to account for why their domestic arrangement is other than they expected (Hochschild, 1989).

The clearest example of the workings of such myths can be found among a couple Hochschild calls Evan and Nancy Holt. After repeated but unsuccessful efforts on Nancy's part to get husband Evan to share more, Nancy considered the possibility

The family is a "unity of interacting personalities," according to symbolic interaction theory. How family members interact with each other is partly determined by their social roles as husband/wife, mother/father, son/daughter, and sibling. Parents and older siblings often fulfill the role of teacher for younger family members.

of a divorce. Unwilling to end her marriage "over a dirty frying pan," she and Evan arrived at a "solution," which Hochschild calls the "upstairs/downstairs" myth. Under this version of domestic reality, Nancy notes that while she does the "upstairs," Evan has taken responsibility for, and freed her from, the "downstairs." Hochschild points out that although portrayed as "sharing," this solution leaves much unequal. The "upstairs" included the living room, dining room, kitchen, two bedrooms, and two bathrooms; whereas the "downstairs" amounted to the garage, which included responsibility for the car and the dog. Nevertheless, by constructing and believing in the idea that they "share," Nancy was able to live with their arrangement. Thus, the meanings she attached to their arrangement ("I do the upstairs; he does the downstairs") became more important than their actual division of responsibilities.

Family myths were used in the opposite direction as well. In other words, couples who believed that traditional divisions of labor (male breadwinner, female homemaker) were better but who could not financially afford such an arrangement, often explained away their failure to achieve them. Such was the case with Frank and Carmen Delacorte. Although Carmen earned one-third of the household income by providing child care in her home, Frank considered her an "at-home wife." Because Carmen could not do all of the housework herself, she used "incompetence" as a strategy to get Frank's help. This way, both were able to believe that they were what they believed they should be, a traditional couple.

Critique

Although symbolic interaction theory focuses on the daily workings of the family, it suffers from several drawbacks. First, the theory tends to minimize the role of power in relationships. If a conflict exists, it may reveal more than differences in meaning and it may take more than simply communicating to resolve it. If one partner strongly wants to pursue a career in Los Angeles and the other just as strongly wants to pursue a career in Boston, no amount of communication and role adjustment may be sufficient to resolve the conflict. Ultimately, the partner with the greater power in the relationship may prevail.

Second, symbolic interaction does not fully account for the psychological aspects of human life, especially personality and temperament. It sees us only as the sum of our roles, thus neglecting the self that exists independently of our roles and limiting our uniqueness as human beings.

Third, the theory emphasizes individualism. It encourages competence in interpersonal relationships, and it values individual happiness and fulfillment over stability, duty, responsibility, and other familial values. As Jay Schvaneveldt (1981) observes, "The welfare and happiness of marital partners are held above the belief that the marital union or family union should stay intact. The happiness of the individual family members appears to be the dominant value."

Fourth, the theory does not place marriage or family within a larger social context. It thereby disregards or minimizes the forces working on families from the outside, such as economic or legal discrimination against minorities and women.

SOCIAL EXCHANGE THEORY

According to **social exchange theory,** we measure our actions and relationships on a cost-benefit basis. People maximize their rewards and minimize their costs by employing their resources to gain the most favorable outcome. An outcome is basically figured by the equation Reward – Cost = Outcome.

How Exchange Works

At first glance, exchange theory may be the least attractive theory we use to study marriage and the family. It seems more appropriate for accountants than for lovers. But all of us use a cost-benefit analysis to some degree to measure our actions and relationships.

The reason why many of us do not recognize our use of this interpersonal accounting is that we do much of it unconsciously. If a friend is unhappy with a partner, you may ask, "What are you getting out of this relationship? Is it *worth it?*" Your friend will start listing pluses and minuses: "On the plus side, I get company and a certain amount of security; on the minus side, I don't get someone who really understands me." When the emotional costs outweigh the benefits of the relationship, your friend will probably end it. This weighing of costs and benefits is social exchange theory at work.

One problem many of us have in recognizing our own exchange activities is that we think of rewards and costs as tangible objects, like money. In personal relationships, however, resources, rewards, and costs are more likely to be things such as love, companionship, status, power, fear, loneliness, and so on. As people enter into relationships, they have certain resources—either tangible or intangible—that others consider valuable, such as intelligence, warmth, good looks, or high social status. People consciously or unconsciously use their various resources to obtain what they want, as when they "turn on" the charm. Most of us have had friends, for example, whose relationships are a mystery to us. We may not understand what our friend sees in his or her partner; our friend is so much better looking and more intelligent than the partner. (Attractiveness and intelligence are typical resources in our society.) But it turns out that the partner has a good sense of humor, is considerate, and is an accomplished musician, all of which our friend values highly.

REFLECTIONS

Think about the benefits you are receiving from your current intimate, marital, or family relationships. What are the costs? Make a list to compare the benefits and costs. Assign a value from 1 to 10 for the various items on your list, with 10 being the highest value and 1 being the lowest. Based on the equation, Reward – Cost = Outcome, how would you predict the outcome? Think about the last time you made a trade-off in a relationship. Was it fair? If it wasn't, how did you feel? How did the other person feel?

Equity

A corollary to exchange is **equity:** exchanges that occur between people have to be fair, to balance out. In the everyday world, we are always exchanging favors: You do the dishes tonight and I'll take care of the kids. Often we do not even articulate these exchanges; we have a general sense that ultimately they will be reciprocated. If, in the end, we feel that the exchange was not fair, we are likely to be resentful and angry. Some researchers suggest that people are most happy when they get what they feel they deserve in a relationship (Hatfield and Walster, 1981). Oddly enough, both partners feel uneasy in an inequitable relationship:

> While it is not surprising that deprived partners (who are, after all, getting less than they deserve) should feel resentful and angry about their inequitable treatment, it is perhaps not so obvious why their overbenefited mates (who are getting more than they deserve) feel uneasy too. But they do. They feel guilty and fearful of losing their favored position.

When partners recognize that they are in an inequitable relationship, they generally feel uncomfortable, angry, or distressed. They try to restore equity in one of three ways:

- They attempt to restore actual equity in the relationship.
- They attempt to restore psychological equity by trying to convince themselves and others that an obviously inequitable relationship is actually equitable.
- They decide to end the relationship.

Society regards marriage as a permanent commitment. Because marriages are expected to endure, exchanges take on a long-term character. Instead of being calculated on a day-to-day basis, outcomes are judged over time.

An important ingredient in these exchanges is whether the relationship is fundamentally cooperative or competitive. In cooperative exchanges, both husbands and wives try to maximize their "joint profit" (Scanzoni, 1979). These exchanges are characterized by mutual trust and commitment. Thus a husband might choose to work part-time and also care for the couple's infant so that his wife may pursue her education. In a competitive relationship, however, each is trying to maximize his or her own individual profit. If both spouses want the freedom to go out whenever or with whomever they wish, despite opposition from the other, the relationship is likely to be unstable.

Ralph and Maureen LaRossa used social exchange ideas as part of their attempt to depict what couples experience in their transitions to new parenthood (LaRossa and LaRossa, 1981). Noting that infants require **continuous coverage** (24 hours a day, 7 days a week supervision), the LaRossas observed that this need introduced a new temporal dynamic into the household. New (first-time) parents must establish some division of responsibility for the care of their newborn and find themselves in one of three modes of relating to their infants: (1) *primary* or direct one-on-one contact or caregiving; (2) *secondary* or less demanding interaction, where one can participate with less than full attention or involvement; or (3) *tertiary* involvement, which is in fact no involvement, or what is called "*down time*." In the LaRossa's analysis, down time becomes the resource that couples bargain for, negotiate over, and argue about. It is perceived as a reward that is increasingly scarce, and each parent attempts to maximize her or his share of it. For a variety of reasons that we will later examine, fathers retain more free time while most primary parenting is done by mothers.

Critique

Social exchange theory assumes that we are all rational, calculating individuals, weighing the costs and rewards of their relationships. In reality, sometimes we are rational, and sometimes we are not. Sometimes we act altruistically without expecting any reward. This is often true of love relationships and parent-child interaction.

Social exchange theory also has difficulty ascertaining the value of costs, rewards, and resources. If

© Laura Dwight/PhotoEdit

The arrival of a child introduces new family dynamics that are differently understood from a variety of theoretical perspectives.

you want to buy eggs, you know they are a certain price per dozen and you can compare buying a dozen eggs with spending the same amount on a notebook. But how does the value of an outgoing personality compare with the value of a compassionate personality? Is a pound of compassion equal to 10 pounds of enthusiasm? Compassion may be the trait most valued by one person but may not be important to another. The values that we assign to costs, rewards, and resources are highly individualistic.

FAMILY DEVELOPMENT THEORY

Of all of the theories discussed here, **family development theory** is the only one that is exclusively directed at families (White and Klein, 2002). It emphasizes the patterned changes that occur in families through stages and across time. In its earliest formulations, family development theory borrowed from theories of individual development and identified a set number of stages that all families pass through as they are formed: growth with the birth of children; change during the raising of children; and contract as children leave and spouses die. Such stages comprised the **family life cycle.** Eventually other concepts were introduced to replace the idea of a family life cycle. Rogers (1973) and Aldous (1978, 1996) proposed the notion of the **family career,** which was said to consist of subcareers like the marital or parental career, which themselves were affected by one's educational or occupational career. Most recently, the idea of the **family life course** has been used to examine the dynamic nature of family experience. The family life course consists of "all the events and periods of time (stages) between events traversed by a family" (White and Klein, 2002). Because all of these concepts emphasize the change and development of families over time, they are complementary and overlapping.

Family development theory looks at the changes in the family that typically commence in the formation of the premarital relationship, proceed through marriage, and continue through subsequent sequential stages. The specification of stages may be based on family economics, family size, or the developmental tasks that families encounter as they move from one stage to the next. The stages are identified by the primary or orienting event characterizing a period of the family history. An eight-stage family life cycle might consist of the following: (1) beginning family, (2) childbearing families, (3) families with preschool children, (4) families with schoolchildren, (5) families with adolescents, (6) families as launching centers, (7) families in middle years, and (8) aging families.

As we grow, each of us responds to certain universal developmental challenges (Person, 1993). For example, all people encounter normative age-graded influences, such as the biological processes of puberty and menopause or sociocultural markers such as the beginning of school and the advent of retirement. Normative history-graded influences come from historical facts that are common to a particular generation, such as the political and economic influences of wars and economic depressions that are similar for individuals in a particular age group (Santrock, 1995).

The life-cycle model give us insights into the complexities of family life and of the different tasks that families perform. This model describes the interacting influences of changing roles and circumstances through time. It provides awareness of ways in which these changes produce corresponding changes in family responsibilities and needs. Planning that utilizes the developmental model alerts the family to seek resources appropriate to the upcoming needs and to be aware of vulnerabilities associated with each family stage (Higgins, Duxbury, and Lee, 1994).

There are a variety of developmental theories that examine specific family phenomena such as "falling in love," choosing a spouse, or experiencing divorce. Instead of attempting to depict all the stages families might encounter, these theories look at the unfolding of specific aspects of family life across stages. You will find such approaches in a number of later chapters.

Critique

An important criticism sometimes made of family development theory is that it assumes the sequential processes of intact, nuclear families. It further assumes that all families go through the same process of change across the same stages. Thus the theory downplays both the diversity of family experience as well as the experiences of those who divorce, remain childless, or bear children but never marry (Winton, 1995). For example, lesbian-headed families are likely to experience a life-cycle pattern that is quite different from the traditional one (Slater,

1995). Similarly, stepfamilies experience different stages and tasks (Ahrons and Rogers, 1987). Nevertheless, the universality of the family life cycle may transcend the individuality of the family form. Single-parent and two-parent families go through many of the same development tasks and transitions. They may differ, however, in the timing and length of those transitions.

A second related criticism points out that gender, race, ethnicity, and social class all create variations in how one experiences family dynamics. The very sequence of stages may reflect a more middle to upper class family reality. Many lower and working class families do not have lengthy periods of early childless marriage. The transitions to marriage and parenthood may be encountered simultaneously or in reverse of what the stages specify. In neglecting these sorts of variations, the developmental model can appear overly simplistic.

STRUCTURAL FUNCTIONALISM

Structural functionalism is a theory that explains how society works, how families work, and how families relate to the larger society and to their own members. The theory is used largely in sociology and anthropology, disciplines that focus on the study of society rather than individuals. When structural functionalists study the family, they look at three aspects: (1) what functions the family serves for society (discussed in Chapter 1), (2) what functional requirements family members perform for the family, and (3) what needs the family meets for its individual members.

Many of us are structural functionalists without knowing it. Those who believe social stability is in the best interest of society share this assumption with structural functionalists. Those who believe families must be intact to fulfill their functions share a common view with structural functionalists. Those who feel that changes in gender roles and the increase in single-parent and dual-earner families threaten the family fear social instability.

Society as a System

Structural functionalism is deeply influenced by biology. It treats society as if it were a living organism, like a person, animal, or tree. In fact, the theory sometimes uses the analogy of a tree in describing society. In a tree, there are many substructures or parts, such as the trunk, branches, roots, and leaves. Each structure has a function. The roots gather nutrients and water from the soil, the leaves absorb sunlight, and so on. Society is like a tree insofar as it has different structures that perform functions for its survival. These structures are called **subsystems.** The subsystems are the major institutions, such as the family, religion, government, and the economy. Each of these structures has a function in maintaining society, just as the different parts of a tree serve a function in maintaining the tree. Religion, for example, gives spiritual support, the government ensures order, and the economy produces goods. The family provides new members for society through procreation and socializes its members so that they fit into society. In theory, all institutions work in harmony for the good of society and one another.

The Family as a System

Families themselves may also be regarded as systems. In looking at families, structural functionalists examine (1) how the family organizes itself for survival, and (2) what functions the family performs for its members. For the family to survive, its members must perform certain functions, which are traditionally divided along gender lines. Men and women have different tasks: Men work outside the home to provide an income, whereas women perform household tasks and child rearing.

According to structural functionalists, the family molds the kind of personalities it needs to carry out its functions. It encourages different personality traits for men and women to ensure its survival. Men develop instrumental traits, and women develop expressive traits. **Instrumental traits** encourage competitiveness, coolness, self-confidence, and rationality—qualities that will help a person succeed in the outside world. **Expressive traits** encourage warmth, emotionality, nurturing, and sensitivity—qualities appropriate for someone caring for a family and a home.

Such a division of labor and differentiation of temperaments is seen as efficient because it allows each spouse to specialize, thus minimizing competition and reducing ambiguity or uncertainty over such things as who should work outside the home or whose outside employment is more important. For these reasons such role allocation may be deemed functional.

Critique

Although structural functionalism has been an important theoretical approach to the family, it has declined in significance in recent decades for several reasons. First, because the theory cannot be empirically tested, we'll never know if it is "right" or "wrong." We can only discuss it theoretically, arguing whether it accounts for what we know about the family.

Second, it is not always clear what function a particular structure serves. "The function of the nose is to hold the *pince-nez* [eyeglasses] on the face," remarked the eighteenth-century philosopher François Voltaire. What is the function of the traditional division of labor along gender lines? Efficiency, survival, or the subordination of women?

If interdependence, specialization, and clarity of role responsibilities are what make breadwinner-homemakers most functional, those same objectives could be met by household arrangements wherein men stay home, rear kids, and tend house, while women earn incomes. In fact, in some relationships these role reversals might be more functional. There are women who earn higher incomes than their husbands, are in jobs with greater opportunities for advancement, and are more dedicated to their careers than are their husbands. If their husbands are frustrated by or stagnated at work but have developed or discovered a deeper than anticipated fulfillment from children, a reversal of the male-provider–female-homemaker household would be most functional for them.

Third, how do we know which family functions are vital? The family, for example, is supposed to socialize children, but much socialization has been taken over by the schools, peer groups, and the media. Is this "functional"?

Fourth, structural functionalism has a conservative bias against change. Aspects that reflect stability are called functional, and those that encourage instability (or change) are called dysfunctional. Traditional roles are functional, but nontraditional ones are dysfunctional. Employed mothers are viewed as undermining family stability because they should be home caring for the children, cleaning house, and providing emotional support for their husbands. But in reality, employed mothers may be contributing to family stability by earning money; their income often pushes their families above the poverty line.

Finally, structural functionalism looks at the family abstractly. It looks at it formally, from a distance far removed from the daily lives and struggles of men, women, and children. It views the family in terms of functions and roles. Family interactions, the very lifeblood of family life, are absent. Because of its formalism, structural functionalism often has little relevance to real families in the real world.

CONFLICT THEORY

Whereas structural functionalists tend to believe that "what is, is good," conflict theorists believe that "what is, is bad, for at least someone. Where structural functionalists assert that existing structures benefit society, conflict theorists ask, "Who benefits?" **Conflict theory** holds that life involves discord. Conflict theorists see society not as basically cooperative but as divided, with individuals and groups in conflict with each other. They try to identify the competing forces.

Sources of Conflict

How can we analyze marriages and families in terms of conflict and power? Marriage and family relationships are based on love and affection, aren't they? Conflict theorists agree that love and affection are important elements in marriages and families, but they believe that conflict and power are also fundamental. Marriages and families are composed of individuals with different personalities, ideas, values, tastes, and goals. Each person is not always in harmony with every other person in the family. Imagine that you are living at home and want to do something your parents don't want you to do, such as spend the weekend with a friend they don't like. They forbid you to carry out your plan. "As long as you live in this house, you'll have to do what we say." You argue with them, but in the end you stay home. Why did your parents win the disagreement? They did so because they had greater power, according to conflict theorists.

Conflict theorists do not believe that conflict is bad; instead, they think it is a natural part of family life. Families always have disagreements, from small ones, such as what movie to see, to major ones, such as how to rear children. Families differ in the number of underlying conflicts of interest, the degree of underlying hostility, and the nature and extent of the expression of conflict. Conflict can take the form of competing goals, such as a husband's wanting to buy a new CD player and a wife's wanting to pay off credit cards. Conflict can also occur because of different role expectations: An employed mother wants to divide housework 50–50, but her husband insists that household chores are "women's work."

REFLECTIONS

Using conflict theory, examine the recurring conflicts in your relationship, marriage, or family. Who wins these various conflicts? What resources do the winners have? The losers? Do they differ according to the type of conflict? What are your resources in relationships? How do you use them?

© Michael Newman/PhotoEdit

According to conflict theory, family life is characterized by much tension and disputes over such things as the division of responsibilities, allocation of resources, and levels of commitment.

Sources of Power

When conflict occurs, who wins? Family members have different resources and amounts of power. There are four important sources of power: (1) legitimacy, (2) money, (3) physical coercion, and (4) love. When arguments arise in a family, a man may want his way "because I'm the head of the house" or a parent "because I'm your mother." These appeals are based on legitimacy—that is, the belief that the person is entitled to prevail by right. Money is a powerful resource in marriages and families. "As long as you live in this house . . ." is a directive based on the power of the purse. Because men tend to earn more than women, they have greater economic power; this economic power translates into marital power. Physical coercion is another important source of power. "If you don't do as I tell you, you'll get a spanking" is one of the most common forms of coercion of children. But physical abuse of a spouse is also common, as we will see in a later chapter. Finally, there is the power of love. Love can be used to coerce someone emotionally, as in "If you really loved me, you'd do what I ask." Or love can be a freely given gift, as in the case of a person's giving up something important, such as a plan, desire, or career, to enhance a relationship.

Everyone in the family has power, although the power may be different and unequal. Adolescent children, for example, have few economic resources, so they must depend on their parents. This dependency gives the parents power. But adolescents also have power through the exercise of personal charm, ingratiating habits, temper tantrums, wheedling, and so on.

Families cannot live comfortably with much open conflict. The problem for families, as for any group, is how to encourage cooperation while allowing for differences. Because conflict theory sees conflict as normal, the theory seeks to channel it and to seek solutions through communication, bargaining, and negotiations. We return to these items in Chapter 5 in our discussion of conflict resolution.

Critique

A number of difficulties arise in conflict theory. First, conflict theory derives from politics, in which self-interest, egotism, and competition are dominant elements. Yet is such a harsh judgment of human nature justified? People's behavior is also characterized by self-sacrifice and cooperation. Love is an important quality in relationships. Conflict theorists do not often talk about the power of love or bonding; yet the presence of love and bonding may distinguish the family from all other groups in society. We often will make sacrifices for the sake of those we love. We will defer our own wishes to another's desires; we may even sacrifice our lives for a loved one. Second, conflict theorists assume that differences lead to conflict. Differences can also be accepted, tolerated, or appreciated. Differences do not necessarily imply conflict. Third, conflict in families is not easily measured or evaluated. Families live much of their lives privately, and outsiders are not always aware of whatever conflict exists or how pervasive it is. Also, much overt conflict is avoided because it is regulated through family and societal rules. Most children obey their parents, and most spouses, although they may argue heatedly, do not employ violence.

FAMILY SYSTEMS THEORY

Family systems theory combines two of the previous sociological theories, structural functionalism and symbolic interaction, to form a psychotherapeutic theory. Mark Kassop (1987) notes that family systems theory creates a bridge between sociology and family therapy.

Structure and Patterns of Interaction

Like functionalist theory, family systems theory views the family as a structure of related parts or subsystems. Each part carries out certain functions. These parts include the spousal subsystem, the parent/child subsystem, the parental subsystem (husband and wife relating to each other as parents), and the personal subsystem (the individual and his or her relationships). One of the important tasks of these subsystems is maintaining their boundaries. For the family to function well, the subsystems must be kept separate (Minuchin, 1981). Husbands and wives, for example, should prevent their conflicts from spilling over into the parent/child subsystem. Sometimes a parent will turn to the child for the affection that he or she ordinarily receives from a spouse. When the boundaries of the separate subsystems blur, as in incest, the family becomes dysfunctional.

As in symbolic interaction, interaction is important in systems theory. A family system consists of more than simply its members. It also consists of the pattern of interactions of family members: their communication, roles, beliefs, and rules. Marriage is more than a husband and wife; it is also their pattern of interactions. The structure of marriage is determined by how the spouses act in relation to each other over time (Lederer and Jackson, 1968). Each partner influences, and in turn is influenced by, the partner. And each interaction is determined in part by the previous interactions. This emphasis on the pattern of interactions within the family is a distinctive feature of the systems approach.

Virginia Satir (1988) compared the family system to a hanging mobile. In a mobile, all the pieces, regardless of size and shape, can be grouped together and balanced by changing the relative distance between the parts. The family members, like the parts of a mobile, require certain distances between one another to maintain their balance. Any change in the family mobile—such as a child leaving the family, family members forming new alliances, hostility distancing the mother from the father—affects the stability of the mobile. This disequilibrium often manifests itself in emotional turmoil and stress. The family may try to restore the old equilibrium by forcing its "errant" member to return to his or her former position, or it may adapt and create a new equilibrium with its members in changed relations to one another.

Analyzing Family Dynamics

In looking at the family as a system, researchers and therapists believe the following:

- *Interactions must be studied in the context of the family system.* Each action affects every other person in the family. The family exerts a powerful influence on our behaviors and feelings, just as we influence the behaviors and feelings of other family members. On the simplest level, an angry outburst by a family member can put everyone in a bad mood. If the anger is constant, it will have long-term effects on each member of the family, who will cope with it by avoidance, hostility, depression, and so on.
- *The family has a structure that can only be seen in its interactions.* Each family has certain preferred patterns of acting that ordinarily work in response to day-to-day demands. These patterns become strongly ingrained "habits" of interactions that make change difficult. A warring couple, for example, may decide to change their

ways and resolve their conflicts peacefully. They may succeed for a while, but soon they fall back into their old ways. Lasting change requires more than changing a single behavior; it requires changing a pattern of relating.

- *The family is a purposeful system; it has a goal.* In most instances, the family's goal is to remain intact as a family. It seeks **homeostasis,** or stability. This goal of homeostasis makes change difficult, for change threatens the old patterns and habits to which the family has become accustomed.
- *Despite resistance to change, each family system is transformed over time.* A well-functioning family constantly changes and adapts to maintain itself in response to its members and the environment. The family changes through the family life cycle—for example, as partners age and as children are born, grow older, and leave home. The parent must allow the parent/child relationship to change. A parent must adapt to an adolescent's increasing independence by relinquishing some parental control. The family system adapts to stresses to maintain family continuity while making restructuring possible. If the primary wage earner loses his or her job, the family tries to adapt to the loss in income; the children may seek work, recreation may be cut, or the family may be forced to move.

Although it has been applied to a variety of family dynamics, systems theory has been particularly influential in studying *family communication* (White and Klein, 2002). As applied by systems theorists, interaction and communication between spouses are the kinds of systems wherein a husband's (next) action or communication toward his wife depends on her prior message to him. But thanks to research in family communications, we recognize that marital communication is more complex than a simple *quid pro quo* or reciprocity expectation, such as "if she is nasty, he is nasty." John Gottman has explored marital communication patterns that differentiate distressed from nondistressed couples. He identifies the importance of nonverbal communication over verbal messages spouses send (White and Klein, 2002). As we will see in later chapters, certain nonverbal messages are especially useful predictors of the eventual success or failure of a relationship (Gottman, Coan, Carrere, and Swanson, 1998; Gottman & Levenson, 1992).

Critique

It is difficult for researchers to agree on exactly what family systems theory is. Many of the basic concepts are still in dispute, even among the theory's adherents, and the theory is sometimes accused of being so abstract as to lose any real meaning (Melito, 1985; White and Klein, 2002).

Family systems theory originated in clinical settings in which psychiatrists, clinical psychologists, and therapists tried to explain the dynamics of dysfunctional families. Although its use has spread beyond clinicians, its greatest success is still in the analysis and treatment of dysfunctional families. As with clinical research, however, the basic question is whether its insights apply to healthy families as well as to dysfunctional ones. Do healthy families, for example, seek homeostasis as their goal, or do they seek individual and family well-being?

FEMINIST PERSPECTIVE

As a result of the feminist movement of the past two decades, new questions and ways of thinking about the meaning and characteristics of families have arisen. Although there is not a unified "feminist family theory," **feminist perspectives** share a central concern regarding family life.

Blending some central ideas of conflict theory with those of interactionist theory, feminists critically examine the ways in which family experience is shaped by **gender**—the social aspects of being female or male. This is the orienting focus that unifies most feminist writing, research, and advocacy. Feminists maintain that family and gender roles have been constructed by society and do not derive from biological or absolute conditions. They believe that family and gender roles have been created by men in order to maintain power over women. Basically, the goals of the feminist perspective are to work to accomplish changes and conditions in society that remove barriers to opportunity and oppressive conditions and are "good for women" (Thompson and Walker, 1995).

Gender and Family—Concepts Created by Society

Who or what constitutes a family cannot be taken for granted. The "traditional family" is no longer the predominant family lifestyle. Today's families have great diversity. What we think family should be is influenced by our own values and family experiences. Research demonstrates that couples actually may construct gender roles in the ongoing interactions that make up their marriages (Zvonkovic et al., 1996).

Are there any basic biological or social conditions that require the existence of a particular form of family? Some feminists would emphatically say no. Some object to the fact that we even try to study the family because to do so accepts as "natural" the inequalities built into the traditional concept of family life. Feminists urge a more extended view of family to include all kinds of sexually interdependent adult relationships regardless of the legal, residential, or parental status of the partnership. For example, families may be formed of committed relationships between lesbian or gay individuals, with children obtained through adoption, from previous marriages, or through artificial insemination.

Feminist Agenda

Feminists strive to raise society's level of awareness regarding the oppression of women. Furthermore, they make the point that all groups defined on the basis of age, class, race, ethnicity, disability, or sexual orientation are oppressed; they extend their concern for greater sensitivity to all disadvantaged groups (Allen and Baber, 1992). Feminists assume that the experiences of individuals are influenced by the social system in which they live. Therefore, the experiences of each individual must be analyzed in order to form the basis for political action and social change. The feminist agenda is to attend to the social context as it impacts on personal experience and to work to translate personal experience into community action and social critique.

Feminists believe that it is imperative to speak out and challenge the system that exploits and devalues women. They are aware of the dangers of speaking out but feel their own integrity will be threatened if they fail to speak out. Some feminists have described themselves as having "double vision"— the ability to be successful in the existing social system while simultaneously working to change oppressive practices and institutions.

Men as Gendered Beings

Inspired and influenced by the writing and research of feminist scholars, many social scientists now focus on how men's experiences are shaped by cultural ideas about masculinity and by their efforts to either live up to or challenge those ideas (Kimmel and Messner, 1998; Cohen, 2001). Instead of assuming that gender only matters to or includes women, this perspective looks at men as men, or as "gendered beings," whose experiences are shaped by the same kinds of forces that shape women's lives (Kimmel and Messner, 1998).

With an increased attention to gender courtesy of feminist scholars, and a more recent refocusing of attention to men as "gendered beings," we now

© David Young-Wolff/PhotoEdit

Men's participation in household tasks is a major concern of both feminist and men's studies perspectives.

have a greatly enlarged and still growing body of literature about men as husbands, fathers, sexual partners, ex-spouses, abusers, and so on (for example, see Cohen, 1987; Coltrane, 1996; Daly, 1993; Gerson, 1993; LaRossa, 1988; Marsiglio, 1998; Johnson, 1996). Throughout this book, we will explore how gender shapes women's and men's experiences of the family issues we examine.

Critique

The feminist perspective is not a unified theory; rather, it represents thinking across the feminist movement. It includes a variety of viewpoints that have, however, an integrating focus relating to the inequity of power between men and women in society and especially in family life (MacDermid et al., 1992).

Some family scholars who conceptualize family life and work as a "calling" have taken issue with feminists' focus on power and economics as a description of family. This has created a moral dialogue concerning the place of family life and work in "the good society" (Ahlander and Bahr, 1995; Sanchez, 1996). Feminists today recognize considerable diversity within their ranks, and the ideas of feminist theorists and other family theorists often overlap.

Applying Theories to Long-Distance Relationships

Although the theories above were illustrated with numerous examples, it is worthwhile to return to our opening scenario and look at some of the questions each of the theories might pose about long-

TABLE 2.1 Applying Theories to Long-Distance Relationships

THEORY	ASSUMPTIONS ABOUT FAMILIES	APPLICATION TO LONG-DISTANCE RELATIONSHIPS
Ecological	Families are influenced by and must adapt to environments.	How do the characteristics of each partner's different living environments affect their abilities to maintain their commitments to the relationship? How does the physical separation place the partners in somewhat different ecological niches, which in turn may be more or less conducive to maintaining the relationship? How does the cultural exosystem impose certain beliefs or expectations that might influence the stability of these relationships?
Symbolic interaction	Family life acquires meaning for members and depends upon the meanings they attach.	What meaning do couples attach to being separated? How does this alter their perceptions of the relationship itself? Does separation prevent or inhibit the construction of a shared definition of the relationship?
Social exchange	Individuals seek to maximize rewards, minimize costs, and achieve equitable relationships.	How do both partners define the costs and rewards associated with their relationship? If the rewards of continuing the relationship are felt to be greater than the costs associated with their physical separation, they will maintain their relationship. If either perceives the costs of being apart as too great, or finds another more rewarding relationship, the long-distance relationship will end.
Family development	Families undergo predictable changes over time across stages.	How do couples handle the transition to a long-distance relationship? What are the stages or phases that couples encounter as they adjust to being separated? What are the key tasks that must be accomplished at each stage in order for the relationship to survive?

distance relationships, given their major assumptions. These examples, illustrated in Table 2.1, are not intended to exhaust all possible questions suggested by each theory, nor do they necessarily favor *either* "absence makes the heart grow fonder" or "out of sight out of mind." They are meant merely as examples of how each theory's core ideas might apply to long-distance relationships.

Conducting Research on Families

In gathering their data, researchers use a variety of techniques. Some researchers ask the same set of questions of great numbers of people. They collect information from people of different age, sex, living situation, and ethnic backgrounds. This is known as "representative sampling." In this way researchers can discover whether age or other background characteristics influence people's responses. This approach to research is called **quantitative research** because it deals with large quantities of information. Furthermore, the information is analyzed and presented statistically. Quantitative family research often uses very sophisticated statistical techniques to assess the relationships between variables. Survey research and, to a lesser extent, experimental research (discussed in the following sections) are examples of quantitative research.

Other researchers study smaller groups or sometimes individuals in a more in-depth fashion. They may place observers in family situations, conduct intensive interviews, do case studies involving information provided by several people, or analyze letters or diaries or other records of people whose experiences represent special aspects of family life. This form of research is known as **qualitative research**

TABLE 2.1 ■ ***(continued)***

THEORY	ASSUMPTIONS ABOUT FAMILIES	APPLICATION TO LONG-DISTANCE RELATIONSHIPS
Structural functionalism	The institution of the family contributes to the maintenance of society. On a familial level, roles and relationships within the family contribute to its continued well-being.	How does physical separation function to maintain or threaten the stability of the relationship? What benefits does separation have for the individual partners and for the couple relationship?
Conflict	Family life is shaped by social inequality. Within families, as within all groups, members compete for scarce resources (e.g., attention, time, power, space).	To what extent does one partner benefit more from being apart? Assuming that one partner has a greater commitment to the relationship, how does physical separation create inequality between partners? How does separation prevent couples from effectively managing and resolving conflict?
Family systems	Families are systems which function and must be understood on that level.	How does being physically separated make it difficult for the couple to communicate effectively? What difficulties does separation create for maintaining the equillibrium of the relationship? How are boundaries between the couple system and the wider society altered by being separated?
Feminist	Gender affects one's experiences of and within families. Gender inequality shapes how women and men experience families. Families perpetuate gender differences.	How are women and men differently affected by separation? Do long-distance heterosexual relationships create a gender unequal relationship? Does separation lead men to exploit women by expecting women—but not men—to remain faithful or monogamous? Do women bear more of the burden of managing and maintaining the relationship?

because it is concerned with a detailed understanding of the object of study. The sections on case methods and observational research are examples of qualitative research (Ambert, Adler, and Detzner, 1995).

In addition to using information provided specifically by people participating in a research project, researchers also utilize information from public sources. This research is called **secondary data analysis.** It involves the reanalyzing of data originally collected for another purpose. Examples might include analyzing U.S. Census data and official statistics, such as state marriage, birth, and divorce records. Secondary data analysis also includes content analysis of various communication media such as newspapers, magazines, letters, and television programs.

Family science researchers conduct their investigations using **ethical guidelines,** agreed upon by professional researchers. These guidelines protect the privacy and safety of people who provide information in the research. For example, any research conducted with college students requires the investigator to present the plan and method of the research to a "human subjects review committee." This ensures that students are participating voluntarily and that their privacy is protected. To protect the privacy of participants, researchers promise them either anonymity or confidentiality. **Anonymity** requires that no one, including the researcher can connect particular responses to the individuals who provided them. Much questionnaire research is of this kind, providing that no identifying information is to be found on the questionnaires. According to the rules of **confidentiality,** the researcher knows the identities of participants and can connect what was said to who said it but promises not to reveal such information publicly.

To protect the safety of research participants, researchers design their studies with the intent to minimize any possible and controllable harm that might come from participation. Such harm is not typically physical harm but rather embarrassment or discomfort. Much of what family researchers study is ordinarily kept very private. Talking about personal matters with an interviewer or answering a series of survey questions may create unintended anxiety on the part of the participants. At best, researchers carefully design their studies so as to reduce the extent and likelihood of such reactions. Unfortunately, they cannot always be completely prevented (Babbie, 2002).

Research ethics also require researchers to conduct their studies and report their findings in ways that assure readers of the accuracy, originality, and trustworthiness of their reports. Falsifying data, misrepresenting patterns of findings, and plagiarizing the research of others are all unethical.

What researchers know about marriage and the family comes from four basic research methods: (1) survey research, (2) clinical research, (3) observational research, and (4) experimental research. There is a continual debate as to which method is best for studying marriage and the family. But such arguments may miss an important point: Each method may provide important and unique information that another method may not (Cowan and Cowan, 1990).

SURVEY RESEARCH

The **survey research** method, using questionnaires or interviews, is the most popular data-gathering technique in marriage and family studies. Surveys may be conducted in person, over the telephone, or by written questionnaires. Typically, the purpose of survey research is to gather information from a smaller, representative group of people and to infer conclusions that are valid for a larger population. Questionnaires offer anonymity, may be completed fairly quickly, and are relatively inexpensive to administer.

Quantitative questionnaire research is an invaluable resource for gathering data that can be generalized to the wider population. Because researchers who use such techniques typically draw or use **probability-based random samples,** they can estimate the likelihood that their sample data can be safely inferred to the population in which they are interested. Furthermore, using preestablished response categories or existing scales or indexes that are used by all respondents allows for more comparability across a particular sample and between the sample data and other related research.

For example, Chloë Bird's 1997 study examines the psychological distress associated with the burdens of parenting, as they vary by gender. Using data from 1,601 men and women under age 60 who participated in the U.S. Survey of Work, Family, and

Well-Being, she contrasted the levels of distress experienced by parents with that of nonparents, and—among parents—compared mothers with fathers. Although the details of her analysis are too complex to be dealt with here, she determined that, on average, parents report higher levels of distress than do people without children, and mothers report higher levels of distress than did fathers (Bird, 1997). In fact, women with children under age 18 living at home reported experiencing the highest levels of distress. From her carefully controlled analysis, Bird determined that it is not children but rather the increased social and economic burdens that accompany children that seem to create the psychological outcomes she identified.

Questionnaires usually do not allow for in-depth responses, however; a person must respond with a short answer, a *yes*, a *no*, or a choice on a scale of, for example, 1 to 10, from *strongly agree* to *strongly disagree*, from *very important* to *unimportant*, and so on. Unfortunately, marriage and family issues are often too complicated for questionnaires to explore in depth.

Interview techniques avoid some of this shortcoming of questionnaires because interviewers are able to probe in greater depth and follow paths suggested by the interviewee. They are also typically better able to capture the particular meanings or the depth of feeling people attach to their family experiences. Consider these two examples, each from research on parents' reactions to the life changes associated with becoming or being parents. The first comes from Sharon Hays's interview study of 38 mothers of 2- to 4-year-old children (Hays, 1996). In describing how priorities get restructured when one becomes a mother, one of Hays' informants offered this comment:

> I think the reason people are given children is to realize how selfish you have been your whole life—you are just totally centered on yourself and what you want. And suddenly here's this helpless thing that needs you constantly. And I kind of think that's why you're given children, so you kinda think, okay, so my youth was spent for myself. Now, you're an adult, they come first. . . . Whatever they need, they come first.

The second example comes from research conducted by one of the authors on men's experiences becoming and being fathers (Cohen, 1993). Here, a 33-year-old municipal administrator describes how becoming a father changed his life:

> I think everything in a personal relationship a baby changes. . . . It's just fantastic . . . it knocked me for a loop. Something creeps into your life and then all of a sudden it dominates your life. It changes your relationship to everybody and everything, and you question every value you ever had. . . . And you say to yourself, "This is a miracle. . . ."

These examples of narrative data convey much about the experience of parenthood that quantitative questionnaire data cannot. By having respondents circle or check the appropriate preestablished response categories to a researcher's questions, we may never identify what that response means to the respondent or how it fits within the wider context of her or his life. However, interviewers are less able to determine how commonly such experiences or attitudes are found. Interviewers may also occasionally allow their own preconceptions to influence the ways in which they frame their questions and to bias their interpretation of responses.

There are problems associated with survey research, whether done by questionnaires or interviews. First, how representative is the sample (the chosen group) that volunteered to take the survey? In the case of a probability-based sample this is not a concern. Self-selection (volunteering to participate) also tends to bias a sample. Second, how well do people understand their own behavior? Third, are people underreporting undesirable or unacceptable behavior? They may be reluctant to admit that they have extramarital affairs or that they are alcoholics, for example. If for any reason people are unable or unwilling to answer questions honestly, the survey technique will produce misleading or inaccurate data.

Nevertheless, surveys are well suited for determining the incidence of certain behaviors or for discovering traits and trends. Much of the research that family scientists conduct and use—on topics as far-reaching as the division of housework and child care, the frequency of and satisfaction with sex, or the impact of divorce on children or adults—is derived from interview or questionnaire data. Surveys are more commonly used by sociologists than by psychologists, because they tend to deal on a general or societal level rather than on a personal or

Understanding Yourself

WHAT DO SURVEYS TELL YOU ABOUT YOURSELF?

Survey questionnaires are the leading source of information about marriage and the family. The questionnaire below, called "The Marriage and Family Life Attitude Survey," was developed by Don Martin to gain information about attitudes toward marriage and the family. On a scale of 1 to 5, indicate for each statement below whether you strongly agree (1), slightly agree (2), neither agree nor disagree (3), slightly disagree (4), or strongly disagree (5).

The Marriage and Family Life Attitude Survey	Strongly agree	Slightly agree	Neither agree nor disagree	Slightly disagree	Strongly disagree
I COHABITATION AND PREMARITAL SEXUAL RELATIONS					
1 I have or would engage in sexual intercourse before marriage.	1	2	3	4	5
2 I believe it is acceptable to experience sexual intercourse without loving one's partner.	1	2	3	4	5
3 I want to live with someone before I marry him/her.	1	2	3	4	5
4 If I lived intimately with a member of the opposite sex, I would tell my parents.	1	2	3	4	5
II MARRIAGE AND DIVORCE					
5 I believe marriage is a lifelong commitment.	1	2	3	4	5
6 I believe divorce is acceptable except when children are involved.	1	2	3	4	5
7 I view my parents' marriage as happy.	1	2	3	4	5
8 I believe I have the necessary skills to make a good marriage.	1	2	3	4	5
III CHILDHOOD AND CHILD REARING					
9 I view my childhood as a happy experience.	1	2	3	4	5
10 If both my spouse and I work, I would leave my child in a day-care center while at work.	1	2	3	4	5
11 If I have a child, I feel only one parent should work so the other can take care of the child.	1	2	3	4	5
12 The responsibility for raising a child is divided between both spouses.	1	2	3	4	5
13 I believe I have the knowledge necessary to raise a child properly.	1	2	3	4	5
14 I believe children are not necessary in a marriage.	1	2	3	4	5
15 I believe two or more children are desirable for a married couple.	1	2	3	4	5
IV DIVISION OF HOUSEHOLD LABOR AND PROFESSIONAL EMPLOYMENT					
16 I believe household chores and tasks should be equally shared between marital partners.	1	2	3	4	5
17 I believe there are household chores that are specifically suited for men and others for women.	1	2	3	4	5
18 I believe women are entitled to careers equal to those of men.	1	2	3	4	5
19 If my spouse is offered a job in a different locality, I will move with my spouse.	1	2	3	4	5
V MARITAL AND EXTRAMARITAL SEXUAL RELATIONS					
20 I believe sexual relations are an important component of a marriage.	1	2	3	4	5
21 I believe the male should be the one to initiate sexual advances in a marriage.	1	2	3	4	5
22 I do not believe extramarital sex is wrong for me.	1	2	3	4	5

	Strongly agree	Slightly agree	Neither agree nor disagree	Slightly disagree	Strongly disagree
VI PRIVACY RIGHTS AND SOCIAL NEEDS					
23 I believe friendships outside of marriage with the opposite sex are important in a marriage.	1	2	3	4	5
24 I believe the major social functioning in a marriage should be with other couples.	1	2	3	4	5
25 I believe married couples should not argue in front of other people.	1	2	3	4	5
26 I want to marry someone who has the same social needs as I have.	1	2	3	4	5
VII RELIGIOUS NEEDS					
27 I believe religious practices are important in a marriage.	1	2	3	4	5
28 I believe children should be made to attend church.	1	2	3	4	5
29 I would not marry a person of a different religious background.	1	2	3	4	5
VIII COMMUNICATION EXPECTATIONS					
30 When I have a disagreement in an intimate relationship, I talk to the other person about it.	1	2	3	4	5
31 I have trouble expressing what I feel toward the other person in an intimate relationship.	1	2	3	4	5
32 When I argue with a person in an intimate relationship, I withdraw from that person.	1	2	3	4	5
33 I would like to learn better ways to express myself in a relationship.	1	2	3	4	5
IX PARENTAL RELATIONSHIPS					
34 I would not marry if I did not get along with the other person's parents.	1	2	3	4	5
35 If I do not like my spouse's parents, I should not be obligated to visit them.	1	2	3	4	5
36 I believe each spouse's parents should be seen an equal amount of time.	1	2	3	4	5
37 I feel parents should not intervene in any matters pertaining to my marriage.	1	2	3	4	5
38 If my parents did not like my choice of a marriage partner, I would not marry this person.	1	2	3	4	5
X PROFESSIONAL COUNSELING SERVICES					
39 I would seek premarital counseling before I got married.	1	2	3	4	5
40 I would like to attend marriage enrichment workshops.	1	2	3	4	5
41 I will seek education and/or counseling in order to learn about parenting.	1	2	3	4	5
42 I feel I need more education of what to expect from marriage.	1	2	3	4	5
43 I believe counseling is only for those couples in trouble.	1	2	3	4	5

After you have completed this questionnaire, ask yourself the questions below:

- Were the questions correctly posed so that your responses adequately portrayed your attitudes?
- Were questions omitted that are important for you regarding marriage and the family? If so, what were they?
- Do your attitudes reflect your actual behavior?

small-group level. But surveys are not able to measure very well how people interact with each other or what they actually do. For researchers and therapists interested in studying the dynamic flow of relationships, surveys are not as useful as clinical, experimental, and observational studies.

As mentioned earlier, many researchers use a technique known as secondary data analysis. Because of the various costs associated with conducting surveys on large, nationally representative samples, researchers often turn to one of the available survey data sets such as the *General Social Survey* conducted by the National Opinion Research Center at the University of Chicago. The GSS includes many social science variables of interest to family researchers. Family researchers also often use data issued by the U.S. Census Bureau, which include many descriptive details about the U.S. population, including characteristics of families and households. Additional examples of available survey data that are of particular value to family researchers include the National Survey of Families and Households (NSFH) and the National Health and Social Life Survey (NHSLS). The NSFH has provided much information about a wide range of family behaviors including the division of housework, the frequency of sexual activity, and the relationships between parents and their adult children. The NHSLS is based on a representative sample of 3,432 Americans, aged 18 to 59, and contains much useful data about sexual behavior (Christopher and Sprecher, 2000).

The major difficulty associated with secondary data analysis is that the material collected in the original survey may "come close to" but not be exactly what you wanted to examine. Perhaps you would have worded it a bit differently to more completely capture the essence of what you are interested in. Likewise, perhaps you would have asked additional related questions to further or more deeply explore your topical interest (Babbie, 2002). This disadvantage does not negate the benefits associated with secondary analysis.

CLINICAL RESEARCH

Clinical research involves in-depth examination of a person or a small group of people who come to a psychiatrist, psychologist, or social worker with psychological or relationship problems. The **case-study method,** consisting of a series of individual interviews, is the most traditional approach of all clinical research; with few exceptions, it was the sole method of clinical investigation through the first half of the twentieth century (Runyan, 1982).

In fact, clinical researchers gather a variety of additional kinds of data including direct, first-hand observation or analysis of records. Rather than a specific technique of data collection, clinical research is distinguished by its examination of individuals and families that have sought some kind of professional help. The advantage of clinical approaches is that they offer long-term, in-depth study of various aspects of marriage and family life. The primary disadvantage is that we cannot necessarily make inferences about the general population from them. People who enter psychotherapy are not a representative sample. They may be more motivated to solve their problems or have more intense problems than the general population (Kitson et al., 1996).

One of the more widely cited and celebrated clinical studies is Judith Wallerstein's longitudinal study of 60 families who sought help from her divorce clinic. Wallerstein has published three books, *Surviving the Breakup: How Children and Parents Cope With Divorce; Second Chances: Men, Women, and Children a Decade After Divorce;* and *The Unexpected Legacy of Divorce: The 25 Year Landmark Study*, following the experiences of most of the children in these families (she has retained 93 of the original 131 children that she first interviewed in 1971) at 5, 10, and 25 years postdivorce (Wallerstein, 1980, 1989, 2000). All three books are sensitively written and richly convey the multitude of short and long-term effects of divorce in the lives of her sample. Her critics have questioned whether findings based on such a clinically drawn sample (60 families from Marin County, California, who sought help as they underwent divorce) apply to divorced families more generally (Coontz, 1998).

Clinical studies, however, have been very fruitful in developing insight into family processes. Such studies have been instrumental in the development of family systems theory, discussed earlier in this chapter. By analyzing individuals and families in therapy, psychiatrists, psychologists, and therapists such as R. D. Laing, Salvador Minuchin, and Virginia Satir have been able to understand how families create roles, patterns, and rules that family members follow without being aware of them.

Exploring Diversity

MEN'S INVOLVEMENT IN HOUSEWORK

The comparative data included here about the division of household labor is representative of the kind of data family researchers gather through survey instruments. It is presented to you in table format, as some other data will be throughout this book, and it can be used for many purposes. Let's look closely at the table. What do the numbers in the table actually represent?

These data represent married men's share of five different household tasks in dual-earner households, as reported by men and women in five different countries (Baxter, 1997). The bottom line indicates women's and men's estimates of the total share of housework dual-earner husbands contribute in each of the five countries. The column heads identify the countries; within each column are males' and females' estimates, whereas the rows identify each of five tasks and the total male contribution. Study the table. What interesting things do you notice?

We can use these table data to note a number of different things. First, in each country men and women report somewhat different estimates of what proportion of each task men contribute, with men always indicating somewhat greater contributions than women credit men with. Perhaps this doesn't surprise you. Second, the gender gap in estimates is greatest among males and females in the United States and smallest among Australian and Swedish women and men. We can see this by adding the differences between women's and men's estimates for each task within each country. In the United States, there was a total difference between male and female estimates of 31 percentage points, or an average gender gap of 6 percent. In Australia, the total difference was 16 percent; the average was 3.2 percent. Although these cross-national differences are not great, they do indicate closer consensus among Australian (followed by Swedish) women and men than among females and males in the United States, Norway, or Canada. Third, we can see which tasks men are most involved in (grocery shopping in four of the five countries, after-meal cleanup in Australia) and where they are least involved (laundry, in all five countries). Observe whether and how that varies by country. Finally, we can note that across these five countries men estimate that they contribute between 25 percent and 28 percent of the total housework, whereas women credit men with between 19 percent and 25 percent of the total housework. Women in the United States and Norway estimate men's contributions at less than 20 percent.

Together, these data reveal that in dual-earner households in the United States and abroad, responsibility for domestic work rests heavily on women's shoulders. The data report only what people say; we do not have behavioral indicators of what they actually do. Furthermore, from these data alone we do not know why these chores are divided as they are. Nor do we know whether women and/or men object to this allocation of responsibilities; for that we would need more and different data. Different theories raised in this chapter offer different explanations of why tasks become gender divided and what implications such divisions have. These sorts of issues will be raised, and data such as these will be used throughout this book.

Husband's Contribution as a Percentage of Domestic Labor, as Reported by Dual-Earner Couples

	UNITED STATES		SWEDEN		NORWAY		CANADA		AUSTRALIA	
	MALE	FEMALE	MALE	FEMALE	MALE	FEMALE	MALE	FEMALE	MALE	FEMALE
Cooking	22	17	28	23	23	17	24	19	25	23
After-meal cleanup	29	22	26	25	32	26	31	24	38	33
Laundry	20	11	19	17	13	8	16	12	16	14
Housekeeping	26	21	31	26	22	17	25	20	21	19
Grocery shopping	30	25	37	32	37	32	34	28	29	24
Total	**25**	**19**	**28**	**25**	**25**	**19**	**26**	**21**	**26**	**23**

SOURCE: From Baxter (1997).

OBSERVATIONAL RESEARCH

Observational and experimental studies (discussed in the next section) account for less than 5 percent of recent research articles (Nye, 1988). In **observational research,** scholars attempt to study behavior systematically through direct observation while remaining as unobtrusive as possible. To measure power in a relationship, for example, an observer-researcher may sit in a home and videotape exchanges between a husband and wife. The obvious disadvantage of this method is that the couple may hide unacceptable ways of dealing with decisions, such as threats of violence, while the observer is present. Individuals within families, as well as families as groups, are concerned with appearances and the impressions they make.

Another problem with observational studies is that a low correlation often exists between what observers see and what the people observed report about themselves (Bray, 1995). Researchers have suggested that self-reports and observations really measure two different views of the same thing: A self-report is an insider's view, whereas an observer's report is an outsider's view (Jacob et al., 1994). Some observational research involves family members being given structured activities to carry out. These activities will involve interaction that can be observed between family members (Milner and Murphy, 1995). They may include problem-solving tasks, putting together puzzles or games, or responding to a contrived family dilemma. Different tasks are intended to elicit different types of family interaction, which will provide the researchers with opportunities to observe behaviors of interest.

A third problem that observational researchers encounter involves the essentially private nature of most family relationships and experiences. Because we experience most of our family life "behind closed doors," researchers typically cannot see what goes on "inside," without being granted access. For more public family behavior, observational data can be effectively utilized.

For example, a 1989 study by sociologist Paul Amato examined the question, "Who takes care of children in public places?" Amato suggested that using "naturalistic observations," wherein people are unaware that they are being watched, eliminated the concern about potential face-saving or impression-making distortions to people's "real behavior." Using researchers strategically stationed in a variety of public places (for example, parks, shopping malls, restaurants) in San Diego, California, and Lincoln, Nebraska, Amato compiled 2,500 observa-

There are aspects of family life that can be easily observed, such as care for children in public.

What goes on at home, behind closed doors, may not be easily accessible to observational researchers.

tions of children with their male and/or female caretakers. He used such observations to test five hypotheses about adult-male-to-child interaction (Amato, 1989).

Overall, Amato found that 43 percent of the young children who were observed were cared for by men. His specific findings indicated that boys were more likely than girls to be looked after by a man; preschool children were most likely, and infants least likely, to have male caretakers; male caretaking was highest in recreational settings and lowest in restaurants; male caretaking rates were higher among men who were accompanied by women than among men by themselves; and there were only modest differences between the California and Nebraska locations. In addition to its substantive contributions, Amato's research showed that though not widely utilized by family researchers, observational research can be used to study certain family phenomena .

REFLECTIONS

Pay attention to the people you see caring for young children in public settings, such as shopping malls, parks, and restaurants. What patterns can you identify? How do those patterns compare to what was reported by Amato more than a decade ago?

Because a major limitation of strictly observational data is identifying meanings people attach to their behavior or attributing motives for why people are doing what they are observed doing, researchers often combine observational data with other sorts of data in a process known as **triangulation.** Gerris, Dekovic, and Janssens (1997) examined whether social class affected the child-rearing values and behaviors of a sample of 237 Dutch mothers and fathers. Researchers interviewed participants, administered a 25-minute "family interaction task" (puzzle-solving between both parents and a target child), which they observed, and asked participants to complete a questionnaire detailing their child-rearing techniques. The observational data were tape-recorded and later analyzed, along with the interview and questionnaire data, to identify a variety of ways in which social class effects surfaced in child rearing.

EXPERIMENTAL RESEARCH

In **experimental research,** researchers isolate a single factor under controlled circumstances to determine its influence. Researchers are able to control their experiments by using variables, aspects or factors that can be manipulated in experiments. Recall the earlier discussion of types of variables, especially independent and dependent variables. In experiments, independent variables are factors manipulated or changed by the experimenter; dependent variables are factors that are affected by changes in the independent variable.

Because it controls variables, experimental research differs from the previous methods we have examined. Clinical studies, surveys, and observational research are correlational in nature. Correlational studies measure two or more naturally occurring variables to determine their relationship to each other. Because correlational studies do not manipulate the variables, they cannot tell us which variable causes the other to change. But because experimental studies manipulate the independent variables, researchers can reasonably determine which variables affect the other variables.

Experimental findings can be very powerful because such research gives investigators control over many factors and enables them to isolate variables. Researchers believing that stepmothers and stepfathers are stigmatized, for example, tested their hypothesis experimentally (Ganong, Coleman, and Kennedy, 1990). They devised a simple experiment in which subjects were asked to evaluate 20 traits of a person in a family who was described in a short paragraph. The person was variously identified as a father or mother in a nuclear family, a biological father or mother in a stepfamily, or a stepfather or stepmother in a stepfamily. When identified as a biological parent in either a nuclear family or a stepfamily, the individual was rated more favorably than when identified as a stepfather or stepmother. This paper-and-pencil experiment confirmed the researchers' hypothesis that stepparents are stigmatized.

The obvious problem with such studies is that we respond differently to people in real life than we do in controlled situations, especially in paper-and-pencil situations. We may not stigmatize a stepparent at all in real life. Experimental situations are usually faint shadows of the complex and varied situations we experience in the real world.

Differences in sampling and methodological techniques help explain why studies of the same phenomenon may arrive at different conclusions. They also help explain a common misperception many of us hold regarding scientific studies. Many of us believe that because studies arrive at different conclusions, none are valid. What conflicting studies may show us, however, is that researchers are constantly exploring issues from different perpectives as they attempt to arrive at a consensus.

Researchers may discover errors or problems in sampling or methodology that lead to new and different conclusions. They seek to improve sampling and methodologies in order to elaborate on or disprove earlier studies. In fact, the very word *research* is derived from the prefix *re-*, meaning "over again," and *search*, meaning "to examine closely." And that is the scientific endeavor: searching and re-searching for knowledge.

Researching Long-Distance Relationships

We have come almost full-circle. At the beginning of this chapter we posed a scenario in which "common sense" failed to resolve contradictory advice to a friend about the likely outcome of her long-distance relationship. We have since revisited this issue from time to time throughout the chapter. It is now time to look at some of what researchers who have approached this phenomenon have learned.

Unfortunately, most research supports a pessimistic view of how long-distance relationships fare (Knox, et al., 2002; Van Horn, et al., 1997; Guldner, 1996; Stafford and Reske, 1990; Schwebel et al., 1992). Although they may not fail because once "out of sight, (we are) out of mind," they don't appear to hold up especially well over time. Using survey research techniques with samples of college students, Van Horn et. al. found that partners in long-distance romantic relationships reported less companionship, less disclosure, less satisfaction, and less certainty about the future together compared to partners in geographically close relationships (Van Horn, et. al., 1997). Comparing 164 students in long-distance relationships with 170 in geographically proximal relationships, Guldner reported that the separated partners showed more depressive symptoms (Guldner, 1996). Schwebel et al. (1992), studied 34 men and 55 women who were in relationships in which they were separated by at least 50 miles from their partners. Within 9 weeks, nearly a quarter of the relationships had ended. Finally, Knox et al. (2002) surveyed 438 undergraduates, in part to test their belief in the "out of sight, out of mind" idea as well as to gauge their experiences of such relationships. Nearly 20 percent of the sample reported being in a long-distance relationship, here meaning separated by 200 miles or more. Thirty-seven percent of the sample reported having ever been in a long-distance relationship that ended. Although more than half of the sample with experience of long-distance relationships had phoned and/or e-mailed several times a week, more than 40 percent felt that the distance had worsened (20 percent) or ended (21.5 percent) their relationships. Conversely, 18 percent said that it "improved" their relationship. Having experienced a long-distance relationship made respondents more likely to believe "out of sight, out of mind."

The most optimistic findings suggest that long-distance relationships are not especially different from more proximal relationships (Guldner and Swensen, 1995). Comparing 194 students in long-distance relationships with 190 who were in geographically close relationships Guldner and Swensen found that the two types of relationships were rated with about the same levels of self-reported relationship satisfaction, and similar levels of intimacy, trust, and degree of relationship progress. This hardly constitutes "making the heart grow fonder," but it is less negative than the other research.

Was there *any* research found to support "absence makes the heart grow fonder?" The answer to that is yes, and yet that itself may be problematic for healthy relationship development. Comparing 34 "geographically close" couples and 37 long-distance couples (separated by an average of 421.6 miles), Stafford and Reske found that long-distance couples were more satisfied with their relationships and with the level of communication they had. They also were by their assessments "more in love." Acknowledging the possibility that the long-distance relationships were "better" than the geographically closer relationships, Stafford and Reske go on to suggest that a process of *idealization* occurs in long-distance relationships, due largely to their more restricted communication (more phone calls and letters as opposed to face-to-face interaction, and less overall interaction). As a consequence of this idealization, couples set themselves up for later problems that couples with less restricted communication (i.e., more geographically close couples) avoid. They "may have little idea of how idealized and inaccurate their images (of their relationships) are (Stafford and Reske, 1990).

It is worth noting that even if *all existing research* painted a negative picture of what happens in long-distance relationships, that does not mean that any particular relationship (i.e., yours, or your friend's) is destined to meet that unfortunate outcome. Family scientists seek to identify and account for *patterns* in social relationships. There are always going to be exceptions to any identified pattern. This is important for two reasons. First, don't assume that patterns reported in this book *will happen* in your life. Your experience may constitute an exception to the more general pattern. Second, and equally important, don't dismiss findings reported here because they don't fit your experiences or those of people you may know. Instead, try to account for why your experience departs from the more generally observed social regularities.

By using our critical thinking skills and by understanding something about the methods and theories used by family researchers, we are in a position to more effectively evaluate the information we receive about families. We are also better able to step outside our personal experience, go beyond what we've always been told, and begin to view marriage and family from a sounder and broader perspective. In Chapters 3 and 4, we will take such steps and explicitly examine the factors and forces that create differences in family experience.

SUMMARY

- We need to be alert to maintain *objectivity* in our consideration of different forms of family lifestyle. The *values* we have about what makes a successful family can cause us to decide ahead of time that certain family lifestyles are "abnormal," because they differ from our own experience or preferences. *Opinions, biases,* and *stereotypes* are ways of thinking that lack objectivity.
- *Fallacies* are errors in reasoning. Two common types of fallacies are *egocentric fallacies* and *ethnocentric fallacies.* These reflect errors in our basic presuppositions that come either from thinking all people are, or should be, the same as we are, or that our way of living is superior to all others.
- Theories attempt to provide frames of reference for the interpretation of data. Theories of marriage and families include ecological theory, symbolic interaction, social exchange theory, structural functionalism, conflict theory, and family systems theory.
- *Ecological theory* examines how families are influenced by and, in return, influence the wider environments in which they function.
- The *symbolic interaction theory* examines how people interact and how we interpret or define others' actions through the symbols they communicate (their words, gestures, and actions). Symbolic interactionists study how social roles and personality interact. Drawbacks to the approach include (1) the tendency to minimize power in relationships, (2) the failure to account fully for the psychological aspects of human life, (3) an emphasis on individualism and personal fulfillment at the expense of the marital or family unit, and (4) inadequate attention to the social context.
- *Social exchange theory* suggests that we measure our actions and relationships on a cost-benefit basis. People seek to maximize their rewards and minimize their costs to gain the most favorable outcome. A corollary to exchange is *equity:* Exchanges must balance out, or hard feelings are likely to ensue. Exchanges in marriage can be either cooperative or competitive. Exchange theory has been criticized because (1) it assumes that individuals are rational and calculating in relationships, and (2) it says that the value of costs, rewards, and resources can be gauged.
- *Structural functionalism* looks at society and families as an organism containing different structures. Each structure has a function. Structural functionalists study three aspects of the

family: (1) the functions the family serves for society, (2) the functional requirements performed by the family for its survival, and (3) the needs of individual members that are met by the family. Family functions are usually divided along gender lines. Criticisms include (1) the inability to test the theory empirically, (2) the difficulty in ascertaining what function a particular structure serves, and (3) the theory's conservative bias against viewing change as functional.

- *Conflict theory* assumes that individuals in marriages and families are in conflict with each other. Power is often used to resolve the conflict. Four important sources of power are (1) legitimacy, (2) money, (3) physical coercion, and (4) love. Criticism of conflict theory includes (1) the theory's politically based view of human nature, (2) its assumption that differences lead to conflict, and (3) the difficulty in measuring and evaluating conflict.
- *Family systems theory* approaches the family in terms of its structure and pattern of interactions. Systems analysts believe that (1) interactions must be studied in the context of the family, (2) family structure can be seen only in the family's interactions, (3) the family is a system purposely seeking *homeostasis* (stability) and (4) family systems are transformed over time. Family systems theory is not a coherent, systematic theory, and it has been criticized because it is based on clinical studies of nonrepresentative families.
- The *feminist perspective* provides an orienting focus for considering *gender* differences relating to family and social issues. In the writing, research, and advocacy of the feminist movement, the goals are to help clarify and remove oppressive conditions and barriers to opportunities for women. The postmodern feminist position has been expanded to include constraints affecting black/white and gay/straight dichotomies.
- The feminist perspective's attention to gender gave rise to men's studies, a field in which scholars examine how masculinity and male socialization shape men's experiences, including their family lives.
- *Family development theory* looks at the changes in the family, beginning with marriage and proceeding through seven sequential stages. This life-cycle model gives us insights into the complexities of family life and the interacting influences of changing roles and circumstances through time. Family development theory has been criticized for assuming that all families follow the sequence of traditional two-parent families. However, family life-cycle models have been developed to include nontraditional forms such as single-parent families, stepfamilies, and families headed by lesbian or gay parents.
- Family researchers use the *scientific method*—well-established procedures that are used to collect information. They may ask the same questions of large numbers of persons, using the *quantitative* method of research, or they may study smaller groups or individuals in more depth, using *qualitative* methods of research. Analyzing information from public sources (newspapers, court records, and media) is termed *secondary data analysis.*
- Research data come from surveys, clinical studies, and direct observation, which are all *correlational* types of studies, in which naturally occurring variables are measured against each other. Data are also obtained from experimental research.
- *Surveys* use questionnaires and interviews. They are most useful for dealing with societal or general issues rather than personal or small-group issues. Inherent problems with the survey method include (1) volunteer bias or an unrepresentative sample, (2) individuals' lack of self-knowledge, and (3) underreporting of undesirable or unconventional behavior.
- *Clinical research* involves in-depth examinations of individuals or small groups who have entered a clinical setting for the treatment of psychological or relationship problems. The primary advantage of clinical studies is that they allow in-depth *case studies;* their primary disadvantage is that the people coming into a clinic are not representative of the general population.
- *Observational research* consists of studies in which interpersonal behavior is examined in a natural setting, such as the home, by an unobtrusive observer.
- In *experimental research,* the researcher manipulates variables. Such studies are of limited use in marriage and family research because of the difficulty of controlling behavior and duplicating real-life conditions.

KEY TERMS

adaptation 35
anonymity 50
bias 32
case-study method 54
clinical research 54
conceptualization 34
confidentiality 50
conflict theory 43
continuous coverage 40
dependent variable 34
egocentric fallacy 32
environment 35
environmental influences 34
equity 39
ethical guidelines 50
ethnocentric fallacy 32
exosystem 35
experimental research 57
fallacy 32
family development theory 41
family myth 37
family systems theory 45
feminist perspective 46
gender 46
homeostasis 46
independent variable 34
interaction 37
objective statement 32
objectivity 32
observational research 56
opinion 32
qualitative research 49
quantitative research 49
scientific method 33
secondary data analysis 50
social exchange theory 39
social role 37
stereotype 32
structural functionalism 42
subsystem 42
survey research 50
symbolic interaction 36
theory 33
value judgment 32
variable 34

SUGGESTED READINGS

Copeland, Ann, and Kathleen White. *Studying Families.* Newbury Park, CA: Sage, 1991. Research issues in studying diverse families, such as cohabiting parents, gay and lesbian families, and childless marriages.

Real, Michael R. *Explaining Media Culture: A Guide.* Thousand Oaks, CA: Sage, 1996. A scholarly, insightful, and thought-provoking examination of the role of media in modern life.

Sussman, Marvin, Suzanne Steinmetz, and Gary Peterson, eds. *Handbook of Marriage and the Family,* 2nd ed. New York: Plenum Press, 1999. The definitive reference book on marriage and the family.

White, James, and David Klein. *Family Theories,* 2nd ed.). Thousand Oaks, CA: Sage, 2002. An in-depth survey of seven major theoretical frameworks.

Winton, C. *Frameworks for Studying Families.* Guilford, CT: Dushkin, 1995.

Zillman, D., J. Bryant, and A. C. Huston, eds. *Media, Children, and the Family: Social, Scientific, Psychodynamic, and Clinical Perspectives.* Hillsdale, NJ: Erlbaum, 1994. One of the key books on TV reality, with essays on gender, social behaviors, and family portrayals.

RESOURCES ON THE INTERNET

Companion Web Site for This Book

http://sociology.wadsworth.com/strong/marriage9e
Gain an even better grasp on this chapter by going to the companion Web site to take one of the Tutorial Quizzes, use the Flash Cards to master key terms, or check out the many other study aids you'll find there. Visit the Marriage and Family Resource Center on the site. You'll also find special features such as GSS Data and Census 2000 information that will put data and resources at your fingertips to help you with that special project or to do some research on your own.

InfoTrac College Edition: Search Word Summary

skepticism
social roles
conflict theorists
moralizing
feminist perspective
minority group

To learn more about these central topics in the study of the family, you can conduct an electronic search using InfoTrac College Edition. To aid in your search and to gain useful tips, see the Student Guide to InfoTrac College Edition that you can access through the companion Web site for this book.

Preview

To gain a sense of what you already know about the material covered in this chapter, answer "True" or "False" to the statements below.

1. Compared with contemporary families, colonial family life was considered a more private matter. True or False?
2. Industrialization transformed the role families played in society as well as the roles women and men played in families. True or False?
3. Slavery destroyed the African-American family system. True or false?
4. Compared with what came both before and after, families of the 1950s were unusually stable. True or false?
5. Within upper-class families, husbands and wives are relatively equal in their household roles and authority. True or false?
6. Lower-class families are the most likely to be single-parent families. True or false?
7. Family relationships can suffer as a result of either downward or upward mobility. True or false?
8. Compared with white families, relationships between African-American husbands and wives are much more traditional. True or false?
9. Looking at Asian-American or Latino families shows much variation within each group, depending upon the country from which they came, why they left, and when they arrived in the United States. True or false?
10. European ethnic groups are as different from each other as they are from African Americans, Latinos, Asian Americans, or Native Americans. True or false?

"If, in fact, the family is a product of its time and place in the hierarchy of social institutions, then American families will be both similar and different—similar in that they share some common experiences, some elements of a common culture . . . different in that class, race, and ethnic differences give a special cast to the shared experience as well as a unique and distinctly different set of experiences."

LILLIAN BRESLOW RUBIN

CHAPTER 3

Dynamics and Diversity of Families

Outline

Get people into a conversation about families and, before too long, it is likely that someone will utter the familiar phrase, "Well, all families are different." Of course, there is a lot of truth to that sentiment. For example, your family is not like your best friend's family in every way. Furthermore, assuming your best friend is someone a lot like you (which is frequently the case among people who become best friends), the differences between your families likely understate how much more richly variable family experience actually is.

While it is true that in some ways *every family is different*, in this and the next chapter we will look at some of the patterned variations that separate and diversify family experiences. Although there are a number of factors that one could include as sources of such variation, we will concern ourselves with the five we consider the most important: time, social class, race, and ethnicity in this chapter, and gender in Chapter 4. We will begin by detailing the historical development of the kinds of families that predominate in the United States today, noting key transformations and the forces that created them. This will accomplish two things: It will give you a better sense of where today's American families have come from; it will also enable you to see how different family life has been across generations, even within the same families. We will then shift our attention to the major racial, ethnic, and economic variations that diversify contemporary American families.

American Families across Time

PREVIEW ANSWERS

1 False
2 True
3 False
4 True
5 False
6 True
7 True
8 False
9 True
10 False

American marriages and families are dynamic and must be understood as the products of wider cultural, demographic, and technological developments (Mintz and Kellogg, 1988). Although we tend to emphasize the changes that have occurred over the past half century (post–World War II), those changes represent only more recent instances of the 300 years of change that comprise the history of American family life. An overview of families in America from the colonial period through the twentieth century may provide a historical perspective of the functions of tradition, culture, and values and the relationship that exists among them. Armed with this brief history, we can recognize and make connections between changes in society and changes in families. Additionally, we will be better positioned to assess the meaning associated with some of the more dramatic changes that have occurred more recently in American family life. Finally, on a more personal level, we can better understand our own genealogies and family histories by recognizing the shifting stage on which they were played out.

THE COLONIAL ERA

The colonial era is marked by differences among cultures, family roles, customs, and traditions. These families were the original crucible from which our contemporary families were formed.

Perspective

HISTORICAL PATTERNS OF MARRIAGE AND THE FAMILY

One difficulty we face in tracing historical patterns of family living is where to draw the boundaries. Specifically, how far back should we reach for our starting point? What societies are we going to consider? In what characteristics of family living are we most interested? Although most of this chapter, like the remainder of the book, focuses on patterns found among American families, it is important to consider a somewhat longer view of their history. Clearly, not all American families trace their ancestry back to European origins. However, Colonial era family life was heavily influenced by the patterns that characterized European family life. Based upon Sheila McIsaac Cooper's fine review of prevalent themes in historical analysis of families, we can note the following.

Until the last few decades of the twentieth century, historians mostly ignored the history of family life. Beginning in the 1960s, historians ceased to view family history as "peripheral" to history, and devoted considerable energies toward filling the void in our historical knowledge (Cooper, 1999). Despite the attention that family history has since received, there is still disagreement among historians as to how to characterize European families of the past. The disagreement extends to issues such as the structure of households and families, the relative importance of emotional ties, and the nature and meaning of childhood (Cooper, 1999).

Historians have disagreed about how common nuclear and extended families were in pre-industrial Europe. Some, most notably Peter Laslett, co-founder of the Cambridge Group for the History of Population and Social Structure, suggest that the nuclear family was the dominant family form both in preindustrial England and throughout western Europe (Cooper, 1999). Others suggest that a form of extended family, the **stem family,** actually characterized family life in many parts of Europe. The stem family consists of a three generational household in which one son and his family continued to reside with his aging parents, ultimately inheriting their property. Though split over the relative prevalence of each type, and of the more familiar extended family, historians "generally agree" that nuclear families were more compatible with such demographic characteristics as age at marriage, life expectancy, and number of children per family (Cooper, 1999).

On the question of the importance of emotional ties, there has been disagreement between the "sentiments school" and their critics. According to historians such as Edward Shorter and Lawrence Stone, between the late seventeenth and early twentieth centuries, families became more affective and love-based rather than rational and calculating. Emotional ties (i.e., *sentiments*) became the distinctive feature of marriages and families. Critics of this perspective argue that romantic love and companionate marriage can be found centuries earlier than credited by the sentiments school (Cooper, 1999).

Finally, historians have disagreed about when modern concepts of childhood and of children's needs first surfaced. The "sentiments" school, in keeping with their overall perspective on families, suggests that the concept of childhood emerged in the late seventeenth and eighteenth centuries. Prior to that, emotional relations between parents and children are depicted as being weak. More optimistic interpretations suggest that the concept of childhood and emotional attachments between parents and children date back at least as far as the Middle Ages (Cooper, 1999).

The disagreements among historians make it difficult to offer definitive characterizations of the kinds of families that existed in Europe in the period before the founding of Colonial America.

Native American Families

The greatest diversity in American family life probably existed during our country's earliest years, when two million Native Americans inhabited what is now the United States and Canada. There were over 240 groups with their own distinct family and kinship patterns. Many groups were **patrilineal:** Rights and property flowed from the father. Others, such as the Zuni and Hopi in the Southwest and the Iroquois in the Northeast, were **matrilineal:** Rights and property descended from the mother.

Native-American families tended to share certain characteristics, although it is easy to overgeneralize. Most families were small. There was a high child mortality rate, and mothers breastfed their infants; during breastfeeding, mothers abstained from sexual intercourse. Children were often born in special birth huts. As they grew older, the young were rarely physically disciplined. Instead, they were taught by example. Their families praised them when they were good and publicly shamed them when they were bad. Children began working at an early stage. Their play, such as hunting or playing with dolls, was modeled on adult activities. Ceremonies and rituals marked transitions into adulthood. Girls underwent puberty ceremonies at first menstruation. For boys, events such as getting the first tooth and killing the first large animal when hunting signified stages of growing up. A vision quest often marked the transition to manhood.

Marriage took place early for girls, usually between the age of 12 and 15 years; for boys, it took place between the age of 15 and 20 years. Some tribes arranged marriages; others permitted young men and women to choose their own partners. Most groups were monogamous, although some allowed two wives. Some tribes permitted men to have sexual relations outside of marriage when their wives were pregnant or breastfeeding.

Colonial Families

From earliest colonial times, America has been an ethnically diverse country. In the houses of Boston, the mansions and slave quarters of Charleston, the maisons of New Orleans, the haciendas of Santa Fe, and the Hopi dwellings of Oraibi (the oldest continuously inhabited place in the United States, dating back to 1150 A.D.), American families have provided emotional and economic support for their members.

THE FAMILY. Colonial America was initially settled by waves of explorers, soldiers, traders, pilgrims, servants, prisoners, farmers, and slaves. In 1565, in St. Augustine, Florida, the Spanish established the first permanent European settlement in what is now the United States. But the members of these first groups came as single men—as explorers, soldiers, and exploiters.

In 1620, the leaders of the Jamestown colony in Virginia, hoping to promote greater stability, began importing English women to be sold in marriage. The European colonists who came to America attempted to replicate their familiar family system. This system, strongly influenced by Christianity, emphasized **patriarchy** (rule by father or eldest male), the subordination of women, sexual restraint, and family-centered production.

The family was basically an economic and social institution, the primary unit for producing most goods and caring for the needs of its members. The family planted and harvested food, made clothes, provided shelter, and cared for the necessities of life. Each member was expected to contribute economically to the welfare of the family. Husbands plowed, planted, and harvested crops. Wives supervised apprentices and servants, kept records, cultivated the family garden, assisted in the farming, and marketed surplus crops or goods, such as grain, chickens, candles, and soap. Older children helped their parents and, in doing so, learned the skills necessary for later life.

As a social unit, the family reared children and cared for the sick, infirm, and aged. Its responsibilities included teaching reading, writing, and arithmetic because there were few schools. The family was also responsible for religious instruction: It was to join in prayer, read scripture, and teach the principles of religion.

Unlike New Englanders, the planter aristocracy that came to dominate the Southern colonies did not give high priority to family life; hunting, entertaining, and politics provided the greatest pleasure. The planter aristocracy continued to idealize gentry ways until the Civil War destroyed the slave system upon which the planters based their wealth.

MARITAL CHOICE. Romantic love was not a factor in choosing a partner; one practical seventeenth-century marriage manual advised women that "this boiling affection is seldom worth anything . . ." (Fraser, 1984). Because marriage had profound economic and social consequences, parents often selected their children's mates. Such choices, however, were not as arbitrary as it may seem. Parents tried to choose partners whom their children already knew and with whom they seemed compatible. Children were expected to accept the parents' choices.

Love came after marriage. In fact, it was a person's duty to love his or her spouse. The inability to desire and love a marriage partner was considered a defect of character.

Although the Puritans prohibited premarital intercourse, they were not entirely successful. **Bundling,** the New England custom in which a young man and woman spent the night in bed together, separated by a wooden bundling board, provided a courting couple with privacy; it did not, however, encourage restraint. An estimated one-third of all marriages in the eighteenth century took place with the bride pregnant (Smith and Hindus, 1975).

FAMILY LIFE. The colonial family was strictly patriarchal. The authority of the husband/father rested in his control of land and property. In an agrarian society such as colonial America, land was the most precious resource. The manner in which the father decided to dispose of his land affected his relationships with his children. In many cases, children were given land adjacent to the father's farm, but the title did not pass into their hands until the father died. This power gave fathers control over their children's marital choices as well as keeping them geographically close.

This strongly rooted patriarchy called for wives to submit to their husbands. The wife was not an equal, but was instead a helpmate. This subordination was reinforced by traditional religious doctrine. Like her children, the colonial wife was economically dependent on her husband. Upon marriage, she transferred to her husband many of the rights she had held as a single woman, such as the right to inherit or sell property, to conduct business, and to attend court.

For women, marriage marked the beginning of a constant cycle of childbearing and child rearing. On the average, colonial women had six children and were consistently bearing children until around age 40. In addition to their maternal responsibilities, colonial women were expected to cook; wash; sew; milk; clean; garden; brew beer; churn butter; harvest fruit; keep chickens; spin wool; bake bread; make cheese; boil laundry; stitch shirts, petticoats, and other garments; salt, pickle, and preserve vegetables, fruits, and meats; and make clothing and soap (Mintz and Kellogg, 1988).

CHILDHOOD AND ADOLESCENCE. The colonial conception of childhood was radically different from ours. First, children were believed to be evil by nature. The community accepted the traditional Christian doctrine that children were conceived and born in sin.

Second, childhood did not represent a period of life radically different from adulthood. Such a conception is distinctly modern (Aries, 1962; Meckel, 1984; Vann, 1982). In colonial times, a child was regarded as a small adult. When children were 6 or 7, childhood ended for them. From that time on, they began to be part of the adult world, participating in adult work and play.

Third, when children reached the age of 10, they were often "bound out" for several years as apprentices or domestic servants. They lived in the home of a relative or stranger, where they learned a trade or skill, were educated, and were properly disciplined.

Adolescence—the separate life stage between childhood and adulthood—did not exist. One went from a shorter childhood (than what we are accustomed to) to adulthood (Mintz and Kellogg, 1988). Thus, our twentieth-century notions of a rebellious life stage filled with inner conflicts, youthful indiscretions, and developmental crises does not fit well with the historical record of Plymouth Colony (Demos, 1970). Evidence indicates that there were different standards of behavior expected and accepted from children under or over the age of 16, which was considered the "age of discretion." However, our concept of an identifiable and unique life stage between childhood and adulthood was absent until the turn of the twentieth century.

African-American Families

In 1619, a Dutch man-of-war docked at Jamestown in need of supplies. As part of its cargo were 20 Africans who had been captured from a Portuguese slaver. The captain quickly sold his captives as indentured servants. Among these Africans was a woman known by the English as Isabella and a man known as Antony; their African names are lost. In Jamestown, Antony and Isabella married. After several years, Isabella gave birth to William Tucker, the first African-American child born in what is today the United States. William's birth marked the beginning of the African-American family, a unique family system that grew out of the African adjustment to slavery in America.

During the eighteenth century and later, West African family systems were severely repressed throughout the New World (Guttman, 1976). At first, some slaves tried unsuccessfully to continue polygamy, which was strongly rooted in many African cultures. Enslaved African Americans were more successful in continuing the traditional African emphasis on the extended family, in which aunts, uncles, cousins, and grandparents played important roles. Although slaves were legally prohibited from marrying, they created their own marriages. Despite

Strong family ties endured in enslaved African-American families. The extended family, important in West African cultures, continued to be a source of support and stability.

© Hulton Archive/Getty Images

the hardships placed on them, they developed strong emotional bonds and family ties. Slave culture discouraged casual sexual relationships and placed a high value on marital stability. On the large plantations, most enslaved people lived in two-parent families with their children. To maintain family identity, parents named their children after themselves or other relatives or gave them African names. In the harsh slave system, the family provided strong support against the daily indignities of servitude. As time went on, the developing African-American family blended West African and English family traditions (see McAdoo, 1996).

NINETEENTH-CENTURY MARRIAGES AND FAMILIES

In the nineteenth century, the traditional colonial family form gradually vanished and was replaced by the modern family.

REFLECTIONS

How far back can you trace your family's history? What would you like to know about it? What values, traits, or memories do you wish to pass on to your descendants?

Industrialization and the Shattering of the Old Family

In the nineteenth century, **industrialization** transformed the face of America. It also transformed American families from self-sufficient farm families to wage-earning urban families. As factories began producing gigantic harvesters, combines, and tractors, significantly fewer farm workers were needed. Looking for employment, workers migrated to the cities, where they found employment in the ever-expanding factories and businesses. Because goods were now bought rather than made in the home, the family began its shift from being primarily a production unit to being a consumer and service-oriented unit. With this shift, a radically new division of labor arose in the family. Men began working outside the home in factories or offices for wages to purchase the family's necessities and other goods. Men became identified as the family's sole provider or breadwinner. Their work was given higher status than women's work because it was paid in wages. Men's work began to be identified as "real" work.

At the same time that industrialization made husbands the breadwinner in the family, it also undercut much of their power over their children as fathers. Children were no longer dependent on their fathers. Opportunities existed outside of and away from one's parental household.

Industrialization also created the housewife, the woman who remained at home attending to household duties and caring for children. Because much of what the family needed had to be purchased with the husband's earnings, the wife's contribution in terms of unpaid work and services went unrecognized, much as it continues today.

Marriage and Families Transformed

Without its central importance as a work unit, and less and less the source of other important societal functions (for example, education, religious worship, protection, recreation), the family became the focus and abode of feelings. The emotional support and well-being of adults and the care and nurturing of the young became the two most important family responsibilities.

THE POWER OF LOVE. This new affectionate foundation of marriage brought love to the foreground. Love as the basis of marriage represented the triumph of individual preference over family, social, or group considerations. Parents had little power in selecting their children's partners, and their children were no longer as economically dependent on them. Women now had a new degree of power; they were able to choose whom they would marry. Women could rule out undesirable partners during courtship; they could choose mates with whom they believed they would be compatible. Mutual esteem, friendship, and confidence became guiding ideals. Without love, marriages were considered empty shells.

CHANGING ROLES FOR WOMEN. The two most important family roles for middle-class women in the nineteenth century were that of housewife and mother. As there was a growing emphasis on domesticity in family life, the role of the housewife increased in significance and status. Home was the center of life, and the housewife was responsible for making family life a source of fulfillment for everyone.

Women also increasingly focused their identities on motherhood. The nineteenth century witnessed the most dramatic decline in fertility in American history. Between 1800 and 1900, fertility dropped by 50 percent, falling from an average of 7 to about 3.5 children per woman. Women reduced their childbearing by insisting that they, not men, control the frequency of intercourse. Child rearing rather than childbearing became one of the most important aspects of a woman's life. Having fewer children allowed more time to concentrate on mothering and opened the door to greater participation in the world outside the family. This outside participation manifested itself in women's heavy involvement in the abolition, prohibition, and women's emancipation movements.

CHILDHOOD AND ADOLESCENCE. A strong emphasis was placed on children as part of the new family. The belief in childhood innocence replaced the idea of childhood corruption. A new sentimentality surrounded the child, who was now viewed as born in total innocence. Protecting children from experiencing or even knowing about the evils of the world became a major part of child rearing.

The nineteenth century also witnessed the beginning of adolescence. In contrast to colonial youths, who participated in the adult world of work and other activities, nineteenth-century adolescents were kept economically dependent and separate from adult activities and often felt apprehensive when they entered the adult world. This apprehension sometimes led to the emotional conflicts associated with adolescent identity crises.

Education also changed as schools, rather than families, became responsible for teaching reading, writing, and arithmetic as well as educating students about ideas and values. Conflicts between the traditional beliefs of the family and those of the impersonal school were inevitable. At school, the child's peer group increased in importance.

The African-American Family: Slavery and Freedom

Although there were large numbers of free African Americans—100,000 in the North and Midwest and 150,000 in the South—most of what we know about the African-American family prior to the Civil War is limited to the slave family.

THE SLAVE FAMILY. By the nineteenth century, the slave family had already lost much of its African heritage. Under slavery, the African-American family lacked two key factors that helped give free African-American and white families stability: autonomy and economic importance. Slave marriages were not recognized as legal. Final authority rested with the owner in all decisions about the lives of slaves. The separation of families was a common occurrence, spreading grief and despair among thousands of slaves. Furthermore, slave families worked for their masters, not themselves. It was impossible for the slave husband/father to become the provider for his family. The slave women worked in the fields beside the men. When an enslaved woman was pregnant, her owner determined her care during pregnancy and her relation to her infant after birth. The age at which children recognized themselves as slaves depended on whether they were children of field or house slaves. If their parents were field slaves, they would be subject to slave discipline from the beginning. But if the children were offspring of house slaves (or the slave owner), they were often playmates of their master's children. But the day would come when such a child would know that he or she was a slave, and that time would be filled with grieving, anger, and humiliation. The knowledge created a deep crisis in the child's concept of self.

Still, it is important to reiterate that slavery did not destroy slave families. Despite the very real oppression and hardship to which they were subjected, slaves survived by reliance on their families and by adapting their family system to the conditions of their lives (Mintz and Kellogg, 1988). This included, for example, relying upon extended kinship networks and, where necessary, on unrelated adults to serve as surrogates for parents absent due to the forced breakup of families.

Furthermore, enslavement did not destroy the African-American family system. In fact, "the stable, two-parent nuclear family" was the norm among slaves (Mintz and Kellogg, 1988). Recognizing this does not diminish the horrors of slavery. Instead, it acknowledges the resilience of those who survived enslavement, and it illustrates how family systems may be pivotal sources of support and key mechanisms of surviving extraordinary distress.

AFTER FREEDOM. When freedom came, the formerly enslaved African-American family had strong emotional ties and traditions forged from slavery and from their West African heritage (Guttman, 1976; Lantz, 1980). Because they were now legally able to marry, thousands of former slaves formally renewed their vows. The first year or so after freedom was marked by what was called "the traveling time," in which African Americans traveled up and down the South looking for lost family members who had been sold. Relatively few families were reunited, although many continued the search well into the 1880s.

African-American families remained poor, tied to the land, and segregated. Despite poverty and continued exploitation, the Southern African-American family usually consisted of both parents and their children. Extended kin continued to be important.

Immigration: The Great Transformation

THE OLD AND NEW IMMIGRANTS. In the nineteenth and early twentieth centuries, great waves of immigration swept over America. Between 1820 and 1920, 38 million immigrants came to the United States. Historians commonly divided them into "old" immigrants and "new" immigrants. The old immigrants, who came between 1830 and 1890, were mostly from western and northern Europe. During this period, Chinese also immigrated in large numbers to the West Coast. The new immigrants, who came from eastern and southern Europe, began to arrive in great numbers between 1890 and 1914 (when World War I virtually stopped all immigration). Japanese also immigrated to the West Coast and Hawaii during this time. Today, Americans can trace their roots to numerous ethnic groups.

National Park Service: Statue of Liberty National Monument

Except for Native Americans, most of us have ancestors who came to America voluntarily or involuntarily. Between 1820 and 1920, more than 38 million immigrants came to the United States.

As the United States expanded its frontiers, surviving Native Americans were incorporated. The United States acquired its first Latino population when it annexed Texas, California, New Mexico, and part of Arizona after its victory over Mexico in 1848.

THE IMMIGRANT EXPERIENCE. Most immigrants were uprooted; they left only when life in the old country became intolerable. The decision to leave their homeland was never easy. It was a choice between life or death and meant leaving behind ancient ties.

Most immigrants arrived in America without skills. Although most came from small villages, they soon found themselves in the concrete cities of America. Again, families were key ingredients in overcoming and surviving extreme hardship. Because families and friends kept in close contact even when separated by vast oceans, immigrants seldom left their native countries without knowing where they were going—to the ethnic neighborhoods of New York, Chicago, Boston, San Francisco, Vancouver, and other cities. There they spoke their own tongues, practiced their own religions, and ate their customary foods. In these cities, immigrants created great economic wealth for America by providing cheap labor to fuel growing industries.

In America, kinship groups were central to the immigrants' experience and survival. Passage money was sent to their relatives at home, information was exchanged about where to live and find work, families sought solace by clustering close together in ethnic neighborhoods, and informal networks exchanged information about employment locally and in other areas.

The family economy, critical to immigrant survival, was based on cooperation among family members. For most immigrant families, as for African-American families, the middle-class idealization of motherhood and childhood was a far cry from reality. Because of low industrial wages, many immigrant families could survive only by pooling their resources and sending mothers and children to work in the mills and factories.

Most groups experienced hostility. Crime, vice, and immorality were attributed to the newly arrived ethnic groups; ethnic slurs became part of everyday parlance. Strong activist groups arose to prohibit immigration and promote "Americanism." Literacy tests required immigrants to be able to read at least 30 words in English. In the early 1920s, severe quotas were enacted that slowed immigration to a trickle.

It is interesting to note what crucial roles families played in enabling people to survive the oppression of enslavement, the difficulties of immigration, and the impoverishment induced by industrialization. In comparison to these more extreme conditions, today one often sees family breakdown, instability, or even abuse in the face of circumstances such as unemployment or underemployment (Rubin, 1976; 1994; Newman, 1988). As we look at some important twentieth-century cultural and social changes, we can begin to make sense of why such different outcomes follow these kinds of despair.

REFLECTIONS

As you read through these historical perspectives, what are your feelings about such struggles and triumphs? How does knowledge of your family history affect you, your values, and your behavior?

TWENTIETH-CENTURY MARRIAGES AND FAMILIES

The Rise of Companionate Marriages: 1900–1960

By the beginning of the twentieth century, the functions of American middle-class families had been dramatically altered from earlier times. Families had lost many of their traditional economic, educational, and welfare functions. Food and goods were produced outside the family, children were educated in public schools, and the poor, aged, and infirm were increasingly cared for by public agencies and hospitals. The primary focus of the family was becoming even more centered on meeting the emotional needs of its members. In time, cultural emphasis would shift from self-sacrificing familism to more self-centered individualism, and one's sense of their connections and obligations to their families would be greatly transformed.

THE NEW COMPANIONATE FAMILY. Beginning in the 1920s, a new ideal family form was beginning to emerge that rejected the "old" family based on male authority and sexual repression. This new family form was based on the **companionate marriage.**

There were four major features of this companionate family (Mintz and Kellogg, 1988): (1) men and women were to share household decision making and tasks; (2) marriages were expected to provide romance, sexual fulfillment, and emotional growth; (3) wives were no longer expected to be guardians of virtue and sexual restraint; (4) children were no longer to be protected from the world but were to be given greater freedom to explore and experience the world; they were to be treated more democratically and encouraged to express their feelings.

Through the Depression and World Wars

The history of twentieth-century family life cannot be told without considering how profoundly family roles and relationships were affected by the Great Depression and two world wars. Although many different connections could be drawn, two seem particularly significant: changes in the relationship between the family and the wider society, and changes in women's and men's roles in and outside of the family.

LINKING PUBLIC AND PRIVATE LIFE. The economic crisis during the Depression was staggering in its scope. Unemployment jumped from under 3 million in 1929 to over 12 million in 1932, while the rate of unemployment rose from 3.2 percent to 23.6 percent. Over that same span of time average family income dropped 40 percent (Mintz and Kellogg, 1988). To cope with this economic disaster, families turned inward, modifying their spending, increasing the numbers of wage earners to include women and children, and pooling their incomes. Often it was a broadened "inward" to which they turned, as people often took in relatives or relied on kinship ties for economic assistance (Mintz and Kellogg, 1988).

Ultimately, these more personal, intrafamilial efforts proved insufficient. President Franklin Roosevelt's New Deal social programs attempted to respond to the social and economic despair that more localized efforts were unable to alleviate. Farm relief, rural electrification, Social Security, and a variety of social welfare provisions were all implemented in the hope of doing what local communities and individual families could not. Such federal initiatives reflected a dramatic ideological shift wherein government now bore responsibility for the lives and well-being of families (Mintz and Kellogg, 1988).

World War II posed its own "policy" problems for families. Precipitated by the mass entrance into the workforce of millions of previously unemployed women, including many with young chil-

© CORBIS

© Lambert/Hulton Archive/Getty Images

During Word War II women were urged to enter the labor force and especially to enter nontraditional occupations left vacant by the deployment of men overseas. The images above illustrate the kinds of messages women received and the kinds of jobs they helped fill.

dren, there was a clear need and opportunity for public resources to be committed to child care. Unfortunately, the federal government's response was slow and inadequate, given the sudden and dramatic increase in need and demand (Filene, 1986; Mintz and Kellogg, 1988). Unlike some of our European allies who invested more heavily in policies and services to accommodate employed mothers (Mintz and Kellogg, 1988), it took the federal government two years to "appropriate funds to build and staff day-care centers, and the funds were sufficient for only one-tenth of the children who needed them" (Filene, 1986). Despite having engineered a propaganda campaign to entice women into jobs vacated by the 16 million men who entered the service, the government remained ambivalent about welcoming mothers of young children into those positions. However inadequate or slow their efforts were, they were still more ambitious than what followed for most of the rest of the century.

GENDER CRISES: THE GREAT DEPRESSION AND WORLD WARS. Both the Depression and the two world wars (especially the World War II) reveal much about the gender foundation on which twentieth-century families rested. During the Depression, it was men whose gender identities and family statuses were threatened by their lost status as providers. During each world war, women were the ones who faced challenges that required them to abandon their gender socialization and step into roles and situations that fell outside their traditional familial roles. In each instance, the familial gender roles and identities had to be altered to match extraordinary circumstances. By looking at how people reacted to these alterations, we see the important familial meanings they attached to their expected roles.

Recall that a consequence of industrialization was the development of the male-as-economic-provider role and the breadwinner-homemaker family. Wage earning became men's primary activity and a major source of their self-identities and family statuses. Men's wages were supposed to be sufficient to support their households. During the Depression, as male unemployment skyrocketed so too did their despair over their sense of having failed as men (Filene, 1986; Mintz and Kellogg, 1988).

What is especially striking about men's reactions to their job loss is their internalization of fault for what was, in fact, a societywide economic crisis. Given how widespread unemployment was, one might think that men would take some comfort in knowing that the predicaments they faced were not of their own making. Yet they had so deeply internalized their sense of themselves as providers that their identities, family statuses, and sense of manhood were all invested in wage earning and providing. When unable to provide, many men were deeply shaken. Some were even driven to the point of emotional breakdown or suicide by their sense of economic failure (Filene, 1986).

For many families, survival depended upon the efforts of wives or the combination of women's earnings, children's earnings, assistance from kin, or some kind of public assistance. For those who depended at least somewhat on women's earnings, there were other gender consequences of running the household. Sometimes, men were pressed by their wives to contribute domestically in the women's "absence." While some did, many others resisted (Filene, 1986). Sometimes, women themselves displayed ambivalence about the meaning of male unemployment and male housework. Whereas 80 percent of the women who were surveyed in 1939 by the *Ladies' Home Journal* thought an unemployed husband should do the domestic work in the absence of his employed wife, 60 percent reported they would lose respect for men whose wives out-earned them (Filene, 1986).

If the Depression illustrates male anxiety about their familial roles as providers, we see in women's experiences during World Wars I and II that gender crises were not limited to men. Both wars share in common the fact that, in the absence of millions of men, women were pressed to step into their vacant shoes and participate in wartime production. One and a half million women entered the wartime labor force during the First World War, many in jobs previously held largely by men (Filene, 1986). During World War II, the number of employed women rose dramatically. Between 1941 and 1945 the numbers of employed women increased by more than 6 million, to a wartime high of 19 million (Degler, 1980; Lindsey, 1997). Furthermore, "nearly half of all American women held a job at some time during the war" (Mintz and Kellogg, 1988). Whereas single women had long worked, and poor or minority women had worked even after marriage, the biggest change in women's labor force participation during World War II was among married, middle-class women. Thus, despite the strong and widely held cultural emphasis on the special nurturing role of women and the belief that the home was a woman's "proper place," American society needed women to take over for the absent men.

As discussed above, to overcome women's potential reluctance, a carefully pitched appeal went out to American women suggesting that it was their patriotic duty to fill the void left by the men overseas. The Office of War Information had to convince women that defense work was desirable, lucrative, and something they were capable of doing well (Lindsey, 1997). Likewise, private industry became more open to the hiring of women, regardless of age or marital status (Filene, 1986), and more strategically and aggressively recruited them into jobs from which they were otherwise excluded.

Once enticed into nontraditional female employment, women received both material and nonmaterial benefits that were hard for many to surrender once the war ended and men returned. Materially, women in traditionally male occupations received higher wages than they had in their past, more sex-segregated work experiences. As important, they also found a sense of gratification and enhanced self-esteem that were often missing from the jobs they were more accustomed to. However reluctant they may have been to take on such work, many were clearly more than a little ambivalent to leave it.

To assist women in their departures from these jobs, pro-family rhetoric and a new ideology extolling the value and importance of women's roles as mothers and caregivers were broadly conveyed by a variety of sources (for example, popular media, social workers, and educators). Many women left the traditionally male occupations they held during the war to marry and have children. Others left their wartime jobs to return to more typical women's jobs. Overall, women's labor-force participation continued to rise through the postwar decades (Fox and Hesse-Biber, 1984), as did the more specific participation of married women with children

The *"Leave It to Beaver"* Families of the 1950s

In the long history of American family life, no other decade has come to symbolize so much about, despite representing relatively little of, that period (Mintz and Kellogg, 1988; Coontz, 1997). In many ways, the 1950s appear to be a period of unmatched family stability. Marriage and birthrates were unusually high, divorce rates were uncharacteristically low, and the economy enabled many to afford to buy houses with only one wage-earning spouse. Drawing upon firsthand experiences or popular media constructions (for example, *Father Knows Best, Ozzie and Harriet, Leave It to Beaver, The Donna Reed Show*), many look at this period as a kind of "Golden Age" of American family life, after which the quality and stability of our families began to steadily decline (Coontz, 1997; Mintz and Kellogg, 1988).

During this decade, marriage and family seemed to be central to American lives. It was a time of youthful marriages, increased birthrates, and a stable divorce rate. Most families were comprised of male breadwinners and female homemakers. For the most part, traditional gender and marital roles prevailed. Man's place was in the world and woman's place was in the home. Women were expected to place motherhood first and to sacrifice their own opportunities for outside advancement for the success of their husbands and the well-being of their children.

Given the meaning that is often invested in this era, it is important to understand that the 1950s were unique. Compared to both what came before and what followed, families of the 1950s were exceptional. This is important: It means that anyone who uses this period as a baseline against which to compare more recent trends in such family characteristics as birth, marriage, or divorce rates starts with a faulty assumption about the representativeness of this decade. Looking at those same trends with a longer view reveals that the changes that followed in the 1950s were more consistent with some of the patterns evident in the nineteenth and earlier part of the twentieth century (Mintz and Kellogg, 1988).

For example, the trend since the Civil War had been an increase in the divorce rate of about 3 percent per decade, until the 1950s. During the 1950s, the divorce rate increased less than in any other decade of the twentieth century. Similarly, after more than one hundred years of declining birthrates and shrinking family sizes, during the 1950s "women of childbearing age bore more children, spaced . . . closer together, and had them earlier and faster" than had previous generations (Mintz and Kellogg, 1988). After all, this was the height of the baby boom; married couples had more children than either those that preceded them or those that followed.

Much of the familial experience of the 1950s was created and sustained by the unprecedented economic growth and prosperity of the postwar economy (Coontz, 1997). The combination of suburbanization and economic prosperity allowed many married couples to achieve the middle-class family dream of home ownership while raising their children under the loving attention of full-time caregiving mothers. Ignoring for the moment the fact that some men and women found the reality associated with such households other than what they had expected and less than they desired, economic vitality enabled many Americans to experience this social ideal (Coontz, 1997).

When we look at family changes that occurred in subsequent decades, we need to recognize that economic factors, again, were important determinants of some of the more dramatic departures from the 1950s model. This especially pertains to the emergence of the dual-earner household. As Stephanie Coontz points out, "By the mid-1970s, maintaining the prescribed family lifestyle meant for many couples giving up the prescribed family form. They married later, postponed children, and curbed their fertility; the wives went out to work" (Coontz, 1997). They did this not in rejection of the family lifestyle of the 1950s but in the pursuit of central features of that lifestyle, such as home ownership.

Of course, we must be careful not to oversimplify family experience of the 1950s. Americans did not all benefit equally from the economic prosperity and opportunity of the decade. Thus, overgeneralizations would leave out the experiences of poor and working-class families and racial minorities, for whom neither full-time mothering nor home ownership were commonplace (Coontz, 1997). Additionally, many women found that the ideal lifestyle of the period left them longing for something more (Friedan, 1963).

Suburbanization especially affected women and families. Most suburbanites were young couples and their children. Few suburban housewives were employed outside the home. Indeed, children dominated the suburban landscape. In subordinating themselves to meeting the needs of their children, some women found themselves isolated from other interests and people. This isolation, coupled with a transient lifestyle occasioned by company-related moves, made loneliness one of suburbia's compelling problems.

Aspects of Contemporary Marriages and Families

The remaining chapters of this book look most closely at families of the latter decades of the twentieth century. Characteristics of these families did not emerge suddenly but were established over a number of years. Beginning with the latter years of the 1950s and escalating through and then beyond the 1960s and 1970s, some dramatic family trends surfaced. These trends more or less persisted through the end of the twentieth century, leaving marriages and families reshaped and the meaning and experience of family life significantly altered.

Birthrates dropped, people delayed and departed marriage as almost never before, and increasingly were drawn to cohabitation. The median age for marriage began to climb in the 1960s, reaching a point (in 1996) that was the highest it had been in more than one hundred years. Even with a slight drop in the last few years of the 1990s, the age at entering first marriage remains four years older for men and nearly five years for women than the 1960 ages (see Table 3.1).

TABLE 3.1 ■ Median Age at First Marriage, 1960–2002

YEAR	MALES	FEMALES
1960	22.8	20.3
1970	23.2	20.8
1980	24.7	22.0
1990	26.1	23.9
2000	26.8	25.1
2002	26.9	25.3

SOURCE: U.S. Census Bureau, 2003. Annual Demographic Supplement to the March 2002 Current Population Survey, Current Population Reports, Series P20–547.

TABLE 3.2 ▪ Trends in Family Characteristics: 1960–2001

	1960	1970	1980	1990	2001
Marriage rate	8.5	10.6	10.6	9.8	8.4
Divorce rate	2.2	3.5	5.2	4.7	4.0
Birthrate[a]	23.7	18.4	15.9	16.7	14.5
Cohabiting couples[b]	439,000	523,000	1.6 million	2.8 million	4.5 million

SOURCE: U.S. Census Bureau, 2002.
[a]Rate per 1000 people.
[b]Total number of cohabiting couples.

As we saw in Chapter 1, marriage and divorce rates rose and fell, and the prevalence of cohabitation substantially increased, while birthrates dropped. But even across this shorter historical span we see that marriage and family trends are not linear; they go up and then drop again (see Table 3.2). Thus it appears that, even in the short term, the only constant in family life is change (Mintz and Kellogg, 1988).

Although the trends shown in Table 3.2 are not the only dimensions of family life that have seen major change (look again at Table 1.1), they are important indicators that the family is indeed a dynamic institution. They are also often the source of much controversy over their larger meaning. It is obvious that such trends depict change, but what is less clear is what those changes say about the vitality of the family.

As we suggested in Chapter 2, some argue that changes such as these are worrisome signs of family decline (Popenoe, 1993). With fewer people marrying, more people divorcing, more people living together (or by themselves) outside of marriage, the importance of the family—as reflected in the stability or desirability of marriage—appears to be declining, and the future of family life is in some doubt. Others take the more liberal position that, in and of itself, change is not a bad thing, and that with these changes come more choices for people about the kinds of families they wish to create and experience (Mintz and Kellogg, 1988; Coontz, 1997). And certainly, today's families do reflect considerable diversity of structure. In painting a picture of today's families we would include many categories: breadwinner-homemaker families with children, two-earner couples with children, single-parent households with children, marriages without children, cohabiting couples with or without children, blended families, role-reversed marriages, and gay and lesbian couples with or without children. Whereas American families have from their beginnings been diverse entities, with varying cultural and economic backgrounds (Mintz and Kellogg, 1988), what distinguishes contemporary families is the diversity represented by the range and spread of people across these varying chosen lifestyles.

FACTORS PROMOTING CHANGE

Marriages and families are shaped by a number of different forces in society. In looking over the description of major changes to American families, we can identify four important factors that initiated these changes: (1) economic changes, (2) technological innovations, (3) demographics, and (4) gender roles and opportunities for women.

Economic Changes

As noted earlier, the family has over time moved from being an economically productive unit to a consuming, service-oriented unit. Where families once met most of the needs of their members —including providing food, clothing, household goods, and occasionally surplus crops which it bartered or marketed—today's families purchase what they need.

Economic factors have been responsible for major change in the familial roles played by women and men. Inflation, economic hardship, and an expanding economy have led to married women entering the labor force in unprecedented numbers. Even married women with preschool-aged children are typically employed outside the home (see Table 3.3). As a result, the dual-earner marriage and the employed mother have become commonplace features of contemporary families.

TABLE 3.3 ■ **Labor Force Participation of Married Women with Children 6 Years Old or Younger**

1960	1970	1980	1990	2000
19%	30%	45%	59%	65%

SOURCE: U.S. Census Bureau.

As women have increased their participation in the paid labor force, other familial changes have occurred. Women are less economically dependent upon either men or marriage. This gives women greater legitimacy in exercising marital power. It has also increased the tension around the division of household chores and raised anxiety and uncertainty over who will care for children.

Technological Innovations

The family has been affected by most major technological developments and innovations—from automobiles, telephones, cell phones, televisions, DVD players, and microwaves to personal computers and the Internet. These devices were not designed to transform families but to enhance transportation, communication, entertainment, and efficiency. Nevertheless, they have had major repercussions in how family life is experienced. For example, such older devices as automobiles and telephones, as well as more recent innovations such as personal computers have aided families in maintaining contact across greater distances, thus allowing extended families to maintain closer relations and nuclear family members to stay available to each other through school- and job-related travel or relocation. The proliferation of automobiles also altered the residential and relationship experiences of many Americans, making it possible for people to live greater distances from where they work—thus contributing to the suburbanization of America—and to experience premarital relationships away from more watchful adult supervision.

Televisions, and more recently the Internet, have altered the recreation and socialization activities in which families engage, with both beneficial and negative consequences. Sitting and watching television programs together gives family members the opportunity for shared experiences. As important as the entertainment function of both television and the Internet are, they also operate as additional socialization agents, beyond parents and other relatives. What we watch on television, or view and read on the Internet, helps shape our values and beliefs about the world around us. Cell phones, e-mail, and instant messaging have altered the ways in which parents monitor children and family members remain in contact with each other.

The range of domestic appliances—from washing machines and dishwashers to microwaves—has altered how the tasks of housework are done. Although we might be tempted to conclude that such devices free people from some of the time- and labor-intensive burdens associated with maintaining homes, historical research has shown that this is not automatically so. For instance, as technology made it possible to more easily wash clothes, the standards for cleanliness increased. In the case of microwaves, the time needed for tasks associated with meal preparation has been reduced, freeing people to spend more time in other activities (not necessarily as families, and often away from their families—at work, for example).

Finally, revolutions in contraception and biomedical technology have had great impact on reshaping the meaning and experience of sexuality and parenthood. Much of the "sexual revolution" in the 1970s and beyond was fueled in part by safer and more certain methods of preventing pregnancy, such as the birth control pill. Regarding parenthood, people who in the past would have been unable to become parents have the opportunity to enjoy childbearing and rearing as a result of **assisted reproductive technologies**—including medical advances such as in vitro fertilization as well as surrogate motherhood and sperm donors. Such developments have thus altered the meaning of parenthood, as multiple individuals may be involved in any single conception, pregnancy, and eventual birth. Sperm and/or egg donors, surrogate mothers, and the parent(s) who do the actual nurturing and raising of the child all can claim in some way to have parented the child in question. Such changes have also complicated the social and legal meanings of parenthood at the same time that they have opened

Understanding Yourself

CONNECTING YOUR FAMILY HISTORY TO THE HISTORY OF FAMILIES

What we experience in our family relationships is partly a product of when we are born and live. This, like your race, ethnicity, and social class, is something over which you neither have control nor exercise choice, yet it limits or offers you choices and constrains or opens up opportunities.

One way to illustrate this is to gather information on your own family's history. If you carefully map out your family's history and compare it with some of the historical indicators discussed in this chapter, you will likely see connections between these broader patterns and your own family's story. You will then be better able to both see the larger picture and understand your family's unique experiences across generations.

What do you know of your family's history? How did your parents first meet? What about their parents? How old were your parents and grandparents when they married? How many siblings do you have? How many did your parents and grandparents have? Did your mother work outside the home when you were younger? Were your grandmothers employed when your parents were children? Comparing across generations, how many divorces have occurred in your family? When was the first one?

There are many ways to explore your family's history. You can examine family photographs, read over letters and diaries, or interview living members to learn what happened, when, and why. Interviews need not be formal but we highly recommend learning as much as you can about your families from surviving members of your families. These may be opportunities to hear family stories and corerect any misunderstanding you have about your families that might be lost as people age and pass away. Family photographs can reveal much about the relationships between members. Gather together photographs of your immediate family and your forebears—grandparents, great-grandparents, and so on. With the help of knowledgeable relatives, identify who is in these pictures. Then look closely at such details as facial expressions, body language, and positioning of family members relative to one another. Are family members clustered closely together or far apart? Is someone standing off from the others? Is someone looking gloomy amidst others who are smiling? What family resemblances do you see? Do you look like a great-great-grandparent, for example?

After you gather information about your pictured relatives, see what aspects of the family discussed in this chapter apply to your family members. Was a great-great-great-grandmother a slave? How did your family weather the Depression? How did relatives go about their daily household tasks? If you can, interview members of your family about what they know. Try to find out stories about the oldest family members. Where did they come from? What important historical events occurred during their lifetimes? In which ones did they actually participate? What were their own experiences of joy and sorrow? What did they pass down—love of learning, ambition, money, pride?

Such family histories will better enable you to understand where you come from and what factors have shaped the family experiences you have had. Connecting our own family history to the wider history of American families is a first step toward a better understanding of both.

up the possibility of parenthood to previously infertile couples or same-sex couples.

REFLECTIONS

As you stop and think about your own family routines and relationships, how much do they seem to be affected by the kinds of technological innovations discussed here? How would your experiences be different *without* these devices?

Demographics

The family has undergone dramatic demographic changes in areas that include family size, life expectancy, divorce, and death. Three important changes have emerged:

- *Increased longevity:* As people live longer they are experiencing aspects of family life that few experienced before. In colonial times, because of a relatively short life expectancy, husbands and wives could anticipate a marriage lasting twenty-five years. Today, couples can remain married for fifty or sixty years. Today's couples can also anticipate living many years together after their children are grown; they can also look forward to grandparenthood or great-grandparenthood. Since many men marry women younger than themselves and die younger than women do, American women can anticipate a prolonged period of widowhood.

- *Increased divorce rate:* The increased divorce rate, beginning in the late nineteenth century (even before 1900 the United States had the highest divorce rate in the world), has led to the rise of single-parent families and stepfamilies. In this way, it has dramatically altered the experience of both childhood and parenthood and has altered our expectations of married life.
- *Decreased fertility rate:* As women bear fewer children, they have fewer years of child-rearing responsibility. With fewer children, partners are able to devote more time to each other and expend greater energy on each child. Children from smaller families benefit in a variety of ways from the greater levels of parental attention, though they lack the advantages of having multiple siblings. Smaller families also afford women greater opportunity for entering the workforce.

Gender Roles/Opportunities for Women

Changes in gender roles are a third force contributing to alterations in American marriages and families. The history summarized above indicated some major changes that took place in women's and men's responsibilities and opportunities. These gender shifts then directly or indirectly led to changes in both the ideology surrounding and the reality confronting families.

In colonial times, men were primarily responsible for tilling fields, harvesting crops, and manufacturing work implements; women produced family goods and necessities. As production was transferred to factories and large-scale farms, men's and women's roles changed and families were no longer self-sufficient. Instead, they relied on wages earned and brought home by men. Women concentrated on mothering and household management and were further expected to shape their children's character and meet the needs of their husbands. Such changes translated into the "invention" of the male breadwinner and female housewife, neither of which existed in the more side-by-side division of labor in colonial families.

The emphasis on child rearing and housework as women's proper duties lasted until World War II, when, as we saw, there was a massive influx of women into factories and stores to replace the men fighting overseas. This initiated a trend in which women increasingly entered the labor force, became less economically dependent on men, and gained greater power in marriage.

The feminist movement of the 1960s and 1970s led many women to reexamine their assumptions about women's roles. Betty Friedan's *The Feminine Mystique* challenged head-on the traditional assumption that women found their greatest fulfillment in being mothers and housewives. The publication of this one book thus tapped into and symbolized a fundamental discontent that many suburban housewives were experiencing. Despite doing what they were supposed to and having all their material needs met, many remained dissatisfied. The women's movement emerged to challenge the female roles of housewife, helpmate, and mother, appealing to some women at the same time that it alienated others.

More recently, the dual-earner marriage made the traditional division of roles an important and open question for women. Today, contemporary women have dramatically different expectations of male/female roles in marriage, child rearing, housework, and the workplace than did their mothers and grandmothers. Changes in marriage, birth, and divorce rates, and in the ages at which people enter marriage, have all been affected by women's enlarged economic roles.

We have also witnessed changes in what men expect and are expected to do in marriage and parenthood. Although it may still be assumed that men will be "good providers," that is no longer enough. Married men face greater pressure to share housework and participate in child care. Although they have been slow to increase the amount of housework they do, there has been a more acceptance of the idea that greater father involvement benefits both children and fathers. New standards of paternal behavior and more participation by fathers in raising children helps explain the ongoing changes—from how dual-earner households function to why we have seen increases in the numbers of single, custodial fathers.

DID YOU KNOW?

The U.S. Census Bureau estimates that there are 105,000 married fathers with children age 15 or younger who function as the primary, at-home caregivers of their children. These fathers take care of approximately 189,000 children. Furthermore, 2 million preschoolers are cared for by their fathers more than any other caregivers while their mothers are at work (U.S. Census Bureau, 2003).

Cultural Changes

We can in conclusion point to a shift in American values from an emphasis on obligation and self-sacrifice to individualism and self-gratification (Bellah et al., 1985; Mintz and Kellogg, 1998; Coontz, 1997). The once strong sense of familism, in which individual self-interest was subordinated to family well-being, has given way to more open and widespread individualism, in which even families can be sacrificed for the sake of one's happiness and personal fulfillment.

This values shift has had consequences for how people weigh and choose among alternative lifestyle paths. For example, complex decisions—about whether and how much to work, whether to stay married or get a divorce, how much time and attention to devote to children or to spouses—are increasingly made against a backdrop of pursuing self-gratification and individual happiness. Values alone have not changed families, but such shifts in values have contributed to the choices people make, out of which new family forms come to predominate (Coontz, 1997).

Social Class Variations in Family Life

A **social class** is a category of people who share a common economic position in the stratified (i.e., unequal) society in which they live. We typically identify classes by using economic indicators such as ownership of property or wealth, amount of income earned, the level of prestige accorded to one's work, and so forth. Social class has both a structural and cultural dimension. Structurally, social class reflects the occupations we hold (or depend upon), the income and power they give us, and the opportunities they present or deny us. The cultural dimension of social class refers to any class-specific values, attitudes, beliefs, and motivations that distinguish classes from each other. Cultural aspects of social class are somewhat controversial, especially when applied to supposed "cultures of poverty"—an argument holding that poor people become trapped in poverty because of the values they hold and the behaviors in which they engage (Harrington, 1962; Lewis, 1966). What is unclear regarding "cultures of class" is how much difference there is in the values and beliefs of different classes and whether such differences cause or follow the more structural dimensions that separate one class from another.

To an extent, there is also a psychological aspect to social class. By this we refer to the internalization of one's economic status in the self-images we form and the self-esteem we possess. These may also be seen as consequences of other aspects of one's class position, such as the self-identity that results from the prestige accorded to one's work or the respect paid to one's accomplishments. Like the structural and cultural components of social class, these are brought home and affect our experiences in our families.

Clearly, many facets of our lives (often referred to as **life chances**) are affected by our **socioeconomic status,** including our health and well-being, safety, longevity, religiosity, and politics. A host of family experiences also vary up and down the socioeconomic ladder. For instance, class variations can be found in such family characteristics as age at marriage, age at parenthood, timing of marriage and parenthood, division of household labor, ideologies of gender, socialization of children, meanings attached to sexuality and intimacy, and the likelihood of violence or divorce.

Conceptualizations of social class vary in how class is defined and how many classes are identified and counted in American society. In some formulations of social class, it is a person's relationship to the means of production that defines class position. In other models, people are grouped into classes depending on having similar incomes, amounts of wealth, degrees of occupational status, and years of education. Whether we claim that the United States has two (owners and workers), three (upper, middle, lower), four (upper, middle, working, lower), six (upper-upper, lower-upper, upper-middle, lower-middle, upper-lower, lower-lower) or more classes, the important point is that life is differently experienced by individuals across the range of identified classes and similarly experienced by people within any one of the class categories.

Using a fairly common model, we can describe these classes as follows.

Upper Classes

Roughly 7 to 10 percent of the population occupies an "upper class" position. The uppermost level of this class represents approximately 3 percent of the population (Renzetti and Curran, 1998; Curry, Jiobu, and Schwirian, 2002). They own 25 to 30 percent of all private wealth and 60 to 70 percent of all corporate wealth. They also receive as much as 25 percent of all yearly income. They are sometimes referred to as the "upper-upper class," or as the "ruling class" or "elite." Their "extraordinary wealth" often takes them into the hundreds of millions if not billions of dollars (Curry, Jiobu, and Schwirian, 2002).

The rest of the upper class live on yearly incomes ranging from hundreds of thousands to billions of dollars, own considerable amounts of wealth, and enjoy much prestige. In fact, some members of the lower-upper class may be wealthier than their elite counterparts, living well in large private homes in exclusive communities and enjoying considerable privilege. The major distinction between the elite and the lower-upper class is between "old" and "new" money (Steinmetz, Clavan, and Stein, 1992; Langman, 1988). In other words, the clearest distinction one can draw is in how they achieved and how long they have enjoyed their affluence.

Middle Classes

In some analyses, the middle class is considered the largest class, representing between 45 to 50 percent of the population (Curry, Jiobu, and Schwirian, 2002; Renzetti and Curran, 1998). Here, too, we can subdivide the middle class into two groupings: the **upper-middle class** and the **lower-middle class.** The former consists of highly paid professionals (for example, lawyers, doctors, engineers) who have annual incomes that may reach into the hundreds of thousands of dollars (Renzetti and Curran, 1998). They are typically college educated although they may not have attended the same elite colleges as the upper-upper class (Curry, Jiobu, and Schwirian, 2002). Women and men of the upper-middle class have incomes that allow them luxuries such as home ownership, vacations, and college educations for their children. The lower-middle class comprises a larger portion of the population, made up of white-collar service workers with incomes between $25,000 and $50,000. They own or rent more modest homes, purchase more affordable automobiles, and hope to send their children to college (Renzetti and Curran, 1998).

Working Class

About a third of the U.S. population is **working class.** They tend to work as skilled laborers, earn between $15,000 and $25,000, and have high school or vocational educations. The working class lives somewhat precariously, with little savings and few liquid assets should illness or job loss occur (Rubin, 1994). They also have difficulty buying their own homes or sending their children to college (Curry, Jiobu, and Schwirian, 2002).

Lower Class

Despite an official estimate of 15 percent of the population being below the "poverty line," a more accurate assessment might indicate that closer to 20 percent of Americans are poor (Seccombe, 2000). As originally established, the poverty line was determined by calculating the annual costs of a "minimal food budget" multiplied by three, since 1960s survey data estimated that families spent one-third of their budgets on food (Seccombe, 2000). Families whose incomes are just *one dollar above* this threshold are not officially classified as poor. Poor families are characterized by irregular employment or chronic underemployment. Individuals work at unskilled jobs that pay minimum wage and offer little security or opportunity for advancement (Renzetti and Curran, 1998). Although many lower-class individuals rent substandard housing, we also find a homelessness problem among poor families. Karen Seccombe (2000) effectively describes the problems: "Poverty affects one's total existence. It can impede adults' and children's social, emotional, biological,

© Bill Bachmann/PhotoEdit

© A. Ramey/Stock Boston, LLC

Family experiences are affected by such variables as social class and race.

and intellectual growth and development." She further notes that over a year's time, the majority of poor families experience one or more of the following: "eviction, utilities disconnected, telephone disconnected, housing with upkeep problems, crowded housing, no refrigerator, no stove, or no telephone" (Seccombe, 2000).

CLASS AND FAMILY LIFE

Working within this framework, we can note some ways in which family life is differently experienced by each of the four classes. Although there are a number of family characteristics one could look at (including divorce, domestic violence, or the division of labor, to name a few), let's look briefly at class-based differences in marriage relationships, parent-child relationships, and ties between nuclear and extended families.

Marriage Relationships

Within upper-class families we tend to find sharply sex-segregated marriages, in which women are subordinated to their husbands. Upper-class women often function as supports for their husbands' successful economic and political activities, thus illustrating the **two-person career** (Papanek, 1973). Although their supportive activities may be essential to the husbands' success, such wives are neither paid nor widely recognized for their efforts. Rather than having their own careers, they often do volunteer work within charitable organizations or their communities. They are free to pursue such activities because they have many servants—from cooks to chauffeurs and nannies—who do the domestic work and some of the child care or supervision .

Middle-class marriages tend to be *ideologically* more egalitarian and are frequently two-career marriages. In fact, middle-class lifestyles increasingly require two incomes. This creates both benefits and costs for middle-class women. The benefits include having more say in family decision making and greater legitimacy in asking for help with domestic and child-rearing tasks. The costs include the failure to receive the help they request. Because they likely earn less than their husbands, the strength of their role in family decision making may thus still be less than that of their husbands. Middle-class couples also highly value and more readily accept the ideal of marriage as a sharing, communicating relationship in which spouses function as "best friends."

Once more explicitly traditional, working-class marriages are becoming more like their middle-class counterparts. Whereas such marriages in the past were clearly more traditional in both rhetoric and their division of responsibilities, in recent years they have moved more toward a model of sharing both roles and responsibilities (Komarovsky, 1962; Rubin, 1976, 1994). The sharply segregated, traditional marriage roles that were evident even just two decades ago have given way to two-earner households, increasingly driven by the need for two incomes. Among those working-class couples who work "opposite" shifts, we find higher levels of sharing domestic and child-care responsibilities, as well as greater male involvement in home life (Rubin, 1994). The reality of being the only parent home forces men to take on tasks that otherwise might be done by wives. Necessity, not ideology, creates this outcome. The meaning of male participation in home life may vary more than actual behavior or vary differently than levels of actual involvement. Male involvement may have greater "value" in the circles in which middle-class men live and work but be more of a practicality or necessity for working-class men. Thus, working-class men may understate, and middle-class men may exaggerate, what and how much they do.

Lower-class marriages are the least stable marriages. In fact, men are often entirely absent from day-to-day family life, as is suggested by the idea of the **feminization of poverty.** Resulting from the combination of high divorce rates and more widespread nonmarital childbearing, almost 40 percent of those living in poverty are single women and their children (O'Hare, 1996). The association of men's wage earning with fulfillment of their family responsibilities subjects lower-class men to harsher experiences within families. They are less likely to marry. If married, they are less likely to remain married, and when married they derive fewer of the benefits that supposedly accrue in marriage.

Parents and Children

The relationships between parents and children vary across social lines, but most research has focused on the middle and working classes (Kohn, 1990). Among the upper class, some of the actual hands-on child rearing may be done by nannies or au pairs. Certainly mothers are involved, and relationships between parents and children are loving, but parental involvement in economic and civic activities may sharply curtail their actual time with children (Langman, 1987). For upper-class parents, an important objective is to see that children acquire the appropriate understanding of their social standing and that they cultivate the right connections with others like themselves. They may attend private and exclusive boarding schools and later join appropriate clubs and organizations. Their eventual choice of a spouse receives especially close parental scrutiny.

Research indicates that working- and middle-class parents socialize their children differently and have different objectives for child rearing. Although all parents want to raise happy and caring children, middle-class parents tend to emphasize autonomy and self-discipline, while working-class parents tend to stress compliance (Kohn, 1990). In a recent study of mothers, Hays (1996) identified differences between what middle- and working-class mothers believed made for a "good mother" and what they thought children most need. Like others, Hays found that working-class mothers tended to stress obedience; middle-class mothers negotiated more with their children. Whereas working-class mothers saw education as essential for their children's later life chances, middle-class mothers took for granted that their children would receive good quality educations and emphasized, instead, the importance of building children's self-esteem. Finally, although both classes of mothers acknowledged using spanking to discipline their children, middle-class mothers spank more selectively and favored other methods of discipline (for example, "time-out") (Hays, 1996).

Lower-class families are the most likely to be single-parent families. Single parents, in general, may suffer stresses and experience difficulties that parents in two-parent households do not, but this situation is exacerbated for low-income single parents (McLanahan and Booth, 1989). Parent-child relationships suffer from a variety of characteristics of lower-class life: unsteady, low-pay employment; substandard housing; uncertainty about obtaining even the most basic necessities (food, clothing, and so forth), all of which can affect the quality of parent-child relationships and the ability of parents to supervise and control what happens to and with their children.

Extended Family Ties

Linkages between nuclear family households and extended kin vary in kind and meaning across social class. By some measures, the least closely connected group may be the middle class, who, due to the

Perspective

SOCIAL CLASS AND TEEN-PARENT COMPARISON

Although there are many ways in which a higher social class eases aspects of family life, that is not always the case. Affluence brings with it its own pressures and problems. Across social classes parents may desire and aspire to make their children's lives better than their own, but there is interesting research that shows significant, and possibly unexpected, social class differences.

Contrary to many other cross-class family comparisons, in some ways *affluent* families create certain problems for their children that are less likely to occur or are less severe in families who struggle economically. An October 1999 *New York Times*/CBS News Poll of 1,038 thirteen- to seventeen-year-olds found that children from more prosperous homes expressed anxiety about matching their parents' successes and expectations. They were also more likely than their less affluent peers to report that, compared to their parents, their lives were harder (Lewin, 1999). As shown in the table below, comparing those whose household incomes were under $30,000 with those from families with incomes over $75,000, shows a notable difference in how each group believes their lives compare with their parents' lives.

Anthropologist Gina Bria, commenting in the *New York Times*, explained the results as follows: "Wealthy kids often have less of a perspective on the struggles of life. And in some strata, there is a lot of pressure to keep the family's new-found status, and children feel the pressure to get into elite schools, to be sure they don't slide back. Falling back is the American nightmare." (Bria, 1999). Those who come from more modest economic origins, and whose lifestyles are more deprived than their higher-income peers, tend to believe that their lot is easier than that of their parents. Thus, they are more likely to believe that they have no right to complain (Rubin, 1976).

How Teens Think Their Lives Compare with Parents' Lives

	INCOME <$30,000	INCOME >$75,000
Life is harder	38%	50%
Life is about the same	18%	21%
Life is easier	41%	28%

SOURCE: *New York Times*, November 7, 1999.

geographic mobility that accompanies their economic status, may find themselves the most physically removed from their kin. Of course, middle class families do visit kin or phone regularly and are available to exchange aid when needed. Still, the emphasis is on the conjugal family of spouses and children.

Closer connections may be found among both the working and upper classes though the reasons differ. In the case of working-class families, there are often both the opportunity and the need for extensive familial involvement. Opportunity results from lesser levels of geographic mobility, which results in closer proximity and allows for more continuous contact to result. The need for involvement is created by the pooling of resources and exchange of services (for example, child care) that often result between adults and their parents or among adult siblings. Intergenerational upward mobility may lessen the reliance on extended families (see discussion below).

Upper-class families, especially among the "old" upper class, highly value the importance of family name and ancestry. They tend to maintain strong and active kinship groups that exert influence in the mate selection processes of members and monitor the behavior of members. Inheritance of wealth gives the kin group more than symbolic importance in their ability to influence behavior of individual members.

Among the lower class, kin ties—both real and fictive—may be essential resources in determining economic and social survival. Grandparents, aunts, and uncles may fill in for or replace absent parents, and multigenerational households (for example, children living with their mothers and grandmothers) are fairly common. **Fictive kin ties** refer to the extension of kinshiplike status to neighbors and friends, thus symbolizing both an intensity of commitment and a willingness to help each other meet needs of daily life (Stack, 1974; Liebow, 1967).

THE DYNAMIC NATURE OF SOCIAL CLASS

Like other aspects of family life, one's social class position is not set in stone. Individuals may experience **social mobility,** movement up *or* down the social class ladder. Either kind of social mobility can affect family relationships, especially intergenerational relationships (Newman, 1988; Sennett and Cobb, 1972). For example, children who see their parents "fall from grace," through job loss and dwindling assets come to look differently at those parents. Fathers who once seemed heroic may become the source of concern, and even resentment, as their job loss threatens the lifestyle of the family on which children depend (Newman, 1988). Children who in their adulthoods climb upward, occasionally find their relationships with parents suffering as a result. As they are exposed to new values and ideas that differ from those held by their parents, generational tension and social distance may follow. Furthermore, as they move into a new social circle, parents (as well as less mobile siblings) may appear to fit less well with their new life circumstances. The more they fit into new circles and circumstances, the less well they may fit comfortably within their ongoing family relationships.

In similar ways, for comparable reasons, marital relationships may be altered by either downward or upward mobility. Research indicates that some men who lose their jobs and "slide downward" react to their economic misfortune by abusing their spouses, turning to alcohol or other substances, withdrawing emotionally, or leaving the home (Rubin, 1994; Newman, 1988). Changes in the marriage are not entirely of men's doing; after an initial period of sympathy and support, wives may grow impatient with their husbands' unemployment, or alter their positive views of the husbands' dedication as a worker or job seeker. Additionally, as couples are forced to scale back their accustomed lifestyle, tensions may rise and resentment and distance may grow.

Upward mobility may also transform marriage relationships. We are familiar with the situation faced by women who, after sacrificing to help launch their husbands' careers by supporting them through school, are left by those same husbands once they have achieved success. With their own increasing economic opportunity, some women find that marriage itself becomes less desirable because of the constraints it continues to impose upon their career development.

Racial and Ethnic Diversity

According to recent census data (U.S. Census Bureau, 2000), 29 percent of the U.S. population are people of color: 13 percent are African American, 11 percent are Hispanic, 4 percent are Asian/Pacific Islander, and 1 percent are Native American. Projections of population change suggest that by the year 2025, non-Hispanic whites will constitute a little over 60 percent of the American population, Hispanics 18.2 percent, African Americans 12.9 percent, Asian/Pacific Islanders 6.5 percent, and Native Americans 1 percent. By 2050, the population is expected to be just over 50 percent white, 24 percent Hispanic, 13 percent black, 9 percent Asian, and 1 percent Native American.

It is important to be aware of the danger of thinking in terms of **ethnocentric fallacies,** beliefs that one's own ethnic group, nation, or culture is innately superior to others. In this section we will consider briefly some of the distinctive characteristics and strengths of families from various ethnic and cultural groups.

We begin our discussion of diversity by noting several important terms. A **racial group** is a group of people, such as whites, blacks, and Asians, classified according to their **phenotype**—their anatomical and physical characteristics. Racial groups share common phenotypical characteristics, such as skin color and facial structure. The concept of race is often misused and misunderstood. We should neither assume a purity or homogeneity within racial groupings (in skin color, facial features, and so on), nor treat racial groups as superior or inferior in comparison to each other. In either of those biological applications, the concept of race is a myth

(Henslin, 2000). Socially, however, the fact that we perceive ourselves within racial classifications and are treated and act toward others on the basis of race, makes it a highly significant factor in shaping our life experiences.

An **ethnic group** is a group of people distinct from other groups because of cultural characteristics. Such things as language, religion, and customs are shared within and differentiate between ethnic groups. These cultural characteristics are transmitted from one generation to another and may shape how each person thinks and acts—both in and outside of families.

Either racial or ethnic groups can be considered **minority groups** depending on their social experience. Minority groups are so designated, not because of their numerical size in the wider population, but because their status (position in the social hierarchy) places them at an economic, social, and political disadvantage (Taylor, 1994b). These terms often overlap. For example, African Americans are simultaneously an ethnic, racial, and minority group in the United States. The term *African American,* used increasingly instead of *black,* reflects the growing awareness of the importance of ethnicity (culture) in contrast to race (skin color) (Smith, 1992; but see Taylor, 1994b).

Ethnic and/or racial differences are often difficult to untangle from social class differences. It may be that some differences in family patterns reflect cultural background factors or distinctive values. However, it is equally plausible that ethnic or racial differences in family patterns reflect the different socioeconomic circumstances under which different groups live (Aponte, Beal, and Jiles, 1999).

CHANGING PERSPECTIVES ON ETHNICITY AND FAMILY

Until the last 25 years, most research about American marriages and families was limited to the white, middle-class family. The nuclear family was the norm against which all other families were evaluated. As we have seen, such a perspective distorts our understanding of other family forms, such as single-parent families and stepfamilies. These family forms were often viewed as pathological because they differed from the traditional norm. A similar distortion also has influenced our understanding of African-American, Latino, Asian-American, and Native-American families. Instead of recognizing the strengths of diverse ethnic family systems, we viewed these families as "tangles of pathology" for failing to meet the model of the traditional nuclear family (Moynihan, 1965).

DID YOU KNOW?

According to Census 2000, 47 million Americans (18 percent) speak languages other than English at home. Of those, 28.1 million speak Spanish. The next most frequently spoken language is Chinese, spoken by more than 2 million people in the United States (Shin and Bruno, 2003).

Part of this distortion resulted from the long-term scarcity of studies on families from African-American, Latino, Asian-American, Native-American, and other ethnic groups. Furthermore, many of the earlier studies were flawed or distorted by a focus on weaknesses rather than strengths, giving the impression that all families from a particular ethnic group were riddled by problems (Dilworth-Anderson and McAdoo, 1988; Taylor, 1994a, 1994b; Taylor et al., 1991). Two of the most prominent examples of ethnocentric distortions are the "culture of poverty" approach to studying African-American families and the "machismo syndrome" used in studying Latino families (Demos, 1990; Mirandé, 1985).

The *culture of poverty* approach, for example, sees African-American families as being deeply enmeshed in illegitimacy, poverty, and welfare as a result of their slave heritage. As one scholar (Demos, 1990) notes, the culture of poverty approach "views black families from a white middle-class vantage point and results in a pejorative analysis of black family life." This approach ignores the majority of families that are intact or middle-class. It also fails to see African-American family strengths, such as strong kinship bonds, role flexibility, love of children, commitment to education, and care for the elderly.

America is a pluralistic society. Thus, it is important that students and researchers alike reexamine diversity among our different ethnic groups as possible sources of strength rather than pathology (DeGenova, 1997). For instance, cultures may vary widely in how the best interests of the child are defined (Murphy-Berman, Levesque, and Berman, 1996). Differences may not necessarily be problems but solutions to problems; they may be signs of adaptation rather than weakness (Adams, 1985). As two family scholars (Dilworth-Anderson and McAdoo, 1988) point out, "Whether a phenomenon is viewed as a problem or a solution may not be objective reality at all but may be determined by the observer's values."

AFRICAN-AMERICAN FAMILIES

According to the 2000 Census, the nearly 35 million African Americans in the United States represented 12.3 percent of the population (U.S. Census Bureau, 2000). African Americans are less likely than the general population to marry, more likely to divorce and more likely to live in single-parent, mostly mother-headed, families. These patterns continued to increase throughout the past decade but even more so among the general population than among African Americans (McLoyd, Cauce, Takeuchi, and Wilson, 2000). Because of high rates of divorce and of births to unmarried women, in 2002 more than half (53 percent) of African-American children lived in households headed by single mothers (48 percent) or single fathers (5 percent) (U.S. Census Bureau, 2003).

There are several other noteworthy features of African-American families. First, African-American families, in contrast to white families, have a long history of being dual-earner families as a result of economic need. As a consequence, employed women have played important roles in the African-American family. They also have more egalitarian family roles. Black men have more positive attitudes toward working wives, take on a slightly larger share of household labor, and spend more time on domestic tasks and childcare activities (McLoyd, Cauce, Takeuchi, and Wilson, 2000). Second, marital relations more often show signs of greater distress than is true of the general population. Some evidence indicates a greater likelihood of spousal violence and lower levels of reported marital happiness among African American marriages (McLoyd, Cauce, Takeuchi, and Wilson, 2000). Third, kinship bonds are especially important, for they provide economic assistance and emotional support in times of need (Taylor, 1994c; Taylor et al., 1991). Fourth, African Americans have a strong tradition of **familism** (emphasis on family and family loyalty), with an important role played by intergenerational ties. Fifth, the African-American community values children highly. Finally, African Americans are much more likely than whites to live in **extended households,** households that contain several different families (Taylor, 1994c). Black children are more likely than other children to live in their grandparent's household or to have a grandparent living with them in their parent's household. Typically, this grandparent is a grandmother (U.S. Census Bureau, 2003).

Many of these characteristics are often associated with poverty and thus may not be problems inherent in African-American families. In fact, when divorce rates are adjusted according to socioeconomic status, racial differences are minimal. Poor African Americans have divorce rates similar to poor whites, and middle-class African Americans have divorce rates similar to middle-class whites (Raschke, 1987). Thus, understanding socioeconomic status, especially poverty, is critical in examining African-American life (Bryant and Coleman, 1988; Julian, McKenry, and McKelvey, 1994; Wilkinson, 1997).

Economically, African-American families are at a clear disadvantage relative to white families. They have more than twice the unemployment rate, nearly three times the poverty rate, and less than two-thirds the median income of whites (see Table 3.4). These economic indicators point out the potential difficulty of comparing black and white family characteristics. Combined with the fact that upper-status African-American families (that is, middle- and upper-middle-class) tend to be as stable as white families of comparable status, these economic indicators suggest that much of what we may assume to be race differences may actually be social class differences masquerading as race.

This more economic argument pertains especially well to an understanding of race differences in marriage rates, divorce rates, and the numbers of single-mother headed families. It appears as though the most widely applied argument is that blacks "marital prospects" have shifted dramatically, especially among the poor (Aponte, Beal and Jiles, 1999). Wilson's notion of the "male marriageable pool index" emphasizes the importance of male employment to their "marriageability" (Wilson, 1987). Downward shifts in male employment

TABLE 3.4 ■ A Comparison of Race, Ethnicity, and Socioeconomic Status in 2000

	WHITES	AFRICAN AMERICANS	LATINOS	ASIANS
Median family income	$54,411	$33,755	$35,504	$56,316
Percentage unemployed	3.5	7.6	7.5	2.6
Percentage of families below poverty	7.3	21.9	19.4	10.8

SOURCE: U.S. Census Bureau, *Statistical Abstract of the United States*, 2003, Tables 39, 40, 44, 46, 598.

patterns would then account for some of the decline in marriage rates and also the increase in single-mother headed families. Not only are African Americans unlikely to devalue marriage, they may actually more highly value marriage than do other groups.

Despite the advantages that come from linking class and race in our attempts to understand family diversity, we cannot simply interpret all race differences as economic in nature. Don't forget that a major feature of race in American society is that it determines much of the treatment we receive from others. Thus, the opportunities we are offered or refused, and whether others insult, avoid, or think less of us, are all affected by race. The interpretation of race differences as only (or even largely) class differences minimizes or ignores the effects of racism and discrimination and fails to acknowledge patterns that may have cultural origins to them—such as greater emphasis on extended family ties or gender equality.

LATINO FAMILIES

Latinos (or Hispanics) are now the largest ethnic group in the United States as well as the fastest growing. The 2000 Census reported 35 million Hispanics, representing 12.5 percent of the U.S. population. Furthermore, it is projected that by 2025 Hispanics

FIGURE 3.1 ■ Marital Status by Ethnicity: 1970 and 2000

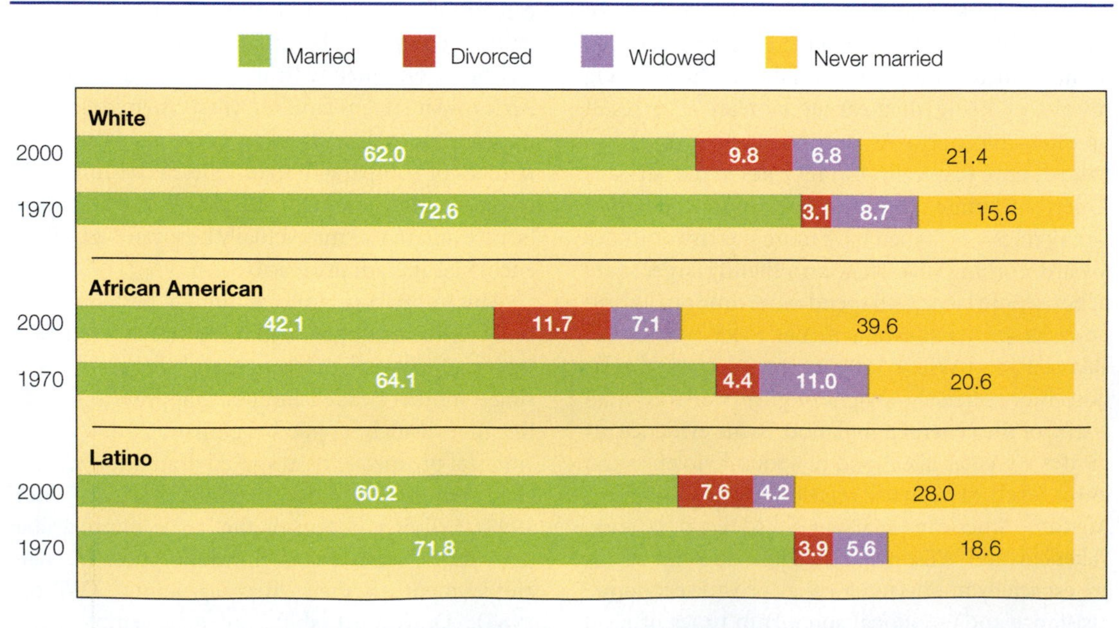

SOURCE: U.S. Census Bureau, *Statistical Abstract of the United States*. Washington, DC: U.S. Government Printing Office, 2002, Table 49.

© Tony Freeman/PhotoEdit

Latino culture emphasizes the family as a basic source of emotional support for children.

will represent 18 percent of the population, and by 2050, 24 percent of the population will be of Hispanic origin. These increases result from both immigration and a higher birthrate among Latinos (U.S. Census Bureau, 1996; Vega, 1991).

Currently, 58.5 percent of Latinos are of Mexican descent, 9.6 percent are Puerto Rican, and another 3.5 percent are Cuban. The remaining 28 percent includes 4.8 percent from Central American countries, 3.8 percent from South American countries, and 2.2 percent from the Dominican Republic (see Figure 3.2). Overall, more than three-fourths of Hispanics live in Western (43.5 percent) and Southern states (32.8 percent), with California and Texas, together, accounting for more than half the Hispanic population in the United States. Hispanics account for 24 percent of the population in the Western United States, a proportion that is nearly twice their national level (12.5 percent). Latinos, mostly of Mexican and Central American descent, are concentrated in California and the Southwest. Latinos of Puerto Rican descent are concentrated in the Northeast, especially New York. The greatest numbers of Cuban Americans are found in Florida. There are significant Latino populations also in Illinois, New Jersey, and Massachusetts (U.S. Census Bureau, 2001).

Continued immigration has transformed the nature of Latino culture in the United States. First, immigration makes both Latino culture and the larger society a "permanently unfinished" society. The newer immigrants are urban and overwhelmingly workers and laborers rather than professionals. Second, in some areas, immigration is changing the proportion of U.S.-born and foreign-born Latinos. In 1960 in California, for example, four out of five Mexicans were born in the United States; today, because of the massive influx of immigrants, only about half are U.S. born (Zinn, 1994).

FIGURE 3.2 ■ Hispanic Population for Selected Groups

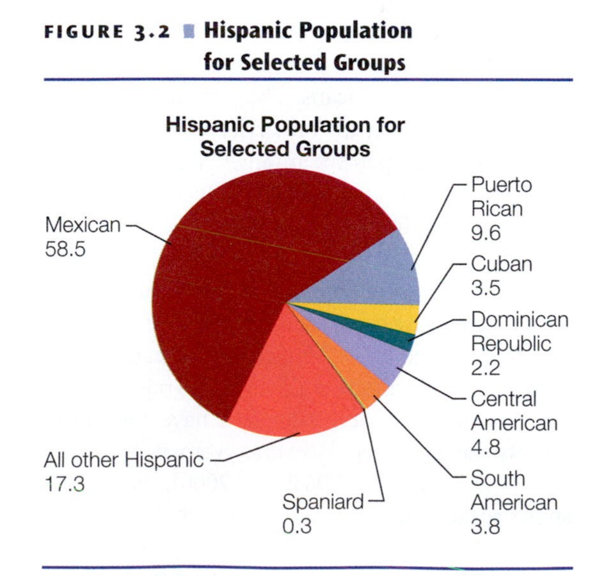

SOURCE: Guzman, Betsy. "The Hispanic Population: Census 2000 Brief." U.S. Census Bureau, May 2001.

TABLE 3.5 ■ Selected Socioeconomic Characteristics of the Hispanic Population

	HISPANIC TOTAL	MEXICAN	PUERTO RICAN	CUBAN
Median income	$31, 663	$31,123	$30,129	$38,312
Families in poverty	20.2%	21.2%	23%	15%
Persons in poverty	22.8%	24.1%	25.8%	17.3%
Living in owner-occupied housing	45.5%	47.8%	35%	58.7%
Persons 25 and over				
high school graduates	57%	51%	64.3%	73%
college graduates	10.6%	6.9%	13%	23%
Unemployment rate	5.7%	5.9%	6.4%	4.4%

SOURCE: U.S. Census Bureau, *Statistical Abstract of the United States*, 2001, Table 41.

It is important to remember that there is considerable diversity among Latinos, in terms of ethnic heritage (such as Mexican, Cuban, or Puerto Rican), socioeconomic status (Sanchez, 1997; Walker, 1993), and family characteristics. As Table 3.5 shows, Cuban Americans have the highest socioeconomic status, as indicated by incomes, poverty rates, home ownership, and educational attainment. As reflected in Table 3.5, Puerto Ricans and Mexican Americans have more similar characteristics, with Mexican Americans being somewhat less likely to be poor, and more likely to own their own homes. They also have a slightly higher median income. Puerto Ricans, on the other hand, are more likely to have graduated from either high school or college.

Furthermore, there are two times as many single-parent families among Puerto Ricans as there are among Cuban Americans. The percentage of children who are born to unmarried mothers ranges from a low of 27 percent among Cubans to 41 percent among Mexicans to a high of 60 percent among Puerto Ricans. Similarly, the percentage of births to teenage mothers ranges from 7.5 percent among Cubans to 20 percent among Puerto Ricans. Puerto Rican women are more likely to have their first child before marriage while Mexican-American women tend to have their first child after marriage. Cuban women tend to marry later and have the lowest fertility rates among Hispanic women (McLoyd, Cauce, Takeuchi, and Wilson, 2000). Figure 3.3 shows some of the variation in household structure among Hispanic families. (For an overview of Mexican-American families, see Sanchez, 1997; for Cuban-American families, see Suarez, 1997; and for Puerto Rican families, see Carrasquillo, 1997). This diversity is further accentuated by the varying proportions of U.S.-born and foreign-born Latinos in each group. Finally, keep in mind that characterizations of Mexican, Puerto Rican, Cuban, or any other Latino family types must avoid overgeneralization. More specifically, in Mexico, Cuba, or Puerto Rico, one finds much diversity, some of which results from socioeconomics, some from rural versus urban living, some from religion, and so on (Aponte, Beal, and Jiles, 1999).

Traditional Mexican and Puerto Rican families can be characterized by two distinctive cultural traits: devotion to family (i.e. *familism*) and male dominance (i.e. *machismo*). *La familia* is based on the nuclear family, but it also includes the extended family of grandparents, aunts, uncles, and cousins, who tend to live close by, often in the same block or neighborhood. There is close kin cooperation and mutual assistance, especially in times of need, when the family bands together. Family unity and interdependence, sometimes extended to include fictive kin (for example, Cuban *compadres* and *comadres*—godparents), reflect the importance of extended kin ties. Male dominance, as suggested though often exaggerated in the misuse or misunderstanding of *machismo*, is part of traditional Latino family systems but has declined, as has familism, especially among dual-earner couples. We will look at both *machismo* and *Marianismo* (the emphasis placed on self-sacrificing motherhood) in Chapter 4. Migration and mobility disrupt traditional Latino family forms and lead to change. This change can be seen as part of a wider process of "convergence," in which

FIGURE 3.3 ■ **Families by Type for Selected Hispanic Groups: 2000**

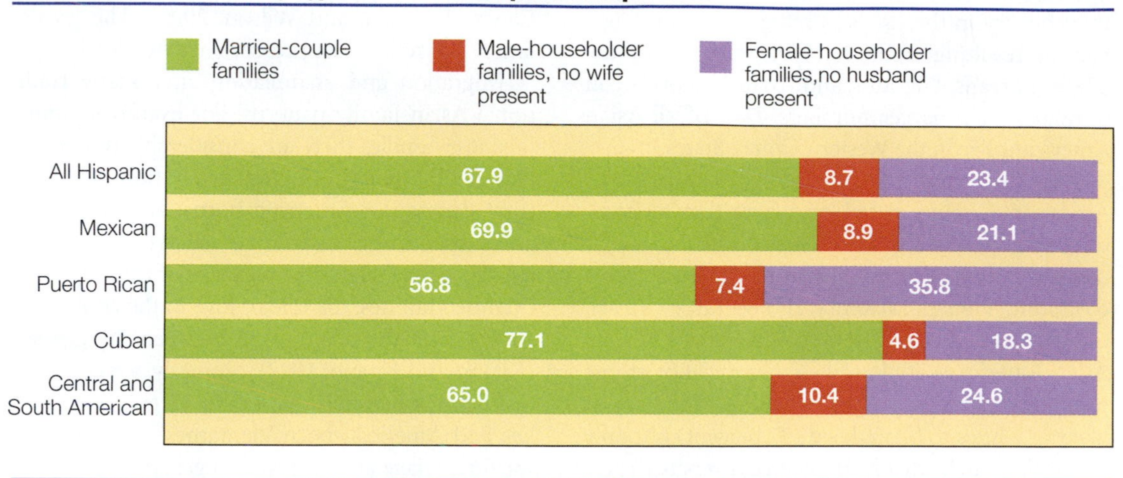

SOURCE: U.S. Census Bureau, *Statistical Abstract of the United States*, 2002, Table 40. Washington, DC: U.S. Government Printing Office.

distinctive ethnic traits diminish over time (Aponte, Beal, and Jiles, 1999).

Children are especially important. Fertility rates are still relatively higher among Hispanics than among the general U.S. population though they are dropping. Because Spanish is important in maintaining ethnic identity, many Latinos, as well as educators, support bilingualism in schools and government. Catholicism is also an important factor in Latino family life. Although there has been a tradition of male dominance, current day-to-day living patterns suggest noteworthy change has occurred. Women have gained power and influence in the family as they have increased their participation in paid employment. When wives are co-providers, Hispanic men spend more time on household tasks (Aponte, Beal, and Jiles, 1999; McLoyd, Cauce, Takeuchi, and Wilson, 2000).

ASIAN-AMERICAN FAMILIES

Asian Americans make up approximately 4 percent of the U.S. population. As revealed in Figure 3.4, Asian Americans are an especially diverse group, comprised of Chinese, Filipino, Japanese, Vietnamese, Cambodian, Hmong, and other groups. In the 2000 U.S. Census, questions about race were modified to allow for individuals to identify whether they were Asian alone, or Asian in combination with some other category. In 2000, the Census reported 10.2 million people identifying themselves as "Asian alone" and an additional 1.7 million who reported themselves as Asian, in combination with some other racial group. Figure 3.4 represents the population of selected Asian groups that results from combining the "Asian alone" and "Asian, in combination" categories into a population numbering nearly 11.9 million people."

The largest Asian-American groups are Chinese Americans, Filipino Americans, Asian Indians, Koreans, Vietnamese, and Japanese Americans. Other groups, such as Cambodians, Laotians, and

FIGURE 3.4 ■ **Selected Asian Groups, Asian Alone or in Combination, 2000**

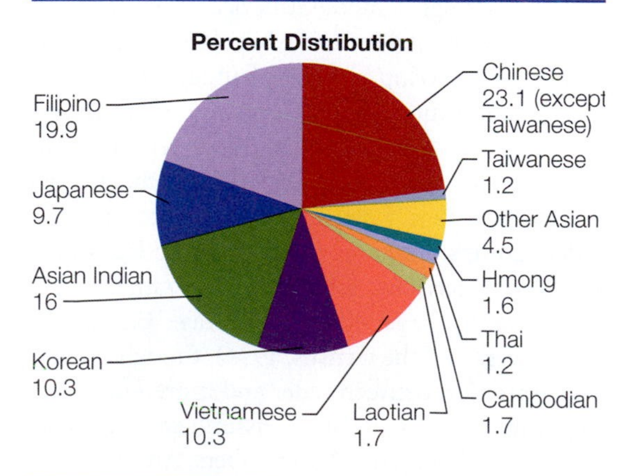

SOURCE: U.S. Census Bureau, "The Asian Population: 2000," Feb. 2002.

Hmong, are more recent arrivals, first coming to this country in the 1970s as refugees from the upheavals resulting from the Vietnam War. In the 1980s, Koreans, Filipinos, and Asian Indians began immigrating in larger numbers. Half of all Asian Americans live in the Western United States.

Much diversity can be observed within Asian-American families, based on where they're from, time of arrival in the United States, and reasons for coming to this country (for example, political versus economic) (Julian, McKenry, and McKelvey, 1994). General comparisons show that in many key ways Asian Americans are less like other racial or ethnic minorities than they are like whites. They are as likely as whites to be married (57 percent) but only half as likely to be divorced (5 percent versus 10 percent). Almost three-fourths of Asian-American households are family households, a level greater than found among whites (66 percent). Almost 90 percent of Asian Americans graduated from high school, and more than half of Asian-American men and nearly half of women are college graduates (Reeves and Bennett, 2002).

Typically, Asian Americans have fewer children, have them within marriage, and have them later than do other ethnic groups. Where 10 percent of European-American, 18 percent of Hispanic, and 23 percent of African-American births occur to women under age 20, only 6 percent of Asian-American births occur to teenaged mothers (McLoyd, Beal, Takeuchi, and Jiles, 2000).

Values that continue to be important to Asian Americans in general include a strong sense of importance of family over the individual, self-control to achieve societal goals, and appreciation of one's cultural heritage. Chinese Americans tend to exercise strong parental control while encouraging their children to develop a sense of independence and strong motivation for achievement (Ishii-Kuntz, 1997; Lin and Fu, 1990). The more recent immigrants retain more culturally distinct characteristics, such as family structure and values, than do older groups, such as Chinese Americans and Japanese Americans. Asian-American families tend to be slightly larger than the average U.S. family (U.S. Bureau of the Census, 1996), though there is wide variation between older and more recent immigrants. Among the more assimilated Japanese, the average family has 2.5 members. Among more recent Asian immigrants (for example, Cambodians, Laotians, Vietnamese, and Hmong), families average between 4 and 5.1 members (McLoyd, Cauce, Takeuchi, and Wilson, 2000). The greater family size reflects the presence of extended kin.

Migration and assimilation alter many traditional Asian family patterns. For example, among Japanese families there are considerable differences between the *Issei* (immigrant generation), the *Nisei* (first generation American-born), and the *Sansei* and subsequent generations on such family characteristics as the relative importance of marriage over extended kin ties, the role of love in the choice of a spouse, and the relationship between the genders (Kitano and Kitano, 1998). Similarly, we can draw distinctions between traditional Vietnamese families and American-born Vietnamese. Attitudes toward marriage and family, changes in familial gender roles, increased prevalence of divorce, and single-parent households all separate the generations. One can see marked change between parents' and children's attitudes on individualism and self-fulfillment versus family obligation and self-sacrifice (Tran, 1998).

The most dramatic change affecting Chinese Americans has been their sheer increase in numbers over the last 30 years. The Chinese-American population increased from 431,000 to 2.7 million between 1970 and 2000. More recent immigrants tend to be from Taiwan or Hong Kong rather than mainland China (Glenn and Yap, 1994). Because of the large numbers of new immigrants, it is important to distinguish between American-born and foreign-born Chinese Americans; little research is available concerning the latter. Contemporary American-born Chinese families continue to emphasize familism, although filial piety and strict obedience to parental authority have become less strong. Chinese Americans tend to be better educated, have higher incomes, and have lower rates of unemployment than the general population. Their sexual values and attitudes toward gender roles tend to be more conservative. Chinese-American women are expected to be employed and to contribute to the household income. Over 1.2 million speak Chinese at home.

NATIVE-AMERICAN FAMILIES

Almost 2.5 million Americans identify themselves as being of native descent, a 26 percent increase between 1990 and 2000. Those who continue to be deeply involved with their own traditional culture give themselves a tribal identity, such as Dine (Navajo), Lakota, or Cherokee (Kawamoto and Cheshire, 1997). The largest tribal groups include the Cherokee (281,069), Navajo (269,202), Sioux (108,272) and Chipewa (105,907). Those who are more acculturated, such as urban dwellers, tend to give themselves an ethnic identity as Native Americans or Indians. Most Americans of native descent consider themselves members of a tribal group rather than an ethnic group. According to John Price (1981), "Specific tribal identities are almost universally stronger and more important than identity as a Native American."

There has been a considerable migration of Native Americans to urban areas since World War II because of poverty on reservations and pressures toward acculturation. Today, 1.2 million Americans of native descent live outside tribal lands; most live in cities,where they are separated from their traditional tribal cultures and may experience great cultural conflict as they attempt to maintain traditional values. Not surprisingly, those in the cities are more acculturated than those remaining on the reservations. Urban Native Americans may attend powwows, intertribal social gatherings centering around drumming, singing, and traditional dances. Powwows are important mechanisms in the development of the Native-American ethnic identity in contrast to the tribal identity. Urban Native Americans, however, may visit their home reservations regularly.

Because of the importance of tribal identities and practices, there is no single type of Native-American family. Although there is considerable variation among different tribal groups, two aspects of Native-American families are important. First, extended families are significant. These extended families may be different from what the larger society regards as an extended family (Wall, 1993). They often revolve around complex kinship networks based on clan membership rather than birth, marriage, or adoption. Concepts of kin relationships may also differ. A child's "grandmother" may be an aunt or great-aunt in European-based conceptualization of kin (Yellowbird and Snipp, 1994).

Second, increasingly large numbers of Native Americans are marrying non-Indians. Among married Native Americans, 53 percent have non-Indian spouses. With such high rates of intermarriage, a key question is whether Native Americans can sustain their ethnic identity. Michael Yellowbird and Matthew Snipp (1994) wonder if "Indians through their spousal choices, may accomplish what disease, western civilization, and decades of Federal Indian policy failed to achieve."

EUROPEAN ETHNIC FAMILIES

The sense of ethnicity among Americans of European descent grew in recent decades. This is especially true among working-class Germans, Italians, Greeks, Poles, Irish, Croats, and Hungarians. This increasing awareness seems to be part of a general rise in ethnic identification over the last thirty years (Rubin, 1994). Earlier, members of European ethnic groups sought to assimilate—to adopt the attitudes, beliefs, and values of the dominant culture. Most white ethnic groups have assimilated to a considerable degree—they have learned English, moved away from their ethnic neighborhoods, and married outside their group, but many continue to be bound emotionally to their ethnic roots. These roots are psychologically important, giving them a sense of community and a shared history. This common culture is manifested in shared rituals, feast days, and saint's days, such as St. Patrick's Day.

Except for some West Coast enclaves, such as Little Italy in San Francisco, white ethnicity is strongest in the East and Midwest. The Irish neighborhoods of Boston, the Polish areas of Chicago, and the Jewish sections of Brooklyn, for example, have strong ethnic identities. Common languages and dialects are spoken in the homes, stores, and parks. Traditional holidays are celebrated; the foods are prepared from recipes passed down through generations. Elders speak of the old country and their villages—even if it was their parents or grandparents who immigrated.

As children grow up and move away from their neighborhoods, their ethnic identity often becomes weaker in terms of language and marriage to others within their group—but they may retain some of the elements of ethnic pride. Their ethnicity is what Herbert Gans (1979) calls symbolic ethnicity—an ethnic identity that's used only when the individual chooses. Symbolic ethnicity has little impact on one's day-to-day life. It is not linked to neighborhoods, accents, the use of a foreign language, or one's working life. Others cannot easily identify the person's ethnicity; he or she "looks" American. Nevertheless, ethnicity has emotional significance. A

THE OHLONE: NATIVE-AMERICAN MARRIAGE AND FAMILY WAYS

In many ways, the Ohlone of the San Francisco and Monterey Bay region in present-day California reflected many basic characteristics of Native-American groups (Margolin, 1978). When an Ohlone boy sought marriage, for example, he presented his request to his parents. He and his prospective bride knew and liked each other but custom called for the marriage to be arranged by both sets of parents. His parents conferred with other important relatives; the family as a whole decided whether the match was good. A relative informally approached the girl's parents to sound out their feelings. Her family called in other members to discuss the proposal. If the proposal was accepted, the boy's family made a formal visit. At this visit, baskets, beads, feathers, furs, and other valuable gifts were brought. Valuable gifts conferred status on the girl and her family because they indicated the family's high standing in the village.

If the marriage proposal was accepted, the girl's family prepared a wedding feast that lasted for several days. Following the feast, the newly married couple wandered outside the village to a thicket of willows, where they consummated their marriage. After a while they returned, with the groom's face scratched and clawed. The marks were a sign that the marriage had been consummated; they showed that the bride was modest. After the marriage, the youth stayed with the wife's family for a few months. Then, if the girl's parents approved of their son-in-law, she left to live with her husband.

Sometimes men had two wives, often in separate villages. Within the village the wives were often sisters, as they preferred sharing husband and housework with a family member. In marriage, fidelity was expected and divorce was discouraged. If divorce was necessary, however, the man or woman simply moved out of the family dwelling. They could then remarry.

The Ohlone restricted their sexual activities because they believed sexuality interfered with their relationship with the spirit world; it weakened their powers. Men were sometimes overcome with desire, however, and ignored the various taboos surrounding sexuality, such as sexual restrictions during menstruation and breastfeeding (which lasted about two years). If a man or woman violated these taboos and later fell ill or had bad luck, it was said that he or she was being punished for weakness.

Although the Ohlone restricted their sexuality, they accepted same-sex relationships. If a boy began imitating women and began wearing women's clothes as he grew older, he was allowed to marry another man. Sometimes he became the second "wife" of a wealthy man whose other wives were women. Women attracted to other women were also accepted, but they were not allowed to adopt male roles; their same-sex orientation was restricted to sexual activities.

person is Irish, Jewish, Italian, or German, for example—not only an American.

European ethnic groups differ from one another in many ways. However, a major study of contemporary American ethnic groups (Lieberson and Waters, 1988) found that European ethnic groups are much more similar to one another than they are to African Americans, Latinos, Asian Americans, and Native Americans. The researchers concluded that a European–non-European distinction remains a central division in our society. There are several reasons for this. First, most European ethnic groups no longer have **minority status**—that is, unequal access to economic and political power. Some scholars suggest that what separates ethnic groups into distinctive lifestyles is their social placement. As groups become more similar in their access to opportunities, their family lifestyles may "converge" toward a common pattern, one that includes smaller families, increased divorce, less interdependent ties with extended families, and less male dominance (Aponte, Beal, and Jiles, 1999). Second, because most European ethnic groups are not physically distinguishable from other white Americans, they are not discriminated against racially.

AMISH FAMILIES

Not all European ethnic groups have blended into mainstream American patterns of family life. The Amish are a religously based group that arrived in North America during colonial times and established distinctive communities in which they main-

When a woman became pregnant, both she and her husband followed prescribed rituals. They abstained from sex and were both careful to avoid meat, fish, and salt, fearing that these would cause a difficult birth. The husband curbed his anger for fear of injuring the baby by disturbing the harmony of nature. He hunted little because it was bad to hurt living things during pregnancy. When the woman went into labor, she went to her hut and was attended by the old women, who caressed, massaged, and encouraged her. Following birth, the mother was led to a stream, where she splashed cold water on herself and the baby. A few days later, she began breastfeeding. From then on, until she weaned her child two years later, both she and her husband refrained from physical contact.

As the Ohlone raised their children, they sought to strengthen the bonds that linked the child to his or her family, clan, and tribe. The child's identity, strength, and fulfillment were found in belonging to family and clan. Selfishness and extreme individualism were discouraged because they weakened the bonds on which the family and community depended.

The child was watched over by the immediate and extended family. Children were not physically punished; instead, good behavior was taught by example. By age five, children were expected to engage in useful work, such as gathering berries and carrying wood. At age eight, boys and girls entered separate worlds. Boys began hunting, working rope and nets, and attending rituals in the sacred sweathouse, where saunalike heat purified their bodies and spirits. During puberty, the boys passed into manhood through a series of ceremonies.

As girls grew older, they helped to grind acorns and gather roots and herbs; they also learned to weave intricate baskets. A girl's passage into womanhood was marked by menarche, her first menstrual bleeding. Menarche was one of the most important events for the Ohlone girl, marking the beginning of her spiritual power. At menarche, she retired to a menstrual hut and began fasting in order to gather her spiritual power. The village women visited her and shared their secrets of female power. In the night, both male and female members of her family performed sacred menstrual dances. From then on, whenever she menstruated, she withdrew to her menstrual hut, where she communed with the spirit world through dreams and fasting.

The Ohlone flourished until the late eighteenth century and the arrival of the Catholic missionaries led by Junipero Serra (who was beatified in 1988 in preparation for sainthood). The padres, seeking to convert and "civilize" the Ohlones, uprooted and herded them into missions and destroyed their culture. After reducing the Ohlones to servitude, the padres became their "defenders" against the demands of Spanish and Mexican settlers and soldiers. Under the mission system, an estimated 100,000 native Californians—half the population—died (Fogel, 1988). Another quarter perished within 10 years of the arrival of the Forty-Niners during the California gold rush.

tained their traditional ways of life across the past three centuries (Hostetler, 1993; Aponte, Beal and Giles, 1999). Familiar to many because of their distinctive styles of dress and their rejection of many modern conveniences, such as automobiles, the Amish have "deliberately and systematically" resisted pressures to assimilate. They reject and avoid not just modern technology but, more importantly, many contemporary cultural patterns (Aponte, Beal and Giles, 1999). Numbering between 150,000 to 200,000, the Amish live an agrarian lifestyle in their isolated farming communities. Their social and familial characteristics reflect their continued adherence to their religious customs.

The Amish family occupies a central role in Amish communities. Individuals subordinate their own self-intrerests to what is defined as best for the family. Loyalty to parents, grandparents, and other relatives is assumed to be lifelong (Hostetler, 1993).

Gender roles are traditional, with the male being "head of the woman, just as Christ is head of the church" (Hostetler, 1999). The subordination of wives to husbands, however, is not total. Women perform key economic roles in Amish households, especially in farming households, where it is essential that women contribute, and where families are dependent upon women's contributions. The marital relationship, itself, is less openly affectionate than would be expected by American cultural ideals. Displays of affection and terms of endearment are "conspicuously absent" (Hostetler, 1999).

The Amish consider their children to be blessings from God. Infants and toddlers are well attended and the recipients of affectionate attention from parents, siblings, grandparents and other kin. Because the Amish live a restrictive, religously based lifestyle, as children age they must learn to live within the restrictions and "exacting disciplines" imposed upon them by parents. In fact, what they learn is how different the Amish lifestyle is from that found among their non-Amish ("English") neighbors and to respect and maintain that differentness. Children and adolescents are taught and expected to display strict obedience to their parents (Hostetler, 1993). In general, respect for elders, like respect for parents, is expected. Amish farms typically contain a separate structure which houses the grandparents who then can remain active within the family, perhaps assisting in the raising of children and in the running of the farm.

Unlike many other ethnic groups, the Amish have sustained a separate identity from that of the wider society. They willfully separate themselves from non-Amish, keeping as much distance as possible in their day-to-day affairs. Most young Amish remain within the religious communities but it is estimated that as many as 20 percent leave, either by choice or expulsion (Aponte, Beal and Jiles, 1999). The success of the Amish in maintaining their way of life may be most threatened by the economic difficulties they face. So much of the characteristic Amish family patterns—the availability of both parents to work together, share responsibility, and care for children and farm—depend upon farming. If forced to accommodate daily life to non-farm work schedules, many of these features would not be possible (Aponte, Beal and Jiles, 1999).

As we have now seen in a host of ways, American families are diverse. They vary across time and, within any given period, between racial, ethnic, and socioeconomic groups. Family diversity will be reflected throughout subsequent more specialized chapters, as relevant variations by race, class, or ethnicity are discussed. Thus, our ultimate goal of understanding American families will be made more complete and representative.

Acknowledging the diversity that exists across families has personal consequences as well. It ought to make us a bit more cautious in generalizing from our own particular set of family experiences to what others "must also experience." Additionally, in noting how historical, economic, and cultural factors shape our families, we link our personal experiences to broader societal forces. In that way, we are better able to apply "sociological imaginations" to family experiences, identifying how our private family worlds are, in large measure, products of when, where, and how we live. Simply put, if we come of age during a period of great economic upheaval, we may put off marrying, bearing children, or divorcing, because of the opportunities and constraints we face. Similarly, the kinds of family experiences we are able to have are limited or enhanced by the economic resources at our disposal, regardless of what we might otherwise choose to do.

Of course, despite the extent to which the factors discussed in this and the next chapter may limit your opportunities or narrow your range of choices, you do choose what kinds of families you wish to create. You decide whether or not to marry, whether or not to bear children, how to rear your children, whether to stay married, and so on. A major goal of this book is to equip you with a foundation of accurate information about family issues from which you can make sound choices more effectively.

SUMMARY

- In the early years of colonization, there were 2 million Native Americans in what is now called the United States. Many of the families were *patrilineal;* rights and property flowed from the father; some other tribal groups were *matrilineal.* Most families were small.
- Diverse groups settled America, including English, Germans, and Africans. In colonial America, marriages were arranged. Marriage was an economic institution and the marriage relationship was *patriarchal* and companionate. The family was self-sufficient. Women's economic

contributions were recognized. Children were also economically important.

- African-American families began in the United States in the early seventeenth century. They continued the African tradition that emphasized kin relations. Most slaves lived in two-parent families that valued marital stability.
- In the nineteenth century, industrialization revolutionized the family's structure; men became wage earners, and women, once they married, became housewives. Childhood was sentimentalized, and adolescence was invented. Marriage was increasingly based on emotional bonds.
- The stability of the African-American enslaved family suffered because it lacked autonomy and had little economic importance. Enslaved families were broken up by slaveholders, and marriage between slaves was not legally recognized. African-American families formed solid bonds nevertheless.
- Thirty-eight million people immigrated to the United States between 1820 and 1920. Most immigrants were uprooted and experienced hostility. Kinship groups were important for survival. The family economy focused on family survival rather than individual success.
- Beginning in the twentieth century, *companionate marriage* became an ideal. Men and women shared household decision making and tasks, marriages were expected to be romantic, wives were expected to be sexually active, and children were to be treated more democratically.
- The 1950s, the golden age of the companionate marriage, was an aberration. It was an exception to the general trend of rising divorce and nontraditional gender roles. Prosperity was unusually high; suburbanization led to increased isolation.
- Contemporary families are culturally diverse. The terms *ethnic group*, *racial group*, and *minority group* are conceptually distinct. An ethnic group is a group of people distinct from other groups because of cultural characteristics, such as language, religion, and customs that are transmitted from one generation to another. A *racial group* is a group of people, such as whites, blacks, and Asians, classified according to phenotypeas well as anatomical and physical characteristics. A *minority group* is a group whose status (position in the social hierarchy) places its members at an economic, social, and political disadvantage.
- African Americans are the largest ethnic group in the United States. *Socioeconomic status* is an important element in understanding African-American families. Many of the problems African-American families experience are the result of low socioeconomic status rather than family structure.
- Several features of African-American families are notable. Because of economic necessity, women traditionally have been employed, which has given them important economic roles in the family and more egalitarian relationships. Kinship bonds are especially important; they provide emotional and economic assistance in times of need. African Americans have a strong tradition of *familism*, with the emphasis on intergenerational ties. The African-American community values children highly. African Americans are much more likely than whites to live in *extended households*.
- Latinos are the fastest growing and second largest ethnic group, as a result of immigration and a higher birthrate than the general population. Important factors in understanding Latino culture are (1) ethnic diversity within the culture, and (2) the role of socioeconomic status. Latinos emphasize extended kin relationships, cooperation, and mutual assistance. *La familia* includes not only the nuclear family but also the extended family. Bilingualism helps maintain ethnic identity.
- Asian Americans are the third largest ethnic group in the United States. Immigration has contributed heavily to the dramatic recent increase in the Asian-American population. Sixty-six percent of Asian Americans are foreign born. The largest Asian-American groups are Chinese Americans, Filipino Americans, and Japanese Americans. The more recent immigrants retain more culturally distinct characteristics, such as family structure and values, than do older groups.
- Native Americans include nearly 2 million Americans. Tribal identity is their key identity. Powwows are social gatherings of diverse tribes that center around drumming, singing, and dancing. Over half of Native Americans live in cities, although many remain in contact with their home

reservation. Extended families are important and are often based on clan membership. About 53 percent of Native Americans are married to non-Indians.

- Ethnic identity among Americans of European descent has been growing, especially among working-class families. For many, their ethnicity is symbolic and has little impact on day-to-day life. Most members of European ethnic groups are physically indistinguishable from other white Americans and no longer have minority status.
- The Amish represent a unique European ethnic group that has systematically avoided assimilation into mainstream American society and has rejected most characteristics of "modern society." Their family relationships are products of their traditional values.

KEY TERMS

adolescence 67
assisted reproductive technologies 77
bundling 66
companionate marriage 71
ethnic group 86
ethnocentric fallacies 85
extended households 87
familism 87
feminization of poverty 83
fictive kin ties 84
life chances 80
lower-middle class 81
matrilineal 65
minority group 86
minority status 94
patriarchy 66
patrilineal 65
phenotype 85
racial group 85
social class 80
social mobility 85
socioeconomic status 80
stem family 65
two-person career 82
upper-middle class 81
working class 81

SUGGESTED READINGS

Billingsley, Andrew. *Climbing Jacob's Ladder.* New York: Simon and Schuster, 1992. A leading family scholar examines the African-American family in historical context.

Chow, Esther Ngan-Ling, Doris Wilkinson, and Maxine Baca Zinn, eds. *Race, Class, and Gender: Common Bonds, Different Voices.* Newbury Park, CA: Sage, 1996. An interdisciplinary look at the intersecting patterns that occur among peoples from different backgrounds and experiences.

Coontz, Stephanie. *The Way We Never Were: American Families and the Nostalgia Trap.* New York: Basic Books, 1992. A historian's view that contrasts the pop images conveyed through the media with the realities of American families since the 1950s.

DeGenova, Mary Kay. *Families in Cultural Context: Strengths and Challenges in Diversity.* Mountain View, CA: Mayfield, 1997. A look at the experiences of 11 American ethnic groups—their cultures, family systems, strengths, challenges, and the myths and stereotypes they confront.

McAdoo, Harriette, ed. *Black Families,* 3rd ed. Thousand Oaks, CA: Sage Publications, 1996. Examines conceptual, historical, demographic, and economic aspects of African-American family systems.

Mindel, C. R. Habenstein, and R. Wright, eds. *Ethnic Families in America: Patterns and Variations,* 4th ed. Upper Saddle River, NJ: Prentice-Hall, 1998. A collection of readings looking at family patterns among a variety of ethnic groups.

Mintz, Steven, and Susan Kellogg. *Domestic Revolutions: A Social History of American Family Life.* New York: Free Press, 1989. A comprehensive history of American family life from the colonial period through the late twentieth century.

Taylor, Ronald, ed. *Minority Families in the United States: A Multicultural Perspective.* Upper Saddle River, NJ: Prentice-Hall, 1998. An excellent collection of readings examining diverse groups and their characteristic family patterns.

RESOURCES ON THE INTERNET

Companion Web Site for This Book

http://sociology.wadsworth.com/strong/marriage9e
Gain an even better grasp on this chapter by going to the companion Web site to take one of the Tutorial Quizzes, use the Flash Cards to master key terms, or check out the many other study aids you'll find there. Visit the Marriage and Family Resource Center on the site. You'll also find special features such as GSS Data and Census 2000 information that will put data and resources at your fingertips to help you with that special project or to do some research on your own.

InfoTrac College Edition: Search Word Summary

parenthood
temperament
self-esteem
fatherhood
gay parents

To learn more about these central topics in the study of the family, you can conduct an electronic search using InfoTrac College Edition. To aid in your search and to gain useful tips, see the Student Guide to InfoTrac College Edition that you can access through the companion Web site for this book.

Preview

To gain a sense of what you already know about the material covered in this chapter, answer "True" or "False" to the statements below.

1. Gender roles reflect the instinctive nature of males and females. True or false?
2. Gender roles are influenced by ethnicity. True or false?
3. The only universal feature of gender is that all societies sort people into only two categories. True or false?
4. Parents are not always aware that they treat their sons and daughters differently. True or false?
5. Peers are the most important influence on gender-role development from adolescence through old age. True or false?
6. Both boys and girls suffer from gender-related problems in school. True or false?
7. For African Americans, the traditional female gender role includes both employment and motherhood. True or false?
8. Research shows it is possible for women and men to establish work or family roles that are counter to their socialization. True or false?
9. Compared with traditional roles, contemporary male gender roles place more emphasis on the expectation that men will be actively involved with their children. True or false?
10. The men's and women's movements have consistently stressed the importance of the family. True or false?

"When I was younger and saw the world in black and white, I believed the woman my mother was was determined by her character, not by social conditions. Now that I see only shades of gray, I know that that is nonsense."

ANNA QUINDLEN

CHAPTER 4

Contemporary Gender Roles

Outline

Did you ever stop to consider how similar or different your life might be if you had been born the opposite sex? Would you be the kind of person you are? Participate in the same activities? Have the same friends? Have the same roles and relationships within your families? Would your goals be the same as they are now? Would you be enrolled in the same college? Take the same courses? Be reading this book? Asking ourselves such questions serves to remind us that much of what we do, who we are, and what happens to us is influenced by gender. In this chapter we will examine how interconnected gender and family experience are. It is no exaggeration to say that we cannot fully understand one without taking into account the other.

The traditional view of masculinity and femininity sees men and women as polar opposites. Our gender stereotypes fit this pattern of polar differences: we believe that men are aggressive, and women are passive; men are instrumental (task oriented), and women are expressive (emotion oriented); men are rational, and women are irrational; men want sex, and women want love (Duncombe and Marsden, 1993; Lips, 1997).

As you will see in this chapter, this *perception* of male-female differences is greater than the actual differences themselves (Hare-Mustin and Murecek, 1990b). We may be accustomed to thinking that we are as different as Martians would be from Venusians, but in fact both women and men inhabit planet Earth (Kimmel, 2000). At the same time, our family experience *is* highly "gendered" (i.e., differently experienced for women and men). Marriages might be said to consist of "two marriages, his and hers," that are not entirely the same (Bernard, 1982). Similarly, we could argue that there are actually "two courtships," "two parenthoods," "two divorces," and so on. In fact, in each area of marriage and family life we often observe differences in what women and men experience. Men and women may define and experience love differently, enter marriage with different emphases and expectations, react to the onset of parenthood and relate to their children differently, divorce for different reasons and with different consequences, and so on. The chapters that follow will identify some of these gender differences.

In this chapter we examine some gender and socialization theories and illustrate how much our families influence how we learn to act masculine and feminine. Next, we explore some areas of family experience that have been and remain differently experienced by women and men. Finally, we discuss changing gender roles by considering gender-based social movements.

PREVIEW ANSWERS

1 False
2 True
3 False
4 True
5 False
6 True
7 True
8 True
9 True
10 False

Understanding Gender and Gender Roles

STUDYING GENDER

Before we continue, we need to define several key terms that are useful in building an understanding of the importance of gender in family life. These terms include *sex, gender, role, gender role, gender-role stereotype, gender-role attitude,* and *gender-role behavior.* **Sex** refers to male or female in a biological sense; it includes chromosomal, hormonal, and anatomical characteristics that differentiate females from males. **Gender** refers to social, cultural, and psychological aspects of being male or female. In general, a **role** consists of culturally defined expectations that an individual is expected to fulfill in a given situation in a particular culture. A **gender role** is the role that a person is expected to perform as a result of being male or female in a particular culture. (The term *gender role* is a more recent concept that has largely replaced the traditional term *sex role.*) A **gender-role stereotype** is a rigidly held and oversimplified belief that all males and females, as a result of their sex, possess distinct psychological and behavioral traits. Stereotypes tend to be false not only for the group as a whole but also for any individual member of the group. The "men are aggressive" stereotype not only is untrue for men as a group, but it also cannot be applied randomly to any individual man: Matthew or Danny, in fact, may not be aggressive. Even if the generalization is statistically valid in describing a group average (males are taller than females), such generalizations do not necessarily predict whether Jason will be taller than Tanya. **Gender-role attitude** refers to the beliefs we have regarding appropriate male and female personality traits and activities. **Gender-role behavior** refers to the actual activities or behaviors we engage in as males and females. When we discuss gender roles, it is important not to confuse stereotypes with reality or to confuse attitudes with behavior.

Historically, most gender-role studies focused on the white middle class. Thus, until recently, less was known about gender roles among African Americans, Latinos, Asian Americans, and other ethnic groups. Even at present, we continue to know less about them (Binion, 1990; Reid and Comas-Diaz, 1990; True, 1990). Students and researchers must be careful to avoid projecting onto other groups the gender-role concepts or aspirations based on their own groups. Too often such projections can lead to distortions or moral judgments. In fact, although we may come to accept one particular standard of behavior as "appropriate" masculinity or femininity, there are actually **multiple masculinities and femininities,** out of which emerges a version that is expected or accepted (Connell, 1995; Kimmel 2000; Messerschmidt, 1993).

These dominant or **hegemonic models of gender** are held up as the standards for all women and men to emulate (Kimmel, 2000). They are also dynamic and culturally variable. They change over time (Kimmel, 1996), differ across space (Gilmore, 1990), and—within a given time and place—are challenged for cultural dominance by those who advocate other versions of masculinity or femininity (Kimmel, 1994; Connell, 1995).

GENDER AND GENDER ROLES

Gender is simultaneously experienced on both the most personal and political levels. At birth, we are identified as either male or female. This identification, based essentially on inspection of one's genitalia, typically leads to the self-identity or **gender identity** we form of ourselves as females or males. We say "typically" because there are individuals who for a combination of reasons are categorized as *transsexuals*—males and females who develop self-identities that differ from the gender category into which they have been placed. They opt for reconstructive surgery to bring their biology into line with the identity they have developed.

We acquire our gender identities at very young ages. Furthermore, gender identity may well be the deepest concept we hold of ourselves. The psychology of insults reveals this depth, for few things offend a person, especially a male, as much as to be tauntingly characterized as a member of the "opposite" sex. Gender identity determines many of the

Generally, the only limit on women's performing traditional male work is social custom rather than individual ability. Most jobs can be done equally well by women or men.

directions our lives will take—for example, whether we will fulfill the role of husband or wife, father or mother. When the scripts are handed out in life, the one you receive depends largely on your gender.

At the same time, gender is a basis for the assignment of social roles, distribution of rewards, and the exercise of power. The majority of societies are **patriarchal systems,** in which males dominate political and economic institutions and exercise power in interpersonal relationships. Although many societies have been identified as more *egalitarian,* truly **matriarchal societies** have not been evident. Within patriarchal societies, families tend to be male dominated. That is to say, in daily decision making and the division of responsibilities men have privileges that women do not (for example, freedom from domestic work). The familial power that men have stems from a variety of sources, including the marriage contract and their wage-earning roles. Later chapters will explore in more detail how gender and power are connected within households and families.

Each culture determines the content of gender roles in its own way. In some cultures, there are more than two gender categories. Among some Asian and Native-American societies, for example, men or women become *berdaches.* They then live as members of the opposite sex. The *Hua* of Papua, New Guinea, perceive gender as fluid, capable of changing over one's life span. In other societies, alternative categories (for example, the *Hjira* of India) are socially recognized for individuals who are *neither* male nor female (Renzetti and Curran, 1999; Nanda, 1990).

We can identify less extreme cultural variations in conceptualizations of gender. Among the Arapesh of New Guinea, both males and females possess what we consider feminine traits. Men and women alike tend to be passive, cooperative, peaceful, and nurturing. The father is said to "bear a child" as well as the mother; only the father's continual care can make a child grow healthily, both in the womb and in childhood. Eighty miles away, the Mundugumor live in remarkable contrast to the peaceful Arapesh. Margaret Mead (1975) offered this observation:

> Both men and women are expected to be violent, competitive, aggressively sexed, jealous, and ready to see and avenge insult, delighting in display, in action, in fighting. . . . Many, if not all of the personality traits which we have called masculine or feminine are as lightly linked to sex as are the clothing, the manners, and the form of head-dress that a society at a given period assigned to either sex. . . .

Biology creates males and females, but culture creates masculinity and femininity.

MASCULINITY AND FEMININITY: OPPOSITES OR SIMILAR?

Until the last generation, a **bipolar gender role** was the dominant model used to explain male-female differences. In this model, males and females are seen as polar opposites, with males possessing exclusively instrumental traits and females possessing exclusively expressive ones. Sandra Bem (1993) describes the culture of the United States as one which looks at gender through a series of "lenses," including the belief that males and females are fundamentally different. She calls this assumption *gender polarization.* Our entire society is organized around such supposed differences (Renzetti and Curran, 2003). In light of the widespread acceptance of this viewpoint and its popularization through John Gray's, *Men Are from Mars, Women Are from Venus,* Michael Kimmel (2000) cleverly calls this viewpoint the *"interplanetary theory of gender."*

Traditional views of masculinity and femininity as opposites have several implications. First, if a person differs from the male or female stereotype, he or she is seen as being more like the other gender. If a woman is sexually assertive, for example, not only is she less feminine but she is believed to be more masculine. Similarly, if a man is nurturing, not only is he less masculine but he is also more feminine. Second, because males and females are perceived as opposites, they cannot share the same traits or qualities. A "real man" possesses exclusively masculine traits and behaviors, and a "real woman" possesses exclusively feminine traits and behaviors. A man is assertive, and a woman is receptive; in reality, men and women are both assertive and receptive. Third, because males and females are viewed as opposites, they are believed to have little in common with each other; and a "war of the sexes" is the norm. Men and women can't understand each other, nor can they expect to do so. Difficulties in their relationships are attributed to their "oppositeness."

The fundamental problem with the view of men and women as opposites is that it is erroneous. Men and women are significantly more alike than they are different.

Our culture, however, has encouraged us to look for differences and, when we find them, to exaggerate their degree and significance. It has taught us to ignore the single most important fact about males and females: that we are both human. As human beings, we are significantly more alike biologically and psychologically than we are different. As men and women, we share similar respiratory, circulatory, neurological, skeletal, and muscular systems. (Even the penis and the clitoris evolved from the same undifferentiated embryonic structure.) Hormonally, both men and women produce androgens and estrogen (but in different amounts). Where men and women biologically differ most significantly is in terms of their reproductive functions: Men impregnate, whereas women menstruate, gestate, and nurse. Beyond these reproductive differences, biological differences are not great. In terms of social behavior, studies suggest that men are more aggressive both physically and verbally than women; the gender difference, however, is not large. Most differences can be traced to gender-role expectations, male-female status, and gender stereotyping.

DID YOU KNOW?

Recent reviews of literature concerning gender find that *neither* men *nor* women are more likely to dominate, more susceptible to influence, or more nurturing, altruistic, or empathetic (Lips, 1997).

Although we are more similar than different in our attributes and abilities, large and meaningful differences do exist in the statuses we occupy and the privileges and responsibilities they carry. Although either gender may have the ability to nurture children, support families, clean, or cook, these tasks are assumed to be more appropriate for one gender than the other. Although women and men may possess the ability to do many kinds of jobs, the labor force is sex-segregated into jobs that are disproportionately male or female. Men's jobs typically carry more prestige, earn higher salaries, and offer more opportunity for advancement than do women's jobs.

We often refer to these differences as "gaps." The "wage gap" refers to the difference between what men tend to earn and what women tend to earn. Recent data indicate that when we compare the median weekly earnings of women and men employed full-time, women earn approximately three-fourths (75.7 percent) of what men earn (Renzetti and Curran, 2003). We can also speak of "prestige gaps" or "mobility gaps." Jobs that tend to be among the most highly respected jobs (typically, jobs such as physician, attorney, engineer) tend to be held disproportionately by men. Jobs that are held largely by women (such as clerical work, elementary and preschool teaching, household service, and nursing) are often undervalued (and, not surprisingly, underpaid). We should not assume that "men's jobs" are highly paid and highly respected, and "women's jobs" are devalued and underpaid. There are many jobs, overwhelmingly held by men, that are not accorded high levels of prestige, such as construction workers, truck drivers, and mechanics, to name a few. Instead, the point is that in those jobs that are rewarded with higher levels of prestige and higher salaries, we tend to find mostly men. Finally, compared to jobs in which we find mostly men, jobs that are typically held by women may lack adequate opportunities for any significant advancement and offer only limited levels of upward mobility (movement "up" in income and position).

Despite possessing traits of both genders, most of us feel either masculine or feminine; we usually do not doubt our gender (Heilbrun, 1982). Unfortunately, when people believe that individuals should *not* have the attributes of the other gender, males suppress their **expressive traits,** and females suppress their **instrumental traits.** As a result, the range of human behaviors is limited by a person's gender role. As psychologist Sandra Bem (1975) points out, "Our current system of sex role differentiation has long since outlived its usefulness, and . . . now serves only to prevent both men and women from developing as full and complete human beings."

When we initially meet a person, we unconsciously note whether the individual is male or female (a process called *gender attribution*) and respond accordingly (Skita and Maslach, 1990). But what happens if we cannot immediately classify a person as male or female? Many of us feel uncomfortable because we don't know how to act if we don't know the gender. This is true even if gender is irrelevant, as in a bank transaction, walking past someone on the street, or answering a query about the time. ("Was that a man or woman?" a person may ask in exasperation, although it really makes no difference.) An inability to tell a person's gender may provoke a hostile response. As Hilary Lips (1997) writes:

> It is unnerving to be unsure of the sex of the person on the other end of the conversation. The labels *female* and *male* carry powerful associations about what to expect from the person to whom they are applied. We use the information the labels provide to guide our behavior toward other people and to interpret their behavior toward us.

Our need to classify people as male or female and its significance is demonstrated in the well-known Baby X experiment (Condry and Condry, 1976). In this experiment, three groups played with an infant known as Baby X. The first group was told that the baby was a girl, the second group was told that the baby was a boy, and the third group was not told what gender the baby was. The group that did not know what gender Baby X was felt extremely uncomfortable, but the group participants then made a decision based on whether the baby was "strong" or "soft." When the baby was labeled a boy, its fussing behavior was called "angry"; when the baby was labeled a girl, the same behavior was called "frustrated." Once the baby's gender was determined (whether correctly or not), a train of responses followed that could have profound consequences in his or her socialization. The study was replicated numerous times with the same general results. Even birth congratulations cards reflect gender stereotyping of newborns (Bridges, 1993). A review of studies on infant labeling found that gender stereotyping is strongest among children, adolescents, and college students (Stern and Karraker, 1989). Stereotyping diminishes among adults, especially among infants' mothers (Vogel et al., 1991).

GENDER AND SEXUAL ORIENTATION

Often we assume that the way an individual acts out his or her gender (gender display) is a sign of their sexual orientation. In other words, we link characteristics of gender with assumptions of sexual preference. Although we often dichotomize sexual preference into a duality of homosexuality and heterosexuality, the universe of sexual preference is more diverse and wide ranging (encompassing bisexuality and situational sexuality). We need to sever this almost automatic assumption that gets made. We assume that women who depart from the variety of behavioral norms associated with femininity and female roles must not be heterosexual; we assume that men who depart from masculinity and reject male roles ("feminine" men) must be gay men. This is not so. One's sexual preference cannot be "read" by ones demeanor or role behavior. Men who fit within norms of "masculine behavior" may be heterosexual, bisexual, or homosexual. Men whose behavior seems "feminine" by wider cultural standards, may be gay, bisexual, or heterosexual. The same holds true for women. There are "feminine" heterosexual women, bisexual women, and lesbians. Women who seem more "masculine" may be heterosexual, bisexual, or lesbians.

On a second level, sexual preference and gender consistently get connected through the ways in which we raise doubts and suspicions about sexual orientation to bolster and reinforce gender norms. Men, especially, may monitor and restrict their behavior so as to avoid the damning and homophobic sort of question, "What are you anyway, a fag?!" These potential doubts serve to accomplish the feat of keeping people conforming to gender roles and expectations.

As we will see, in a variety of ways gender transcends sexual orientation. The similarities that exist between heterosexual and gay men (e.g., in areas like acceptance of non-monogamous relationships) are due to their being men, and men tend to be more tolerant of and interested in infidelity than are women.

Gender and Gender Socialization

There are several prominent theories used to explain the significance of gender in our culture and how we learn what is expected of us. These include gender theory, social learning theory, and cognitive development theory.

GENDER THEORY

In studying gender, feminist scholars begin with two assumptions: (1) that male-female relationships are characterized by power issues and (2) that society is constructed in such a way that males dominate females. They argue that on every level, male-female relationships—whether personal, familial, or societal—reflect and encourage male dominance, putting females at a disadvantage. Male dominance is neither natural nor inevitable, however. Instead, it is created by social institutions, such as religious groups, government, and the family (Acker, 1993; Ferree, 1991). The question is, how is male-female inequality created?

Social Construction of Gender

In the 1980s, gender theory emerged as an important model explaining inequality. According to this theory, gender is a **social construct,** an idea or concept created by society through the use of social power. The theory asserts that society may be best understood by how it is organized according to gender, and that social relationships are based on the *socially perceived* differences between females and males that are used to justify unequal power relationships (Scott, 1986; White 1993). Imagine, for example, an infant crying in the night. In the mother/father parenting relationship, which parent gets up to take care of the baby? In most cases, the mother does because (1) women are socially perceived to be nurturing and (2) it's the woman's "responsibility" as mother (even if she hasn't slept in two nights and is employed full-time).

Gender theory focuses on (1) how specific behaviors (such as nurturing or aggression) or roles (such as child rearer, truck driver, or secretary) are defined as male or female; (2) how labor is divided into man's work and woman's work, both at home and in the workplace; and (3) how different institutions bestow advantages on men (such as male-only clergy in many religious denominations or women receiving less pay than men for the same work).

The key to the creation of gender inequality is the belief that men and women are "opposite" sexes—that they are opposite each other in personalities, abilities, skills, and traits. Furthermore, the differences between the genders are unequally valued: Reason and aggression (defined as male traits) are considered more valuable than sensitivity and compliance (defined as female traits). Making men and women appear to be opposite and of unequal value requires the suppression of natural similarities by the use of social power. The exercise of social power might take the form of greater societal value being placed on looks than on achievement for women, sexual harassment of women in the workplace or university, patronizing attitudes toward women, and so on.

"Doing Gender"

Some gender scholars emphasize how gender is reproduced or constructed in everyday social situations. They argue that more than what we *are,* gender is something that we *do* (West and Zimmerman, 1987; Risman, 1998). As Fox and Murry (2000) explain it, "men and women not only vary in their degree of masculinity or femininity but have to be constantly persuaded or reminded to be masculine and feminine. That is, men and women have to "do" gender rather than "be" a gender." We "do gender" whenever we take into account the gendered expectations in social situations and act accordingly. We don't so much perform an internalized role as tailor our behaviors to convey our suitability as a woman or a man in the particular situation in which we find ourselves (West and Zimmerman, 1987). To fail to conform to the expectations for someone of our gender in a given situation exposes us to potential criticism, ridicule, or rejection as an incompetent or immoral man or woman (Risman, 1998). But in living up to or within those social expectations, we help create and sustain the idea of gender difference. According to Kimmel (2000, p. 104), "successfully being a man or a woman simply means convincing others that you are what you appear to be."

Although we see the social construction or "doing" of gender in all kinds of social settings, the family is a particularly gendered domain (Risman, 1998). There are cultural expectations about how wage earning, housework, child care, and sexual intimacy should be allocated and performed between women and men. Thus, much of the experience that people have in their families is understandable as both an exercise in and a consequence of how they and others "do gender."

Gender as Social Structure

Another key idea shared by many gender theorists is the notion that gender itself is a social structure that constrains behavior by the opportunities it offers or denies us (Risman, 1987, 1998; Lorber, 1994; Connell, 1987). The consequences of the different opportunities afforded women and men can be seen at the *individual* level in the development of gendered selves, at the *interactional* level in the cultural expectations and situational meanings that shape how we "do gender," and at the *institutional* level in such things as sex-segregated jobs, a wage gap, and other economic and institutional realities that differentiate women's and men's experiences (Risman, 1998). Although we may more often focus on individuals making choices that reflect their internalization of gender expectations, situations and institutions also shape behavior.

GENDER SOCIALIZATION VIA SOCIAL LEARNING THEORY

Many theorists see gender like any other socially acquired role. We have to be socialized to act according to the expectations attached to our status as female or male. The emphasis on socialization has been considerable although consensus on the process of socialization has not. Social learning theory is derived from behaviorist psychology. In explaining our actions, behaviorists emphasize observable events and their consequences rather than internal feelings and drives. According to behaviorists, we learn attitudes and behaviors as a result of social interactions with others (hence, the term *social learning*).

The cornerstone of social learning theory is the belief that consequences control behavior. Acts that are regularly followed by a reward are likely to occur again; acts that are regularly followed by a punishment are less likely to recur. Girls are rewarded for playing with dolls ("What a nice mommy!"), but boys are not ("What a sissy!").

This behaviorist approach has been modified recently to include cognition—that is, mental processes (such as evaluation and reflection) that intervene between stimulus and response. The cognitive processes involved in social learning include our ability to (1) use language, (2) anticipate consequences, and (3) make observations. These cognitive processes are important in learning gender roles. By using language, we can tell our daughter that we like it when she does well in school and that we don't like it when she hits someone. A person's ability to anticipate consequences affects behavior. A boy does not need to wear lace stockings in public to know that

Playing "dress up" is one way children model the characteristics and behaviors of adults. It is part of the process of learning what is appropriate for someone of their gender.

such dressing will lead to negative consequences. Finally, children observe what others do. A girl may learn that she "shouldn't" play video games by seeing that the players in video arcades are mostly boys.

We also learn gender roles by imitation, according to social learning theory. Learning through imitation is called **modeling.** Most of us are not even aware of the many subtle behaviors that make up gender roles—the ways in which men and women use different mannerisms and gestures, speak differently, use different body language, and so on. We don't "teach" these behaviors by reinforcement. Children tend to model friendly, warm, and nurturing adults; they also tend to imitate adults who are powerful in their eyes—that is, adults who control access to food, toys, or privileges. Initially, the most powerful models that children have are their parents. Reflecting on your own family, you might examine the division of labor in your household. How is housework divided? How is unpaid household work valued in comparison with employment in the workplace?

As children grow older and their social world expands, so do the number of people who may act as their role models: siblings, friends, teachers, media figures, and so on. Children sift through the various demands and expectations associated with the different models to create their own unique selves.

COGNITIVE DEVELOPMENT THEORY

In contrast to social learning theory, **cognitive development theory** focuses on the child's active interpretation of the messages he or she receives from the environment. Whereas social learning theory assumes that children and adults learn in fundamentally the same way, cognitive development theory stresses that we learn differently, depending on our age. Swiss psychologist Jean Piaget (1896–1980) showed that children's abilities to reason and understand change as they grow older.

Lawrence Kohlberg (1969) took Piaget's findings and applied them to how children assimilate gender-role information at different ages. At age 2, children can correctly identify themselves and others as boys or girls, but they tend to base this identification on superficial features, such as hair and clothing. Girls have long hair and wear dresses; boys have short hair and never wear dresses. Some children even believe they can change their sex by changing their clothes or hair length. They don't identify sex in terms of genitalia, as older children and adults do. No amount of reinforcement will alter their views because their ideas are limited by their developmental stage.

When children are 6 or 7, they begin to understand that gender is permanent; it is not something they can change as they can their clothes. They acquire this understanding because they are capable of grasping the idea that basic characteristics do not change. A woman can be a woman even if she has short hair and wears pants. Oddly enough, although children can understand the permanence of sex, they tend to insist on rigid adherence to gender-role stereotypes. Even though boys can play with dolls, children of both sexes believe they shouldn't because "dolls are for girls." Researchers speculate that children exaggerate gender roles to make the roles "cognitively clear."

According to social learning theory, boys and girls learn appropriate gender-role behavior through reinforcement and modeling. But according to cognitive development theory, once children learn that gender is permanent, they independently strive to act like "proper" girls or boys. They do this on their own because of an internal need for congruence, the agreement between what they know and how they act. Also, children find performing the appropriate gender-role activities rewarding in itself. Models and reinforcement help show them how well they are doing, but the primary motivation is internal.

How Family Matters: Learning Gender Roles

Although biological factors, such as hormones, clearly are involved in the development of male and female differences, the extent of biological influences is not well understood. Moreover, it is difficult to analyze the relationship between biology and behavior, for learning begins at birth. In this section, we'll explore gender-role learning from infancy through adulthood, emphasizing the influence of our families in the construction of our ideas about gender.

CHILDHOOD AND ADOLESCENCE

In our culture, infant girls are usually held more gently and treated more tenderly than boys, who are ordinarily subjected to rougher forms of play. The first day after birth, parents tend to describe their daughters as soft, fine featured, and small, and their sons as hard, large featured, big, and attentive. Fathers tend to stereotype their sons more extremely than mothers do (Fagot and Leinbach, 1987). Although it is impossible for strangers to know the gender of a diapered baby, once they learn the baby's gender, they respond accordingly. Such gender role socialization occurs throughout our lives. By middle childhood, although conforming to gender-role behavior and attitudes becomes increasingly important, there is still considerable flexibility (Absi-Semaan, Crombie, and Freeman, 1993). It is not until late childhood and adolescence that conformity becomes most characteristic. The primary agents forming our gender roles are parents. Eventually, teachers, peers, and the media also play important roles.

Parents as Socialization Agents

During infancy and early childhood, a child's most important source of learning is the primary caretaker—often both parents, but also often just the mother, father, grandmother, or someone else. Most parents may not be aware of how much their words and actions contribute to their children's gender-role socialization (Culp et al., 1983). Nor are they aware that they treat their sons and daughters differently because of their gender. Although parents may recognize that they respond differently to sons than to daughters, they usually have a ready explanation—the "natural" differences in the temperament and behavior of girls and boys. Parents may also believe that they adjust their responses to each particular child's personality. In an everyday living situation that involves changing diapers, feeding babies, stopping fights, and providing entertainment, it may be difficult for harassed parents to recognize that their own actions may be largely responsible for the differences they attribute to nature.

The role of nature cannot be ignored completely, however. Temperamental characteristics may be present at birth. Also, many parents who have conscientiously tried to raise their children in a nonsexist way have been frustrated to find their toddler sons shooting each other with carrots or their daughters primping in front of the mirror.

Children are socialized in gender roles in many ways. Children's literature, for example, typically depicts girls as passive and dependent, whereas boys are instrumental and assertive (Kortenhaus and Demarest, 1993). In the more than four thousand children's books published annually, females are rarely portrayed as brave or independent, and are typically presented in supporting roles (Renzetti and Curran, 2003). Children's toys and clothing also reinforce gender differences. In general, children are socialized by their parents through four very subtle processes: manipulation, channeling, verbal appellation, and activity exposure (Oakley, 1985):

- *Manipulation:* Parents manipulate their children from infancy onward. They tend to treat a daughter more gently, tell her she is pretty, and advise her that nice girls do not fight. They treat a son roughly, tell him he is strong, and advise him that big boys do not cry. Eventually, children incorporate their parents' views in such matters as integral parts of their personalities. It may well be that differences in girls' and boys' behaviors result from the fact that parents expect their children to behave differently (Connors, 1996 cited in Renzetti and Curran, 2003).
- *Channeling:* Children are channeled by having their attention directed to specific objects and away from others. Toys, for example, are differentiated by gender. Dolls are considered appropriate for girls and cars, for boys. Toy companies market their products with gender themes, as can be seen in toy ads and displays in retail stores. Parents purchase different toys for their daughters and sons. Children, influenced by advertising, the reinforcement by their parents, and the enthusiasm of their peers, are attracted to gendered toys (Renzetti and Curran, 2003).
- *Verbal appellation:* Parents use different words with boys and girls to describe the same behavior. A boy who pushes others may be described as "active," whereas a girl who does the same may be called "aggressive."
- *Activity exposure:* The activity exposure of boys and girls differs markedly. Although both are usually exposed to feminine activities early in life, boys are discouraged from imitating their mothers, whereas girls are encouraged to be "mother's little helpers." Even the chores chil-

One way children learn gender roles is through activity exposure. Girls may be encouraged to help their mothers inside the house, whereas boys may participate with their fathers in outdoor work.

dren do are categorized by gender. Girls may wash dishes, make beds, and set the table; boys are assigned to carry out trash, rake the yard, and sweep the walk (Blair, 1992). The boy's domestic chores take him outside the house, whereas the girl's keep her in it—another rehearsal for traditional adult life. A study of fourth and fifth graders found that television reinforced these stereotypic views (Signorielli and Lears, 1992).

It is generally accepted that parents socialize their children differently according to gender (Fagot and Leinbach, 1987). Fathers more than mothers pressure their children to behave in gender-appropriate ways. Fathers set higher standards of achievement for their sons than for their daughters, play more interactive games with them and encourage them to explore their environments (Renzetti and Curran, 2003). Fathers emphasize the interpersonal aspects of their relationships with their daughters and encourage closer parent-child proximity. Mothers also reinforce the interpersonal aspect of their parent-daughter relationships (Block, 1983). They typically engage in more frequent "emotion talk" with their daughters than their sons, and—unsurprisingly—as early as first grade girls are more adept at monitoring emotion and social behavior (Renzetti and Curran, 2003).

Both parents of teenagers and the teenagers themselves believe that parents treat boys and girls differently. It is not clear, however, whether parents are reacting to differences or creating them (Fagot and Leinbach, 1987). It is probably both, although by that age, gender differences are fairly well established in the minds of adolescents.

Various studies have indicated that ethnicity and social class are important in socialization (Renzetti and Curran, 2003; Zinn, 1990; see Wilkinson, Chow, and Zinn, 1992, for scholarship on the intersection of ethnicity, class, and gender). Among whites, working-class families tend to differentiate more sharply than middle-class families between boys and girls in terms of appropriate behavior; they tend to place more restrictions on girls. African-American families tend to socialize their children toward more egalitarian gender roles (Taylor, 1994c). There is evidence that African-American families socialize their daughters to be more independent than white families do. Indeed, among African Americans, the "traditional" female role model may never have existed. The African-American female role model in which the woman is both wage earner and homemaker is more typical and more accurately reflects the African-American experience (Lips, 1997).

Other Sources of Socialization

Families are not the only influences on the ideas we acquire about gender. Our early lives are lived in the company of many others who also shape our ideas about men and women, femininity and masculinity. As children grow older, their social world expands and so do their sources of learning.

SCHOOL. Around the time children enter day-care centers or kindergarten, teachers (and peers, discussed next) become important influences. Day-care centers, nursery schools, and kindergartens are often a child's first experience in the wider world outside the family. Teachers become important role models for their students. Because most day-care, nursery school, kindergarten, and elementary school teachers are women, children tend to think of child-adult interactions as primarily taking place with women. Teachers also monitor children's behavior, reinforcing gender differences along the way.

Girls generally excel over boys in all areas during grade school, but by middle school boys are better in math, science, history, geography, reading, and spelling. (Sadker and Sadker, 1994). Classroom observations find that boys are louder, are more demanding, and receive a disproportionate amount of the teacher's attention. Teachers call on boys more often, are more patient with boys in their explanations, and are more generous in their praise to boys. Girls are praised for their appearance and neatness of work. As a result, girls become more tentative and hesitant as they enter middle school. By high school, they preface their answers with disclaimers: "I'm probably wrong, but . . ." or "I'm not sure, but . . ." Intelligent girls are often devalued by boys. Only in all-girl schools, write Myra Sadker and David Sadker (1994), do female students tend to assert themselves vigorously in class. They argue that in gender-segregated schools and classes, girls do not have to compete with boys for the teacher's attention. In such schools, girls are not overly concerned with their appearance, and they do not fear that their intelligence will make them undesirable as dates.

Increased attention and concern have also been directed at what boys experience in school (Pollack, 1998; Sadker and Sadker, 1994). For example, the attention boys receive is not always positive—they are subject to more discipline and receive more of the teacher's anger than do girls, even when the disruptiveness of their behavior is similar. Furthermore, their academic performance often suffers, as indicated by their rates of failing, acting up, and/or dropping out (Sadker and Sadker, 1994; Renzetti and Curran, 1995; Pollack, 1998). Thus, both girls and boys "pay" for what school does to them, and both would benefit from greater gender equity.

Gender doesn't operate alone in shaping school experiences. Race and class matter too. In schools, black males face especially difficult circumstances and receive the most unfavorable teacher treatment when compared with white males, white females, or black females (Sadker and Sadker, 1994; Basow, 1992). They receive the most recommendations for special education and are subjected to low expectations by teachers. Teachers describe black males as having the worst work habits, and they predict lower levels of academic success for them, *regardless of their actual behavior* (Basow, 1992).

Schools that have developed nonsexist curricula show that ordinary males and females can act and work in nonstereotypical ways (Bigler and Liben, 1992). Research suggests that if schools consciously structured activities involving both girls and boys, there would be considerably more interaction between the two (Mead and Ignicio, 1992). Furthermore, both girls and boys would benefit by acquiring the skills ordinarily restricted to one sex. Boys could learn homemaking skills, such as cooking, whereas girls could learn more assertiveness skills, such as political debate.

PEERS. A child's age-mates, or **peers,** become especially important when the child enters school. By granting or withholding approval, friends and playmates influence what games children play, what they wear, what music they listen to, what television programs they watch, and even what cereal they eat. Peer influence is so pervasive that it is hardly an exaggeration to say that in some cases, children's peers tell them what to think, feel, and do.

Peers also provide standards for gender-role behavior in several ways (Carter, 1987b):

- Peers reinforce gender-role norms through play activities and toys. With their friends, girls play with dolls that cry and wet or with glamorous dolls with well-developed figures and expensive tastes. Boys play together with dolls known as "action figures," such as G.I. Joe and Power Rangers, equipped with automatic weapons and bigger-than-life biceps.

- Peers react with approval or disapproval to others' behavior. Smiles encourage a girl to play with makeup or a boy to play with a football.
- Peers influence the adoption of gender-role norms through verbal approval or disapproval. "That's for boys!" or "Only girls do that!" discourages girls from playing with footballs or boys from playing house.
- Children's perceptions of their friends' gender-role attitudes, behaviors, and beliefs encourage them to adopt similar ones in order to be accepted. If a girl's female friends play soccer, she is more likely to play soccer. If a boy's male friends display feelings, he is more likely to display feelings.

During adolescence, peers continue to have a strong influence, but research indicates that parents can be more influential than peers (Gecas and Seff, 1991). Parents influence their adolescent's behavior primarily by establishing norms, whereas peers influence others through modeling behavior. Even though parents tend to fear the worst from their children's peers, peers provide important positive influences. It is within their peer groups, for example, that adolescents learn to develop intimate relationships (Gecas and Seff, 1991). Also, adolescents tend to be more egalitarian in gender roles than do parents, especially fathers (Thornton, 1989).

POPULAR CULTURE AND MASS MEDIA. In all its forms, the mass media depicts females and males quite differently. In fact, we can safely assert that the media typically have "ignored, trivialized, or condemned women," a process known as symbolic annihilation (Renzetti and Curran, 2003; Tuchman, et al. 1978).

Much of television programming promotes or condones negative stereotypes about gender, ethnicity, age, and gay men and lesbians. Women are significantly underrepresented on television (MRTW, 1999; Signorielli, 1997). Through the 1970s, men outnumbered women on prime-time television three to one; today, women account for approximately 39 percent of major characters (MRTW, 1999). Even on *Sesame Street,* 84 percent of the characters were male in 1992, compared with 76 percent five years earlier ("Muppet Gender Gap," 1993).

The women depicted on television are less representative of women than are the men depicted representative of men. Nearly two-thirds of all female prime-time characters are in their twenties and thirties. Although 22 percent of male characters are in their forties, only 12 percent of female characters are in their forties (MRTW, 1999). Almost half of female characters are "thin and attractive;" only 16 percent of men are "thin or very thin" (Renzetti and Curran, 2003; Signiorelli, 1997). Television women are portrayed as emotional and needing emotional support; they are also sympathetic and nurturing. Not surprisingly, women are usually portrayed as wives, mothers, or sex objects (Vande Berg and Strekfuss, 1992).

On television male characters are shown as more aggressive and constructive than female characters. They solve problems and rescue others from danger. Only in the last few years in prime-time series have males been shown in emotional, nurturing roles. Although things have improved, ethnic and sexual stereotypes continue to be commonly found in television.

GENDER DEVELOPMENT IN ADULTHOOD

Although more attention has been directed at early experiences and socialization in childhood and adolescence, gender development doesn't stop there. Many life experiences that we have in adulthood alter our ideas about and actions as males and females. Once again, families loom large in reshaping our gendered ideas and behaviors. From a 1970s perspective known as *role transcendence,* an individual goes through three stages in developing his or her gender-role identity: (1) undifferentiated stage, (2) polarized stage, and (3) transcendent stage (Hefner, Rebecca, and Oleshansky, 1975).

Young children have not clearly differentiated their activities into those considered appropriate for males or females. As children enter school, however, they begin to identify behaviors as masculine or feminine. They tend to polarize masculinity and femininity as they test the appropriate roles for themselves. As they enter young adulthood, they slowly begin to shed the rigid male-female polarization as they are confronted with the realities of relationships. As they mature and grow older, men and women transcend traditional masculinity and

Perspective

GENDER ROLES AND VIDEOS

Recreation and entertainment for young people increasingly encompass video images and technologies. Much as popular music was revolutionized by the "invention" of the music video (Twitchell, 1992), the video game industry has revolutionized "play" for millions of young Americans, especially males. Billions of dollars and countless hours have been spent on arcade or home video games (Dietz, 1998). Together these media have also altered the experience of gender socialization.

In video games, when females are present, they are most often either victims ("damsels in distress") or sex objects ("visions of beauty with large breasts and thin hips"), but rarely heroes or action characters. In the typical video game, females are entirely absent (Dietz, 1998).

There is considerable verbal or physical aggressiveness against both genders in music videos and video games (Kalis and Neuendorf, 1989). In music videos, female aggression is often provoked by jealousy. Male aggression is frequently unprovoked. Aggression is often a part of male swagger—the assertion of power and status—especially in heavy metal and rap videos. Critic James Twitchell (1992) writes of these two forms: "Both are rife with adolescent misogyny, homophobia, and threats of violence. They are rude, bawdy, boastful, with a kind of 'in your face' aggression . . . characteristic of insecure masculinity."

Many female vocalists appear in music videos. However, the majority of music videos are dominated by male singers or male groups, and women provide erotic backdrop or vocal backup (Seidman, 1992; Sommers-Flanagan, Sommers-Flanagan, and Davis, 1993). Most women are depicted as sex objects (Kaloff, 1993).Women are typically pictured condescendingly, are provocatively dressed, or both.

In video games, aggression and violence are major components of many games. Most of the time, male characters are the perpetrators of video game violence and their targets are generally other male characters or some non-human characters (such as monsters, aliens, creatures, or animals). Occasionally (20 percent in one study of a sample of 33 Nintendo or Sega games), violence is directed at female characters, though that is not typical (Dietz, 1998). Overall, violent themes and aggressive action are commonplace.

Because television prohibits the explicit depiction of sexual acts, music videos use sexual innuendos and suggestiveness to impart their sexual meanings (Baxter et al., 1985; Sherman and Dominick, 1987). One study found that adolescent or male viewers generally rated music videos, especially sexually provocative ones, more positively than did older or female viewers (Greeson, 1991). Another study found that both male and female undergraduates responded with positive emotions to music videos with sexual content; they responded negatively to those with violence. The music videos declined in appeal when sex and violence were combined (Hansen and Hansen, 1991). Females have stronger positive responses to soft rock, while males seem to enjoy hard rock (Toney and Weaver, 1994).

Although listening to popular music continues to be important in our lives—adolescents, for example, listen to about 10,500 hours of it between seventh and twelfth grade—now music is "seen" as well as heard. Visual images are as important as are the music and the lyrics; indeed, the images may even be more important than the music.

Music videos and video games present popular images of masculinity and femininity: the people in music videos are youthful, attractive, fashionably attired, hip, and sexual.

According to a study by Steven Seidman (1992), men are more adventuresome, generally aggressive, violent, and domineering than women. Women are more affectionate, dependent, and fearful. Men are often construction workers, mechanics, firefighters, or in the military. Women are often secretaries, librarians, or cheerleaders. Women are systematically excluded from most white-collar professions. In Seidman's study, three times as many women were in blue-collar jobs, such as waitress and hairstylist, as in white-collar jobs. Not a single woman was portrayed as a politician, business executive, or manager. The occupational stereotyping is similar to that found in prime-time television and in commercials. Such stereotyping serves to reinforce the status quo (Signorielli, 1989).

Cumulatively, video games and music videos become part of the gender socialization process. Their themes—male as aggressive and violent, females as sex objects and victims—fit, both with each other and with other popular media content (such as television, film, cartoons, and advertising). Clearly, no single game or video will determine a person's attitudes toward women or his or her propensity toward violence. Collectively, however, such images help shape and reinforce traditional gender attitudes and make aggressive outcomes more likely.

femininity. They combine masculinity and femininity into a more complex role.

More recent research details some of the ways adult life experiences can transform how we act as a male or female (Gerson, 1985, 1993; Risman, 1986, 1987, 1988, 1998). These more structural analyses have shown how adult life experiences both inside and outside of families have the potential to restructure our identity, redefine our role responsibilities, and take us in directions that are quite different from those suggested by our early gender socialization. In adulthood, new or different sources of gender-role learning may include marriage and parenthood as well as college and experiences in the workplace.

COLLEGE. The college and university environment, which encourages critical thinking and independent behavior, contrasts markedly with high school. In the college setting, many young adults learn to think critically, to exchange ideas, and to discover the bases for their actions. It is in college that many young adults first encounter alternatives to traditional gender roles, either in their personal relationships or in their courses. A longitudinal study of gender roles found that traditional and egalitarian gender-role attitudes affected dating relationships in college but had little impact on later life (Peplau, Hill, and Rubin, 1993).

MARRIAGE. Marriage is an important source of gender-role learning, for it creates the roles of husband and wife. For many individuals, no one is more important than a partner in shaping gender-role behaviors through interaction. Our partners have expectations of how we should act as a husband or wife, and these expectations are important in shaping behavior.

Husbands tend to believe in innate gender roles more than wives do. This should not be especially surprising, because men tend to be more traditional and less egalitarian about gender roles (Thornton et al., 1983). Husbands stand to gain more in marriage by believing that women are "naturally" better at cooking, cleaning, shopping, and caring for children.

PARENTHOOD. For most men and women, motherhood alters life more significantly and visibly than fatherhood does. For some men, fatherhood may mean little more than providing for their children. It is unlikely to find many who would associate motherhood only with providing. As parents, mothers do more, are expected to do more, and are expected to juggle the more that they do along with their paid employment. As a consequence, fatherhood does not create the same work-family conflict that motherhood does. A man's work role allows him to fulfill much of his perceived parental obligation.

Yet contemporary fatherhood has become more complex. Not only have our expectations shifted toward a more nurturing model of fatherhood, but where traditional fatherhood was tied to marriage, today a third of all current births occur outside of marriage and half of all current marriages end in divorce. What, then, is the father's role for a man who is not married to his child's mother or who is divorced and does not have custody? What are his role obligations as a single father as distinguished from those of married fathers? For many men, the answers are painfully unclear, as evidenced by the low rates of contact between unmarried or divorced fathers and their children.

Women today have somewhat greater latitude as wives. It is now both accepted and expected that women will work outside the home at least until they become mothers, and more than likely that they will continue or return to paid employment sometime after they have children. Even with increases in the numbers of women who remain childless, women are still under considerable pressure to become mothers. Once children are born, roles tend to become more traditional, even in previously nontraditional marriages. Often, the wife remains at home, at least for a time, and the husband continues full-time work outside the home. The woman must then balance her roles as wife and mother against her own needs and those of her family.

THE WORKPLACE. It is well established that men and women are psychologically affected by their occupations (Menaghan and Parcel, 1991; Schooler, 1987). Work that encourages self-direction, for example, makes people more active, flexible, open, and democratic; restrictive jobs tend to lower self-esteem and make people more rigid and less tolerant. If we accept that female occupations are usually low status with little room for self-direction, we can understand why women are not as achievement oriented as men. Because men and women have different opportunities for promotion, they have different attitudes toward achievement. Women typically downplay any desire for promotion, suggesting that promotions would interfere with their family responsibilities. But this really may be related to a need to protect themselves from frustration because many women are in dead-end jobs in which promotion to management positions is unlikely.

Household work affects women psychologically in many of the same ways that paid work does in female-dominated occupations such as clerical and service jobs (Schooler, 1987). Women in both situations feel greater levels of frustration owing to the repetitive nature of the work, time pressures, and being held responsible for things outside their control. Such circumstances do not encourage self-esteem, creativity, or a desire to achieve.

Remaking Women and Men

Focusing on adulthood is important because it reveals the gaps that often exist between earlier gender socialization and adult experiences. The lives we ultimately lead are often very different from those we were raised to lead or expected to lead (Gerson, 1993; Risman, 1987, 1998). To some scholars, this diminishes the importance of socialization and discredits theories that deterministically link early socialization to later life outcomes (Gerson, 1993). In some ways, those theories are no better than *biological determinism,* in which we are ultimately limited to those behaviors that our genetic or hormonal characteristics allow. They simply substitute socialization for biology (Risman, 1989).

Socialization is important, especially in affecting our expectations and offering us role models for lives we might live. But life is more circuitous than linear. Unanticipated twists and turns often take us in directions we neither expected nor intended. Research on women's and men's career and family experiences bear this out. For example, Kathleen Gerson's research on women's and men's career and family choices reveals that many people may develop commitments to either careers or parenting that stem from their experiences in jobs and relationships (Gerson, 1985, 1993). Some women and men who anticipate "traditional" adult outcomes move in nontraditional directions based on the levels of fulfillment and opportunity at work, the experiences and aspirations of their partners, and their experiences with children. Similarly, men and women who aspire to nontraditional outcomes (career attachment for women, involved fatherhood for men) may "reluctantly" abandon those directions as a result of firsthand experiences at home and work.

Barbara Risman's research on single custodial fathers pointed to similar adult development. Men, who reluctantly found themselves as lone, custodial parents developed nurturing abilities that their socialization had not included. More important than how they were raised was how they interacted with their children, as well as the fact that there was no female in their lives to whom these tasks could be assigned. Thus, these single fathers "mothered" their children in ways that were more like women's relationships with children than what one would have predicted (Risman, 1986). Importantly, socialization contributes to but does not guarantee any particular family outcome.

Gender Matters in Family Experiences

Within the past generation, there has been a significant shift from traditional toward more egalitarian gender roles (Brewster and Padavic, 2000). Women have changed more than men, but men also are changing. These changes seem to affect all classes, though not to the same extent. Also, there is still resistance to change, as those from conservative religious groups, such as Mormons, Catholics, and fundamentalist and evangelical Protestants, continue to adhere more strongly to traditional roles (Jensen and Jensen, 1993; Spence et al., 1985).

Contemporary gender role attitudes have changed, in part, as a response to the steady increase in women's participation in the labor force. Although a large share of this increase occurred in the 1970s and 1980s, as did the move toward more egalitarian attitudes, it continued through the 1990s. College-educated women and men, especially, are considerably less likely to hold traditional ideas about gender and work and family roles (Brewster and Padavic, 2000).

Within the family, although attitudes toward gender roles have become more liberal, in practice, gender roles continue to place women at a disadvantage, especially by making them responsible for housekeeping and child-care activities (Atkinson, 1987; Coltrane, 2000; Hochschild, 1989). Some of the most important changes affecting men's and women's roles in the family are briefly described in the following sections.

MEN'S ROLES IN FAMILIES

As noted earlier, according to traditional gender-role stereotypes, many of the traits ascribed to one gender are not ascribed to the other. Men show instrumental traits, women display expressive ones (Hort, Fagot, and Leinbach, 1990). Theoretically, these traits complemented women's and men's traditional roles in the family. Because of assumed basic gender differences, men were expected to participate in the world of work and politics. Their central male role was worker; in the family, this role translated to breadwinner. Because women were thought to be primarily expressive, they were expected to remain in the home as wives and mothers.

Central features of the traditional male role, whether among whites, African Americans, Latinos, or Asian Americans, include dominance, work, and family. Males are generally regarded as being more power oriented than females. Statistically, men demonstrate higher degrees of aggression, especially violent aggression (such as assault, homicide, and rape), seek to dominate and lead, and show greater competitiveness. Although aggressive traits are thought to be useful in the corporate world, politics, and the military, such characteristics are rarely helpful to a man in fulfilling marital and family roles requiring understanding, cooperation, communication, and nurturing.

Traditionally, the centrality of men's work identity affected their family roles as husbands and fathers. When men see their primary family function as that of provider, it takes precedence over all other family functions, such as nurturing and caring for children, doing housework, preparing meals, and being intimate. Because of this focus, traditional men may get confused by their spouses' expectations of intimacy; they believe that they are good husbands simply because they are good providers (Rubin, 1983). When circumstances render them unable to provide, the blow to their self-identities can be quite powerful (Rubin, 1994).

Across ethnic and racial lines traditional male roles have centered around providing. Thus the somewhat traditional gender rhetoric of the 1990s Million Man March on Washington, D.C., by African-American men, was not that far from the more explicitly traditional rhetoric espoused by the Christian Promise Keepers. Both groups implored men to live up to their responsibilities to their families and communities, and central to the familial responsibilities was to lead and provide.

However, given the overlap between race, ethnicity, and economic status discussed in the previous chapter, certain categories of men face more difficulty meeting the expectations of the traditional provider role. Because African Americans and Latinos often fare less well economically, men frequently are left unable to lay claim to the household status and power that traditional masculine roles promise.

Occasionally, characterizations of Latino families have exaggerated the extent of male dominance, as suggested by the notion of *machismo.* Although such a notion may have been somewhat more accurate in depicting gender ideologies of rural Mexico and the Caribbean in the first half of the twentieth century, it is inaccurately applied to contemporary Latino families (McLoyd, Cauce, Takeuchi, and Wilson, 2000). In fact, both African-American and Chicano men have more positive attitudes toward employed wives.

At the same time, however, they are also more likely to support the traditional male economic provider role. However, cultural beliefs about men's roles as providers may not be achievable under limited economic resources. Ethnic differences in traditional notions of masculinity and men's roles are more evident among older and less educated African Americans and among Mexican Americans not born in the United States (McLoyd, Cauce, Takeuchi, Wilson, 2000).

Because the key assumption about male gender roles has been the centrality of work and economic success, many earlier researchers failed to look closely at how men interacted within their families. Over the past two decades, as part of a closer examination of men's lives, we witnessed a dramatic increase in the popular and scholarly attention paid to men's family lives (see, for example, Cohen, 1987, 1993; Coltrane, 1996; Gerson, 1993; Daly, 1993, 1996; LaRossa, 1988; Popenoe, 1996; Marsiglio, 1998).

Researchers finally began to ask about men's lives some of the same questions that were previously asked about women, looking at whether and how men juggle paid work and family and maintain sufficient involvement in each (Gerson, 1993; Daly, 1996; Coltrane, 1996). Although we may not yet treat working fathers with the same concern we bring to working mothers, we have made strides in examining how men experience conflicts between work and parenting.

Additionally, research indicates that men consider their family role to be much broader than that of family breadwinner. As researchers have pointed out, economic responsibilities are only one part of men's family lives (Cohen, 1993; Gerson, 1993; Coltrane, 1996). Other dimensions of men's experiences include emotional, psychological, community, and legal dimensions; they also include housework and child-care activities (Goetting, 1982). Later chapters will look at men's experiences of marriage, parenting, and the division of household labor.

Still, even with enlarged emphasis on men's more nurturing qualities, men continue to be expected to work and to support or help support their families. The male provider role remains one of men's central roles in marriage and family life; the provider role is not central for married women. Although their financial contributions may be no less essential to maintain their family standard of living or even remain out of poverty, women are not judged as successful wives and mothers based on whether they succeed at paid employment. As a result, men have less role freedom to choose whether to work than women have, though there are marriages in which men stay home and women act as providers (Russell, 1987; Cohen and Durst, 2001). When a man's roles of worker and father come into conflict, usually it is the father role that suffers. A factory worker may want to spend time with his children, but his job does not allow flexibility. Because he must provide income for his family, he will not be able to be more involved in parenting. In a familiar scene, a child comes into the father's home office to play, and the father says, "Not now. I'm busy working. I'll play with you later." When the child returns, the "not now, I'm busy" phrase is repeated. The scene recurs as the child grows up, and one day, as his child leaves home, the father realizes that he never got to know him or her.

Many men strive to avoid this all too familiar nightmare and prevent the father-child estrangement that they may remember experiencing as children. They go out of their way to be more involved and more nurturing with their children. However, what they learn is that the complexity of juggling work and family is not restricted to women. Men who attempt the same juggling act often experience similar role strain and role overload (Gerson, 1993).

Additionally, men continue to have greater difficulty expressing their feelings than do women (Real, 1997). Men tend to cry less and show love, happiness, and sadness less. When men do express their feelings, they are more forceful, domineering, and boastful; women, in contrast, tend to express their feelings more gently and quietly. When a woman asks a man what he feels, a common response is "I don't know," or "Nothing." Such men have lost touch with their inner lives because they have repressed feelings that they have learned are inappropriate. This male inexpressiveness often makes men strangers to both themselves and their partners.

Men continue to expect and, in many cases, are expected to be the dominant member in a relationship. Unfortunately, the male sense of power and command often does not facilitate personal relationships. Without mutual respect and equality gen-

As contemporary male gender roles allow increasing expressiveness, men are encouraged to nurture their children.

uine intimacy is difficult to achieve. One cannot control another person and at the same time be intimate with that person.

WOMEN'S ROLES IN FAMILIES AND WORK

Although the main features of traditional male gender roles vary more by class than ethnicity, there are more striking ethnic differences in traditional female roles.

Traditional white female gender roles center around women's roles as wives and mothers. When a woman leaves adolescence, she is expected to either go to college or to get married and have children. Although a traditional woman may work prior to marriage, she is not expected to defer marriage for work goals, and soon after marriage she is expected to be "expecting." Once married, she is expected to devote her energies to her husband and family and work but to find her meaning as a woman by fulfilling her roles as wife and mother. Within the household, she is expected to subordinate herself to her husband. Often this subordination is sanctioned by religious teachings.

We still know relatively less about the lives of married African-American women, as most research focuses more upon unmarried mothers and the poor (Wyche, 1993). Yet we do know that the traditional white female gender role does not extend to African-American women. This may be attributed to a combination of the African heritage, slavery (which subjugated women to the same labor and hardships as men), and economic discrimination that pushed women into the labor force. Karen Drugger (1988) notes:

> A primary cleavage in the life experiences of Black and White women is their past and present relationship to the labor process. In consequence, Black women's conceptions of womanhood emphasize self-reliance, strength, resourcefulness, autonomy, and the responsibility of providing for the material as well as emotional needs of family members. Black women do not see labor-force participation and being a wife and mother as mutually exclusive; rather, in Black culture, employment is an integral, normative, and traditional component of the roles of wife and mother.

One study (Leon, 1993) found that African-American women appear more instrumental than either white or Latina women; they also have more flexible gender and family roles. African-American men are generally more supportive than white or Latino men of egalitarian gender roles.

In traditional Latina gender roles, the notion of *Marianismo* has been the cultural counterpart to *machismo.* Drawn from the Catholic ideal of the Virgin Mary, Marianismo stresses women's roles as self-sacrificing mothers, suffering for their children (McLoyd, Cauce, Takeuchi, and Wilson, 2000). Thus traditional Latino women are expected to subordinate themselves to males (Vasquez-Nuthall, Romero-Garcia, and De Leon, 1987). But this subordination is based more on respect for the male's role as provider than on subservience (Becerra, 1988). It also appears to be waning. Latina women are increasingly adopting values that are incompatible with a belief in male dominance and female subordination. They also display higher levels of marital satisfaction and less depression when their husbands share more of the domestic work (McLoyd, Cauce, Takeuchi, and Wilson, 2000). Wives have greater equality if they are employed; they also have more rights in the family if they are educated (Baca Zinn, 1994).

Latino gender roles, unlike those of Anglos, are strongly affected by age roles in which the young subordinate themselves to the old. In this dual arrangement, notes Becerra (1988), "females are viewed as submissive, naive, and somewhat childlike. Elders are viewed as wise, knowledgeable, and deserving of respect." As a result of this intersection of gender and age roles, older women are treated with greater deference than younger women.

Even though the traditional roles for women have typically been those of wife and mother, increasingly over the past few decades an additional role has been added: employed worker or professional. (This change has most notably affected white women, as African-American women have traditionally combined wife, mother, and worker roles.) It is now generally expected that most women will be employed at various times in their lives. For most women, entrance into work or a career does not exclude their more traditional roles as wives and mothers. Such work, however, may conflict with their family roles. Women generally attempt to reduce the conflict between work and family roles by giving family roles precedence. As a result, they tend

BECOMING "AMERICAN": IMMIGRANTS AND GENDER ROLES IN TRANSITION

The enormous rise in immigration beginning in the 1980s brought great numbers of Latinos and Asians to the United States. Today, more than 750,000 immigrants arrive annually; a large percentage are children and adolescents. The process of adaptation to a new environment, culture, and language places individuals under considerable stress. In adjusting to life in the United States, women tend to experience greater stress than men. Mental health workers attribute women's greater stress to conflicting gender role expectations and family roles (Aneshensel and Pearling, 1987).

The acquisition of "American" gender roles may be difficult and complicated. New immigrants must deal with several problems, including those resulting from (1) their immigrant status, including issues involved in being a newcomer and a member of an ethnic group (which is often a minority group as well); (2) sorting out what it means to be female or male both in American culture and in their original cultures; and (3) integrating their ethnic identity and their gender role (Goodenow and Espín, 1993).

For young immigrants, acquiring a meaningful gender role is complicated by the conflict between the norms of their new and old cultures. Their old social environment has disappeared and a new, alien one taken its place. Most adolescents are able to try out their different roles with peers whose language they understand, but most new immigrants do not have this advantage. English leaves them bewildered. In addition, different meanings are given to everyday things and events. What does a certain gesture or action mean? Whom should I trust? What are appropriate choices? What kinds of clothes should I wear?

Immigrants must adjust to two gender-role cultures: their original culture and the American culture. One notable problem is that many cultures are far less egalitarian than American culture. This creates strains between men and women as they become acculturated. Furthermore, immigrant males and females face different problems in acculturation. Families expect females to maintain traditional roles and values, whereas males are often encouraged to become Americanized quickly. For females, conflict is most likely to arise over appropriate gender-role behavior and sexuality. Conflict is most acute for females who were reared in traditional cultures. Young women often feel pulled in different directions at the same time. Adolescents must become American without alienating their families or losing their cultural heritage.

Carol Goodenow and Olivia Espín (1993) interviewed adolescent Latina

to work outside the home in greatest numbers before motherhood and after divorce, when single mothers generally become responsible for supporting their families. After marriage, most women are employed even after the arrival of the first child. Regardless of whether a woman is working full-time, she almost always continues to remain responsible for housework and child care. In spite of the fact that employment is often a necessity for mothers, most feel great emotional conflict because they want to be or believe they should be at home with their children.

The cultural expectations attached to mothering impose high standards of devotion and labor-intensive, self-sacrifice on women who become mothers, what is described as the **intensive mothering ideology**—the belief that children need full-time, unconditional attention from mothers in order to develop into healthy, well-adjusted people. Obviously, this puts all mothers in a demanding position, but it creates a particularly difficult dilemma for mothers who also choose or need to work outside the home (Hays, 1996). It leads increasing numbers of women to question whether or not to have children, and, if they do become mothers, how much of their time and attention their children actually need.

Women from ethnic and minority groups, however, are less likely than whites to view motherhood as an impediment. African-American women and Latinas tend to place greater value on motherhood than the white or Anglo majority. For African Americans, tradition has generally combined work and motherhood; the two are not viewed as necessarily antithetical (Basow, 1993). For Latinas, the cultural and religious emphasis on family, the higher status conferred on motherhood, and their own familial attitudes have contributed to high birthrates (Jorgensen and Adams, 1988).

immigrants to see how they adjusted to American gender roles. They found that although the girls did not challenge their original culture's traditional sexual norms, they did challenge its gender roles. The girls said that if they had remained in their native countries, they would have felt pressured to marry and have children early. Each wanted to marry someday, but first each wanted to "make something of myself" or "have a career, not just to get married." They wanted adventure, "to be somebody," to "do things." For the girls, "the personal freedom offered by life in the United States, especially the relative freedom for women, was exhilarating. . . .Their vision of the future focused on careers or jobs—very different from the limited domestic roles that might easily have been theirs." As Latinas become more acculturated, they adopt more egalitarian gender roles (Vasquez-Nuthall, Romero-Garcia, and De Leon, 1987). First-generation Latinas have held significantly more traditional gender-role attitudes than second- and third-generation Latinas. This traditionalism is reflected in part by the reluctance of first-generation Latinas to work outside the home (Ortiz and Cooney, 1984). Researchers found that the shift from a culture emphasizing definite male-female gender roles to one with greater freedom has affected the self-identities of immigrant Latinas (Salgado de Snyder, Cervantes, and Padilla, 1990).This shift has been a potential source of both personal and family conflict. As a Latina demands greater input in making decisions or asks her partner's assistance in child care—behavior more typical among Anglos, she is likely both to encounter resistance from her husband and to feel guilty for violating traditional Latino norms. Reiko True (1990) describes similar findings among Asian-American women. Struggling to establish themselves in American society, they "often find themselves trapped within restrictive roles and identities as defined by Asian cultures or by general American stereotypes."

Although immigrant males also experience stress resulting from changed male-female relationships, their greatest stress tends to be occupational. This difference results from the priority of work roles over family roles for males. Latino males have experienced occupational stress, for example, reflecting their difficulty in finding good jobs or job advancement as a result of language differences, education, or discrimination (Salgado de Snyder, Cervantes, and Padilla, 1990). Although many Asian male immigrants tend to be from white-collar backgrounds, they also experience occupational stress resulting from discrimination and language difficulties.

Gender-role stress increases for both men and women as their children adopt more typically American gender roles. Parents and children may come into conflict about appropriate gender-related behavior because parents tend to be more culturally traditional and their children tend to be more Americanized.

Although husbands were once the final authority, wives have greatly increased their power in decision making. They are no longer expected to be submissive but to have significant, if not equal, input in marital decision making. This trend toward equality is limited in practice by an unspoken rule of marital equality: "Husbands and wives are equal, but husbands are more equal." In actual practice, husbands may continue to have greater power than wives, becoming what sociologist John Scanzoni (1982) once described as the "senior partner" of the marriage.

This is not absolute or inevitable, however. Interesting exceptions to this pattern exist, especially among some dual-earner couples. Some couples develop and act upon an ideology of sharing and fairness, valuing and pursuing such relationship characteristics as equality and equity (Schwartz, 1994; Risman and Johnson-Sumerford, 1997). Although these "peer" and **post-gender relationships** (i.e., relationships lived outside the constraints of gender expectations) are not yet the norm, they do reflect the most concerted efforts to establish greater equality in marriage.

BREAKDOWN OF INSTRUMENTAL/ EXPRESSIVE DICHOTOMY

The identification of masculinity with instrumentality and femininity with expressiveness appears to be breaking down. As a group, men perceive themselves to be more instrumental than do women, and women perceive themselves as being more expressive than do men. A substantial minority of both genders is relatively high in both instrumentality and expressiveness or is low in both. It is interesting that the instrumental/expressiveness ratings men

MAKING GENDER MATTER LESS

Research on intimate relationships, marriage, and family consistently reveals the importance of gender in dividing up responsibilities and shaping personal experiences. The familiar notion of the "two marriages" inside every marriage highlights the differences between "his" and "her" marriage (Bernard, 1982). Data on housework and childcare further validate the importance of gender. Women perform two to three times as much housework as men, and employed wives experience greater stress and enjoy less leisure than their husbands (Coltrane, 2000). The consistency with which such inequalities are reported may give the impression of inevitability. Let's look a little more closely at some research that offers a more hopeful scenario to those who might wish to someday create more equal partnerships.

Sociologists Barbara Risman and Danette Johnson-Sumerford interviewed 15 couples who explicitly reject conventional conceptions of gender, opting instead for more gender-neutral relationships. That is, they carefully and intentionally share responsibility for paid work and share responsibility as caregivers for their children. At minimum, they "changed how gender works in their families." Furthermore, "in the negotiation of marital roles and responsibilities, they have moved beyond using gender as their guidepost" (Risman and Johnson-Sumerford, 1998, p. 24).

In recruiting their sample, rather than relying only on couples' characterizations of their relationships as equal or "fair," Risman and Johnson-Sumerford included only those who split their household labor 40/60 or better (i.e., closer to 50/50), and who agreed that they were equally responsible for breadwinning and childrearing. Couples were interviewed and subsequently observed in their homes. Interviews explored life histories of each member, histories of the development of their relationship, how they arrived at their equitable division of labor, how they felt about and divided child care, and their experiences in paid work. Here, we look briefly at the different paths couples took to construct their "postgender" marriages.

Dual-career couples: The most common path to "postgender marriage" begins with a marriage of two career-oriented professionals, in which at least the wife, but preferably the husband as well, values equality. They are ideologically committed to sharing. Both partners retain strong commitments to their careers although both scale back in order to achieve the lifestyle they desire. There was hardly any use of gender as a principle around which

and women give each other have very little to do with how they rate themselves as masculine or feminine (Spence and Sawin, 1985).

Constraints of Contemporary Gender Roles

Even though substantially more flexibility is offered to men and women today, contemporary gender roles and expectations continue to limit our potential. Indeed, there is considerable evidence that some stereotypes about gender traits are still very much alive. Men are perceived as having more undesirable self-oriented traits (such as being arrogant, self-centered, and domineering) than women. Women are viewed as having more traits reflecting a lack of a healthy sense of self (such as being servile and spineless).

Research suggests that the traditional female gender role does not facilitate self-confidence or mental health. Both men and women tend to see women as being less competent than men. A study by Lyn Brown and Carol Gilligan (1992), revealed that the self-esteem of adolescent girls plummeted between the age of 9 and the time they started high school.

Traditional women married to traditional men experience the most symptoms of stress (feeling tired, depressed, or worthless) and express the most dissatisfaction about life as a whole (Spence et al., 1985). The combination of gender-role stereotypes and racial/ethnic discrimination tends to encourage feelings of both inadequacy and lack of physical daily life was to be organized. This lifestyle is the most common path to postgender marriage.

Dual nurturers: Dual nurturer couples place their priorities on home and family, not careers. Rather than being attached to careers, their work is for money to enable them to spend their time together and with their children. In one dual nurturer couple, neither spouse had consistently held a full-time job. Instead, they piece together part-time work, seeing to it that they weren't both working simultaneously each day.

Posttraditionalists: As the label suggests, this path begins with a traditional, though not necessarily male-breadwinner, female-homemaker, arrangement. Instead, "traditional" means a gender-based division of household roles and labor. These couples found themselves dissatisfied in gender-based arrangements, whether in their current or a former marriage. They were strongly motivated to avoid the sort of unfairness that often plagues dual-earner couples.

External forces: This path consists of couples being "pushed" by circumstance toward more equal domestic arrangements. They used gender neutral-language in describing their arrangements and in explaining the forces that took them there (e.g., economic factors such as a wife's higher salary and less flexible work schedule than her husband, or illness). Whatever the circumstances, they came to recognize and appreciate the gender equality that resulted.

Regardless of the route couples took to arrive at their postgender family arrangements, they used criteria other than gender to organize their daily activities. In fact, they have "rejected *hegemonic* notions of gender" and the idea that "wifehood involves a script of domestic service or that breadwinning is an aspect of successful masculinity."

Risman and Johnson-Sumerford acknowledge that such couples are still rare and their lifestyles may require high levels of female income and professional autonomy if women are to be able to move beyond male dominance or privilege. They also note that all but two couples employ paid help with domestic tasks such as cleaning, dusting, bathrooms, and yard work. Such outside help certainly made life easier, and perhaps made fairness more achievable. However, couples who used paid help reported that domestic responsibilities had been shared even before they started paying for housekeeping services.

The importance of research such as Risman and Johnson-Sumerford's is that it reveals a wider range of possible marital outcomes than most literature reports. There is no inevitable inequality that engulfs married couples. Equality and fairness take work and persistence but are possible for those who seek them.

attractiveness among African-American women, Latinas, and Asian-American women (Basow, 1993).

The situation of contemporary women in dual-earner households imposes its own constraints on women's lives. Because they continue to shoulder the bulk of responsibility for housework and child care *on top of full-time jobs*, they often experience fatigue, stress, resentment, and a lack of leisure (Hochschild, 1989). Especially for women who try to be "supermoms," the volume and complexity of work and family can force them to cut back on their aspirations or compromise their expectations for marriage and motherhood (Hochschild, 1989, 1997). Significantly, despite the ongoing stresses, women who "juggle" are less distressed and more fulfilled than full-time homemakers (Crosby, 1991).

Finally, there is still a "double standard of aging" that treats men and women differently. As women get older, they tend to be regarded as more masculine and as unattractive. Also, our culture treats aging in men and women differently: As men age, they become distinguished; as women age, they simply get older. Masculinity is associated with independence, assertiveness, self-control, and physical ability; with the exception of physical ability, none of these traits necessarily decreases with age. Because older women are considered to have lost their attractiveness and because they have fewer potential partners, they are less likely to marry.

RESISTANCE TO CHANGE

We may think that we want change, but both men and women reinforce traditional gender-role stereotypes among themselves and each other (Hort, Fagot, and Leinbach, 1990). Women's self-help books often reflect a conservative bias by encouraging women to discover their "inner" feminine selves, which mirror traditional gender roles (Schrager,

1993). Both genders react more negatively to men displaying so-called female traits (such as crying easily or needing security) than to women displaying male traits (such as assertiveness or worldliness), and both define male gender-role stereotypes more rigidly than they do female stereotypes. Men, however, do not define women as rigidly as women do men. And both men and women describe the ideal female in very androgynous terms (Hort, Fagot, and Leinbach, 1990).

Despite the limitations that traditional gender roles may place on us, changing them is not easy. Gender roles are closely linked to self-evaluation. Our sense of adequacy often depends on gender-role performance as defined by parents and peers in childhood ("You're a good boy/girl"). Because gender roles often seem to be an intrinsic part of our personality and temperament, we may defend these roles as being natural, even if they are destructive to a relationship or to ourselves. To threaten an individual's gender role is to threaten his or her gender identity as male or female because people do not generally make the distinction between gender role and gender identity. Such threats are an important psychological mechanism that keeps people in traditional but dysfunctional gender roles.

Finally, the social structure itself works to reinforce traditional gender norms and behaviors and make change more difficult. Some religious groups, for example, strongly support traditional gender roles. The Catholic Church, conservative Protestantism, Orthodox Judaism, and fundamentalist Islam, for example, view traditional roles as being divinely ordained. Accordingly, to violate these norms is to violate God's will. The marketplace also helps enforce traditional gender roles. The wage disparity between men and women (remember, women earn about 75 percent of what men earn) is a case in point. Such a significant difference in income makes it "rational" that the man's work role take precedence over the woman's work role. If someone needs to remain at home to care for the children or an elderly relative, it makes "economic sense" for a heterosexual woman to quit her job because her male partner probably earns more money than she does.

Gender Movements and the Family

Gender issues have been the source of much collective action and the focus of a number of social movements that press for change. These movements include the range of perspectives within the con-

© Paul Conklin/PhotoEdit

© M. Greenlar/The Image Works

The National Organization for Women (NOW) and the Promise Keepers are two examples of organized gender movements. In the rhetoric and rallies that comprise such movements, family issues loom large.

temporary women's movement but also a variety of "men's movements," that, although less visible, have organized to change aspects of men's lives. We look briefly here at some of the ways these movements have framed and acted on family matters.

A complete history of American feminism is beyond the scope of this book. In the eighteenth, nineteenth, and twentieth centuries, women organized around issues such as economic justice, abolition of slavery, and women's suffrage. In their antislavery activity during the nineteenth century, many women were sensitized to the extent of their own oppression and disadvantage, which helped energize their pursuit of voting rights (Renzetti and Curran, 1999; Lindsey, 1997). After gaining the right to vote with the passage of the Nineteenth Amendment in 1919, many women withdrew from active feminist involvement as they thought they had reached equality with men (Renzetti and Curran, 1999).

During the 1960s, feminism resurfaced dramatically. Catalyzed by the publication of Betty Friedan's *The Feminine Mystique,* many women began to look critically at the sources of their "problems with no names," and the family was seen as a major culprit. Additionally, wage inequality was made a public issue through President Kennedy's Commission on the Status of Women in 1961 and the passage of the Equal Pay Act of 1963. Then in 1966, the National Organization for Women (NOW) was established. In its 38-year history, this liberal, reform-oriented feminist organization has grown to include more than half a million members in its more than 500 chapters throughout the United States. It is the largest, though not the only, organized plank of the Women's Movement, and its philosophy represents one of a number of "feminisms" (Renzetti and Curran, 2003; Lindsey, 1997). Contemporary feminist positions range across a spectrum that includes a host of perspectives, including *liberal feminism, socialist feminism, radical feminism, lesbian feminism, multiracial feminism,* and *postmodern feminism* (Lorber, 1998; Renzetti and Curran, 1999). Each has its own specific emphasis on issues and advocates different strategies to improve women's lives.

Judith Lorber sorts the various feminist perspectives into three broader categories: **Gender-reform feminism** is geared toward giving women the same rights and opportunities that men enjoy; **gender-resistant feminism** advocates more radical, separatist strategies for women out of the belief that their subordination is too embedded in the existing social system; and **gender-rebellion feminism** tends to emphasize overlapping and interrelated inequalities of gender, sexual orientation, race, and class (Lorber, 1998; Renzetti and Curran, 1999).

Given this diversity of opinion, it is difficult to characterize a "feminist" position on families. Furthermore, such attempts occasionally exaggerate or simplify more complex positions. In her critique of American feminism, economist Sylvia Hewlett notes that neither liberal feminism ("equal rights" feminism) nor more radical feminist positions have recognized the commitments women feel toward their families and the consequences of those commitments. By stressing equal rights and full equality with men, liberal feminism may have downplayed the responsibilities women carry within families and denied that women may need different supports than those needed by men (Hewlett, 1986). Some of the more radical feminist positions articulated in the 1960s and 1970s may have been antimarriage or antimotherhood, as either or both have at times been seen as relationships that oppress women and keep them from achieving their full capabilities.

Hewlett compares both approaches to a movement more characteristic of European feminist activity: **social feminism**—the belief that workplace and family supports are essential if women are to experience a high quality of life (Hewlett, 1986). Feminist critics of Hewlett rightly point out that the pressure for public support for families has and continues to come most strongly from women; thus her characterization is said to be unfair. Although American feminists have been active at the forefront of pushing for parental leave, child care, and so forth, organizations such as NOW still more heavily stress abortion rights, reproductive freedom, opposing bigotry against lesbians and gays, and ending violence against women rather than more specifically family-focused issues.

Divisions of opinion and multiple perspectives on gender inequality constitute a basic similarity between women and men. Just as there is no one perspective on how women should be or what they should do, neither is there unanimity about men's lives. Just as there are multiple feminisms, each with its own agenda, so, too, are there different viewpoints on whether, in what direction, and how men ought to change (Clatterbaugh, 1997; Messner, 1997; Renzetti and Curran, 1995).

In recent years, at one time or another, we have witnessed the crowning of each of the following as "the men's movement": the mythopoetic men's movement, the men's rights/fathers' rights advocates, the Christian men's movement (for example, Promise Keepers), and the pro-feminist, gay-affirmative men's organization, NOMAS (National Organization of Men Against Sexism). Each represents just a part among many or just one of a number of movements. Many of these movements differ in what they see as men's roles in and responsibilities to their families.

Central to a **pro-feminist men's movement** is the issue of *fairness.* Pro-feminist men believe that men ought to share responsibilities within their households and that women and men ought to be equal partners. Also, pro-feminists argue that men and children would both benefit from closer connections between fathers and their children.

Both the Christian Promise Keepers and the organizers of the 1995 Million Man March and rally in Washington, D.C., by African-American men, also stressed the idea of men's *responsibilities* to their families, though their versions of responsibilities included somewhat more traditional notions of men's roles as the heads of their households. They also argued that men needed to be more accountable to spouses and children. Finally, the men's rights movement has stressed the discrimination that men face in and out of family matters. They note, for example, that only men can be subject to compulsory military service. They also look at what they believe to be inequalities in areas of divorce settlements and custody or visitation arrangements.

It is interesting to note the different positions taken on the family by the various feminist and men's movements. Although it is inaccurate and overstated to suggest that feminists are *antifamily,* the resurgent women's movement of the 1960s did grow in part out of the articulation of discontent by Betty Friedan in *The Feminine Mystique.* Similarly, early "second wave" feminists (1960s–1970s) attempted to sever the automatic connections typically made between women, children, and families as a way of liberating women to pursue other aspirations.

Conversely, across most of the positions of the men's movement is a sense that men need to enlarge their family role, live up to or "honor" their commitments to their families, and/or share in caring for children and households. Such involvement is often seen as potentially "liberating" for men, as it reconnects them to their emotional sides and broadens their lives beyond wage earning.

Looked at more closely, these movements are really not as different as they seem. What feminists railed against was not the *family* but the *gendered family.* They were less antagonistic to what women felt toward and did in the family than what men did not. Because of the differential burden carried by women in households, family life imposed constraints on women's opportunities for outside involvements in ways it did not on men's. More recently, the various men's movements have acknowledged men's lack of involvement or weaker commitments and opposed defining men solely in terms of what they do away from the family.

Contemporary gender roles are still in flux. Few men or women are entirely egalitarian or traditional. Even those who are androgynous or who have egalitarian attitudes, especially males, may be more traditional in their behaviors than they realize. Few with egalitarian or androgynous attitudes, for example, divide all labor along lines of ability, interest, or necessity rather than gender. Also, marriages that claim to be traditional rarely have wives who submit to their husbands in all things. Among contemporary men and women, women find that their increasing access to employment puts them at odds with their traditional (and personally valued) role as mother. Women continue to feel conflict between their emerging equality in the workplace and their continued responsibilities at home. Within marriages and families, the greatest areas of gender inequality continue to be the division of housework and child care. But change continues to occur in the direction of greater gender equality, and this equality promises greater intimacy and satisfaction for both men and women in their relationships.

SUMMARY

- A *gender role* is the role a person is expected to perform as a result of being male or female in a particular culture. *Gender-role stereotypes* are rigidly held and oversimplified beliefs that males and females, as a result of their sex, possess distinct psychological and behavioral traits. *Gender-role attitudes* are beliefs that we have about ourselves and others regarding appropriate male and female personality traits and activities. *Gender-role behaviors* are the actual activities or behaviors that we or others engage in as males and females. *Gender identity* refers to being male or female.
- Although our culture encourages us to think that men and women are "opposites," they are actually more similar than different. Innate gender differences are generally minimal; differences are encouraged by socialization. *Gender schemas* reflect our tendency to divide objects, activities, and behaviors into masculine and feminine categories.
- Within any given society at a particular point in time, there are multiple versions of masculinity and femininity, one of which comes to dominate our thinking about gender. Across societies, much variation exists in how gender is perceived, including the perception of how many gender categories there are.
- Gender relations are also power relations. *Patriarchal societies* are societies in which men dominate. Logically, *matriarchal societies* would be societies in which women dominate political and economic life. Researchers have not found any society that truly embodies a matriarchal social structure.
- According to *gender theory,* social relationships are based on the socially perceived differences between males and females that justify unequal power relationships. The key to creating gender inequality is the belief that men and women are opposite in personalities, abilities, skills, and traits. Furthermore, the differences between the sexes are unequally valued in society.
- Symbolic interactionists view gender as something we actively create or "do" in everyday situations and relationships, not an internalized set of behavioral and personal attributes.
- The two most important socialization theories are social learning theory and cognitive development theory. *Social learning theory* emphasizes learning behaviors from others through rewards and punishments and *modeling.* This behaviorist approach has been modified to include cognitive processes, such as the use of language, the anticipation of consequences, and observation. *Cognitive development theory* asserts that once children learn that gender is permanent, they independently strive to act like "proper" boys or girls because of an internal need for congruence.
- Children learn their gender roles through manipulation, channeling, verbal appellation, and activity exposure. Parents, teachers, and *peers* (age-mates) are important agents of socialization during childhood and adolescence. Ethnicity and social class also influence gender roles. Among African Americans, strong women are important female role models.
- During adolescence, parents can be more important influences than peers. Peers, however, have more egalitarian gender-role attitudes than do adults. The media tend to portray traditional stereotypes of men and women as well as ethnic groups. For students, colleges and universities are important sources of gender-role learning, especially for nontraditional roles. Marriage, parenthood, and the workplace also influence the development of adult gender roles.
- Research indicates that in any individual's life, gender is dynamic. The roles we play in adulthood may be very different from roles we expected or started out to play. Situations, opportunities and constraints can alter the path established by socialization.
- Traditional male roles, whether for whites, African Americans, Latinos, or Asian Americans, emphasize dominance and work. A man's central family role has been viewed as being the provider. For women, there is greater role diversity according to ethnicity. Traditional female roles among middle-class whites emphasize being a wife and mother. Among African Americans, women are expected to be instrumental; there is no conflict between work and motherhood. Among Latinos, women are deferential to men generally from respect rather than subservience; elders, regardless of gender, are afforded respect.

- Contemporary gender roles are more egalitarian than the traditional ones of the past. Important changes affecting contemporary gender roles include (1) the acceptance of women as workers and professionals; (2) increased questioning, especially among white women, of motherhood as a core female identity; (3) greater equality in marital power; (4) the breakdown of the instrumental/expressive dichotomy; and (5) the expansion of male family roles.
- Limitations of contemporary gender roles for men include the primacy of the provider role, which limits men's father and husband roles; difficulty in expressing feelings; and a sense of dominance that precludes intimacy. Limitations for women include diminished self-confidence and mental health; another limitation is the association of femininity with youth and beauty, which creates a disadvantage as women age. Ethnic women may suffer both racial discrimination and gender-role stereotyping, which compound each other. Expectations and limitations reduce our potential as humans and create characteristic stresses.
- Changing gender-role behavior is often difficult because (1) each sex reinforces the traditional roles of its own and the other sex; (2) we evaluate ourselves in terms of fulfilling gender-role concepts; (3) gender roles have become an intrinsic part of ourselves and our roles; and (4) the social structure reinforces traditional roles.
- There has been a variety of social movements dedicated to challenging or changing women's or men's roles. These include a variety of feminisms, including those oriented around gaining women equal rights with men (*gender-reform feminism*), more radical and separatist feminist movements (*gender-resistant feminism*), and movements designed to emphasize how gender overlaps with other bases of oppression, like age, race or class. Among men, there are also a variety of "movements" and perspectives, including the *profeminist men's movement* and a few more male-oriented viewpoints. Ironically, early 1960s and 1970s feminists often rallied against women being associated with family responsibilities. Most of the current men's movements attempt to reconnect men with families.

KEY TERMS

SUGGESTED READINGS

Bem, Sandra. *The Lenses of Gender: Transforming the Debate on Sexual Inequality.* New Haven, CT: Yale University Press, 1993. A leading gender-role scholar examines gender concepts.

Gerson, Kathleen. *No Man's Land: Men's Changing Commitments to Work and Family.* New York: Basic Books, 1993. An interview study with more than 100 men about their work and family commitments and involvements. Gerson sorts men into three camps— family oriented men, breadwinning men, and autonomy-seeking men.

Gerson, Kathleen. *Hard Choices: How Women Decide About Work, Career, and Motherhood.* Berkeley, CA: University of California Press, 1985. Gerson's earlier book, in which she examined the different orientations toward work and motherhood of a sample of working-class and middle-class women, noting how those orientations differ and develop.

Hochschild, Arlie. *The Second Shift.* New York: Viking, 1989. A classic study of the division of household labor among dual-earner couples. Hochschild looks not only at how couples divide up chores but also at the processes whereby they reach their arrangements and the strategies they use to deal with unsatisfying arrangements.

Kimmel, Michael S. *The Gendered Society.* New York: Oxford University Press, 2000. A text on gender from the leading scholar on men and masculinity.

Pollack, William. *Real Boys: Rescuing Our Sons from the Myths of Boyhood.* New York: Holt, 1998. Psychologist William Pollack, an authority on masculinity and male development, explores some of the ways in which conventional ideas about masculinity lead to personal and academic problems for boys and young men.

Risman, Barbara. *Gender Vertigo.* New Haven: Yale University Press, 1998. Risman draws on her own original research on custodial single fathers, egalitarian couples, and baby boom mothers to build her argument that social experience is partly derived form society's gender structure, which is interconnected with other social institutions.

Schwartz, Pepper. *Love Between Equals: How Peer Marriage Really Works.* New York: Free Press, 1994. Schwartz's study of couples whose relationships are unconventional. They share more of the domestic and wage-earning responsibilities, believe in the importance of fairness, and have a level of intimacy ("deep friendship") that is greater than reported of most marriages.

Tavris, Carol. *The Mismeasure of Woman.* New York: Simon and Schuster, 1992. An examination of various misconceptions and biases that affect our understanding of women.

RESOURCES ON THE INTERNET

Companion Web Site for This Book

http://sociology.wadsworth.com/strong/marriage9e
Gain an even better grasp on this chapter by going to the companion Web site to take one of the Tutorial Quizzes, use the Flash Cards to master key terms, or check out the many other study aids you'll find there. Visit the Marriage and Family Resource Center on the site. You'll also find special features such as GSS Data and Census 2000 information that will put data and resources at your fingertips to help you with that special project or to do some research on your own.

InfoTrac College Edition: Search Word Summary

gender	marital power
egalitarian	social construct

To learn more about these central topics in the study of the family, you can conduct an electronic search using InfoTrac College Edition. To aid in your search and to gain useful tips, see the Student Guide to InfoTrac College Edition that you can access through the companion Web site for this book.

Preview

To gain a sense of what you already know about the material covered in this chapter, answer "True" or "False" to the statements below.

1 A high value on romantic love is unique to the United States. True or false?

2 The development of mutual dependence is an important factor in love. True or false?

3 Love and commitment are inseparable from each other. True or false?

4 Friendship and love share many of the same characteristics. True or false?

5 Men fall in love more quickly than women fall in love. True or false?

6 Heterosexuals, gay men, and lesbians are equally as likely to fall in love. True or false?

7 In many ways, love is like the attachment an infant experiences for his or her parent or primary caregiver. True or false?

8 In stalking relationships, men stalk women but women do not stalk men. True or false?

9 A high degree of jealousy is a sign of true love. True or false?

10 Partners with different styles of loving are likely to have more satisfying relationships because their styles are complementary. True or false?

"The meeting of two personalities is like the contact of two chemical substances; if there is any reaction, both are transformed."

CARL JUNG

CHAPTER 5

Friendship, Love, and Commitment

Outline

Ahhhh, love . . . Americans have an obsession with love. Listen to popular music, go to the movies, or travel through bookstores and you quickly discover that in what we listen to, watch, and read love stands out as a central theme. Popular song titles and lyrics are often testimonies to the power, pleasure, and pain associated with falling in and out of love (for example, recent hits such as *Because You Loved Me, Love Will Keep Us Alive, In Your Eyes, Love Hurts, How Do I Live Without You?*). Films are just as love-obsessed, although there is more competition from other genres, too. The phenomenal success of *My Big Fat Greek Wedding*, a feel good love story, reflects our fixation with love. Other recent motion pictures, such as *Sweet Home Alabama, Shakespeare in Love, Jerry Maguire, Titanic,* and *Sleepless in Seattle*, are all contemporary examples of a long-standing movie fascination with love that is also seen in older or classic films such as *Gone With the Wind, Casablanca, From Here to Eternity*, and *Love Story*. Romance novels sell widely, accounting for nearly half of all mass-market paperback books sold in the United States. They are read by an estimated 45 million Americans. Obviously, Americans can't get enough of love.

American families, like the culture that surrounds them, place high value on love. Decisions about entering or exiting a marriage, assessments of the quality and success of any particular marriage, and devotion between spouses or parents and children all come down to love. On both an individual and familial level, then, it is important to consider the role love plays in our lives.

The Importance of Love

Love is essential to our lives. Love binds us together as partners, husbands and wives, parents and children, and friends and relatives. The importance of romantic love cannot be overrated, for we make major life decisions, such as marrying, on the basis of love (Simpson et al., 1986). Love creates bonds that enable us to endure the greatest hardships, suffer the severest cruelty, overcome any distance. Because of its significance, we may torment ourselves with the question: Is it really love? Many of us may find scenes such as the following familiar and potentially humorous (Greenberg and Jacobs, 1966):

YOU: Do you love me?
MATE: Yes, of course I love you.
YOU: Do you really love me?
MATE: Yes, I really love you.
YOU: You are sure you love me—you are absolutely sure?
MATE: Yes, I'm absolutely sure.
YOU: Do you know the meaning of love?
MATE: I don't know.
YOU: Then how can you be sure you love me?
MATE: I don't know. Perhaps I can't.
YOU: You can't, eh? I see. Well, since you can't even be sure you love me, I can't really see much point in our remaining together. Can you?
MATE: I don't know. Perhaps not.
YOU: You've been leading up to this for a pretty long time, haven't you?

PREVIEW ANSWERS
1 False
2 True
3 False
4 True
5 True
6 True
7 True
8 False
9 False
10 False

Love is both a feeling and an activity. We feel love for someone and act in a loving manner. But we can also be angry with the person we love as well as frustrated, bored, or indifferent. This is the paradox of love; it encompasses opposites. Love includes affection and anger, excitement and boredom, stability and change, bonds and freedom. Its paradoxical quality makes some ask whether they are really in love when they are not feeling "perfectly" in love or when their relationship is not going smoothly. Love does not give us perfection; however, it does give us meaning. In fact, as sociologist Ira Reiss (1980a) suggests, an important question to ask is this: Is the love I feel the kind of love on which I can build a lasting relationship or marriage?

We can look at love in many ways besides through the eyes of lovers, although other ways may not be as entertaining. Whereas love was once the province of lovers, madmen, poets, and philosophers, social scientists have begun to appear on the scene. Although there is something to be said for the mystery of love, understanding how love works in the day-to-day world may help us keep our love vital and growing.

Love and American Families

Love is the basis for family formation in the United States, as it has been for most of the last two centuries (Mintz and Kellogg, 1988). Although American marriages were never quite as formally arranged as they have been in other places in the world, throughout the eighteenth century they were guided by more practical considerations and subject to more parental control. By the end of the nineteenth century most active parental involvement in their children's marriage choices had dissipated (Murstein, 1986). Economic developments had decreased the dependency of adult children on their parents; increasingly, economic opportunity could be found on one's own, which allowed people to more confidently choose their own mates without worrying as much about the consequences of parental disapproval. With increasing economic activity among women, a spread of legal and social recognition of women's rights, and enhanced opportunity for young people to meet and mingle, American courtship was transformed (Mintz and Kellogg, 1988; Murstein, 1986). Love, as experienced, perceived, and pursued by individuals, became the vehicle that drove mate selection.

In the early decades of the twentieth century, new ideals about marriage and family emerged. Although American family life had already shifted from an economic to an emotional emphasis with the appearance of the democratic family, this was extended even further with the emergence and celebration of companionate marriage. In this newer model of marriage, spouses were to be each other's best friends, confidants, and romantic partners (Mintz and Kellogg, 1988). Love was the foundation upon which marriage was built and the criterion by which spouses were chosen.

Neither "falling in love" nor the experience of romantic love are uniquely American; 90 percent of the 166 societies examined by Jankowiak and Fischer recognize and value love as an important element in building intimate relationships (Jankowiak and Fischer, 1992). But love does appear to have a more central role in American mate selection than in other Western societies (Goode, 1982). It fits well with and helps reinforce other features of American families and society. Love-based marriage validates the importance of individual autonomy and freedom from parental intervention and control, establishes the relative independence of the **conjugal family** from the extended family, and fits with the wider social freedoms granted to adolescents and young adults (Goode, 1982). Conversely, in societies where nuclear families are embedded in extended families, or where it is important for economic or political reasons to create alliances and exchanges through marriages, romantic love is not the central factor in mate selection.

The high emphasis we place on love as the basis for spousal choice contributes to the American patterns of divorce and remarriage. The qualities we "fall in love" with may not be easy to sustain across the lifelong duration of a marriage. Thus we are more likely to perceive our marriages as "failures" when we sense that those qualities are gone or diminished. We then seek those same idealized qualities from subsequent marriage partners.

Within our marriage practices we find at least two distinct but related cultural beliefs about the character and place of love: (1) that love is the

While it is difficult to come up with a formal definition of love, we usually know what we mean when we tell someone we love them. Such feelings are important at the individual, relationship, and institutional level.

© Esbin Anderson/The Image Works

criterion for choosing a spouse, and (2) that love is uncontrollable and irrational (*love is blind*). As much data show, Americans tend to follow a pattern known as **homogamy**—the tendency to marry people much like themselves. The prevalence of homogamy casts some doubt on some of our ideas about love and marriage.

Perhaps love is more controllable and rational than we pretend (and therefore not blind), as we seem to fall in love with people like ourselves. On the other hand, if love *is* blind (that is, uncontrollable and irrational), it must not be the only determinant of mate selection. In other words, if—as the song lyrics suggest—love and marriage go together like a horse and carriage, we are selective in which horses we harness to our carriages. We don't marry simply because we've fallen in love (love and marriage don't necessarily go together) and probably recognize some "loves" as unwise marriages. Finally, the social circles within which we live and move limit love. Thus, our "one and only" is drawn from a smaller pool than what the romantic mystique surrounding love suggests. With these qualifications in mind, we should still remember that the vast majority of Americans who marry say they are marrying because they are in love.

Friendship, Love, and Commitment

Love is closely linked to friendship and commitment in our intimate relationships. Each helps to create and sustain the other.

REFLECTIONS

What are the ideas you associate with love? With friendship? With commitment? How do they overlap? Have you ever mistaken love for commitment? Friendship for love?

Friendship is the foundation for love and commitment. Love reflects the positive factors, such as caring and attraction, that draw two people together and sustain them in a relationship. Commitment reflects the stable factors, including not only love but also obligations and social pressure, that help maintain the relationship for better or for worse. Although love and commitment are related, they are not necessarily connected. One can exist without the other. It is possible to love someone without being committed. It is also possible to be committed to someone without loving that person. Yet, when all is said and done, most of us long for a love that includes commitment and a commitment that encompasses love.

FRIENDSHIP AND LOVE

Friendship and love breathe life into humanity. They bind us together, provide emotional sustenance, buffer us against stress, and help to preserve our physical and mental well-being. The loss of a friend and especially a loved one can lead to illness and even suicide.

What distinguishes love from friendship? Should marriage satisfy all of our needs (including friendship)? What is this thing called friendship?

Two researchers (Todd and Davis, 1985) set out to distinguish the differences between characteristics of love and friendship. In their study of 250 college students and community members, they found that though love and friendship were alike in many ways, some crucial differences make love relationships both more rewarding and more vulnerable. Best friends were similar to spouse/lover relationships in several ways: levels of acceptance and of confiding, trust, respect, understanding, spontaneity, and mutual acceptance. Levels of satisfaction and happiness with the relationship were also found to be similar for both groups. What separated friends from lovers was that lovers had much more fascination and a sense of exclusiveness with their partners than did friends. Though love had a greater potential for distress, conflict, and mutual criticism, it ran deeper and stronger than friendship.

Friendship appears to be the foundation for a strong love relationship. Shared interests and values, acceptance, trust, understanding, and enjoyment are at the root of friendship and form a basis for love. Adding passion and emotional intimacy alters the nature of the friendship.

While some believe that marriage should satisfy all their needs, it is important to remember that when people marry they do not cease to be separate individuals. Friendships and patterns of behavior continue, so the effect that maintaining friendship has on marital satisfaction must be understood.

With men and women marrying later than ever before and women being an integral part of the workforce, close friendships, including other-sex friends, are more likely to be a part of the tapestry of relationships in people's lives. Partners need to communicate and to seek understanding regarding the nature of activities and the degree of emotional closeness they find acceptable in their spouses' friendships. Boundaries should be clarified and opinions shared. Many couples have found cross-sex friendships to be acceptable and even desirable. Like other significant issues that interface with marriage, keys to their success are the ability to communicate concerns and the maturity of all individuals involved.

GENDER, LOVE, AND FRIENDSHIP

The research on intimate relationships from friendship through love and marriage reveals important gender differences. In most of the scientific literature, there is a recurring theme highlighting men's supposed shortcomings as friends and partners. Unlike women, who are said to relate more easily and deeply with others and who develop a greater capacity for disclosing and sharing their inner selves, men maintain greater emotional distance, even as they experience their closest relationships.

Francesca Cancian (1985) argued that there is a gender bias in our cultural constructions of love that may distort our understanding of how both men and women love. Defining or "seeing" love in largely expressive terms (telling each other how you feel) ignores other aspects of women's and men's intimacy. For example, much of what women do as expressions of love (for spouses and children, especially) consists of more instrumental tasks associated with nurturing and caregiving. Though done out of love, such activities may not be seen as displays of love. Likewise, if men believe they "show" or express love by *what they do* more than by *what they say,* conceptualizing or recognizing love in terms of things said renders men's sincere attempts to show intimacy invisible and leaves them looking especially inadequate as intimate partners.

This critique of love pertains equally well to friendship. We tend to conceptualize "real" or "true" friendship by such qualities as emotional support and self-disclosure. Friends share their inner lives with each other; they tell each other how they feel, including how they feel about each other. The closer the friend, the more personal and more frequent the disclosures. This conceptualization measures friendship against a female standard and may underestimate the "real" intimacy that men's friendships contain, especially if such closeness is expressed in other, more covert ways (Swain, 1989).

There are differences in the number and nature of men's and women's friendships. Men reportedly have more friendships but lack the kind of "close friendships" women have. Whereas women spend much time with friends sharing disclosures, men spend their time doing things but revealing little. Additionally, men display less affection, either via words or touch, than women do toward their friends (Dolgin, 2001).

Men are more open and intimate in cross-sex relationships than in their friendships with other men (Dolgin, 2001). Wives or romantic partners are often the closest confidants in men's lives. In those relationships, men find themselves reaping the benefits that come from greater disclosure, even if the levels at which they disclose don't always match what their partners desire. Certainly, the tendency to funnel their intimacy into one relationship, especially marriage, is consistent with the cultural expectations of marriage as best-friendship. But even outside marriage, the depth of men's disclosure to women stands in contrast to the male-male style, suggesting not so much an inability as an unwillingness or a discomfort at male-male intimacy.

With regard to love, the genders differ in a number of ways. Men fall in love more quickly than women, describe more instrumental styles of love, and are more likely to see sex as an expression of love. Because men have fewer deeply intimate, self-disclosing friendships, when they find this quality in a relationship they are more likely to perceive that relationship as special. Having more intimates with whom they can share their feelings, women are less likely to be as quick to characterize a particular relationship as love. In addition, traditionally, women could less safely do so unless other, economic, criteria were also met. Thus, while men could afford to be more romantic, women needed to be more realistic (Knox and Schact, 2000). Other differences surface in the connection between love and sex. Although men are often depicted as easily separating sex and love, there is also evidence that within relationships men see sex as a means of expressing or showing love (Rubin, 1983; Cancian, 1987). Women's experiences of love and sexuality are different. Although sexual scripts have been changed in the direction of more open and acceptable expressions of female sexuality, to feel loved requires more than sexual expression.

Gender differences may be more exaggerated in *what people say* than in *what they do*. This is certainly the case with friendship, where bigger differences show up in how the genders talk about friendship than in what they experience as friends (Walker, 1994). In researching real versus stated gender differences in friendship, Karen Walker discovered that although her male and female informants validated the more common characterizations of how women's and men's friendships differ, on talking about *their friends*, they revealed more complex, often gender "inappropriate" patterns of relating. Thus men had male friends to whom they disclosed personal information, and women had some relationships that resembled men's friendship patterns (Walker, 1994).

In identifying the factors that shape men's and women's intimate relationships, most researchers point to aspects of gender socialization (McGill, 1985; Basow, 1992). Some emphasize the elements of the dominant cultural constructions of masculinity and femininity, wherein men are inexpressive, competitive, rational, and uncomfortable with revealing their innermost feelings, especially feelings of vulnerability or of affection toward other males (Bell, 1981; Rubin, 1985; McGill, 1985; Stein, 1986). Women are allowed and encouraged to express a wider range of feelings without concern for the consequences.

Other researchers suggest that gender-specific relationship styles emerge because of differences in how males and females resolve the developmental task of early childhood identity formation (Chodorow, 1978; Rubin, 1985). As a result of being "mothered," and having one's closest early relationship be with a female, the genders develop different ways of relating. Females develop "permeable ego boundaries" that are open to relationships with others, and retain a strong connection with their mothers. Males are forced to separate from their mothers, identify with absent or less present fathers, and build boundaries around themselves in relation to their most nurturant caregivers. This haunts them throughout their later relationships, as it makes them less able to "connect" intimately with others

(Rubin, 1985). Women experience themselves in the context of relationships, whereas men—depicted as "selves in separation"—remain oriented more toward independence and task completion (Kilmartin, 1994).

We might also emphasize the role-model consequences of being "mothered" but not "fathered." Without a loving, attentive, nurturing presence from fathers or other male role models, boys come to inhibit their own emotional expressiveness, identifying such behavior as typical of mothers (and women, in general) and to be avoided. Because of the relative involvement of mothers versus fathers in caring for young children, and the greater prevalence of single-mother versus single-father households, boys have fewer available role models for intimacy. Furthermore, what role models they have are also products of gender socialization and carry a style of relating that results from that socialization. Girls have the opportunity to observe up close a caring, loving female role model from which they learn how to relate and express love.

GENDER EXCEPTIONS: "LOVE BETWEEN EQUALS"

The gender differences depicted above, though common in the literature on love and intimacy, are not inevitable. As noted in the previous chapter, there are marriages and intimate heterosexual relationships that depart from traditional gender patterns (Schwartz, 1994; Risman and Johnson-Sumerford, 1997). Pepper Schwartz's research on *peer couples* is an excellent illustration of loving relationships that avoid the aforementioned gender patterns. Schwartz conceptualized **peer marriage** as a relationship built on principles of equity, equality, and **deep friendship.** The emphases on equity and equality made peer couples strive toward fairness and equality in what each person contributed to and enjoyed within their relationship. Thus they shared chores, had equal say in decision making, and participated equally in child rearing. More important for the present discussion is the element of deep friendship. By this, Schwartz meant that these couples most valued an intense companionship, "a collaboration of love and labor in order to produce profound intimacy and mutual respect" (Schwartz, 1994).

In peer marriages spouses become more alike over time, and thus both husbands and wives are more likely to display and value a blend of female and male styles of intimacy. Women value and appreciate the instrumental displays of love from their partners (for example, finding her husband has had her car serviced) because they know what it is like to make or take time from one's demanding daily routines to attend to such things. Because husbands are more involved in daily domestic and child-rearing routines, they share interests and concerns with their wives that traditional spouses do not. With enlarged identities outside the marriage and home, peer wives also need less of the conventional, conversational demonstrations of love and affection. They and their husbands have "learned love on each other's terms" (Schwartz, 1994).

Of course, Schwartz's peer couples, like Risman and Johnson-Sumerford's post-gender couples, are not very common. They represent what is possible in marriage, but creating such lifestyles requires both an ideological commitment to sharing and equality and an ability to withstand scrutiny and curiosity from more typical couples. For the most part, such lifestyles also require each spouse to have a job or career that the other recognizes as equal in importance to his or her own.

GENDER, LOVE, AND SEXUALITY

One area in which persistent gender differences have been long observed is in the relationship between love and sex. Although love and sex are often thought of as being separate phenomena, recent research suggests that for both men and women, sex includes intimacy and caring, key aspects of love (Aron and Strong, n.d.; Strong and DeVault, 1997). Nevertheless, gender differences do exist, especially in terms of casual relationships. (See Chapter 8 for a further discussion of sexuality.)

Men, Sex, and Love

Men and women who are not in an established relationship have different expectations. Men are more likely than women to separate sex from affection. Studies consistently demonstrate that for the majority of men, sex and love can be easily separated (Blumstein and Schwartz, 1983; Carroll et al., 1985; Laumann et al., 1994).

Although men are more likely than women to separate sex and love, Linda Levine and Lonnie

Barbach (1985) found in their interviews that men indicated that their most erotic sexual experiences took place in a relational context. The authors quote one man as follows:

> Emotions are everything when it comes to sex. There's no greater feeling than having an emotional attachment with the person you're making love to. If those emotions are there, it's going to be fabulous. . . . They don't call it "making love" for nothing.

Most men in the study responded that it was primarily the emotional quality of the relationship that made their sexual experiences special. Furthermore, for men more than women sex is a way of sharing or expressing love. Some of the couples Lillian Rubin interviewed for her Book *Intimate Strangers: Men and Women Together* illustrate the different ways men and women link love and sex. As Rubin notes, for men, the sexual "sparks the emotional," and sex becomes the "one arena where it is legitimate for men to contact their deeper feeling states and express them" (Rubin, 1983). One of her male informants offered this explanation:

> Having sex with her makes me feel much closer so it makes it easier to bridge the emotional gap so to speak. It's like the physical sex opens up another door, and things and feelings can get expressed that I couldn't before.

Women, Sex, and Love

Women generally view sex from a relational perspective. In the decision to have sexual intercourse, the quality and degree of intimacy of a relationship were more important for women than men (Christopher and Cate, 1984). Women were more likely to report feelings of love if they were sexually involved with their partners than if they were not sexually involved (Peplau et al., 1977). Love is also more closely related to feelings of self-esteem (Walsh, 1991).

Women generally seek emotional relationships, but men may initially seek physical relationships. This difference in intentions can place women in a bind. Carole Cassell (1984) suggests that today's women face a "damned if you do, damned if you don't" dilemma in their sexual relationships. A woman might have sexual intercourse with a man only to have him turn around and say goodbye. On the other hand, if she doesn't have sex with him, he might say he respects her but still say good-bye. A young woman Cassell interviewed related the following:

> I really hate the idea that, because I'm having sex with a man whom I haven't known for a long, long time, he'll think I don't value myself. But it's hard to know what to do. If you meet a man and date him two or three times and don't have sex, he begins to feel you are either rejecting him or you have serious sex problems. . . . But I really dread feeling that I could turn into, in his eyes, an easy lay, a good-time girl. I want men to see me as a grown-up woman who has the same right as they do to make sexual choices.

Traditionally, women were labeled "good" or "bad" on the basis of their sexual experience and values. "Good" women were virgins, sexually naive, or passive; "bad" women were sexually experienced, independent, and passionate. According to Lillian Rubin (1990), this attitude has not entirely changed. Rather, we are ambivalent about sexually experienced women. One exasperated woman leaped out of her chair and began to pace the floor, exclaiming to Rubin: "I sometimes think what men really want is a sexually experienced virgin. They want you to know the tricks, but they don't like to think you did those things with someone else."

Gay Men, Lesbians, and Love

Love is equally important for heterosexuals, gay men, lesbians, and bisexuals (Patterson, 2000; Aron and Aron, 1991; Keller and Rosen, 1988; Kurdek, 1988; Peplau and Cochran, 1988). Many heterosexuals, however, perceive lesbian and gay love relationships as less satisfying and less loving than heterosexual ones. In a study of 360 heterosexual undergraduates (Testa et al., 1987), students were presented with identical information about a couple that was variously described as heterosexual, gay, and lesbian. When the couple was identified as heterosexual, it ranked high on love and satisfaction. When the couple was identified as lesbian or gay, it was ranked significantly lower on love and satisfaction. Because the couples were identical except for sexual orientation, the researchers concluded that there was a heterosexual bias in the perception of gay and lesbian love relationships. It is well documented that love is important for lesbians and gay men; like heterosexual relationships, theirs have multiple emotional dimensions (Patterson, 2000; Adler, Hendrick, and Hendrick, 1989).

Men in general are more likely than women to separate love and sex; gay men are especially likely to make this separation. Although gay men value love, they also tend to value sex as an end in itself. Furthermore, they place less emphasis on sexual exclusiveness in their relationships (Patterson, 2000). Researchers suggest, however, that heterosexual males are not very different from gay males in terms of their acceptance of casual sex. Lesbians and heterosexual couples tend to be more supportive than gay men of monogamy and sexual fidelity. This is probably due to gender more than sexual orientation, as heterosexual males would be as likely as gay males to engage in casual sex if women were equally interested. Women, however, are not as interested in casual sex; as a result, heterosexual men do not have as many willing partners available as do gay men (Foa et al., 1987; Symons, 1979).

For lesbians, gay men, and bisexuals, love has special significance in the formation and acceptance of their identities. Although significant numbers of women and men have had sexual experiences with members of the same sex or both sexes, relatively few identify themselves as lesbian or gay. An important element in solidifying such an identity is loving someone of the same sex. Love signifies a commitment to being gay or lesbian by unifying the emotional and physical dimensions of a person's sexuality (Troiden, 1988). For the gay man or lesbian, it marks the beginning of sexual wholeness and acceptance. In fact, some researchers believe that the ability to love someone of the same sex, rather than having sex with him or her, is the critical element that distinguishes being gay or lesbian from being heterosexual (Money, 1980).

Sex without Love

Is love necessary for sex? Most of us might assume that it is or it *should be*, but our assumption is based on our values. The question cannot be answered by reference to empirical or statistical data. A more fundamental question is this: Is sexual activity legitimate in itself, or does it require justification? John Crosby (1985) observes:

> Sexual pleasure is a value in itself and hence capable of being inherently meaningful and rewarding. The search for extrinsic justification and rationalization simply belies our reluctance to believe that sex is a pleasurable activity in and of itself.

To believe that sex does not require love as a justification, argues Crosby, does not deny the significance of love and affection in sexual relations. In fact, love and affection are important and desirable for enduring relationships. They are simply not necessary, Crosby believes, for affairs in which erotic pleasure is the central feature.

Ironically, although sex without love may violate social norms, it is the least threatening form of extrarelational sex. Even those who accept their partners' having sex outside the relationship find it especially difficult to accept their partners' having a meaningful affair. "They believe," note Philip Blumstein and Pepper Schwartz (1983) "that two intense romantic relationships cannot coexist and that one would have to go."

LOVE, MARRIAGE, AND SOCIAL CLASS

In many ways, our romantic view of love-based marriage represents a middle-class version of marriage. Among upper-class families, there is a greater urgency in assuring that one's children marry the "right kind," as considerable wealth and social position may be at stake. Furthermore, upper-class families have more ability to exercise such control by the threat of withholding inheritance from the maverick child who dares act without consideration of parental preference (Goode, 1982). Among the working class, marriage was often entered as a means to escape economic instability, parental authority, and to be seen as an adult (Rubin, 1976, 1992). This may now be less true, as working-class marriages have taken on more characteristics of the middle-class ideal (for example, expecting more sharing, communication) (Rubin, 1995). Still, the economic circumstances that define one's life may induce different ways of linking love and marriage.

PROTOTYPES OF LOVE AND COMMITMENT

Despite centuries of discussion, debate, and complaint by philosophers and lovers, no one has succeeded in finding a definition of love on which all can agree. Ironically, such discussions seem to engender conflict and disagreement rather than love and harmony.

Because of the unending confusion surrounding definitions of love, some researchers wonder whether such definitions are even possible (Fehr, 1988; Kelley, 1983). In the everyday world, however, we do seem to have something in mind when we tell someone we love him or her. We may not have formal definitions of love, but we do have **prototypes** (that is, models) of what we mean by love stored in the backs of our minds. Researchers suggest that instead of looking for formal definitions of love and commitment, we examine people's prototypes; that is, we consider what people mean by the concept of love. In some ways these prototypes may be more important than formal definitions. For example, when we say "I love you," we are referring to our prototype of love rather than its definition. If we find ourselves thinking about our partners all the time, feeling happy when we are with them and sad (or less happy) when we are apart, and spending all our available time together, we will compare these thoughts, feelings, and behaviors to our mental models or prototypes of love (Regan, 2003). If our experiences match the different characteristics of love, we will define ourselves as in love. By thinking in terms of prototypes, we can study how people actually use the words *love* and *commitment* in real life and how their meanings of love and commitment help define the progress of their intimate relationships.

To discover people's prototypes, researcher Beverly Fehr (1988) asked 172 respondents to rate the central features of love and commitment. The 12 central attributes of love they listed are given in order:

- Trust
- Caring
- Honesty
- Friendship
- Respect
- Concern for the other's well-being
- Loyalty
- Commitment
- Acceptance of the other, the way he or she is
- Supportiveness
- Wanting to be with the other
- Interest in the other

The 12 central attributes of commitment, listed in order, were as follows:

- Loyalty
- Responsibility
- Living up to your word
- Faithfulness
- Trust
- Being there for the other in good and bad times
- Devotion
- Reliability
- Giving your best effort
- Supportiveness
- Perseverance
- Concern about the other's well-being

There are many other characteristics identified as features of love (euphoria, thinking about the other all the time, butterflies in the stomach) or commitment (putting the other first, contentment). These, however, tend to be peripheral. As relationships progress, the central aspects of love and commitment become more characteristic of the relationship than the peripheral ones. According to Fehr (1988), the central features "act as true barometers of a move toward increased love or commitment in a relationship." Similarly, violations of central features of love and commitment are considered to be more serious than violations of peripheral ones. A loss of caring, trust, honesty, or respect threatens love, while the disappearance of butterflies in the stomach does not. Similarly, lack of responsibility or faithfulness endangers commitment, whereas discontent is not perceived as threatening. Researchers have found that love *and* commitment are correlated to satisfaction in romantic relationships (Hendrick, Hendrick, and Adler, 1988).

ATTITUDES AND BEHAVIORS ASSOCIATED WITH LOVE

A review of the research on love finds a number of attitudes, feelings, and behaviors associated with love (Kelley, 1983).

Positive Attitudes and Feelings

Positive attitudes and feelings toward the other bring people together. Zick Rubin (1973) found that there were four feelings identifying love:

- Caring for the other—wanting to help him or her
- Needing the other—having a strong desire to be in the other's presence and to have the other care for one
- Trusting the other—mutually exchanging confidences
- Tolerating the other—accepting his or her faults

Of these, caring appears to be the most important, followed by needing, trusting, and tolerating (Steck et al., 1982). J. R. Davitz (1969) identified similar feelings associated with love but noted in addition that respondents reported feeling an inner glow, optimism, and cheerfulness. They felt harmony and unity with the person they loved. They were intensely aware of the other person, feeling that they were fully concentrated on him or her.

Loving Behaviors

Love is also expressed in certain behaviors. One study (Swensen, 1972) found that romantic love was expressed in several ways, with the expression of love often overlapping thoughts of love:

- Verbally expressing affection, such as saying "I love you"
- Self-disclosing, such as revealing intimate facts about oneself
- Giving nonmaterial evidence, such as offering emotional and moral support in times of need and showing respect for the other's opinion
- Expressing nonverbal feelings such as happiness, contentment, and security when the other is present
- Giving material evidence, such as providing gifts or small favors or doing more than one's share of something
- Physically expressing love, such as by hugging, kissing, and making love
- Tolerating and accepting the other's idiosyncrasies, peculiar routines, or annoying habits such as forgetting to put the cap on the toothpaste

These behavioral expressions of love are consistent with the prototypical characteristics of love. In addition, research supports the belief that people "walk on air" when they are in love. Researchers have found that those in love view the world more positively than those who are not in love (Hendrick and Hendrick, 1988a).

Though little research exists on ethnicity and attitudes and behaviors associated with love, one study of Mexican American college students suggests that they share many of the same attitudes and behaviors described above (Castaneda, 1993). Both females and males valued communication/sharing, trust, mutual respect, shared values and attitudes, and honesty. Data from white, middle-class adults indicate that men and women are quite similar in their love attitudes across adulthood (Montgomery and Sorell, 1997).

DID YOU KNOW?

Those in committed romantic or marital relationships have higher levels of satisfaction than those in noncommitted relationships. (Hecht, Martson, and Larken, 1994).

FACTORS AFFECTING COMMITMENT

Although we generally make commitments to a relationship because we love someone, love alone is not sufficient to make a commitment last. Our commitments seem to be affected by several factors that can strengthen or weaken the relationship. Ira Reiss (1980a) believes that there are three important factors in commitment to a relationship:

1. *The balance of costs to benefits:* Whether we like it or not, human beings have a tendency to look at romantic and marital relationships from a cost-benefit perspective. Most of the time, when we are satisfied, we are unaware that we judge our relationships in this manner. But as we saw in our discussion of social exchange theory in Chapter 2, when there is stress or conflict, we might ask ourselves, "Just what am I getting out of this relationship?" Then we add up the pluses and minuses. If the result is on the plus side, we are encouraged to continue the relationship; if the result is negative, we are more likely to discontinue it.
2. *Normative inputs:* Normative inputs for relationships are the values that you and your partner hold about love, relationships, marriage, and family. These values can either sustain or detract

from a commitment. How do you feel about a love commitment? A marital commitment? Do you believe that marriage is for life? Does the presence of children affect your beliefs about commitment? What are the values that your friends, family, and religion hold regarding your type of relationship?

3. *Structural constraints:* The structure of a relationship will add to or detract from commitment. Depending on the type of relationship—whether it is dating, living together, or marriage—different roles and expectations are structured in. In marital relationships, there are partner roles (husband/wife) and economic roles (employed worker/homemaker). There may also be parental roles (mother/father).

These different factors interact to increase or decrease the commitment.

Commitments are more likely to endure in marriage than in cohabiting or dating relationships, which tend to be relatively short lived. They are more likely to last in heterosexual relationships than in gay or lesbian relationships (Testa et al., 1987). Ethnicity may also be the greatest predictor of satisfaction and commitment to a friendship (deVries, Jacoby, and Davis, 1996). The reason commitments tend to endure in marriage may or may not have anything to do with a couple being happy. Marital commitments may last because norms and structural constraints compensate for the lack of personal satisfaction.

For most people, love seems to include commitment and commitment seems to include love. Beverly Fehr (1988) found that if a person violated a central aspect of love, such as caring, that person was also seen as violating the couple's commitment, and if a person violated a central aspect of commitment, such as loyalty, it called love into question.

Because of the overlap between love and commitment, we can mistakenly assume that someone who loves us is also committed to us. As one researcher (Kelley, 1983) points out: "Expressions of love can easily be confused with expressions of commitment. . . . Misunderstandings about a person's love versus commitment can be based on honest errors of communication, on failures of self-understanding." Or a person can intentionally mislead the partner into believing that there is a greater commitment than there actually is. Even if a person is committed, it is not always clear what the commitment means: Is it a commitment to the person or to the relationship? Is it for a short time or a long time? Is it for better and for worse?

The Development of Love: The Wheel Theory

Sociologist Ira Reiss (1980a) suggests that love develops and is maintained through four processes: (1) rapport, (2) self-revelation, (3) mutual dependency, and (4) fulfillment of the need for intimacy. Reiss calls the processes the **wheel theory of love** to emphasize their interdependence. A reduction in any one affects the development or maintenance of a love relationship (see Figure 5.1). If a couple habitually argues, for example, the arguments will affect the partners' mutual dependency and their need for intimacy; this in turn will weaken their rapport.

Rapport

When two people meet, they quickly sense if rapport exists between them. This rapport is a sense of ease, the feeling that they understand each other in some special way. Feelings of rapport are dependent on social environment in two ways. First, we tend to feel rapport with those who share the same social and cultural background. If one person has only a grade school education and the other a college education, it is not likely that they will share many of the same values. If one person is upper class and the other is working class, their life experiences have probably been quite different. Second, we tend to feel rapport with those who share our role conceptions. Two people who believe in egalitarian gender roles, for example, are more likely to feel rapport than are a radical feminist and a male chauvinist.

Self-Revelation

If you feel rapport with someone, you are likely to feel relaxed and confident. As a result, self-revelation—the disclosure of intimate feelings—is likely to occur. But self-revelation also depends on what is considered proper by your socialization group. For example, upper-class people have a tendency to be

FIGURE 5.1 ■ **Graphical Representation of Reiss's Wheel Theory of Love**

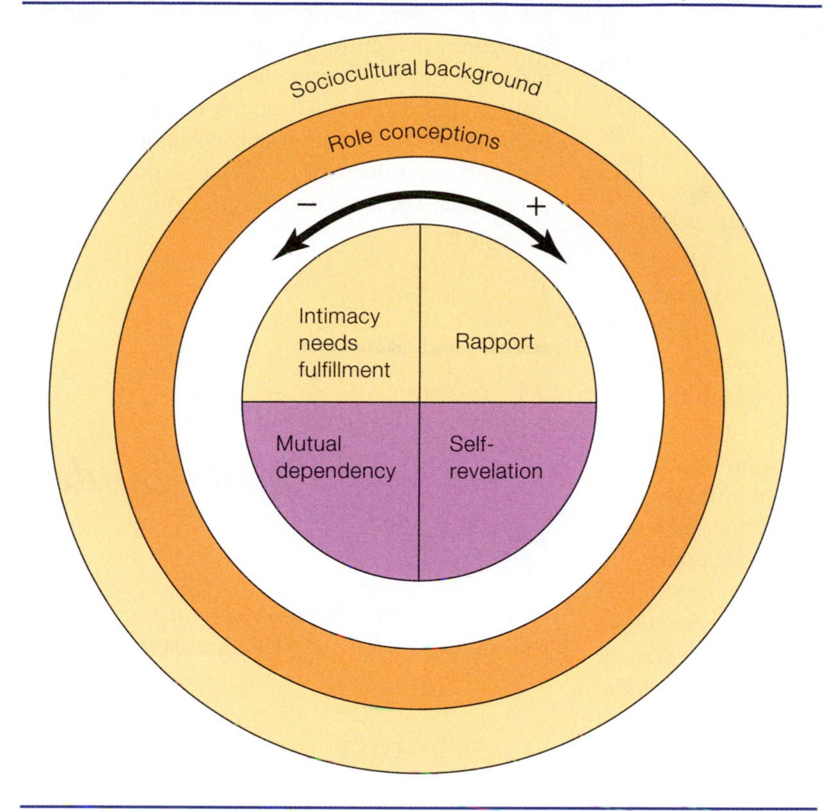

According to this theory, the development of intimacy is most likely to take place between those who share the same sociocultural background and role conceptions. Intimacy develops from a feeling of rapport, which leads to self-revelation; self-revelation leads to mutual dependency, which in turn may lead to intimacy need fulfillment.

reserved about themselves. Middle-class people feel more comfortable in revealing intimate aspects of their lives and feelings.

REFLECTIONS

As you examine the wheel theory of love diagram, ask yourself whether your love relationships follow the course Reiss suggests. What creates rapport for you? What factors increase or decrease self-revelation? When self-revelation increases, does mutual dependency also increase? If mutual dependency decreases, do self-revelation and rapport decrease? What impact have social background and role conceptions had on the development of your relationships?

Mutual Dependency

After two people feel rapport and begin revealing themselves to each other, they may become mutually dependent. Each needs the other to share pleasures, fears, and jokes, as well as sexual intimacies; each becomes the other's confidant. Each person develops ways of acting and being that cannot be fulfilled alone. Going for a walk is no longer something you do by yourself; you do it with your partner. Sleeping no longer takes place in a single bed but in a larger one with your partner. You are a couple.

Here, too, social and cultural background is important. The forms of mutually dependent behavior that develop are influenced by each person's conception of the role of courtship. Interdependency may develop through dating, getting together, or living together. Premarital intercourse may or may not be acceptable.

Fulfillment of Intimacy Needs

According to Reiss (1980a), we all have a basic need for intimacy—"the need for someone to love, the need for someone to confide in, and the need for sympathetic understanding." These needs are important for fulfilling our roles as a partner or parent. Reiss describes the relationship among the four processes, which culminate in intimacy, as follows:

> By virtue of rapport, one reveals oneself and becomes dependent, and in the process of carrying out the relationship one fulfills certain basic intimacy needs. To the extent that these needs are fulfilled, one finds a love relationship developing. In fact, the initial rapport that a person feels on first meeting someone can be presumed to be a dim awareness of the potential intimacy need fulfillment of this other person for one's own needs. If one needs sympathy and support, and senses these qualities in a date, rapport will be felt more easily; one will reveal more and become more dependent, and if the hunch is right, and the person is sympathetic, one's intimacy needs will be fulfilled.

Conceptualized as a "wheel," and not steps or stages, the processes flow into one another in one direction (as spokes do as a wheel spins) to develop and maintain love, but flow in the opposite direction tends to weaken it. If we feel less comfort (that is, rapport) with the other, we may reveal fewer thoughts or feelings, feel less dependent upon the other for a sense of happiness or contentment, and seek and fulfill our intimacy needs elsewhere. This seems to approximate what happens through the process of divorce or the ending of intimate relationships as well (Vaughan, 1986). Thus the model can depict falling in or out of love. The "+" and "–" in Figure 5.1 indicate the directions in which the processes can increase or decrease love. The outer ring on the diagram, "sociocultural background," produces the next ring, "role conceptions." All four processes are influenced by role conceptions, which define what a person should expect and do in a love relationship.

How Do I Love Thee? Approaches to the Study of Love

Researchers have developed a number of ways to study love (Hendrick and Hendrick, 1987).

STYLES OF LOVE

Sociologist John Lee (1973, 1988;) describes six basic styles of love:

- **Eros:** love of beauty
- **Ludus:** playful love
- **Storge:** companionate love
- **Mania:** obsessive love
- **Agape:** altruistic love
- **Pragma:** practical love

TABLE 5.1 ■ The Full Range of Possible Combinations of Commitment, Passion, and Intimacy

TYPE	COMMITMENT	PASSION	INTIMACY
Liking	–	–	+
Infatuation	–	+	–
Empty love	+	–	–
Romantic love	–	+	+
Companionate love	+	–	+
Fatuous love	+	+	–
Consummate love	+	+	+

These styles, Lee cautions, are relationship styles, not individual styles. The style of love may change as the relationship changes or when individuals enter different relationships. In addition to these pure forms there are mixtures of the basic types: storgic-eros, ludic-eros, and storgic-ludus. The six basic types can be described in greater detail.

EROS. Erotic lovers delight in the tactile, the sensual, the immediate; they are attracted to beauty (though beauty may be in the eye of the beholder). They love the lines of the body, its feel and touch. They are fascinated by every detail of their beloved. Their love burns brightly but soon flickers and dies.

LUDUS. For ludic lovers, love is a game, something to play at rather than to become deeply involved in. Love is ultimately ludicrous. Love is for fun; encounters are casual, carefree, and often careless. "Nothing serious" is the motto of ludic lovers.

STORGE. Storge (pronounced *STOR-gay*) is the love between companions. It is, writes Lee, "love without fever, tumult, or folly, a peaceful and enchanting affection." It begins usually as friendship and then gradually deepens into love. If the love ends, it also occurs gradually, and the couple often become friends once again. Of such love Theophile Gautier wrote, "To love is to admire with the heart; to admire is to love with the mind."

MANIA. The word *mania* comes from the Greek word for madness. The Russian poet Mikhail Lermontov aptly described a manic lover:

> He in his madness prays for storms
> And dreams that storms will bring him peace.

For manic lovers, nights are marked by sleeplessness and days by pain and anxiety. The slightest sign of affection brings ecstasy for a short while, only to have it disappear. Satisfactions last but a moment before they must be renewed. Manic love is roller-coaster love.

AGAPE. Agape (pronounced *ah-GA-pay*) is love that is chaste, patient, selfless, and undemanding; it does not expect to be reciprocated. Agape emphasizes nurturing and caring as their own rewards. It is the love of monastics, missionaries, and saints more than that of worldly couples.

PRAGMA. Pragmatic lovers are, first and foremost, logical in their approach toward looking for someone who meets their needs. They look for a partner who has the background, education, personality, religion, and interests that are compatible with their own. If they meet a person who meets their criteria, erotic or manic feelings may develop. But, as Samuel Butler warned, "Logic is like the sword—those who appeal to it shall perish by it."

According to Lee, a person must thus find a partner who shares the same style and definition of love in order to have a mutually satisfying love affair. The more different two people are in their styles of love, the less likely it is that they will understand each other's love. Love styles are also linked to gender and ethnicity (Hendrick and Hendrick, 1986).

Research indicates that heterosexual and gay men have similar attitudes toward eros, mania, ludus, and storge, and that gay male relationships have multiple emotional dimensions (Adler, Hendrick, and Hendrick, 1989). As to cultural differences, different styles tend to characterize Asians, African Americans, Latinos, and whites. Asian Americans have a more pragmatic style of love than do Latinos, African Americans, or Caucasians, and place a high value on affection, trust, and friendship (pragma and storge). Latinos frequently score higher on the ludic characteristics (Regan, 2003).

DYNAMICS AND DIFFERENCES: THE TRIANGULAR THEORY OF LOVE

The **triangular theory of love** is composed of three elements that can be visualized as the points of a triangle: intimacy, passion, and decision/commitment (see Figure 5.2). Each can be enlarged or diminished in the course of a love relationship, and their changes will affect the quality of the relationship. They can also be combined in different ways. Each combination offers a different type of love—for example, romantic love, infatuation, empty love, and liking. The components may be combined differently at different times in the same love relationship (see Figure 5.3). Table 5.1 displays the combinations of commitment, passion, and intimacy in each type of love. There has been little research in this theory (Acker and Davis, 1992).

FIGURE 5.2 ■ The Triangular Theory of Love

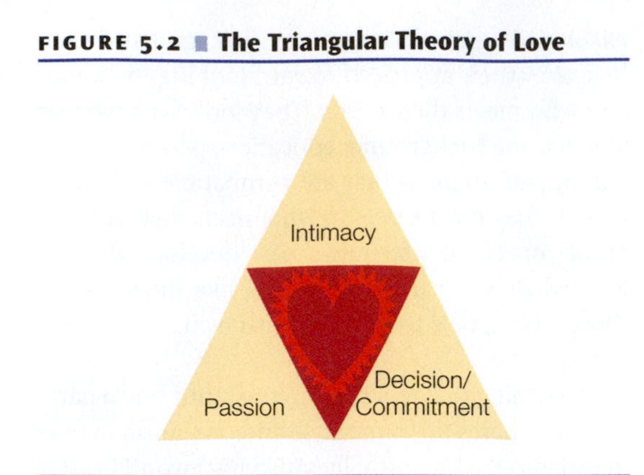

The passion, intimacy, and decision/commitment components of love can be combined in a variety of ways to form differently shaped triangles. These shapes may change over time.

THE INTIMACY COMPONENT. Intimacy refers to the warm, close feelings of bonding you get when you love someone. According to Sternberg and Grajek (1984), there are ten signs of intimacy:

- Wanting to promote your partner's welfare
- Feeling happiness with your partner
- Holding your partner in high regard
- Being able to count on your partner in time of need
- Being able to understand each other
- Sharing yourself and your possessions with your partner
- Receiving emotional support from your partner
- Giving emotional support to your partner
- Being able to communicate with your partner about intimate things
- Valuing your partner's presence in your life

THE PASSION COMPONENT. The passion component refers to the elements of romance, attraction, and sexuality in a relationship. These may be fueled by the desire to increase self-esteem, to be sexually active or fulfilled, to affiliate with others, to dominate, or to subordinate.

THE DECISION/COMMITMENT COMPONENT. The decision/commitment component consists of two parts, one short term and one long term. The short-term part refers to your decision that you love someone. You may or may not make the decision consciously, but it usually occurs before you decide to make a commitment to that person. The commitment represents the long-term aspect; it is the maintenance of love, but a decision to love someone does not necessarily entail a commitment to maintaining that love.

Kinds of Love: The Different Combinations

According to Sternberg (1988), the intimacy, passion, and decision/commitment components can be combined in eight basic ways. These combinations form the basis for classifying love:

- Liking (intimacy only)
- Romantic love (intimacy and passion)
- Infatuation (passion only)
- Fatuous love (passion and commitment)
- Empty love (decision/commitment only)

FIGURE 5.3 ■ The Triangles of Love

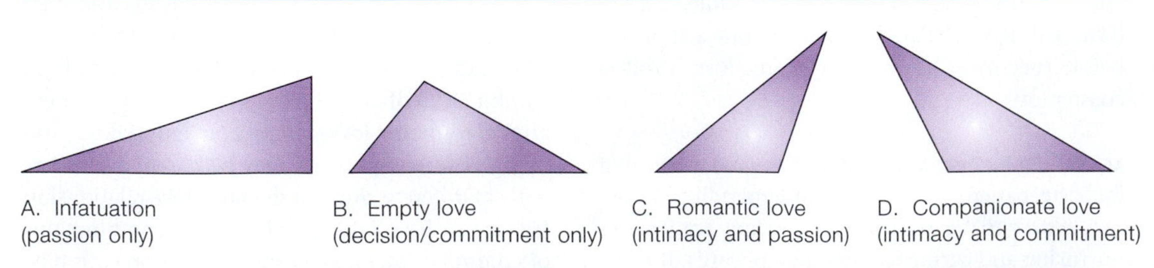

The greater the intensity of love one experiences, the greater will be one's love triangle in area. The greater a given component of love, the further the point from the center of the triangle. Thus A, B, C, and D experience different kinds of love; as a result, their triangles are differently shaped.

- Companionate love (intimacy and commitment)
- Consummate love (intimacy, passion, and commitment)
- Nonlove (absence of intimacy, passion, and commitment)

These types represent extremes that probably few of us experience. Not many of us, for example, experience infatuation in its purest form, in which there is absolutely no intimacy. The categories are nevertheless useful for examining love. We will discuss the various categories below (except for empty love which is not really love at all).

LIKING: INTIMACY ONLY. Liking represents the intimacy component alone. It forms the basis for close friendships but is neither passionate nor committed. As such, liking is often an enduring kind of love. Boyfriends and girlfriends may come and go, but good friends remain.

ROMANTIC LOVE: INTIMACY AND PASSION. Romantic love combines intimacy and passion. It is similar to liking, but it is more intense as a result of physical or emotional attraction. It may begin with an immediate union of the two components—with friendship that intensifies with passion or with passion that also develops intimacy. Although commitment is not an essential element of romantic love, it may develop.

INFATUATION: PASSION ONLY. Infatuation is love at first sight. It is the kind of love that idealizes its object; it rarely sees the other as a "real" person who sometimes has bad breath—but is worthy of being loved nonetheless. Infatuation is marked by sudden passion and a high degree of physical and emotional arousal. It tends to be obsessive and all consuming. One has no time, energy, or desire for anything or anyone but the beloved (or thoughts of him or her). To the dismay of the infatuated individual, infatuations are usually asymmetrical: One's passion (or obsession) is rarely returned equally, and the greater the asymmetry, the greater the distress in the relationship.

FATUOUS LOVE: PASSION AND COMMITMENT. Fatuous, or deceptive, love is whirlwind love; it begins the day a couple meet and quickly results in cohabitation or engagement, and then marriage. It goes so fast one hardly knows what has happened. Often enough, nothing much really did happen that will permit the relationship to endure. As Sternberg (1988) observes, "It is fatuous in the sense that a commitment is made on the basis of passion without the stabilizing element of intimate involvement—which takes time to develop." Passion fades soon enough, and all that remains is commitment. But commitment that has had relatively little time to deepen is a poor foundation on which to build an enduring relationship. With neither passion nor intimacy, the commitment wanes.

COMPANIONATE LOVE: INTIMACY AND COMMITMENT. Companionate love is essential to a committed relationship. It often begins as romantic love, but as the passion diminishes and the intimacy increases, it is transformed. Some couples are satisfied with such love; others are not. Those who are dissatisfied in companionate love relationships may seek extrarelational affairs to maintain passion in their lives. They may also end the relationship to seek a new romantic relationship in the hope that it will remain romantic.

CONSUMMATE LOVE: INTIMACY, PASSION, AND COMMITMENT. Consummate love is born when intimacy, passion, and commitment combine to form their unique constellation. It is the kind of love we dream about but do not expect in all our love relationships. Many of us can achieve it, but it is difficult to sustain over time. To sustain it, we must nourish its different components, for each is subject to the stress of time.

NONLOVE: NONEXISTENT LOVE. Nonlove can take many forms, such as attachment for financial reasons, fear, or the fulfillment of neurotic needs.

The Geometry of Love

The shape of the love triangle depends on the intensity of the love and the balance of the parts. Intense love relationships create triangles with greater area; such triangles occupy more of one's life. Just as love relationships can be balanced or unbalanced, so can love triangles. The balance determines the shape of the triangle (see again Figure 5.3). A relationship in which the intimacy, passion, and commitment components are equal forms an equilateral triangle. But if the components are not equal, unbalanced triangles form. The size and shape of a person's triangle give a good pictorial sense of how that person

Understanding Yourself

YOUR STYLE OF LOVE

John Lee, who developed the idea of styles of love, also developed a questionnaire that allows men and women to identify their styles of love. Complete the questionnaire to identify your style , then ask yourself to which style of love you find yourself drawn in others. Is it the same or different?

To diagnose your style of love, look for patterns across characteristics. If you consider your childhood less happy than that of your friends, were discontent with life when you fell in love, and very much want to be in love, you have "symptoms" that are rarely typical of eros and almost never true of storge but that do suggest mania. Where a trait does not especially apply to a type of love, the space in that column is blank. Storge, for instance, is not the presence of many symptoms of love but precisely their absence; it is a cool, abiding affection.

Graph Your Own Style of Loving

Consider each characteristic as it applies to a current relationship that you define as love, or to a previous one if that is more applicable. For each, note whether the trait is almost always true (AA), usually true (U), rarely true (R), or almost never true (AN).

	Characteristic	EROS	LUDUS	STORGE	MANIA	LUDIC EROS	STORGIC EROS	STORGIC LUDUS	PRAGMA
1	You consider your childhood less happy than the average of peers.	R		AN	U				
2	You were discontented with life (work, etc.) at the time your encounter began.	R		AN	U	R			
3	You have never been in love before this relationship.				U	R	AN	R	
4	You want to be in love or have love as security.	R	AN		AA		AN	AN	U
5	You have a clearly defined ideal image of your desired partner.	AA	AN	AN	AN	U	AN	R	AA
6	You felt a strong gut attraction to your beloved on the first encounter.	AA	R	AN	R		AN		
7	You are preoccupied with thoughts about the beloved.	AA	AN	AN	AA			R	
8	You believe your partner's interest is at least as great as yours.		U	R	AN			R	U
9	You are eager to see your beloved almost every day; this was true from the beginning.	AA	AN	R	AA		R	AN	R
10	You soon believed this could become a permanent relationship.	AA	AN	R	AN	R	AA	AN	U
11	You see "warning signs" of trouble but ignore them.	R	R		AA		AN	R	R
12	You deliberately restrain frequency of contact with partner.	AN	AA	R	R	R	R	U	

feels about another. The greater the match between each person's triangle in a relationship, the more likely each is to experience satisfaction in the relationship.

LOVE AS ATTACHMENT

Attachment theory of love maintains that the degree and quality of attachments we experience in early life influence our later relationships. It has been increasingly used in recent years in studying personal relationships, including love. It examines love as a form of attachment that finds its roots in infancy (Hazan and Shaver, 1987; Shaver, Hazan, and Bradshaw, 1988). Research suggests that romantic love and infant/caregiver attachment have similar emotional dynamics. Phillip Shaver and his associates (1988) suggest that "all important love relationships—especially the first ones with parents and later ones with lovers and spouses—are attachments." On the basis of infant/caregiver work by John Bowlby (1969, 1973, 1980), some researchers suggest numerous similarities between attachment

	EROS	LUDUS	STORGE	MANIA	LUDIC EROS	STORGIC EROS	STORGIC LUDUS	PRAGMA
13 You restrict discussion of your feelings with beloved.	R	AA	U	U	R		U	U
14 You restrict display of your feelings with beloved.	R	AA	R	U	R		U	U
15 You discuss future plans with beloved.	AA	R	R				AN	AA
16 You discuss wide range of topics, experiences with partner.	AA	R			U	R	AA	
17 You try to control relationship, but feel you've lost control.	AN	AN	AN	AA	AN	AN		
18 You lose ability to be first to terminate relationship.	AN	AN		AA	R	U	R	R
19 You try to force beloved to show more feeling, commitment.	AN	AN		AA		AN	R	
20 You analyze the relationship,weigh it in your mind.			AN	U		R	R	AA
21 You believe in the sincerity of your partner.	AA			U	R	U	AA	
22 You blame partner for difficulties of your relationship.	R	U	R	U	R	AN		
23 You are jealous and possessive but not to the point of angry conflict.	U	AN	R		R	AN		
24 You are jealous to the point of conflict, scenes, threats, etc.	AN	AN	AN	AA	R	AN	AN	AN
25 Tactile, sensual contact is very important to you.	AA		AN		U	AN		R
26 Sexual intimacy was achieved early, rapidly in the relationship.	AA		AN	AN	U	R	U	
27 You are willing to work out sex problems, improve technique.	U	R		R	U		R	U
28 You have a continued high rate of sex, tactile contact throughout the relationship.	U		R	R	U	R		R
29 You declare your love first, well ahead of partner.		AN	R	AA		AA		
30 You consider love life your most important activity, even essential.		AN	R	AA		AA		
31 You are prepared to "give all" for love once under way.	U	AN	U	AA	R	AA	R	R
32 You are willing to suffer abuse, even ridicule, from partner.		AN	R	AA			R	AN
33 Your relationship is marked by frequent differences of opinion, anxiety.	R	AA	R	AA	R	R		R
34 The relationship ends with lasting bitterness, trauma for you.	AN	R	R	AA	R	AN	R	R

SOURCE: From J. A. Lee, "The Styles of Love," *Psychology Today*, 1974. Reprinted by permission.

and romantic love (Downey, Bonica, and Rincon, 1999; Bringle and Bagby, 1992; Shaver et al., 1988). These include the following:

ATTACHMENT

- Attachment bond's formation and quality depend on attachment object's responsiveness and sensitivity.
- When attachment object is present, infant is happier.
- Infant shares toys, discoveries, objects with attachment object.
- Infant coos, talks baby talk, "sings."
- There are feelings of oneness with attachment object.

ROMANTIC LOVE

- Feelings of love are related to lover's interest and reciprocation.
- When lover is present, person feels happier.
- Lovers share experiences and goods, give gifts.
- Lovers coo, talk baby talk, and sing.
- There are feelings of oneness with lover.

According to research by Downey(1996), rejection by parents of their children's needs can lead to the development of **rejection sensitivity,** or the tendency to anticipate and overreact to rejection. Individuals who develop rejection sensitivity seek to avoid rejection by their partners and closely monitor, even overanalyze the relationship dynamics for signs of potential rejection. As Regan (2003) notes, even "minimal or ambiguous" rejection cues may lead to feelings of rejection and to anger, jealousy, and despondency. Rejection-sensitive people tend to be less satisfied with their relationships, and more likely to see them end.

Studies conducted by Mary Ainsworth and colleagues (1978, cited in Shaver et al., 1988) indicate that there are three different styles of infant attachment: (1) secure, (2) anxious/ambivalent, and (3) avoidant. In *secure attachment,* the infant feels secure when the mother is out of sight. He or she is confident that the mother will offer protection and care. In *anxious/ambivalent attachment,* the infant shows separation anxiety when the mother leaves. He or she feels insecure when the mother is not present. In *avoidant attachment,* the infant senses the mother's detachment and rejection when he or she desires close bodily contact. The infant shows avoidance behaviors with the mother as a means of defense. In Ainsworth's study, 66 percent of the infants were secure, 19 percent were anxious/ambivalent, and 21 percent were avoidant.

Some researchers (Feeney and Noller, 1990; Shaver et al., 1988) believe that the styles of attachment developed during infancy continue through adulthood. Others, however, question the validity of applying infant research to adults as well as the stability of attachment styles throughout life (Hendrick and Hendrick, 1994). Still others found a significant association between attachment styles and relationship satisfaction (Brennan and Shaver, 1995).

Secure Adults

Secure adults find it relatively easy to get close to others. They are comfortable depending on others and having others depend on them. They believe they are worthy of love and support and expect to receive them in their relationships (Regan, 2003). They generally do not worry about being abandoned or having someone get too close to them. More than avoidant and anxious/ambivalent adults, they feel that others generally like them; they believe that people are generally well intentioned and good hearted. In contrast to others, secure adults are less likely to believe in media images of love and more likely to believe that romantic love can last. Their love experiences tend to be happy, friendly, and trusting. They are more likely to accept and support their partners. Reportedly, compared to others, secure adults find greater satisfaction and commitment in their relationships (Pistol, Clark, and Tubbs, 1995).

Anxious/Ambivalent Adults

Anxious/ambivalent adults feel that others do not or will not get as close as they themselves want. They worry that their partners do not really love them or that they will leave them. They feel unworthy of love and need approval from others (Regan, 2003). They also want to merge completely with the other person, which sometimes scares that person away. More than others, anxious/ambivalent adults believe that it is easy to fall in love. Their experiences in love are often obsessive and marked by a desire for union, high degrees of sexual attraction and jealousy, and emotional highs and lows.

Avoidant Adults

Avoidant adults feel discomfort in being close to others; they are distrustful and fearful of becoming dependent (see Bartholomew, 1990). Thus, to avoid the pain they expect to come from eventual rejection, they maintain distance and avoid intimacy (Regan, 2003). More than others, they believe that romance seldom lasts but that at times it can be as intense as it was at the beginning. Their partners tend to want more closeness than they do. Avoidant lovers fear intimacy and experience emotional highs and lows and jealousy.

In adulthood, the attachment styles developed in infancy combine with sexual desire and caring behaviors to give rise to romantic love. Comparing across these three types of attachment styles indicates that women and men with secure attachment styles tend to be the preferred type of romantic partner by women and men alike. They also tend to find more satisfaction in their relationships, experience more happiness, hold more positive views of their partners, and display fewer negative emotions (Regan, 2003).

Unrequited Love

As most of us know from painful experience, love is not always returned. We may reassure ourselves that, as Tennyson wrote 150 years ago, " 'Tis better to have loved and lost / Than never to have loved at all." Too often, however, such words sound like a rationalization. Who among us does not sometimes think, " 'Tis better never to have loved at all"? **Unrequited love**—love that is not returned—is a common experience.

Several researchers (Baumeister, Wotman, and Stillwell, 1993) accurately captured some of the feelings associated with unrequited love in the title of their research article: "Unrequited Love: On Heartbreak, Anger, Guilt, Scriptlessness, and Humiliation." They found that unrequited love was distressing for both the would-be lover and the rejecting partner. Would-be lovers felt both positive and intensely negative feelings about their unlucky attempt at a relationship. Nearly half of them (44) reported that the unreciprocated love caused them pain and suffering, jealousy and anger. Almost a quarter of them (22 percent) experienced fears about rejection. However, positive feelings were more common than negative feelings. More than half looked back on the experience positively (Regan, 2003). The rejectors, however, felt uniformly negative about the experience. Unlike the rejectors, the would-be lovers felt that the attraction was mutual, that they had been led on, and that the rejection had never been clearly communicated. Rejectors, by contrast, felt that they had not led the other person on; moreover, they felt guilty about hurting him or her. Nevertheless, many found the other person's persistence intrusive and annoying; they wished the other would have simply gotten the hint and gone away. Approximately half (51 percent) felt annoyed by the unwanted attention, 61 percent felt badly about having to reject the other, and 70 percent felt a range of negative emotions such as frustration, and resentment (Regan, 2003). Whereas rejectors saw would-be lovers as self-deceptive and unreasonable, would-be lovers saw their rejectors as inconsistent and mysterious.

Unrequited love presents a paradox: If the goal of loving someone is an intimate relationship, why should we continue to love a person with whom we could not have such a relationship? Arthur Aron and his colleagues addressed this question in a study of almost 500 college students (Aron et al., 1989b). The researchers found three different attachment styles underlying the experience of unrequited love:

- *The Cyrano style:* The desire to have a romantic relationship with a specific person regardless of how hopeless the love is. In this style, the benefits of loving someone are so great that it does not matter how likely the love is to be returned. Being in the same room with the beloved—because he or she is so wonderful—may be sufficient. This style is named after Cyrano de Bergerac, the seventeenth-century French poet and musketeer, whose love for Roxanne was so great that it was irrelevant that she loved someone else.
- *The Giselle style:* The misperception that a relationship is more likely to develop than it actually is. This might occur if one misreads the other's cues, such as in mistakenly believing that friendliness is a sign of love. This style is named after Giselle, the tragic ballet heroine who was misled by Count Albrecht to believe that her love was reciprocated.
- *The Don Quixote style:* The general desire to be in love, regardless of whom one loves. In this style, the benefits of being in love—such as being viewed as a romantic or the excitement of extreme emotions—are more important than actually being in a relationship. This style is named after Cervantes's Don Quixote, whose love for the common Dulcinea was motivated by his need to dedicate knightly deeds to a lady love. "It is as right and proper for a knight errant to be in love as for the sky to have stars," Don Quixote explained.

Using attachment theory, the researchers found that some people were predisposed to be Cyranos, others Giselles, and still others Don Quixotes. Anxious/ambivalent adults tended to be Cyranos, avoidant adults often were Don Quixotes, and secure adults were likely to be Giselles. Those who were anxious/ambivalent were most likely to experience unrequited love; those who were secure were least likely to experience such love. Avoidant adults experienced the greatest desire to be in love in general; yet they had the least probability of being in a specific relationship. Anxious/ambivalent adults showed the greatest desire for a specific relationship; they also had the least desire to be in love in general.

REFLECTIONS

Have you experienced unrequited love? How did it differ from requited love? Do you have a "style" of unrequited love? Have you been the object of someone's unrequited love? How did you handle it?

Stalking as Extreme Unrequited Love

When unrequited love is joined by obsessive thinking, the stage is set for what has come to be known as **stalking** or *obsessive relational intrusion* (Regan, 2003). In such instances, one person pursues another seeking an intimate relationship. The victim is not interested in a relationship of this nature. The pursuer may send unwanted letters or gifts, leave notes, make phone calls, and visit, watch, or follow the target of his or her affection. Women and men are reportedly equally likely to be victimized in such a way. Among college students, surveys report between 20 and 30 percent have been targets of such behavior.

One can differentiate between milder and more extreme forms of stalking behavior (Spitzberg and Cupach, 2001; Regan, 2003). The milder forms include the following behaviors:

- Repeated phone calls in which the pursuer argues with the target
- Calling the victim and hanging up without speaking
- Begging the victim for another chance
- Watching or staring from a distance
- Gossiping to others about a supposed (but nonexistent) relationship

Some stalkers may resort to more extreme and threatening behaviors:

- Threatening to physically harm the victim
- Following the victim
- Damaging the property or possessions of the victim
- Sexually exposing oneself
- Forcing the victim to engage in sexual relations
- Recording conversations or taking photos of the victim without his or her knowledge

Although the more extreme forms are less common, perhaps as many as 30 percent of victims report such behaviors (Regan, 2003). The consequences experienced by the targets of these various forms of relational stalking may include fear, depression, anger, self-blame, sleeplessness, curtailed lifestyle, distrust of others, and physical symptoms including illness. Targets may try a variety of strategies to deal with the unwanted attention. Avoidance (ignoring, not responding, not accepting gifts, etc.) is common. Other strategies include direct confrontation, retaliation, and the seeking of formal protection. These may not achieve the desired outcome of lessening or stopping the behavior. As Regan notes, avoidance strategies may be too ambiguous and therefore not correctly understood by pursuers. Direct confrontation may actually give the pursuer what she or he is seeking, more contact. Both retaliation and the use of formal protection may serve to anger not stop the pursuer (2003).

Jealousy: The Green-Eyed Monster

Does jealousy prove love? Many of us think it does. We may try to test someone's interest or affection by attempting to make him or her jealous by flirting with another person. If our date or partner becomes jealous, the jealousy is taken as a sign of love (Salovey and Rodin, 1991; White, 1980a). But provoking jealousy proves nothing except that the other person can be made jealous. Making jealousy a litmus test of love is dangerous, for jealousy and love are not necessarily related. Jealousy may be a more accurate yardstick for measuring insecurity or possessiveness than love (see Mullen [1993] for a discussion of changing cultural attitudes toward jealousy).

It's important to understand jealousy for several reasons. First of all, jealousy is a painful emotion filled with anger and hurt. Its churning can turn us inside out and make us feel out of control. If we can

Comstock

Jealousy is not necessarily a sign of love.

understand jealousy, especially when it is irrational, then we can eliminate some of its pain. Second, jealousy can help cement or destroy a relationship. It helps maintain a relationship by guarding its exclusiveness, but in its irrational or extreme forms, it can destroy a relationship by its insistent demands and attempts at control. We need to understand when and how jealousy is functional and when and how it is not. Third, jealousy is often linked to violence (Follingstad et al., 1990; Laner, 1990; Riggs, 1993). It is a factor in precipitating violence or emotional abuse in dating relationships among both high school and college students (Burcky, Reuterman, and Kopsky, 1988; Stets and Pirog-Good, 1987). Marital violence and rape are often provoked by jealousy (Russell, 1990). It is often used by abusive partners to justify their violence (Adams, 1990). Rather than being directed at a rival, jealous aggression is often used against one's partner (Paul and Galloway, 1994).

WHAT IS JEALOUSY?

Jealousy is an aversive response that occurs because of a partner's real, imagined, or likely involvement with a third person (Bringle and Buunk, 1985; Sharpsteen, 1993). It functions as a boundary-making mechanism. Jealousy sets the boundaries for what an individual or group feels are important relationships; others cannot trespass these limits without evoking it (Reiss, 1986).

Jealousy: The Psychological Dimension

Jealousy is a painful experience. It is an agonizing compound of hurt, anger, depression, fear, and doubt. We feel less attractive and acceptable to our partner when we are jealous (Bush, Bush, and Jennings, 1988). Jealous responses are most intense in committed or marital relationships because both partners assume "specialness." This specialness occurs because our intimate partner is different from everyone else: It is with him or her that we are most confiding, revealing, vulnerable, caring, and trusting. There is a sense of exclusiveness. To have sex outside the relationship violates that sense of exclusiveness because sex symbolizes "specialness." Words such as *unfaithfulness, cheating,* and *infidelity* reflect the sense that an unspoken pledge has been broken. This pledge is the normative expectation that serious relationships, whether dating or marital, will be sexually exclusive (Lieberman, 1988).

As our lives become more and more intertwined, we become less and less independent. For some, this loss of independence increases the fear of losing the partner. But it takes more than simple dependency to make a person jealous. The relationship must be

Exploring Diversity

JEALOUSY IN POLYGAMOUS CULTURES: TIBETAN SOCIETY

In traditional Tibetan society, polyandry (the less common variety of polygamy in which a woman has more than one husband) was the cultural ideal. This, however, was a special sort of polyandry—fraternal polyandry. For Tibetans in Tibet and Nepal, the best form of marriage was that of one woman married to several men who were brothers. Fraternal polyandry was what every household sought to accomplish, but of course a family had to have more than one son to arrange this kind of marriage. With only one son, the best a family could do was arrange a monogamous marriage, which was considered a less fortunate match.

For Westerners, polyandry has always needed more explaining than polygyny. Having several wives seems, even to Westerners, more natural. Certainly co-wives may be jealous of one another—novels and histories have described the hostility felt by one wife toward another in traditional Chinese society, for example—but they must carry on. To conceive of men overcoming their jealousy at sharing one wife takes more imagination, because co-husbands would "naturally" be jealous of one another, and jealousy would inevitably lead to fighting or killing. How, therefore, could polyandry ever work?

Anthropologists and historians have explored several hypotheses to explain the practice of polyandry among Tibetans. The leading hypothesis is that polyandry is particularly adaptive to certain kinds of ecological and economic conditions. Traditionally, Tibetans, using plow and oxen, farmed very difficult terrain. Their primary crop was buckwheat. The most economically successful households were those that maximized the number of adult males, who not only did the plowing but also herded sheep and engaged in long-distance salt trading. According to this hypothesis, polyandry created households composed of more than one adult male to handle these various economic tasks.

But what about "natural" jealousy? Tibetan boys were raised to place supreme value on loyalty and cooperation among brothers. Households built on the marriage of several brothers to one wife not only were more successful economically but were also prestigious. These households were living proof of the solidarity of brothers. Of course, there were tensions of various kinds with Tibetan families, but the brothers sought to resolve them in order to preserve the common residence and marriage.

threatened or we must suspect, rightly or wrongly, that it is being threatened. Those who are most jealous lack self-esteem and feel insecure, either about themselves or about their relationships (Berscheid and Frei, 1977; McIntosh and Tate, 1990). This is true for both whites and African Americans (McIntosh, 1989).

Social psychologists suggest that there are two types of jealousy: suspicious and reactive (Bringle and Buunk, 1991). **Suspicious jealousy** is jealousy that occurs where there is either no reason to be suspicious or only ambiguous evidence to suspect that a partner is involved with another. **Reactive jealousy** is jealousy that occurs when a partner reveals a current, past, or anticipated relationship with another person.

Suspicious jealousy generally occurs when a relationship is in its early stages. The relationship is not firmly established, and the couple is unsure about its future. The smallest distraction, imagined slight, or inattention can be taken as evidence of interest in another person. Even without any evidence, a jealous person may worry ("Is she seeing someone else but not telling me?"). This person may engage in vigilance, watching the partner's every move ("I'd like to audit your marriage and family class"). He may snoop, unexpectedly appearing in the middle of the night to see if someone else is there ("I was just passing by and thought I'd say hello"). The partner may try to control the other's behavior ("If you go to your friend's party without me, we're through"). Suspicious jealousy may have both legitimate and negative functions in a relationship. While it may be a reasonable response to circumstantial evidence and warn the partner what will happen if there are serious transgressions, if unfounded, it can be self-defeating.

Reactive jealousy occurs when one partner learns of the other's present, past, or anticipated sexual involvement with another. This usually provokes the most intense jealousy. If the affair occurred in the early part of the present relationship, the unknowing partner may feel that the primary relationship has been based on a lie. Trust is ques-

The eldest brother chose the woman whom he and his brothers would marry. He looked for a woman who was attractive to him, who came from a good family, and who was known to be a hard worker. Upon marriage, the woman left her parents' home and settled with her husbands in theirs. Initially, she slept mostly with the eldest brother. Some of the younger brothers might be awkward adolescents, or even younger. However, over time, all brothers took turns sleeping with their wife. The guiding spirit in polyandrous marriage was that the eldest brother was to defer to his younger brothers, so that they too might form an attachment to their wife. According to anthropologist Nancy Levine (1988), who studied Tibetan polyandry for many years, the younger husbands, who at first might be too young to be of any interest to the wife, would with time become increasingly attractive to her. However, the wife, too, was concerned about promoting equal time and good feeling among the brothers. Protocol required that during breakfast the wife would signal the husband with whom she would sleep that night. Thus it would be clear to all whose turn it was.

The wife in a polyandrous marriage kept track of her menstrual cycles and believed she knew which of the brothers fathered which children. Generally, of course, the eldest brother fathered the first children. But after that, it was the wife who assigned paternity at the birth of a child. Although the brothers believed she knew who the father was, the wife also knew that it was important that each brother have a son.

Levine found in her study that statistically, households experiencing the most domestic difficulty—for example, households whose problems of equal time and equal attraction were chronic—were those with three or more brothers. In some of these households, problems of time and attraction could be resolved only by marriage to an additional wife. This situation was considered to be a great failure by both the family and the community. Ideally, one wife was best, as all the brothers understood. Families struggled to resolve the troublesome issues and avoid the second marriage. However, in some cases, a second marriage was inescapable and the ideal second wife was a sister of the first. Anthropologists refer to this kind of plural marriage as sororal polygynous fraternal polyandry.

Fraternal polyandry as practiced in Tibetan society (and elsewhere, among the Todaz of India and the Sinhalese of Sri Lanka) raises several questions: What is a family? What is marriage? Who is a parent? Among the most interesting issues, however, is the "natural fact" of sexual jealousy. What role does culture play in its creation?

tioned. Every word and event must be reevaluated in light of this new knowledge: "If you slept with him when you said you were going to the library, did you also sleep with him when you said you were going to the laundromat?" Or "How could you say you loved me when you were seeing her?" The damage can be irreparable.

Boundary Markers

As we noted earlier, jealousy represents a boundary marker. It points out what the boundaries are in a particular relationship. It determines how, to what extent, and in what manner others can interact with members of the relationship. It also shows the limits within which the members of the relationship can interact with those outside the relationship. Culture prescribes the general boundaries of what evokes jealousy, but individuals adjust them to the dynamics of their own relationships.

Boundaries may vary, depending on the type of relationship, gender, sexual orientation, and ethnicity. Sexual exclusiveness is generally important in serious dating relationships and cohabitation; it is virtually mandatory in marriage (Blumstein and Schwartz, 1983; Buunk and van Driel, 1989; Hansen, 1985; Lieberman, 1988). Men are generally more restrictive toward their partners than women; heterosexuals are more restrictive than gay men and lesbians. Although we know very little about jealousy and ethnicity, traditional Latinos and new Latino and Asian immigrants appear to be more restrictive than Anglos and African Americans (Mindel, Habenstein, and Wright, 1988). Despite variations on where the boundary lines are drawn, jealousy functions to guard those lines.

Although our culture sets down general marital boundaries, each couple evolves its own boundaries. For some, it is permissible to carve out an area of individual privacy. In some relationships, partners may have few or many friends of their own (of the same or other sex), activities, and interests apart from the couple. In others there are no separate spheres because of jealousy or a lack of interest. But wherever a married couple draws its boundaries, each member understands where the line is drawn. The partners implicitly or explicitly know what behavior will evoke a jealous response (Bringle and Buunk, 1991). For some, it is having lunch with a member of the other sex (or same sex, if they are gay or lesbian); for others, it is having dinner; for still others, it is having dinner and seeing a movie. It is often disingenuous for a married partner to say that he or she didn't know that a particular action (a flirtatious suggestion, a lingering touch, or dinner with someone else) would provoke a jealous response.

GENDER DIFFERENCES IN JEALOUSY

Both men and women fear that their partner might be attracted to someone else because of dissatisfaction with the relationship and the desire for sexual variety. Women, however, feel especially vulnerable to losing their partner to an attractive rival (Nader and Dotan, 1992; White, 1981). Men and women become jealous about different matters. Men tend to experience jealousy when they feel their partner is sexually involved with another man. Women, by contrast, tend to experience jealousy over intimacy issues (Buss et al., 1992). Women feel the most jealousy when they believe their partner is both emotionally and physically involved with another woman (White, 1981).

Both men and women react to jealousy with anger. But men are more likely to express anger, and women are more likely to suppress it and feel depressed. Ira Reiss (1986) suggests that this difference in expressing jealousy is consistent with cultural restraints prohibiting women from displaying anger. At the same time, it reflects their greater powerlessness vis-à-vis men. As a result, women may turn their anger inward, transforming it into depression.

There may be some truth in the belief that women are more jealous than men, but such jealousy is not inherent in being female. It is more likely related to the greater sexual freedom that men are permitted. (Even in marriage, it is more acceptable for men to "roam" because they are believed to be "naturally" more sexual than women.) If women appear to be more jealous than men, it may be because men, granted greater autonomy than women, have more opportunities to evoke jealous responses from women (Reiss, 1986).

MANAGING JEALOUSY

Jealousy can be unreasonable or a realistic reaction to genuine threats. Unreasonable jealousy can become a problem when it interferes with an individual's well-being or that of the relationship. Dealing with irrational suspicions can often be difficult, for such feelings touch deep recesses in ourselves. As noted earlier, jealousy is often related to personal feelings of insecurity and inadequacy. The source of such jealousy lies within a person, not within the relationship.

If we can work on the underlying causes of our insecurity, then we can deal effectively with our irrational jealousy. Excessively jealous persons may need considerable reassurance, but at some point they must confront their own irrationality and insecurity. If they do not, they emotionally imprison their partner. Their jealousy may destroy the very relationship they were desperately trying to preserve.

But jealousy is not always irrational. Sometimes there are real reasons for it, such as the violation of relationship boundaries. In this case, the cause lies not within person but within the relationship. Gordon Clanton and Lynn Smith (1977) write:

> Jealousy cannot be treated in isolation. *Your* jealousy is not *your* problem alone. It is also a problem for your partner and for the person whose interest in your partner sparks your jealousy. Similarly, when your partner feels jealous, you ought not to dismiss the matter by pointing a finger and saying "That's your problem." Typically, three or more persons are involved in the production of jealous feelings and behaviors. Ideally, all three should take a part of the responsibility for minimizing the negative consequences.

Managing jealousy requires the ability to communicate, the recognition by each partner of the feelings and motivations of the other, and a willingness to reciprocate and compromise (Ridley and Crowe, 1992). If the jealousy is well founded, the partner may need to modify or end the relationship with the "third party" whose presence initiated the jealousy. Modifying the third-party relationship reduces the jealous response and, more important, symbolizes the partner's commitment to the primary relationship. If the partner is unwilling to do this— because of lack of commitment, unsatisfied personal needs, or other problems in the primary relationship—the relationship is likely to reach a crisis. In such cases, jealousy may be the agent for profound change.

The Transformation of Love: From Passion to Intimacy

Intense, passionate love does not last forever at the same high level. Instead, it fades or transforms itself into a more enduring love based on intimacy.

THE INSTABILITY OF PASSIONATE LOVE

Ultimately, romantic love may be transformed or replaced by a quieter, more lasting love. In fact, those in secure companionate love relationships, according to one study, experience the highest levels of satisfaction; they are much more satisfied than those in traditional romantic relationships (Hecht et al., 1994).

The Passage of Time: Changes in Intimacy, Passion, and Commitment

According to researcher Robert Sternberg (1988), time affects our levels of intimacy, passion, and commitment.

INTIMACY OVER TIME. When we first meet someone, intimacy increases rapidly as we make critical discoveries about each other, ranging from our innermost thoughts of life and death to our preference for strawberry or chocolate ice cream. As the relationship continues, the rate of growth decreases and then levels off. After the growth levels off, the partners may no longer consciously feel as close to each other. This may be because they are beginning to drift apart, or it may be because they are becoming intimate at a different, less conscious, deeper level. This kind of intimacy is not easily observed. It is a latent intimacy that nevertheless is forging stronger, more enduring bonds between the partners.

PASSION OVER TIME. Passion is subject to habituation. What was once thrilling—whether love, sex, or roller coasters—becomes less so the more we get used to it. Once we become habituated, more time with a person (or more sex or more roller coaster rides) does not increase our arousal or satisfaction.

If the person leaves, however, we experience withdrawal symptoms (fatigue, depression, anxiety), just as if we were addicted. In becoming habituated, we have also become dependent. We fall beneath the emotional baseline we were at when we met our partner. Over time, however, we begin to return to that original level.

COMMITMENT OVER TIME. Unlike intimacy and passion, time does not necessarily diminish, erode, or alter commitments. Our commitment is most affected by how successful our relationship is. Even initially, commitment grows more slowly than intimacy or passion. As the relationship becomes long term, the growth of commitment levels off. Our commitment will remain high as long as we judge the relationship to be successful. If the relationship begins to deteriorate, after a time the commitment will probably decrease. Eventually, it may disappear and an alternative relationship may be sought.

Disappearance of Romance as Crisis

The disappearance (or transformation) of passionate love is often experienced as a crisis in a relationship. A study of college students (Berscheid, 1983) found that half would seek divorce if passion disappeared from their marriage. But intensity of feeling does not necessarily measure depth of love. Intensity, like the excitement of toboggan runs, diminishes over time. It is then that we begin to discover if the love we experience for each other is one that will endure.

Our search for enduring love is complicated by our contradictory needs. Elaine Hatfield and William Walster (1981) offer this observation:

> What we really want is the impossible—a perfect mixture of security and danger. We want someone who understands and cares for us, someone who will be around through thick and thin, until we are old. At the same time, we long for sexual excitement, novelty, and danger. The individual who offers just the right combination of both ultimately wins our love. The problem, of course,

is that, in time, we get more and more security—and less and less excitement—than we bargained for.

The disappearance of passionate love, however, enables individuals to refocus their relationship. They are given the opportunity to move from an intense one-on-one togetherness that excludes others to a togetherness that includes family, friends, and external goals and projects. They can look outward on the world together.

The Reemergence of Romantic Love

Contrary to what pessimists believe, many people find that they can have both love and romance and that the rewards of intimacy include romance.

Romantic love may be highest during the early part of marriage and decline as stresses from child rearing and work intrude on the relationship. Most studies suggest that marital satisfaction proceeds along a U-shaped curve, with highest satisfaction in the early and late periods (see Chapter 10). Romantic love may be affected by the same stresses as general marital satisfaction. In fact, romantic love begins to increase as children leave home. In later life, romantic love may play an important role in alleviating the stresses of retirement and illness.

New research on the differences in love attitudes across family life stages reveals some unexpected and perhaps encouraging news for older romantics. Montgomery and Sorell (1997) write:

> The love attitudes endorsed by the broad age-range sample contradicts notions that romantic, passionate love is the privilege of youth and young relationships, functioning to bring partners together. Instead, individuals throughout the life-stages of marriage consistently endorse the love attitudes involving passion, romance, friendship, and self -giving love, and these results indicate that any popularization of young single adulthood as the enviable passionate idea is erroneous.

Being physically limited does not necessarily inhibit love and sexuality any more than being able-bodied guarantees them.

Perspective

IN SEARCH OF TRUE LOVE

What we expect and experience of love varies across the life span of a relationship. The romantic mystique that defines the early stages of committed relationships may be difficult to sustain across many years together. It also may become less definitive of the kind of lifelong relationship that is implied by "till death do us part." Consider the following story of love's "final days," poignantly told by journalist Mike Harden. It captures what is meant when we exchange our vows and promise to love each other forever.

In the end, real love means in sickness and in health

When Frank Steger pushed himself into an upright position in the hospital bed, the heart monitor's fluid cursive line disintegrated into an erratic scribble.

"I told the doctor," he said, peeking at the edge of the curtain to make sure wife, Mary, was not within earshot, "that I felt like I was drowning. He said, 'This is how it happens with congestive heart disease.' I told him I'd rather he throw me off the roof instead."

Mary returned to the room, drawing a chair to his bedside.

"Thirsty," he complained.

She lifted the straw to his lips as he pulled the oxygen mask aside.

The medicine was making him sick. She fetched the basin, wrapped a firm arm around his spasm-wracked shoulders, mopped the sweat from his forehead.

In sickness and in health, I thought. They were supposed to be preparing for a Florida vacation, not holding on to one another in the cardiac care unit at Mount Carmel East Hospital.

"Help me sit up," he whispered hoarsely.

In the end, love comes down to this; not Gable's devilish first appraisal of Leigh, not Lancaster and Kerr rolling in the surf. But, "Help me sit up."

A late December rain spattered against the pane. Christmas had come and gone in the half-darkened room, a blur of canned carols punctuated by beeps and buzzes, lit by the winking light on the intravenous monitor.

"Merry Christmas," the cardiologist hailed, parting the curtains.

Christmas had always been a festive time for them. Standing rib roast, all the trimmings. Lift the glasses to the new year. To your health and the health of all who sleep beneath your roof.

When breath came harder, he slept sitting up in the chair next to the bed. By then, the body had turned against itself, the mutinous kidneys loosing their slow poison on the weakened heart.

Mary paused in the waiting room to remove her street shoes and put on her slippers. She did not want to wake him now that sleep was such a rationed luxury. Soundlessly, she slipped into the chair next to his.

In the end, love is not the smoldering glance across the dance floor, the clink of crystal, a leisurely picnic spread upon summer's clover. It is the squeeze of a hand. I'm here. I'll be here, no matter how long the fight, even when what you want most is to close your eyes and be done with it all. Water? You need water? Here. Drink. Let me straighten your pillow.

"Help me into bed," he said, he who had once been warrior triumphant in the business world. He was tough, demanding, but never as much on others as himself. If you gave him your best, no one could hurt you. If you gave him less, no one could hide you. He was never accused of being a yes man. She had been beside him when the future was golden, beside him when health sent his career into eclipse.

Mary. Faithful Mary.

"I'm thirsty," he said.

"Here," she said, "let me get you something."

Along the road they had once traveled so often to visit family, the hearse wound its way past stubbled fields, shuttered roadside markets. The minister, clutching his Bible against his chest as though it alone was sufficient cloak against the wind whipping across Pickaway County, passed final benediction:

"Ashes to ashes, dust to dust."

He stopped to pick up his hat as the funeral director placed the folded flag in Mary's lap.

When all is said and done, love is not rapture and fire. It is a hand steadier than one's own squeezing harder than a heartbeat. Wine changes back to water. Roses no longer come with love messages, but best wishes for a quick recovery. Endearment is exhibited by what once might have been considered insignificant kindnesses, but which, in the end, become the tenderest of ministrations.

On the day after the funeral, trying to busy herself with chores that could easily wait, she plopped the laundry basket in front of her granddaughter. The child tugged out the end of the sheet her Frank had always held when they did the wash. When the child brought the folded end to meet the corners her grandmother held, she kissed her playfully, just as he had once done.

I'm thirsty, Grandma.

Here, let me get you something.

SOURCE: Reprinted with permission, from *The Columbus Dispatch. Mike Harden, "In the End, Real Love Means in Sickness and in Health."*

As we age the dynamics that characterize our intimate relationships change even when the relationships, themselves, endure.

© David Young-Wolff/PhotoEdit

So it is that, among those whose marriages survive, passion and romance do not necessarily decline over time.

INTIMATE LOVE: COMMITMENT, CARING, AND SELF-DISCLOSURE

Perhaps one of the most profound questions we can ask about love is how to make it stay. The key to making love stay seems to be not in love's passionate intensity but in the transformation of that intensity into intimate love. Intimate love is based on commitment, caring, and self-disclosure.

Commitment

Commitment is an important component of intimate love because it is a "determination to continue" a relationship or marriage in the face of bad times as well as good (Reiss, 1980a). It is based on conscious choice rather than on feelings, which, by their very nature, are transitory. Commitment is a promise of a shared future, a promise to be together come what may.

Commitment has become an important concept in recent years. We seem to be as much in search of commitment as we are in search of love or marriage. We speak of "making a commitment" to someone or to a relationship. (Among singles, commitment is sometimes referred to as "the C-word.") A committed relationship has become almost a stage of courtship, somewhere between dating and being engaged or living together.

Caring

Caring is placing another's needs before your own. As such, caring requires treating your partner as valued for simply being himself or herself. It requires what the philosopher Martin Buber called an I-Thou relationship. Buber described two fundamental ways of relating to people: I-Thou and I-It. In an I-Thou relationship, each person is treated as a Thou—that is, as a person whose life is valued as an end in itself. In an I-It relationship, each person is treated as an It; a person has worth only as someone who can be used. When a person is treated as a Thou, his or her humanity and uniqueness are paramount.

Self-Disclosure

When we self-disclose, we reveal ourselves—our hopes, our fears, our everyday thoughts—to others. Self-disclosure deepens others' understanding of us. It also deepens our own understanding, for we discover unknown aspects as we open ourselves to others. (Self-disclosure is discussed in detail in Chapter 6.) Without self-disclosure, we remain opaque and hidden. If others love us, such love leaves us with anxiety. Are we loved for ourselves or for the image we present to the world?

Together, commitment, caring, and self-disclosure help transform love. But in the final analysis, perhaps the most important means of sustaining love is our words and actions. Caring words and deeds provide the setting for maintaining and expanding love (Byrne and Murnen, 1988).

While we increasingly understand the dynamics and varied components of love, the experience of love itself remains ineffable, the subject of poetry rather than scholarship. A journal article is not a love poem, and romantics should not forget that love exists in the everyday world. Researchers have helped us increasingly understand love in the light of day—its nature, its development, its varied aspects—so that we may better be able to enjoy it in the moonlight.

SUMMARY

- *Prototypes* of love and commitment are models of how people define these two ideas in everyday life. The central aspects of the love prototype include trust, caring, honesty, friendship, respect, and concern for the other; central aspects of the commitment prototype include loyalty, responsibility, living up to one's word, faithfulness, and trust.
- Attitudes and feelings associated with love include caring, needing, trusting, and tolerating. Behaviors associated with love include verbal, nonverbal, and physical expressions of affection; self-disclosure; giving of nonmaterial and material evidence; and tolerance.
- Commitment is affected by the balance of costs to benefits, normative inputs, and structural constraints.
- Though friendship and love share many of the same characteristics, love contains passion, fascination, and instability. Friendship is the foundation for loving relationships.
- The *wheel theory of love* emphasizes the interdependence of four processes: (1) rapport, (2) self-revelation, (3) mutual dependency, and (4) fulfillment of intimacy needs.
- According to John Lee, there are six basic styles of love: *eros, ludus, storge, mania, agape,* and *pragma.*
- The *triangular theory of love* views love as consisting of three components: (1) intimacy, (2) passion, and (3) decision/commitment.
- The *attachment theory of love* views love as being similar in nature to the attachments we form as infants. The attachment (love) styles of both infants and adults are secure, anxious/ambivalent, and avoidant.
- *Unrequited love* is a common experience. There are three styles of unrequited love: (1) the Cyrano style—the desire to have a relationship with another, regardless of how hopeless it is; (2) the Giselle style—the misperception that a relationship is more likely to develop than it actually is; and (3) the Don Quixote style—the general desire to be in love. Anxious/ambivalent adults are most likely to be Cyranos, avoidant adults to be Don Quixotes, and secure adults to be Giselles.
- Sometimes, unrequited love is expressed through obsessive relational intrusion, or stalking. Between 20 and 30 percent of college students

report having been stalked, and women and men are equally likely to be victimized.

- *Jealousy* is an aversive response that occurs because of a partner's real, imagined, or likely involvement with a third person. Jealousy acts as a boundary marker for relationships. Jealous responses are most likely in committed or marital relationships because of the presumed "specialness" of the relationship. Specialness is symbolized by sexual exclusiveness. As individuals become more interdependent, there is a greater fear of loss. Fear of loss, coupled with insecurity, increases the likelihood of jealousy.
- Time affects romantic relationships. The rapid growth of intimacy tends to level off, and we become habituated to passion. Commitment tends to increase, provided that the relationship is judged to be rewarding.
- Romantic love tends to diminish. It may either end or be replaced by intimate love. Many individuals experience the disappearance of romantic love as a crisis. Romantic love seems to be most prominent in adolescence and in early and later stages of marriage. Intimate love is based on commitment, caring, and self-disclosure.

KEY TERMS

agape 144
attachment theory of love 148
conjugal family 133
deep friendship 137
eros 144
homogamy 134
jealousy 153
ludus 144
mania 144
peer marriage 137
pragma 144
prototype 140
reactive jealousy 154
rejection sensitivity 150
stalking 152
storge 144
suspicious jealousy 154
triangular theory of love 145
unrequited love 151
wheel theory of love 142

SUGGESTED READINGS

Ackerman, Diane. *A Natural History of Love.* New York: Random House, 1994. An historical and cultural perspective on love.

Cancian, Francesca. *Love in America: Gender and Self-Development.* New York: Cambridge University Press, 1987. A look at love as it relates to gender: feminine, masculine, and androgynous.

Clark, Don. *The New Loving Someone Gay.* Rev. and exp. ed. Berkeley, CA: Celestial Arts, 1987. An exploration of the many aspects of gay and lesbian love, including love in gay and lesbian relationships, love between gay men—or lesbians—and their families and friends, by a gay therapist.

Regan, Pamela. *The Mating Game: A Primer on Love, Sex, and Marriage.* Thousand Oaks, CA: Sage Publications, 2003. An excellent overview of research and theory about love, sexuality, and mate selection.

Schwartz, Pepper. *Love Between Equals: How Peer Marriage Really Works.* New York: Free Press, 1994. Schwartz describes unique features of egalitarian relationships built on deep friendship and intimacy.

Solomon, Robert C. *Love: Emotion, Myth, & Metaphor.* Buffalo, NY: Prometheus Books, 1990. A book that seeks to separate love as an emotion from the many myths and illusions surrounding it.

RESOURCES ON THE INTERNET

Companion Web Site for This Book

http://sociology.wadsworth.com/strong/marriage9e
Gain an even better grasp on this chapter by going to the companion Web site to take one of the Tutorial Quizzes, use the Flash Cards to master key terms, or check out the many other study aids you'll find there. Visit the Marriage and Family Resource Center on the site. You'll also find special features such as GSS Data and Census 2000 information that will put data and resources at your fingertips to help you with that special project or to do some research on your own.

InfoTrac College Edition: Search Word Summary

love	friendship
attachment	commitment
jealousy	romance

To learn more about these central topics in the study of the family, you can conduct an electronic search using InfoTrac College Edition. To aid in your search and to gain useful tips, see the Student Guide to InfoTrac College Edition that you can access through the companion Web site for this book.

Preview

To gain a sense of what you already know about the material covered in this chapter, answer "True" or "False" to the statements below.

1 Conflict and intimacy go hand in hand in intimate relationships. True or false?

2 Touching is one of the most significant means of communication.True or false?

3 Always being pleasant and cheerful is the best way to avoid conflict and sustain intimacy. True or false?

4 Studies suggest that those couples with the highest marital satisfaction tend to disclose more than those who are unsatisfied. True or false?

5 Negative communication patterns before marriage are a poor predictor of marital communication because people change once they are married. True or false?

6 Good communication is primarily the ability to offer excellent advice to your partner to help him or her change. True or false?

7 Physical coercion is the method men use most frequently when disagreement arises between them and their partners. True or false?

8 The party with the least interest in continuing a relationship generally has the power in it. True or false?

9 Latinos and Asian Americans tend to rely on the nonverbal expression of intense feelings in contrast to direct verbal expressions. True or false?

10 Wives tend to give more negative messages than husbands. True or false?

"The greatest thing in family life is to take a hint when a hint is intended and not to take a hint when a hint isn't intended."

ROBERT FROST

CHAPTER 6

Communication, Power, and Conflict

Outline

Think about the kinds of relationships that are the focus of this book. What is it that we expect from marriages, families, and other intimate relationships? Chances are, if you list the many characteristics or qualities you desire in such relationships, somewhere on that list will be "communication." We want our loved ones to understand us, to share their feelings and ideas with us and to understand the ideas or feelings that we voice to them. In other words, we want to be able to communicate effectively. Chances are that "conflict" will not be on such a list. We tend to see conflict as a negative to be avoided. Yet, conflict is a basic feature of intimate relationships. As long as we value, care about, and live with others, we will experience occasions where we disagree. How we resolve those disagreements tells us much about the health of our relationships.

Both communication and conflict are inextricably connected to intimacy. When we speak of communication, we mean more than just the ability to discuss problems and resolve conflicts. We also mean communication for its own sake: the pleasure of being in each other's company, the excitement of conversation, the exchange of touches and smiles, the loving silences. Through communication we disclose who we are, and from this self-disclosure, intimacy grows.

One of the most common complaints of married partners, especially unhappy partners, is that they don't communicate. But it is impossible not to communicate—a cold look may communicate anger as effectively as a fierce outburst of words. What these unhappy partners mean by "not communicating" is that their communication drives them apart rather than bringing them together, feeds conflict rather than resolving it. Communication patterns are strongly associated with marital satisfaction (Noller and Fitzpatrick, 1991).

In this chapter, we explore how communication brings people together: how to develop communication skills, how to self-disclose, how to give feedback and affirm your partner. We also discuss the relationship between conflict and intimacy, exploring the types of conflict and the role of power in marital relationships. We look at common conflicts about sex, money, and housework. Finally, we explore some ways of resolving conflicts.

PREVIEW ANSWERS

1 True
2 True
3 False
4 True
5 False
6 False
7 False
8 True
9 True
10 True

Verbal and Nonverbal Communication

When we communicate, the messages we send and receive contain both a verbal and a nonverbal component. The verbal part expresses the *basic content* of the message, whereas the nonverbal part expresses the *relationship* part of the message. The relationship part conveys the attitude of the speaker (friendly, neutral, or hostile) and indicates how the words are to be interpreted (as a joke, request, or command). The full content of any message has to be understood according to both the verbal and nonverbal parts.

For a message to be most effective, both the verbal and nonverbal components must be in agreement. If you are angry and say "I'm angry," and your facial expression and voice both show anger, the message is clear. But if you say "I'm angry" in a neutral tone of voice and smile, your message is ambiguous. If you say "I'm not angry" but clench your teeth and use a controlled voice, your message is also unclear.

Nonverbal Communication

There is no such thing as *not communicating*. Even when you are not talking, you are communicating by your silence (for example, an awkward silence, a hostile silence, or a tender silence). You are communicating by the way you position your body and tilt your head and through your facial expressions, your physical distance from the other person, and so on. In fact, take a moment, right now, and look around you. If there are other people in your presence, how are they communicating nonverbally?

One of the problems with nonverbal communication, however, is the imprecision of its messages. Is a person frowning or squinting? Does the smile indicate friendliness or nervousness? A person may be in reflective silence, but we may interpret the silence as disapproval or distance.

FUNCTIONS OF NONVERBAL COMMUNICATION

An important study of nonverbal communication and marital interaction found that nonverbal communication has three important functions in marriage (Noller, 1984): (1) conveying interpersonal attitudes, (2) expressing emotions, and (3) handling the ongoing interaction.

Conveying Interpersonal Attitudes

Nonverbal messages are used to convey attitudes. Gregory Bateson describes nonverbal communication as revealing "the nuances and intricacies of how two people are getting along" (quoted in Noller, 1984). Holding hands can suggest intimacy; sitting on opposite sides of the couch can suggest distance. Not looking at each other in conversation can suggest awkwardness or a lack of intimacy.

According to psychologist John Gottman (1994), even the simple act of rolling one's eyes in response to a statement or complaint made by one's spouse can convey *contempt*, a feeling that the target of the expression is undesirable. In fact, Gottman suggests that rolling one's eyes is one of five significant reactions that signal serious flaws in a relationship and expose it to a high likelihood of a later breakup.

Expressing Emotions

Our emotional states are expressed through our bodies. A depressed person walks slowly; a happy person walks with a spring. Smiles, frowns, furrowed brows, tight jaws, tapping fingers—all express emotion. Expressing emotion is important because it lets our partner know how we are feeling so that he or she can respond appropriately. It also allows our partner to share our feelings—to laugh or weep with us.

Handling the Ongoing Interaction

Nonverbal communication helps us handle the ongoing interaction by indicating interest and attention. An intent look indicates our interest in the conversation; a yawn indicates boredom. Posture and eye contact are especially important. Are we leaning toward the person with interest or slumping back, thinking about something else? Do we look at the person who is talking, or are we distracted, glancing at other people as they walk by or watching the clock?

PROXIMITY, EYE CONTACT, AND TOUCH

Three forms of nonverbal communication are of particular importance: (1) proximity, (2) eye contact, and (3) touch.

Proximity

Nearness in terms of physical space, time, and so on, is referred to as **proximity.** Where we sit or stand in relationship to another person signifies levels of intimacy or the type of relationship. Many of our words conveying emotion relate to proximity, such as feeling "distant" or "close," or being "moved" by someone. We also "make the first move," "move in" on someone else's partner, or "move in together."

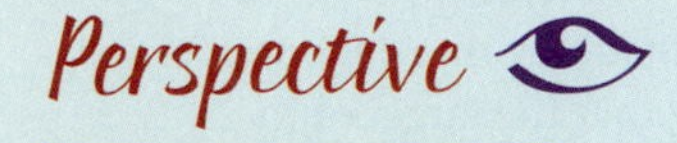

FAMILY RULES AND COMMUNICATION

Family Rules

According to family systems theory (discussed in Chapter 2), **rules** are patterned or characteristic responses. Family rules are generally unwritten; in fact, most of the time we probably don't even know we have them. They are formed from habit; like any habit, they are difficult to change.

All families have a **hierarchy of rules,** the ranking of rules in order of significance. Individuals have their rules; so does each family member in his or her roles as mother, father, son, daughter, and so on. Above individual and member rules are **family rules**—the combined members' rules. They are arrived at either consensually or through power struggles—both of which may be unconscious. These family rules are "policies" that evolve over time, such as "No one will discuss Daddy's drinking." The rules may be overt (openly recognized) or they may be covert (hidden and unrecognized).

Superior to family rules in the hierarchy are **meta-rules**—abstract, general, unarticulated rules at the apex of the hierarchy of rules. Meta-rules are different from individual and family rules primarily because they are more abstract. If not discussing Daddy's drinking is a family rule, the meta-rule is "Don't discuss anything that will cause a problem." When we talk of conspiracies of silence, we are usually talking about meta-rules. Meta-rules are much more difficult to change.

Feedback

The way each person responds influences every other person through feedback. Let's say the family rule is that no one confronts the father about his yelling at other family members.

The wife responds by withdrawing affection, the son goes into his room and shuts the door, the daughter tries to placate her father. These responses become new input. The father feels that his daughter is supportive of him; he is angry at his wife for withdrawing. The mother is angry at her daughter for responding to her father's yelling and ignoring her. The son is an absent member of the family.

Feedback is the basic principle for processing information. Feedback permits the system to alter its activities, structure, and direction to further its own goals. Information is processed through a feedback loop. The input becomes the system's output; the output in turn becomes new input. There are two types of feedback: negative feedback and positive feedback.

Negative feedback tells the system that change is unnecessary, that things are normal or returning to normal. In the example above, the feedback the father received was basically negative feedback because no one told him that his behavior was unacceptable. The feedback did not tell him that he had to change.

Positive feedback tells the system that things are changing, that the system is deviating from its normal state. Positive feedback *amplifies* the original input. If the mother had responded to her husband's yelling behavior by telling him to stop, he would have received positive feedback. But because the family rule is to avoid confrontation, the father received negative feedback—that is, none. Imagine, however that the family rule was to meet outburst with outburst. If the family rule called for anger, then the output would be quarreling and discussion in the family. This angry output then would

In a social situation, the face-to-face distances between people when starting a conversation are clues to how the individuals wish to define the relationship. All cultures have an intermediate distance in face-to-face interactions that are neutral. In most cultures, decreasing the distance signifies an invitation to greater intimacy or a threat. Moving away denotes the desire to terminate the interaction. When you stand at an intermediate distance from someone at a party, you send the message that intimacy is not encouraged. If you want to move closer, however, you risk the chance of rejection. Therefore, you must exchange cues, such as laughter or small talk, before moving closer in order to avoid facing direct rejection. If the person moves farther away during this exchange ("Excuse me, I think I see a friend"), he or she is signaling disinterest. But if the person moves closer, there is the "proposal" for greater intimacy. As relationships develop, there is close gazing into each other's eyes, holding hands, and walking with arms around each other—all of which require close proximity.

But because of cultural differences, there can be misunderstandings. The neutral intermediate dis-

become new input into the family system. If the family rule allows the unbridled expression of anger, then the anger will be amplified until it goes out of control. This process causes polarization and escalation. It is known as the *positive feedback spiral.*

Family rules and meta-rules are also known as *rules of transformation* because they transform or interpret input as it becomes output. If the husband loses his job, a family rule might call for blaming. As a result of the family rule, the job loss might generate hostility, family disintegration, and a major crisis. A different family could have different rules dealing with similar input. If the husband loses his job, the family rule might call for unity. In this case, the output would be family solidarity, which might reduce the crisis (Broderick and Smith, 1979).

Changing Family Rules

To show how the family processes information, let's take an extended example used by Mary Hicks and her colleagues (1983). As you read this example, remember that we are discussing an interactive process.

The family reacts to input, processes the input, and then sends it back to the environment as output, which becomes additional input. Here is an illustration concerning a woman working outside the home:

- *Environmental input:* "Men and women are equal."
- *Wife's rule:* "I should be able to work outside the home and get help from my husband in household work."
- *Husband's rule:* "I should give my full energy to my career; my wife should support me in this."
- *Family rule:* "Husband should work and wife should be responsible for child care and housework."
- *Meta-rule:* "Husband is primary and wife is subordinate."
- *System output:* "Husband is breadwinner; wife supports husband through child care, housekeeping, and additional income."

Look at these rules carefully. This family system is experiencing stress because the wife is out of step with the rest of the system. The wife is providing positive feedback. Through a process of negotiation between husband and wife, the rules in the hierarchy will change, although it will probably take considerable time and negotiation.

New rules evolve as society continues to change and move toward more egalitarian norms." After many cycles and over an extended period of time," note Hicks and her colleagues, "the combination of inputs, rule transformations, and outputs may result in a situation such as the one that exists in contemporary Western societies, in which an egalitarian norm is evolving, but with husbands' and families' rules still slow to change." If the family evolves these norms, the rules for the preceding example may be somewhat like the following:

- *Environmental input:* "Men and women are equal."
- *Wife's rule:* "I should be able to work and receive help from my husband."
- *Husband's rule:* "My wife should be able to work, but my career is primary."
- *Family rule:* "Wife can work; husband will share in household work and child care as long as it does not interfere with his career."
- *Meta-rule:* "Husband is senior partner; wife is junior partner."
- *Family system output:* "Wife will increase financial contribution and husband will increase child care and household work."

As Hicks and her co-workers point out, some changes have been made in all components, but the husband's, family's, and meta-rules are still resistant to an egalitarian ethic. Because disparity still exists, stress will continue.

tance for Latinos, for example, is much closer than for Anglos, who may misinterpret the distance as close (too close for comfort). In social settings, this can lead to problems. As Carlos Sluzki (1982) points out, "A person raised in a non-Latino culture will define as seductive behavior the same behavior that a person raised in a Latin culture defines as socially neutral. "Because of the miscue, the Anglo may withdraw or flirt, depending on his or her feelings. If the Anglo flirts, the Latino may respond to what he or she believes is the other's initiation. Additionally, the neutral responses of people in cultures that have greater intermediate distances and less overt touching, such as Asian-American culture, may be misinterpreted negatively by people with other cultural backgrounds.

Eye Contact

Much can be discovered about a relationship by watching how people look at each other. Making eye contact with another person, if only for a split second longer than usual, is a signal of interest. Brief and extended glances, in fact, play a significant role in women's expression of initial interest (Moore, 1985). (The word *flirting* is derived from the old English word *fliting,* which means "darting back and forth," as so often occurs when one flirts with his or her eyes.) When you can't take your eyes off another person, you probably have a strong attraction to him or her. In fact, you can often distinguish people in love by their prolonged looking into each other's eyes. In addition to eye contact, dilated pupils may be an indication of sexual interest (or poor lighting).

Research suggests that the amount of eye contact between a couple having a conversation can distinguish between those who have high levels of conflict and those who don't. Those with the greatest degree of agreement have the greatest eye contact with each other (Beier and Sternberg, 1977). Those in conflict tend to avoid eye contact (unless it is a daggerlike stare). As with proximity, however, the level of eye contact may differ by culture.

We convey feelings via a variety of nonverbal means—proximity, touch, and eye contact.

REFLECTIONS

Think about your own nonverbal communication. In instances where you and another person had significant eye contact, what did the eye contact mean? As you think about touch, what are the different kinds of touch you do? What meanings do you ascribe to the touch you give and the touch you receive?

Touch

A review of the research on touch finds it to be extremely important in human development, health, and sexuality (Hatfield, 1994). It is the most basic of all senses; it contains receptors for pleasure and pain, hot and cold, rough and smooth. "Skin is the mother sense and out of it, all the other senses have been derived," writes anthropologist Ashley Montagu (1986). Touch is a life-giving force for infants. If babies are not touched, they may fail to thrive and may even die. We hold hands with small children and those we love. Many of our words for emotion are derived from words referring to physical contact: *attraction, attachment,* and *feeling.* When we are emotionally moved by someone or something, we speak of being "touched."

But touch can also be a violation. A stranger or acquaintance may touch you in a way that is too familiar. Your date or partner may touch you in a manner you don't like or want. And sexual harassment is yet another form of unwelcome touching.

Touch often signals intimacy, immediacy, and emotional closeness. In fact, touch may very well be the most intimate form of nonverbal communication. One researcher (Thayer, 1986) writes, "If intimacy is proximity, then nothing comes closer than touch, the most intimate knowledge of another." Touching seems to go "hand in hand" with self-disclosure. Those who touch seem to self-disclose more; in fact, touch seems to be an important factor in prompting others to talk more about themselves (Heslin and Alper, 1983; Norton, 1983).

The amount of contact, from almost imperceptible touches to "hanging all over" each other, helps differentiate lovers from strangers. How and where a person is touched can suggest friendship, intimacy, love, or sexual interest. Levels of touching differ between cultures and ethnic groups. Members of Latin cultures (both European and American) and Jews touch more than do Anglo-Americans, whereas Asian Americans touch less (Henley, 1977). June Dobbs Butt (1981) describes the importance of touching among African Americans:

> Perhaps it is in the touching and the enjoyment of contact with human bodies that black culture is most alive, and is introduced into the life of the growing child. The fondness for touch permeates black culture from cradle to grave.

Despite stereotypes about women touching and men avoiding touch, studies suggest that there are no consistent differences between males and females in the amount of overall touching (Andersen, Lustig, and Andersen, 1987). Men do not seem to initiate touch with women any more than women do with men. Women are markedly unenthusiastic, however, about receiving touches from strangers and express greater concern about being touched.

Sexual behavior relies above all else on touch: the touching of self and others, and the touching of hands, faces, chests, arms, necks, legs, and genitals. Sexual behavior is skin contact. In sexual interactions, touch takes precedence over sight, as we close our eyes to caress, kiss, and make love. In fact, we shut our eyes to focus better on the sensations aroused by touch; we shut out visual distractions to intensify the tactile experience of sexuality.

The ability to interpret nonverbal communication correctly appears to be an important ingredient in successful relationships. The statement "I can tell when something is wrong " reveals the ability to read nonverbal clues, such as body language or facial expressions. This ability is especially important in ethnic groups and cultures that rely heavily on nonverbal expression of feelings, such as Latino and Asian-American cultures. Although the value placed on nonverbal expression may vary between groups and cultures, the ability to communicate and understand nonverbally remains important in all cultures. A comparative study of Chinese and American romantic relationships, for example, found that shared nonverbal meanings were important for the success of relationships in both cultures (Gao, 1991).

Gender Differences in Communication

The idea that women and men communicate differently has been the subject of much research and writing (Rubin, 1983; Tannen, 1990; Gray, 1993), including best sellers bemoaning our lack of understanding and inabilities to communicate with each other. Gender differences surface whether we examine nonverbal or verbal communication, and they become especially pronounced in cross-sex interaction.

Compared with men's nonverbal communication patterns, women smile more; express a wider range of emotions through their facial expressions; occupy, claim, and control less space; and maintain more eye contact with others with whom they are interacting (Borisoff and Merrill, 1985; Lindsey, 1997). In their use of language and their styles of speaking, further differences emerge (Lakoff, 1975; Tannen, 1990; Lindsey, 1997). Women use more qualifiers (for example, "It's *sort of* cold out"), use more tag questions ("It's sort of cold out, *don't you think?*"), use a wider variety of intensifiers ("It was *awfully* nice out yesterday, now it's sort of cold out, don't you think?"), and speak in more polite and less insistent tones. Male speech contains fewer words for such things as color, texture, food, relationships, and feelings, but men use more and harsher profanity (Lindsey, 1997). In cross-sex interaction, men talk more and interrupt women more than women interrupt men. In same-gender conversation, men disclose less personal information and restrict themselves to safer topics, such as sports, politics, or work (Lindsey, 1997).

The male styles of both verbal and nonverbal communication fit more with positions of dominance, women's with positions of subordination. At the same time, women's style is one of cooperation and consensus; thus, it is also situationally appropriate and advantageous to relationship building and maintenance (Tannen, 1990; Lindsey, 1997). In light of these facts, researchers differ in their interpretations of these gender patterns, between those who see women's style as artifacts of subordination versus those who see gender patterns as reflecting difference.

Communication Patterns and Marriage

Communication occupies an important place in marriage. When couples have communication problems they often fear that their marriages are seriously flawed. Some research indicates that the most frequent complaint of couples seeking therapy is about their communication problems (Burleson and Denton, 1997).

There has been an explosion of research on premarital and marital communication in the last decade. Researchers are finding significant correlations between the nature of communication and satisfaction, as well as finding differences in male versus female communication patterns in marriage.

PREMARITAL COMMUNICATION PATTERNS AND MARITAL SATISFACTION

"Drop dead, you creep!" is hardly the thing to say when trying to resolve a disagreement in a dating relationship. But it may be an important clue as to whether such a couple should marry. Many couples who communicate poorly before marriage are likely to continue the same way after marriage, and the result can be disastrous for future marital happiness. Researchers have found that how well a couple communicates before marriage can be an important predictor of later marital satisfaction (Cate and Lloyd, 1992). If communication is poor before marriage, it is not likely to get significantly better after marriage—at least not without a good deal of effort and help.

Self-disclosure—the revelation of deeply personal information about one's self—prior to or soon after marriage is related to relationship satisfaction later. In one study (Surra, Arizzi, and Asmussen, 1988), men and women were interviewed shortly after marriage and four years later. The researchers found that self-disclosure was an important factor for increasing each other's commitment later. Talking about your deepest feelings and revealing yourself to your partner builds bonds of trust that help cement a marriage.

Whether a couple's interactions are basically negative or positive can also predict later marital satisfaction. In a notable experiment by John Markham (1979), 14 premarital couples were evaluated using "table talk," sitting around a table and simply engaging in conversation. Each couple talked about various topics. Using an electronic device, each partner electronically recorded whether the message was positive or negative. Markham found that the negativity or positivity of the couple's communication pattern had little impact on their marital satisfaction during their first year. This protective quality of the first year is known as the **honeymoon effect**—which means you can say almost anything during the first year and it will not have a serious impact on marriage (Huston, McHale, and Crouter, 1986). But after the first year, couples with negative premarital communication patterns were less satisfied than those with positive communication patterns. A later study (Julien, Markman, and Lindahl, 1989) found that those premarital couples who responded more to each other's positive communication than to each other's negative communication were more satisfied in marriage four years later.

Cohabitation and Later Marital Communication

As we will see in the next chapter, researchers have revealed a cohabitation effect on marriage. Specifically, couples who live together before marrying are more likely to separate and divorce than couples who don't live together before marriage. For now, that may seem counterintuitive. Wouldn't couples who live together first find it easier to adjust to marriage? Doesn't cohabitation weed out the unsuccessful matches before marriage? Later, we will consider the range of explanations for this cohabitation effect. At this point we will simply look at how communication patterns might contribute to later marital failure.

Among the possible explanations for the cohabitation effect, Cohan and Kleinbaum (2002) hypothesized that spouses who live together before marrying display more negative problem solving and support behavior compared with their counterparts who marry without first living together. Why would cohabitation lead to poorer marital communication? Cohan and Kleinbaum suggest four possible reasons:

1. Couples who live together and then marry have a longer relationship history. The longer union

duration places couples who cohabited further along into the period when marital satisfaction may drop off. This leaves open the possibility that it is not communication as much as duration that plagues these couples after marriage (i.e., cohabitants who marry have been together longer).

2. Couples who live together come from backgrounds that may predispose them to poorer communication abilities. Compared with couples who don't cohabit, cohabitants tend to be younger, less religious, and more likely to come from divorced homes. Cohan and Kleinbaum point out that this translates into them being less mature, less traditional, and less likely to have had good parental role models for effective communication.
3. People who cohabit may be more accepting of divorce and less committed to marriage. Thus, they may expend less effort or energy at developing good marital communication skills because they are less sure they will stay married.
4. Cohabitation is associated with factors such as alcohol use, infidelity, and lower marital satisfaction, which in turn are correlated with less effective communication.

In studying 92 couples who were in their first two years of marriage, they found that premarital cohabitation was associated with poorer marital communication. Couples with one or more cohabitation experiences displayed poorer, more divisive and destructive communication behaviors than did couples with no prior cohabitation experience (Cohan and Kleinbaum, 2002).

MARITAL COMMUNICATION PATTERNS AND SATISFACTION

Researchers have found a number of patterns that distinguish the communication patterns in satisfied and dissatisfied marriages (Gottman, 1995; Hendrick, 1981; Noller and Fitzpatrick, 1991; Schaap, Buunk, and Kerkstra, 1988). Couples in satisfied marriages tend to have the following characteristics:

- Willingness to accept conflict but to engage in conflict in nondestructive ways.
- Less frequent conflict and less time spent in conflict. Both satisfied and unsatisfied couples, however, experience conflicts about the same topics, especially about communication, sex, and personality characteristics.

© Scott Barrow

Touch is one of our primary means of communication. It conveys intimacy, immediacy, and emotional closeness.

- The ability to disclose or reveal private thoughts and feelings, especially positive ones, to one's partner. Dissatisfied spouses tend to disclose mostly negative thoughts to their partners.
- Expression by both partners of more or less equal levels of affection, such as tenderness, words of love, and touch.
- More time spent talking, discussing personal topics, and expressing feelings in positive ways.
- The ability to encode (send) verbal and nonverbal messages accurately and to decode (understand) such messages accurately from their spouses. This is especially important for husbands. Unhappy partners may actually decode the messages of strangers more accurately than those from their partners.

GENDER DIFFERENCES IN PARTNERSHIP COMMUNICATION

Researchers have identified several gender differences in marital communication (Klinetob and Smith, 1996; Noller and Fitzpatrick, 1991; Thompson and Walker, 1989). First, wives tend to send clearer messages to their husbands than their husbands send to them. Wives are often more sensitive and responsive to their husbands' messages, both during conversation and during conflict. They are more likely to reply to either positive messages ("You look great") or negative messages ("You look awful") than are their husbands, who may not reply at all.

Second, wives tend to give more positive or negative messages; they tend to smile or laugh when they send messages, they send fewer clearly neutral messages. Husbands' neutral responses make it more difficult for wives to decode what their partners really are trying to say. If a wife asks her husband if they should go to dinner or see a movie and he gives a neutral response, such as, "Whatever," does he really not care, or is he pretending he doesn't care in order to avoid possible conflict?

Third, although communication differences in arguments between husbands and wives are usually small, they nevertheless follow a typical pattern. Wives tend to set the emotional tone of an argument. They escalate conflict with negative verbal and nonverbal messages ("Don't give me that!") or de-escalate arguments by setting an atmosphere of agreement ("I understand your feelings"). Husbands' inputs are less important in setting the climate for resolving or escalating conflicts. Wives tend to use emotional appeals and threats more than husbands, who tend to reason, seek conciliation, and find ways to postpone or end an argument. A wife is more likely to ask, "Don't you love me?" whereas a husband is more likely to say, "Be reasonable."

A prominent type of marital communication is referred to as **demand-withdraw communication**—a pattern in which one spouse makes an effort to engage the other spouse in a discussion of some issue of importance. The spouse raising the issue may criticize, complain, or suggest a need for change in his or her spouse's behavior. The other spouse, in response to such overtures, withdraws by either leaving the discussion, failing to reply, or changing the subject (Klinetob and Smith, 1996).

The demand-withdraw pattern has been found by researchers to be associated with gender. In 60 percent of couples, wives "demand" and husbands "withdraw." In 30 percent of couples , these roles are reversed. In the remaining 10 percent, spouses demand and withdraw about equally (Klinetob & Smith, 1996). Researchers have considered a variety of explanations for the gender differences in demanding and withdrawing, including psychological, biological, and structural factors (Christensen and Heavey, 1990). Research conducted by Klinetob and Smith suggests that the demand and withdraw roles vary according to whose issue is being discussed: "During discussions of a wife-generated topic, she was the demander and her husband withdrew. During discussions of a husband-generated topic, he demanded and she withdrew" (1996:954). They further suggest that because marriage relationships often favor husbands, husbands will be less likely to bring up issues for discussion, because the relationship *as is* is more acceptable to them. On the other hand, because wives may be more discontented with aspects of the relationship and bring them up for discussion, they more frequently occupy the "demand" position (Klinetob & Smith, 1996).

Although the demand-withdraw pattern is fairly common, it is not a particularly healthy style of communication and conflict resolution. It is associated with less marital satisfaction and a higher likelihood of relationship failure (Regan, 2003).

Sexual Communication

In order to have a satisfying sexual relationship a couple must be able to communicate effectively with each other about expectations, needs, attitudes, and preferences (Regan, 2003). Both the frequency with which couples engage in sexual relations and the quality of their involvement depend on such communication.

Among heterosexuals, in both married and cohabiting relationships, women and men follow a sexual script that leaves the initiation of sex (i.e., the communication of desire and interest) to men, with women then in a position of accepting or refusing men's overtures. Reviewing the literature on sexual communication, Regan observes that regardless of who takes the role of initiating, their efforts are usually met with positive responses. Both attempts to initiate and positive responses are rarely communicated explicitly and verbally:

> A person who desires sexual activity might turn on the radio to a romantic soft rock station, pour his or her partner a glass of wine, and glance suggestively in the direction of the bedroom. The partner . . . might smile, put down his or her book, and engage in other nonverbal behaviors

that continue the sexual interaction without explicitly acknowledging acceptance. (Regan, 2003, p. 84)

Interestingly, lack of interest or refusal of sexual initiations is communicated directly and verbally (e.g., "Not tonight, I have a lot of work to do"). By framing one's refusal in terms of some kind of account, the refusing partner allows the rejected partner to save face (Regan, 2003).

Effective sexual communication may be difficult but it is important if couples hope to construct and keep mutually satisfying sexual relationships. One must trust one's partner enough to express one's feelings about sexual needs, desires, and dislikes and one must be able to hear the same from one's partner without feeling judgmental or defensive (Regan, 2003).

Developing Communication Skills

Studies suggest that poor communication skills precede the onset of marital problems (Gottman, 1994; Markman, 1981; Markman et al., 1987). Even family violence has been seen by some as the consequence of deficiencies in one's abilities to communicate (Burleson and Denton, 1997).

While we cannot *not* communicate, we can enhance the quality of our communication so that we can understand each other and enhance our relationships. We can learn to communicate constructively rather than destructively. What follows, we hope, will help you develop good communication skills so that your relationships will be mutually rewarding.

STYLES OF MISCOMMUNICATION

Virginia Satir noted in *Peoplemaking* (1988), her classic work on family communication, that people can be classified according to four styles of miscommunication:

Placaters: Always agreeable, placaters are passive, speak in an ingratiating manner, and act helpless. If a partner wants to make love when a placater does not, the placater will not refuse, because that might cause a scene. No one knows what placaters really want or feel—and they themselves often do not know either.

Blamers: Acting superior, blamers are tense, they are often angry, and they gesture by pointing. Inside, they feel weak and want to hide this from

© Tony Freeman/PhotoEdit

How partners express and handle conflict verbally as well as nonverbally says much about the direction in which the relationship is heading.

Perspective

ETHNICITY AND COMMUNICATION

Different ethnic groups within our culture have different language patterns that affect the way they communicate. African Americans, for instance, have distinct communication patterns (Hecht, Collier, and Ribeau, 1993). Language and expressive patterns are characterized by emotional vitality, realness, and valuing direct experience, among other things (White and Parham, 1990). Emotional vitality is expressed in the animated use of words. Realness refers to "telling it like it is" using concrete nonabstract words. Direct experience is valued because "there is no substitute in the Black ethos for actual experience gained in the course of living" (White and Parham, 1990). "Mother wit"—practical or experiential knowledge—may be valued over knowledge gained from books or lectures.

Latinos, especially traditional Latinos, assume that intimate feelings will not be discussed openly (Guerrero Pavich, 1986). One researcher (Falicov, 1982) writes this about Mexican Americans: "Ideally, there should be a certain formality in the relationship between spouses. No deep intimacy or intense conflict is expected. Respect, consideration, and curtailment of anger or hostility are highly valued." Confrontations are to be avoided; negative feelings are not to be expressed. As a consequence, nonverbal communication is especially important. Women are expected to read men's behavior for clues to their feelings and for discovering what is acceptable. Because confrontations are unacceptable, secrets are important. Secrets are shared between friends but not between partners.

Asian-American ethnic groups are less individualistic than the dominant American culture. Whereas the dominant culture views the ideal individual as self-reliant and self-sufficient, Asian-American subcultures are more relationally oriented. Researchers Steve Shon and Davis Ja (1982) note the following about Asian Americans:

> They emphasize that individuals are the products of their relationship to nature and other people. Thus, heavy emphasis is placed on their relationship with other people, generally with the aim of maintaining harmony through proper conduct and attitudes.

Asian Americans are less verbal and expressive in their interactions than are both African Americans and whites; instead, they rely to a greater degree on indirect and nonverbal communication, such as silence and the avoidance of eye contact as signs of respect (Del Carmen, 1990). Because harmonious relationships are highly valued, Asian Americans tend to avoid direct confrontation if possible. Japanese Americans, for example, "value implicit, nonverbal intuitive communication over explicit, verbal, and rational exchange of information" (Del Carmen, 1990).To avoid conflict, verbal communication is often indirect or ambiguous; it skirts around issues instead of confronting them. As a consequence, in interactions Asian Americans rely on the other person to interpret the meaning of a conversation or nonverbal clues.

everyone (including themselves). If a blamer runs short of money, the partner is the one who spent it; if a child is conceived by accident, the partner should have used contraception. The blamer does not listen and always tries to escape responsibility.

Computers: Very correct and reasonable, computers show only printouts, not feelings (which they consider dangerous). "If one takes careful note of my increasing heartbeat," a computer may tonelessly say, "one must be forced to come to the conclusion that I'm angry." The partner who is interfacing, also a computer, does not change expression and replies, "That's interesting."

Distractors: Acting frenetic and seldom saying anything relevant, distractors flit about in word and deed. Inside, they feel lonely and out of place. In difficult situations, distractors light cigarettes and talk about school, politics, business— anything to avoid discussing relevant feelings. If a partner wants to discuss something serious, a distractor changes the subject.

MISCOMMUNICATION AND THE LIKELIHOOD OF DIVORCE

Psychologist John Gottman, a leading authority on communication and miscommunication in marriage, conducted a longitudinal observational study of couple interaction in which he identified five reactions to conflict that are particularly problematic. Along with *contempt* (see earlier discussion of nonverbal communication), Gottman suggested that *criticism, defensiveness, stonewalling* (resisting a partner's complaint), and *belligerence* (a defiant challenge to one's partner) are all signs of serious risk of eventual divorce (Gottman, 1994; Gottman et al., 1998). Conversely, couples who communicate with affection and interest, and manage to maintain humor in the midst of conflict, can use such *positive affect* to diffuse potentially threatening conflict (Gottman et al., 1998).

REFLECTIONS

Do you find that any of the stated styles of miscommunication characterize your own communication patterns? Your partner's style? What happens if you and your partner have similar styles? Different styles?

WHY PEOPLE DON'T COMMUNICATE

We can learn to communicate, but it is not always easy. Traditional male gender roles, for example, work against the idea of expressing feelings. This role calls for men to be strong and silent, to ride off into the sunset alone. If men talk, they talk about things—cars, politics, sports, work, money—but not about feelings. Also, both men and women may have personal reasons for not expressing their feelings. They may have strong feelings of inadequacy: "If you really knew what I was like, you wouldn't like me." They may feel ashamed of, or guilty about, their feelings: "Sometimes I feel attracted to other people, and it makes me feel guilty because I should only be attracted to you." They may feel vulnerable: "If I told you my real feelings, you might hurt me." They may be frightened of their feelings: "If I expressed my anger, it would destroy you." Finally, people may not communicate because they are fearful that their feelings and desires will create conflict: "If I told you how I felt, you would get angry."

OBSTACLES TO SELF-AWARENESS

Before we can communicate with others, we must first know how we ourselves feel. Though feelings are valuable guides for actions, we often place obstacles in the way of expressing them. First, we suppress "unacceptable" feelings, especially feelings such as anger, hurt, and jealousy. After a while, we may not even consciously experience them. Second, we deny our feelings. If we are feeling hurt and our partner looks at our pained expression and asks us what we're feeling, we may reply, "Nothing." We may actually feel nothing because we have anesthetized our feelings. Third, we project our feelings. Instead of recognizing that we are jealous, we may accuse our partner of being jealous; instead of feeling hurt, we may say our partner is hurt.

Becoming aware of ourselves requires us to become aware of our feelings. Perhaps the first step toward this self-awareness is realizing that feelings are simply emotional states—they are neither good nor bad in themselves. As feelings, however, they need to be felt, whether they are warm or cold, pleasurable or painful. They do not necessarily need to be acted on or expressed. It is the acting out that holds the potential for problems or hurt.

SELF-DISCLOSURE

Self-disclosure creates the environment for mutual understanding (Derlega et al., 1993). We live much of our lives playing roles—as student, worker, husband, wife, son, or daughter. We live and act these roles conventionally. They do not necessarily reflect our deepest selves. If we pretend that we are only these roles and ignore our deepest selves, we have taken the path toward loneliness and isolation. We may reach a point at which we no longer know who we are. In the process of revealing ourselves to others, we discover who we are. In the process of our sharing, others share themselves with us. Self-disclosure is reciprocal.

Keeping Closed

Having been taught to be strong, men may be more reluctant to express feelings of weakness or tenderness than women. Many women find it easier to disclose their feelings, perhaps because from earliest childhood they are more often encouraged to express them (see Notarius and Johnson, 1982).

If distinct differences exist, they can drive wedges between men and women. One sex does not understand the other. The differences may plague a marriage until neither partner knows what the other wants; sometimes partners don't even know what they want for themselves. One woman described her predicament this way:

> I'm not sure what I want. I keep talking to him about communication, and he says, "Okay, so we're talking; now what do you want?" And I don't know what to say then, but I know it's not what I mean. I sometimes get worried because I think maybe I want too much. He's a good husband; he works hard; he takes care of me and the kids. He could go out and find another woman who would be very happy to have a man like that, and who wouldn't be all the time complaining at him because he doesn't feel things and get close (Rubin, 1976).

What is missing is the intimacy that comes from self-disclosure. People live together, or are married, but they feel lonely. There is no contact, and the loneliest loneliness is to feel alone with someone with whom we want to feel close.

How Much Openness?

Can too much openness and honesty be harmful to a relationship? How much should intimates reveal to each other? Some studies suggest that less marital satisfaction results if partners have too little *or too much* disclosure; a happy medium offers security, stability, and safety. But a review of studies on the relationship between communication and marital satisfaction finds that a linear model of communication is more closely related to marital satisfaction than the too-little/too much curvilinear model (Boland and Follingstad, 1987). In the linear model of communication, the greater the self-disclosure, the greater the marital satisfaction, provided that the couple are highly committed to the relationship and willing to take the risks of high levels of intimacy. High self-disclosure can be a highly charged undertaking. Studies suggest that high levels of negativity are related to marital distress (Noller and Fitzpatrick, 1991). It is not clear whether the negativity reflects the marital distress or causes it. Most likely, the two interact and compound each other's effects.

Research by Burleson and Denton suggests that the relationship between communication skill and marital success and satisfaction is "quite complex" (1997, p. 889). In a study of 60 couples, researchers explored the importance of four communication skills in determining marital satisfaction:

- *Communication effectiveness:* producing messages that have their intended effect
- *Perceptual accuracy:* correctly understanding the intentions underlying a message
- *Predictive accuracy:* accurately anticipating the effect of one's message on another

Self-disclosure is reciprocal.

© Laurie DeVault Photography

- *Interpersonal cognitive complexity:* the capacity to process social information

Prior research had indicated that each of the above were important in differentiating satisfied from dissatisfied couples or nondistressed from distressed couples. Based on their research, Burleson and Denton suggest that *communication skill* may not adequately explain levels of distress or dissatisfaction. In fact, the intentions and feelings being communicated were more important factors separating distressed from nondistressed couples. Spouses in distressed couples had "more negative intentions" toward each other. "The negative communication behaviors frequently observed in distressed spouses may result more from ill will than poor skill" (1997, p. 897). Burleson and Denton also observe that *good communication skills* can worsen marital relationships when spouses have "negative intentions toward one another" (1997, p. 900).

TRUST

When we talk about intimate relationships, the two words that most frequently pop up are *love* and *trust.* As we saw in our discussion of love in Chapter 5, trust is an important part of love. But what is trust? **Trust** is the belief in the reliability and integrity of a person.

When a person says, "Trust me," he or she is asking for something that does not easily occur. For trust to develop, three conditions must exist (Book et al., 1980). First, a relationship has to exist and have the likelihood of continuing. We generally do not trust strangers or people we have just met with information that makes us vulnerable, such as our sexual anxieties. We trust people with whom we have a significant relationship.

Second, we must be able to predict how a person will likely behave. If we are married or in a committed relationship, we trust that our partner will not do something that will hurt us, such as having an affair. In fact, if we discover that our partner is involved in an affair, we often speak of our trust being violated or destroyed. If trust is destroyed in this case, it is because the predictability of sexual exclusiveness is no longer there.

Third, the person must also have other acceptable options available to him or her. If we were marooned on a desert island alone with our partner, he or she would have no choice but to be sexually monogamous. But if a third person, who was sexually attractive to our partner, swam ashore a year later, then our partner would have an alternative. Our partner would then have a choice of being sexually exclusive or nonexclusive; his or her behavior would then be evidence of trustworthiness—or the lack of it.

DID YOU KNOW?

The happiest couples are those who balance autonomy with intimacy and negotiate personal and couple boundaries through supportive communication (Scarf, 1995)

Trust is critical in close relationships for two reasons (Book et al., 1980). First, self-disclosure—which is vital to closeness—makes a person vulnerable and thus requires trust. A person will not self-disclose if he or she believes that the information may be misused—for example, by a partner who resorts to mocking behavior or revealing a secret. Second, the degree to which you trust a person influences the way you are likely to interpret ambiguous or unexpected messages. If your partner says he or she wants to study alone tonight, you are likely to take the statement at face value if you have a high trust level. But if you have a low trust level, you may believe your partner is going to meet someone else while you are studying in the library.

Trust in personal relationships has both a behavioral and a motivational component (Book et al., 1980). The behavioral component refers to the probability that a person will act in a trustworthy manner. The motivational component refers to the reasons a person engages in trustworthy actions. Whereas the behavioral element is important in all types of relationships, the motivational element is important in close relationships. One has to be trustworthy for the "right" reasons. As long as you trust your mechanic to charge you fairly for rebuilding your car's engine, you don't care why he or she is trustworthy. But you do care why your partner is trustworthy. For example, you want your partner to be sexually exclusive to you because he or she loves you or is attracted to you. Being faithful because of duty or because your partner can't find anyone better is the wrong motivation. Disagreements about the motivational bases for trust are often a source of conflict. "I want you because you love me, not because you need me" or "You don't really love me; you're just saying that because you want sex" are typical examples of conflict about motivation.

GIVING FEEDBACK

Self-disclosure is reciprocal. If we self-disclose, we expect our partner to self-disclose as well. As we self-disclose, we build trust; as we withhold self-disclosure, we erode trust. To withhold ourselves is to imply that we don't trust the other person, and if we don't, he or she will not trust us.

A critical element in communication is **feedback,** the ongoing process in which participants and their messages create a given result and are subsequently modified by the result (see Figure 6.1). If someone self-discloses to us, we need to respond to his or her self-disclosure. The purpose of feedback is to provide constructive information to increase self-awareness of the consequences of our behaviors toward each other.

If your partner discloses to you his or her doubts about your relationship, for example, you can respond in a number of ways:

- You can remain silent. Silence, however, is generally a negative response, perhaps as powerful as saying outright that you do not want your partner to self-disclose this type of information.
- You can respond angrily, which may convey the message to your partner that self-disclosing will lead to arguments rather than understanding and possible change.
- You can remain indifferent, responding neither negatively nor positively to your partner's self-disclosure.

FIGURE 6.1 ■ Communication Loop

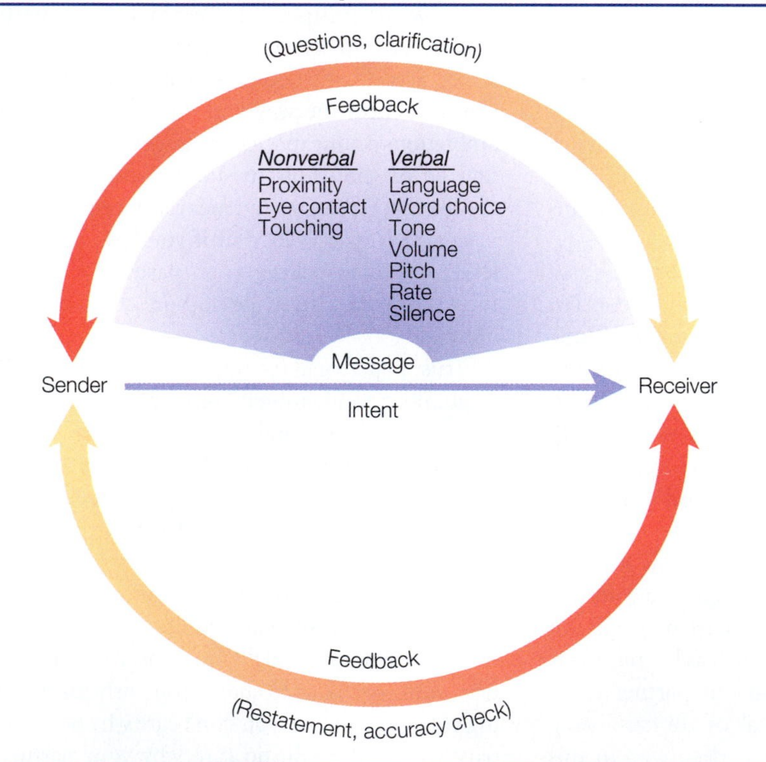

In successful communication, feedback between the sender and receiver ensures that both understand (or are trying to understand) what is being communicated. For communication to be clear, the message and the intent behind the message must be congruent. Nonverbal and verbal components must also support the intended message. Verbal aspects of communication include not only language and word choice but also characteristics such as tone, volume, pitch, rate, and periods of silence.

- You can acknowledge your partner's feelings as being valid (rather than right or wrong) and disclose how you feel in response to his or her statement. This acknowledgment and response is constructive feedback. It may or may not remove your partner's doubts, but it is at least constructive in that it opens up the possibility for change, whereas silence, anger, and indifference do not.

Some guidelines, developed by David Johnston for the Minnesota Peer Program, may help you engage in dialogue and feedback with your partner:

1. *Focus on "I" statements.* An "I" statement is a statement about your feelings: "I feel annoyed when you leave your dirty dishes on the living room floor." "You" statements tell another person how he or she is, feels, or thinks: "You are so irresponsible. You're always leaving your dirty dishes on the living room floor." "You" statements are often blaming or accusatory. Because "I" messages don't carry blame, the recipient is less likely to be defensive or resentful.
2. *Focus on behavior rather than on the person.* If you focus on a person's behavior rather than on the person, you are more likely to secure change. A person can change behaviors but not himself or herself. If you want your partner to wash his or her dirty dishes, say, "I would like you to wash your dirty dishes; it bothers me when I see them gathering mold on the living room floor." This statement focuses on behavior that can be changed. If you say, "You are such a slob; you never clean up after yourself," then you are attacking the person. He or she is likely to respond defensively: "I am not a slob. Talk about slobs, how about when you left your clothes lying in the bathroom for a week?"
3. *Focus feedback on observations rather than on inferences or judgments.* Focus your feedback on what you actually observe rather than on what you think the behavior means. "There is a towering pile of your dishes in the living room" is an observation. "You don't really care about how I feel because you are always leaving your dirty dishes around the house" is an inference that a partner's dirty dishes indicate a lack of regard. The inference moves the discussion from the dishes to the partner's caring. The question "What kind of person would leave dirty dishes for me to clean up?" implies a judgment: only a morally depraved person would leave dirty dishes around.
4. *Focus feedback on observations based on a more-or-less continuum.* Behaviors fall on a continuum. Your partner doesn't *always* do a particular thing. When you say that he or she does something sometimes or even most of the time, you are actually measuring behavior. If you say that your partner always does something, you are distorting reality. For example, there were probably times (however rare) when your partner picked up the dirty dishes. "Last week I picked up your dirty dishes three times" is a measured statement. "I always pick up your dirty dishes" is an exaggeration that will probably provoke a hostile response.
5. *Focus feedback on sharing ideas or offering alternatives rather than on giving advice.* No one likes being told what to do. Unsolicited advice often produces anger or resentment because advice implies that you know more about what a person needs to do than the other person does. Advice implies a lack of freedom or respect. By sharing ideas and offering alternatives, however, you give the other person the freedom to decide based on his or her own perceptions and goals. "You need to put away your dishes immediately after you are done with them" is advice. To offer alternatives, you might say, "Having to walk around your dirty dishes bothers me. What are the alternatives other than my watching my step? Maybe you could put them away after you finish eating, clean them up before I get home, or eat in the kitchen. What do you think?"
6. *Focus feedback according to its value to the recipient.* If your partner says something that upsets you, your initial response may be to lash back. A cathartic response may make you feel better for the time being, but it may not be useful for your partner. If, for example, your partner admits lying to you, you can respond with rage and accusations, or you can express hurt and try to find out why he or she didn't tell you the truth.
7. *Focus feedback on the amount the recipient can process.* Don't overload your partner with your response. Your partner's disclosure may touch deep, pent-up feelings in you, but he or she may not be able to comprehend all that you say. If you respond to your partner's revelation of doubts

with a listing of all the doubts you have ever experienced about yourself, your relationship, and relationships in general, you may overwhelm your partner.

8. *Focus feedback at an appropriate time and place.* Choose a time when you are not likely to be interrupted. Turn the television off and the phone answering machine on. Also, choose a time that is relatively stress free. Talking about something of great importance just before an exam or a business meeting is likely to sabotage any attempt at communication. Finally, choose a place that will provide privacy; don't start an important conversation if you are worried about people's overhearing or interrupting you. A dormitory lounge during the soaps, Grand Central Station, a kitchen teeming with kids, or a car full of friends is an inappropriate place.

MUTUAL AFFIRMATION

Good communication in an intimate relationship involves mutual affirmation, which includes three elements: (1) mutual acceptance, (2) liking each other, and (3) expressing liking in both words and actions. Mutual acceptance consists of people accepting each other as they are, not as they would like each other to be. People are who they are, and they are not likely to change in fundamental ways without a tremendous amount of personal effort, as well as a considerable passage of time. The belief that an insensitive partner will somehow magically become sensitive after marriage, for example, is an invitation to disappointment and divorce.

If you accept people as they are, you can like them for their unique qualities. Liking someone is somewhat different from being romantically involved. It is not rare for people to dislike those with whom they are romantically linked.

We also need to express our feelings of warmth, affection, and love. To one's partner, unexpressed words, actions, thoughts, kindnesses, deeds, touches, caresses, and kisses can be the same as nonexistent or unfelt ones. "You know that I love you" without the expressions of love is a meaningless statement. A simple rule of thumb for communicating love is: If you love, show love.

Mutual affirmation entails our telling others that we like them for who they are, that we appreciate the little things as well as the big things that they do. Think about how often you say to your partner, your parents, or your children, "I like you," "I love you," "I appreciate your doing the dishes," or "I like your smile." Affirmations are often most frequent during dating or the early stages of marriage or living together. As you get to know a person better, you may begin noting things that annoy you or are different from you. Acceptance turns into negation and criticism: "You're selfish," "Stop bugging me," "You talk too much," or "Why don't you clean up after yourself?"

If you have a lot of negatives in your interactions, don't feel too bad. Many of our negations are habitual. When we were children, our parents may have been negating: "Don't leave the door open," "Why can't you get better grades?" "Stand straight and pull in your stomach." How often did they affirm? Once you become aware that negations are often automatic, you can change them. Because negative communication is a learned behavior, you can unlearn it. One way is to make the decision consciously to affirm what you like; too often we take the good for granted and feel compelled to point out only the bad.

Power, Conflict, and Intimacy

The politics of family life—who has the power, who makes the decisions, who does what—can be every bit as complex and explosive as politics at the national level. **Power** is the ability or potential ability to influence another person or group. Most of the time we are not aware of the power aspects of our relationships. One reason for this is that we tend to believe that intimate relationships are based on love alone. Another reason is that the exercise of power is often subtle. When we think of power, we tend to think of coercion or force; as we shall see, however, marital power takes many forms. A final reason why we are not always aware of power is that power is not constantly exercised. It comes into play only when an issue is important to both people and they have conflicting goals.

CHANGING SOURCES OF MARITAL POWER

Traditionally, husbands have held authority over their wives. In Christianity, the subordination of wives to their husbands has its basis in the New Testament. Paul (Colossians 3:18–19) states: "Wives, submit yourselves unto your husbands, as unto the Lord." Such teachings reflected the dominant themes of ancient Greece and Rome. Western society continued to support wifely subordination to husbands. English common law stated, "The husband and wife are as one and that one is the husband." A woman assumed her husband's identity, taking his last name on marriage and living in his house.

The U.S. courts have institutionalized these power relationships. The law, for example, supports the traditional division of labor in many states, making the husband legally responsible for supporting the family and the wife legally responsible for maintaining the house and rearing the children. She is legally required to follow her husband if he moves; if she does not, she is considered to have deserted him. But if she moves and her husband refuses to move with her, she is also considered to have deserted him (Leonard and Elias, 1990).

Legal and social support for the husband's control of the family has declined since the 1920s and especially since the 1960s. An egalitarian standard for sharing power in families has taken much of its place (Sennett, 1980). The wife who works has especially gained more power in the family. She has greater influence in deciding family size and how money is to be spent.

The formal and legal structure of marriage makes the male dominant, but the reality of marriage may be quite different. Sociologist Jessie Bernard (1982) makes an important distinction between authority and power in marriage. Authority is based in law, but power is based in personality. A strong, dominant woman is likely to exercise power over a more passive man simply by the force of her personality and temperament.

If we want to see how power really works in marriage, we must look beneath the stereotypes. Women have considerable power in marriage, although they often feel that they have less than they actually do. They may fail to recognize the extent of their power; because cultural norms theoretically put power in the hands of their husbands, women may look at norms rather than at their own behavior. A woman may decide to work, even against her husband's wishes, and she may determine how to discipline the children. Yet she may feel that her husband holds the power in the relationship because he is *supposed* to be dominant. Similarly, husbands often believe that they have more power in a relationship than they actually do because they see only traditional norms and expectations.

BASES OF MARITAL POWER

Power is not a simple phenomenon. Researchers generally agree that family power is a dynamic, multidimensional process (Szinovacz, 1987). Generally speaking, no single individual is always the most powerful person in every aspect of the family. Nor is power necessarily always based on gender, age, or relationship. Power often shifts from person to person, depending on the issue.

According to J. P. French and Betram Raven (1959), there are six bases of marital power:

1. *Coercive power* is based on the fear that one partner will punish the other. Coercion can be emotional or physical. A pattern of belittling, threatening, or being physical can intimidate and threaten another. This is the least common form of power but is used in partner rape or abuse.
2. *Reward power* is based on the belief that the other person will do something in return for agreement. If, for example, your partner attempts to understand your feelings about a specific issue, he or she may expect you to do the same.
3. *Expert power* is based on the belief that one partner has greater knowledge than the other. If you believe that your partner has more wisdom about child rearing, for instance, you may defer the rewards, incentives, and discipline to him or her.
4. *Legitimate power* is based on acceptance of roles giving the other person the right to demand compliance. Gender roles are an important part of legitimacy as they give an aura to rights based on gender. Traditional gender roles legitimize male initiation in dating and female acceptance or refusal rights. Sociologists refer to legitimate power as authority.
5. *Referent power* is based on identifying with the partner and receiving satisfaction by acting

similarly. If you have great respect in your partner's communication skills and his or her ability to actively listen, provide feedback, and disclose in an honest manner, you are more likely to model yourself after him or her.

6. *Informational power* is based on the partner's persuasive explanation. If, for example, your partner refuses to use a condom, you can provide information about the prevalence and danger of STDs and AIDS.

RELATIVE LOVE AND NEED THEORY

Another way of looking at the sources of marital power is through the **relative love and need theory,** which explains power in terms of the individual's involvement and needs in the relationship. Each partner brings certain resources, feelings, and needs to a relationship. Each may be seen as exchanging love, companionship, money, help, and status with the other. What each gives and receives, however, may not be equal. One partner may be gaining more from the relationship than the other. The person gaining the most from the relationship is the one who is most dependent. Constantina Safilios-Rothschild (1970) offers this observation:

> The relative degree to which the one spouse loves and needs the other may be the most crucial variable in explaining the total power structure. The spouse who has relatively less feeling for the other may be the one in the best position to control and manipulate all the "resources" that he has in his command in order to effectively influence the outcome of decisions.

Love is a major power resource in a relationship. Those who love equally are likely to share power equally (Safilios-Rothschild, 1976). Such couples are likely to make decisions according to referent, expert, and legitimate power.

PRINCIPLE OF LEAST INTEREST

Akin to relative love and need as a way of looking at power is the **principle of least interest.** Sociologist Willard Waller (Waller and Hill, 1951) coined this term to describe the curious (and often unpleasant) situation in which the partner with the least interest in continuing a relationship has the most power in it. At its most extreme form, it is the stuff of melodrama. "I will do anything you want, Charles," Laura says pleadingly, throwing herself at his feet. "Just don't leave me." "Anything, Laura?" he replies with a leer. "Then give me the deed to your mother's house." Quarreling couples may unconsciously use the principle of least interest to their advantage. The less involved partner may threaten to leave as leverage in an argument: "All right, if you don't do it my way, I'm going." The threat may be extremely powerful in coercing a dependent partner. It may have little effect, however, if it comes from the dependent partner because he or she has too much to lose to be persuasive. The less involved partner can easily call the other's bluff.

RESOURCE THEORY OF POWER

In 1960, sociologists Robert Blood and Donald Wolfe studied the marital decision-making patterns as revealed by their sample of 900 wives. Using "final say" in decision making as an indicator of relative power, Blood and Wolfe inquired about a variety of decisions (for example, whether the wife should be employed, what type of car to buy, where to live) and who "ultimately" decided what couples should do. They noted that men tended to have more of such decision-making power and attributed this to their being the sole or larger source of the financial resources on which couples depended. They further observed that as wives share of resources increased, so did their roles in decision making (Blood and Wolfe, 1960).

The resource theory has been met with both criticism and some empirical support. By focusing so narrowly on resources, the theory overlooks other sources of gendered power. Specifically, it fails to explain the power men continue to enjoy when they are outearned by their wives (Thompson and Walker, 1989), or when they are "househusbands," and thus completely dependent on wives' incomes (Cohen and Durst, 2000). The theory has also been criticized for equating power with decision making and for ignoring that power occasionally frees one from having to make decisions. Although resources alone don't account for power, they may, in combination with other factors, influence it, especially among heterosexual couples (Blumstein and Schwartz, 1983; Schwartz, 1994).

RETHINKING FAMILY POWER

Even though women have considerable power in marriages and families, it would be a serious mistake to overlook the inequalities between husbands and wives. As feminist scholars have pointed out, major aspects of contemporary marriage point to important areas where women are clearly subordinate to men: the continued female responsibility for housework and child rearing, inequities in sexual gratification (sex is often over when the male has his orgasm), the extent of violence against women, and the sexual exploitation of children are examples.

Feminist scholars suggest several areas that require further consideration (Szinovacz, 1987). First, they believe that too much emphasis has been placed on the marital relationship as the unit of analysis. Instead, they believe that researchers should explore the influence of the larger society on power in marriage —specifically, the relationship between the social structure and women's position in marriage. Researchers could examine, for example, the relationship of women's socioeconomic disadvantages, such as lower pay and fewer economic opportunities than men, to female power in marriage.

Second, these scholars argue that many of the decisions that researchers study are trivial or insignificant in measuring "real" family power. Researchers cannot conclude that marriages are becoming more egalitarian on the basis of joint decision making about such things as where a couple goes for vacation, whether to buy a new car or appliance, or which movie to see. The critical decisions that measure power are such issues as how housework is to be divided, who stays home with the children, and whose job or career takes precedence.

Some scholars suggest that we shift the focus from marital power to family power. Researcher Marion Kranichfeld (1987) calls for a rethinking of power in a family context. Even if women's marital power may not be equal to men's, a different picture of women in families may emerge if we examine power within the entire family structure, including power in relation to children. The family power literature has traditionally focused on marriage and marital decision making. Kranichfeld, however, feels that such a focus narrows our perception of women's power. Marriage is not family, she argues, and it is in the larger family matrix that women exert considerable power. Their power may not be the same as male power, which tends to be primarily economic, political, or religious. But if *power* is defined as the ability to change the behavior of others intentionally, "women in fact have a great deal of power, of a very fundamental and pervasive nature, so pervasive, in fact, that it is easily overlooked," according to Kranichfeld (1987). She further observes:

> Women's power is rooted in their role as nurturers and kinkeepers, and flows out of their capacity to support and direct the growth of others around them through their life course. Women's power may have low visibility from a nonfamily perspective, but women are the lynchpins of family cohesion and socialization.

Recent feminist scholarship has revealed that even among self-professed "equal couples," power processes seem to favor men. Kudson-Martin and Mahoney's 1998 study of "equal couples"—where each spouse perceives the relationship to be characterized by mutual accommodation and mutual attention to each other, and where each spouse has the same ability to get cooperation from the other in meeting one's needs or wants—is a case in point. Despite the fact that couples described their relationships as equal and their roles as "non-gender specific," men wielded more power than women. Wives more than husbands made concessions to fit their daily lives around their husbands' schedules. Women were also more likely than their husbands to report worrying about upsetting or offending their spouses, to do what their spouses wanted, and to attend to their spouses' needs (Fox and Murry, 2000). It appears as if characterizing an unequal marriage as equal allows a couple to ignore real if covert power differences that might otherwise threaten their relationships (Fox and Murry, 2000).

POWER VERSUS INTIMACY

The problem with power imbalances or the blatant use of power is the negative effect on intimacy. As Ronald Sampson (1966) observes in his study of the psychology of power, "To the extent that power is the prevailing force in a relationship—whether between husband and wife or parent and child, between friends or between colleagues—to that extent love is diminished." If partners are not equal, self-disclosure may be inhibited, especially if the powerful person believes his or her power will be lessened by sharing feelings (Glazer-Malbin, 1975). Genuine intimacy appears to require equality in power relationships. Decision making in the happiest marriages seems to be based not on coercion or tit for tat but on caring, mutuality, and respect for the other person. Women or men who feel vulnerable to their mates may withhold feelings or pretend to feel what they do not. Unequal power in marriage may encourage power politics. Each partner may struggle with the other to keep or gain power.

It is not easy to change unequal power relationships after they become embedded in the overall structure of a relationship; yet they can be changed. Talking, understanding, and negotiating are the best approaches. Still, in attempting changes, a person may risk estrangement or the breakup of a relationship. He or she must weigh the possible gains against the possible losses in deciding whether change is worth the risk.

Intimacy and Conflict

Conflict between people who love each other seems to be a mystery. The coexistence of conflict and love has puzzled human beings for centuries. An ancient Sanskrit poem reflected this dichotomy:

> In the old days we both agreed
> That I was you and you were me.
> But now what has happened
> That makes you, you
> And me, me?

We expect love to unify us, but sometimes it doesn't. Two people do not become one when they love each other, although at first they may have this feeling. Their love may not be an illusion, but their sense of ultimate oneness is. In reality, they retain their individual identities, needs, wants, and pasts while loving each other—and it is a paradox that the more intimate two people become, the more likely they may be to experience conflict. But it is not conflict itself that is dangerous to intimate relationships; it is the manner in which the conflict is handled.

Conflict is natural in intimate relationships. If this is understood, the meaning of conflict changes, and it will not necessarily represent a crisis in the relationship. David and Vera Mace (1979), prominent marriage counselors, observe that on the day of marriage, people have three kinds of raw material with which to work. First, there are things they have in common—the things they both like. Second are the ways in which they are different, but the differences are complementary. Third, unfortunately, are the differences between them that are not at all complementary and that cause them to meet head-on with a big bang. In every relationship between two people, there are a great many of those kinds of differences. So when we move closer together to each other, those differences become disagreements. The presence of conflict within a marriage or family does not necessarily indicate that love is going or gone. It may mean just the opposite.

DID YOU KNOW?

When the communication patterns of newly married African Americans and whites were examined, couples who believed in avoiding marital conflict were less happy two years later than those who confronted their problems (Crohan, 1992).

BASIC VERSUS NONBASIC CONFLICTS

Relationships suffer from two types of conflict—basic and nonbasic—which have different effects on relationship quality and stability. Basic conflicts challenge the fundamental assumptions or rules of a relationship, thus leading to the possible end of the relationship. Nonbasic conflicts are more common and less consequential; couples learn to live with them.

Basic Conflicts

Basic conflicts revolve around carrying out marital roles and the functions of marriage and the family, such as providing companionship, working, and rearing children. It is assumed, for example, that a husband and wife will have sexual relations with each other. But if one partner converts to a religious sect that forbids sexual interaction, a basic conflict is likely to occur because the other spouse considers sexual interaction part of the marital premise. No room for compromise exists in such a matter. If one partner cannot convince the other to change his or her belief, the conflict is likely to destroy the relationship. Similarly, despite recent changes in family roles, it is still expected that the husband will work to provide for the family. If he decides to quit work altogether and not function as a provider in any way, he is challenging a basic assumption of marriage. His partner is likely to feel that his behavior is unfair. Conflict ensues. If he does not return to work, his wife is likely to leave him.

Nonbasic Conflicts

Nonbasic conflicts do not strike at the heart of a relationship. The husband wants to change jobs and move to a different city, but the wife may not want to move. This may be a major conflict, but it is not a basic one. The husband is not unilaterally rejecting his role as a provider. If a couple disagree about the frequency of sex, the conflict is serious but not basic because both agree on the desirability of sex in the relationship. In both cases, resolution is possible.

Experiencing and Managing Conflict

Differences and conflicts are part of any healthy relationship. If we handle conflicts in a healthy way, they can help solidify our relationships. But conflicts can go on and on, consuming the heart of a relationship, turning love and affection into bitterness and hatred. In the following section, we will look at ways of resolving conflict in constructive rather than destructive ways. In this manner, we can use conflict as a way of building and deepening our relationships.

DEALING WITH ANGER

Differences can lead to anger, and anger transforms differences into fights, creating tension, division, distrust, and fear. Most people have learned to handle anger by either venting or suppressing it. David and Vera Mace (1980) suggest that many couples go through a love-anger cycle. When a couple comes close to each other, they may experience conflict, and they recoil in horror, angry at each other because just at the moment they were feeling close, their intimacy was destroyed. Each backs off; gradually they move closer again until another fight erupts, driving them away from each other. After a while, each learns to make a compromise between closeness and distance to avoid conflict. They learn what they can reveal about themselves and what they cannot.

Another way of dealing with anger is to suppress it. Suppressed anger is dangerous because it is always there, simmering beneath the surface. Ultimately, it leads to resentment, that brooding, low-level hostility that poisons both the individual and the relationship.

Anger can be dealt with in a third way; when conflict escalates into violence. Especially in a culture that cloaks families in privacy, surrounds people with beliefs that legitimize violence, and gives them the sense that they have a right to influence what their loved ones do, escalating anger can result in assault, injury, and even death. Given the relative power of men over women and adults over children, threats against one person's supposed advantage may provoke especially harsh reactions. We will look closely at the causes, context, and consequences of family violence in Chapter 13.

Finally and most constructively, anger can be recognized as a symptom of something that needs to be changed. If we see anger as a symptom, we realize that what is important is not venting or suppressing the anger but finding its source and eliminating it. David and Vera Mace (1980) offer this suggestion:

> When your disagreements become conflict, the only thing to do is to take anger out of it, because when you are angry you cannot resolve a conflict. You cannot really hear the other person because you are just waiting to fire your shot. You cannot be understanding; you cannot be empathetic when you are angry. So you have to take the anger out, and then when you have taken the anger out, you are back again with a disagreement. The disagreement is still there, and it can cause another disagreement and more anger unless you clear it up. The way to take the anger out of disagreements is through negotiation.

We can learn to use conflict as a way to build and deepen our relationships.

© Laurie DeVault Photography

CONFLICT RESOLUTION AND MARITAL SATISFACTION

Happy couples tend to act in positive ways to resolve conflicts, such as changing behaviors (putting the cap on the toothpaste rather than denying responsibility) and presenting reasonable alternatives (purchasing toothpaste in a dispenser). Unhappy or distressed couples, in contrast, use more negative strategies in attempting to resolve conflicts (if the cap off the toothpaste bothers you, then *you* put it on). A study of happily and unhappily married couples found distinctive communication traits as these couples tried to resolve their conflicts (Ting-Toomey, 1983). The communication behaviors of happily married couples displayed the following traits:

- *Summarizing:* Each person summarized what the other said: "Let me see if I can repeat the different points you were making."
- *Paraphrasing:* Each put what the other said into his or her own words: "What you are saying is that you feel bad when I don't acknowledge your feelings."
- *Validation:* Each affirmed the other's feelings: "I can understand how you feel."
- *Clarification:* Each asked for further information to make sure that he or she understood what the other was saying: "Can you explain what you mean a little bit more to make sure that I understand you?"

In contrast, unhappily married couples displayed the following reciprocal patterns:

- *Confrontation:* Both partners confronted each other: "You're wrong!" "Not me, buddy. It's you who's wrong."
- *Confrontation and defensiveness:* One partner confronted while the other defended himself or herself: "You're wrong!" "I only did what I was supposed to do."
- *Complaining and defensiveness:* One partner complained while the other was defensive: "I work so hard each day to come home to this!" "This is the best I can do with no help."

Attachment style (discussed in Chapter 5) seems to influence the way conflict is expressed in relationships (Pistole, 1989). In contrast to anxious/ambivalent and avoidant adults, secure adults are more satisfied in their relationships and use conflict strategies that focus on maintaining the relationship. Helping the relationship stay cohesive is more important than "winning" the battle. Secure adults are more likely to compromise than are anxious/ambivalent adults, and anxious/ambivalent adults are more likely than avoidant adults to give in to their partners' wishes, whether they agree with them or not.

Conflict Resolution across Relationship Types

All couple relationships experience conflict. Using self-report and partner-report data, Kurdek (1994a) explored how conflicts were handled by 75 gay, 51 lesbian, 108 married nonparent, and 99 married parent couples. Essentially, the differences across couple type were less impressive than were the similarities. The four types of couples did not significantly differ in their level of ineffective arguing, and there were no noteworthy differences in their styles of conflict resolution as measured by the Conflict Resolution Styles Inventory (CRSI). The CRSI includes four styles of conflict resolution: (1) *positive problem solving* (including negotiation and compromise), (2) *conflict engagement* (e.g., personal attacks), (3) *withdrawal* (refusing to further discuss an issue), and (4) *compliance* (e.g., giving in). Ratings were obtained from both partners about themselves and the other partner. There was little indication that the frequency with which conflict resolution styles were utilized varied across couple type. As Kurdek (1994a) notes, there is similarity in relationship dynamics across couple types.

COMMON CONFLICT AREAS: SEX, MONEY, AND HOUSEWORK

Even if, as the Russian writer Leo Tolstoy suggested, every unhappy family is unhappy in its own way, marital conflicts still tend to center around certain issues, especially communication, children, sex, money, personality differences, how to spend leisure time, in-laws, infidelity, and housekeeping. In this section, we focus on three areas: sex, money, and housework. Then we discuss general ways of resolving conflicts.

Fighting about Sex

Fighting and sex can be intertwined in several different ways (Strong and DeVault, 1997). A couple can have a specific disagreement about sex that leads to a fight. One person wants to have sexual intercourse and the other does not, so they fight. A couple can have an indirect fight about sex. The woman does not have an orgasm, and after intercourse, her partner rolls over and starts to snore. She lies in bed feeling angry and frustrated. In the morning she begins to fight with her partner over his not doing his share of the housework. The housework issue obscures why she is really angry. Sex can also be used as a scapegoat for nonsexual problems. A husband is angry that his wife calls him a lousy provider. He takes it out on her sexually by calling her a lousy lover. They fight about their lovemaking rather than about the issue of his provider role. A couple can fight about the wrong sexual issue. A woman may berate her partner for being too quick during sex, but what she is really frustrated about is that he is not interested in oral sex with her. She, however, feels ambivalent about oral sex ("Maybe I smell bad"), so she cannot confront her partner with the real issue. Finally, a fight can be a cover-up. If a man feels sexually inadequate and does not want to have sex as often as his partner, he may pick a fight and make his partner so angry that the last thing she would want to do is to have sex with him.

In power struggles, sexuality can be used as a weapon, but this is generally a destructive tactic (Szinovacz, 1987). A classic strategy for the weaker person in a relationship is to withhold something that the more powerful one wants. In male-female struggles, this is often sex. By withholding sex, a woman gains a certain degree of power. A small minority of men also use sex in its most violent form: They rape (including date rape and marital rape) to overpower and subordinate women. In rape, aggressive motivations displace sexual ones.

It is hard to tell during a fight if there are deeper causes than the one about which a couple is currently fighting. Are you and your partner fighting because you want sex now and your partner doesn't? Or are there deeper reasons involving power, control, fear, or inadequacy? If you repeatedly fight about sexual issues without getting anywhere, the ostensible cause may not be the real one. If fighting does not clear the air and make intimacy possible again, you should look for other reasons for the fights. It may be useful to talk with your partner about why the fights do not seem to accomplish anything. Step back and look at the circumstances of the fight; what patterns occur; and how each of you feels before, during, and after a fight.

Money Conflicts

An old Yiddish proverb addresses the problem of managing money quite well: "Husband and wife are the same flesh, but they have different purses." Money is a major source of marital conflict. Intimates differ about spending money probably as much as, or more than, any other single issue.

WHY PEOPLE FIGHT ABOUT MONEY. Couples disagree or fight over money for a number of reasons. One of the most important has to do with power. Earning wages has traditionally given men power in families. A woman's work in the home has not been rewarded by wages. As a result, full-time homemakers have been placed in the position of having to depend on their husbands for money. In such an arrangement, if there are disagreements, the woman is at a disadvantage. If she is deferred to, the old cliché "I make the money but she spends it" has a bitter ring to it. As women increasingly participate in the workforce, however, power relations within families are shifting. Studies indicate that women's influence in financial and other decisions increases if they are employed outside the home.

Another major source of conflict is allocation of the family's income. Not only does this involve deciding who makes the decisions but it also includes setting priorities. Is it more important to pay a past-due bill or to buy a new television set to replace the broken one? Is a dishwasher a necessity or a luxury? Should money be put aside for long-range goals, or should immediate needs (perhaps those your partner calls "whims") be satisfied? Setting financial priorities plays on each person's values and temperament; it is affected by basic aspects of an individual's personality. A miser probably cannot be happily married to a spendthrift. Yet we know so little of our partner's attitudes toward money before marriage that a miser might very well marry a spendthrift and not know it until too late.

Dating relationships are a poor indicator of how a couple will deal with money matters in marriage. Dating has clearly defined rules about money: Either the man pays, both pay separately, or they take turns paying. In dating situations, each partner is financially independent of the other. Money is not pooled, as it usually is in a committed partnership or marriage. Power issues do not necessarily enter spending decisions because each person has his or her own money. Differences can be smoothed out fairly easily. Both individuals are financially independent before marriage but financially interdependent after marriage. Even cohabitation may not be an accurate guide to how a couple would deal with money in marriage, as cohabitors generally do not pool all (or even part) of their income. It is the working out of financial interdependence in marriage that is often so difficult.

TALKING ABOUT MONEY. Talking about money matters is difficult. People are very secretive about money. It is considered poor taste to ask people how much money they make. Children often do not know how much money is earned in their families; sometimes spouses don't know either. One woman remarked that it is easier to talk with a partner about sexual issues than about money matters: "Money is the last taboo," she said. But, as with sex, our society is obsessed with money.

We find it difficult to talk about money for several reasons. First of all, we don't want to appear to be unromantic or selfish. If a couple is about to marry, a discussion of attitudes toward money may lead to disagreements, shattering the illusion of unity or selflessness. Second, gender roles make it difficult for women to express their feelings about money because women are traditionally supposed to defer to men in financial matters. Third, because men tend to make more money than women, women feel that their right to disagree about financial matters is limited. These feelings are especially prevalent if the woman is a homemaker and does not make a financial contribution, but they devalue her child-care and housework contributions.

Housework and Conflict

The division of responsibility for housework is one of the most significant issues faced by dual-earner couples (Kluwer, Heesink, and Van De Vliert, 1997). It can become a source of tension and conflict within marriage (Hochschild, 1989). Part of this is an understandable consequence of the inequality in each spouse's contribution; most men do not do much housework. Whether or not they are employed outside the home, and whether there are children in the home or not, wives bear the bulk of housework responsibility. A husband's lack of involvement can create resentment and affect the levels of both conflict and happiness in a marriage. In fact, in her acclaimed study of the division of housework among

50 dual-earner couples, Arlie Hochschild (1989) argued that men's level of sharing "the second shift" (that is, unpaid domestic work and child care) influenced the levels of marital happiness couples enjoyed and their relative risk of divorce. This held true whether couples were traditional or egalitarian in their views of marriage.

In a study of 54 Dutch couples, Kluwer, Heesink, and Van De Vliert (1996) found that conflict about household work was related to wives' dissatisfactions with their and their husbands' relative contributions and expenditures of time. They note that 72 percent of the wives preferred to do less than they actually did; that is, when they spent more time on housework than they preferred to, they were dissatisfied. They also tended to be dissatisfied if they perceived their husbands spending less time than they preferred them to on housework. Just over half (52 percent) of the wives wished their husbands would do more housework than they actually did (Kluwer, Heesink, and Van De Vliert, 1996).

How much each spouse contributes to the household is only the more observable aspect of the "politics of housework." Additionally, couples must reach agreements about *standards, schedules,* and *management* of housework. Conflicts about standards are struggles over whose standards will predominate: Who gets to decide whether things are "clean enough"? Similarly, disputes about schedules reflect whose time is more valuable, which partner works around the other's sense of priorities. Who waits for whom? Finally, arguments about who bears responsibility for organizing, initiating, or overseeing housework tasks are also disputes about who will have to ask the other for help, carry more responsibility in his or her head, and risk refusal from an uncooperative partner.

Thus, housework conflicts have both practical and symbolic dimensions. Practically speaking, there are things that somehow must get done in order for households to run smoothly and families to function efficiently. Couples must decide who shall do them, and how and when they should be done. On a more symbolic level, disputes over housework may be experienced as conflicts about the level of commitment each spouse feels toward their marriage. Because marriage symbolizes the union of two people who share their lives, work together, consult each other, and take each other's feelings and needs into consideration, resisting housework or doing it only under duress may be seen as a less than equal commitment. We will look more in detail at the dynamics surrounding the division of housework in Chapter 12.

The absence of overt conflict over the allocation of tasks and time does not mean that there is no conflict. It means only that the conflict is not openly expressed. Wives in more traditional marriages are more likely than wives in egalitarian relationships to avoid conflict over housework even if they are dissatisfied with their domestic arrangements. They may withdraw from discussions of the division of labor as a way of avoiding the issue. Because egalitarian couples may engage in more open discussion and conflict over housework responsibities, such conflict gives them more opportunity to establish a solution (Kluwer, Heesink, and Van De Vliert, 1997).

Unequal involvement in housework and childcare can breed resentment.

RESOLVING CONFLICTS

There are a number of ways to end conflicts and solve problems. You can give in, but unless you believe that the conflict ended fairly, you are likely to feel resentful. You can try to impose your will through the use of power, force, or the threat of force, but using power to end conflict leaves your partner with the bitter taste of injustice. Less productive conflict resolution strategies include *coercion* (e.g., threats, blame, sarcasm), *manipulation* (e.g., attempting to make your partner feel guilty) and *avoidance* (Regan, 2003).

More positive strategies for resolving conflict, include *supporting your partner* (through active listening, compromise, or agreement), *assertion* (e.g., clearly stating your position, keeping the conversation on topic), and *reason* (e.g., the use of rational argument and the consideration of alternatives) (Regan, 2003). Finally, you can end the conflict through negotiation. In negotiation, both partners sit down and work out their differences until they come to a mutually acceptable agreement (see Figure 6.2). Conflicts can be solved through negotiation in three primary ways: (1) agreement as a gift, (2) bargaining, and (3) coexistence.

FIGURE 6.2 ■ Family Problem-Solving Loop

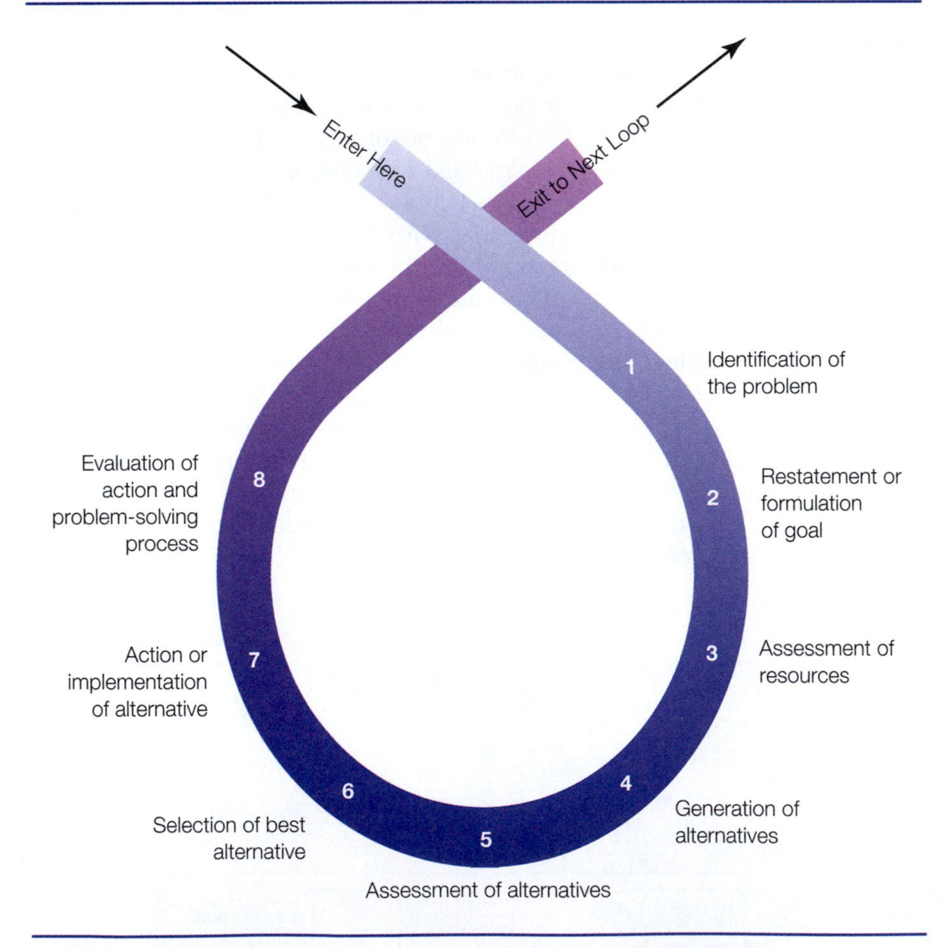

Most family problem solving occurs in the ebb and flow of daily family events. Though family dynamics and transition take various forms, it is interesting to note which types might have relevance for family issues.

SOURCE: D. Kieren, T. O. Maguire, and N. Hurlbut. "A Marker Method to Test a Phasing Hypothesis in Family Problem-Solving Interaction." *Journal of Marriage and the Family* 58, 2 (May 1996): 442–455. Copyright © 1996 by the National Council on Family Relations. Used by permission.

Understanding Yourself

IN TIMES OF TROUBLE: SEEKING OUTSIDE HELP

In spite of good intentions and communication skills, we may not be able to resolve our relationship problems on our own. Accepting the need for professional assistance may be a significant first step toward reconciliation and change. Experts advise counseling when communication is hostile, conflict goes unresolved, individuals cannot resolve their differences, and/or a partner is thinking about leaving.

Marriage and partners counseling are professional services whose purpose is to assist individuals, couples, and families gain insight into their motivations and actions within the context of a relationship while providing tools and support to make positive changes. A skilled counselor offers objective, expert, and discreet help. Much of what counselors do is crisis or intervention oriented.

It may be more valuable and perhaps more effective to take a preventive approach and explore dynamics and behaviors before they cause more significant problems. This may occur at any point in relationships: during the engagement, prior to an anticipated pregnancy, or at the departure of a last child.

Each state has its own degree and qualifications for marriage counselors. The American Association for Marital and Family Therapy (AAMFT) is one association that provides proof of education and special training in marriage and family therapy. Graduate education from an accredited program in either social work, psychology, psychiatry, or human development coupled with a license in that field ensures both education and training as well as offering the consumer recourse if questionable or unethical practices occur. This recourse is, however, only available if the practitioner holds a valid license issued by the state in which he or she practices. Mental health workers belong to any one of several professions:

- *Psychiatrists* are licensed medical doctors who, in addition to completing at least six years of post-baccalaureate medical and psychological training, can prescribe medication.
- *Clinical psychologists* have usually completed a Ph.D., which requires at least six years of postbaccalaureate course work. A license requires additional training and the passing of state boards.
- *Marriage and family counselors* typically have a master's degree and additional training to be eligible for state board exams.
- *Social workers* have master's degrees requiring at least two years of graduate study plus additional training to be eligible for state board exams.
- *Pastoral counselors* are clergy who have special training in addition to their religious studies.

Financial considerations may be one consideration when selecting which one of the above to see. Typically, the more training a professional has, the more he or she will charge for services.

Finding a therapist can be done via a referral from a physician, school counselor, family, friend, clergy, or by the state department of mental health. In any case, it is important to meet personally with the counselor in order to decide if he or she is right for you. Besides inquiring about his or her basic professional qualifications, it is important to feel comfortable with this person, to decide whether your value and belief systems are compatible, and to assess his or her psychological orientation. Shopping for the right counselor may be as important a decision as deciding to enter counseling in the first place.

Marriage or partnership counseling has a wide variety of approaches: Individual counseling focuses on one partner at a time; joint marital counseling involves both people in the relationship; and family systems therapy includes as many family members as possible. Regardless of the approach, all share the premise that in order to be effective, those involved should be willing to cooperate. Additional logistical questions, such as the number and frequency of sessions, depend on the type of therapy.

At any time during the therapeutic process, you have the right to stop or change therapists. Before you do, however, ask yourself whether your discomfort is personal or if it has to do with the techniques or personality of the therapist. Discuss this issue with the therapist prior to making a change. Finally, if you believe that your therapy is not benefiting you, change therapists.

Agreement as a Gift

If you and your partner disagree on an issue, you can freely agree with your partner as a gift. If you want to go to the Caribbean for a vacation and your partner wants to go backpacking in Alaska, you can freely agree to go to Alaska. An agreement as a gift is different from giving in. When you give in, you do something you don't want to do. When you agree without coercion or threats, the agreement is a gift of love, given freely without resentment. As in all exchanges of gifts, there will be reciprocation. Your partner will be more likely to give you a gift of agreement. This gift of agreement is based on referent power, discussed earlier.

Bargaining

Bargaining in relationships—the process of making compromises—is different from bargaining in the marketplace or in politics. In relationships, you want to get the most equitable deal for both you and your partner, not just the best deal for yourself. At all points during the bargaining process, you need to keep in mind what is best for the relationship as well as for yourself, and you need to trust your partner to do the same. In a marriage, both partners need to win. The result of conflict in a marriage should be to solidify the relationship, not to make one partner the winner and the other the loser. To achieve your end by exercising coercive power or withholding love, affection, or sex is a destructive form of bargaining. If you get what you want, how will that affect your partner and the relationship? Will your partner feel you are being unfair and become resentful? A solution has to be fair to both, or it won't enhance the relationship.

Coexistence

Sometimes differences can't be resolved, but they can be lived with. If a relationship is sound, differences can be absorbed without undermining the basic ties. All too often we regard a difference as a threat rather than as the unique expression of two personalities. Rather than being driven mad by the cap left off the toothpaste, perhaps we can learn to live with it.

If you can't talk about what you like and what you want, there is a good chance that you won't get either one. Communication is the basis for good relationships. Communication and intimacy are reciprocal: Communication creates intimacy, and intimacy in turn helps create good communication.

If we fail to communicate, we are likely to turn our relationships into empty facades, with each person acting a role rather than revealing his or her deepest self. But communication is learned behavior. If we have learned *how not to* communicate, we can learn *how to* communicate. Communication will allow us to maintain and expand ourselves and our relationships.

SUMMARY

- Communication includes both verbal and nonverbal communication. The functions of nonverbal communication are to convey interpersonal attitudes, express emotions, and handle the ongoing interaction. For communication to be clear, verbal and nonverbal messages must agree. *Proximity,* eye contact, and touch are important forms of nonverbal communication. Levels of touching differ between cultures and ethnic groups.
- There are gender differences in nonverbal, verbal, and partnership communication. Wives tend to send clearer messages. Husbands may give neutral messages or withdraw, whereas wives tend to give more positive or negative messages. Also, wives tend to set the emotional tone and escalate arguments more than husbands do.
- Researchers are finding that how well a couple communicates before marriage can be an important predictor of later marital satisfaction. *Self-disclosure* prior to marriage is related to relationship satisfaction later. Whether a couple's premarital interactions are basically negative or positive can also predict later marital satisfaction.

- Research indicates that happily married couples (1) are willing to engage in conflict in nondestructive ways, (2) have less frequent conflict and spend less time in conflict, (3) disclose private thoughts and feelings to partners, (4) express equal levels of affection, (5) spend more time together, and (6) accurately encode and decode messages.
- Virginia Satir identified four styles of miscommunication; (1) placaters, (2) blamers, (3) computers, and (4) distractors. Placaters are passive, helpless, and always agreeable; blamers act superior, are often angry, do not listen, and try to escape responsibility; computers are correct, reasonable, and expressionless; and distractors are frenetic and tend to change the subject.
- Barriers to communication include the traditional male gender role (because it discourages the expression of emotion); personal reasons, such as feelings of inadequacy; and the fear of conflict. To express ourselves, we need to be aware of our own feelings. We prevent self-awareness through suppressing, denying, and projecting feelings. A first step toward self-awareness is realizing that our feelings are neither good nor bad but simply emotional states.
- According to some researchers, both low and high levels of self-disclosure may be related to lower levels of marital satisfaction. This is referred to as the curvilinear model. Other researchers take a more linear approach, maintaining that high levels of self-disclosure result in higher levels of marital satisfaction.
- *Trust* is the belief in the reliability and integrity of a person. In order for trust to develop, certain things must occur: (1) a relationship has to exist and have the likelihood of continuing, (2) we must be able to predict how our partner will likely behave, and (3) our partner must also have other acceptable options available to him or her. Trust is critical in close relationships because self-disclosure requires trust; how much you trust a person influences the way you are likely to interpret ambiguous or unexpected messages from him or her.
- *Feedback* is the ongoing process in which participants and their messages create a given result and are subsequently modified by the result. Constructive feedback includes (1) focusing on "I" statements, (2) focusing on behavior rather than on the person, (3) focusing feedback on observations rather than on inferences or judgments, (4) focusing feedback on the observed incidence of behavior, (5) focusing feedback on sharing ideas or offering alternatives rather than on giving advice, (6) focusing feedback according to its value to the recipient, (7) focusing feedback on the amount the recipient can process, and (8) focusing feedback at an appropriate time and place.
- The basis of good communication in a relationship is mutual affirmation. Mutual affirmation includes mutual acceptance, mutual liking, and expressing liking in words and actions.
- A common pattern among married couples is what's known as *demand-withdraw communication.* One partner will raise an issue for discussion and the other partner withdraws from the conversation instead of attempting to communicate.
- *Power* is the ability or potential ability to influence another person or group. Traditionally, legal as well as de facto power rested in the hands of the husband. Recently, wives have been gaining more actual power in relationships, although the power distribution still remains unequal. The six bases of marital power are (1) coercive, (2) reward, (3) expert, (4) legitimate, (5) referent, and (6) informational. Other theories of power include the *relative love and need theory,* the *principle of least interest,* and the *resource theory of power.*
- Conflict is natural in intimate relationships. Types of conflict include basic versus nonbasic conflicts and situational versus personality conflicts. Basic conflicts may threaten the foundation of a marriage because they challenge fundamental rules; nonbasic conflicts do not threaten basic assumptions and may be negotiable. Situational conflicts are based on specific issues. Personality conflicts are unrealistic conflicts based on the need of the partner or partners to release pent-up feelings or on their fundamental personality differences.
- Major sources of conflict include sex, money, and housework. Conflicts about sex can be specific disagreements about sex, indirect disagreements in which a partner feels frustrated or angry and

takes it out in sexual ways, disagreements about the wrong sexual issue, or arguments that are ostensibly about sex but that are really about nonsexual issues. Money conflicts occur because of power issues, disagreements over the allocation of resources, or differences in values. Conflict over housework is often about how much each person does. It can also focus on other issues, such as time, standards, or responsibility for managing what gets done.

- Conflict resolution may be achieved through negotiation in three ways: (1)agreement as a freely given gift, (2) bargaining, and (3) coexistence.
- People usually handle anger in relationships by suppressing or venting it. Anger, however, makes negotiation difficult. When anger arises, it is useful to think of it as a signal that change is necessary.
- Happily married couples use certain techniques to resolve conflict, including summarizing, paraphrasing, validation, and clarification. Unhappy couples use confrontation, confrontation and defensiveness, and complaining and defensiveness.

KEY TERMS

demand-withdraw communication 174
family rules 168
feedback 180
hierarchy of rules 168
honeymoon effect 172
meta-rules 168
power 182
principle of least interest 184
proximity 167
relative love and need theory 184
rules 168
self-disclosure 172
trust 178

SUGGESTED READINGS

Cupach, W. R., and B. H. Spitzberg, eds. *The Darkside of Interpersonal Communication.* Hillsdale, NJ: Lawrence Erlbaum, 1994. A discussion of conversational dilemmas, distressed marital relationships, and other issues that stress families.

Gottman, John. *Why Marriages Succeed or Fail.* New York: Simon and Schuster, 1994. A leading authority examines factors associated with positive or negative marital outcomes.

Notarius, Clifford, and Howard Markman. *We Can Work It Out: Making Sense of Marital Conflict.* New York: Putnam, 1993. Tips on improving communication and resolving conflicts from leading researchers on marital communication.

Satir, Virginia. *The New Peoplemaking.* Rev. ed. Palo Alto, CA: Science and Behavior Books, 1988. One of the most influential (and easy-to-read) books of the last 25 years on communication and family relationships.

Tannen, Deborah. *You Just Don't Understand: Women and Men in Conversation.* New York: Morrow, 1990. A bestselling, intelligent, and lively discussion of how females use communication to achieve intimacy and males use communication to achieve independence.

Ting-Toomey, Stella, and Felipe Korzenny, eds. *Cross-Cultural Interpersonal Communication.* Newbury Park, CA: Sage Publications, 1991. A groundbreaking collection of scholarly essays on communication and relationships among different ethnic and cultural groups, including African-American, Latino, Korean, and Chinese ethnic groups and cultures.

RESOURCES ON THE INTERNET

Companion Web Site for This Book

http://sociology.wadsworth.com/strong/marriage9e
Gain an even better grasp on this chapter by going to the companion Web site to take one of the Tutorial Quizzes, use the Flash Cards to master key terms, or check out the many other study aids you'll find there. Visit the Marriage and Family Resource Center on the site. You'll also find special features such as GSS Data and Census 2000 information that will put data and resources at your fingertips to help you with that special project or to do some research on your own.

InfoTrac College Edition: Search Word Summary

communication in marriage
conflict
power
conflict management
interpersonal confrontation
self-disclosure

To learn more about these central topics in the study of the family, you can conduct an electronic search using InfoTrac College Edition. To aid in your search and to gain useful tips, see the Student Guide to InfoTrac College Edition that you can access through the companion Web site for this book.

Preview

To gain a sense of what you already know about the material covered in this chapter, answer "True" or "False" to the statements below.

1. Looking for a mate can be compared to shopping for goods in a market. True or false?
2. Generally, the most important factor in judging someone at the first meeting is how he or she looks. True or false?
3. There is a significant shortage of single eligible African-American men that makes marriage less likely for African-American women. True or false?
4. If a woman asks a man out on a first date, it is generally a sign that she wants to have sex with him. True or false?
5. The lesbian subculture values being single and unattached more than being involved in a stable relationship. True or false?
6. Singles, compared to their married peers, tend to be more dependent on their parents. True or false?
7. An important dating problem that men cite is their own shyness. True or false?
8. Cohabitation has become part of the courtship process among young adults. True or false?
9. Compared to married couples, cohabiting couples have a more accepting attitude toward infidelity. True or False?
10. Previously married cohabitants are more likely than never-married cohabitors to view living together as a test of marital compatibility. True or false?

"The opposite of loneliness is not togetherness. It is intimacy."

RICHARD BACH

CHAPTER 7

Singlehood, Pairing, and Cohabitation

Outline

Do you know what this is?

Real Life Juliet Seeks Romeo: I have been searching the world over, looking for my true love. I am a friendly, ambitious, compassionate, hardworking female, who enjoys music, dancing, travel, and the beach. Looking for someone who wants to share a movie, dinner, a laugh, and maybe a lifetime. I know you're out there somewhere.

Of course, we all know this is a personal ad, one of the many found each day in newspapers and magazines, or on the Internet. Along with dating services, computer matchmakers, or singles clubs, such personal ads represent some of the more recent ways Americans attempt to find their "one and only." Recent television programs have pushed such attempts into previously uncharted water. On February 15, 2000, the Fox Network aired *Who Wants to Marry a Multi-Millionaire?* In this special two-hour program, 50 women, selected from a pool of thousands, competed with each other in beauty pageant-like competitions to see who would be selected by, proposed to, and married, on camera, to an anonymous, wealthy bachelor. Although the newlyweds Rick and Darva quickly went their separate ways and annulled their marriage, one wondered, what could be next? Now we know: two *Joe Millionaires, Who Wants to Marry My Dad?, Bachelor,* and *Married by America*. Each of these "reality" shows has tried to capitalize on our age-old fascination with how people get together.

There is considerable social science interest, too, in understanding how people find their spouses or partners. In addition, researchers have studied *who* we choose and *why* we choose those particular individuals. In this chapter, we not only look at the general rules by which we choose partners but also examine dating, romantic relationships, living together, and the singles world. Over the last several decades, many aspects of pairing, such as the legitimacy of premarital intercourse and cohabitation, have changed considerably, radically affecting marriage. Today, large segments of American society accept and approve of both premarital sex and cohabitation. Marriage has lost its exclusiveness as the only legitimate relationship in which people can have sex and share their everyday lives. Increasing numbers of Americans experience both premarital sex and cohabitation in their lives. Additionally, more people are forgoing marriage altogether and are instead remaining single. These issues, too, will be examined in this chapter.

PREVIEW ANSWERS

1 True
2 True
3 True
4 False
5 False
6 False
7 True
8 True
9 True
10 True

Choosing Partners

How do we choose the people we date, live with, or marry? At first glance, it seems that we choose them on the basis of love. In theory, we are free to select those people with whom we fall in love but other factors enter into the process and our choices are somewhat limited by rules of mate selection.

There is a game you can play if you understand some of the principles of mate selection in our culture. Without ever having met a friend's new boyfriend or girlfriend, you can deduce many things about him or her, using basically the same method of deductive reasoning that Sherlock Holmes used to astound Dr. Watson. For example, if a female friend at college has a new boyfriend, it is safe to guess that he is about the same age or a little older, probably taller, and a college student. Furthermore, he is probably about as physically attractive as your friend (if not, they will probably break up within six months); his parents probably are of the same ethnic group and social class as hers; and most likely, he is about as intelligent as your friend. If a male friend has a new girlfriend, many of the same things apply, except that she is probably the same age or younger and shorter than he is. After you have described your friend's new romantic interest, don't be surprised if he or she exclaims, "Good grief, Holmes, how did you know that?" Of course, not every characteristic may apply, but you will probably be correct in most instances. These are not so much guesses as deductions based on the principle of homogamy, discussed later in this chapter.

THE MARKETPLACE OF RELATIONSHIPS

Bargaining and exchange affect the process of choosing partners. We select each other in a kind of *marketplace of relationships.* We use the notion of a "marketplace" to convey the fact that, as in a commercial marketplace, when we form relationships we exchange goods. Unlike a real marketplace, however, the relationship marketplace is a process, not a place, in which *we* are the goods that are exchanged. Each of us has certain resources—such as socioeconomic status, looks, and personality—that determine our marketability. We bargain with these resources. We size ourselves up and rank ourselves as a good deal, an average package, or something to be "remaindered"; we do the same with potential dates and, ultimately, mates.

PHYSICAL ATTRACTIVENESS: THE HALO EFFECT, RATING, AND DATING

The Halo Effect

Imagine yourself at a party, unattached. You notice that someone is standing next to you as you reach for some chips. He or she says hello. In that moment, you have to decide whether to engage him or her in conversation. On what basis do you make that decision? Is it looks, personality, style, sensitivity, intelligence, or something else?

If you're like most of us, you consciously or unconsciously base this decision on appearance. If you decide to talk to the person, you probably formed a positive opinion about his or her appearance. In other words, he or she looked "cute," looked like a "fun person," gave a "good first impression," or seemed "interesting." Elaine Hatfield and Susan Sprecher (1986) explain the process:

> Appearance is the sole characteristic apparent in every social interaction. Other information may be more meaningful but far harder to ferret out. People do not have their IQs tattooed on their foreheads, nor do they display their diplomas prominently about their persons. Their financial status is a private matter between themselves, their bankers, and the Internal Revenue Service.

Physical attractiveness is particularly important during the initial meeting and early stages of a relationship. If you don't know anything else about a person, you tend to judge him or her on appearance.

REFLECTIONS

How important are looks to you? Have you ever mistakenly judged someone by his or her looks? How did you discover your error? Have you ever made trade-offs in a relationship? What did you and your partner trade?

Most people would deny that they are attracted to others just because of their looks. We do so unconsciously, however, by inferring qualities based on looks. This inference is based on a **halo effect**—the assumption that good-looking people possess more desirable social characteristics than unattractive people. In a well-known experiment (Dion et al., 1972), students were shown pictures of attractive people and asked to describe what they thought these people were like. Attractive men and women were assumed to be more sensitive, kind, warm, sexually responsive, strong, poised, and outgoing than others; they were assumed to be more exciting and to have better characters than "ordinary" people. Research indicates that overall, the differences between perceptions of very attractive and average people are minimal. It is when attractive and average people are compared to those considered to be unattractive that there are pronounced differences, with those perceived as unattractive being rated more negatively (Hatfield and Sprecher, 1986).

The Rating and Dating Game

In casual relationships, the physical attractiveness of a romantic partner is especially important. Hatfield and Sprecher (1986) suggest three reasons why people prefer attractive people over unattractive ones. First, there is an "aesthetic appeal," a simple preference for beauty. Second, there is the "glow of beauty," in which we assume that good-looking people are more sensitive, kind, warm, modest, self-confident, sexual, and so on. Third, there is the status we achieve by dating attractive people.

Several studies (cited in Hatfield and Sprecher, 1986) have demonstrated that good-looking companions increase our status. In one study, men were asked their first impressions of a man seen alone, arm-in-arm with a beautiful woman, and arm-in-arm with an unattractive woman. The man made the best impression with the beautiful woman. He ranked higher alone than with an unattractive woman. In contrast to men, women do not necessarily rank as high when seen with a handsome man. A study in which married couples were evaluated found that it made no difference to a woman's ranking if she was unattractive but had a strikingly handsome husband. If an unattractive man had a strikingly beautiful wife, it was assumed that he had something to offer other than looks, such as fame or fortune.

Are Looks Important to Everyone?

For all of us ordinary-looking people, it's a relief to know that looks aren't everything.

Looks are most important to certain types or groups of people and in certain situations or locations (for example, in classes, at parties, and in bars, where people do not interact with one another extensively on a day-to-day basis). Looks are less important to those in ongoing relationships and to those older than young adults. Looks are also less important if there are regular interactions between individuals—for example, those who work together or commute in the same automobile (Hatfield and Sprecher, 1986). Looks tend to be especially important in adolescence and youth because of our need to conform. It is at this time that we are most vulnerable to pressure from our peers to go out with handsome men and beautiful women.

Men tend to care more about how their partners look than do women (Regan, 2003). This may be attributed to the disparity of economic and social power. Because men tend to have more assets (such as income and status) than women, they do not have to be concerned with their potential partner's assets. Therefore, they can choose partners in terms of their attractiveness. Because women lack the earning power and assets of men, they have to be more practical. They have to choose a partner who

Trade-Offs

People don't necessarily gravitate to the most attractive person in the room. Instead, they tend to gravitate to those who are about as attractive as themselves. Sizing up someone at a party or dance, a man may say, "I'd never have a chance with her; she's too good-looking for me." Even if people are allowed to specify the qualities they want in a date, they are hesitant to select anyone notably different from themselves in social desirability.

We tend to choose people who are our equals in terms of looks, intelligence, education, and so on (Hatfield and Walster, 1981). However, if two people are different in looks or intelligence, usually the individuals make a trade-off in which a lower-ranked trait is exchanged for a higher-ranked trait. A woman who values status, for example, may accept a lower level of physical attractiveness in a man if he is wealthy or powerful.

People tend to choose partners who are about as attractive as themselves.

can offer security and status. Unsurprisingly, then, women are more likely than men to emphasize the importance of socioeconomic factors (Regan, 2003).

Most research on attractiveness has been done on first impressions or early dating. Researchers are finding, however, that attractiveness is important in established relationships as well as in beginning or casual ones. Most people expect looks to become less important as a relationship matures, but Philip Blumstein and Pepper Schwartz (1983) found that the happiest people in cohabiting and married relationships thought of their partners as attractive. People who found their partners attractive had the best sex lives. Physical attractiveness continues to be important throughout marriage.

Bargains and Exchanges

Likening relationships to markets or emphasizing exchange as a basis for choosing partners may not seem romantic, but both are deeply rooted in marriage and family customs. In some cultures, for example, arranged marriages take place after extended bargaining between families. The woman is expected to bring a dowry in the form of property (such as pigs, goats, clothing, utensils, or land) or money, or a woman's family may demand a bride-price if the culture places a premium on women's productivity. Traces of the exchange basis of marriage still exist in our culture in the traditional marriage ceremony when the bride's parents pay the wedding costs and "give away" their daughter.

Gender Roles

Traditionally, relationship exchanges have been based on gender. Men have used their status, economic power, and role as protector in a trade-off for women's physical attractiveness and nurturing, childbearing, and housekeeping abilities; women, in return, have gained status and economic security in the exchange.

As women enter careers and become economically independent, the terms of bargaining change. When women achieve their own occupational status and economic independence, what do they ask from men in the marriage exchange? Clearly, many women expect men to bring more expressive, affective, and companionable resources into marriage. An independent woman does not have to settle for a man who brings little more to the relationship than a paycheck; she wants a man who is a companion, not simply a provider.

But even today, a woman's bargaining position is not as strong as a man's. As we noted earlier, women earn only about three-fourths of what men earn. Women are still significantly underrepresented in the professions. Furthermore, many of the things that women traditionally used to bargain with in the marital exchange—children, housekeeping services, or sexuality—are today devalued or available elsewhere. Children are not the economic assets they once were. A man does not have to rely on a woman to cook for him, sex is often accessible in the singles world, and someone can be paid to do the laundry and clean the apartment.

Women are at a further disadvantage because of the double standard of aging. Physical attractiveness is a key bargaining element in the marital marketplace, but the older a woman gets, the less attractive she is considered. For women, youth and beauty are linked in most cultures. As women get older, their field of eligible partners declines because men tend to choose younger women as mates.

The Marriage Squeeze and Mating Gradient

An important factor affecting the marriage market is the ratio of men to women. Researchers Guttentag and Secord (1983) argue that whenever there is a shortage of women in society, marriage and monogamy are valued; when there is an excess of women, marriage and monogamy are devalued. The scarcer sex is able to weight the rules in its favor. It gains bargaining power in the marriage marketplace.

The **marriage squeeze** refers to the gender imbalance reflected in the ratio of available unmarried women and men. Because of this imbalance, members of one gender tend to be "squeezed" out of the marriage market. The marriage squeeze is distorted, however, if we look at overall figures of men and women without distinguishing between age and ethnicity. Overall, there are significantly more unmarried women than men: 87 single men for every 100 single women (U.S. Census Bureau, 2003). This figure, however, is somewhat deceptive. From ages 18 to 44, the prime years for marriage, there are significantly more unmarried men than women, reversing the overall marriage squeeze. Combining widowed, divorced, and never married together, in 2002 there were 113 unmarried men, aged 18 to 44, for every 100 unmarried women (U.S. Census Bureau, 2003). Thus, women in this age group have greater bargaining power and are able to demand marriage and monogamy. But once ethnicity is taken into consideration, the marriage squeeze "squeezes" many African-American women of all ages out of the marriage market. With eligible males scarcer, African-American men have greater bargaining power and are less likely to marry because of more attractive alternatives. (See Figure 7.1 and the Perspective box on the African-American male shortage.)

"All the good ones are taken" is a common complaint of women in their mid-thirties and beyond, even if there are still more men than women in that age bracket. The reason for this is the **mating gradient,** the tendency for women to marry men of higher status. Sociologist Jessie Bernard (1982) comes to this conclusion:

> In our society, the husband is assigned a superior status. It helps if he actually is superior in ways—in height, for example, or age or education or occupation—for such superiority, however slight, makes it easier for both partners to conform to the structural imperatives. The [woman] wants to be able to "look up" to her husband, and he, of course, wants her to. The result is a situation known sociologically as the marriage gradient.

Although we tend to marry those with the same socioeconomic status and cultural background, men tend to marry women slightly below them in age, education, and so on. Bernard continues:

> The result is that there is no one for the men at the bottom to marry, no one to look up to them. Conversely, there is no one for the women at the top to look up to; there are no men superior to them. . . . [T]he never-married men . . . tend to be "bottom-of-the-barrel" and the women . . . "cream-of-the-crop."

The marriage gradient puts high-status women at a disadvantage in the marriage marketplace.

FIGURE 7.1 ■ **Ratio of Unmarried Men to Unmarried Women by Age and Ethnicity, 2002**

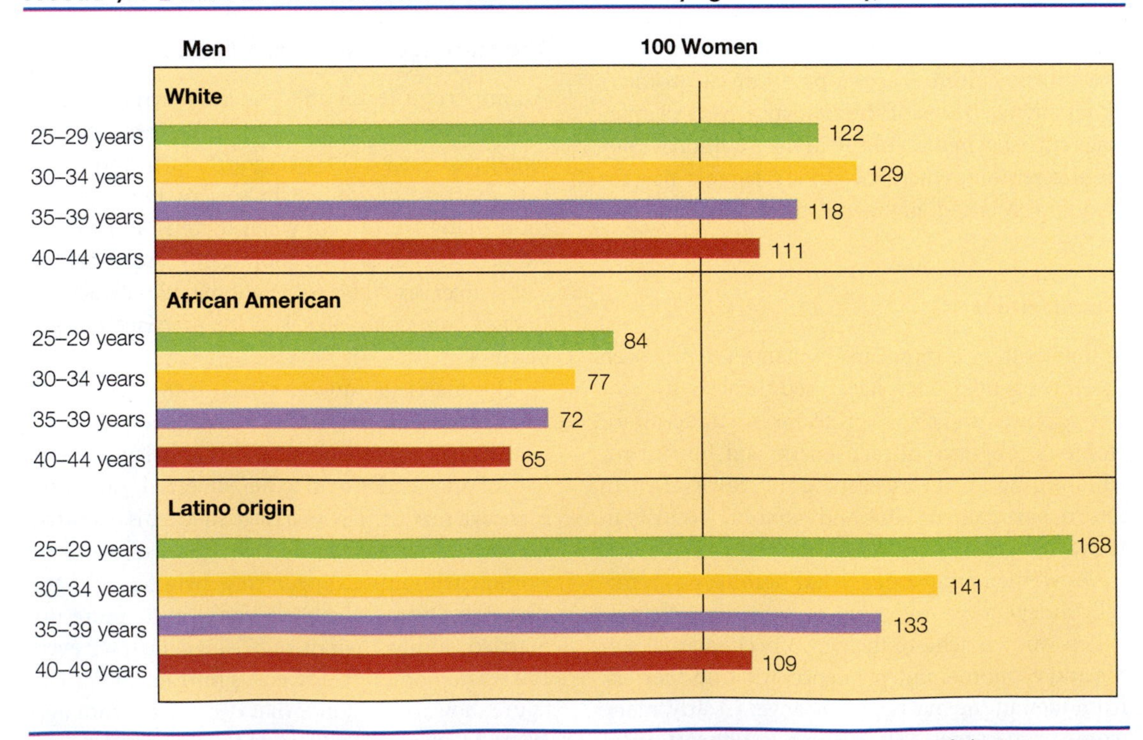

SOURCE: "Married Status of Persons 15 Years and Over by Family Status, Age, Sex, Race, and Hispanic Origin," *Current Population Reports 2002*, unpublished Table 7.

Perspective

THE AFRICAN-AMERICAN MALE SHORTAGE

The ratio of males to females is an important factor affecting relationships and marriage rates. For most groups, there is an abundance of available males in the age groups most likely to marry. Among African Americans, however, there is a significant shortage of eligible males and an excess of women. This abundance of women may affect male/female relationships as well as the likelihood of many women marrying, having fulfilling relationships, bearing children outside of marriage, and raising children without a partner. In Chapter 3, we referred to the male marriageable pool index, (MMPI), a developed by Wilson (1987). As the researchers intended it, the MMPI helps account for many marriage and family patterns observed among African Americans. Let's look more closely at the shortage of eligible black men and its consequences.

Consider the following: Among whites aged 25–29, there are 132 single men per 100 women; among Latinos, there are 163 men per 100 women. Among whites aged 30–34, there are 128 single men per 100 single women; among Latinos, the ratio is 130 single men for 100 women. By the simple law of supply and demand, white and Latina women can be more selective about their partners because the pool of single men is so large. The percentage of married individuals is 63 percent among whites and 59 percent among Hispanics (U.S. Census Bureau, 1998).

But the story is different for African Americans: Single African-American men are in short supply. Among blacks aged 25–29, there are 87 single men per 100 women; among those aged 30–34, there are 76 men per 100 women. By age 40–44, there are 84 single men per 100 women. There are an estimated 1.5 million more African-American women than men (Staples and Johnson, 1993).

There are several important consequences resulting from this. First, the percentage of married African Americans has declined significantly, from 64 percent in 1971 to 42.1 percent in 2000 (U.S. Census Bureau, 2002). Second, single African-American women have sex less often than single white women (Tanfer and Cubbins, 1992). Third, many black women give birth while single and raise their children with the assistance of their extended family and the biological father (Staples, 1988). Births to single mothers are associated with enduring poverty. Fourth, interracial marriages have risen dramatically since 1980, especially between blacks and whites. Currently, 12.1 percent of all new marriages involving an African American are interracial, with black men more likely to marry white women than the reverse (Besharov and Sullivan, 1996).

Inner-city African Americans have higher rates of singlehood than middle-class blacks. Among college-educated African Americans, the gender ratio is more extreme as people tend to marry those with similar educational backgrounds. Overall, the ratio of single college-educated black women is 2 to 1. For divorced black women over 35 with more than 5 years of college, the ratio of comparable men is 38 to 1 (Staples, 1991).

Furthermore, significantly fewer African Americans marry because of the lack of eligible men. Not only are there *fewer* males because of the gender ratio, but because of lack of jobs or skills, they are often unemployed. Marriage among blacks is often a function of the male's being employed (Tucker and Taylor, 1989). More African Americans than whites are single because of social problems, such as the gender ratio and high unemployment, rather than a rejection of marriage (Tucker and Taylor, 1989).

The proportion of single males to females decreased sharply between 1970 and 1995 among African Americans, while it rose among both whites and Latinos (U.S. Census Bureau, 1996). Why the decline in African-American males? Sociologist Robert Staples (1988) points to the effects of institutional racism: high infant mortality, premature death, devastating homicide rates, poor healthcare access, HIV/AIDS, and illegal drugs. High unemployment, disproportionate incarceration rates, increasing school dropout rates, and drug abuse further make the affected young African-American men less desirable as mates. Staples (1988) summarizes the situation: "Due to the operational effects of institutional racism, large numbers of black males are incarcerated, unemployed, narcotized, or fall prey to early death."

It is important to establish social policies to reverse the devastation visited upon African Americans by discrimination and poverty. Such policies are important in reversing the gender imbalance and the consequences that have developed over the past 25 to 30 years.

THE FIELD OF ELIGIBLES

The men and women we date, live with, or marry usually come from the **field of eligibles**—that is, those whom our culture approves of as potential partners. The field of eligibles is defined by two principles: **endogamy** (marriage within a particular group) and **exogamy** (marriage outside a particular group).

Endogamy

People usually marry others from a large group—such as the nationality, ethnic group, or socioeconomic status with which they identify—because they share common assumptions, experiences, and understandings. Endogamy strengthens group structure. If people already have ties as friends, neighbors, work associates, or fellow church members, a marriage between such acquaintances solidifies group ties. To take an extreme example, it is easier for two Americans to understand each other than it is for an American and a Fula tribesperson from Africa.

Americans are monogamous and urban, whereas the Fula are polygamous wandering herders. But another, darker force may lie beneath endogamy: the fear and distrust of outsiders, those who are different from ourselves. Both the need for commonality and the distrust of outsiders urge people to marry individuals like themselves.

Exogamy

The principle of exogamy requires us to marry outside certain groups—specifically, outside our own family (however defined) and outside our same sex. Exogamy is enforced by taboos that are deeply embedded within our psychological makeup. The violation of these taboos may cause a deep sense of guilt. A marriage between a man and his mother, sister, daughter, aunt, niece, grandmother, or granddaughter is considered incestuous; women are forbidden to marry their corresponding male relatives. Beyond these blood relations, however, the definition of incestuous relations changes. One society defines marriages between cousins as incestuous, whereas another may encourage such marriages.

Some states prohibit marriages between stepbrothers and stepsisters as well as cousins; others do not. As we noted in Chapter 1, the longstanding cultural presumption and legal precedent that we must marry someone of the other sex has met with numerous challenges by some lesbian and gay couples and gay rights organizations. With the Massachusetts Court having declared the prohibition of same sex marriages unconstitutional, the Commonwealth of Massachusetts is scheduled to become the first state to legalize gay marriage.

Heterogamy and Homogamy

Endogamy and exogamy interact to limit the field of eligibles. (See this chapter's Understanding Yourself box, on page 213, which discusses Internet personals, computer dating, and homogamy.) The field is further limited by society's encouragement of **homogamy**, the tendency to choose a mate whose personal or group characteristics are similar to ours. This is also known as positive assortative mating (Blackwell, 1998). **Heterogamy** refers to the tendency to choose a mate whose personal or group characteristics differ from our own. The strongest pressures are toward homogamy. We may make homogamous choices regarding any number of characteristics, including age and race, but also such characteristics as height (Blackwell, 1998). As a result, our choices of partners tend to follow certain patterns. These homogamous considerations generally apply to heterosexuals, gay men, and lesbians alike in their choice of partners.

There has been a growing tendency toward allowing individuals choice of partners without state interference. In 1966, the U.S. Supreme Court, for example, declared unconstitutional laws prohibiting marriage between individuals of different races (*Loving vs. Virginia*). About 20 percent of whites and 8 percent of African Americans continue to believe intermarriage should be illegal (Wilkerson, 1991). Although at present only Massachusetts is in line to legalize same-sex marriage, other states, such as Vermont, have moved in the direction of guaranteeing equal protection and equal rights that heterosexuals receive when they marry. Denial of legal marriage rights and its many protections and benefits is otherwise unconstitutional. Although Vermont *could* have become the first state to legalize gay marriage, instead it enacted "civil union" legislation, thus creating a fully equivalent domestic partnership alternative to which gay men and lesbians have access.

The most important elements of homogamy are race and ethnicity, religion, socioeconomic status, age, and personality characteristics. These elements are strongest in first marriages and weaker in second and subsequent marriages (Glick, 1988). They also strongly influence our choice of sexual partners, as our sexual partners are often potential marriage partners (Michael, Gagnon, Laumann, and Kolata, 1994).

RACE AND ETHNICITY. Most marriages are between members of the same race. Of the nearly 55 million married couples in the United States in 2000, 98 percent of them consisted of husbands and wives of the same race. Of the more than a million interracial couples, one-fourth were marriages between blacks and whites (Fields and Casper, 2001). By 1993, 12 percent of all new marriages involving African Americans were interracial. This is nearly double the percentage in 1980 (6.6 percent), and four times the percentage (2.6 percent) from 1970 (Besharov and Sullivan, 1996). It is suggested that the reasons why both black groom/white bride and black bride/white groom are increasing is the rise of a black middle class, making both African-American men and women more attractive to middle-class whites. Black women still face obstacles to marriage of any kind; they are more than twice as likely to have children born out of wedlock. About 1.2 percent of marriages consist of one partner who is white and one from an Asian, Native American, or other nonwhite group (Fields and Casper, 2001).

The degree of intermarriage between ethnic groups is of concern to some members of these groups because it affects the rate of assimilation and continued ethnic identity (Stevens and Schoen, 1988). Almost half of all Japanese Americans marry outside their ethnic group (Takagi, 1994). Over half of all Native Americans are married to non-Native Americans (Yellowbird and Snipp, 1994). For both Japanese Americans and Native Americans, intermarriage leads to profound questions about their continued existence as distinct ethnic groups in the twenty-first century. Among European ethnic groups in this country, such as Italians, Poles, Germans, and Irish, only one in four marries within his or her ethnic group. The ethnic identity of these groups has decreased considerably since the beginning of this century. Interestingly, Louisiana Cajuns have very high rates of ethnic homogamy, especially for a group of their size and considering the length of time they have been in the United States. Among married Cajun women, more than 75 percent were married to Cajun men; among Cajun men, over 70 percent were also homogamous (Bankston III and Henry, 1999).

RELIGION. Until the late 1960s, religion was a significant factor in marital choice. Today, most religions still oppose interreligious marriage because they believe it weakens individual commitment to the faith. Nonetheless, interreligious dating and marriage have been increasing. Almost half of all Catholics marry outside their faith (Maloney, 1986). Almost 40 percent of Jews choose a non-Jewish partner, up from 6 percent in the early 1960s (Mindel, Haberstein, and Wright, 1988). Those who marry from different religious backgrounds are at slightly greater risk of divorce than those from similar backgrounds (Bumpass, Martin, and Sweet, 1991; Lehrer and Chiswick, 1993; Sander, 1993).

Religious groups tend to discourage interfaith marriages, believing that such marriages, in addition to weakening individual beliefs, lead to children's being reared in a different faith or secularize the family. Such fears, however, may be overstated. Among Catholics who marry Protestants, for example, there seems to be little secularization by those who feel themselves to be religious (Petersen, 1986). Some who are from different religious backgrounds, however, do convert to their spouses' religions.

SOCIOECONOMIC STATUS. Most people marry others of their own socioeconomic status. They also more often marry those with the same or similar educational backgrounds as themselves. Even if a person marries outside their ethnic, religious, or age group, the selected spouse will probably be from the same socioeconomic level. Furthermore, some ethnic or racial homogamy may actually be due more to tendencies toward socioeconomic homogamy (Bankston III and Henry, 1999).

Socioeconomic homogamy results from the combination of choice-shaping factors, such as shared values, tastes, goals, and expectations, and opportunity-determining factors such as residential neighborhood, school, and/or occupation. Additionally, control is exerted by affluent families to ensure that their children marry at the "right" level. Of course, not everyone marries homogamously. Men more than women will sometimes marry below their socioeconomic level (**hypogamy**);

women more often "marry up" (a practice known as **hypergamy**).

AGE. Reflecting the data in Chapter 3 on trends in age at marriage, Americans have long tended to marry those of roughly the same or similar ages. Typically, the man is slightly older than the woman. Age is important because we view ourselves as members of a generation, and each generation's experience of life leads to different values and expectations. Furthermore, different developmental and life tasks confront us at different ages. A 20-year-old woman wants something different from marriage and from life than a 60-year-old man does. By marrying people of similar ages, we often ensure congruence for developmental tasks. The gap between grooms' and brides' ages has narrowed in recent years, as the ages at which both men and women enter marriage have climbed.

RESIDENTIAL PROPINQUITY. An additional homogamous factor is based on the principle of **residential propinquity**—the tendency we have to select partners (for relationships and for marriages) from a geographically limited locale. Put differently, the likelihood of marriage decreases as the distance between two people's residences increases. The obvious explanation behind this is one of opportunity. In most instances, in order to start dating or get together with someone you have to first meet. Our chances of meeting are greater when our daily activities (shopping, commuting, eating out, and so forth) overlap.

Although it is easy to trivialize this tendency as too obvious to be meaningful, consider the implications it has for the other patterns of homogamy. American communities are often segregated by class and/or race. In some towns, they may even have religious splits (for example, the Catholic side or Protestant side of town, a Jewish neighborhood). Public schools, being neighborhood-based, further the tendency for us to associate with others like ourselves. Thus, the types of people we are most likely to come into contact with and with whom we might develop intimate relationships or eventually marry are a lot like ourselves. Thus, residential propinquity explains some of the other homogamous tendencies by how it limits our opportunity. At the same time, the cultural beliefs that homogamous marriages are better or more likely to be stable, might reinforce people's tendencies to "look locally," where they are more likely to be surrounded by people like themselves.

These factors in the choice of partner interact with one another. Ethnicity and socioeconomic status, for example, are often closely related because of discrimination. Many African Americans and Latinos are working class and are not as well educated as whites. Whites generally tend to be better off economically and are usually better educated. Thus, a marriage that is endogamous in terms of ethnicity is also likely to be endogamous in terms of education and socioeconomic status.

REFLECTIONS

Keeping heterogamy and homogamy in mind, think about those who are or have been your romantic or marital partners. In what respects have your partners shared the same racial, ethnic, religious, socioeconomic, age, and personality characteristics with you? In what respects have they not? Have shared or differing characteristics affected your relationships? How?

Are homogamous relationships "better" or "stronger" relationships? Data on intermarriages by religion, race, and/or class are inconsistent on this question. Some studies reveal greater difficulties in nonhomogamous relationships and higher likelihood of divorce among those who intermarry. Others fail to substantiate the negative outcome (Eshelman, 1997). The most consistent findings are related to those risks associated with religious intermarriages, though these risks are not great.

There are three possible explanations as to why heterogamous marriages might be less stable than homogamous marriages (Udry, 1974):

1. Heterogamous couples may have considerably different values, attitudes, and behaviors, which may create a lack of understanding and promote conflict.
2. Heterogamous marriages may lack approval from parents, relatives, and friends. Couples are then cut off from important sources of support during crises.
3. Heterogamous couples are probably less conventional and therefore less likely to continue an unhappy marriage for the sake of appearances.

THE STAGES OF MATE SELECTION

Let's say that you meet someone who fits all the criteria of homogamy: same ethnic group, religion, socioeconomic background, age, and personality traits. Homogamously speaking, he is "Mr. Right," she is "Ms. Right," and your children would be "Little Rights." He or she is the person your parents dream of your marrying. But, unfortunately, you can't stand this person. Homogamy by itself doesn't work. A range of theories has been suggested to address the question of why we select the particular individuals that we do. Do "*opposites attract*"? Do "*birds of a feather flock together*"? Do we unconsciously select people like our parents? What is more important: finding someone who seems to think as we do about things or someone whose behavior fits what we expect in a partner?

Each of the preceding questions illustrates an existing theory of mate selection. The commonsense notion that "opposites attract" is in keeping with **complementary needs theory,** the belief that people select as spouses those whose needs are different. Thus, a very assertive person who has difficulty compromising will be drawn to a less outgoing and highly adaptable person. The notion that "birds of a feather flock together" is more in keeping with theories such as value theory or role theory, in which gratification follows from finding someone who feels and/or thinks like we do. Having someone who shares our view of what's important in life or who acts in ways that we desire in a partner validates us, and this sense of validation leads to an intensification of what we feel toward that other person. **Parental image theory** suggests that we seek partners who are similar to our opposite-sex parent. Some versions of parental image theory draw on Freudian concepts such as the *Oedipus complex,* whereas others point more toward the lasting impressions made by our parents (Eshelman, 1997; Murstein, 1986).

Bernard Murstein developed a multifactor, sequential theory known as **stimulus-value-role theory** to depict what happens between the "magic moment" with its mysterious chemistry of attraction and the decision to maintain a long-term relationship such as marriage. Murstein's theory is based on exchange and identifies three stages of romantic relationships. At each stage, if the exchange seems more or less equitable, the two will progress to the next stage and ultimately remain together (Murstein, 1986)

In the *stimulus* stage, each person is drawn or attracted to the other before actual interaction. This attraction can be physical, mental, or social; it can be based on a person's reputation or status. The stimulus stage, according to Murstein, is most prominent on the first encounter, when one has little information with which to evaluate the other person.

In the next stage, the *value* stage, partners weigh each other's basic values, trying to determine if they are compatible. Each person tries to figure out the other's philosophy of life, politics, sexual values, religious beliefs, and so on. If both highly value rap music, it is a plus for the relationship. If they disagree on religion—one is a dedicated fundamentalist, the other is a goddess follower— it is a minus for the relationship. Each person adds or subtracts the pluses and minuses along value lines. Based on the outcome, the couple will either disengage or go on to the next stage. Values are usually determined between the second and seventh meetings.

Eventually, in the *role* stage, each person analyzes the other's behaviors, or how the person fulfills his or her roles as lover, companion, friend, worker— and potential husband or wife, mother or father. Are the person's behaviors consistent with marital roles? Is he or she emotionally stable? This aspect is evaluated in the eighth and subsequent encounters.

Although the stimulus-value-role theory has been one of the most prominent theories explaining relationship development, some scholars have criticized it. Most notably, scholars ask if men and women really test their degree of fit (Huston et al., 1981). For example, religious fundamentalists and goddess worshippers may sometimes believe that they are compatible. They may not discuss religion; instead they might focus on the "incredible" physical attraction in their relationship. Or they may make errors in arithmetic; they may mistakenly believe that religion is not that important, only to discover after they are married that it is very important.

Dating and Romantic Relationships

As more and more people delay marriage, never marry, or seek to remarry after divorce or widowhood, romantic relationships will, according to

Surra (1991), "take different shapes at different points in time, as they move in and out of marriage, friendship, romance, cohabitation, and so on." As a result, researchers are shifting from the traditional emphasis on mate selection toward the study of the formation and development of romantic relationships, such as the dynamics of heterosexual dating, cohabitation, postdivorce relationships, and gay and lesbian relationships. The field of personal relationships is developing a broad focus that explores relationship dynamics (Duck, 1993; Kelley et al., 1983; Perlman and Duck, 1987).

BEGINNING A RELATIONSHIP: SEEING, MEETING, AND DATING

Although the general rules of mate selection are important in the abstract, they do not tell us how relationships begin. The actual process of beginning a relationship is discussed below.

Seeing

On a typical day, we may see dozens, hundreds, or thousands of men and women. But seeing isn't enough; masses of people blur into one another. We must become aware of someone for a relationship to begin. It may only take a second from the moment of noticing to meeting, or sometimes it may take days, weeks, or months. Sometimes "noticing" occurs between two people simultaneously; other times it may take considerable time; sometimes it never happens.

REFLECTIONS

What are some of the settings in which you "see" people? How do the settings affect the strategies you use to meet others? How do you move from meeting to "going out" with someone? What are your feelings at each stage of seeing, meeting, and dating?

The setting in which you see someone can facilitate or discourage meeting each other (Murstein, 1976, 1987). **Closed fields,** such as small classes or seminars, dormitories, parties, and small workplaces, are characterized by a small number of people who are likely to interact whether they are attracted or not. In such settings, you are likely to "see" and interact more or less simultaneously. In contrast, **open fields,** such as beaches, shopping malls, bars, raves, amusement parks, and large university campuses, are characterized by large numbers of people who do not ordinarily interact with one another.

Meeting

How is a meeting initiated? Among heterosexuals, does the man initiate it? On the surface, the answer appears to be yes, but in reality, the woman often "covertly initiates . . . by sending nonverbal signals of availability and interest" (Metts and Cupach, 1989). A woman will glance at a man once or twice and catch his eye; she may smile or flip her hair. If the man moves into her physical space, the woman then relies on nodding, leaning close, smiling, or laughing (Moore, 1985).

Regardless of who initiates contact, there are a variety of verbal and nonverbal signals that are used to convey attraction and interest to a potential partner. Smiling, moving closer to, gazing at, laughing, and displaying "positive facial expressions" are all gestures to convey interest or "flirt" (Regan, 2003). Touch is also an important element in flirting, whether the touch consists of lightly touching the arm or hand or the face or hair of the target of interest, or rubbing one's fingers across the other's arm (Regan, 2003).

If a man believes a woman is interested, he often initiates a conversation using an opening line. The opening line tests the woman's interest and availability. You have probably used or heard an array of opening lines. According to women, the most effective are innocuous, such as "I feel a little embarrassed, but I'd like to meet you" or "Are you a student here?" The least effective are sexual come-ons, such as "You really turn me on. Do you want to have sex?" Women, more than men, prefer direct but innocuous opening lines over cute, flippant ones, such as "What's a good-looking babe like you doing in a college like this?"

A recent Web-search for "pick-up lines" identified more than *two million* sites. There were sites specializing in math pick-up lines, astrology pick-up lines, *X-Files* pick-up lines, *Dr. Who* pick-up lines, Christian pick-up lines, Jewish pick-up lines, Gothic pick-up lines, as well as "cheesy," humorous, and bad pick-up lines. There were also lines for women to use with men, men to use with women, men to use with men, and women to use with women. The following list is a sampling of some actual opening lines that men or women have used to initiate contact: To many of us, these lines seem silly and unlikely to generate the kind of impression that might lead to forming a relationship. Readers of this text can probably identify other such lines that they have used or received.

"You must be tired, because you've been running through my mind all day."

"You know, if you held up eleven roses in front of a mirror, you would be looking at twelve of the most beautiful things in the world."

"You must be from Tennessee, because you're the only 10 I see."

Do you have a quarter? I promised my mother I'd call her when I met the girl/guy of my dreams."

"Did they just turn on a fan in here or was that you blowing me away?"

"You're like a great song. I can't get you out of my head."

"Shhhh, not so loud. You don't want to wake up the other angels in heaven or they'll realize you snuck out."

"Hold still, there's something in your eye. Never mind, it's just a sparkle."

"Did the sun just come out or did you smile at me?"

"Remember this moment so that we can tell our children how we first met."

"If I had a nickel for every time I met someone as beautiful as you, I'd have a nickel."

Men are more likely to initiate a meeting directly, whereas women are more likely to wait for the other person to introduce himself or herself or to be introduced by a friend (Berger, 1987). About a third or half of all relationships rely on introductions (Sprecher and McKinney, 1993). An introduction has the advantage of a kind of prescreening, as the mutual acquaintance may believe that both may hit it off. Parties are the most common settings in which young adults meet, followed by classes, work, bars, clubs, sports settings, or events centered around hobbies, such as hiking (Marwell et al., 1982; Shostak, 1987; Simenauer and Carroll, 1982).

Technology via the Internet continues to gain popularity as a way for people to "meet" a potential partner. Online, people can introduce themselves in fantasy-like images. A growing number of people first "meet" in cyberspace, find common interests, and form relationships that develop and intensify prior to ever actually meeting. According to a recent *New York Times* article, more than 45 million Americans visited online dating sites just in the month of May 2003. Subscribers to such sites are spending approximately $100 million a quarter through this year (Harmon, 2003).

Single men and women also increasingly rely on personal classified ads, where men tend to advertise themselves as "success objects" and women advertise themselves as "sex objects" (Davis, 1990). Their ads typically reflect stereotypical gender roles. Men advertise for women who are attractive and deemphasize intellectual, work, and financial aspects. Women advertise for men who are employed, financially secure, intelligent, emotionally expressive, and interested in commitment. Men are twice as likely as women to place ads. Other alternative forms of meeting others include video dating services, introduction services, computer bulletin boards, and 900 party-line phone services. (One video dating service claims to have over 150,000 members, who pay an average of $2,000 for its services; over 9,000 of its members have married each other.)

Single men and women often rely on their churches and church activities to meet other singles. Black churches are especially important for middle-class African Americans, as they have less chance of meeting other African Americans in integrated work and neighborhood settings. They also attend concerts, plays, film festivals, and other social gatherings oriented toward African Americans (Staples, 1991).

For lesbians and gay men, the problem of meeting others is exacerbated by the fact that they cannot necessarily assume that the person in whom they are interested shares their orientation. Instead, they must rely on identifying cues, such as meeting at a gay or lesbian bar, wearing a gay/lesbian pride button, participating in lesbian/gay events, or being introduced by friends to others identified as being gay or lesbian (Tessina, 1989). Once a like orientation is established, gay men and lesbians usually engage in nonverbal processes to express interest. Lesbians and gay men both tend to prefer innocuous opening lines. To prevent awkwardness, the opening line usually does not make an overt reference to orientation unless the other person is clearly lesbian or gay.

Dating

For many of us, asking someone out for the first time is not easy. Shyness, fear of rejection, and traditional gender roles that expect women to wait to be asked may fill us with anxiety and nervousness. (Sweaty palms and heart palpitations are not uncommon when asking someone out the first time.) Both men and women contribute, although sometimes differently, to initiating a first date. Men are more likely to ask directly for a date: "Want to go see a movie?" Women are often more indirect. They hint or "accidentally on purpose" run into the other person: "Oh, what a surprise to see you *here* studying for your marriage and family midterm!" Although women may initiate dates, they do so less frequently than do men (Berger, 1987).

Additionally, research indicates that both women and men believe that men *should* initiate first dates, that men display a greater willingness to do so, and that men have a higher frequency of actual "first moves." Interestingly, men also express a desire for women to more actively participate in initiating relationships, either by asking directly for a date or at least hinting. Men report that the most passive stance, in which women wait for men to ask/initiate, is less preferred (Regan, 2003).

Costs and Benefits of Romantic Relationships

As anyone who has had a romantic relationship can attest, relationships bring positive and negative experiences. In other words, when asked, people identify both rewards (e.g., companionship, sexual gratification, feeling loved and loving another, intimacy, expertise in relationships, and enhanced self-esteem) and costs (loss of freedom to socialize or date, investment of time and effort, loss of identity, feeling worse about oneself, stress and worry about the health or durability of the relationship, as well as other nonsocial costs like lower grades) of romantic relationships (Sedikedes, Oliver, and Campbell, 1994, cited in Regan, 2003).

Males and females differ some in what costs and rewards they identify. More males than females identify sexual gratification as a benefit they get from romantic relationships, while women are much more likely than men to identify the benefit of enhanced self-esteem. More women than men mention loss of identity, feeling worse about themselves or growing too dependent on their partners as relationship costs. Males, on the other hand, stress perceived loss of freedom (to socialize or date) and financial costs more than women do (Regan, 2003).

POWER IN DATING RELATIONSHIPS

Power doesn't seem to be a concern for most people in dating relationships. In fact, one study of dating couples found that both men and women thought that they had about equal power in their relationships (Sprecher, 1985). If power does become a source of conflict, it may not become as intense an

Dating is a source of pleasure as well as problems. It is also the process through which most Americans find their spouses.

© Michael Newman/PhotoEdit

Understanding Yourself

INTERNET PERSONALS, COMPUTER DATING, AND HOMOGAMY

Many search engines and Web sites allow users to search through personal ads to find a suitable match. For example, Yahoo and America Online each have sections of "personals." Sites specifically designed for matchmaking and finding dates are abundant. In fact, a Web search for "Personals and Dating Services" netted nearly a million sites. There are free sites and pay sites, sites for finding Christian partners or Jewish partners, sites for smokers and for nonsmokers, sites for people in the military and for single parents, "shy folks over 30," and nondrinkers as well as sites that specialize in interracial matches and some that specialize in matching gay men, lesbians, or bisexuals with potential partners. The more popular sites include Match.com, DreamMates.com, MatchDoctor.com, e-mates.com, kiss.com, and americansingles.com.

Although users can often peruse the ads and photos without registering, most sites ask users to register and answer a series of questions that describe themselves and their "ideal matches." Upon completing these initial screening questions, the database is then made accessible, and users can search through the ads and e-mail those who interest them.

Typically, screening questions ask about basic characteristics including height, weight (or body type), hair color, occupation, education, marital status (i.e., never married, separated, divorced, or widowed), whether one has (or wants) children, religious preferences, political views, drinking habits, and smoking habits. Additionally, on most sites, users indicate their interests and what they enjoy doing (e.g., movies, music, outdoor activities, reading, dancing, travel, theater).

Some sites do the matching themselves, after asking members to complete extensive questionnaires or personality profiles. In compiling profiles, some sites seek detailed self-assessments with close to 100 personality characteristics including warmth, cleverness, ambitiousness, intelligence, compassion, loyalty, modesty, submissiveness, aloofness, arrogance, impulsiveness, spirituality, humor, perfectionism, sensitivity, and generosity. People rate themselves on such characteristics and indicate their importance in a potential partner.

As you think about the processes described above, to what extent do you find homogamy operating? Is there an assumption that people ought to be paired with people like themselves? Furthermore, what do you think would be your chances of liking a person who was your "ideal match"? What other characteristics would be important to you?

issue as in a marriage. Strong power conflicts can be avoided by dissolving the relationship in question.

In marriage, the person with the most economic resources often has the most power. Generally, this means the man because he is more likely to be employed and to earn more money than the woman. In dating relationships, however, economic resources do not appear to be as important a power resource for men as their ability to date other women (Sprecher, 1985). The easier men think it will be for them to go out with other women, the more power they have in a dating relationship. The fact that men have more power, prestige, and status than do women cannot help but to affect heterosexual interactions. (Henley and Freeman, 1995).

The question of who initiates, who touches, and who terminates sexual advances prescribes male leadership and dominance. Even though many people do not wish to have unequal sexual relationships, modes of expression and resistance and difficulty in changing communication patterns help to maintain an edge of inequality and imbalance among women. For equality to occur, women need to determine what they wish to express and how they wish to keep those behaviors that give them strength.

PROBLEMS IN DATING

Dating is often a source of both fun and intimacy, but a number of problems may be associated with it. Think about your own romantic relationships. When a disagreement occurs, who generally wins? Does it depend on the issue? When one person wants to go to the movies and the other wants to go to the beach, where do you end up going? If one wants to engage in sexual activities and the other doesn't, what happens?

LOVE VERSUS ARRANGED MARRIAGE IN CROSS-CULTURAL PERSPECTIVE

Is love the natural or ideal basis for marriage? Certainly in American society, marriage without love is considered somewhat shocking, even scandalous. It is the subject of soap operas and whispered gossip: "He married her for her money." "She married his family name." Although we might consider marriage without love an exceptional case, anthropologists tell us that in traditional cultures most people do not consider love a particularly sound basis for marriage.

Marriage customs vary dramatically across cultures, and marriage means very different things in different cultures. If we consider how marriages come about—how they are "arranged"—we find that it is usually not the bride and groom themselves who have decided to marry, as is the case in our own society today. Typically, the elders have done the matchmaking. In most cultures, the parents of the bride and groom were charged with arranging the marriage of their children. In some cultures, mothers were the primary matchmakers, as in traditional Iroquois culture. In others, fathers had a dominant voice in arranging marriage, as in traditional Chinese society. In still other cultures, the pool of elders involved in matchmaking was more extensive, including grandparents, aunts, uncles, and even local political and religious authorities, such as tribal chiefs and clan leaders.

In traditional societies, marriage was important family business. In fact, marriage was a major event in the life of two families—both the bride's and the groom's—as well as for the clan, tribe, and community to which each family belonged. Because marriage united two families, there were important matters to be taken into account before agreeing to any particular match. Typically, the economic or class standing of both families, as well as their social reputation or honor, was a paramount concern. Families needed to know how a particular marriage would affect the family as a whole. Will this marriage create ties to the chief's family that will bring political advantage to our family? How many cattle will the groom's family send us? Will the bride bring a large dowry of household goods and furniture? Will this marriage bring trade with our tribe? Can we count on his or her clan for additional workers at harvest time and additional strength in times of war?

With issues of this magnitude at stake, marriage could not be left up to the young people themselves. Sentimental feelings of love would certainly cloud their judgment. Marriage was not primarily a personal or intimate event focused on a young couple alone. The feelings and love between an individual bride and groom were subordinate to the greater interests and welfare of the family, clan, and community.

Anthropologists report that in most traditional cultures, newly married couples did not live separately from their parents. Part of the experience of getting married in our society is striking out on our own and setting up house together, independent of our parents. However, in traditional societies the new couple was expected to reside with the parents of either the bride or the groom, depending on the culture. In some cultures, aunts and uncles also lived in the extended family households into which the newly married couple settled. For instance, in many matrilineal cultures—where related females formed the backbone of families and clans—it was customary for the bride and groom to reside with the bride's family. In traditional Iroquois society, it was the groom who moved into the bride's longhouse, which included the bride's mother and father, her maternal grandparents, her mother's sisters, and their husbands and children. By contrast, in patrilineal cultures—where related males formed the basis of families and clans—the newly married couple settled in the home of the groom's family. In tradi-

Female/Male Differences

Divergent gender-role conceptions may complicate dating relationships. Often when two people lack complementary gender-role conceptions, the woman is more egalitarian and the man, more traditional.

As we saw earlier, it is difficult for women to initiate dates directly. They usually wait to be asked; and when they are asked, it may be by the wrong person. Another problem is who pays when going out on a date. Some women in Mirra Komarovsky's (1985) study feared that male acquaintances would be put off if they offered to pay their share. Other women who offered to pay, whether traditional or egalitarian, found themselves mocked by their dates. Some men who allowed their dates to pay still tional Chinese society, the bride moved in with her husband, his parents and paternal grandparents, his father's brothers, and their wives and children. In traditional societies like these, the newly married couple was not independent but rather dependent on family elders, remaining under their authority and protection.

Let us look more closely at the meaning of marriage and the role of love in one culture: Bedouin society in northern Egypt along the Libyan border. Traditionally, the Bedouin were nomadic pastoralists whose livelihood depended on herding sheep and goats and on trade with neighboring peoples for agricultural products. Bedouin culture is patrilineal. Bedouin camps are composed of extended families formed around related males—typically a grandfather, his adult sons and their wives, and his grandsons and unmarried granddaughters. On marriage, the bride leaves her family and goes to live with her husband and his relatives in their camp.

If marriage in American culture is based on cultural notions about individualism and love, marriage in Bedouin society is based on notions of family honor and duty. The elder males in the Bedouin family arrange a marriage to maximize family honor. They try to establish a link through marriage with an honorable family—one that is strong in numbers of men; has large herds of sheep and established business connections; is independent but able to support an impressive number of wives, children, and clients; and is of good blood and overall reputation. Moreover, the bride herself must have personal qualities that will enhance her husband's family. She is first under the authority of her husband's parents and his older male relatives, and she must know her place as a daughter-in-law. She must work hard and obey and serve her in-laws. She must show proper respect for all of her husband's relatives, conducting herself modestly through proper veiling and deferential behavior. In so doing, she will reflect honor on her husband and his family.

As in other patrilineal societies, an important concern of the male elders in Bedouin society is the protection of their authority and control over junior family members. They are also concerned with the promotion of a sense of loyalty and duty among the next generation of brothers, who will one day form the core of the family. This demands that the relationship between husband and wife be subordinated to family interests. A wife must not cause trouble between her husband and his father or his brothers. She should not win her husband's heart to the extent that he will forget his primary loyalties to his father, elders, and brothers. Thus, romantic attachment or love is viewed by Bedouin elders as potentially threatening to their own control over junior family members. Excessive love and passion can make a man disobedient, disrespectful, and even dependent, lessening both his and his family's honor.

Bedouin marriages are usually arranged between a young man and woman who belong to different camps, thus creating blind marriages. That is, the bride and groom typically have not met prior to their engagement and marriage. The practice of arranging blind marriages further enhances the control and authority of the older generation over the young couple. Without ever having met, two people can hardly be in love at the time of their marriage.

In the less frequent case where bride and groom have perhaps made slight acquaintance—on the occasion of other weddings or celebrations—it is still very unlikely that they will be in love at the time of their own marriage. Indeed, if family elders were to learn that affection or attraction existed between a young man and woman, they would not agree to the marriage. That would be asking for trouble. Of course, the emotion, attraction, and commitment that we mean by the word *love* may in time develop between husband and wife. However, in Bedouin society, as in most traditional societies, love is neither a necessary nor an advisable condition in arranging a good marriage.

insisted on choosing where they went, whether the women wanted to go there or not. Still other men allowed their dates to pay but not publicly; instead, for example, the women secretly slipped money to them under the dinner table.

In a random study of 227 women and 107 men in college, nearly 25 percent of the women said that they received unwanted pressure to engage in sex, usually before the establishment of an emotional bond between the couple. Women rated places to go (23 percent) and communication (22 percent) almost as high on their list of dating problems as sexual pressure (Knox and Wilson, cited in Knox, 1991). Komarovsky (1985) cites male sexual pressure as a major barrier between men and women. A woman who wants to see a man again faces a dilemma: how to encourage him to ask her out again without engaging in more sexual activity than she wants. In Komarovsky's study, men whose sexual advances were rejected by their dates often salved their hurt egos by accusing the women of having sexual hang-ups or being lesbians.

The number one problem cited by 35 percent of men was communicating with their dates (Knox and Wilson, cited in Knox, 1991). Men often felt that they didn't know what to say, or they felt anxious about the conversation dragging. Communication may be a particularly critical problem for men because traditional gender roles do not encourage the development of intimacy and communication skills among males. A second problem, shared by almost identical numbers of men and women, was where to go. A third problem, named by 20 percent of the men but not mentioned by women, was shyness. Although men can take the initiative to ask for a date, they also face the possibility of rejection. For shy men, the fear of rejection is especially acute. A final problem—and again one not shared by women—was money, cited by 17 percent of the men. Men apparently accept the idea that they are the ones responsible for paying for a date.

Because of the variety of problems that plague relationships, many couples ultimately break up. In the process of breaking up, both the initiator and the rejected partner suffer.

Extrarelational Sex in Dating and Cohabiting Relationships

You don't have to be married to be unfaithful (Blumstein and Schwartz, 1983; Hansen, 1987; Laumann et al., 1994). Both cohabiting couples and couples in committed relationships usually have expectations of sexual exclusiveness. But, like some married men and women who take vows of fidelity, they do not always remain sexually exclusive. Blumstein and Schwartz (1983) found that those involved in cohabiting relationships had similar rates of extrarelational involvement as did married couples, except that cohabiting males had somewhat fewer partners than husbands did. Gay men had more partners than did cohabiting and married men, and lesbians had fewer partners than any other group.

Large numbers of both men and women have sexual involvements outside dating relationships that are considered exclusive. One study of college students (Hansen, 1987) indicated that over 60 percent of the men and 40 percent of the women had been involved in erotic kissing outside a relationship; 35 percent of the men and 11 percent of the women had had sexual intercourse with someone else. Of those who knew of their partner's affair, a large majority felt that it had hurt their own relationship. When both partners had engaged in affairs, each believed that their partner's affair had harmed the relationship more than their own had. Both men and women seem to be unable to acknowledge the negative impact of their own outside relationships. It is not known whether those who tend to have outside involvement in dating relationships are also more likely to have extramarital relationships after they marry.

BREAKING UP

"Most passionate affairs end simply," Hatfield and Walster (1981) note. "The lovers find someone they love more." Love cools; it changes to indifference or hostility. Perhaps the relationship ends because one partner shows a side of him or herself that the other partner decides is undesirable. Or couples disclose *too much*, revealing negative feelings or ideas that lead to unhappiness and ultimately to the demise of the relationship (Regan, 2003).

Relationships are also susceptible to outside influences. Perhaps, some new opportunity for greater fulfillment appears in the person of someone else or in the opportunity to return to a more autonomous and independent state. Even satisfying relationships may end under these circumstances (Regan, 2003).

Breaking up is typically painful because few relationships end by mutual consent. For college students, breakups are more likely to occur during vacations or at the beginning or end of the school year. Such timing is related to changes in the person's daily living schedule and the greater likelihood of quickly meeting another potential partner.

Research indicates that relationships often begin to sour as one partner grows quietly dissatisfied (Duck, 1982; Vaughan, 1990). Duck (1982) calls this the *intrapsychic* phase, Vaughan (1990) talks about "keeping secrets." The point of both ideas is that often one partner decides that something is wrong with the relationship, considers the possibility of ending the relationship, weighs the likely outcomes associated with being out of the relationship, and begins to build an identity as a "single." All of this may happen before even informing one's partner about what is going on. By the time the "initiator" informs the partner, the partner is forced to play "catch-up," in that the initiator is a few steps ahead in the exiting process. This will be further discussed in Chapter 14.

Breaking up is rarely easy, whether you are on the initiating or "receiving" end. In fact, as Regan (2003) summarizes, the more satisfied you are with your partner, the closer you feel to your partner, and the more difficult you believe it will be to find another relationship, the harder it is to experience a breakup. Social support and self-esteem appear to be important factors in helping one recover more quickly and completely (Regan, 2003).

If you initiate a breakup, thinking about the following may help:

- *Be sure that you want to break up.* If the relationship is unsatisfactory, it may be because conflicts or problems have been avoided or have been confronted in the wrong way. Conflicts or problems, instead of being a reason to break up, may be a rich source of personal development if they are worked out. Sometimes people erroneously use the threat of breaking up as a way of saying, "I want the relationship to change."
- *Acknowledge that your partner will be hurt.* There is nothing you can do to erase the pain your partner will feel; it is only natural. Not breaking up because you don't want to hurt your partner may actually be an excuse for not wanting to be honest with him or her or with yourself.
- *Once you end the relationship, do not continue seeing your former partner as "friends" until considerable time has passed.* Being friends may be a subterfuge for continuing the relationship on terms wholly advantageous to yourself. It will only be painful for your former partner because he or she may be more involved in the relationship than you. It may be best to wait to become friends until your partner is involved with someone else (and by then, he or she may not care if you are friends or not).
- *Don't change your mind.* Ambivalence after ending a relationship is not a sign that you made a wrong decision; neither is loneliness. Both indicate that the relationship was valuable for you.

If your partner breaks up with you, keep the following in mind:

- *The pain and loneliness you feel are natural.* Despite their intensity, they will eventually pass. They are part of the grieving process that attends the loss of an important relationship, but they are not necessarily signs of love.
- *You are a worthwhile person, whether you are with a partner or not.* Spend time with your friends; share your feelings with them. They care. Do things that you like; be kind to yourself.
- *Keep a sense of humor.* It may help ease the pain. Repeat these clichés: No one ever died of love. (Except me.) There are other fish in the ocean. (Who wants a fish?)

Singlehood

A quick observation of demographics in this country point to a new and increasing way of life: singlehood. The trend, which has taken root and grown substantially since 1960, includes divorced, widowed, and never-married individuals. Each year more and more adult Americans are single.

According to the 2000 Census, 24 percent of the U.S. population, 18 and older, had never married. This percentage varies by race and ethnicity; 20.6 percent of non-Hispanic whites, 39.4 percent of African Americans, 28 percent of Hispanics, and 28.5 percent of Asian and Pacific Islanders had never married. Furthermore, an additional 10.1 percent of non-Hispanic whites, 11.6 percent of African Americans, 7.7 percent of Hispanics, and 4.6 percent of Asian and Pacific Islanders were divorced (Fields, 2000). Thus, the population of singles is quite large.

Over 68 million adult Americans (18 or older) are unmarried (divorced or never married). If one includes the 13.6 million widows and widowers, the number of "unmarrieds" swells to over 80 million Americans (Fields, 2000). The varieties of unmarried lifestyles in the United States are too complex to examine under any single category. They represent a diverse group: never married, divorced, young, old, single parents, gay men, lesbians, widows, widowers, and so on, and they live in diverse situations that affect how they experience their singleness. In research on singles, however, those who are generally regarded as "single" are young or middle-aged, heterosexual, not living with someone, and working rather than attending school or college. Although there are, of course, numerous single lesbians and gay men, they have not traditionally been included in such research.

SINGLES: AN INCREASING MINORITY

The growth in the percentage of never-married adults, from 20.3 percent in 1980 to 24 percent in 2000, has occurred across all population groups. This increase reflects a change in the way in which society views this way of life. Singles appear to be postponing marriage to an age which, for many, makes better economic and social sense (U.S. Census Bureau, 2000). The growing divorce rate is also contributing to the numbers of singles. In 2000, 8.8 percent of men and 10.8 percent of women 18 and over were divorced (Fields, 2000). The proportion of widowed men and women has declined somewhat but still remains similar to past numbers. Among older people, singlehood most often occurs because of the death of a spouse rather than by choice. Nevertheless, as society moves toward valuing individualism and choice, the numbers composing singlehood will most likely continue to grow.

DID YOU KNOW?

In 2000, nearly 27 million Americans aged 15 and older lived alone. Of these, 58 percent were females, 42 percent were males.

The number of single adults is rising as a result of several factors (Buunk and van Driel, 1989; Macklin, 1987):

- Delayed marriage, with a median age at first marriage of 26.8 for men and 25.1 for women in 2000 (U.S. Census Bureau, 2001). The longer one postpones marriage, the greater the likelihood of never marrying. As shown in Table 7.1, the percent of never-married men and women of typical "marrying ages" has dramatically increased in the past 30 years. It is estimated that between 8 and 9 percent of men and women now in their twenties will never marry.

TABLE 7.1 ■ Percent of Never-Married Women and Men, 1970–2000

	MALE		FEMALE	
AGE	1970	2000	1970	2000
20–24	35.8	83.7	54.7	72.8
25–29	10.5	51.7	19.1	38.1
30–34	6.2	30.1	9.4	21.9
35–39	5.4	20.3	7.2	14.3
40–44	4.9	15.7	6.3	11.8

SOURCE: Fields, 2000.

- Expanded lifestyle and employment options currently open to women.
- Increased rates of divorce and decreased likelihood of remarriage, especially among African Americans.
- Increased number of women enrolled in colleges and universities.
- More liberal social and sexual standards.
- Uneven ratio of unmarried men to unmarried women.

RELATIONSHIPS IN THE SINGLES WORLD

When people form relationships within the singles world, both the man and woman tend to remain highly independent. Singles work, and as a result, the individuals tend to be economically independent of each other. They may also be more emotionally independent because much of their energy may already be heavily invested in their work or careers. The relationship that forms consequently tends to emphasize autonomy and egalitarian roles. The fact that single women work is especially important. Single women tend to be more involved in their work, either from choice or necessity, but the result is the same: They are accustomed to living on their own without being supported by a man. The various factors that draw people to singlehood or marriage are illustrated in Table 7.2.

The emphasis on independence and autonomy blends with an increasing emphasis on self-fulfillment, which, some critics argue, makes it difficult for some to make commitments. Commitment requires sacrifice and obligation, which may conflict with ideas of "being oneself." A person under obligation can't necessarily do what he or she "wants" to do; instead, a person may have to do what "ought" to be done (Bellah et al., 1985).

According to Barbara Ehrenreich (1984), men are more likely to flee commitment, because they need women less than women need men. They feel oppressed by their obligation to be the family breadwinner. In the marital exchange, argues Ehrenreich, men need women less than women need men, because men make more money than women and can obtain many of the "services" provided by wives—such as cooking, cleaning, intimacy, and sex—outside marriage without being tied down by family demands and obligations. Thus, men may not have a strong incentive to commit, marry, or stay married.

Nevertheless, a recent study (Marks, 1996) revealed that "flying solo at mid-life" appears to be

TABLE 7.2 ■ Pushes and Pulls Toward Marriage and Singlehood

	PUSHES	PULLS
Toward Marriage	Cultural norms	Love and emotional security
	Loneliness	Physical attraction and sex
	Parental pressure	Desire for children
	Economic pressure	Desire for extended family
	Social stigma of singlehood	Economic security
	Fear of independence	Peer example
	Media images	Social status as "grown-up"
	Gult over singlehood	Parental approval
Toward Singlehood	Fundamental problems in marriage	Freedom to grow
	Stagnant relationship with spouse	Self-sufficiency
	Feelings of isolation with spouse	Expanded friendships
	Poor communication with spouse	Mobility
	Unrealistic expectations of marriage	Career opportunities
	Sexual problems	Sexual exploration
	Media images	

SOURCE: Adapted from Peter J. Stein. "Singlehood: An Alternative to Marriage." *The Family Coordinator* 24, 4 (1975).

more problematic for men than for women. Single women appear to have better psychological well-being than do single men. For those who were socialized during an era of traditional gender roles and family values with marriage as the norm, there seemed to be a degree of mental health risk associated with singlehood, especially for men.

CULTURE AND THE INDIVIDUAL VERSUS MARRIAGE

Despite the importance of intimate relationships for our development as human beings, our culture is ambivalent about marriage. This is nothing new. Paul of Tarsus (1 Corinthians 7:7–9) declared that it was best for people to remain chaste, as he himself had done. "But if they cannot," he wrote, "let them marry: For it is better to marry than to burn."

The tension between the alternatives of singlehood and marriage is diminishing as society begins to view singlehood as an option rather than a deviant lifestyle. The singles subculture is glorified in the mass media; the marriages portrayed on television are situation comedies or soap operas abounding in extramarital affairs. Yet many are rarely fully satisfied with being single and yearn for marriage. They are pulled toward the idea of marriage by their desires for intimacy, love, children, and sexual availability. They are also pushed toward marriage by parental pressure, loneliness, and fears of independence. Married persons, at the same time, are pushed toward singlehood by the limitations they feel in married life. They are attracted to singlehood by the possibility of creating a new self, having new experiences, and achieving independence.

TYPES OF NEVER-MARRIED SINGLES

Much depends on whether a person is single by choice and whether he or she considers being single a temporary or permanent condition (Shostak, 1987). If one is voluntarily single, his or her sense of well-being is likely to be better than that of a person who is involuntarily single. Arthur Shostak (1981, 1987) has divided singles into four types:

- *Ambivalents:* Ambivalents are voluntarily single and consider their singleness temporary. They are not seeking marital partners, but they are open to the idea of marriage. These are usually younger men and women who are actively pursuing education, career goals, or "having a good time." Ambivalents may be included among those who are cohabiting.
- *Wishfuls:* Wishfuls are involuntarily and temporarily single. They are actively seeking marital partners but have been unsuccessful so far. They consciously want to be married.
- *Resolveds:* Resolved individuals regard themselves as permanently single. A small percentage are priests, nuns, or single parents who prefer rearing their children alone. The largest number, however, are "hard-core" singles who simply prefer the state of singlehood.
- *Regretfuls:* Regretful singles prefer to marry but are resigned to their "fate." A large number of these are well-educated, high-earning women over 40 who find a shortage of similar men as a result of the marriage gradient.

Singles may shift from one type to another at different times. All but the resolveds share an important characteristic: They want to move from a single status to a romantic couple status. "The vast majority of never-married adults," writes Shostak (1987), "work at securing and enjoying romance." Never-married singles share with married Americans "the high value they place on achieving intimacy and sharing love with a special one."

SINGLES: MYTHS AND REALITIES

There are many longstanding myths about singles (Cargan and Melko,1982; Waehler, 1996). Although first identified more than 20 years ago, notice how familiar these notions still sound:

- *Singles are dependent on their parents.* Few differences exist between singles and marrieds in their perceptions of their parents and relatives. They do not differ in perceptions of parental warmth or openness and differ only slightly in the amount and nature of parental conflicts.
- *Singles are self-centered.* Singles value friends more than do married people.
- Singles are also more involved in community service projects.
- *Singles have more money.* Fewer than half the singles interviewed made more than $20,000 a year (adjusted to 1995). Married couples were better

off economically than singles, in part because both partners often worked.

- *Singles are happier.* Singles tend to believe that they are happier than marrieds, whereas marrieds believe that they are happier than singles. Single men, however, exhibited more signs of stress than did single women.
- *Singles view singlehood as a lifetime alternative.* The majority of singles expected to be married within five years. They did not view singlehood as an alternative to marriage but as a transitional time in their lives.

Cargan and Melko also determined that the following statements characterize singlehood more accurately:

- *Singles don't easily fit into married society.* Singles tend to socialize with other singles. Married people think that if they invite singles to their home, they must match them up with an appropriate single member of the other sex. Married people tend to think in terms of couples.
- *Singles have more time.* Singles are more likely to go out twice a week and much more likely to go out three times a week compared with their married peers. Singles have more choices and more opportunities for leisure activities.
- *Singles have more fun.* Although singles tend to be less happy than marrieds, they engage more in sports and physical activities, and have more sexual partners than do marrieds. Apparently, fun and happiness are not equated.
- *Singles are lonely.* Singles tend to be lonelier than married people; the feeling of loneliness is more pervasive for the divorced than for the never married.

GAY AND LESBIAN SINGLEHOOD

In the late nineteenth century, groups of gay men and lesbians began congregating in their own clubs and bars. There, in relative safety, they could find acceptance and support, meet others, and socialize. By the 1960s, some neighborhoods in the largest cities (such as Christopher Street in New York and the Castro district in San Francisco) became identified with gay men and lesbians. These neighborhoods feature not only openly lesbian or gay bookstores, restaurants, coffee houses, and bars, but also clothing stores, physicians, lawyers, hair salons—even driver's schools. They have gay churches, such as the Metropolitan Community Church, where gay men and lesbians worship freely; they have their own political organizations, newspapers, and magazines (such as *The Advocate*). They have family and child-care services oriented toward the needs of the gay and lesbian communities; they have gay and lesbian youth counseling programs.

In these neighborhoods men and women are free to express their affection as openly as heterosexuals; they experience little discrimination or intolerance; they are more involved in lesbian or gay social and political organizations. More recently, with increasing acceptance in some areas, many middle-class lesbians and gay men are moving to suburban areas. In the suburbs, however, they remain more discreet than in the larger cities (Lynch, 1992).

The urban gay male subculture that emerged in the 1970s emphasized sexuality. Although relationships were important, sexual experiences and variety were more important (Weinberg and Williams, 1974). This changed with the rise of the HIV/AIDS epidemic. Beginning in the 1980s, the gay subculture placed an increased emphasis on the relationship context of sex (Carl, 1986; Isensee, 1990). Relational sex has become normative among large segments of the gay population (Levine, 1992). Most gay men have sex within dating or love relationships. (In fact, some AIDS organizations are giving classes on gay dating in order to encourage safe sex.) One researcher (Levine, 1992) says of the men in his study: "The relational ethos fostered new erotic attitudes. Most men now perceived coupling, monogamy, and celibacy as healthy and socially acceptable."

Beginning in the 1950s and 1960s, young and working-class lesbians developed their own institutions, especially women's softball teams and exclusively female gay bars as places to socialize (Faderman, 1991). During the late 1960s and 1970s, **lesbian separatists,** lesbians who wanted to create a separate "womyn's" culture distinct from heterosexuals *and* gay men, rose to prominence. They developed their own music, literature, and erotica; they had their own clubs and bars. But by the middle of the 1980s, according to Lilian Faderman (1991), the lesbian community underwent a "shift to moderation." The community became more diverse, including Latina, African-American, Asian-American, and older women. It has developed closer ties with the gay community. They now view gay men as sharing much in common with them because of the common prejudice directed against both groups.

In contrast to the gay male subculture, the lesbian community centers its activities around couples. Lesbian therapist JoAnn Loulan (1984) writes: "Being single is suspect. A single woman may be seen as a loser no one wants. Or there's the 'swinging single' no one trusts. The lesbian community is as guilty of these prejudices as the world at large."

Lesbians tend to value the emotional quality of relationships more than the sexual components. Lesbians usually form longer-lasting relationships than gay men (Tuller, 1988). Lesbians' emphasis on emotions over sex and the more enduring quality of their relationships reflects their socialization as women. Being female influences a lesbian more than being gay.

Cohabitation

Few changes in patterns of marriage and family relationships have been as dramatic as changes in cohabitation. Over the past 40 years cohabitation has increased *tenfold.* Its increase across all socioeconomic, age, and racial groups makes it no longer a moral issue but rather a family lifestyle. It also appears to be a lifestyle that is here to stay

THE RISE OF COHABITATION

There are an estimated 5.5 million cohabiting couples in the United States, including 4.9 million heterosexual couples and, in addition, nearly 600,000 gay and lesbian couples (U.S. Census Bureau, 2002, Table 49). Forty years ago there were only approximately 400,000 such couples. Thus, we can see how steep an increase has occurred, especially since 1970 (see Figure 7.2).

In the United States, cohabiting couples still lack most of the rights that married couples enjoy. According to Judith Seltzer, children of cohabiting couples may also be disadvantaged unless they have legally identified fathers (Seltzer, 2000). This situation differs greatly in many other parts of the world. In Sweden, for instance, the law treats unmarried cohabitants and married couples the same in such areas as taxes and housing. In many Latin American countries, cohabitation has a long and socially accepted history as a substitute for formal marriage (Seltzer, 2000).

TYPES OF COHABITATION

There is no single reason to cohabit, just as there is no single type of person who cohabits. At least seven different reasons to cohabit are described below:

- *Temporary casual convenience.* Two people share the same living quarters because it is expedient and convenient to do so.
- *Affectionate dating or going together.* If two people live together because they enjoy being with each other, they will continue living together as long as it is mutually satisfying.
- *Economic advantage or necessity.* Even though approximately 80 percent of cohabitants are under age 45, older people find that they can retain their financial benefits by cohabiting if they are not married.
- *Trial marriage.* This includes persons who are "engaged to be engaged," as well as couples who are trying to discover if they want to marry each other.
- *Respite from being single.* For many, cohabiting offers a comfortable and oftentimes more familiar domestic lifestyle.
- *Temporary alternative to marriage.* Cohabiting may offer an alternative when the timing of a marriage must be postponed.
- *Permanent alternative to marriage.* Living together is an acceptable alternative for those who reject traditional marriage.

Living together has become more widespread and accepted in recent years for several reasons:

- *The general climate regarding sexuality is more liberal than it was a generation ago.* Sexuality is more widely considered to be an important part of a person's life, whether or not he or she is married. The moral criterion for judging sexual intercourse has shifted; love rather than marriage is now widely regarded as making a sexual act moral.
- *The meaning of marriage is changing.* Because of the dramatic increase in divorce over the last 25 years, marriage is no longer thought of as a nec-

FIGURE 7.2 ■ Cohabitation: 1960 to 2001

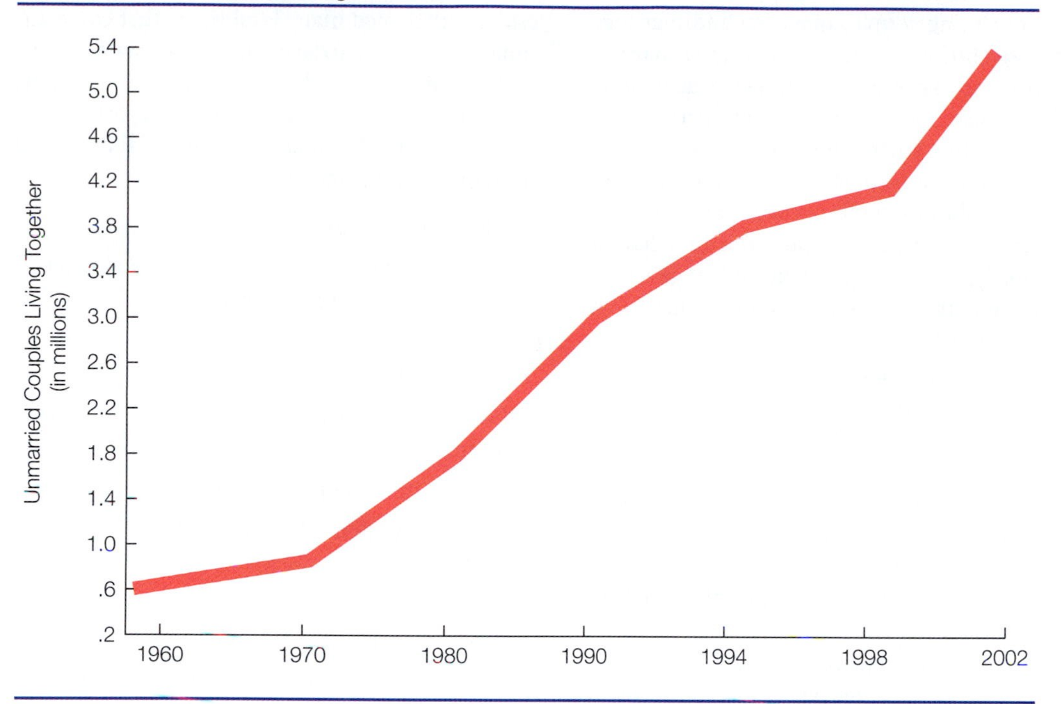

SOURCE: U.S. Bureau of the Census, 2001.

essarily permanent commitment. Permanence is increasingly replaced by serial monogamy—a succession of marriages. Because the average marriage now lasts only seven years, the difference between marriage and living together is losing its sharpness.

- *Men and women are delaying marriage longer.* More than half of cohabiting couples eventually marry (Smock, 2000). As long as children are not desired, living together offers advantages for many couples. When children are wanted, however, the couple will usually marry so that the child will be "legitimate."

Cohabitation does not seem to threaten marriage. Eleanor Macklin (1987), one of the major researchers in the field, notes:

> Non-marital cohabitation in the United States serves primarily as a part of the courtship process and not as an alternative to marriage. The great majority of young persons plan to marry at some point in their lives . . . and most cohabiting relationships either terminate or move into legal marriage within a year or two.

The most notable social impact of cohabitation is that it delays the age of marriage for those who live together. Ideally, cohabitation could actually encourage more stable marriages because the older a person is at the time of marriage, the less likely he or she is to divorce.

DID YOU KNOW?

Although cohabitation has increased for all educational groups and for whites, Latinos and African Americans, it is more common among those with lower levels of education and income (Seltzer, 2000)

Concomitantly, as the age of marriage continues to increase, cohabitation may become the preferred form for premarital bonding among those who are beyond the early adult years (Cate and Lloyd, 1992).

Although there may be a number of advantages to cohabitation, there are also disadvantages. Parents may refuse to provide support for school as long as their child is living with someone, or they may not welcome their child's partner into their home. Cohabiting couples may also find that they cannot easily buy houses together, as banks may not count their income as joint; they also usually don't qualify for insurance benefits. If one partner has children, the other partner is usually not as involved with the children as he or she would be if they were married. Cohabiting couples who live together may find themselves socially stigmatized if they have a child. Finally, cohabiting relationships generally don't last more than two years; couples either break up or get married.

Living together takes on a different quality among those who have been previously married. About 40 percent of cohabiting relationships have at least one previously married partner. Remarriage rates have dipped, in part, as postmarital cohabitation has increased (Seltzer, 2000). About a third of all cohabiting couples have children from their earlier relationships. As a result, the motivation in these relationships is often colored by painful marital memories and the presence of children (Bumpass and Sweet, 1990). In these cases, men and women tend to be more cautious about making their commitments. The majority of remarriages are preceded by cohabitation (Ganong and Coleman, 1994). Even though cohabiting couples are less likely to stay together compared to married couples, having children in the household somewhat stabilizes the couples (Wu, 1995).

DOMESTIC PARTNERSHIPS

Domestic partners—cohabiting heterosexual, lesbian, and gay couples in committed relationships—are gaining some legal rights. Domestic partnership laws, which grant some of the protection of marriage to cohabiting partners, are increasing the legitimacy of cohabitation. In 1997, San Francisco extended health insurance and other benefits to their employees' domestic (which includes same-sex) partners. It was the first such city ordinance in the nation. Individual employers, such as the Gap, Levi Strauss, and Walt Disney Company already have started domestic partner policies. Sweden, Iceland, and Denmark legally recognize domestic partnerships. In Denmark and Iceland, lesbians and gay men have the same rights and responsibilities in domestic partnerships as do married couples, except in matters involving adoption and child custody.

Domestic partners, whether heterosexual, gay, or lesbian, are denied many legal rights that come automatically with marriage. Thus, as we saw in Chapter 1, cohabiting couples live without many rights and benefits that their married counterparts obtain upon marrying. Recalling only some of those noted in Chapter 1, these include the right to

- File joint tax returns
- Automatically make medical decisions if your partner is injured or incapacitated
- Automatically inherit your partner's property if he or she dies without a will
- Enter hospitals, jails, and other places restricted to "immediate family"
- Create a marital life estate trust
- Claim the unlimited marital deduction from estate taxes
- Receive survivor's benefits
- Obtain health and dental insurance, bereavement leave, and other employment benefits
- Collect unemployment benefits if you quit your job to move with a partner who has obtained a new job
- Live in neighborhoods zoned "family only"
- Get residency status for a noncitizen partner to avoid deportation

Of course, we need to keep in mind that heterosexual domestic partnerships and same-sex domestic partnerships may be motivated for different reasons. Among heterosexuals, domestic partnership is a deliberately chosen alternative to marriage. For at least some gay and lesbian couples, domestic partnerships are the closest approximation to legal marriage available to them. Many same-sex couples would prefer to be married if marriage was an option.

GAY AND LESBIAN COHABITATION

The 2000 U.S. Census reported nearly 600,000 gay or lesbian couples living together. Other estimates put the number at over 1.5 million same-sex cohabiting couples. The relationships of gay men and lesbians have been stereotyped as less committed than heterosexual couples because (1) lesbians and gay men cannot legally marry, (2) they may not emphasize as strongly sexual exclusiveness, and (3) heterosexuals misperceive love between gay and lesbian couples as being somehow less "real" than love between heterosexual couples.

Numerous similarities exist between gay and heterosexual couples, according to Ann Peplau (1981, 1988). Regardless of their sexual orientation, most people want a close, loving relationship with another person. For lesbians, gay men, and heterosexuals, intimate relationships provide love, romance, satisfaction, and security. There is one important difference, however. Heterosexual couples tend to adopt a traditional marriage model, whereas gay couples tend to have a "best friend" model. Peplau (1988) :

> A friendship model promotes equality in love relationships. As children, we learn that the husband should be the "boss" at home, but friends "share and share alike." Same-sex friends often have similar interests, skills, and resources—in part because they are exposed to the same gender-role socialization in growing up. It is easier to share responsibilities in a relationship when both partners are equally skilled or inept at cooking, making money, and disclosing feelings.

With this model, tasks and chores are often shared, alternated, or done by the person who has more time. Usually, both members of the couple support themselves; rarely does one financially support the other (Peplau and Gordon, 1982).

Few lesbian and gay relationships are divided into the traditional heterosexual provider/homemaker roles. Among heterosexuals, these divisions are gender linked as male or female. But in cases in which the couple consists of two men or two women, these traditional gender divisions make no sense. As one gay male remarked, "Whenever I am asked who is the husband and who is the wife, I say, 'We're just a couple of happily married husbands.'" Tasks are often divided pragmatically, according to considerations such as who likes cooking more (or dislikes it less) and work schedules (Marecek, Finn, and Cardell, 1988). Most gay couples are dual-earner couples; neither partner supports or depends on the other economically. Furthermore, because gay and lesbian couples are the same gender, the economic discrepancies based on greater male earning power are absent. One partner does not necessarily have greater power than the other based on income. Although gay couples emphasize egalitarianism, if there are differences in power they are attributed to personality; if there is a significant age difference, the older partner is usually more powerful (Harry, 1988).

COHABITATION AND MARRIAGE COMPARED

Different Commitments

A lesser level of commitment characterizes cohabiting couples when compared to married couples. When a couple lives together, their primary commitment is to each other but it is a more transitory commitment. As long as they feel they love each other, they will stay together. In marriage, the couple makes a commitment not only to each other but to their marriage. Cohabitants are less committed both to the institution of marriage and to the certainty of a future together (Waite and Gallagher, 2001; Forste and Tanfer, 1996; Schwartz, 1983).

Living together tends to be a more temporary arrangement than marriage (Seltzer, 2000; Teachman and Polonko, 1990). Half of cohabiting relationships end within a year because the couple either marries or breaks up. Cohabiting couples are three times as likely as married couples (29 percent versus 9 percent for married couples) to break up within two years (Seltzer, 2000). A man and woman who are living together may not work as hard to save their relationship. They are less certain of a lifetime together and live somewhat more autonomous lives. In marriage, each partner will do more to save their marriage; they may give up dreams, work, ambitions, and extramarital relationships to make a marriage work.

Although society encourages married couples to make sacrifices to save their marriage, unmarried couples rarely receive the same support. Parents may even urge their "living together" children to split up rather than give up plans for school or a career. If the couple is beginning to encounter sexual difficulties, it is more likely that they will split up if they are cohabiting than if they are married. It may be easier to abandon a problematic relationship than to change it. Among cohabitors who intend to marry, relationships are not significantly different from marriages. The intention to marry is highest among cohabiting couples with high incomes (Brown and Booth, 1996).

Sex

There are differences in the sexual relationships and attitudes of cohabiting and married couples. Waite and Gallagher (2001) suggest that married couples experience more fulfilling sexual relationships because of their long-term commitment to each other and their emphasis on *exclusivity.* Because they expect to remain together, married couples have more incentive to work on their sexual relationships and discover what most pleases their partners.

Cohabitants have more frequent sexual relations. Whereas 43 percent of married men reported that they had sexual relations at least twice a week, 55 percent of cohabiting men said they had sex two or three times a week or more. Among married women, 39 percent said that they had sex at least twice a week compared with 60 percent of never-married cohabiting women. Sex may also be a more important aspect of cohabiting relationships than it is of marriages. Waite and Gallagher (2001) go as far as calling it the "defining characteristic" of cohabitants' relationships.

Married men and women are also more likely to be sexually monogamous. According to data from the National Sex Survey (see next chapter), 4 percent of married men said they had been unfaithful over the past 12 months; four times as many, 16 percent, of the cohabitants reported infidelity. Among women, the equivalent comparison shows that 1 percent of married women compared with 8 percent of cohabiting women expressed having had sex outside of their relationship. Similar findings were obtained by Treas and Giesen (2000):

> We found cohabitors more likely than married people to engage in infidelity even when we controlled for permissiveness of personal values regarding extramarital sex. This finding suggests that cohabitors' lower investments in their unions, not their unconventional values, accounted for their greater risk of infidelity. (p. 59)

Finances

A striking difference between cohabiting and married couples is the pooling of money as a symbol of commitment (Waite and Gallagher, 2001; Blumstein and Schwartz, 1983). People generally assume that in marriage, the couple will pool their money. This arrangement suggests a basic trust or commitment to the relationship: the individual is willing to sacrifice his or her particular economic interests to the interests of the relationship. Among most cohabiting couples, money is not pooled. In fact, one of the reasons couples cohabit rather than marry is to maintain a sense of financial independence. One man said, "A strong factor in the success of the rela-

Heterosexuals, gay men, and lesbians cohabit. A significant difference between heterosexual and gay cohabitation is that many gay men and lesbians who would like to marry are prohibited by law from doing so.

© David Young-Wolff/PhotoEdit

tionship is the fact that we're economically independent of each other. We make no decisions that involve joint finances and that simplifies life a great deal" (Waite and Gallagher, 2001; Blumstein and Schwartz, 1983).

Only if the couple expects to be living together for a long time or to marry do they pool their income. As Blumstein and Schwartz (1983) point out:

> Since the majority of cohabitors do not favor pooling, these facts say important things about pooling and commitment. When couples begin to pool their finances, it usually means they see a future for themselves. The more a couple pools, the greater the incentive to organize future financial dealings in the same way. As a "corporate" sense of the couple emerges, it becomes more difficult for the partners to think of themselves as unattached individuals.

Relationship Quality and Mental Health

Research by Brown and Booth (1996) indicates that cohabiting couples have poorer relationship quality than do married couples. Specifically, cohabiting couples report lower levels of happiness with their relationships, more fighting, and more violence than do married couples. However, these differences disappear or greatly diminish when one considers only cohabitants who have expressed the intention to marry (Brown and Booth, 1996). In fact, Brown and Booth point out that among the more than 75 percent of cohabitors who plan to marry their partners, their relationships are not qualitatively different from marriage (Brown and Booth, 1996).

Researchers have looked at the mental health characteristics of cohabitants as they compare to singles and married couples (Ross, 1995; Horwitz and White, 1998). Some research suggests that cohabitants are more like married people. Married and cohabiting individuals have similar levels of depression, and both are less depressed than those without partners (Ross, 1995). However, Horwitz and White's research comparing rates of depression and alcohol problems among unmarried cohabiting, married, and single people found cohabitants to have higher rates of both depression and alcohol problems than married people. The mental health of cohabitors was more like single people than married ones. Furthermore, cohabiting men had the highest rate of alcohol problems of the three groups, suggesting that something about cohabitation (for example, unconventionality or financial pressures among those wishing to marry) may cause high rates of alcohol problems. Among married men, those who cohabited before marriage were no different in their level of alcohol problems than those who had not first cohabited (Horwitz and White, 1998).

Work

Traditional marital roles call for the husband to work; it is left to the discretion of the couple whether the woman works. The husband is basically responsible for supporting his wife and family. Of course, contemporary families often cannot afford the luxury of a one-wage earner household. Still, gender roles in marriage have emphasized men's economic provision as a major component of men's family responsibilities. In cohabiting relationships, the man is not expected to support his partner (Blumstein and Schwartz, 1983). If the woman is not in school, she is expected to work. If she is in school, she is nevertheless expected to support herself.

Some married couples may fight about the wife's going to work; such fights do not generally occur among cohabiting couples. With less certainty about the future of their relationships, cohabiting women may be less willing to restrict their outside employment or to spend time and energy that could be spent on paid work on housework.

Married couples often disagree about the division of household work. Both married and cohabiting women tend to do more of the domestic work than their male partners. But cohabiting women spend less time on housework than do married women, as the homemaker role is not as significant for them. Married women may spend as many as six hours more per week on housework tasks than cohabiting women spend (Waite and Gallagher, 2001; Seltzer, 2000; Shelton and John, 1993).

Societal Support

Compared with marriage, cohabitation receives less societal support, except from peers. It may be considered an inferior or immoral relationship—inferior to marriage because it does not symbolize lifetime commitment, and immoral because it involves a sexual relationship without the sanction of marriage.

This lack of social reinforcement is an important factor in the greater instability of living-together relationships. Parents usually do not support cohabitation with the same enthusiasm as they would marriage. (If they do not like their son's or daughter's partner, however, they may console themselves with the thought that "at least they are not married.") Research indicates that parents' attitudes toward cohabitation in general become more positive if their own children cohabit (Axinn and Thornton, 1993).

Unmarried couples often find the greatest amount of social support from their friends, especially other couples who are living together. They are able to share similar problems with fellow cohabitants, such as whether to tell parents, how to handle visiting home together, difficulties in obtaining housing, and so forth. Because unmarried couples tend to have similar values, commitments, and uncertainties, they are able to give one another support in the larger noncohabiting world.

IMPACT OF COHABITATION ON MARITAL SUCCESS

Although it may seem surprising and goes against the logic used by cohabiting couples who think that cohabitation helps prepare them for marriage, as we've noted before, such couples are actually more likely to divorce than those who do not live together before marriage (Bumpass & Lu, 2000). In marriages that were previously cohabiting relationships, there are higher levels of disagreement and instability, lower levels of commitment, and, thus, a greater likelihood of divorce.

What is still unclear is what it is about cohabitation that causes later marital difficulties. Is it the *types of people* who choose to live together before marrying? Is it something about the *experience of living together* that causes problems later? Brown and Booth (1996) suggest that these outcomes may result more from the characteristics of people who cohabit rather than from the cohabiting experience itself. People who live together before marriage tend to be more liberal, more sexually experienced, and more independent than people who do not live together before marriage. They also tend to have slightly lower incomes and are slightly less religious than noncohabitants (Smock, 2000).

At the same time, there is evidence that cohabitation itself may affect individual partners and their relationships. Compared with married couples, cohabiting partners tend to have more similar incomes and divide household tasks more equally. These arrangements may be harder to sustain once married, and strain or conflict may occur (Seltzer, 2000).

DID YOU KNOW?

Children often turn cohabitation into marriage. Cohabiting couples in which the woman becomes pregnant have a greater likelihood of marrying than cohabiting couples where no pregnancy occurs. Also, cohabiting couples who already have children from previous relationships are more likely to marry than couples who don't have children (Seltzer, 2000).

As more and more people across different backgrounds enter cohabitation relationships, we will be better positioned to see whether the *experiences of cohabitation* or *characteristics of cohabitors* have greater impact on later marriage. As cohabitation grows in number and acceptability, its effects on marriage may also change. One thing we can suggest is that at least some poorly chosen relationships break up at the cohabitation stage. Thus, although it may not protect couples from later marital failure, it does show some high-risk couples that they were not meant for each other. This spares them the later experience of a divorce (Seltzer, 2000).

More and more individuals experience a longer period of singlehood as the age of marriage increases. For many people, being single is a transitional state before cohabitation or marriage. For others singlehood also occurs following divorce or at the end of a cohabiting relationship. Still others choose to remain single throughout their lives. Many of those who are single are lesbian or gay, but they often form cohabiting relationships.

As we have seen, whom we choose as a partner is a complex matter. Our choices are governed by rules of homogamy and exogamy as much as by the heart. But the process of dating or cohabiting helps us determine how well we fit with each other. While these relationships may sometimes be viewed as a prelude to marriage, they are important in their own right. Whatever their outcome, these relationships provide a context for love and personal development.

SUMMARY

- The marriage marketplace refers to the selection activities of men and women when sizing up someone as a potential date or mate. In this marketplace, each person has resources, such as social class, status, age, and physical attractiveness. The marital exchange is based on gender roles. Traditionally, men offer status, economic resources, and protection; women offer nurturing, childbearing, homemaking skills, and physical attractiveness. Recent changes in women's economic status have given women more bargaining power; the decline in the value of housekeeping and children and the increased availability of sexual relations in the singles world have given men more bargaining power.
- The ratio of unmarried men to women also may affect bargaining power in the marriage marketplace. The *marriage squeeze* refers to the gender imbalance reflected in the ratio of available unmarried women to men. Overall, there are significantly more unmarried women than men, but in the age group of 15 to 39 years, there are significantly more unmarried men than women (except among African Americans, where women significantly outnumber men). Marital choice is also affected by the *mating gradient,* the tendency for women to marry men of higher status.
- Initial impressions are heavily influenced by physical attractiveness. A *halo effect* surrounds attractive people, from which we infer that they have certain traits, such as warmth, kindness, sexiness, and strength. The "rating and dating" game is the evaluation of men and women by their appearance. Most people, however, choose equals in terms of looks, intelligence, and education. If there is an appreciable difference in looks, there is some kind of trade-off in which a lower-ranked trait is exchanged for a higher-ranked one.
- The *field of eligibles* consists of those of whom our culture approves as potential partners. It is limited by the principles of *endogamy* (marriage within a particular group) and *exogamy* (marriage outside a particular group). The field of eligibles is further limited by *homogamy* (the tendency to choose a mate whose individual or group characteristics are similar to ours) and by *heterogamy* (the tendency to choose a mate whose individual or group characteristics are different from ours). Homogamy is especially powerful in our culture, particularly in terms of race and ethnicity, religion, socioeconomic status, and age.
- The theories that attempt to explain mate selection include parental image, complementary needs, value, and role theories.
- Bernard Murstein explained romantic relationships in terms of the three-stage *stimulus-value-role theory.* In the stimulus stage, each person is attracted to the other before the actual interaction. In the value stage, each weighs the other's basic values for compatibility. In the role stage, each person analyzes the other's behaviors in roles as lover, companion, and so on.
- Beginning a relationship typically depends on seeing, meeting, and dating another person. The setting in which you see someone can facilitate or discourage a meeting. A *closed field* allows you to see and interact more or less simultaneously. An *open field,* characterized by large numbers of people who do not ordinarily interact, makes meeting more difficult. When meetings do occur, women often covertly initiate them by sending nonverbal signals of availability and interest. Men then initiate conversation with an opening line. African-American churches are especially important for middle-class African Americans as meeting places. For lesbians and gay men, the problem of meeting is accentuated because they cannot assume that the person in whom they are interested shares their sexual orientation. Instead, they must rely on identifying cues.
- Men tend to initiate dates directly, whereas women tend to initiate dates indirectly. Power tends to be more equal in dating relationships than in marriage. Having dating alternatives is an important element in male power; women must determine what they wish to express and how they wish to keep those behaviors that give them strength. Strong self-esteem is essential for healthy relationships. Divergent gender-role conceptions are problems for both genders in dating. For women, problems in dating include sexual pressure, communication, and where to go

on the date; for men, problems include communication, where to go, shyness, and money.

- Relationships in the singles world tend to stress independence and autonomy. The emphasis on individual self-fulfillment works against the making of commitments. The marital exchange favors men and may also work against their making a commitment. Singles may be classified into four categories: (1) ambivalents, who are voluntarily and temporarily single; (2) wishfuls, who are involuntarily and temporarily single; (3) resolveds, who are permanently single by choice; and (4) regretfuls, who are permanently and involuntarily single.
- Women and men in the gay world often gather in gay neighborhoods where they may openly socialize. Single gay men increasingly focus on relationships rather than sex alone. Lesbians emphasize relationships and are less supportive of women outside relationships.
- *Domestic partnership* laws grant some legal rights to cohabiting couples, including gay and lesbian couples. Cohabitation has become increasingly accepted because of a more liberal sexual climate, the changed meaning of marriage, and delayed marriage. Reasons for cohabitation include: temporary casual convenience, affectionate dating or going together, trial marriage, temporary alternative to marriage, and permanent alternative to marriage.
- Between 600,000 and 1.5 million gay men and lesbians cohabit. Whereas heterosexual cohabiting couples tend to adopt a traditional marriage model, lesbians and gay men utilize a "best friend" model that promotes equality in roles and power.
- Compared with marriage, cohabitation is more transitory, has different commitments, lacks economic pooling, and has less social support.

KEY TERMS

closed field 210
complementary needs theory 209
domestic partner 224
endogamy 206
exogamy 206
field of eligibles 206
halo effect 201
heterogamy 206
homogamy 206
hypergamy 208
hypogamy 207
lesbian separatist 221
marriage squeeze 203
mating gradient 204
open field 210
parental image theory 209
residential propinquity 208
stimulus-value-role theory 209

SUGGESTED READINGS

Duck, Steve, ed. *Dynamics of Relationships.* Thousand Oaks, CA: Sage, 1994. A collection of scholarly essays exploring the interpersonal skills necessary to build and maintain successful relationships.

———. *Meaningful Relationships: Talking, Sense, and Relating.* Thousand Oaks, CA: Sage, 1994. An examination of relationship dynamics from a symbolic interactionist perspective: Relationships are based on shared meanings conveyed through our everyday conversations and symbols.

Goss, Robert. *Our Families, Our Values: Snapshots of Queer Kinship.* Binghamton, NY: Haworth Press, Inc., 1997. An exploration of the various ongoing efforts to give religious pride to the various configurations of gay relationships, families, and values.

Hatfield, Elaine, and Susan Sprecher. *Mirror, Mirror: The Importance of Looks in Everyday Life.* Albany, NY: State University of New York Press, 1985. An important survey concerning the significance of looks in relationships and our daily lives.

Hendrick, Susan, and Clyde Hendrick. *Liking, Loving, and Relating,* 2nd ed. Pacific Grove, CA: Brooks/Cole, 1992. An excellent academic introduction to the field of close relationships, including the close-relationships perspective, an overview of current research, research methodology, and ethical issues in research.

Miell, Dorothy, and Rudi Dallos, eds. *Social Interactions and Personal Relationships.* Newbury Park, CA: Sage, 1996. A clearly written book that explores the interactions between people.

Regan, Pamela. *The Mating Game: A Primer on Love, Sex, and Marriage.* Newbury Park, CA: Sage, 2002. A well-written compilation of research on couple relationships.

RESOURCES ON THE INTERNET

Companion Web Site for This Book

http://sociology.wadsworth.com/strong/marriage9e
Gain an even better grasp on this chapter by going to the companion Web site to take one of the Tutorial Quizzes, use the Flash Cards to master key terms, or check out the many other study aids you'll find there. Visit the Marriage and Family Resource Center on the site. You'll also find special features such as GSS Data and Census 2000 information that will put data and resources at your fingertips to help you with that special project or to do some research on your own.

InfoTrac College Edition: Search Word Summary

dating	endogamy
lesbians	unmarried couples
commitment	arranged marriage

To learn more about these central topics in the study of the family, you can conduct an electronic search using InfoTrac College Edition. To aid in your search and to gain useful tips, see the Student Guide to InfoTrac College Edition that you can access through the companion Web site for this book.

Preview

To gain a sense of what you already know about the material covered in this chapter, answer "True" or "False" to the statements below.

1 Unlike most human behavior, sexual behavior is instinctive. True or false?

2 A significant number of women require manual or oral stimulation of the clitoris to experience orgasm. True or false?

3 It is normal for children to engage in sexual experimentation with other children of both sexes. True or false?

4 Both men and women report feelings of obligation or pressure to engage in sexual intercourse. True or false?

5 A decline in the frequency of intercourse almost always indicates problems in the marital relationship. True or false?

6 Most married women and men have had an extramarital sexual relationship. True or false?

7 The American Psychiatric Association has rejected the idea of homosexuality as a form of mental disorder. True or false?

8 Latinos are generally less permissive about sex than African Americans or Anglos. True or false?

9 Because of their knowledge, college students rarely put themselves at risk for HIV/AIDS. True or false?

10 Condoms are not very effective as contraceptive devices. True or false?

"To approach sex carelessly, shallowly, with detachment, and without warmth is to dine night after night in erotic greasy spoons. In time, one's palate will become insensitive, one will suffer (without knowing it) emotional malnutrition, the skin of the soul will fester with scurvy, the teeth of the heart will decay."

TOM ROBBINS

CHAPTER 8

Understanding Sexuality

Outline

It is now time to consider sex. For many of you, that must seem like a silly statement. Quite apart from this book, we often consider, think about, or take steps to pursue—or avoid—sexual encounters. Our popular culture is heavilly sexualized. Advertising, in particular, uses sexual innuendo and image to sell us any number of products. Furthermore, being sexual is an essential part of being human. Through our sexuality we are able to connect with others on the most intimate levels, revealing ourselves and creating deep bonds and relationships. Sexuality is a source of great pleasure and profound satisfaction. It is the means by which we reproduce, bringing new life into the world and transforming ourselves into mothers and fathers. Paradoxically, sexuality also can be a source of guilt and confusion, a pathway to infection, and a means of exploitation and aggression. Examining the multiple aspects of sexuality helps us understand our own sexuality and that of others. It provides the basis for enriching our relationships.

In this chapter, we offer an overview of sexuality and sexual issues. We begin by considering the sources of our sexual learning and proceed through psychosexual development in young, middle, and later adulthood, including the gay/lesbian/bisexual identity process. We will consider the shifts in sexual scripts from traditional to modern and the social control of sexuality, When we examine sexual behavior, we will cover the range of activities and relationships in which people engage. Ultimately, we will also look at nonconsensual sexual relations, sexual problems and dysfunctions; birth control; sexually transmissible diseases and HIV/AIDS; and sexual responsibility. We hope that this chapter will help you make sexuality a positive element in your life and relationships.

Psychosexual Development in Young Adulthood

PREVIEW ANSWERS

1 False
2 True
3 True
4 True
5 False
6 False
7 True
8 True
9 False
10 False

At each period in our psychosexual development, we are presented with different challenges. Adolescents are sexually mature (or close to it) in a physical sense, but they are still learning their gender and social roles; they may also be struggling to understand the meaning of their sexual feelings for others and their sexual orientation. During young adulthood—from the late teens through mid-thirties—many of the same tasks continue and new ones are added.

SOURCES OF SEXUAL LEARNING

Before we examine the developmental tasks of young adulthood, let's look at some of the sources of our sexual learning.

Parental Influence

Children learn a great deal about sexuality from their parents. For the most part, however, they learn not because their parents set out to teach them but because they are avid observers of their parents' behavior. Much of what they learn concerns the hidden nature of sexuality (Roberts, 1983): "The silence that surrounds sexuality in most families and in most communities carries its own important messages. It communicates that some of the most important dimensions of life are secretive, off limits, bad to talk about or think about."

Parents further convey sexual attitudes when they react to their children's sexual curiosity or expression. How parents react, for example, to children who touch their "private parts" or try to touch either their mother's breasts or some other woman's breasts, conveys meanings about sex to a child. Parents who overreact may create a sense that sex is wrong. On the other hand, parents who acknowledge sexuality rather than ignoring or condemning it, help children develop positive body images, comfort with sexual matters, and higher self-esteem (Miracle, Miracle, and Baumeister, 2003).

As young people enter adolescence, they are especially concerned about their own sexuality, but they are often too embarrassed or distrustful to ask their parents directly about these "secret" matters, and most parents are ambivalent about their children's developing sexual nature. They are often fearful that their children (daughters especially) will become sexually active if they have too much information. They tend to indulge in wishful thinking: "I'm sure Jenny's not really interested in boys yet;" "I know Joey would never do anything like that." Parents may put off talking seriously with their children about sex, waiting for the "right time," or they may bring up the subject once, say their piece, breathe a sigh of relief, and never mention it again. Sociologist John Gagnon calls this the "inoculation" theory of sex education: "Once is enough" (cited in Roberts, 1983). But children need frequent "boosters" where sexual knowledge is concerned. When a parent does undertake to educate a child about sex, it is usually the mother. Thus, most children grow up believing that sexuality is an issue that men don't deal with unless they have a specific problem.

Although parental norms and beliefs are generally influential, they do not appear to have a strong effect on an adolescent's decision to become sexually active. (Peers seem to be the more important factor.) But a lack of rules and structure also seems to be related to more permissive sexual attitudes and premarital sex among adolescents (Forste and Heaton, 1988; Hovell et al., 1994). Parental communication does have an impact on adolescent sexual behavior, as does the presence of two parents in the home (Miracle, Miracle, and Baumeister, 2003). A strong bond with parents also appears to lessen teens' dependence on the approval of their peers and to lessen the need for interpersonal bonding that may lead to sexual relationships (DiBlasio and Benda, 1992; Miller and Fox, 1987)

Parents also have an impact on whether an adolescent will use contraception (Baker, Thalberg, and Morrison, 1988). Research indicates that parental concern and involvement with sons and daughters is a key factor in preventing teenage pregnancy (Hanson et al., 1987). Instilling values such as respect for others and responsibility for one's actions provides a context in which young people can use their knowledge about sex. A study of African-American adolescents (Scott-Jones and Turner, 1988) found that the majority of parents not only gave their daughters information about sex but also instructed them about contraception. Mothers and grandmothers are especially important sources of sex information for African-American adolescents (Tucker, 1989). The strategy of advising "Don't have sex; but if you do, use a condom" may actually be effective in helping prevent adolescent pregnancies and sexually transmissible diseases. Sex educator Sol Gordon (1984) recommends this "double standard of sex education."

Peer Influence

Adolescents garner a wealth of information and *misinformation* from one another about sex. They also put pressure on one another to carry out traditional gender roles. Boys encourage other boys to be sexually active even if the others are unprepared or uninterested. Those who are pressured must camouflage their inexperience with bravado, which increases misinformation; they cannot reveal sexual ignorance. Comedian Bill Cosby (1968) recalled the pressure to have sexual intercourse as an adolescent: "But how do you find out how to do it without blowin' the fact that you don't know how to do it?" On his way to his first sexual encounter he realized that he didn't have the faintest idea of how to proceed:

> So now I'm walkin', and I'm trying to figure out what to do. And when I get there, the most embarrassing thing is gonna be when I have to take my pants down. See, right away, then, I'm buck naked . . . buck naked in front of this girl. Now, what happens then? Do . . . do you just . . . I don't even know what to do . . . . I'm gonna just stand there and she's gonna say, "You don't know how to do it." And I'm gonna say, "Yes, I do, but I forgot." I never thought of her showing me, because I'm a man and I don't want her to show me. I don't want nobody to show me, but I wish somebody would kinda slip me a note.

Even though many teenagers find their early sexual experiences less than satisfying, many still seem to feel a great deal of pressure to conform, which means continuing to be sexually active (DiBlasio and Benda, 1992). The following are typical statements from a group of teenagers in one study (De Armand, 1983): "I had to do it. I was the only virgin" (from a 15-year-old girl); "If you want a boyfriend, you have to put out" (from a 13-year-old girl); "You do what your friends do or you will be bugged about it" (from a 12-year-old boy). The students interviewed also said that "everyone" has to make the decision about having sexual intercourse by age 14. Researcher Charlotte De Armand (1983) comments that although, of course, "not 'everyone' is involved, certainly a large number of young people are making this decision at 14 or younger." Encouragingly, four large national probability samples from the Youth Risk Behavior Survey indicate that the 1990s saw a shift in adolescent sexual activity. According to data collected between 1991 and 1997, there was an 11 percent increase in the "incidence of virgin adolescents." This shift occurred mostly among males and among blacks and whites (but not Hispanics). Still, it represents a "significant reversal" from the 1970s and 1980s (Christopher and Sprecher, 2000)

DID YOU KNOW?

Although most research and public discussion of adolescent sexuality focuses on sexual intercourse, noncoital sexual activity (e.g., oral sex, mutual masturbation, etc.) is common among those who have and those who haven't had sexual intercourse (Miracle, Miracle, and Baumeister, 2003).

Media Influence

The media have a profound impact on our sexual attitudes (Wolf and Kielwasser, 1991; McMahon, 1990). Anthropologist Michael Moffatt (1989) noted that about a third of the students he studied at Rutgers University mentioned the impact of college and college friends on their sexual development and another third mentioned their parents and religious values. But he found that the major influence on their sexuality was contemporary American pop culture:

> The direct sources of the students' sexual ideas were located almost entirely in mass consumer culture: the late-adolescent/young-adult exemplars displayed in movies, popular music, advertising, and on TV; Dr. Ruth and sex manuals; *Playboy, Penthouse, Cosmopolitan, Playgirl,* etc.; Harlequins and other pulp romances (females only); the occasional piece of real literature; sex education and popular psychology as it had filtered through these sources, as well as through public schools, and as it continued to filter through the student-life infrastructure of the college; classic soft-core and hard-core pornographic movies, books, and (recently) videocassettes.

To these we can add the Internet, with its numerous sexually oriented Web sites, and DVDs, which increasingly are replacing videocassettes.

REFLECTIONS

Think about your different sources of sexual learning—parents, peers, the media, and partners. How did each influence your sexual development? Did their messages complement or compete with each other? Which influences were strongest? Why?

Partner Influence

Parents, peers, and the media become less important in our sexual learning as we get older, being replaced by our sexual partners. The experience of interpersonal sexuality is ultimately the most important source of modifying traditional sexual scripts. Linda Levine and Lonnie Barbach (1985) describe the sources of men's sexual learning:

> Before their first sexual encounter, men could only rely on secondary sources for information about sex. But once they lost their virginity, women became their primary source of information. It was their continued sexual experience that ultimately expanded and enriched men's sexual repertoire. Their skill at the game of love evolved over time, through trial and error. Each

experience left them with a clearer sense of themselves as sexual men. But until they acquired this self-confidence, many men were reluctant to drop their he-man façades and reveal to their partners that they were less than skilled lovers.

In relationships, men and women learn that the sexual scripts and models they learned from parents, peers, and the media do not necessarily work in the real world. They adjust their attitudes and behaviors in everyday interactions. If they are married, sexual expectations and interactions become important factors in their sexuality.

DEVELOPMENTAL TASKS IN YOUNG ADULTHOOD

Several tasks challenge young adults as they develop their sexuality:

- *Integrating love and sex:* Traditional gender roles call for men to be sex oriented and women to be love oriented. In adulthood, we need to develop ways of uniting sex and love instead of polarizing them as opposites.
- *Forging intimacy and commitment:* Young adulthood is characterized by increasing sexual experience. Through dating, cohabitation, and courtship, we gain knowledge of ourselves and others as potential partners. As our relationships become more meaningful, the degree of intimacy and interdependence increases. Sexuality can be a means of enhancing intimacy and self-disclosure as well as a means of obtaining physical pleasure. As we become more intimate, we need to develop our ability to make commitments.
- *Making fertility/childbearing decisions:* Childbearing is socially discouraged during adolescence but becomes increasingly legitimate for young adults in their twenties, especially if they are married. Fertility issues are often critical but unacknowledged, especially for single young adults. If these adults are sexually active, how important is it for them to prevent or defer pregnancy? What will they do if the woman unintentionally gets pregnant?
- *Establishing a sexual orientation:* As children and adolescents, we may engage in sexual experimentation such as playing doctor, kissing, and fondling members of both sexes. Such activities are not necessarily associated with sexual orientation. But by young adulthood a heterosexual, gay, or lesbian orientation emerges. Most young adults develop a heterosexual orientation. Others find themselves attracted to members of the same sex and begin to develop a gay, lesbian, or bisexual identity.
- *Developing a sexual philosophy:* As we move from adolescence to adulthood, we reevaluate our moral standards, moving from moral decision making based on authority to standards based on our own personal principles of right and wrong and caring and responsibility (Gilligan, 1982; Kohlberg, 1969). We become responsible for developing our own moral code, which includes sexual issues. In doing so, we need to develop a philosophical perspective to give coherence to our sexual attitudes, behaviors, beliefs, and values. We need to place sexuality within the larger framework of our lives and relationships. We need to integrate our personal, religious, spiritual, or humanistic values with our sexuality.

SEXUAL SCRIPTS

Our gender roles are critical in learning sexuality. Gender roles tell us what behavior (including sexual behavior) is appropriate for each gender. Our sexual impulses are organized and directed through culturally shared sexual scripts, which we learn and act out. Scripts are the acts, rules, stereotyped interaction patterns, and expectations associated with a particular role. *Sexual scripts* enable individuals to interpret emotions and sensations as sexually meaningful and provide them with methods of organizing sexual situations (Hynie, Lydon, Cote, and Wiener, 1998).

Thus, a sexual script is like a sexual road map or blueprint in that it gives general directions; it is more a sketch than a detailed picture of how our culture expects us to act as sexual beings. But even though a script is generalized, it is often more important than our own experiences in guiding our actions. Over time, we may modify or change our scripts, but we will not throw them away. A **sexual script** consists of expectations of how one is to behave sexually as a female or male and as a heterosexual, lesbian, or gay male.

The scripts we are "given" for sexual behavior tend to be traditional. These scripts are most powerful during adolescence, when we are first learning to be sexual. Gradually, as we gain experience, we modify and change our sexual scripts. As children and adolescents, we learn our sexual scripts primarily from our parents, peers, and the media. As we get older, interactions with our partners become increasingly important. In adolescence, both middle-class whites and middle-class African Americans appear to share similar values and attitudes about sex and male-female relationships (Howard, 1988).

Female Sexual Scripts

Whereas traditional male sexual scripts focus on sex more than feelings, traditional female sexual scripts focus on feelings more than sex, on love more than passion. In the female sexual script, sex is "relational," a way of "expressing or achieving emotional and psychological intimacy *within certain prescribed relationships*" (Hynie, Lydon, Cote, and Wiener, 1998, emphases added). The traditional female sexual scripts include several assumptions (Barbach, 1982):

- *Sex is both good and bad.* Women must deal with conflicting attitudes about sex. What makes sex good? Marriage or a committed relationship. What makes sex bad? A casual or uncommitted relationship. Sex is so good that you need to save it for your husband (or for someone with whom you are deeply in love). If it is not sanctioned by love or marriage, you'll get a bad reputation.
- *Girls don't want to know about their bodies "down there."* Girls are taught not to look at their genitals, not to touch them, especially not to explore them. As a result, women know very little about their genitals. They are often concerned about vaginal odors, which makes them uncomfortable about cunnilingus (oral sex).
- *Sex is for men.* Men are supposed to want sex; women are supposed to want love. Women are supposed to be sexually passive, waiting to be aroused. Sex is not supposed to be a pleasurable activity as an end in itself; it is something performed by women for men.
- *Men should know what women want.* Men are supposed to know what women want, even if women don't tell them. Women are supposed to remain pure and sexually innocent. It is up to the man to arouse the woman, even if he doesn't know what a particular woman finds arousing. To keep her image of sexual innocence, she does not tell him what she wants.
- *Women shouldn't talk about sex.* Many women cannot talk about sex easily because they are not expected to have strong sexual feelings. Some women may know their partners well enough to have sex with them but not well enough to communicate their needs to them.
- *Women should look like beautiful models.* The media present ideal women as having slender hips, firm and full breasts, and no fat; these women are always young, with never a pimple, wrinkle, or gray hair. As a result of these media images, many women are self-conscious about their physical appearance. They worry that they are too fat, too plain, too old. They often feel awkward without clothes on to hide their imagined flaws.
- *Women are nurturers.* Women are supposed to give and men are supposed to receive. Women give themselves, their bodies, their pleasures to men. His needs come first—his desire over hers, his orgasm over hers. If a woman always puts her partner's enjoyment first, she may be depriving herself of her own enjoyment.
- *There is only one right way to experience orgasm.* Women often "learn" that there is only one right way to experience orgasm: during sexual intercourse as a result of penile stimulation. But there are many ways to reach orgasm: through oral sex; manual stimulation before, during, or after intercourse; masturbation; and so on. Women who rarely or never have an orgasm during heterosexual intercourse but believe intercourse is the only legitimate route to orgasm are deprived of expressing themselves sexually in other ways.

Male Sexual Scripts

In traditional sexual scripts, men are perceived to be highly sexually aggressive. Once set in motion, a male's sexual response is thought to be difficult to control (Denov, 2003). Traditional male sexual scripts also portray sex as "recreational," or pleasure-centered for men (Hynie, Lydon, Cote, and Wiener, 1998). Therapist Bernie Zilbergeld (1993) has identified the following assumptions in the male sexual script:

- *Men should not have (or at least should not express) certain feelings.* Men should not express doubts; they should be assertive, confident, and aggressive. Tenderness and compassion are not masculine feelings.
- *Performance is the thing that counts.* Sex is something to be achieved, to be won. Feelings only get in the way of the job to be done. Sex is not for intimacy but for orgasm.
- *The man is in charge.* As in other things, the man is the leader, the person who knows what is best. The man initiates sex and gives the woman her orgasm. A real man doesn't need a woman to tell him what women like; he already knows.
- *A man always wants sex and is ready for it.* It doesn't matter what else is going on; a man wants sex. He is always able to become erect. He is a machine.
- *All physical contact leads to sex.* Because men are basically sexual machines, any physical contact is a sign for sex. Touching is the first step toward sexual intercourse, not an end in itself. There is no physical pleasure except sexual pleasure.
- *Sex equals intercourse.* All erotic contact leads to sexual intercourse. Foreplay is just that: warming up, getting your partner excited for penetration. Kissing, hugging, erotic touching, oral sex are only preliminaries to intercourse.
- *Sexual intercourse always leads to orgasm.* The orgasm is the proof of the pudding. The more orgasms, the better the sex. If a woman does not have an orgasm, she is not sexual. The male feels that he is a failure because he was not good enough to give her an orgasm. If she requires clitoral stimulation to have an orgasm, she is considered to have a problem.

Common to all these myths is a separation of sex from love and attachment. Sex is seen as a performance.

A study of sexual stereotypes found the following eight traits to be associated with the traditional male role: (1) sexual competence, (2) ability to give partners orgasms, (3) sexual desire, (4) prolonged erection, (5) being a good lover, (6) fertility, (7) reliable erection, and (8) heterosexuality (Riseden and Hort, 1992). The researchers observe that their results "offer the rather sad suggestion that men's gender [role] identity may be heavily dependent on the vagaries of a capricious physiological event (getting and maintaining an erection)."

Contemporary Sexual Scripts

As gender roles have changed, so have sexual scripts. Traditional sexual scripts have been challenged by more liberal and egalitarian ones. Sexual attitudes and behaviors have become increasingly liberal for both white and African-American males and females, but African-American attitudes and behaviors have been and continue to be somewhat more liberal than those of whites (Belcastro, 1985; Weinberg and Wilson, 1988; Wyatt et al., 1988). We do not know as much about how Latino sexuality and Asian-American sexuality have changed, as there is less research on the sexual scripts, values, and behaviors in those cultures.

Many women have made an explicit break with the more traditional scripts, especially the good girl/bad girl dichotomy and the belief that "nice" girls don't enjoy sex (Moffatt, 1989). College age women as well as older, professional women who are single are among those most likely to reject the old images (Davidson and Darling, 1988b).

Contemporary sexual scripts include the following elements for both sexes (Gagnon and Simon, 1987; Reed and Weinberg, 1984; Rubin, 1990; Seidman, 1989):

- Sexual expression is positive.
- Sexual activities are a mutual exchange of erotic pleasure.
- Sexuality is equally involving of both partners, and the partners are equally responsible.
- Legitimate sexual activities are not limited to sexual intercourse but also include masturbation and oral-genital sex.
- Sexual activities may be initiated by either partner.
- Both partners have a right to experience orgasm, whether through intercourse, oral-genital sex, or manual stimulation.
- Nonmarital sex is acceptable within a relationship context.
- Gay, lesbian, and bisexual orientations and relationships are increasingly open and accepted or tolerated, especially on college campuses and in large cities.

These scripts give greater recognition to female sexuality. They are also relationship centered rather than male centered. Society, however, still does not grant women full equality with males. Women who

Understanding Yourself

THE SOCIAL CONTROL OF SEXUALITY AND YOUR SEXUAL SCRIPTS

Our discussion of sexual scripts illustrates some ways in which society influences our sexual attitudes and behavior. This is not unique to the United States; in fact, every society regulates and controls the who, what, when, where, and why of sexuality. Your own script will change over time, depending on your age, sexual experience, and interaction with intimate partners and others. Let's examine some questions you are likely to encounter.

Who

The factors of homogamy and heterogamy are almost as strong in selecting sexual partners as they are in choosing marital partners, in part because sexual and marital partners are often the same. Society tells you to have sex with people who are unrelated, around your age, and of the other sex (heterosexual). Less acceptable is having sex with yourself (masturbation), with members of the same sex (gay or lesbian sexuality), and with members of different ethnic groups. In most societies, extramarital relationships are prohibited. Examine the "whos" in your sexual script, then think about the following questions:

- With whom do you engage in sexual behaviors?
- How do your choices reflect homogamy and heterogamy?
- What social factors influence your choice?
- Does your autoerotic behavior change if you are in a relationship? How? Why?

What

The "whats" in your sexual script are the types of sexual behaviors in which you engage. Society classifies sexual acts as good or bad, moral or immoral, appropriate or inappropriate. Although these designations may seem to be absolute, they are culturally relative.

- What sexual acts are part of your sexual script?
- How are they regarded by society?
- How important is the level of commitment in a relationship in determining your sexual behaviors?
- What level of commitment do you need for kissing? Petting? Sexual intercourse?
- What occurs if you and your partner have different sexual scripts for engaging in various sexual behaviors?

When

The "whens" of your sexual script refer to timing. You might make love when

have several concurrent sexual partners or casual sexual relationships, for example, are much more likely to be regarded as promiscuous than are men in similar circumstances (Williams and Jacoby, 1989). Interestingly enough, the "suppression of female sexuality" is often carried out by women, whether through maternal influence or the judgments of female peer groups (Miracle, Miracle and Baumeister, 2003)

GAY, LESBIAN, AND BISEXUAL IDENTITIES

In contemporary America, people are generally classified as **heterosexual** (sexually attracted to members of the other gender), **homosexual** (sexually attracted to members of the same gender), or **bisexual** (attracted to both genders). Even this three-category approach may not accurately depict the range that exists in our "sexual orientations"—who we are attracted to, who we have relations with, who we fantasize about, the type of lifestyle we live, and how we identify ourselves. On any of these items we may be exclusively oriented toward the other sex or our same sex, mostly drawn to the other sex or our own sex, or oriented to both sexes about equally. Finally, our sexual orientation may change over time. Thus, what was true of past relationships or attractions may not fit with the present or may differ from what we envision for our future (Klein, 1990; Miracle, Miracle, and Baumeister, 2003).

In discussing gay men and lesbians, we (like many researchers) generally do not use the term *homosexual* because it often conveys negative or pathological connotations. Instead, we use the more neutral terms **gay male** to refer to men and **lesbian** to refer to women. Replacing the term *homosexual* helps us see individuals as whole persons by emphasizing the fact that sexuality is not the only aspect of their lives. Their lives also include love, commit-

your parents are out of the house or, if a parent yourself, when your children are asleep. Usually, such timing is related to privacy, but it may also be related to the age at which sexual activity is expected to start and stop, how often people are expected to engage in sexual relations, and when in a relationship sex should begin. Finally, it may pertain to times when sex is considered appropriate or inappropriate. Some societies frown upon a woman engaging in sex during her menstrual flow, for a period after the birth of a child, or while nursing (Miracle, Miracle and Baumeister, 2003).

- When do you engage in sexual activities?
- Are the times related to privacy?
- When did you experience your first erotic kiss?
- At what age did you first have sexual intercourse? If you have not had intercourse, at what age do you think it would be appropriate?
- How was the timing for your first intercourse determined, or how will it be determined?
- What influences (friends, parents, religion) are brought to bear on the age timing of sexual activities?

Where

Where do sexual activities occur with society's approval? In our society, usually in the bedroom, where a closed door signifies privacy. For adolescents, automobiles, fields, beaches, and motels may be identified as locations for sex; churches, classrooms, and front yards usually are not. "Where" may also extend to where it might be considered appropriate or inappropriate to discuss sex, or to expose parts of your body.

- What do you think are the acceptable places to be sexual?
- What makes them acceptable for you?
- Have you ever had conflicts with partners about the "wheres" of sex? Why?

Why

The "whys" are the explanations we give ourselves and others about our sexual activities. There are many reasons for having sex: procreation, love, passion, revenge, intimacy, exploitation, fun, pleasure, relaxation, boredom, achievement, relief from loneliness, exertion of power, and on and on. Some of these reasons are approved by society; others are not. Some we conceal; others we do not.

- What are your reasons for sexual activities?
- Do you have different reasons for different activities, such as masturbation, oral sex, and sexual intercourse?
- Do the reasons change with different partners? With the same partner?
- Which reasons are approved by society and which are disapproved?
- Which reasons do you make known, and which do you conceal? Why?

ment, desire, caring, work, children, religious devotion, passion, politics, loss, and hope. Sex is important, obviously, but it is not the only significant aspect of their lives, just as it is not the only significant aspect of the lives of heterosexuals.

Those with lesbian or gay orientations have been called sinful, sick, or perverse, reflecting traditional religious, medical, and psychoanalytic approaches. Contemporary thinking in sociology and psychology has rejected these older approaches as biased and unscientific. Instead, sociologists and psychologists have focused their work on how women and men come to identify themselves as lesbian or gay, how they interact among themselves, and what impact society has on them (Heyl, 1989). Researchers reject the idea that lesbians and gay men are inherently deviant or pathological. As noted sociologist Howard Becker (1963) has pointed out, "Deviant behavior is behavior that people so label." Deviance is created by social groups that make rules whose violation results in violators being labeled deviant and treated as outsiders. Lesbian and gay behavior, then, is deviant only insofar as it is called deviant.

How does one "become" gay, lesbian, bisexual—or even heterosexual, for that matter? A person's **sexual orientation**—sexual identity as heterosexual, gay, lesbian, or bisexual—is complex. It depends on the interaction of numerous factors. These factors, which may be a combination of social, biological, and personal ones, lead to the unconscious formation of a person's sexual orientation. Two of the most important factors are the gender of one's sexual partner and whether one labels oneself as heterosexual, gay, lesbian, or bisexual.

The actual percentage of the population that is lesbian, gay, or bisexual is not known. Among women, about 13 percent have had orgasms with other women, but only 1 to 3 percent identify themselves as lesbian (Fay et al., 1989; Kinsey, Pomeroy, and Martin, 1948, 1953; Marmor, 1980c). Among males, including adolescents, as many as 20 to 37

© Miriam Grosman

Two significant factors in identifying sexual orientation are (1) the gender of one's partner, and (2) the label one gives oneself (lesbian, gay, bisexual, or heterosexual).

percent have had orgasms with other males, according to Kinsey's studies. Ten percent were predominantly gay for at least three years; 4 percent were exclusively gay throughout their entire lives (Kinsey et al., 1948). A review of studies on male same-sex behavior between 1970 and 1990 estimated that a minimum of 5 to 7 percent of adult men had had sexual contact with other men in adulthood. Based on their review, the researchers suggested that about 4.5 percent of men are exclusively gay (Rogers and Turner, 1991). A large-scale study of 3,300 men aged 20 to 39 reported that 2 percent had engaged in same-sex sexual activities and 1 percent considered themselves gay (Billy et al., 1993). In 1994, the National Health and Social Life Study found that of the participants, 2.8 percent of men and 1.4 percent of women described themselves as homosexual or bisexual, although approximately 6 percent of men and 4 percent of women said they had had a sexual experience with someone of the same sex at least once since puberty (Laumann et al., 1994).

What are we to make of these differences between studies? In part, the variances may be explained by different methodologies, interviewing techniques, sampling, or definitions of homosexuality. Furthermore, sexuality is more than simply sexual behaviors; it also includes attraction and desire. One can be a virgin or celibate and still be gay or heterosexual. Finally, sexuality is varied and changes over time; its expression at any one time is not necessarily its expression at another.

DID YOU KNOW?

Studies from the 1970s indicated that the average gay male did not identify himself as gay until age 19 to 21. With increased societal awareness of homosexuality, lesbian and gay adolescents now recognize their orientation at younger ages (Savin-Williams and Rodriguez, 1993; Troiden and Goode, 1980).

Identifying Oneself as Gay or Lesbian

Many researchers believe that a person's sexual interest or direction as heterosexual, gay, or lesbian is established by age 4 or 5 (Marmor, 1980a, 1980b). But identifying oneself as lesbian or gay takes considerable time and includes several phases, usually beginning in late childhood or early adolescence (Blumenfeld and Raymond, 1989; Troiden, 1988). **Homoeroticism**—erotic attraction to members of the same gender—almost always precedes gay or lesbian activity by several years.

STAGES IN ACQUIRING A LESBIAN OR GAY IDENTITY. The first stage in acquiring a lesbian or gay identity is marked by fear and suspicion that somehow one's desires are different from those of others. At first, the person finds it difficult to label the emotional and physical desires for the same sex. The initial reactions often include fear, confusion, and denial. Adolescents especially fear their family's discovery of their homoerotic feelings. In the second stage, the person labels these feelings of attraction, love, and desire as homoerotic if they recur often enough. The third stage includes the person's self-definition as lesbian or gay. This may take a considerable struggle, for it entails accepting a label that society generally calls deviant. Questions then arise about whether to tell parents or friends, whether to hide one's identity ("to be in the closet") or make the identity known ("to come out of the closet").

Some gay men and lesbians may go through two additional stages. One stage is to enter the gay subculture. A gay person may begin acquiring exclusively gay friends, going to gay bars and clubs, or joining gay activist groups. In the gay world, gay and lesbian identities incorporate a way of being in which sexual orientation is a major part of the identity as a person. As Michael Denneny (quoted in Altman, 1982) says: "I find my identity as a gay man as basic as any other identity I can lay claim to. Being gay is a more elemental aspect of who I am than my profession, my class, or my race." Similarly, Pat Califia (quoted in Weeks, 1985) explains the process: "Knowing I was a lesbian transformed the way I saw, heard, perceived the whole world. I became aware of a network of sensations and reactions that I had ignored all my life."

The final stage begins with a person's first lesbian or gay love affair. This marks the commitment to unifying sexuality and affection. Sex and love are no longer separated. Most lesbians and gay men have had such love affairs, despite the stereotypes of anonymous gay sex.

COMING OUT. Being lesbian or gay is increasingly associated with a total lifestyle and way of thinking. In making the gay or lesbian orientation a lifestyle, **coming out**—publicly acknowledging one's gayness—has become especially important as an affirmation of sexuality. Coming out is a major decision because it may jeopardize many relationships, but it is also an important means of self-validation. By publicly acknowledging a gay or lesbian orientation, a person begins to reject the stigma and condemnation associated with it. Generally, coming out occurs in stages, first involving family members, especially the mother and siblings and later the father. Coming out to the family often creates a crisis, but generally the family accepts the situation and gradually adjusts (Holtzen and Agresti, 1990). Religious beliefs, prejudice, and misinformation about gay and lesbian sexuality, however, often interfere with a positive parental response, initially making adjustment difficult (Borhek, 1988; Cramer and Roach, 1987). After the family, friends may be told and, in fewer cases, employers and coworkers.

Gay men and women are often "out" to varying degrees. Some may be out to no one, not even themselves. Some are out only to their lovers, others to close friends and lovers but not to their families, employers, associates, or fellow students. Still others may be out to everyone. Because of fear of reprisal, dismissal, or public reaction, lesbian and gay school teachers, police officers, members of the military, politicians, and members of other such professions are rarely out to their employers, co-workers, or the public.

Outing refers to the practice of publicly identifying closeted gays or lesbians. There is often a political justification for why people who may wish to keep their sexual orientation hidden should be outed. The rationale behind outing suggests that if gays and lesbians stay quiet about their sexual orientation, negative stereotypes about homosexuals remain unchallenged. Conversely, as heterosexuals discover that some of their friends and family members, or even public figures that they are familiar with and respect, are gay or lesbian, they may modify their attitudes about homosexuality in a more accepting direction (Miracle, Miracle, and Baumesiter, 2003).

DID YOU KNOW?

A large-scale survey conducted by the *New York Times* found that 55 percent of the respondents believed behavior between adult gay men or lesbians was morally wrong. At the same time, 78 percent believed gay men and lesbians should have equal job opportunities; 43 percent supported gay men and lesbians in the military (and an equal percentage opposed it); and 42 percent believed laws should be passed to guarantee equal rights for gay men and lesbians (Schmalz, 1993).

Anti-Gay/Lesbian Prejudice and Discrimination

Anti-gay prejudice is a strong dislike, fear, or hatred of lesbians and gay men because of their homosexuality. **Homophobia** is an irrational or phobic fear of gay men and lesbians. Not all anti-gay feelings are phobic in the clinical sense of being excessive and irrational. They may be unreasonable or biased. (Nevertheless, they may be within the norms of a biased culture.) Because prejudice may not be clinically phobic, the term *homophobia* is being increasingly replaced by the less clinical *anti-gay prejudice* (Haaga, 1991).

As a belief system, anti-gay prejudice justifies discrimination and violence based on sexual orientation. In his classic work on prejudice, Gordon

Allport (1958) states that social prejudice is acted out in three stages: (1) offensive language, (2) discrimination, and (3) violence. Gay men and lesbians experience each stage. They are called *faggot, dyke, queer,* and *homo.* They are discriminated against in terms of housing, equal employment opportunities, insurance, adoption, parental rights, family acceptance, and so on, and they are the victims of violence known as gay bashing or queer bashing. Among college students, anti-gay prejudice extends to heterosexuals who voluntarily choose to room with a lesbian or gay man. They are assumed to have "homosexual tendencies" and to have many of the negative stereotypical traits of gay men and lesbians, such as poor mental health (Sigelman et al., 1991).

The 1990s saw the hate-based murders of Matthew Shepard and Billy Jack Gaither. Both men died because they were gay men whose killers targeted them and brutalized them for that reason. Like other kinds of minorities, gays are frequent targets in hate or bias crimes. FBI data for 1997 indicate that approximately one of every seven reported hate crimes was based on sexual motivation. This makes anti-gay hate crimes the third most common, trailing behind only race-based and religiously directed crimes.

According to one study of anti-gay hate crimes in eight U.S. cities, 19 percent of the gay men and lesbians who responded reported having been punched, kicked, beaten, or hit at least once because of their sexual orientations. Almost half (44 percent) at some point had faced threats of such violence. Most broadly, 94 percent had suffered some anti-gay victimization, including such things as being verbally abused, chased or pelted with objects, spat upon, or assaulted. Lifetime assault victimization rates for homosexuals range from between one in ten to one in five (Renzetti and Curran, 1995; McCaghy and Capron, 1997).

Anti-gay prejudice is derived from several factors (Marmor, 1980): (1) a deeply rooted insecurity concerning the person's own sexuality and gender identity, (2) a strong fundamentalist religious orientation, and (3) simple ignorance concerning homosexuality.

In many states, gay men and lesbians may face criminal prosecution for engaging in oral and anal sex, for which heterosexuals are seldom charged. Medical and public health efforts against HIV/AIDS were inhibited initially because HIV/AIDS was perceived as "the gay plague" and was considered punishment against gay men for their "unnatural" sexual practices (Altman, 1985). The fear of HIV/AIDS has contributed to increased anti-gay prejudice among some heterosexuals (Lewes, 1992). Anti-gay prejudice influences parents' reactions to their lesbian and gay children, often leading to estrangement (Holtzen and Agresti, 1990).

BANG! Inc./Getty Images

© Evan Agostini/Getty Images

The hate-based killing of Matthew Shepherd inspired memorial demonstrations and raised awareness about the extent of homophobia in the United States.

Anti-gay prejudice adversely affects heterosexuals too. First, it creates fear and hatred—aversive emotions that cause distress and anxiety. Second, it alienates heterosexuals from gay family members, friends, neighbors, and co-workers (Holtzen and Agresti, 1990). Third, it limits range of behaviors and feelings, such as hugging or being emotionally intimate with same-sex friends, for fear that such intimacy may be "homosexual" (Britton, 1990). Heterosexuals may restrict displays of affection with their same-sex friends for fear that such displays could be misinterpreted (Garnets et al., 1990). Fourth, anti-gay prejudice may lead to exaggerated displays of masculinity by heterosexual men trying to prove they are not gay (Mosher and Tomkins, 1988).

What can be done to reduce prejudice against lesbians and gay men? Education and positive social interactions appear to be important vehicles for change. Education can affect negative attitudes. Two researchers (Serdahely and Ziemba, 1984) studied the impact of including a unit on homosexuality in their human sexuality course. They found that students who at the beginning of the course scored above the class mean on homophobic attitudes, by the end had a significant decrease in their scores. Other researchers also have reported increased tolerance following human sexuality courses (Stevenson, 1990). Negative attitudes about homosexuality may be reduced by the arranging of positive interactions between heterosexuals and gay men and lesbians. These interactions should be in settings of equal status, common goals, cooperation, and a moderate degree of intimacy. Such interactions may occur when family members or close friends come out. Other interactions may emphasize common group membership (religious, social, ethnic, or political, for example) on a one-to-one basis. Religious volunteers working with people with HIV/AIDS often decrease their homophobia through their caring for and comforting of those with the infection (Kayal, 1992).

Bisexuality

As we noted earlier, bisexuals are attracted to members of both genders. Becoming bisexual requires the rejection of two recognized categories of sexual identity: heterosexual and homosexual. In a nationwide study conducted by Samuel Janus and Cynthia Janus (1993), about 5 percent of the men and 3 percent of the women identified themselves as bisexual. Data from the National Health and Social Life Survey, a comprehensive national survey of sexual behavior in the United States, found a smaller percentage (less than 1 percent) who self-identified themselves as bisexual. If one looks at reports of "sexual attraction," 3.9 percent of men and 4.1 percent of women report themselves attracted to "mostly the opposite gender," both genders, or "mostly the same gender" (Laumann et al., 1994).

Because they reject both heterosexuality and homosexuality, bisexuals often find themselves stigmatized by gay men and lesbians as well as by heterosexuals. Heterosexuals view bisexuals as "really" homosexual. Gay men and lesbians view bisexuals as "fence-sitters" not willing to admit their homosexuality or as people simply "playing" with their orientation. Thus, bisexuality may not be taken seriously by either group. Loraine Hutchins and Lani Kaahumanu (1991b) believe that bisexuality arouses hostility because it "challenges current assumptions about the immutability of people's orientations and society's supposed divisions into discrete groups."

Research on homosexuality is extensive, but there is little research on bisexual identity. (There is considerable HIV/AIDS research, however, on bisexual behavior among men who identify themselves as heterosexual.) It was not until 1994 that the first model of bisexual identity formation was developed (Weinberg, Williams, and Pryor, 1994). According to this model, bisexual women and men go through several stages in developing their identity. The first stage, often lasting years, is *initial confusion.* Many are distressed by being sexually attracted to both sexes; others believe that their attraction to the same sex means an end to their heterosexuality; still others are disturbed by their inability to categorize their feelings as either heterosexual or homosexual. The second stage is *finding and applying the bisexual label.* For many, discovering there is such a thing as bisexuality is a turning point. Some find that their first heterosexual or same-sex experience permits them to view sex with both sexes as pleasurable; others learn of the term *bisexuality* from friends and are able to apply it to themselves. The third stage, *settling into the identity,* is characterized as feeling at home with the bisexual label. For many, self-acceptance is critical. The fourth stage is *continued uncertainty.* Bisexuals don't have a community or social environment that reaffirms their identity. Despite being settled in, many feel persistent pressure from gay men and lesbians to relabel themselves as homosexual and to engage exclusively in same-sex activities.

Psychosexual Development in Middle Adulthood

Psychosexual development and change does not end with young adulthood. It continues throughout our lives. In middle age and old age, our lives, bodies, sexuality, relationships, and environment continue to change. New tasks and new satisfactions arise to replace or supplement older ones.

DEVELOPMENTAL TASKS IN MIDDLE ADULTHOOD

In the middle adult years, some of the tasks in psychosexual development begun in young adulthood may continue. These tasks, including issues of intimacy and childbearing, may have been deferred or only partly completed in young adulthood. Because of separation or divorce, we may find ourselves facing the same intimacy and commitment tasks at age 40 that we thought we completed 15 years earlier (Cate and Lloyd, 1992). But life does not stand still; it moves steadily forward, whether we're ready or not. Other developmental issues appear, including the following:

- *Redefining sex in marital or other long-term relationships:* In new relationships, sex is often passionate, intense; it may be the central focus. But in long-term marital or cohabiting relationships, the passionate intensity associated with sex is often eroded by habituation, competing parental and work obligations, fatigue, and unresolved conflicts. Sex may need to be redefined as a form of intimacy and caring. Individuals may also need to decide how to deal with the possibility, reality, and meaning of extramarital or extrarelational affairs.
- *Reevaluating one's sexuality:* Single men and women may need to weigh the costs and benefits of sex in casual or lightly committed relationships. In long-term relationships, sexuality often becomes less central to relationship satisfaction. Nonsexual elements, such as communication, intimacy, and shared interests and activities, become increasingly important to relationships. Women who have deferred their childbearing begin to reappraise their decision: Should they remain childfree, "race" against their biological clocks, or adopt a child? Some individuals may redefine their sexual orientation. The sexual philosophy of those in middle adulthood continues to be reexamined and to evolve.
- *Accepting the biological aging process:* As we age, our skin wrinkles, our flesh sags, our hair grays (or falls out), our vision blurs—and we become in the eyes of society less attractive and less sexual. By our forties, our physiological responses have begun to slow noticeably. By our fifties, society begins to "neuter" us, especially if we are women who have gone through menopause. The challenges of aging are to accept its biological mandate and to reject the stereotypes associated with it.

SEXUALITY AND MIDDLE AGE

Men and women view aging differently. As men approach their fifties, they fear the loss of their sexual capacity but not their attractiveness; in contrast, women fear the loss of their attractiveness but generally not their sexuality. As both age, purely psychological stimuli, such as fantasies, become less effective for arousal. Physical stimulation remains effective, however.

Among American women, sexual responsiveness continues to grow from adolescence until it reaches its peak in the late thirties or early forties; it is usually maintained at more or less the same level into the sixties and beyond. Men's physical responsiveness is greatest in late adolescence or early adulthood; beginning in men's twenties, responsiveness begins to slow imperceptibly. Changes in male sexual responsiveness become apparent only when men are in their forties and fifties. As a man ages, achieving erection requires more stimulation and time and the erection may not be as firm.

Around the age of fifty, the average American woman begins menopause, which is marked by a cessation of the menstrual cycle. Menopause is not a sudden event. Usually, for several years preceding menopause, the menstrual cycle becomes increasingly irregular. Although menopause ends fertility, it does not end interest in sexual activities. The decrease in estrogen, however, may cause thinning and dryness of the vaginal walls, which makes intercourse painful. The use of vaginal lubricants will remedy much of the problem. There is no male equivalent to menopause. Male fertility slowly declines, but men in their eighties are often fertile.

Because of physical changes, notes Herant Katchadourian (1987), "middle-aged couples may be misled into thinking that this change heralds a sexual decline as an accompaniment to aging." Katchadourian continues:

> Sexual partners who have been together for a long time have the benefits of trust and affection. In the younger years of marriage, sex tends to be a battleground where scores are settled and peace is made, but if a couple has stuck together until middle age, sex should become a demilitarized zone. . . . They continue to enjoy the physical pleasures of sex but do not stop there. . . . [T]he sensual quality of the person, rather than the body as such, becomes the main course.

Psychosexual Development in Later Adulthood

As we leave middle age, new tasks confront us, especially dealing with the process of aging itself. Our health and the presence or absence of a partner are key aspects of this time in our lives.

DEVELOPMENTAL TASKS IN LATER ADULTHOOD

Many of the psychosexual tasks older Americans must undertake are directly related to the aging process:

- *Changing sexuality:* As older men's and women's physical abilities change with age, their sexual responses change as well. A 70-year-old person, though still sexual, is not sexual in the same manner as an 18-year-old. As men and women continue to age, their sexuality tends to be more diffuse, less genital, and less insistent. Chronic illness and increasing frailty understandably result in diminished sexual activity. These considerations contribute to the ongoing evolution of the individual's sexual philosophy.
- *Loss of partner:* One of the most critical life events is the loss of a partner. After age 60, there is a significant increase in spousal deaths. As having a partner is the single most important factor determining an older person's sexual interactions, the death of a partner signals a dramatic change in the survivor's sexual interactions.

The developmental tasks of later adulthood are accomplished within the context of continuing aging. Their resolution helps prepare us for acceptance of our own eventual mortality.

SEXUALITY AND THE AGED

The sexuality of older Americans tends to be invisible; that is, society tends to discount their sexuality (Libman, 1989). In fact, one review of the literature on aging concludes that the decline in sexual activity among aging men and women is more cultural than biological in origin (Kellett, 1991). Several reasons for this exist in our culture (Barrow and Smith, 1992). First, we associate sexuality with the young, assuming that sexual attraction exists only between those with youthful bodies. Interest in sex is considered normal and virile in 25-year-old men, but in 75-year-old men it is considered lecherous. Second, we associate the idea of romance and love with the young; many of us find it difficult to believe that the aged can fall in love or love intensely. Third, we associate sex with procreation, measuring a woman's femininity by her childbearing and motherhood and a man's masculinity by the children he sires. Finally, the aged do not have sexual desires as strong as those of the young and they do not express their desires as openly. Intimacy is especially valued and is important for an older person's well-being (Mancini and Blieszner, 1992).

Sexuality is one of the least understood aspects of life in old age. Many older people continue to adhere to the standards of activity or physical attraction they held when they were young (Creti and Libman, 1989). They need to become aware of the taboos and stereotypes about aging that they held when they were younger so they can enjoy their sexuality in their later years (Kellett, 1991).

Aging lesbians and gay men face a double bias: They are old *and* gay. But like other stereotypes of aging Americans, theirs reflects myths rather than realities. A study of gay men over age 60 found that over 80 percent accepted their gayness and about half worried about growing old (Berger, 1982). A study of aging lesbians found that 71 percent were satisfied with being lesbian and about half were concerned about aging (Kehoe, 1988).

Aged men and women face different sexual problems. Physiologically, men are less responsive than they used to be. The decreasing frequency of intercourse and the increasing time required to attain an erection produce anxieties in many older men about erectile dysfunction (impotence)—anxieties that may very well lead to such dysfunction. When the natural slowing down of sexual responses is interpreted as the beginning of erectile dysfunction, this self-diagnosis triggers a vicious spiral of fears and even greater difficulty in attaining or maintaining an erection. One study (Weitzman and Hart, 1987) found that about 31 percent of elderly male respondents were unable to have an erection. **Viagra,** a relatively recent oral medication for the treatment of erectile dysfunction, helped restore sexual activity to many older men and to quite a few younger men who suffered similarly. It has since been joined by other oral medications that treat erectile dysfunction. These will be discussed later.

Women have different concerns. They face greater social constraints than men. Women are confronted with an unfavorable gender ratio (29 unmarried men per 100 unmarried women over age 65), a greater likelihood of widowhood, and norms against marrying younger men. Grieving over the death of a partner, isolation, and depression also affect their sexuality (Rice, 1989). Finally, there is a double standard of aging. In our culture, as men age, they become distinguished; as women age, they simply get older. Femininity is connected with youth and beauty. But as women age, they tend to be regarded as more masculine. A young woman, for example, is "beautiful," but an older woman is "handsome"—a term ordinarily used for men of any age.

The greatest determinants of an aged individual's sexual activity are health and the availability of a partner. Researchers studied more than 800 married whites and African Americans over age 60 (Marisiglio and Donnelly, 1991). They found that over half the sample (and 24 percent of those older than 76) had sexual intercourse within the previous month. Those who had sex during the month had it an average of four times. Among the sexually active

Keri Pickett

Sexuality among the aged tends to be sensual and affectionate. Older couples may experience an intimacy forged by years of shared joys and sorrows that is as intense as the passion of young love.

older people, there were no differences by gender or ethnicity. Those who do not have partners may turn to masturbation as an alternative to sexual intercourse (Pratt and Schmall, 1989).

After age 75, a significant decrease in sexual activity takes place. This seems to be related to health problems, such as heart disease, arthritis, and diabetes. Often older people indicate that they continue to feel sexual desires; they simply lack the ability to express them because of their health. In a study of men and women in nursing homes whose ages averaged 82 years, 91 percent reported no sexual activity immediately prior to their interviews; 17 percent of these men and women, however, expressed a desire for sexual activity (White, 1982). Unfortunately, most nursing homes make no provision for the sexuality of the aged. Instead, they actively discourage sexual expression—not only sexual intercourse but also masturbation—or try to sublimate their clients' sexual interests into crafts or television.

Sexual Behavior

In this section we examine various sexual behaviors. For a discussion of sexual structure and the sexual response cycle, see Appendix A.

AUTOEROTICISM

Autoeroticism consists of sexual activities that involve only the self. It includes sexual fantasies, masturbation, and erotic dreams. A universal phenomenon in one form or another, autoeroticism is one of our earliest expressions of sexual stirrings. It is also one that traditionally has been condemned in our society. By condemning it, however, our culture sets the stage for the development of deeply negative inhibitory attitudes toward sexuality.

Sexual Fantasies

Erotic fantasizing is probably the most universal of all sexual behaviors. Nearly everyone has experienced erotic fantasies, but because they may touch on feelings or desires considered personally or socially unacceptable, they are not widely discussed. They may also interfere with an individual's self-image, causing a loss of self-esteem as well as confusion.

Sexual fantasies serve a number of important functions in maintaining our psychic equilibrium. First of all, sexual fantasies help direct and define our erotic goals. They take our generalized sexual drives and give them concrete images and specific content. Second, sexual fantasies allow us to plan or anticipate erotic situations that may arise. They provide a form of rehearsal, allowing us to practice in our minds how to act in various situations. Third, sexual fantasies provide escape from a dull or oppressive environment. Routine or repetitive labor often gives rise to sexual fantasies as a way of coping with the boredom in which we are trapped. Fourth, even if our sexual lives are satisfactory, we may indulge in sexual fantasies to bring novelty and excitement into a relationship. Many people fantasize things they would not actually do in real life. Fantasy offers a safe outlet for sexual curiosity. Fifth, sexual fantasies also have an expressive function in somewhat the same manner as dreams do. Our sexual fantasies may offer a clue to our current interests, pleasures, anxieties, fears, or problems.

Various studies report that between 60 and 90 percent of respondents fantasize during sex—the percentage depending on gender, age, and ethnicity (Miracle, Miracle and Baumeisater, 2003; Knafo and Jaffe, 1984; Price and Miller, 1984). A large-scale study (Michael et al., 1994) found that 54 percent of the men and 19 percent of the women thought about sex daily. Twenty-three percent of the men and 11 percent of the women bought X-rated videos. Women's and men's sexual fantasies tend to differ, with a common pattern being that men more often fantasize about *doing something sexual to someone.* Women's fantasies more often consist of *having something sexual done to them* (Leitenberg and Henning, 1995, cited in Miracle, Miracle and Baumeister, 2003). Despite their prevalence, fantasies during intercourse may provoke guilty feelings in a number of people, especially when they involve people other than one's partner (Cado and Leitenberg, 1990).

Erotic Dreams

Almost all of the men and two-thirds of the women in Kinsey and his colleagues' studies (1948, 1953) reported having had overtly sexual dreams. Sexual images in dreams are frequently very intense. Although people tend to feel responsible for fantasies, which occur when they are awake, they are usually less troubled by sexual dreams.

Dreams almost always accompany nocturnal orgasm. The dreamer may awaken, and men usually ejaculate. Approximately 2 to 3 percent of a woman's orgasms, may be nocturnal, while for men the number may be around 8 percent of the total (Kinsey et al., 1948, 1953). Furthermore, research estimates that 80 percent of men and 40 percent of women have experienced a "nocturnal orgasm" (Miracle, Miracle, and Baumeister, 2003). Although the dream content may not be overtly sexual, it is always accompanied by sensual sensations. Women seem to feel less guilt or fear about nocturnal orgasms than do men, accepting them more easily as pleasurable experiences. Men tend to worry about them, perhaps because they emit semen.

Masturbation

Masturbation is the manual stimulation of one's genitals. Individuals masturbate by rubbing, caressing, or otherwise stimulating their genitals to bring themselves sexual pleasure. Masturbation is an important means of learning about our bodies. Girls, boys, women, and men may masturbate during particular periods or throughout their entire lives. An analysis of research articles on gender roles and sexual behavior found that the greatest male-female difference was in masturbation (Oliver and Hyde, 1993). Males had significantly more masturbatory experience than females.

By the end of adolescence, virtually all males and about two-thirds of females have masturbated to orgasm (Knox and Schact, 1992; Lopresto, Sherman, and Sherman, 1985). Masturbation continues after adolescence. In recent years, there seems to have been a slight increase in the incidence and frequency of masturbation. Gender differences, however, continue to be significant (Atwood and Gagnon, 1987; Leitenberg, Detzer, and Srebnik, 1993).

Data from the National Health and Social Life Survey revealed that across age, race, marital status, education, or religion, women were less likely to masturbate, masturbated less frequently, and less often masturbated to orgasm. Interestingly, males tended to report more guilt feelings about masturbating (Laumann et al., 1994).

Although the rate is significantly lower for those who are married, many people, especially men, continue to masturbate even after they marry. There are many reasons for continuing the activity during marriage: Masturbation is a pleasurable form of sexual excitement; a spouse may be away or unwilling to engage in sex; sexual intercourse may not be satisfying; the partners may fear sexual inadequacy; one partner may want to act out fantasies. In marital conflict, masturbation may act as a distancing device, with the masturbating spouse choosing masturbation over sexual intercourse as a means of emotional protection (Betchen, 1991).

Cohabitation has a different effect than marriage on frequency of masturbation. Many cohabiting men masturbate frequently, despite the presence or availability of a sexual partner. Thus, social factors other than the presence of a partner affect masturbation. In fact, in citing reasons for why they masturbate, only a third of women and men list an unavailable partner (Laumann et al., 1994).

Attitudes toward masturbation vary along ethnic lines. Whites are the most accepting of masturbation, for example; African Americans are less accepting. The differences can be explained culturally. Whites tend to begin their coital activities later than African Americans; whites therefore regard masturbation as an acceptable alternative to sexual intercourse. African Americans, by contrast, tend to accept interpersonal sexual activity at an earlier age. They therefore may view masturbation as a sign of personal and sexual inadequacy. As a result, many African Americans see sexual intercourse as normal and masturbation as deviant (Cortese, 1989; Kinsey et al., 1948; Wilson, 1986). Recently, masturbation has become more accepted as a legitimate sexual activity within the African-American community (Wilson, 1986; Wyatt et al., 1988).

Latinos, like African Americans, hold less permissive attitudes than Anglos about masturbation (Cortese, 1989; Padilla and O'Grady, 1987). In Mexican culture, masturbation is not considered an acceptable sexual option for either males or females (Guerrero Pavich, 1986). In part, this is because of the cultural emphasis on sexual intercourse and the influence of Catholicism, which regards masturbation as sinful. As with other forms of sexual behavior, acceptance becomes more likely as Latinos become more assimilated. A study of Mexican-American college students found their attitudes toward masturbation more liberal than those of the Mexican-American community as a whole (Padilla and O'Grady, 1987).

DID YOU KNOW?

A study of college students in a human sexuality class found that 87 percent of the men and 58 percent of the women had masturbated (Knox and Schact, 1992). In a larger study, among adults of all ages, 63 percent of the men and 42 percent of the women had masturbated in the previous year (Michael et al., 1994).

INTERPERSONAL SEXUALITY

We often think that sex is sexual intercourse and that sexual interactions end with orgasm (usually the male's). But sex is not limited to sexual intercourse. Heterosexuals engage in a wide variety of sexual activities, which may include erotic touching, kissing, and oral and anal sex. Except for sexual intercourse, gay and lesbian couples engage in more or less the same sexual activities as heterosexuals.

Touching

Because touching, like desire, does not in itself lead to orgasm, it has largely been ignored as a sexual behavior. Sex researchers William Masters and Virginia Johnson (1970) suggest a form of touching they call **pleasuring**—nongenital touching and caressing. Neither partner tries to stimulate the other sexually; the partners simply explore each other. Such pleasuring gives each a sense of his or her own responses; it also allows each to discover what the other likes or dislikes. We can't assume we know what any particular individual likes, for there is too much variation among people. Pleasuring opens the door to communication; couples discover that the entire body is erogenous, rather than just the genitals.

As we enter old age, touching becomes increasingly significant as a primary form of erotic expression. Touching in all its myriad forms—ranging from holding hands to caressing, massaging to hugging, walking with arms around each other to fondling—becomes the touchstone of eroticism for the elderly. One study found touching to be the primary form of erotic expression for married couples over 80 years old (Bretschneider and McCoy, 1988).

Kissing

Kissing as a sexual activity is probably the most common and acceptable of all premarital sexual activities, occurring in over 90 percent of all cultures (Jurich and Polson, 1985; Fisher, 1992, cited in Miracle, Miracle, and Baumeister, 2003). The tender lover's kiss symbolizes love, and the erotic lover's kiss, of course, simultaneously represents and is passion. Both men and women in one study regarded kissing as a romantic act, a symbol of affection as well as attraction (Tucker, Marvin, and Vivian, 1991). A cross-cultural study of jealousy found that kissing is also associated with a couple's boundary maintenance: In each culture studied, kissing a person other than the partner evoked jealousy (Buunk and Hupka, 1987).

The lips and mouth are highly sensitive to touch and are exquisitely erotic parts of our bodies. Kisses discover, explore, and excite the body. They also involve the senses of taste and smell, which are especially important because they activate unconscious memories and associations. Often we are aroused by familiar smells associated with particular sexual memories: a person's body smells, perhaps, or perfumes associated with erotic experiences. In some cultures—among the Borneans, for example—the word *kiss* literally translates as "smell." In fact, among traditional Eskimos and Maoris there is no mouth kissing, only the nuzzling that facilitates smelling.

Although kissing may appear innocent, it is in many ways the height of intimacy. The adolescent's first kiss is often regarded as a milestone, a rite of passage, the beginning of adult sexuality (Alapack, 1991). Philip Blumstein and Pepper Schwartz (1983) report that many of their respondents found it unimaginable to engage in sexual intercourse without kissing. In fact, they found that those who have a minimal (or nonexistent) amount of kissing feel distant from their partners but engage in coitus nevertheless as a physical release.

The amount of kissing differs according to orientation. Lesbian couples tend to engage in more kissing than heterosexual couples, and gay male couples kiss less than heterosexual couples. As many

Kissing is probably the most acceptable premarital sexual activity.

as 95 percent of lesbian couples, 80 percent of heterosexual couples and 71 percent of gay couples engage in kissing whenever they have sexual relations (Blumstein and Schwartz, 1983).

Oral-Genital Sex

In recent years, oral sex has become part of our sexual scripts. It is engaged in by heterosexuals, gay men, and lesbians. The two types of oral-genital sex are cunnilingus and fellatio. **Cunnilingus** is the erotic stimulation of a woman's vulva by her partner's mouth and tongue. **Fellatio** is the oral stimulation of a man's penis by his partner's sucking and licking. Cunnilingus and fellatio may be performed singly or simultaneously. Oral sex is an increasingly important and healthy aspect of adults' sexual selves (Wilson and Medora, 1990).

Although oral-genital sex is increasingly accepted by white middle-class Americans, it remains less permissible among certain ethnic groups. African Americans, for example, have lower rates of oral-genital sex than do whites; many African Americans consider it immoral (Wilson, 1986; Laumann et al., 1994). Oral sex is becoming increasingly accepted, however, by African-American women (Wyatt, Peters, and Guthrie, 1988). This is especially true if they have a good relationship and communicate well with their partners (Wyatt and Lyons-Rowe, 1990). Among married Latinos, oral sex is relatively uncommon. When it occurs, it is usually at the instigation of men, as women are not expected to be interested in erotic variety (Guerrero Pavich, 1986). Although little is known about older Asian Americans and Asian immigrants, college-age Asian Americans appear to accept oral-genital sex to the same degree as middle-class whites (Cochran, Mays, and Leung, 1991).

Among both sexes, the same percentages report receiving and performing oral sex (Laumann et al., 1994). For both sexes, fellatio is less common than either sexual intercourse or cunnilingus (Newcomer and Udry, 1985). A study of university students of both sexes found that oral sex was regarded as an egalitarian, mutual practice (Moffatt, 1989). Students felt less guilty about it than about sexual intercourse because oral sex was not "going all the way."

DID YOU KNOW?

Between 60 and 95 percent of the men and women in various studies report that they have engaged in oral sex. Among adults of all ages, 27 percent of the men and 19 percent of the women had oral sex in the previous year (Delamater and MacCorquodale, 1979; Michael et al., 1994; Peterson et al., 1983).

Sexual Intercourse

Sexual intercourse or **coitus**—the insertion of the penis into the vagina and subsequent stimulation—is a complex interaction. As with many other types of activities, the anticipation of reward triggers a pattern of behavior. The reward may not necessarily be orgasm, however, because the meaning of sexual intercourse varies considerably at different times for different people. There are many motivations for sexual intercourse; sexual pleasure is only one. Other motivations include showing love, having children, giving and receiving pleasure, gaining power, ending an argument, demonstrating commitment, seeking revenge, proving masculinity or femininity, or degrading someone (including oneself).

Although sexual intercourse is important for most sexually involved couples, its significance is different for men and women. For men, sexual intercourse appears to be only one of several activities, such as fellatio and cunnilingus, that they enjoy. For women, however, intercourse is often central to their sexual satisfaction. More than any other heterosexual sexual activity, sexual intercourse involves equal participation by both partners. Ideally, both partners equally and simultaneously give and receive. Many women report that the sense of sharing during intercourse is important to them.

Men tend to be more consistently orgasmic than women in sexual intercourse. Part of the reason may be that the clitoris frequently does not receive sufficient stimulation from penile thrusting alone to permit orgasm. Many women need manual stimulation during intercourse to be orgasmic. They may also need to be more assertive. A woman can manually stimulate herself or be stimulated by her partner before, during, or after intercourse. But to do so, she has to assert her own sexual needs and move away from the idea that sex is centered around male orgasm. The sexual script has to be redefined.

DID YOU KNOW?

According to a scientific, nationwide study of adults of all ages, about one-third of Americans have sexual intercourse twice a week, one-third a few times a month, and one-third a few times a year or not at all. Married couples are more likely to engage in coitus than singles; married women are more likely to be orgasmic. About 40 percent of married couples and 25 percent of singles report having coitus twice a week (Michael et al., 1994).

Anal Eroticism

Sexual activities involving the anus are known as **anal eroticism.** The male's insertion of his erect penis into his partner's anus is known as **anal intercourse.** Both heterosexuals and gay men may participate in this activity. For heterosexual couples who engage in it, anal intercourse is generally an experiment or occasional activity rather than a common mode of sexual expression. About 10 percent of men and 9 percent of women report engaging in anal sex in the previous year (Michael et al., 1994), and one in four men and one in five women reported having ever experienced anal sex (Laumann et al., 1994). Among gay men, anal intercourse is less common than oral sex. It is, nevertheless, an important ingredient in the sexual satisfaction of many gay men (Blumstein and Schwartz, 1983). From a health perspective, anal intercourse is the riskiest form of sexual interaction. It is the most prevalent sexual means of transmitting the human immunodeficiency virus (HIV) among both gay men and heterosexuals. Because the delicate rectal tissues are easily torn, HIV (carried within semen) can enter the bloodstream. (HIV will be discussed in detail later in the chapter.)

Sexual Enhancement

Sexual behavior cannot be isolated from our personal feelings and relationships. If our sexuality is not a source of personal growth, meaning, and pleasure, we need to think about its role in our lives and our relationships. Sometimes dissatisfaction arises because the relationship itself is unsatisfactory. At other times the relationship itself is good but the erotic fire needs to be lit or rekindled. Such relationships may grow through **sexual enhancement**— improving the quality of a sexual relationship, especially by providing accurate information about sexuality, developing communication skills, fostering positive attitudes, and increasing self-awareness.

CONDITIONS FOR GOOD SEX

According to noted sex therapist Bernie Zilbergeld (1992), enhancing our sexual relationships requires the following:

1. Accurate information about sexuality, especially your own and your partner's.
2. An orientation toward sex based on pleasure (including arousal, fun, love, and lust) rather than on performance and orgasm.
3. Being involved in a relationship that allows each person's sexuality to flourish.
4. An ability to communicate verbally and nonverbally about sex, feelings, and relationships.

5. Being equally assertive and sensitive about your own sexual needs and those of your partner.
6. Accepting, understanding, and appreciating differences between partners.

Being aware of your own sexual needs is often critical to enhancing your sexuality. Gender-role stereotypes and negative learning about sexuality often cause us to lose sight of our own sexual needs. Following these sexual stereotypes may impede our ability to have what therapist Carol Ellison (1985) calls "good sex." Ellison writes that you will know you are having good sex if you feel good about yourself, your partner, your relationship, and what you're doing. It's good sex if, after a while, you still feel good about yourself, your partner, your relationship, and what you did. Good sex does not necessarily include orgasm or intercourse. It can be kissing, holding, masturbating, oral sex, anal sex, and so on. It can be heterosexual, gay, lesbian, or bisexual.

Zilbergeld (1993) suggests that to fully enjoy our sexuality, we need to explore our "conditions for good sex." There is nothing unusual about requiring conditions for any activity. For a good night's sleep, for example, each of us has certain conditions. We may need absolute quiet, no light, a feather pillow, an open window. Others, however, can sleep during a loud dormitory party, curled up in the corner of a stuffy room. Of conditions for good sex, Zilbergeld writes:

> In a sexual situation, a condition is anything that makes you more relaxed, more comfortable, more confident, more excited, more open to your experience. Put differently, a condition is something that clears your nervous system of unnecessary clutter, leaving it open to receive and transmit sexual messages in ways that will result in a good time for you.

Different individuals report different conditions for good sex. Some common conditions include the following:

- *Feeling intimate with your partner:* This is often important for both men and women, despite stereotypes of men wanting only sex. If the partners are feeling distant from each other, they may need to talk about their feelings before becoming sexual. Emotional distance can take the heart out of sex.
- *Feeling sexually capable:* Generally this relates to an absence of anxieties about sexual performance. For men, this includes anxiety about erections or about ejaculating too soon. For women, it includes worry about painful intercourse or lack of orgasm. For both men and women, it includes worry about whether they are good lovers.

© Jerrican/Photo Researchers, Inc.

Common conditions for a satisfying sexual relationship include feelings of intimacy, capability, trust, arousal, alertness, and positiveness about the environment and situation.

- *Feeling trust:* Both partners may need to know they are emotionally safe with the other. They need to feel confident that they will not be judged or ridiculed or talked about.
- *Feeling aroused:* A person does not need to be sexual unless he or she is sexually aroused or excited. Simply because your partner wants to be sexual does not mean that you have to be.
- *Feeling physically and mentally alert:* Both partners should not feel particularly tired, ill, stressed, preoccupied, or under the influence of excessive alcohol or drugs.
- *Feeling positive about the environment and situation:* A person may need privacy, to be in a place where he or she feels protected from intrusion. Each needs to feel that the other is sexually interested and wants to be sexually involved.

INTENSIFYING EROTIC PLEASURE

One of the most significant elements in enhancing one's physical experience of sex is intensifying arousal, which requires us to focus on increasing erotic pleasure rather than on sexual performance. This can be done in many ways, some of which are described here.

Sexual arousal is the physiological responses, fantasies, and desires associated with sexual anticipation and sexual activity. We have different levels of arousal. These levels are not necessarily associated with particular types of sexual activities. Sometimes we feel more sexually aroused when we kiss than when we have sexual intercourse or oral sex. Masturbation may sometimes be more exciting than oral sex or coitus.

The first element in increasing your sexual arousal is having your conditions for good sex met. If you need a romantic setting, go for a walk at the beach by moonlight or listen to music by candlelight. If you want limits on your sexual activities, tell your partner. If you need a certain kind of physical stimulation, show or tell your partner what you like.

A second element in increasing arousal is focusing on the sensations you are experiencing. Once you begin an erotic activity, such as massaging or kissing, do not let yourself be distracted. When you're kissing, don't think about what you're going to do next or about an upcoming test in your marriage and family class. Instead, focus on the sensual experience. Zilbergeld (1993) explains the process:

> Focusing on sensations means exactly that. You put your attention in your body where the action is. When you're kissing, keep your mind on your lips. This is *not* the same as thinking about your lips or the kiss; just put your attention in your lips. As you focus on your sensations, you may want to convey your pleasure to your partner. Let him or her know through your sounds and movements that you are excited.

Sexual Relationships

Sexuality exists in various relationship contexts that may influence our feelings and activities. These include nonmarital, marital, and extramarital contexts.

NONMARITAL SEXUALITY

Nonmarital sex is sexual activities, especially sexual intercourse, that take place outside of marriage. We use the term *nonmarital sex* rather than *premarital sex* to describe sexual behavior among unmarried adults in general. We will use the term **premarital sex** when we are referring to never-married adults under the age of 30. There are several reasons to make premarital sex a subcategory of nonmarital sex. First, because increasing numbers of never-married adults are over 30 years of age, "premarital sex" does not adequately describe the nature of their sexual activities. Second, at least 10 percent of adult Americans will never marry; it is misleading to describe their sexual activities as "premarital." Third, many adults are divorced, separated, or widowed; 30 percent of divorced women and men will never remarry. Fourth, between 3 and 10 percent of the population is lesbian or gay; gay and lesbian sexual relationships cannot be categorized as "premarital," until gays and lesbians are given the right to marry.

Sexuality in Dating Relationships

Over the last several decades, there has been a remarkable increase in the acceptance of premarital sexual intercourse. Over the last three decades there has been a decline in the numbers of people who believe that premarital sex is "always wrong" and an increase in the percentages who feel it is "not wrong at all." This trend has been interpreted as a shift toward "moral neutrality" regarding intercourse before marriage (Christopher and Sprecher, 2000). For adolescents and young adults, the advent of effective birth control methods, changing gender roles that permit females to be sexual, and delayed marriages have played a major part in the rise of premarital sex. For middle-aged and older adults, increasing divorce rates and longer life expectancy have created an enormous pool of once-married men and women who engage in nonmarital sex. Only **extramarital sex**—sexual interactions that take place outside the marital relationship—continues to be consistently frowned upon.

The increased legitimacy of sex outside of marriage has transformed both dating and marriage. Sexual intercourse has become an acceptable part of the dating process for many couples, whereas only petting was acceptable before. Furthermore, marriage has lost some of its power as the only legitimate setting for sexual intercourse (Sprecher and McKinney, 1993). One important result is that many people no longer feel that they need to get married to express their sexuality in a relationship (Scanzoni et al., 1989).

There appears to be a general expectation among students that they will engage in sexual intercourse sometime during their college careers. Although college students expect sexual involvement to occur within an emotional or loving relationship (Robinson et al., 1991), this emotional connection may be relatively transitory.

FACTORS LEADING TO PREMARITAL SEXUAL INVOLVEMENT. What factors lead individual men and women to have premarital sexual intercourse? One study indicated that among men and women who had premarital sex, the most important factors were their love (or liking) for each other, physical arousal and willingness of both partners, and planning and arousal prior to the encounter (Christopher and Cate, 1985). Among nonvirgins, as with virgins, love or liking between the partners was extremely important (Christopher and Cate, 1984), but feelings of obligation or pressure were about as important as actual physical arousal. Women reported affection as being slightly more important than did men. An interesting finding is that men perceived slightly more pressure or obligation to engage in intercourse than women.

Examining the sexual decision process more closely, researcher Susan Sprecher (1989) identifies individual, relationship, and environmental factors affecting the decision to have premarital intercourse:

- *Individual factors:* The individual factors that influence the decision to have premarital intercourse include previous sexual experience, sexual attitudes, personality characteristics, and gender. The more premarital sexual experience a man or woman has had, the more likely he or she is to engage in sexual activities. Once the psychological barrier against premarital sex is broken, sex appears to become less taboo. This seems to be especially true if the earlier sexual experiences were rewarding in terms of pleasure and intimacy. Those with liberal sexual attitudes are more likely to engage in sexual activity than those with restrictive attitudes. In terms of personality characteristics, men and women who do not feel high levels of guilt about sexuality are more likely to engage in sex, as are those who value erotic pleasure. Men tend to initiate sexual activity more frequently than women, but both women and men use similar tactics to initiate sex (implying commitment, increasing attention, and displaying "status cues"). There is a gender difference in "compliance" with partner-initiated sex, such that women are more likely than men to comply, and they do it to maintain their relationships (Christopher and Sprecher, 2000).
- *Relationship factors:* Two of the most important factors determining sexual activity in a relationship are the level of intimacy and the length of time the couple has been together. Even those who are less permissive in their sexual attitudes accept sexual involvement if the relationship is emotionally intimate and long standing. Individuals who are less committed (or not committed) to a relationship are less likely to be sexually involved. Finally, persons in relationships in which power is shared equally are more likely to be sexually involved than those in inequitable relationships.
- *Environmental factors:* In the most basic sense, the physical environment affects the opportu-

nity for sex. Because sex is a private activity, the opportunity for it may be precluded by the presence of parents, friends, roommates, or children (Tanfer and Cubbins, 1992). The cultural environment, too, affects premarital sex. The values of one's parents or peers may encourage or discourage sexual involvement. A person's ethnic group also affects premarital involvement. Generally, African Americans are more permissive than whites, and Latinos are less permissive than non-Latinos (Baldwin, Whitely, and Baldwin, 1992). Furthermore, a person's subculture—such as the university or church environment, the singles world, the gay and lesbian community—exerts an important influence on sexual decision making.

INITIATING A SEXUAL RELATIONSHIP. As we saw in Chapter 7, after we meet someone, we weigh each other's attitudes, values, and philosophy to see if we are compatible. If the relationship continues in a romantic vein, we may include physical intimacy. To signal the transition from nonphysical to physical intimacy, one of us must make the first move. Making the first move marks the transition from a potentially sexual relationship to an actual one.

If the relationship develops according to traditional gender-role patterns, as described above, the male will make the first move to initiate sexual intimacy, whether it is kissing, petting, or engaging in sexual intercourse (O'Sullivan and Byers, 1992). At what point this occurs generally depends on two factors: the level of intimacy and the length of the relationship (Sprecher, 1989). The more emotionally involved the couple are, the more likely it is that they will be sexually involved as well. The duration of the relationship also affects the likelihood of sexual involvement. Initial sexual involvement can occur as early as the first meeting or much later, as part of a well-established relationship. Some people become sexually involved immediately ("lust at first sight"), but the majority begin their sexual involvement in the context of an ongoing relationship. In fact, even in one-night stands or short-term affairs, more couples knew each other at least a year before engaging in sex than knew each other just for a couple or few days (Miracle, Miracle, and Baumeister, 2003).

Strategies for making the first move vary, depending on the motives of each individual and the nature of the relationship. If the people do not know each other well, both are likely to rely on traditional sexual scripts, with the man making the first overt move. If the motive for both partners is sexual pleasure, both may acknowledge their lack of interest in commitment. But if one desires pleasure and the other commitment, different strategies may be used. The pleasure-oriented partner, for example, may cease making overtures, feign commitment, or utilize sexual pressure. The other partner may withhold sex unless a commitment is made.

In new or developing relationships, communication about sexuality is generally indirect and ambiguous. Direct strategies are sometimes used to initiate sexual involvement, but they usually are used when the person is confident in the other's interest or is not concerned about being rejected. A study of sexual initiation among college students found that males and females used similar strategies to initiate sex (O'Sullivan and Byers, 1992).

In new relationships, we communicate indirectly about sex because we want to become sexually involved with the other person but we also want to avoid rejection. By using indirect strategies—such as turning down the lights, moving closer, touching the other's face or hair—we may test the other's interest in sexual involvement. If the other person responds positively to our cues, we can initiate a sexual encounter. At the same time, if the other turns the lights up, moves away, or does not respond to our touching, he or she gives a message of disinterest. Because we have not made direct overtures, we can save face. The sexual cues, innuendos, and signals can pass unacknowledged. Consequently a direct refusal does not occur. We can breathe a sigh of relief because we have avoided rejection.

Because so much of our sexual communication is indirect, ambiguous, or nonverbal, we run a high risk of being misinterpreted. There are four reasons for our communications being misinterpreted (Cupach and Metts, 1991). First, men and women tend to disagree about when sexual activities should take place in a relationship. Men tend to want sexual involvement earlier and with lower levels of intimacy than do women. Second, men may be skeptical about women's refusals, and may often misinterpret women's cues. Men also believe that women often say "no" when they actually mean "coax me," so they interpret it as token resistance. Third, because women communicate indirectly, they may be unclear in signaling their disinterest. They may turn their face aside, move a man's hand back to its proper place, say it's getting late, or try to change the subject. Research indicates that women are most effective when they make strong, direct verbal refusals; men become more compliant if women are persistent in their direct verbal refusals (Christopher and Frandsen, 1990; Murnen, Perot, and Byrne, 1989). Fourth, men are more likely than women to interpret nonsexual behavior or cues as sexual. Although both men and women flirt for fun, men are more likely to flirt with a sexual purpose and to interpret a woman's flirtation as sexual or "teasing."

DIRECTING SEXUAL ACTIVITY. As we begin a sexual involvement, we have several tasks to accomplish. First, we need to practice safe sex. We need to gather information about our partners' sexual history and determine whether he or she practices safe sex, including the use of condoms. Unlike much of our sexual communication, which is nonverbal or ambiguous, we need to use direct verbal discussion in practicing safe sex. Second, unless we are intending a pregnancy, we need to discuss birth control. Although the use of condoms will help prevent the spread of sexually transmissible diseases, condoms alone are only moderately effective as contraception. To be highly effective, they must be used in conjunction with contraceptive foam or jellies or other devices. Responsibility for contraception, like safe sex, generally requires verbal communication.

In addition to communicating about safe sex and contraception, we also need to communicate about what we like and need sexually. What kind of foreplay do we like? What kind of afterplay? Do we like to be orally or manually stimulated during intercourse? If so, how? What does each partner need to be orgasmic? Many of our needs and desires can be communicated nonverbally by our movements or by other physical cues. But if our partner does not pick up our nonverbal signals, we need to discuss them directly and clearly to avoid ambiguity.

Sexuality in Cohabiting Relationships

As we saw in Chapter 7, cohabitation has become a widespread phenomenon in American culture. Despite its increasing importance for men and women of all ages, little research exists on sexuality in such relationships. In contrast to married men and women, cohabitants have sexual intercourse more frequently, are more egalitarian in initiating sexual activities, and are more likely to be involved in sexual activities outside their relationship (Waite and Gallagher, 2001; Blumstein and Schwartz, 1983). The higher frequency of intercourse, however, may be due to the "honeymoon" effect: Cohabitants may be in the early stages of their relationship, the stages when sexual frequency is highest. Blumstein and Schwartz found that 22 percent of female cohabitants but only 9 percent of married women had been involved in extrarelational sex in the previous year; 25 percent of male cohabitants and 11 percent of married men were similarly involved. The differences in frequency of extrarelational sex may result from a combination of two factors: Norms of sexual fidelity may be weaker in cohabiting relationships, and men and women who cohabit tend to conform less to conventional norms.

SEXUALITY IN GAY AND LESBIAN RELATIONSHIPS

Because of their socialization as males, gay men are likely to initiate sexual activity earlier in the relationship than are lesbians. In large part, this is because both partners are free to initiate sex and because men are not expected to refuse sex, as women are (Isensee, 1990). Lesbians do not initiate sex as frequently as do gay or heterosexual men. They often feel uncomfortable because women have not been socialized to initiate sex.

In both gay and lesbian relationships, the more emotionally expressive partner is likely to initiate sexual interaction. The gay or lesbian partner who talks more about feelings and who spontaneously gives the partner hugs or kisses is the one who more often begins sexual activity.

One of the major differences between heterosexuals and gay men and lesbians is in how they handle extrarelational sex. In the gay and lesbian culture, sexual exclusivity is negotiable. Sexual exclusiveness is not necessarily equated with commitment or fidelity among gay men, although it often is among lesbians (Renzetti and Curran, 1995).

As a result of these differing norms, gay men and lesbians must decide early in the relationship whether they will be sexually exclusive (Isensee, 1990). If they choose to have a nonexclusive relationship, they need to discuss how outside sexual interests will be handled. They need to decide whether to tell each other, whether to have affairs with friends, what degree of emotional involvement will be acceptable, and how to deal with jealousy.

MARITAL SEXUALITY

When people marry, they may discover that their sexual life is very different than it was before marriage. Sex is now morally and socially sanctioned. It is in marriage that the great majority of heterosexual interactions take place. Yet as a culture we feel ambivalent about marital sex. On the one hand, marriage is the only relationship in which sexuality is fully legitimized. On the other hand, marital sex is an endless source of humor and ridicule: "Marital sex? What's that?" Two journalists watched prime-time television on the major networks for one week (Hanson and Knopes, 1993) and found that of the 45 sexual scenes depicted, only four were between married couples. There were almost six times as many depictions of sexual activities between unmarried men and women as between married couples. There was four times as much extramarital sex as marital sex. On television, men have sex more often with prostitutes than with their wives. Erotic activity is often linked with violence. Sex research is not much different from popular culture. The empirical research devoted to healthy marital sexuality is virtually nonexistent.

Sexual Interactions

A variety of large scale studies report consistent findings in regard to how often married couples engage in sexual intercourse and in how sexual frequency changes over the course of a marriage. Married couples report engaging in sexual relations about once or twice a week, or about six to seven times a month (Christopher and Sprecher, 2003).

Sexual intercourse tends to diminish in frequency the longer a couple is married. For newly married couples, the average rate of sexual intercourse is about three times a week. Data from over 13,000 respondents in the National Survey of Families and Households, reported that couples under the age of 24 had sex on average 11.7 times per month (or approximately three times per week). (Call, Sprecher and Schwartz, 1995 cited in Christopher and Sprecher, 2000). As couples get older, sexual frequency drops. In early middle age, married couples make love an average of one and a half to two times a week. After age 50, the rate is about once a week or less. Among couples 75 and older, the frequency is a little less than once a month (Christopher and Sprecher, 2000).

This decreased frequency, however, does not necessarily mean that sex is no longer important or that the marriage is unsatisfactory. It often means simply that one or both members are too tired. For dual-worker families and families with children, fatigue and lack of private time may be the most significant factors in the decline of frequency (Olds, 1985). Blumstein and Schwartz (1983) found that most people attributed their decline in sexual intercourse to lack of time or physical energy or to "being accustomed" to each other. Also, activities and interests other than sex engage them. The decline in interest and frequency of sex may begin within the first two years of marriage (Christopher and Sprecher, 2000).

Most married couples don't seem to believe that declining frequency is a major problem if they rate their overall relationship as good (Cupach and Comstock, 1990). Sexual intercourse is only one erotic bond among many in marriage. There are also kissing, caressing, nibbling, stroking, massaging, dining by candlelight, walking hand in hand, looking into each other's eyes, and talking intimately.

Married men continue to initiate sexual encounters overtly more frequently than do women, but women signal their interest or willingness. They pace the frequency of intercourse by showing their interest through nonverbal cues, such as a "certain look" or lighting candles by the bed; they may also overtly suggest "doing the wild thing." Their partners pick up on the cues and "initiate" sexual interactions. In marital relationships, many women feel comfortable about initiating sex. In part this may be related to the decreasing significance of the double standard as relationships continue. In marital relationships, the woman's initiation may be viewed positively, as an expression of love; it may also be the result of couples becoming more egalitarian in their gender-role attitudes.

Positive responses to initiation are usually nonverbal, such as beginning or continuing the sexual interaction by kissing or touching erotically. In most cases, when a partner refuses the sexual initiation, the couple "agree" not to have sex. They may decide to have sex at a different time, or they may "agree to disagree"—that is, they may find disagreement acceptable and nonthreatening. Partners were most satisfied in the way the disagreement was resolved if the initiation was made verbally, such as with the question: "Do you want to make love?" than if the initiation was made physically, such as by erotic touching or kissing. Partners find it easier to say or accept no to a verbal request than to a physical one. Contrary to the common stereotype, it appears that women do not restrict sexual activities any more than men do (Byers and Heinlein, 1989; O'Sullivan and Byers, 1992).

New Meanings to Sex

Sex within the marriage is significantly different from premarital sex in at least three ways. First, sex in marriage is expected to be *monogamous.* Second, procreation is a legitimate goal. Third, such sex takes place in the everyday world. These differences present each person with important tasks.

MONOGAMY. One of the most significant factors shaping marital sexuality is the expectation of monogamy. Before marriage or following divorce a person may have various sexual partners, but within marriage all sexual interactions are expected to take place between the spouses. In fact, approximately 90 percent of Americans believe extramarital sexual relations are "always" or "almost always" wrong (Miracle, Miracle, and Baumeister, 2003; Christopher and Sprecher, 2000). This expectation of monogamy lasts a lifetime; a person marrying at 20 commits himself or herself to 40 to 60 years of sex with the same person. Within a monogamous relationship, each partner must decide how to handle fantasies, desires, and opportunities for extramarital sexuality. Do you tell your spouse that you have fantasies about other people? That you masturbate? Do you flirt with others? Do you have an extramarital relationship? If you do, do you tell your spouse? How do you handle sexual conflicts or difficulties with your partner? How do you deal with sexual boredom or monotony?

SOCIALLY SANCTIONED REPRODUCTION. Sex also takes on a procreative meaning within marriage. Although it is obviously possible to get pregnant before marriage, in most segments of society, marriage remains the only socially sanctioned setting for having children. At marriage, partners are confronted with the task of deciding whether and when to have children. It is one of the most crucial decisions they will make, for having children profoundly alters a relationship. If the couple decide to have a child, their lovemaking may change from simply an erotic activity to an intentionally reproductive act as well.

CHANGED SEXUAL CONTEXT. The sexual context changes with marriage. Because married life takes place in a day-to-day living situation, sex must also be expressed in the day-to-day world. Sexual intercourse must be arranged around working hours and at times when the children are at school or asleep. One or the other partner may be tired, frustrated, or angry.

Two examples from interviews one of this book's authors did illustrate this quite vividly. In the first, a 33-year-old father of one contrasted where he and his wife prioritized sex:

> It's more important to me than to my wife . . . My wife always says, "I can't just have sex like you. Everything's gotta be . . . you know, the dishes gotta be washed, the place has gotta be cleaned. I got a thousand things on my mind." I say, "Yeah, well I got a thousand things on my mind too, but the first thing is sex!" They (women) can't do that.

In another example, a 30-year-old husband describes life before and after marriage.

> You don't think of (this) when you're single. You go out with the guys, you work all day, then you go out and play basketball for a couple of hours, afterward you go out, have a couple of beers, come home exhausted, and just plop into bed. Nobody's there to complain. Do the same thing when you're married, and you come home and your wife says, "Hi sweetheart. How about tonight?" You say, "I'm exhausted, honey, please . . ." And she says, "But that's what you said last night." . . . After we got married, the honeymoon came and went fine, but then you get into your routine. And I'm not one of those guys who can handle that every night. When I go to bed I like to go to sleep. (Cohen, 1986)

In marriage, some of the emotions associated with premarital sex may disappear. Some of the passion of romantic love eventually disappears as well, to be replaced with a love based on intimacy, caring, and commitment. Although we may tend to believe that good sex depends on good techniques, it really depends more on the quality of the marriage. As humorist Garrison Keillor (1994) reminds us:

> Despite jobs and careers that eat away at their evenings and weekends and nasty whiny children who dog their footsteps and despite the need to fix meals and vacuum the carpet and pay bills, [married] couples still manage to encounter each other regularly in a lustful, inquisitive way and throw their clothes in the corner and do thrilling things in the dark and cry out and breathe hard and afterward lie sweaty together feeling *extreme pleasure.*

DID YOU KNOW?

Between 20 and 25 percent of American men and 10 to 15 percent of American women reported having extramarital affairs (Christopher and Sprecher, 2000).

EXTRAMARITAL SEXUALITY

As we saw earlier, extramarital sex consists of sexual interactions that take place outside the marital relationship. A fundamental assumption in our culture is that marriages are monogamous. Each person remains the other's exclusive intimate partner in terms of both emotional and sexual intimacy. Extramarital relationships violate that assumption. Figure 8.1 shows the percentage of Americans who have had extramarital affairs according to a 1994 survey.

Personal characteristics and the quality of the marriage appear to be the most important factors associated with extramarital relationships. The personal characteristics—feelings of alienation, need for intimacy, emotional dependence, and egalitarian gender roles—are stronger correlates of extramarital sex than is the quality of the marriage. Generally, the lower the marital satisfaction and the lower the frequency and quality of marital intercourse, the greater the likelihood of extramarital sexual relationships. Most people become involved in extramarital sex because they feel something is missing in their marriage. They have judged it

FIGURE 8.1 ■ **Lifetime Incidence of Infidelity by Gender and Age**

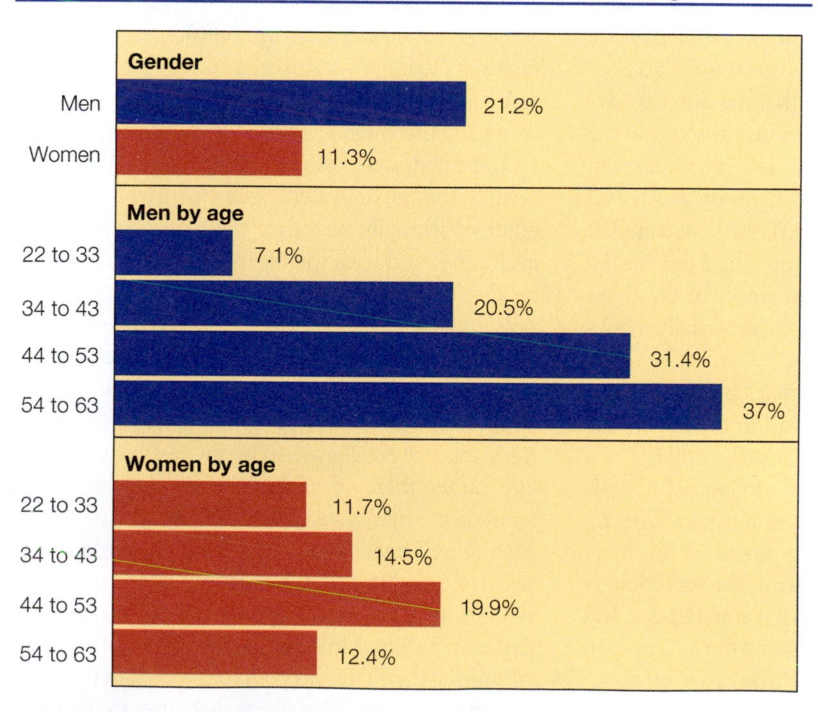

SOURCE: National Opinion Research Center, 1994 survey.

defective, although not defective enough to consider divorce. Extramarital relationships are a compensation for these deficiencies.

We tend to think of extramarital involvements as being sexual, but they may actually assume several forms (Moore-Hirschl et al., 1995; Thompson, 1993). They may be (1) sexual but not emotional, (2) sexual and emotional, or (3) emotional but not sexual (Thompson, 1984). Very little research has been done on extramarital relationships in which the couple are emotionally but not sexually involved. Anthony Thompson's study, however, found that among married and cohabiting individuals, the three types of extrarelational involvement were about equally represented.

As a result of marital assumptions, both sexual and nonsexual extramarital relationships take place without the knowledge or permission of the other partner. If the marital assumptions are violated, David Weiss (1983) points out, we have "guidelines" on how to handle the violation: "These guidelines encourage the 'adulterer' to be secretive and discreet, suggest that guilt will be a consequence, and maintain that the spouse will react with feelings of jealousy and rejection if the [extramarital sex] is discovered."

If the extramarital relationship is discovered, a marital crisis ensues. Many married people believe that the spouse who is unfaithful has broken a basic trust. Sexual accessibility implies emotional accessibility. When one spouse learns that the other is having an affair, the emotional commitment of the spouse having the affair is brought into question. How can the partner prove that he or she is still trustworthy? It cannot be done. Trust is assumed; it can never be proved. Furthermore, the extramarital relationship of one partner may imply to the other (rightly or wrongly) that he or she is sexually inadequate or uninteresting.

People who engage in extramarital affairs have a number of different motivations, and these affairs satisfy a number of different needs (Adler, 1996; Moultrup, 1990). Research by Ira Reiss and his colleagues (1980) suggests that extramarital affairs are related to two variables: unhappiness in the marriage and premarital sexual permissiveness. Generally speaking, in happy marriages, a partner is less likely to seek outside sexual relationships. A person who had premarital sex is more likely to have extramarital sex; once the first prohibition is broken, the second holds less power.

REFLECTIONS

Is extramarital sexuality ever justifiable? What about nonsexual affairs? Should partners be honest with each other about close involvements outside of the relationship?

Characteristics of Extramarital Sex

The majority of extramarital sexual involvements are sporadic. Most extramarital sex is not a love affair; it is generally more sexual than emotional. Affairs that are both emotional and sexual appear to detract more from the marital relationship than do affairs that are only sexual or only emotional (Thompson, 1984). More women than men consider their affairs emotional; almost twice as many men as women consider their affairs only sexual. About equal percentages of men and women are involved in affairs that they view as both sexual and emotional. Research suggests that men are more bothered by the sexual nature of a partner's infidelity where women are disturbed more by the emotional aspect (Christopher and Sprecher, 2000).

An emotionally significant extramarital affair creates a complex system of relationships among the three individuals (Moultrup, 1990). Long-lasting affairs can form a second but secret "marriage." In some ways, these relationships resemble polygamy, in which the outside person is a "junior" partner with limited access to the other. Such relationships form a triangular system. The two involved in the affair continually negotiate their relationship with each other and with the uninvolved partner (whose needs, demands, or possible presence or suspiciousness must always be considered). Meanwhile, the uninvolved partner mistakenly believes he or she is involved in a dyadic (two-person) system. As a result, he or she misinterprets situations. The partner's absence is believed to be the result of working late rather than an affair. The involved partners, who know their system is triadic, must try to meet each other's needs for time, affection, intimacy, and sex while taking the uninvolved partner into consideration. Such extramarital systems are stressful and demanding. Most people find great difficulty in sustaining them. If both people involved in the affair are married, the dynamics become even more complex.

Sexually Open Marriages

In an **open marriage,** the partners agree to allow each other to have openly acknowledged and independent sexual relationships with others. Blumstein and Schwartz (1983) found that 15 to 26 percent of the couples in their sample had "an understanding" that permitted extramarital relations under certain conditions, such as having affairs only out of town, never seeing the same person twice, and never having sex with a mutual friend. Knapp and Whitehurst (1977) found that successful open marriages required (1) a commitment to the primacy of the marriage, (2) a high degree of affection and trust between the spouses, (3) good interpersonal skills to manage complex relationships, and (4) nonmarital partners who did not compete with the married partner.

A study by Rubin and Adams (1986) attempted to measure the impact of sexually open marriages on marital stability. It matched 82 couples in 1978 and followed up 74 of them five years later. It found no significant difference in marital stability related to whether the couples were sexually open or monogamous in their marriages. Among the marriages that broke up, the reasons given were not related to extramarital sex. No appreciable differences were found in terms of marital happiness and jealousy.

Nonconsensual Sexual Behavior

There are many varieties of **nonconsensual sexual behaviors**—sexual behaviors that people experience against their will and without their consent. Less extreme forms include **voyuerism** and **exhibitionism** as well as obscene phone calls. Voyeurs (also called "peepers") are sexually aroused by covertly watching people undress or engage in sexual relations. Exhibitionists (also known as "flashers") expose themselves in public settings, such as parks, libraries, and deserted streets. Obscene phone calls are sometimes referred to as "less extreme forms" of nonconsensual sexual behavior because they tend to be nonviolent. However, they can cause victims to feel unsafe or violated (Miracle, Miracle, and Baumeister, 2003). It is estimated that almost all women receive such phone calls at some point. Rape, sexual abuse of children, and sexual harassment are all examples of **sexual coercion**—the more extreme forms of nonconsensual sex (Miracle, Miracle, and Baumeister, 2003). We will deal with each again in later chapters but for now it is worthwhile to note the following:

- *Sexual harassment* consists of "deliberate, repeated, verbal comments, gestures, or physical contact of a sexual nature that is considered unwelcome by the recipient" (Miracle, Miracle, and Baumeister, 2003). It occurs in schools, workplaces, and the military, though its prevalence is hard to measure definitively. There are two categories of harassment. The first is *quid pro quo harassment,* an explicit bartering and exchange of favors for sex. The harasser may offer higher grades, promotions at work, avoidance of punishment, and so on, in exchange for desired sexual contact. The second category is the creation of a *hostile environment* that creates discomfort and interferes with a victim's abilities to work, study, or perform expected behavior (Kimmel, 2004). Most harassment is unreported, despite consequences that may include psychological distress, educational failure, absenteeism, or quitting one's job.
- *Rape* refers to the use or threat of force to obtain sex. Although we may think more often about rape between strangers, the actual occurrence of "stranger rape" is far less than marital, acquaintance, or date rape. However, rapes occurring within relationships are the most underreported type of rape (Miracle, Miracle, and Baumeister, 2003). As a result, accurate estimates of the prevalence of sexual assault and rape in marriage are difficult to come by. Somewhere between 2 and 14 percent of married women are victims of forced sex by their husbands (Christopher and Sprecher, 2000).
- Acquaintance or date rape refers to instances of rape in which the offender and victim are previously acquainted, often as current or former dating partners. There are certain characteristics that are associated with perpetrators and victims of date rape. Some characteristics that are found more frequently among victims of date rape than among non-victims include poorer peer relationships, having more same-sex friends who

are sexually active, receiving less parental supervision, experiencing parental sexual abuse and/or family violence, and having a history of behavioral problems and delinquency (Christopher and Sprecher, 2000). Sexually coercive men differ from noncoercive men in that they date more frequently, start engaging in sexual behavior at younger ages, have more partners, and have a preference for casual and/or novel sexual encounters. Overall, they are described as having a "predatory" approach to sex (Christopher and Sprecher, 2000).

- *Sexual abuse of children* is defined as any form of sexual contact between a child and an adult. There need not be force applied, the child need not resist, and the behavior in question need not include acts of sexual penetration. In fact, fondling, kissing, taking nude pictures, or showing explicit sexual pictures to a child can be considered abusive if they are undertaken for the sexual gratification of the adult. The most common form of child sexual abuse is touching or fondling of the genitals. This occurs in over 80 percent of the cases in which boys are the victims, and in over 90 percent of the cases in which girls are victimized.
- Incest is a particular form of sexual abuse of children, occurring more often in situations of family disruption or accompanying spousal abuse, alcoholism, and marital problems. Fathers and stepfathers are more frequent perpetrators than mothers, but some suggest that siblings are the most frequent if least reported perpetrators of incest (Miracle, Miracle, and Baumeister, 2003).

Sexual Problems and Dysfunctions

Many of us who are sexually active may sometimes experience sexual difficulties or problems. Those problems that are recurring, causing distress to the individual or the partner, are known as **sexual dysfunctions.** Although some sexual dysfunctions are physical in origin, many are psychological. Some dysfunctions have immediate causes, others originate in conflict within the self, and still others are rooted in a particular sexual relationship.

Both men and women may suffer from hypoactive (low or inhibited) sexual desire (Hawton, Catalan, and Fagg, 1991). Other dysfunctions experienced by women are orgasmic dysfunction (the inability to attain orgasm), arousal difficulties (the inability to become erotically stimulated), and dyspareunia (painful intercourse). The most common dysfunctions among men include erectile dysfunction (the inability to achieve or maintain an erection), premature ejaculation (the inability to delay ejaculation after penetration), and delayed orgasm (difficulty in ejaculating) (Spector and Carey, 1990). Figure 8.2 shows the percentage of heterosexual adults in the general U.S. population who reported experiencing sexual problems during the previous year, in response to a recent survey (Laumann et al., 1994).

ORIGINS OF SEXUAL PROBLEMS

Physical Causes

It is generally believed that between 10 and 20 percent of sexual dysfunctions are structural in nature. Physical problems may be *partial* causes in another 10 or 15 percent (Kaplan, 1983; LoPiccolo, 1991). Various illnesses may have an adverse effect on a person's sexuality (Wise, Epstein, and Ross, 1992). Alcohol and some prescription drugs, such as medication for hypertension, may affect sexual responsiveness (Buffum, 1992; "Drugs," 1992).

Among women, diabetes, hormone deficiencies, and neurological disorders, as well as alcohol and alcoholism, can cause orgasmic difficulties. Painful intercourse may be caused by an obstructed or thick hymen, clitoral adhesions, a constrictive clitoral hood, or a weak pubococcygeus muscle. Coital pain caused by inadequate lubrication and thinning vaginal walls often occurs as a result of decreased estrogen associated with menopause. Lubricants or hormone replacement therapy often resolve the difficulties.

Among males, diabetes and alcoholism are the two leading physical causes of erectile dysfunctions; atherosclerosis is another important factor (LoPiccolo, 1991; Roenrich and Kinder, 1991). Smoking may also contribute to sexual difficulties (Rosen et al., 1991).

FIGURE 8.2 ■ **Heterosexual Sexual Dysfunctions in a Nonclinical Sample**

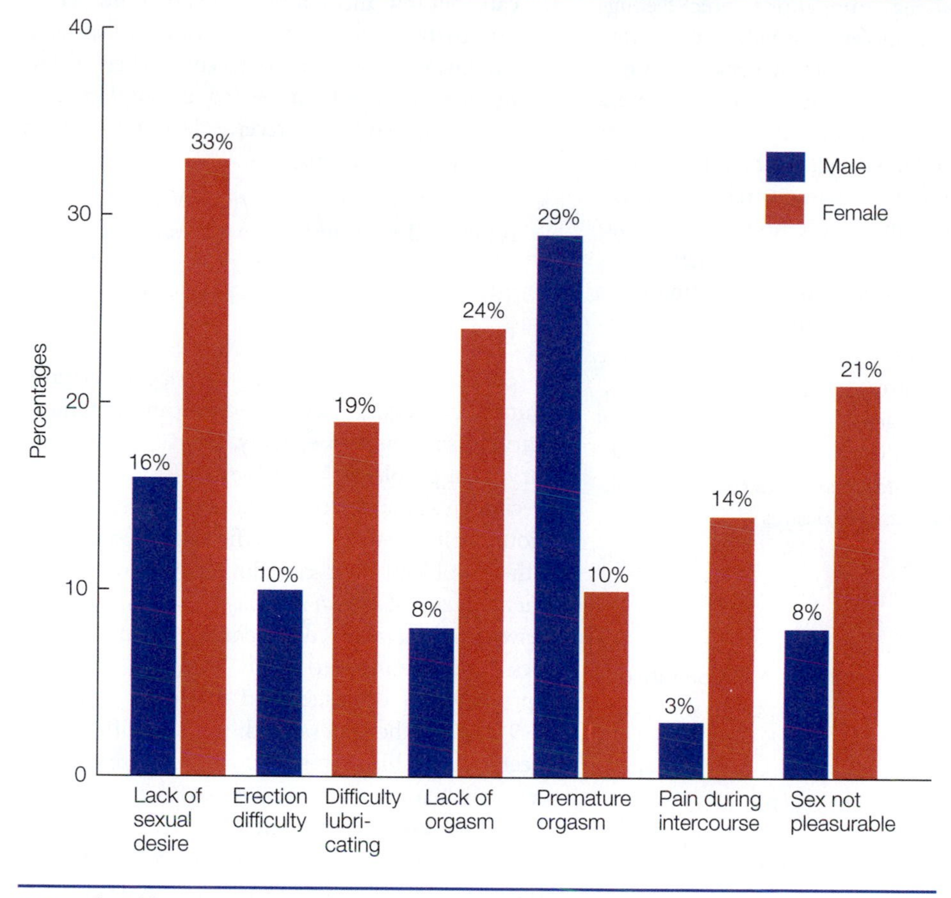

SOURCE: Adapted from Laumann et al., 1994, pp. 370–371.

Psychological/Relationship Causes

Two of the most prominent causes of sexual dysfunctions are performance anxiety and conflicts within the self. *Performance anxiety*—the fear of failure—is probably the most important immediate cause of erectile dysfunctions and, to a lesser extent, of orgasmic dysfunctions in women (H. Kaplan, 1979). If a man does not become erect, anxiety is a fairly common response. Some men experience their first erectile problem when a partner initiates or demands sexual intercourse. Women are permitted to say no, but many men have not learned that they too may say no to sex. Women suffer similar anxieties, but they tend to center around orgasmic abilities rather than the ability to have intercourse. If a woman is unable to experience orgasm, a cycle of fear may arise, preventing future orgasms. A related source of anxiety is an excessive need to please one's partner.

Conflicts within the self are guilt feelings about one's sexuality or sexual relationships. But guilt and emotional conflict do not usually eliminate a person's sexual drive; rather, they inhibit the drive and alienate the person from his or her sexuality. These psychic conflicts often are deeply rooted; they may be unconscious. Among gay men and lesbians, concerns about sexual orientation may be an important cause of such conflicts (George and Behrendt, 1987).

The relationship itself, rather than either individual, sometimes can be the source of sexual problems. Disappointment, anger, or hostility may become integral parts of a deteriorating or unhappy relationship. Such factors ultimately affect sexual interactions, for sex can become a barometer for the whole relationship. Helen Kaplan (1979) suggests that relationship discord affects our sexuality in several ways. First, we may *transfer* or redirect feelings we have about someone else (usually parents, former partners, or other important persons) to our current partner. Second, *power struggles* may be a central theme in a relationship. In such cases, sexuality becomes a tool in struggles for control. Third, we may have *unwritten rules* in our relationships—the usually unconscious assumptions and expectations about how each should act in the relationship, such as reading each other's minds, putting one's partner first, and so on. Fourth, partners may engage in *sexual sabotage* by asking for sex at the wrong time, putting pressure on each other, or frustrating or criticizing each other's sexual desires and fantasies. People most often do this unconsciously. Fifth, *poor communication* may undermine our ability to express our needs and desires.

Sex Between Unequals, Sex Between Equals

Sociologist Pepper Schwartz (1994) identified a number of sexual problems that plague traditional marriages because of the gender heirarchy that defines such marriages. In such relationships partners are "too distant, too different, and too inequitable" to enjoy complete sexual fulfillment. Sexual problems among traditional couples include the following:

- *Failure of timing:* This results when one person is more in charge of the couple's sexual relationship and his or her needs define when the couple has sexual relations. If men more often initiate sex, couples may suffer from a lack of synchronicity.
- *Failure of intimacy:* If traditional couples lack the same depth of intimacy that defines more egalitarian relationships, it is apparent in their sexual relationship. According to Schwartz, this absence of sharing and communicating can prevent them from finding complete fulfillment in their sexual relationship.
- *Failure of sexual empathy:* Some couples fail to realize that what one finds pleasing the other may not. This is particularly true in the most traditional marriages, where "men and women have little experience of each other's lives; they respect each other's sexual needs very little and refuse to take turns learning what to do for each other" (Schwartz, 1994).
- *Failure of reciprocity:* Inequality in relationships can spill into inequality in the bedroom, where one partner, more often the woman, feels as if she gives more than she receives. There is less mutual massage than desired, or she feels that she is touched less or receives less oral sex than she gives or performs.
- *Failure of overromanticization:* Women in more traditional relationships may possess overly romanticized expectations of sexual relations. These are often beyond what most "ordinary" men can live up to.

Schwartz notes that peer marriages, relationships built on deep friendship and commitments to fairness, sharing, and equality, avoid these particular sexual problems. What they suffer, instead, is a decline in sexual intensity. Some of this results just from habituation. More specific to peer couples are other problems that can diminish sexual excitement, most notably an inability to transform themselves from their everyday identities based on sameness and openness to erotic identities based on "principles of opposites and mystery" (Schwartz, 1994). Thus, the very same things that differentiate peer relationships from their more common and less equal counterparts may make it hard for them to sustain sexual energy. These problems are not insurmountable, but they do require special effort on the part of peer couples to create a separate and special sexual environment removed from more mundane life matters.

RESOLVING SEXUAL PROBLEMS

Perhaps the first step in dealing with a sexual problem is to turn to your own immediate resources. Begin by discussing the problem with your partner; find out what he or she thinks. Discuss specific strategies that might be useful. Sometimes simply communicating your feelings and thoughts will resolve the difficulty. Seek out friends with whom you can share your feelings and anxieties. Find out what they think. Ask them whether they have had similar experiences and how they handled them. Try to keep your perspective—and your sense of humor.

Partners, friends, and books may provide permission for you to engage in sexual exploration and discovery. From these sources we may learn that many of our sexual fantasies and behaviors are very common. Such methods are most effective when the dysfunctions arise from a lack of knowledge or mild sexual anxieties.

If you are unable to resolve your sexual difficulties yourself, seek professional assistance. It is important to realize that seeking such assistance is not a sign of personal weakness or failure. Rather, it is a sign of strength, for it demonstrates an ability to reach out and a willingness to change. It is a sign that you care for your partner, your relationship, and yourself.

In March of 1998, the Food and Drug Administration approved Viagra, the first oral treatment for male impotence. Since then, Viagra has become an economic and cultural phenomenon. In just its first year of availability, Viagra had sales of $1 billion, propelling its manufacturer, Pfizer, to the second spot among the world's largest drug companies. In 2002, Viagra had sales in excess of $1.7 billion. It is estimated that 9 tablets are dispensed every second worldwide.

Viagra was originally developed to improve blood flow to the heart as a treatment for chest pain. It was then discovered that the drug also had the effect of enabling men with erectile disfunction to become aroused. With success rates of as much as 80 to 85 percent, Viagra has restored many sexual relationships among mostly middle-aged and older couples. It has not been without costs, however. Some men suffer side effects, typically consisting of headaches, diarrhea, and nausea. More extreme consequences include deaths; a small number of men have died as a result of taking Viagra, and men with cardiac ailments are discouraged from taking this drug.

Since the approval of Viagra, two other drugs have been introduced for men with erectile dysfunction. Levitra, (*vardenafil*) is an erectile dysfunction drug marketed by Bayer and GlaxoSmithKline, that gained approval from the Food and Drug Administration in August 2003. From the same family of drugs as Viagra, its makers claim that Levitra may work for men who have had unsuccessful results from Viagra, and that older men who have higher cardiovascular and other medical risks that recommend against using Viagra may more safely use Levitra. More recently, a third medication—Cialis—developed by Eli Lilly and Icos was approved by the FDA for prescription use in the United States. Cialis, already available in more than 40 other countries, is also from the same family of drugs as Viagra and Levitra. Unlike its two predecessors, Cialis has shown longer effectiveness, lasting up to 36 hours, and leading to its nickname as "the weekend drug."

For those whose problems stem mostly from psychological or relationship causes, therapists can help deal with sexual problems on several levels. Some focus directly on the problem, such as lack of orgasm, and suggest behavioral exercises, such as pleasuring and masturbation, to develop an orgasmic response. Others focus on the couple relationship as the source of difficulty. If the relationship improves, they believe, then sexual responsiveness will also improve. Still others work with the individual to help develop insight into the origins of the problem in order to overcome it. Therapy can also take place in a group setting. Group therapy may be particularly valuable for providing partners with an open, safe forum in which they can discuss their sexual feelings and experience and discover commonalities with others.

Birth Control

Most of us think of sexuality in terms of love, passionate embraces, and entwined bodies. Sex involves all of these, but what we so often forget (unless we are worried) is that sex is also a means of reproduction. Whether we like to think about it or not, many of us (or our partners) are vulnerable to unintended pregnancies. Not thinking about pregnancy does not prevent it; indeed, not thinking about it may even contribute to the likelihood of its occurring. Unless we practice **abstinence,** refraining from sexual intercourse, we need to think about unintended pregnancies and then take the necessary steps to prevent them. Chapter 10 will take up the issues of birth control via contraception and abortion.)

DID YOU KNOW?

An estimated four in ten American women will become pregnant by the time they turn 20 (Alan Guttmacher Institute, 1997).

Sexually Transmitted Diseases and HIV/AIDS

"Do you have chlamydia, gonorrhea, herpes, syphilis, HIV, or any other sexually transmissible disease that I should know about?" is hardly a question you want to ask someone on a first date. But it is a question to which you really need to know the answers before you become sexually involved. Just because a person is nice is no guarantee that he or she does not have a **sexually transmitted disease (STD)**, a disease spread through sexual contact, such as sexual intercourse or oral or anal sex. No one can tell by a person's looks, intelligence, or moral fervor whether he or she has contracted a sexually transmitted disease, and the costs are too great for anyone to become sexually involved with a person without knowing about the presence of any of these diseases.

Americans are in the middle of the worst STD epidemic in our history. There are an estimated 15 million new cases of sexually transmitted infections in the United States each year, the highest rate of infection of any industrialized nation in the world (Miracle, Miracle, and Baumeister, 2003). College students are as vulnerable as anyone else. Untreated chlamydia and gonorrhea can lead to pelvic inflammatory disease (PID) in women, a major cause of infertility.

DID YOU KNOW?

Each year, pelvic inflammatory disease (PID) affects 1.7 million women in the United States, many of whom become sterile (Hilts, 1990a).

Overall, HIV (which will be described in the next section) has infected more than a million Americans. This includes 850,000 diagnosed AIDS cases, and another nearly 200,000 cases of HIV infection that had not progressed to AIDS. Approximately 56 percent of these infections were transmitted through male-male or heterosexual sexual contact (Centers for Disease Control, 2003). HIV and AIDS cases are increasing at a disproportionate rate for African Americans and Latinos; sexually transmitted cases among heterosexuals are increasing at a greater rate than among gay men. These figures suggest that virtually all adults in the United States are or will soon be related to, personally know, work with, or go to school with people who are infected with HIV or will know others whose friends, relatives, or associates test HIV-positive.

PRINCIPAL STDS

The most prevalent STDs in the United States are chlamydia, gonorrhea, genital warts, genital herpes, syphilis, hepatitis, and HIV/AIDS. Conditions that may be sexually transmitted include urethritis (in both women and men) and vaginitis and PID (in women). Table 8.1 (on pages 270–271) briefly describes the symptoms, exposure intervals, treatments, and other information regarding the principal STDs.

HIV/AIDS

The **human immunodeficiency virus (HIV)** is the virus that causes **acquired immunodeficiency syndrome (AIDS)**. The disease is so termed because of its characteristics.

> Acquired—because people are not born with it.
>
> ImmunoDeficiency—because the disease relates to the body's immune system, which is lacking in immunity.
>
> Syndrome—because the symptoms occur together as a group.

Although there is no vaccine to prevent or cure HIV or to prevent the subsequent AIDS symptoms, we have considerable knowledge about the nature of the virus and how to prevent its spread:

- *HIV attacks the body's immune system.* HIV is carried in the blood, semen, and vaginal secretions of infected persons. A person may be HIV-positive (infected with HIV) for many years before developing AIDS symptoms.
- *HIV is transmitted only in certain clearly defined circumstances.* It is transmitted through the exchange of blood (as by shared needles or transfusions of contaminated blood), through sexual contact involving semen or vaginal secretions, and, prenatally, from an infected woman to her fetus through the placenta. It is also possible for

infected mothers to transmit the infection during delivery or via nursing (Miracle, Miracle, and Baumeister, 2003).

- *Heterosexuals, bisexuals, gay men, and lesbians are all susceptible to the sexual transmission of HIV.* Sexual transmission accounts for 68 percent of AIDS and 57 percent of HIV infections among men through 2002. Male-male sexual contact is attributed to 55 percent of all AIDS cases, and 47 percent of all non-AIDS HIV infections among men. The rate of heterosexual HIV transmission is rising faster than the rate of gay transmission. Among women, heterosexual contact accounts for 42 percent of HIV/AIDS cases.
- *There is a definable progression of HIV infection and a range of illnesses associated with AIDS.* HIV attacks the immune system. Once this is impaired, AIDS symptoms occur, as opportunistic diseases—diseases that the body normally resists—infect the individual. The most common opportunistic diseases are pneumocystis carinii pneumonia and Kaposi's sarcoma, a skin cancer. It is an opportunistic disease rather than HIV that kills the person with AIDS.
- *The presence of HIV can be detected through antibody testing.* To date there are no widely available tests to detect HIV itself. The Western Blot and ELISA antibody tests are reasonably accurate blood tests that show whether the body has developed antibodies in response to HIV. The newest antibody test, OraSure, tests for the presence of antibodies in a fluid in the mouth (oral mucosal transudate). It is as accurate as but more expensive than blood testing. Anonymous testing is available at many college health centers and community health agencies. Self-test kits may also be purchased at pharmacies; the results of these are obtained anonymously by telephone after a drop of blood has been sent in for analysis. HIV antibodies develop between one and six months after infection. Antibody testing should take place one month after possible exposure to the virus and, if the results are negative, again six months later. If the antibody is present, the test will be positive. That means that the person has been infected with HIV and an active virus is present. The presence of HIV does not mean, however, that the person necessarily will develop AIDS symptoms in the near future; symptoms generally occur seven to ten years after the initial infection.
- *All those with HIV (whether or not they have AIDS symptoms) are HIV carriers.* They may infect others through unsafe sexual activity or by sharing needles; if they are pregnant, they may infect the fetus.
- *The distribution of HIV and AIDS varies by race and ethnicity.* Blacks and Latinos make up a larger percentage of HIV cases than one would predict based on their population size (see Figure 8.3).

DID YOU KNOW?

According to a Centers for Disease Control study, regular condom use almost tripled among women after they were counseled at clinics—from 13 percent to 36 percent ("Urban Women," 1997).

FIGURE 8.3 ■ **AIDS and Ethnicity, 1996**

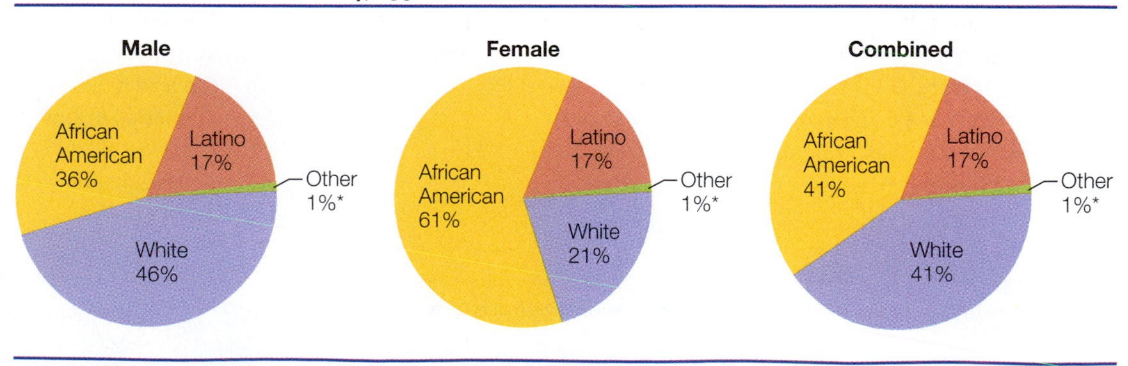

*Includes Asian and Pacific Islander as well as American Indian and Native Alaskan.
SOURCE: Centers for Disease Control, 2003.

TABLE 8.1 ■ Principal Sexually Transmitted Diseases

STD AND INFECTING ORGANISM	SYMPTOMS	TIME FROM EXPOSURE TO OCCURENCE	MEDICAL TREATMENT	COMMENTS
Chlamydia *(Chlamydia trachomatis)*	Women: 80% asymptomatic; others may have vaginal discharge or pain with urination. Men: 30–50% asymptomatic; others may have discharge from penis, burning urination, pain and swelling in testicles, or persistent low fever.	7–21 days	Doxycycline, tetracycline, erythromycin	If untreated, may lead to pelvic inflammatory disease (PID) and subsequent infertility in women.
Gonorrhea *(Neisseria gonorrhoeae)*	Women: 50–80% asymptomatic; others may have symptoms similar to chlamydia. Men: itching, burning, or pain with urination; discharge from penis ("drip").	2–21 days	Penicillin, tetracycline, or other antibiotics	If untreated, may lead to pelvic inflammatory disease (PID) and subsequent infertility in women.
Genital warts (Human papilloma virus)	Variously appearing bumps (smooth, flat, round, clustered, fingerlike, white, pink, brown, etc.) on genitals, usually penis, anus, vulva, vagina, or cervix.	1–6 months (usually within 3 months)	Surgical removal by freezing, cutting, or laser therapy. Chemical treatment with podophyllin (80% of warts eventually reappear).	Virus remains in the body after warts are removed.
Genital herpes (Herpes simplex virus)	Small, itchy bumps on genitals, becoming blisters that may rupture, forming painful sores; possibly swollen lymph nodes; flulike symptoms with first outbreak.	3–20 days	No cure although acyclovir may relieve symptoms. Nonmedical treatments may help relieve symptoms.	Virus remains in the body, and outbreaks of contagious sores may recur. Many people have no symptoms after the first outbreak
Syphilis *(Treponema pallidum)*	Stage 1: Red, painless sore (chancre) at bacteria's point of entry. Stage 2: Skin rash over body, including palms of hands and soles of feet.	Stage 1: 1–12 weeks Stage 2: 6 weeks to 6 months after chancre appears	Penicillin or other antibiotics	Easily cured, but untreated syphilis can lead to ulcers of internal organs and eyes, heart disease, neurological disorders, and insanity.

SOURCE: Bryan Strong and Christine DeVault. *Human Sexuality.* Mountain View, CA: Mayfield Publishing Company, 1997.

TABLE 8.1 ▪ ***(continued)***

STD AND INFECTING ORGANISM	SYMPTOMS	TIME FROM EXPOSURE TO OCCURENCE	MEDICAL TREATMENT	COMMENTS
Hepatitis (Hepatitis A or B virus)	Fatigue, diarrhea, nausea, abdominal pain, jaundice, darkened urine due to impaired liver function.	1–4 months	No medical treatment available; rest and fluids are prescribed until the disease runs its course.	Hepatitis B more commonly spread through sexual contact; can be prevented by vaccination.
Urethritis (various organisms)	Painful and/or frequent urination; discharge from penis; women may be asymptomatic.	1–3 weeks	Penicillin, tetracycline, or erythromycin, depending on organism.	Laboratory testing is important to determine appropriate treatment.
Vaginitis *(Gardnerella vaginalis, Trichomonas vaginalis,* or *Candida albicans)*	Intense itching of vagina and/or vulva; unusual discharge with foul or fishy odor; painful intercourse. Men who carry organisms may be asymptomatic.	2–21 days	Depends on organism; oral medications include metronidazole and clindamycin. Vaginal medications include clotrimazole and miconazole.	Not always acquired sexually. Other causes include stress, oral contraceptives, pregnancy, tight pants or underwear, antibiotics, douching, and dietary imbalance.
HIV infection and AIDS (Human immunodeficiency virus)	Possible flulike symptoms but often no symptoms during early phase. Variety of later symptoms including weight loss, persistent fever, night sweats, diarrhea, swollen lymph nodes, bruiselike rash, persistent cough.	Several months to several years	No cure available, although many symptoms can be treated with medications. Antiviral drugs may strengthen immune system. Good health practices can delay or reduce the severity of symptoms.	Cannot be self-diagnosed; a blood test must be performed to determine the presence of the virus.
Pelvic inflammatory disease (PID) (women only)	Low abdominal pain; bleeding between menstrual periods; persistent low fever.	Several weeks or months after exposure to chlamydia or gonorrhea (if untreated)	Penicillin or other antibiotics; surgery.	Caused by untreated chlamydia or gonorrhea; may lead to chronic problems such as arthritis and infertility.

PROTECTING YOURSELF AND OTHERS

Although many people have changed their behaviors to reduce the risk of STDs and HIV infection, many continue to jeopardize their health and lives—as well as the health and lives of their partners and loved ones—by failing to take adequate precautions. They worry about STDs and HIV but do not take the necessary steps—such as always using condoms—to prevent infections. One study of sexually active young adults found that 44 percent had not changed their behavior in any way to reduce the risk of HIV infection (Cochran, Keidan, and Kalechsteir, 1989). A study of female college students found that except for causing an increase in regular condom use—from 21 percent in 1986 to 41 percent in 1989—public health campaigns have not had a substantial influence on the habits and behavior of these well-educated young adults (DeBuono et al., 1990).

A male university student from Illinois who knew that heterosexuals are at risk for contracting HIV maintained a dismissive position: "I just don't see AIDS as being much of a threat to heterosexuals, and I don't find a lot of pleasure in using a condom" (Johnson, 1990). Another study of college students found that they believe they can "identify" infected men and women (Maticka-Tyndale, 1991). A female student in Ohio explained why she does not insist that her partner use a condom: "I have an attitude—it may be wrong—that any guy I would sleep with would not have AIDS" (Johnson, 1990).

Abstinence is the best protection from STDs and HIV. If you are sexually active, however, the key to protecting yourself and others is to talk with your partner about STDs in an open, nonjudgmental way and to use condoms. The best way of finding out whether your partner has an STD is by asking. If you feel nervous about broaching the subject, you can rehearse talking about it. It may be sufficient to ask in a lighthearted manner, "Are you as healthy as you look?" or because many people are uncomfortable asking about STDs, you can open up the topic by revealing your anxiety: "This is a little difficult for me to talk about because I like you and I'm embarrassed, but I'd like to know whether you have herpes, or HIV, or whatever." If you have an STD, you can say, "Look, I like you, but we can't make love right now because I have a chlamydial infection and I don't want you to get it."

Remember, however, that every person who believes he or she doesn't have an STD may honestly not know. Women with chlamydia and gonorrhea, for example, generally don't exhibit symptoms. Both men and women infected with HIV may not show any symptoms for years, although they are capable of spreading the infection through sexual contact.

If you don't know whether your partner has an STD, use a condom. Even if you don't discuss STDs, condoms are simple and easy to use without much discussion. Both men and women can carry them. A woman can take a condom from her purse and give it to her partner. If he doesn't want to use it, she can say, "No condom, no sex." (Condoms and safer sex practices are discussed in greater detail in the Resource Center.)

Sexual Responsibility

Because we have so many sexual choices today, we need to understand what responsibilities our sexuality entails. Sexual responsibility includes the following:

- *Disclosure of intentions:* Each person needs to reveal to the other whether a sexual involvement indicates love, commitment, recreation, and so on.
- *Freely and mutually agreed-upon sexual activities:* Each individual has the right to refuse any or all sexual activities without the need to justify his or her feelings. There can be no physical or emotional coercion.
- *Use of mutually agreed-upon contraception in sexual intercourse if pregnancy is not intended:* The persons in a sexual relationship are equally responsible for preventing an unintended pregnancy in a mutually agreed-upon manner.
- *Use of "safer sex" practices:* Each person is responsible for practicing safer sex unless both have been monogamous with each other for at least five years or have recently tested negative for HIV. Safer sex practices guard against sexually transmitted diseases, especially HIV/AIDS. Such practices do not transmit semen, vaginal secretions, or blood during sexual activities.

- *Disclosure of infection from or exposure to STDs:* Each person must inform his or her partner about personal exposure to an STD because of the serious health consequences, such as infertility or AIDS, that may follow untreated infections. If you are infected, you must refrain from behaviors—such as sexual intercourse, oral-genital sex, and anal intercourse—that may infect your partner. To help ensure that STDs are not transmitted, use a condom.
- *Acceptance of the consequences of sexual behavior:* Each person needs to be aware of and accept the possible consequences of his or her sexual activities. These consequences can include emotional changes, pregnancy, abortion, and sexually transmitted diseases.

REFLECTIONS

How do you deal with the issue of STDs in your relationships? Do you discuss the topic with a partner prior to becoming sexually involved? How do you or would you bring up the topic of STDs? Have you ever contracted an STD? If so, how did you feel about your partner? Yourself?

Responsibility in many of these areas is facilitated when sex takes place within the context of an ongoing relationship. In that sense, sexual responsibility is a matter of values. Is responsible sex possible outside an established relationship? Are you able to act in a sexually responsible way? Sexual responsibility also leads to the question of the purpose of sex in your life. Is it for intimacy, erotic pleasure, reproduction, or other purposes?

As we consider the human life cycle from birth to death, we cannot help but be struck by how profoundly sexuality weaves its way through our lives. From the moment we are born, we are rich in sexual and erotic potential, which begins to take shape in our sexual experimentations of childhood. As children, we are still unformed, but the world around us haphazardly helps give shape to our sexuality. In adolescence, our education continues as a mixture of learning and yearning. But as we enter adulthood, with greater experience and understanding, we undertake to develop a mature sexuality: we establish our sexual orientation as heterosexual, gay, lesbian, or bisexual; we integrate love and sexuality; we forge intimate connections and make commitments; we make decisions regarding our fertility and sexual health; we develop a coherent sexual philosophy. Then, in our middle years, we redefine sex in our intimate relationships, accept our aging, and reevaluate our sexual philosophy. Finally, as we become elderly, we reinterpret the meaning of sexuality in accordance with the erotic capabilities of our bodies. We come to terms with the possible loss of our partner and our own end. In all these stages, sexuality weaves its bright and dark threads through our lives.

SUMMARY

- Our primary sources of sexual learning are parents, peers, and the media. As we get older, interactions with our partners become increasingly important.
- There are several tasks that we must undertake in developing our sexuality as young adults, including (1) integrating love and sex, (2) forging intimacy and commitment, (3) making fertility/childbearing decisions, (4) establishing a sexual orientation, and (5) developing a sexual philosophy.
- The traditional female *sexual scripts* include the following ideas: Sex is both good and bad (depending on the context); girls don't want to know about their bodies "down there"; sex is for men; men should know what women want; women shouldn't talk about sex; women should look like beautiful models; women are nurturers; and there is only one right way to experience an orgasm. Traditional male sexual scripts include the following: Men should not have (or at least should not express) certain feelings; performance

is the thing that counts; the man is in charge; a man always wants sex and is ready for it; all physical contact leads to sex; sex equals intercourse; and sexual intercourse always leads to orgasm.

- Contemporary sexual scripts are more egalitarian, consisting of the following beliefs: Sexual expression is positive; sexual activities are a mutual exchange of erotic pleasure; sexuality involves both partners equally and the partners are equally responsible; legitimate sexual activities include masturbation and oral-genital sex; sexual activities may be initiated by either partner; both partners have a right to experience orgasm; and nonmarital sex is acceptable within a relationship context.
- Between 1 and 10 percent of American men are *gay,* and between 1 and 3 percent of American women are *lesbian* at one time or another in their lives. Identifying oneself as gay or lesbian occurs over considerable time, usually beginning in late childhood or early adolescence. The first stage is marked by fear and suspicion that one's sexual desires are different from those of others. In the second stage the person labels these feelings as gay or lesbian. The third stage includes the person's self-definition as gay or lesbian. Two additional stages include entering the gay subculture and having a gay or lesbian love affair.
- Lesbian, gay, and bisexual individuals may be subject to *antigay prejudice* or *homophobia,* leading to verbal abuse, discrimination, or violence. Education and positive social interactions can reduce such prejudice.
- *Bisexuals* are attracted to members of both genders. In developing a bisexual identity men and women go through several stages: (1) initial confusion, (2) finding and applying the bisexual label, (3) settling into the identity, and (4) continued uncertainty. Bisexuals don't have a community or social environment that reaffirms their identity.
- Developmental tasks in middle adulthood include (1) redefining sex in marital or other long-term relationships, (2) reevaluating one's sexuality, and (3) accepting the biological aging process. In middle age, women tend to reach their sexual peak, which is often maintained into their sixties and beyond; they also experience menopause. The sexual responsiveness of men declines somewhat, causing men to require greater stimulation and time to become aroused. There is no male equivalent to menopause.
- Many of the psychosexual tasks older Americans must undertake are directly related to the aging process. They include changing sexuality and loss of a partner. The sexuality of older Americans tends to be invisible because (1) we associate sexuality with youth, (2) we associate romance and love with youth, (3) we associate sex with procreation, and (4) the elderly themselves do not have desires as strong as those of the young. The main determinants of sexual activity in old age are health and the availability of a partner.
- *Autoeroticism* consists of sexual activities that involve only the self. It includes sexual fantasies, masturbation, and erotic dreams. Erotic fantasizing is the most universal of all sexual behaviors. Sexual fantasies serve several functions: (1) they help direct and define our erotic goals, (2) they allow us to plan or anticipate erotic situations, (3) they provide escape from a dull or oppressive environment, and (4) they bring novelty and excitement into a relationship. Erotic dreams are widely experienced.
- *Masturbation* is an important means of learning about our bodies. By the end of adolescence, most men and the majority of women have masturbated to orgasm. Most people continue to masturbate during marriage, although married men tend to masturbate to supplement their sexual activities whereas women tend to masturbate as a substitute for such activities.
- Oral-genital sex, which includes *cunnilingus* and *fellatio,* is practiced by heterosexuals, gay men, and lesbians.
- *Sexual intercourse (coitus)* is the insertion of the penis into the vagina and the stimulation that follows. It is a complex interaction, involving more than erotic pleasure or reproduction. It is a form of communication that may have many motivations and express a host of feelings.
- *Anal eroticism* is practiced by both heterosexuals and gay men. From a health perspective, *anal intercourse* is dangerous because it is the most common means of sexually transmitting HIV.
- Nonconsensual sex consists of behaviors ranging from exhibitionism and voyeurism to rape, sexual harassment, and child sexual abuse.

- Sexual enhancement means improving the quality of a sexual relationship. It is based on accurate information about sexuality, developing communication skills, fostering positive attitudes, and increasing self-awareness. Awareness of your own sexual needs is often critical to enhancing your sexuality. Enhancing sex includes the intensification of arousal, which requires focusing on erotic pleasure.
- *Nonmarital sex* is sexual activities, especially sexual intercourse, that take place outside of marriage. *Premarital sex* is sexual activities between younger, never-married adults under the age of 30. Premarital sexual intercourse is widely accepted. In sexual decision making, love or liking is often the most important factor leading men and women to have premarital intercourse. In contrast to married couples, cohabitants have a higher frequency of sex, greater equality in initiating sexual activities, and more extrarelational sex.
- In gay and lesbian relationships, gay men are likely to initiate sexual activity earlier in the relationship than are lesbians. Lesbians do not initiate sex as often as do gay or heterosexual men. In the gay and lesbian culture, sexual exclusivity is negotiable.
- Marital sex tends to decline in frequency over time, but this does not necessarily signify marital deterioration. Sex is only one bond among many in marriage. Sex within marriage is different from premarital sex in the following ways: (1) Sex in marriage is expected to be monogamous, (2) procreation is a legitimate goal, and (3) the sex takes place in the everyday world.
- *Extramarital sex* assumes three basic forms: (1) sexual but not emotional, (2) sexual and emotional, and (3) emotional but not sexual. Women are increasingly engaging in extramarital affairs; male involvement has remained relatively high. Extramarital affairs appear to be related to two variables: unhappiness in the marriage and premarital sexual permissiveness.
- *Sexual dysfunctions* are recurring persistent problems in giving and receiving erotic satisfaction. Sexual dysfunctions may be physiological or psychological in origin. Both men and women are subject to hypoactive (low or inhibited) sexual desire. The most common female problems are orgasmic dysfunction, arousal difficulties, and dyspareunia (painful intercourse). The most common male problems are erectile dysfunction, premature ejaculation, and delayed orgasm. Two of the most prominent causes of sexual dysfunction are performance anxiety and conflicts within the self. Relationship discord affects sexuality through transference, lack of trust, power struggles, contractual disappointments, sexual sabotage, and lack of communication.
- *Sexually transmitted diseases (STDs),* especially chlamydia and gonorrhea, are epidemic. *Acquired immunodeficiency syndrome (AIDS)* is caused by the *human immunodeficiency virus (HIV),* which attacks the body's immune system. HIV is carried in the blood, semen, and vaginal fluid of infected persons. Heterosexuals, bisexuals, and gay men and lesbians are susceptible to the sexual transmission of HIV. If one is sexually active, the keys to protection against STDs, including HIV/AIDS, are communication and condom use.

KEY TERMS

abstinence 267
acquired immunodeficiency syndrome (AIDS) 268
anal eroticism 253
anal intercourse 253
anti-gay prejudice 243
autoeroticism 249
bisexual 240
coitus 252
coming out 243
cunnilingus 252
exhibitionism 263
extramarital sex 256
fellatio 252
gay male 240
heterosexual 240
homoeroticism 242
homophobia 243
homosexual 240
human immunodeficiency virus (HIV) 268
lesbian 240
masturbation 250
nonconsensual sexual behavior 263
nonmarital sex 255
open marriage 263
outing 243
pleasuring 251
premarital sex 255
sexual coercion 263
sexual dysfunction 264
sexual enhancement 253
sexual intercourse 252
sexual orientation 241
sexual script 237
sexually transmitted disease (STD) 268
Viagra 248
voyeurism 263

SUGGESTED READINGS

Laumann, Edward, John Gagnon, Robert Michael, and Stuart Michaels, *The Social Organization of Sexuality: Sexual Practices in the United States.* Chicago: University of Chicago, Press 1994. Large, comprehensive survey of sexual attitudes and behaviors reported by more than 3,400 American women and men. The more scholarly companion to *Sex in America* (see below).

Marsiglio, William. *Procreative Man,* New York: New York University Press, 1998. An insightful sociological overview and analysis of men's experiences in the realm of procreation. Marsiglio looks at men's experiences of contraception, abortion, pregnancy, fatherhood, and assisted reproductive techniques.

Michael, Robert, John Gagnon, Edward Laumann, and Gina Kolata. *Sex in America: The Definitive Survey.* Boston: Little, Brown, 1994. A random, cross-sectional survey, conducted under the supervision of noted sociologists, in which over 3,200 men and women were personally interviewed. Emphasizes the social context of sexuality.

Reiss, Ira. *The End of Shame: Shaping Our Next Sexual Revolution.* Buffalo, NY: Prometheus, 1990. An impassioned plea for transforming sexuality into a positive force in contemporary society—made by a leading sociological researcher in human sexuality.

Rubin, Lillian. *Erotic Wars.* New York: Farrar, Strauss, and Giroux, 1990. A fine portrayal by a thoughtful therapist of the yearnings of and misunderstandings between men and women.

Shilts, Randy. *And the Band Played On: People, Politics, and the AIDS Epidemic.* New York: St. Martin's Press, 1987. A compelling story of the medical, political, and human responses to the AIDS crisis.

Steinberg, David. *The Erotic Impulse: Honoring the Sensual Self.* New York: Jeremy Tarcher, 1992. An outstanding collection of essays and poems by writers, poets, teachers, and psychologists.

Strong, Bryan, and Christine DeVault. *Human Sexuality,* 2nd ed. Mountain View, CA: Mayfield, 1997. A comprehensive introduction to human sexuality.

Weinberg, Martin, Colin Williams, and Douglas Pryor. *Dual Attraction: Understanding Bisexuality.* New York: Oxford University Press, 1994. An important empirical work.

RESOURCES ON THE INTERNET

Companion Web Site for This Book

http://sociology.wadsworth.com/strong/marriage9e
Gain an even better grasp on this chapter by going to the companion Web site to take one of the Tutorial Quizzes, use the Flash Cards to master key terms, or check out the many other study aids you'll find there. Visit the Marriage and Family Resource Center on the site. You'll also find special features such as GSS Data and Census 2000 information that will put data and resources at your fingertips to help you with that special project or to do some research on your own.

InfoTrac College Edition: Search Word Summary

intimacy	sexual orientation
coming out	contraception
monogamy	adultery

To learn more about these central topics in the study of the family, you can conduct an electronic search using InfoTrac College Edition. To aid in your search and to gain useful tips, see the Student Guide to InfoTrac College Edition that you can access through the companion Web site for this book.

Preview

To gain a sense of what you already know about the material covered in this chapter, answer "True" or "False" to the statements below.

1. More women than men tend to live with their parents. True or false?
2. Couples who are unhappy before marriage significantly increase their happiness after marriage. True or false?
3. Marriage, more than parenthood, radically affects a woman's life. True or false?
4. The advent of children generally increases a couple's marital satisfaction. True or false?
5. Age at marriage is a strong indicator of later marital success. True or false?
6. In-law relationships tend to be characterized by low emotional intensity. True or false?
7. Asian, Latino, and African American families are more likely than white families to take in extended family. True or false?
8. The empty nest syndrome, characterized by maternal depression after the last child leaves home, is more a myth than a problem for American women. True or false?
9. The vast majority of long-term marriages involve couples who are blissful and happily in love. True or false?
10. The key to marital satisfaction in the later years is continued good health. True or false?

"Chains do not hold a marriage together. It is threads, hundreds of tiny threads, which sew people together through the years."

SIMONE SIGNORET

CHAPTER 9

Family Processes, Family Life Cycles

Outline

Have you ever looked really closely at a family photo album, say one that belonged to a parent or grandparent? If you have, you know that these albums are fascinating representations of the dynamics inherent in all families. If you get the chance, study one of your family's old albums closely. Typically, you'll find photos of now deceased relatives, which means you can "meet" ancestors that you never got to know in person. Maybe even more interesting might be the photos of your parents as children and adolescents or young adults, looking the way they did before they met and when they first got together. Then, of course, there are wedding photos that capture the excitement and hope that brides and grooms wear on their faces as they embark on a shared married life. Eventually, there are baby pictures with the same spouses now new parents. As you turn the pages and study the photos, you can see the changes as children grow and parents age. You can see the kids growing taller, their bodies taking on more mature shapes. You also start to see more lines on parents' faces, some grey in their hair, perhaps some thickening around their waists. In short, you see a family that has formed, grown up, and changed.

The basic truth conveyed by such visual images is the subject of this chapter. Our families are dynamic. They are always changing to meet new situations, new emotions, new commitments, and new responsibilities. The marriage process may begin informally with cohabitation or formally with engagement. Marriage itself ends with divorce or with the death of a partner. When we enter marriage, we may think we know how to act within it, but we find that the reality of marriage requires us to be more flexible than we had anticipated. We need flexibility to meet our needs, our partners' needs, and the needs of the marriage. We may have periods of great happiness and great sorrow within marriage. We may find boredom, intensity, frustration, and fulfillment. Some of these may occur because of our marriage; others may occur in spite of it. But as we shall see, marriage encompasses constantly evolving changes and possibilities.

In this chapter we use a developmental perspective to examine changes in marriages and families that occur over time and across stages. We look at beginning marriages, including factors predicting marital success, cohabitation, engagement, and weddings, as well as the establishment of marital roles and boundaries. Then we look at youthful marriages, especially the impact of children and the individual changes we may experience. We turn next to middle-aged marriages, examining families with young children and adolescents, families as launching centers of the young, and the process of reevaluation. Then we review later-life marriages, especially the extended family, grandparenting, retirement, caregiving, and widowhood. Finally, we survey the different patterns of lasting marriages.

PREVIEW ANSWERS

1 False
2 False
3 False
4 False
5 True
6 True
7 True
8 True
9 False
10 True

A Developmental Approach

The developmental perspective, or "family development theory," as it was referred to in Chapter 2, is an interdisciplinary approach that unites sociology and related disciplines with child and human development (Duvall, 1988). It sees individual development and family development as interacting with each other.

INDIVIDUAL DEVELOPMENT

Our identity, our sense of who we are, is not fixed or frozen. It changes as we mature. At different points in our lives, we are confronted with different developmental tasks, such as acquiring trust and becoming intimate. Our growth as human beings depends on the way we perform these tasks. Erik Erikson (1963) describes the human life cycle as containing eight developmental stages, as listed below. At each stage, we have an important developmental task to accomplish, and each stage intimately involves the family. As we enter young adulthood, these stages may also involve marriage or other intimate relationships (Nichols and Pace-Nichols, 1993).

The way we deal with these stages, is strongly influenced by our families, marriages, or other intimate relationships. We cannot separate our identity from our relationships. The eight stages can be summarized as follows:

- *Infancy: Trust Versus Mistrust.* In the first year of life, children are wholly dependent on their parenting figures for survival. It is in this stage that they learn to trust by having their needs satisfied and by being loved, held, and caressed. Without loving care, an infant may develop a mistrusting attitude toward others and toward life in general.
- *Toddler: Autonomy Versus Shame and Doubt.* Between ages 1 and 3, children learn to walk and talk; they also begin toilet training. At this stage they need to develop a sense of independence and mastery over their environment and themselves.
- *Early Childhood: Initiative Versus Guilt.* Ages 4 to 5 are years of increasing independence. The family must allow the child to develop initiative while at the same time directing the child's energy. The child must not be made to feel guilty about his or her desire to explore the world.
- *School Age: Industry Versus Inferiority.* Between ages 6 and 11, children begin to learn that their activities pay off and that they can be creative. The family needs to encourage the child's sense of accomplishment. Failing to do so may lead to feelings of inferiority in the child.
- *Adolescence: Identity Versus Role Confusion.* The years of puberty, between ages 12 and 18, may be a time of turmoil as well as discovery and growth. Adolescents try new roles as they make the transition to adulthood. To make a successful transition, they need to develop goals, a philosophy of life, and a sense of self. The family needs to be supportive as the adolescent tentatively explores adulthood. If the adolescent fails to establish a firm identity, he or she is likely to drift without purpose.
- *Young Adulthood: Intimacy Versus Isolation.* In young adulthood, the adolescent leaves home and begins to establish intimate ties with other people through cohabitation, marriage, or other important intimate relationships. A young adult who does not make other intimate connections may be condemned to isolation and loneliness.

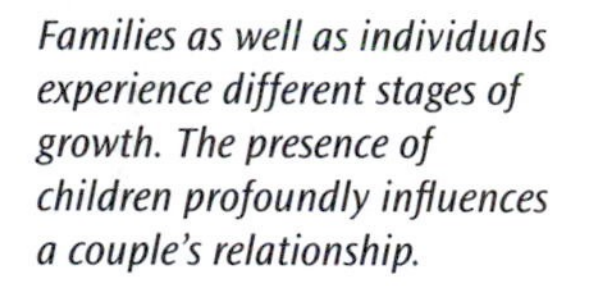

Families as well as individuals experience different stages of growth. The presence of children profoundly influences a couple's relationship.

- *Adulthood: Generativity Versus Self-Absorption.* Generativity is the bearing of offspring, productiveness, or creativity. In adulthood, the individual establishes his or her own family and finds satisfaction in family relationships. It is a time of creativity. Work becomes important as a creative act, perhaps as important as family or an alternative to family. The failure to be generative may lead to self-centeredness and a "what's-in-it-for-me" attitude toward life.
- *Maturity: Integrity Versus Despair.* In old age, the individual looks back on life to understand its meaning—to assess what has been accomplished and to gauge the meaning of his or her relationships. Those who can make a positive judgment have a feeling of wholeness about their lives. The alternative is despair.

Throughout our life cycles, our goals and concerns change. Among young adults, goals include education and family-related goals, such as marriage and having children. Among middle-aged adults, goals shift to concern about children's lives and about property, such as buying or maintaining homes. Among the elderly, health, retirement, leisure activities, and interest in the world predominate (Nurmi, 1992).

FAMILY DEVELOPMENT

Just as individuals change and develop over time, so do marriages and families. Within the stages of the **family life cycle,** each marriage and family has its own unique history (Aldous, 1978). The concept of the family life cycle uses a developmental framework to explain people's behavior in families. According to this framework, families change over time both in terms of the people who are members of the family and the roles they play. At various stages in its life cycle, the family has different developmental tasks to perform. Much of the behavior of family members can be explained in terms of the family's developmental stage. The key factor in such developmental stages is the presence of children. The family organizes itself around its child-rearing responsibilities. A woman's role as wife is different when she is childless than when she has children. A man's role is different when he is the father of a 1-year-old than when he is the father of a 15-year-old.

The developmental approach gives us important insights into the complexities of family life. Not only is the family performing various tasks at different points, but also each family member takes on various developmental tasks during each different stage. The task for families with adolescents is to give their children greater autonomy and independence. While the family is coping with this new developmental task, an adolescent daughter has her own individual task of trying to develop a satisfactory identity. Meanwhile, her older brother is struggling with intimacy issues, her younger sister is developing industry, her parents are dealing with issues of generativity, and her grandparents are confronting issues of integrity. The life cycle or life course emphasis highlights the common experiences families have in the course of their shared lives.

An Eight-Stage Family Life Cycle

One of the most widely used approaches divides the family life cycle into eight stages (Duvall and Miller, 1985), reflecting the model life cycle. It does not encompass the family life cycle of single-parent or remarried families, as we discuss shortly.

- *Stage I: Beginning Families.* During the first stage, the married couple has no children. In the past, this stage was relatively short because children soon followed marriage. Today, however, this stage may last until a couple is in their late twenties or thirties. The average age for women to have their first child was 25.1 in 2002, an all-time high in the United States (http://www.cdc.gov/nchs/data/nvsr/nvsr52/nvsr52_10.pdf). Because this includes births to both married and unmarried women, the average age at which married women have their first child is likely somewhat older than 25. Most studies on marital satisfaction agree that couples experience their greatest satisfaction during this stage (Glenn, 1991).
- *Stage II: Childbearing Families.* Because families tend to space their children about 30 months apart, the family in Stage II is still considered to be forming, as new births are likely during this period. By the second stage, the average family has two children. Although mothers are deeply involved in childbearing and child rearing, 63 percent of married women with children under 6 are in the labor force. Nearly 60 percent of married women with children 3 or younger are also employed (U.S. Census Bureau, 2001, Tables 577 and 578). Stage II lasts around two and a half years. Marital satisfaction begins to

lessen and continues to decline through the stage of families with schoolchildren (Stage IV) or those with adolescents (Stage V).

- *Stage III: Families with Preschool Children.* The family's oldest child is 30 months to 6 years in Stage III. The parents, especially the mother, are still deeply involved in child rearing. This stage lasts about three and a half years.
- *Stage IV: Families with Schoolchildren.* The family's oldest child is between 6 and 13 years old. With the children in school and more free time, the mother has more options available to her. If they hadn't already done so by now, three out of four women have reentered the job market. This stage lasts about seven years.
- *Stage V: Families with Adolescents.* The oldest child is between 13 and 20 years old, and marital satisfaction reaches its nadir. This stage lasts about seven years.
- *Stage VI: Families as Launching Centers.* The first child has been launched into the adult world. This stage lasts until the last child leaves home, a period averaging about eight years. Virtually all studies show that marital satisfaction begins to rise for most couples during this stage.
- *Stage VII: Families in the Middle Years.* Lasting from the time the last child has left home to retirement, this stage is commonly referred to as the "empty nest." It is a distinct and relatively new phase in the family life cycle. Until this century, most parents continued to have children until middle age, and the child-rearing and launching periods were extended into old age. More recently, however, because of deferred marriage and the high cost of housing, many adult children continue to live at home or return home after being away for a number of years. This "not-so-empty nest" phase, argue some scholars, constitutes a more recent variation in Stage VII (Glick, 1989a; Mattessich and Hill, 1987). At this time, many families begin caretaking activities for elderly relatives, especially parents and parents-in-law.
- *Stage VIII: Aging Families.* The working members of the aging family have retired. Usually, the husband retires before the wife because he tends to be older. Ill health begins to take its toll. Eventually, one of the spouses dies, usually the husband. The surviving spouse may live alone or with other family members or be cared for by them.

Transitions from one stage to another in the family life cycle seem to affect relationships. Change is marked by stress as we adjust to new situations and roles. A more positive outcome is likely when couples have other meaningful roles in their lives, such as work, hobbies, or school.

REFLECTIONS

As you examine the different family life cycle stages, determine the stage in which your family of orientation or cohabitation is. What are the tasks your family confronts? In what individual developmental tasks are the different members of your family engaged?

Limitations and Variations

As stated in Chapter 2, a major limitation of the family life cycle approach is its tendency to focus on the intact nuclear family as *the* family. Paul Glick, who coined the term *family life cycle* in 1947, has noted that a number of social changes have affected the traditional family life cycle (Glick, 1989a). Deferred marriages, cohabitation, divorce, remarriage, single parenthood, and gay and lesbian families introduce notable variations in the life course of the family.

As marriages are increasingly deferred, cohabitation rates have skyrocketed. It may be useful, Glick suggests, to include a cohabitational stage that recognizes the significance of living together in family development. In addition, the family life cycle of single adults is considerably different from that of their married counterparts. For childless adults who have never married, the families of orientation may become the central family focus. Single women, for example, may be intimately tied to their parents as caregivers; these women also may be connected to children as aunts (Allen and Pickett, 1987; Allen, 1989).

Families that include children with disabilities must negotiate the family life cycle differently, depending on the child's unique abilities and limitations For example, independence may come later, or the child may never be able to live outside the family (Hanline, 1991). Adoptive families have their own unique life cycles (Hajal and Rosenberg, 1991). They may simultaneously become beginning families and families with schoolchildren with the adoption of a 5-year-old child. Lesbian and gay families also have special family life cycle issues as a result of having two mothers or two fathers, sperm donor fathers, or parents who are the heterosexual partners from earlier marriages or relationships (Slater and Mencher, 1991). Such families often encompass members of their communities as part of their extended families.

Scholars are also recognizing the importance of ethnicity to family life cycles. The timing of the various stages, for example, is affected by ethnicity. African Americans are less likely to marry than are whites, and they are likely to marry at a later age than whites. Single parenthood is more prevalent among African-American families than among white ones. In 1995, for example, 64 percent of African-American families, 36 percent of Hispanic families, and 25 percent of white families were headed by single parents (U.S. Census Bureau, 1996). African Americans are also more likely than whites or Latinos to begin their family life cycles with unmarried single-parent families (U.S. Census Bureau, 1994).

Immigrant families find themselves experiencing unique issues in the family life cycle as old values clash with new ones, such as whether unmarried adult children (especially women) should leave home and whether aged parents should live with their children. Furthermore, if the family is split—with some members living in the country of origin and some in the United States—family members may differ in their expectations of what should occur during various stages of the family life cycle (Hong and Ham, 1992). Religious commitments of parents can also clash with the predominant society, especially over such issues as sex education and discipline.

Of all these variations, researchers are increasingly focusing on two widely prevalent alternatives to the traditional intact family life cycle: the single-parent and the stepfamily life cycles. Today, nearly one out of three families is a single-parent family and one out of six is a stepfamily.

About half of single-parent families are headed by divorced women, who usually divorce between Stages II and IV, during the childbearing through school-age stages. Approximately two-thirds of divorced women will remarry. Divorced mothers are less likely to remarry than divorced women without children. If they remarry, the single-parent stage generally lasts between three and six years. If these divorced women do not remarry, they continue their single parenting until their adolescent children are launched (Mattessich and Hill, 1987).

The second type of single-parent family originates with unmarried mothers. Birth to unmarried parents now rivals divorce as a pathway in which children enter family structures (Acquilino, 1996). Overall, a third of all births in 2002 were to unmarried mothers. Sixty-nine percent of African-American children, 42 percent of Hispanic children, and 27 percent of white children were born to an unmarried mother (U.S. Census Bureau, 2001, Table 76). About 30 percent of unmarried mothers are adolescents, and another 36 percent are between ages 20 and 24 years. They usually have become pregnant unintentionally. Their family life cycle begins with a single-parent family as its first stage. In this stage, the young mother (especially if she is an adolescent) and her child often live with the child's grandmother or cohabit with the child's father. The second stage may be marriage. After marriage, the family life cycle may follow more traditional patterns or diverge back into the single-parent pattern. Most studies indicate

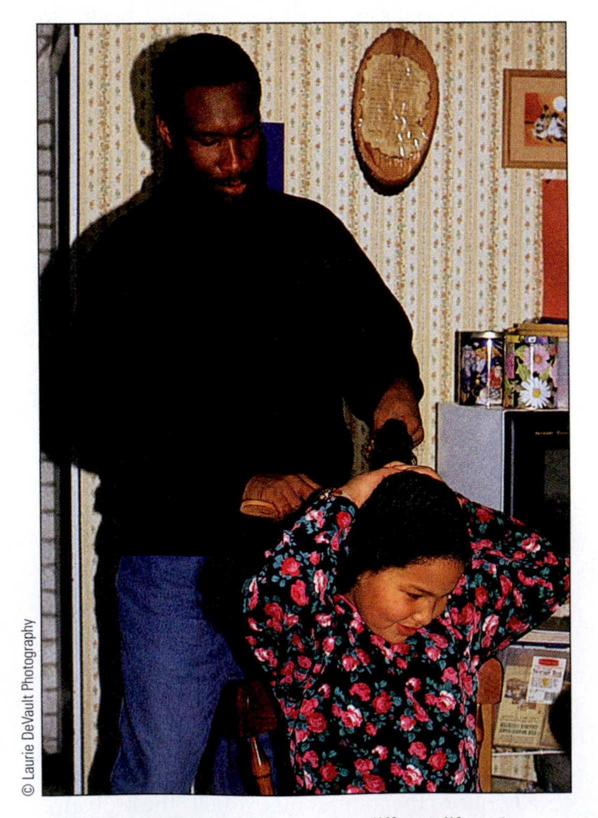

The single-parent family experiences a different life-cycle pattern than the two-parent family.

multiple disadvantages to both mothers and children, including poverty, lack of education, and less productive lives (Polakow, 1993).

Stepfamilies are formed when the husband or the wife (or both) has children from a previous marriage or relationship. After single parenthood, the stepfamily generally forms during the school-age and adolescent family stages. The parents and stepparents, who are usually aged 30 years or older, have a double developmental task: they must simultaneously enter both the beginning family and childrearing stages.

DID YOU KNOW?

Overall, a third of all births in 2002 were to unmarried mothers.

Beginning Marriages

Americans are waiting longer to marry today than in previous generations. Whatever the reason, increasing age at time of marriage probably results in young adults beginning marriage with more maturity, independence, work experience, and education. These are important assets to bring into marriage.

PREDICTING MARITAL SUCCESS

The period before marriage is especially important because couples learn about each other—and themselves. Courtship sets the stage for marriage. Many of the elements important for successful marriages, such as the ability to communicate in a positive manner and to compromise and resolve conflicts, develop during courtship. They are often apparent long before a decision to marry has been made (Cate and Lloyd, 1992).

The time and patterns that precede marriage can often predict how happy a couple will be during marriage. Happy couples are more likely than unhappy couples to be satisfied in their marriages, and couples who are unhappy before marriage are more likely to be unhappy after marriage as well (Olson and DeFrain, 1994). Each individual brings into marriage the same strengths and weaknesses that he or she brought into the earlier relationship. Marriage permits relationships to grow—or to deteriorate.

Whether marriage is an arena for growth or disenchantment depends on the individuals and the nature of their relationship. It is a dangerous myth that marriage will change a person for the better: An insensitive single person simply becomes an insensitive husband or wife. In fact, undesirable traits tend to become magnified in marriage because one has to live with them in close, unrelenting, and everyday proximity.

Family researchers have found numerous premarital factors to be important in predicting later marital happiness and satisfaction. Although they may not necessarily apply in all cases—and when we are in love, we believe we are the exceptions—they are worth thinking about. According to Rodney Cate and Sally Lloyd (1992), these premarital factors include background, personality, and relationship factors.

Background Factors

Age at marriage is important. Adolescent marriages (i.e., marriages when either party is younger than 20) are especially likely to end in divorce. Marriage age seems to have less effect as one waits longer to enter marriage. In other words, differences between those who marry in their thirties versus those who marry in their mid to late twenties are slight. Differences between those who wait until they are at least twenty and those who marry as teens are, however, considerable. Length of courtship is also related to marital happiness. The longer you date and are engaged to someone, the more likely you are to discover whether you are compatible with each other. But you can also date "too long." Those who have long, slow-to-commit, up-and-down relationships are likely to be less satisfied in marriage. They are also more likely to divorce. Such couples may torture themselves (and their friends) with the familiar dilemma of whether to split up or get married—and then get married, to their later regret.

Level of education seems to affect both marital adjustment and divorce. Education may give us additional resources, such as income, insight, or status, that contribute to our ability to carry out our marital roles.

Exploring Diversity

THE LIFE CYCLE IN CROSS-CULTURAL PERSPECTIVE: THE CASE OF THE MASAI

In American society, the concept of the life cycle is tied to our understanding of individual physiological and psychological development, as well as related cultural notions of life stages: childhood, adolescence, marriage and adulthood, and old age. Underlying all these measures of the life cycle is a focus on the individual. In U.S. society, each person progresses through the stages of life—through the life cycle—individually. Our cultural emphasis on the individual is, from a cross-cultural perspective, extreme. To us, of course, the focus on the individual seems natural, precisely because our society is shaped and permeated by notions of the individual: individual rights, responsibilities, opinions, votes, and identities.

Anthropologists and historians have found that persons in other cultures are not seen as "individuals" as we define the term. Because we take the individual as a fundamental and universal identity, our cultural bias leads us to project individual identities onto persons in other cultures. But in most traditional societies, a person's primary identity is a clan identity. Which clan you belong to—whether birth established membership in your mother's clan or your father's—is determined by custom. Thus, you as a person would be known and identified by others first as a son or daughter of a particular clan. Your relationships with other people, both your behavior toward them and their behavior toward you, would be determined by your primary identity, your clan identity.

In addition to kinship, age sets in some traditional societies were also important in shaping a person's identity, as well as his or her position in the life cycle. (An age set consists of members of society who are approximately the same age.) Most societies that have age set systems are found in Africa, but they are also found among the Cheyenne of the Great Plains in North America and the Chavante of Brazil. Let's look at age sets by focusing our cross-cultural lens on the Masai peoples of Kenya and Tanzania (Paul, 1988; Saitoti, 1986).

Traditionally, the Masai based their livelihood almost totally on pastoralism; specifically, the herding of short-horned zebu cattle. The Masai themselves did no farming but instead traded with their neighbors for needed agricultural products. As with many other pastoral peoples, the Masai do not depend on their cattle primarily for meat but in fact derive most of their caloric intake from the blood of their cattle and dairy products. The Masai refer to themselves as "the people of the cattle." All of life revolves in some way around the management and accumulation of cattle. Boys grow up helping their fathers herd cattle; girls grow up assisting their mothers with milking and caring for calves. Anthropologists have found that such pastoral societies as the Masai are in general more androcentric, or male focused, than other kinds of societies. They are usually patrilineal: You inherit your father's clan identity, not your mother's.

Among the Masai, marriages are made of cattle. A man needs cattle to marry, for the bride's father demands many cattle for his daughter. Thus, a young man, without a substantial herd of his own, must borrow heavily from his own father. Herd size both confers status on a man and enables him to marry several wives. The ability to marry wives and accumulate cattle is tied to the age-set system and a man's place in it.

In essence, Masai society is organized into fraternities. When a boy is physically mature and around 15 years of age, he will undergo a public circumcision ceremony along with other local boys of about the same age. (Circumcision ceremonies are held only every few years; schedules are staggered in neighboring Masai areas.)

Childhood environment, such as attachment to family members, parents' marital happiness, and low parent-child conflict, is associated with one's own marital happiness. This is especially true for women: Some studies indicate that it is the woman's relationship with her family of orientation that is crucial to later marital happiness. In fact, it may spell trouble if the man is too close to his family of orientation. Most studies on childhood environment, however, are based on men and women who came of age prior to the 1960s. The social context of marriage has changed dramatically since then, with the rise of divorce, smaller families, and changing gender roles. It is not clear, for example, exactly how today's young adults may be affected by parental divorce. Parental divorce may cause one either to shy away from marriage or to marry with the determination not to repeat the parents' mistakes. Once

Those boys who are initiated in the same "season" will henceforth and for the duration of their lives be members of the same fraternity or age set. They share a basic lifelong identity and loyalty to one another as brothers. Together, they will progress through a fixed series of graduations from one life stage, or grade, to the next, each fifteen years in duration. The first and most important age grade is the warrior grade, into which boys are initiated at their circumcision ceremony.

Traditionally, it was men of the warrior grade who were the first line of defense against hostile incursions and cattle raids by neighboring peoples. Warriors lived separately from the camps of their fathers, together in their own camp in the bush. The warrior identity was marked by ritual—warriors drank milk communally in the forest and were not permitted to drink it alone—as well as by special dress and ornamentation, weapons, dances, songs, and calls. Only warriors could wear the lion headdress, carry a warrior's spear, and give the lion's call at night. Young boys looked forward to the day they would become warriors, and old men fondly recalled their own warrior days. The warrior was the cultural ideal of manhood. Warriors were considered beautiful, virile, and fierce—and were much admired by women.

Generally, men did not marry until around 30 years of age, when they entered the next stage of life, junior elderhood. As warriors, young men owned only a few cattle apiece. Although they might try to add additional head to their herds—even raiding their own fathers' herds in a show of defiance—warriors were occupied with their defense duties and ritual obligations. With the initiation of a new class of warriors, however, and the reluctant graduation of the "old" class into junior elderhood, there came new responsibilities. Junior elders worked hard to build up their cattle herds, married, and began to father children. Elderhood followed junior elderhood at around age 45; senior elderhood commenced at around age 60. Each graduation was marked with ritual and, of course, the initiation of a new group of boys as warriors. For all elders, herd management and marriage were important business. As an elder or senior elder, a man might—if sufficiently wealthy in cattle—marry a third or fourth wife. An extremely wealthy man might have as many as seven wives. However, cattle that could be used to acquire perhaps a fourth or fifth wife for a senior elder could also, of course, be used to acquire a first wife for his son.

Women did not have their own separate age system. Their progress through life was in a sense more individual but was tied to the male age grades in certain important respects. A young girl was initially free to admire and be courted by warrior boyfriends. With her first period, however, a young girl's life changed radically. She immediately underwent a clitoridectomy, making her marriageable, which was celebrated by her female relatives with gifts and feasting. Soon afterward, she was married in an arranged marriage to a much older man. You will recall that most men did not marry until after graduation into junior elderhood, at age 30. Thus, one general effect of the age system was to concentrate wives at the upper or older end of the male age hierarchy, by denying men of the youngest age grade the right to marry. In addition, men were forbidden to take the daughters of men of their own age set as wives. This was considered incestuous because these women were the daughters of "brothers." However, men of an age set could, as brothers, share wives with one another if given permission.

In societies with age-set systems like the Masai, the life cycle was divided into fixed stages through which all males who were initiated together progressed as a group. A young man did not move through the life cycle by meeting individual psychological or physiological criteria, as in our society. Instead, the age-set system created fixed group identities and life stages, through which men of the age set moved together, as had their fathers in their own age sets before them.

married, one's likelihood of success is affected by parental divorce. Those who grew up in households where parents divorced are more likely to experience a divorce themselves.

Personality Factors

How does having a flexible personality affect marital success? How about a contentious personality? A giving one? An obnoxious one? As you can imagine, your partner's personality will affect your life considerably. But there is little research on personality characteristics and marital success. We do know, however, that opposites do not usually attract; instead, they repel. We choose partners who share similar personality characteristics because similarity allows greater communication, empathy, and understanding (Antill, 1983; Buss, 1984; Kurdek and Smith, 1987; Lesnick-Oberstein and Cohen, 1984). It may be that personality characteristics are most significant during courtship. It is then that those with undesirable or incompatible personalities are weeded out—at least in theory.

Because researchers tend to focus on relationship process and change, they are reluctant to examine personality. Personality seems fixed and unchanging. Nevertheless, it does affect marital processes. A rigid personality may prevent negotiation and conflict resolution. A dominating personality may disrupt the give-and-take necessary to making a relationship work.

Relationship Factors

Most existing research has focused on aspects of premarital relationships that might predict marital success. Loving each other did not seem to have much impact on whether couples fought. Couples who had other partners simultaneously prior to marriage or who compared their partners with others had lower levels of satisfaction. Another study on communication and marital satisfaction examined the same couples after one, two and a half, and five and a half years of marriage (Markman, 1981, 1984). During the first year, there was no relationship between communication and marital satisfaction, but after two and a half and five and a half years, the more negative the communication, the less satisfactory the marriage. Researchers suggest that negative interactions did not significantly affect the first year of marriage because of the **honeymoon effect,** the tendency of newlyweds to overlook problems (Huston, McHalle, and Crouter, 1986). Failure to fulfill one's partner's expectations about marital roles, such as intimacy and trust, predicted marital dissatisfaction (Kelley and Burgoon, 1991).

Olson and DeFrain (1994) assert that they can more or less predict an engaged couple's eventual marital satisfaction based on their current relationship. The factors they find significant in reviewing the research literature include the ability to do the following:

- Communicate well with each other
- Resolve conflicts in a constructive way
- Develop realistic expectations about marriage
- Like each other as people
- Agree on religious and ethical issues
- Balance individual and couple leisure activities with each other

In addition, how each person's parents related to each other and to their daughter or son is also an important predictor. It is in our families of orientation that we learn our earliest (and sometimes most powerful) lessons about intimacy and relationships (Larsen and Olson, 1989).

ENGAGEMENT, COHABITATION, AND WEDDINGS

The first stage of the family life cycle may begin with engagement or cohabitation followed by a wedding, the ceremony that represents the beginning of a marriage.

Engagement

Engagement is the culmination of the formal dating process. Today, in contrast to the past, engagement has greater significance as a ritual than as a binding commitment to be married. Engagement is losing even its ritualistic meaning, however, as more couples start out in the less formal patterns of "getting together" or living together. These couples are less likely to become formally engaged. Instead, they announce that they "plan to get married." Because it lacks the formality of engagement, "planning to get married" is also less socially binding.

Engagements currently average between 12 and 16 months (Carmody, 1992). They perform several functions:

- Engagement signifies a commitment to marriage and helps define the goal of the relationship as marriage.
- Engagement prepares couples for marriage by requiring them to think about the realities of everyday married life: money, friendships, religion, in-laws, and so forth. They are expected to begin making serious plans about how they will live together as a married couple.
- Engagement is the beginning of kinship. The future marriage partner begins to be treated as a member of the family. He or she begins to become integrated into the family system.
- Engagement allows the prospective partners to strengthen themselves as a couple. The engaged

pair begin to experience themselves as a social unit. They leave the youth or singles culture and prepare for the world of the married, a remarkably different world.

Men and women typically need to deal with several key psychological issues during engagement (Wright, 1990). The issues include (1) anxiety, a general uneasiness that comes to the surface when you decide to marry; (2) maturation and dependency needs, questions about whether you are mature enough to marry and to be interdependent; (3) losses, or regret over what you give up by marrying, such as the freedom to date and responsibility for only yourself; (4) partner choice, worry about whether you're marrying the right person; (5) gender-role conflict, disagreement over appropriate male/female roles; (6) idealization and disillusionment, the tendency to believe that your partner is "perfect" and to become disenchanted when she or he is discovered to be "merely" human; (7) marital expectations, or beliefs that the marriage will be blissful and conflict free and that your partner will be entirely understanding of your needs; and (8) self-knowledge, an understanding of yourself, including your weaknesses as well as your strengths.

REFLECTIONS

As you look at the factors predicting marital success, look at your own past relationships. Retrospectively, what factors, such as background, personality characteristics, and relationship characteristics, might have predicted the quality of your relationship? Were any particular characteristics especially important for you? Why?

Cohabitation

The rise of cohabitation has led to its becoming an alternative beginning to the contemporary family life cycle (Glick, 1989a; Surra, 1991). More than an alternative to marriage, cohabitation is more typically an alternative way of entering marriage. More than half of first unions result from cohabitation (Seltzer, 2000; London, 1991).

Although cohabiting couples may be living together before marriage, their relationship is not legally recognized until the wedding, nor is the relationship afforded the same social legitimacy. Most relatives do not consider cohabitants, for example, as kin, such as sons-in-law or daughters-in-law. At the same time, as discussed in Chapter 7, there is evidence that marriages that follow cohabitation have a higher divorce rate than do marriages that begin without cohabitation (De Maris and Rao, 1992; Hall and Zhao, 1995). Cohabitation does, however, perform some of the same functions as engagement, such as preparing the couple for some of the realities of marriage and helping them think of themselves in terms of being a couple as well as individuals.

Weddings

Weddings are ancient rituals that symbolize a couple's commitment to each other. The word *wedding* is derived from the Anglo-Saxon *wedd,* meaning "pledge." But it also included a pledge to the bride's father to pay him in money, cattle, or horses for his daughter (Ackerman, 1994; Chesser, 1980). When the father received his pledge, he "gave the bride away." The exchanging of rings dates back to ancient Egypt and symbolizes trust, unity, and timelessness because a ring has no beginning and no end. It is a powerful symbol. To return a ring or take it off in anger is a symbolic act. Not wearing a wedding ring may be a symbolic statement about a marriage. Another custom, carrying the bride over the threshold, was practiced in ancient Greece and Rome. It was a symbolic abduction growing out of the belief that a daughter would not willingly leave her father's house. The eating of cake is similarly ancient, representing the offerings made to household gods; the cake made the union sacred (Coulanges, 1960). The African tradition of jumping the broomstick, carried to America by enslaved tribespeople, has been incorporated by many contemporary African Americans into their wedding ceremonies (Cole, 1993).

The honeymoon tradition can be traced to a pagan custom for ensuring fertility: Each night after the marriage ceremony, until the moon completed a full cycle, the couple drank mead, honey wine. The honeymoon was literally a time of intoxication for the newly married man and woman. Flower girls originated in the Middle Ages; they carried wheat to symbolize fertility. Throughout the world, gifts are exchanged, special clothing is worn, and symbolically important objects are used or displayed in weddings (Werner et al., 1992).

The weddings of today are big business. Not all couples, however, have formal church weddings. Civil weddings now account for almost one-third of all marriage ceremonies (Ravo, 1991). Because of the expense, many couples are opting for civil ceremonies, which sometimes cost no more than $30, in addition to the marriage license. License fees, officiant's fees, and registration fees vary, though, from state to state. Ceremonies for second marriages are more likely to be civil than religious. Fewer religious ceremonies may also be a result of declining religious homogamy; either the bride and groom are from different religious backgrounds or are not religious.

Whether a first, second, or subsequent marriage, the central meaning of a wedding is that it symbolizes a profound life transition. Most significantly, the partners take on marital roles. For young men and women entering marriage for the first time, marriage signifies a major step into adulthood. Some of the apprehension felt by those planning to marry may be related to their taking on these important new roles and responsibilities. Many will have a child in the first year of marriage. Therefore, the wedding must be considered a major rite of passage. Before two people exchange their marriage vows, they are single, and their primary responsibilities are to themselves. Their parents may have greater claims on them than the couple have on each other. But with the exchange of vows, they are transformed. When they leave the wedding scene, they leave behind singlehood. They are now responsible to each other as fully as they are to themselves and more than they are to their parents.

Weddings are important rituals acknowledging the transition from singlehood to couplehood. Tradition plays an important role, as seen in this Korean ceremony.

Of course, the transition to marriage is much more than just a legal one. In fact, if we focus too much on the ceremonial aspect of marriage, we overlook two important points. First, getting married is a *process* that begins well before *and continues after* couples exchanges their vows. Second, the legal or ceremonial aspect of getting married may not be the most profound part of the transition.

THE STATIONS OF MARRIAGE

Past analyses of both divorce and remarriage have used the concept of the stations of marriage to represent both the dynamic and multidimensional nature of transitions out of and back into marriage (Bohannan, 1970; Goetting, 1982). In highlighting the different dimensions of experience that are part of divorce and reemarriage, these analyses depict the multidimensional, complex process of getting married equally well.

First articulated by anthropologist Paul Bohannan, the "six stations of divorce" reflect the following areas of experience that divorcing people encounter (Bohannan, 1970): (1) emotional, (2) psychic, (3) community, (4) economic, (5) legal, and (6) coparental. They may not experience them in the same sequence (thus, they are not *stages*) but somewhere on the path out of marriage they will encounter these stations. (See Chapter 14 for further discussion of Bohannan's stations of divorce). A decade later, Ann Goetting applied this same framework, with the same "six stations," to depict the complexities of remarriage.

Both Bohannan and Goetting stressed that that marital transitions are thick with complexity. They commence prior to and last well beyond the official or legal transition, be it divorce or remarriage. In fact, both Bohannan and Goetting tended to downplay the difficulties associated with the *legal station* even though more societal emphasis tends to be directed there. Their point was that in important social and emotional ways one can be divorced before the legal decree or remain emotionally "married" despite an official divorce. In recognizing the different dimensions of the divorce and remarriage experience, both analyses highlight the multidimensionality of marriage. Applying their notions of "stations" we can say that getting married consists of the following:

- *Emotional marriage:* The experiences associated with falling in love and the intensification of an emotional connection between two people. In the love-based marriages forming our society, as people fall in love they may contemplate an eventual marriage.
- *Psychic marriage:* The change in one's identity from that of an autonomous individual to that of a partner in a couple. As this occurs, one may encounter shifts in one's priorities, sense of self, perceptions of social reality, and expectations for the future (Berger and Kellner, 1970).
- *Community marriage:* The changes in one's social relationships and social network that accompany the shift in priorities and identity described above. It is actually a two-way process of redefining and *being redefined by* others. Friends may perceive themselves as no longer able to make the same claims or hold the same expectations about a formerly single or unattached friend. People may begin to refer to each partner only as a couple. In other words, *Matt and Jen* replaces *Matt* or *Jen.* As relationships become even more serious, the couple will be introduced to each other's family and may also find one's partner being incorporated into one's own family events. This certainly occurs as couples get engaged and proceed toward marriage. Once married, new spouses are unquestionably looked upon differently *because they are married.* They may even find their single friends becoming less interesting to or interested in them.
- *Legal marriage:* The legal relationship that—as we have seen—provides a couple with a host of rights and responsibilities. Clearly, legal marriage also restricts the individual's right to marry again without first ending the current marriage. However, aside from these and some restrictions on whom one may marry (which, granted, are not insignificant matters), there are few legal interventions into marriage as long as both parties remain content with their marriage. On a day-to-day basis, one may not notice changes in one's relationship that are due exclusively to this dimension of marriage.
- *Economic marriage:* The variety of economic changes that a couple experience when they marry. If both are employed, they now have more financial resources that need to be managed and allocated in ways that differ from their single days. For example, they may have to decide whether to pool all their assets or retain some separate finances; they must determine how much they will save or spend; they have to come to some working agreement about essential items that they need versus nonessential items that they want. Whether the decision they face is which overdue bill to pay this month or whether to buy a Lexus or an SUV, they will have to change the way they previously made economic decisions and decide as part of a couple. Typically, there are stylistic differences in one's spending or money management that require some compromise.
- *Coparental marriage:* The changes induced in marriage relationships by the arrival (via birth, remarriage, or adoption) of children. Important in both Bohannan's and Goetting's analyses, the coparental marriage is not really part of getting or becoming married per se. With regard to divorce, the coparental station includes attending to such issues as day-to-day care and custody, financial support, and visitation. In the coparental remarriage, the primary issue is to establish stepparenting roles and relationships (see Chapter 15). As far as a "station of marriage" we might say that if one party has any children, both partners will need to establish routines and share responsibilities. If childless at marriage, the coparental station would refer to those issues that change married relationships once children arrive (see Chapter 11).

Although neither Bohannan nor Goetting described a seventh station, we might include a domestic marriage—all of the negotiating, dividing, managing, and performing of day-to-day household chores. Couples must establish a working division of household labor. Even if they have cohabited before marriage, there is no guarantee that their "cohabiting division of labor" will be sustained in marriage.

By conceptualizing becoming married in these terms, we can state the following important points.

One may indeed feel and function as married prior to being legally married. That in no way guarantees success in marriage, as the research on cohabitants who marry is fairly pessimistic. But it does mean that when people think about the process of getting married, if they think in terms of prior to versus post wedding (essentially the legal station) the transition may seem less sweeping than in fact it is.

Becoming married transforms lives in all of the ways depicted above. However, because one has likely encountered at least the emotional, psychic, and community (or some of it) stations of marriage by the time one enters legal marriage, one has an opportunity to begin to remake one's life for marriage without yet being married. Bear in mind, too, that couples may experience these stations in different sequences. Cohabitants may experience all of these stations of marriage before legally marrying. Marriages entered into because of pregnancy or as escape from one's single lifestyle, will encounter these dimensions in a different order than those who marry out of first dating and falling in love. What's useful though about the concept of stations is how it helps us appreciate how broadly and deeply marriage changes two people.

© Mark Richards/PhotoEdit

Although lesbian and gay marriages have not gained full legal recognition, some same-sex couples nevertheless celebrate their relationships with a marriage ceremony. Because couples in such relationships cannot divide tasks or allocate roles on the basis of gender, they must use other criteria such as expertise or preference.

ESTABLISHING MARITAL ROLES

The expectations that two people have about their own and their spouse's marital roles are based on gender roles and their own experience. There are four traditional assumptions about husband/wife responsibilities: (1) the husband is the head of the household, (2) the husband is responsible for supporting the family, (3) the wife is responsible for domestic work, and (4) the wife is responsible for child rearing. More than mere expectations, these assumptions reflect traditional legal marriage (Weitzman, 1981).

The traditional assumptions about marital responsibilities do not necessarily reflect marital reality, however. For example, the husband traditionally may be regarded as head of the family, but power tends to be more shared, though perhaps not equally. In dual-earner families, both men and women contribute to the financial support of the family. Although responsibility for domestic work still tends to reside largely with women, men are gradually increasing their involvement in household labor, especially child care. The mother is generally still responsible for child rearing, but fathers are participating more.

Our gender-role attitudes and behaviors contribute to our marital roles. They create marital roles that reflect both traditional and nontraditional beliefs about men and women (Huston and Geis, 1993; Thoits, 1992). Although there have been significant changes over the last generation concerning gender and marital roles, many expectations have not significantly changed. According to one study (Ganong and Coleman, 1992), although both men and women expected their future partners to be successful, both expected the husband to be more successful than the wife. In addition, women expected their husbands to make significantly more money than the women did, to be better educated and more intelligent, and to be more competent in general. Even among dual-earner couples, husbands tend to continue holding traditional role expectations for themselves and their wives, such as the idea that wives are responsible for household tasks.

Marital Tasks

Newly married couples need to begin a number of marital tasks in order to build and strengthen their marriages. The failure to complete these tasks successfully may contribute to what researchers identify as the **duration-of-marriage effect**—the accumulation over time of various factors such as unresolved conflicts, poor communication, grievances, role overload, heavy work schedules, and child-rearing responsibilities that might cause marital disenchantment (see the Perspective box in this section that examines marital satisfaction). These tasks are primarily adjustment tasks and include the following:

- *Establishing marital and family roles:* Discuss marital-role expectations for self and partner; make appropriate adjustments to fit each other's needs and the needs of the marriage; discuss childbearing issues; and negotiate parental roles and responsibilities.
- *Providing emotional support for the partner:* Learn how to give and receive love and affection, support the other emotionally, and fulfill one's own identity as both an individual and a partner.
- *Adjusting personal habits:* Adjust to each other's personal ways by enjoying, accepting, tolerating, or changing personal habits, tastes, and preferences, such as differing sleep patterns, levels of personal and household cleanliness, musical tastes, and spending habits.
- *Negotiating gender roles:* Adjust gender roles and tasks to reflect individual personalities, skills, needs, interests, and equity.
- *Making sexual adjustments:* Learn how to physically show affection and love, discover mutual pleasures and satisfactions, negotiate timing and activities, and decide on the use of birth control.
- *Establishing family and employment priorities:* Balance employment and family goals; recognize the importance of unpaid household labor as work; negotiate child-care responsibilities; decide on whose employment, if either, receives priority; and divide household responsibilities equitably.
- *Developing communication skills:* Share intimate feelings and ideas with each other; learn how to talk to each other about difficulties; share moments of joy and pain; establish communication rules; and learn how to negotiate differences to enhance the marriage.
- *Managing budgetary and financial matters:* Establish a mutually agreed-upon budget; make short-term and long-term financial goals, such as saving for vacations or home purchase; and establish rules for resolving money conflicts.
- *Establishing kin relationships:* Participate in extended family and manage boundaries between family of marriage and family of orientation.
- *Participating in the larger community:* Make friends, nurture friendships, meet neighbors, and become involved in community, school, church, or political activities.

As you can see, a newly married couple must undertake numerous tasks as their marriage takes form. Marriages take different shapes according to how different tasks are shared, divided, or resolved. It is no wonder that many newlyweds find marriage harder than they expected. But if the tasks are undertaken in a spirit of love and cooperation, they offer the potential for marital growth, richness, and connection (Whitbourne and Ebmeyer, 1990). If the tasks are avoided or undertaken in a selfish or rigid manner, however, the result may be conflict and marital dissatisfaction.

Identity Bargaining

People carry around idealized pictures of marriage long before they meet their marriage partners. They have to adjust these preconceptions to the reality of the partner's personality and the circumstances of the marriage. The interactional process of role adjustment is called **identity bargaining** (Blumstein, 1975). The process is critical to marriage. A study of African-American and white newlyweds, for example, found that marital interactions that affirmed a person's identity predicted marital well-being (Oggins, Veroff, and Leber, 1993). Mirra Komarovsky (1987) points out that a spouse has a "vital stake" in getting his or her partner to fulfill certain obligations. "Hardly any aspect of marriage is exempt from mutual instruction and pressures to change," she writes.

Identity bargaining is a three-step process. First, a person has to identify with the role he or she is performing. A man must feel that he is a husband, and a woman must feel that she is a wife. The wedding ceremony acts as a catalyst for role change from the single state to the married state.

Perspective

EXAMINING MARITAL SATISFACTION

Because marriage and the family have moved to the very center of people's lives as a source of personal satisfaction, we generally evaluate them according to how well they fulfill emotional needs (although such fulfillment is not the only measurement of satisfaction). Marital satisfaction influences not only how we feel about our marriages and our partners but also how we feel about ourselves. If we have a good marriage, we tend to feel happy and fulfilled (Glenn, 1991).

Considering the various elements that make up or affect a marriage—from identity bargaining to economic status—it should not be surprising that marital satisfaction ebbs and flows. The ebb, however, begins relatively early for many couples. Researchers have found significant declines in the average level of marital satisfaction beginning in the first year of marriage (McHale and Huston, 1985). In some cases the decline in marital satisfaction during the first stage of marriage may mean that we have chosen the wrong partner. In fact, those with low marital satisfaction during the beginning marriage stage are four to five times as likely to divorce as those with high satisfaction (Booth et al., 1986). Satisfaction generally continues to decline during the first ten years of marriage, or perhaps longer (Glenn, 1989). Studies consistently indicate that marital satisfaction changes over the family life cycle, following a U-shape or curvilinear curve (Finkel and Hansen, 1992; Glenn, 1991; Suitor, 1991; but see Vaillant and Vaillant, 1993). Satisfaction is highest during the initial stages and then begins to decline but rises again in the later years. There seems to be little difference in marital satisfaction between first marriages and remarriages (Vemer et al., 1989).

It was once thought that couples with average or higher marital satisfaction would have stable marriages, whereas those with low satisfaction would have divorce-prone marriages. Although low marital satisfaction may make a marriage more likely to end in divorce, marital dissatisfaction alone cannot predict eventual divorce (Kitson and Morgan, 1991). Many unhappy marriages continue to endure in the face of misery and discord; sometimes they outlast much happier marriages. Unhappy marriages continue if there are too many barriers to divorce (such as a potential decline in the standard of living) and if the available alternatives seem less attractive than the current marriage. Happier marriages sometimes end in divorce if there are few barriers and better alternatives (Glenn, 1991).

Decline in Marital Satisfaction

Why does marital satisfaction tend to decline soon after marriage? Two explanations for changes in marital satisfaction have been given by researchers. The first ascribes the changing patterns of satisfaction to the presence of children; the second points to the effects of time on marital satisfaction.

Children and Marital Satisfaction

Traditionally, researchers have attributed decline in marital satisfaction to the arrival of the first child: Children take away from time a couple spends together, are a source of stress, and cost money. The decline reaches its lowest point when the oldest child enters adolescence (or school, according to some studies). When children begin leaving home, marital satisfaction begins to rise again.

It seems paradoxical that children cause marital satisfaction to decline. For many people, children are among the things they value most in their marriages. First, attributing the decline to children creates a single-cause fallacy—that is, attributes a complex phenomenon to one factor when there are probably multiple causes. Second, the arrival of children at the same time that marital satisfaction declines may be coincidental, not causal. Other undetected factors may be at work.

Although many societal factors make child rearing a difficult and sometimes painful experience for some families, it is also important to note that children create parental roles and the family in its most traditional sense. For some, the marital relationship may be less than fulfilling with children present, but many couples may make a trade-off for fulfillment in their parental roles. In times of marital crisis, parental roles may be the glue that holds the relationship together until the crisis passes. Many couples will endure intense situational conflict, not for the sake of the marriage but for the sake of the children. If the crisis can be resolved, the marriage may be even more solid than before. We need to balance marital satisfaction with family satisfaction.

The Duration of Marriage Effect and Marital Satisfaction

More recently, researchers have looked for factors besides children that might explain decline in marital satisfaction. The most persuasive alternative is the duration-of-marriage effect.

The duration-of-marriage effect is most notable during the first stage of marriage rather than during the transition to parenthood that follows (White and Booth, 1985). This early decline may reflect the replacement of unrealistic expectations about marriage by more realistic ones—a challenge to be intimate and loving in the everyday world. Because beginning marriage requires us to undertake numerous relational tasks, the transition from singlehood to marriage is a time filled with challenges. How we handle these challenges may set the tenor for our marriages for years to come. Those who handle the challenges successfully have the tools with which to enhance their marriages; those who fail may see their marriages decline in satisfaction.

Social and Psychological Factors in Marital Satisfaction

Social factors are important ingredients in marital satisfaction. Income level, for example, is a significant factor. Blue-collar workers have less marital satisfaction than white-collar, managerial, and professional workers because their lower income creates financial distress. Unemployment and economic uncertainty as sources of stress and tension also directly affect marital satisfaction. If a couple have an insufficient income or are deeply in debt, how to allocate their resources—for rent, repairing the car, or paying dental bills—becomes critical, sometimes involving conflict-filled decisions.

Psychological factors also affect marital satisfaction (London, Wakefield, and Lewak, 1990). Although it was once believed that marital satisfaction was dependent on a partner's fulfilling complementary needs and qualities (an introvert's marrying an extrovert, for example), research has failed to substantiate this assertion. Instead, marital success seems to depend on partners' being similar in their psychological makeup and personalities. Outgoing people are happier with outgoing partners; tidy people like tidy mates. Furthermore, a high self-concept (how a person perceives himself or herself), as well as how the spouse perceives the person, also contributes to marital satisfaction. Finally, similarity in perception, such as "seeing" events, relationships, and values through the same lenses, may be critical in marital satisfaction (Deal, Wampler, and Halverson, 1992).

Attitudes toward gender and marital roles may have an impact on marital satisfaction. There are numerous studies that approach this relationship from different angles. Together they indicate the significance of social roles in marriage. One study found that the discrepancy between how you expect your partner to behave and his or her actual behavior could predict marital satisfaction. Discrepancies in expectations were particularly significant in terms of intimacy, equality, trust, and dominance. Interestingly, discrepancies were more important in predicting dissatisfaction than was the fulfillment of expectations (Kelley and Burgoon, 1991). This finding is not entirely surprising. We seem to take for granted that our partner will fulfill our expectations, so it may be an unpleasant surprise to discover that our spouse is not interested in (or lacks the ability for) intimacy or that he or she is untrustworthy.

Expressiveness seems to be an important quality in marital satisfaction (L. King, 1993). One study found that expressive traits were more closely related to higher marital satisfaction than were instrumental ones (Juni and Grimm, 1993). Wives whose husbands discussed their relationships tended to be more satisfied with their marriages than other wives (Acitelli, 1992).

A psychological perspective may help explain why many people in middle-years marriages experience low levels of satisfaction. The middle stages of the family life cycle are also when adults enter midlife, a time characterized by psychological crises and reevaluation. The causes of decreased marital satisfaction may be the result of the adults' own psychological concerns and distress. Steinberg and Silverberg (1987) come to this conclusion: "It is reasonable to assume that these feelings may provoke disenchantment in the marital relationship, regardless of changes in the adolescent or parent-adolescent relationship."

Even though much of the literature points to declines in marital satisfaction over time, we must remember that not all marriages suffer a significant decline. For many marriages, the decline is small or satisfaction rises after a relatively short period. Even for those whose marriages decrease significantly in satisfaction, the decline may be offset by other satisfactions, such as pleasure in parental roles or a sense of security. Studies consistently demonstrate, for example, that the psychological and physical health of married men and women tends to be better than that of individuals who are unmarried (Gove, Styles, and Hughes, 1990; Waite, 1995).

It is important to understand that marital satisfaction is not static. It fluctuates over time, battered by stress, enlarged by love. Husband and wife continuously maneuver through myriad tasks, roles, and activities—from sweeping floors to kissing each other—to give their marriages form. Children, who bring us both delight and frustration, constrain our lives as couples but challenge us as mothers and fathers and enrich our lives as a family. Trials and triumphs, laughter and tears punctuate the daily life of marriage. If we are committed to each other and to our marriage, work together in a spirit of flexibility and cooperation, find time to be alone together, and communicate with each other, we lay the groundwork for a rich and meaningful marriage.

Second, a person must be treated by the other as if he or she fulfills the role. The husband must treat his wife as a wife; the wife must treat her husband as a husband. The problem is that a couple rarely agree on what actually constitutes the roles of husband and wife. This is especially true now as the traditional content of marital roles is changing.

Third, the two people must negotiate changes in each other's roles. A woman may have learned that she is supposed to defer to her husband, but if he makes an unfair demand, how can she do this? A man may believe that his wife is supposed to be receptive to him whenever he wishes to make love, but if she is not, how should he interpret her sexual needs? A woman may not like housework (who does?), but she may be expected to do it as part of her marital role. Does she then do all the housework, or does she ask her husband to share responsibility with her? A man believes he is supposed to be strong, but sometimes he feels weak. Does he reveal this to his wife?

Eventually, these adjustments must be made. At first, however, there may be confusion; both partners may feel inadequate because they are not fulfilling their role expectations. While some may fear losing their identity in the give and take of identity bargaining, the opposite may be true: One's sense of identity may actually grow in the process of establishing a relationship. In the process of forming a relationship, we discover ourselves. An intimate relationship requires us to define who we are.

ESTABLISHING BOUNDARIES

When people marry, many still have strong ties to their parents. Until the wedding, their family of orientation has greater claim to their loyalties than their spouse-to-be. Once the marriage ceremony is completed, however, the newlyweds can establish their own family independent of their families of orientation. The couple must negotiate a different relationship with their parents, siblings, and in-laws. Loyalties shift from their families of orientation to their newly formed family. The families of orientation must accept and support these breaks. Indeed, opening themselves to outsiders who have become in-laws places no small stress on families (Carter and McGoldrick, 1989). At the same time, however, many so-called in-law problems may actually be problems between the couple. It's easier to complain about a mother-in-law, for example, than it is to deal with troubling issues in one's own relationship (Silverstein, 1992).

The new family must establish its own boundaries. The couple should decide how much interaction with their families of orientation is desirable and how much influence these families may have. The addition of extended family can bring into contact people who are very different from one another in culture, life experiences, and values. There are often important ties to the parents that may prevent new families from achieving their needed independence. First is the tie of habit. Parents are used to being superordinate; children are used to being subordinate. The tie between mothers and daughters is especially strong; daughters often experience greater difficulty separating themselves from their mothers than do sons. These continuing ties may cause an adult child to feel conflicting loyalties toward parents and spouse (Cohler and Geyer, 1982). Much conflict occurs when a spouse feels that an in-law is exerting too much influence on his or her partner (for example, a mother-in-law's insisting that her son visit every Sunday and the son's accepting despite the protests of his wife; or a father-in-law's warning his son-in-law to establish himself in a career or risk losing his wife). If conflict occurs, husbands and wives often need to put the needs of their spouses ahead of those of their parents.

Another tie to the family of orientation may be money. Newly married couples often have little money or credit with which to begin their families. They may turn to parents to borrow money, co-sign loans, or obtain credit. But financial dependence keeps the new family tied to the family of orientation. The parents may try to exert undue influence on their children because it is their money, not their children's money, that is being spent. They may try to influence their children's purchases, or they may refuse to loan money to buy something of which they disapprove.

A review of research on in-laws found that in-law relationships generally had little emotional intensity (Goetting, 1989). The relationship between married women and their mothers-in-law and mothers seems to change with the birth of a first child (Fischer, 1983). Mother-daughter relationships seem to improve as the mother shifts some of her maternal role onto the grandchild. In-laws gave minimal direct support. Bonding between in-laws tends to be between women, and if

there is a divorce, divorced women are more likely than their ex-husbands to maintain supportive ties with former in-laws (Serovich, Price, and Chapman, 1991).

The critical task is to form a family that is interdependent rather than totally independent or dependent. It is a delicate balancing act as parents and their adult children begin to make adjustments to the new marriage. We need to maintain bonds with our families of orientation and to participate in the extended family network, but we cannot let those bonds turn into chains.

Youthful Marriages

Youthful marriages represent Stages II through IV in the family life cycle: childbearing families (Stage II), families with preschool children (Stage III), and families with schoolchildren (Stage IV).

IMPACT OF CHILDREN

Typically, husbands and wives both work until their first child is born; about half of all working women leave the workplace for at least a short period of time to attend to child-rearing responsibilities after the birth of the first child. The husband continues his job or career. Although the first child makes the husband a father, fatherhood generally does not visibly alter his relationship with his work. For example, it may redefine his motivation for work and the responsibility he feels to provide. Thus, even if he appears to continue at work relatively unaffected, he may be experiencing important changes

The woman's life, however, changes more dramatically and visibly with motherhood. If she continues her outside employment, she is usually responsible for arranging child care and juggling her employment responsibilities when her children are sick, and, if her story is like that of most employed mothers, she continues to have primary responsibility for the household and children. If she withdraws from the workplace, her contacts during most of the day are with her children and possibly other mothers. This relative isolation requires her to make a considerable psychological adaptation in her transition to motherhood, leading in some cases to unhappiness or depression.

Typical struggles in families with young children concern child-care responsibilities and parental roles. The woman's partner may not understand her frustration or unhappiness because he sees her fulfilling her roles as wife and mother. She herself may not fully understand the reasons for her feelings. The partners may increasingly grow apart during this period. During the day they move in different worlds, the world of the workplace and the world of the home; during the night they cannot relate easily because they do not understand each other's experiences. Research suggests that men are often overwhelmed by the emotional intensity of this and other types of conflict (Gottman, 1994). With all

Understanding Yourself

MARITAL SATISFACTION

An important question in studying marital satisfaction is how to measure it (Fincham and Bradbury, 1987). One measure widely used is Graham Spanier's Dyadic Adjustment Scale, which we have included a sample of here. This scale is an example of the type of questionnaire scholars use as they examine marital adjustment. What are the advantages of a questionnaire such as this? The disadvantages

Answer the questions below and then ask yourself if you think they can measure marital satisfaction. (*Hint:* You must first define what marital satisfaction is.) If you are currently involved in a relationship or marriage, you and your partner might be interested in answering the questions separately and comparing your answers. Do you have similar perceptions of your relationship? At the end of this course, answer the questions again without referring to your first set of answers. Then compare your responses. What do you infer from this comparison?

The Marital Satisfaction Survey	Always agree	Almost always agree	Occasionally disagree	Frequently disagree	Almost always disagree	Always disagree
1 Handling family finances	5	4	3	2	1	0
2 Matters of recreation	5	4	3	2	1	0
3 Religious matters	5	4	3	2	1	0
4 Demonstrations of affection	5	4	3	2	1	0
5 Friends	5	4	3	2	1	0
6 Sex relations	5	4	3	2	1	0

that accompanies the transition to parenthood (see next two chapters), it is unsurprising that more frequent conflict and tension ensue and that couples often change the ways in which they handle or resolve conflict (Crohan, 1996).

For adoptive families, the transition to parenthood may differ somewhat from that of biological families (Levy-Shiff, Goldschmidt, and Har-Even, 1991). Adoptive parents report more positive expectations about having a child, as well as more positive experiences in their transition to parenthood. In part this may be explained by adoptive parents' being able to fulfill parental roles that they vigorously sought. For them, parenting is a much more conscious decision than for many biological parents, for whom a pregnancy sometimes just "happens." For adoptive parents to become parents, considerable effort and expense must be undertaken; they are less likely to question their decision to become parents.

INDIVIDUAL CHANGES

Around the time people are in their thirties, the marital situation changes substantially. The children have probably started school and the mother, who usually has the lioness's share of duties, now begins to have more freedom from child-rearing responsibilities. She evaluates her past and decides on her future. The majority of women who left jobs to rear children return to the workplace by the time their children reach adolescence. By working, women generally increase their marital power.

Husbands in this period may find that their jobs have already peaked; they can no longer look forward to promotions. They may feel stalled and become depressed as they look into the future, which they see as nothing more than the past repeated for 30 more years. Their families may provide emotional satisfaction and fulfillment, however, as a counterbalance to workplace disappointments.

Jumpstart. Reprinted by permission of UFS, Inc.

Middle-Aged Marriages

Middle-aged marriages generally represent Stages V and VI in the family life cycle: families with adolescents (Stage V) and families as launching centers (Stage VI). Some parents may continue to raise young children while others, especially if one partner is considerably younger than the other, may choose to start a new family. Couples in these stages are usually in their forties and fifties.

> **DID YOU KNOW?**
>
> Family values, such as support, communication, and respect, along with marital satisfaction, face their greatest challenge in families with adolescent children (Larson and Richards, 1994).

FAMILIES WITH YOUNG CHILDREN

Increasing dramatically since 1970 are the over-35 women who have chosen to postpone childbearing until they are emotionally or financially ready. In 2000, more than 546,000 babies were born to women over age 35 and 94,000 to women over 40 (U.S. Census Bureau, 2002, Table 68). Although there have always been older women having children, in the past these mothers were having their last child, not their first. Because the majority of these over-35 women have a higher education, job status, and income, they also experience a lower divorce rate, are more stable, and are frequently more attentive to their young.

FAMILIES WITH ADOLESCENTS

Adolescents require considerable family reorganization on the part of parents: They stay up late; play loud music; infringe upon their parents' privacy; and leave a trail of empty pizza cartons, popcorn, dirty socks, and Big Gulp cups in their wake. As Carter and McGoldrick (1989) point out:

> Families with adolescents must establish qualitatively different boundaries than families with younger children. . . . Parents can no longer maintain complete authority. Adolescents can and do open the family to a whole array of new values as they bring friends and new ideas into the family arena. Families that become derailed at this stage are frequently stuck at an earlier view of their children. They may try to control every aspect of their lives at a time when, developmentally, this is impossible to do successfully. Either the adolescent withdraws from the appropriate involvements for this developmental stage or the parents become increasingly frustrated with what they perceive as their own impotence.

Although the majority of teenagers do not cause "storm and stress" (Larson and Ham, 1993), increased family conflict may occur as adolescents begin to assert their autonomy and independence. Conflicts over tidiness, study habits, communication, and lack of responsibility may emerge. Adolescents want rights and privileges but have difficulty accepting responsibility. Conflicts are often contained, however, if both parents and adolescents tacitly agree to avoid "flammable" topics, such as how the teenager spends his or her time or money. Such tactics may be useful in maintaining family peace, but in the extreme they can backfire by decreasing family closeness and intimacy. Despite the growing pains accompanying adolescence, parental bonds generally remain strong (Gecas and Seff, 1991).

FAMILIES AS LAUNCHING CENTERS

Some couples may be happy or even grateful to see their children leave home, some experience difficulties with this exodus, and some continue to accommodate their adult children under the parental roof.

The Empty Nest

As children are "launched" from the family (or ejected, as some parents wryly put it), the parental role becomes increasingly less important in daily life. The period following the child's exit is commonly known as the **empty nest.** Most parents make the transition reasonably well (Anderson, 1988). In fact, marital satisfaction generally begins to rise for the first time since the first stage of marriage (Glenn, 1991). For some parents, however, the empty nest is seen as the end of the family. Children have been the focal point of much family happiness and pain, and now they are gone.

Traditionally, it has been asserted that the departure of the last child from home leads to an "empty nest syndrome" among women, characterized by depression and identity crisis. However, there is little evidence that the syndrome is widespread. Rather, it is a myth that reinforces the traditional view that women's primary identity is found in motherhood. Once deprived of their all-encompassing identity as mothers, the myth goes, women lose all sense of purpose. (In reality, however, mothers may be more likely to complain when faced with adult children who have not left home.)

The couple must now re-create their family minus their children. Their parental roles become less important and less stressful on a day-to-day basis (Anderson, 1988). The husband and wife must rediscover themselves as man and woman. Some couples may divorce at this point if the children were the only reason the pair remained together. The outcome is more positive when parents have other more meaningful roles, such as school, work, or other activities to turn to (Lamanna and Riedmann, 1997).

The Not-So-Empty Nest: Adult Children at Home

Just how empty homes actually are after children reach age 18 is open to question. In fact, census data revealed that in 2000 56 percent of 18- to 24-year-old males, and 43 percent of 18- to 24-year-old females were living with one or both parents (Fields and Casper, 2001). Some are not moving out at all before their mid-twenties and many are doing an extra rotation through their family home after a temporary or lengthy absence. This later group is sometimes referred to as the **boomerang generation.**

In a 1995 survey of first-time college freshmen, 19 percent said wanting to get away from home was a very important reason to go to school. A larger share (25 percent) were living at home while they attended school, according to University of California at Los Angeles' Annual American Freshman Study.

Hispanics are more likely than other young adults to take a traditional route of staying home until they marry. Blacks are less likely than whites or Hispanics to leave home before marriage. Though family income may influence nest-leaving, ethnic or racial tradition seems to be more important in determining whether young adults will leave home (American Demographics, 1996). Most, however, move away from home when they marry.

Researchers note that there are important financial and emotional reasons for this trend (Mancini and Bliezner, 1991). High unemployment, housing costs, and poor wages are factors causing adult children to return home. High divorce rates, as well as personal problems, push adult children back to the parental home for social support and child care, as well as cooking and laundry services.

REFLECTIONS

Recall your family of orientation when you were an adolescent. How did you and your parents deal with establishing new family boundaries and with issues of autonomy and independence? What was the process of "launching" like? Has it been completed? If you continue to live at home, what difficulties has it caused you and your parents?

Young adults at home are such a common phenomenon that one of the leading family life cycle scholars suggests an additional family stage: *adult children at home* (Aldous, 1990). This new stage generally is not one that parents have anticipated. Almost half reported serious conflict with their children. For parents, the most frequently mentioned problems were the hours of their children's coming and going and their failure to share in cleaning and maintaining the house. Most wanted their children to be "up, gone, and on their own."

REEVALUATION

Middle-aged people find that they must reevaluate relations with their children, who have become independent adults, and must incorporate new family members as in-laws. Some must also begin considering how to assist their own parents, who are becoming more dependent as they age.

Couples in middle age tend to reexamine their aims and goals (Steinberg and Silverberg, 1987). On the average, husbands and wives have 13 more years of marriage without children than they used to, and during this time their partnership may become more harmonious or more strained. The man may decide to stay at home or not work as hard as before. The woman may commit herself more fully to her job or career, or she may remain at home, enjoying her new child-free leisure. Because the woman has probably returned to the workplace, wages and salary earned during this period may represent the highest amount the couple will earn.

DID YOU KNOW?

Average life expectancy is 74.4 for men and 79.8 for women. By the time individuals reach 65, their life expectancy rises to 81.4 for men and 84.4 for women. If they reach 75, they can anticipate living a decade (men) or dozen years (women) more (National Center for Health Statistics, www.cdc.gov/nchs/data/hus/tables/2003/03hus027.pd).

As people enter their fifties, they probably have advanced as far as they will ever advance in their work. They have accepted their own limits, but they also have an increased sense of their own mortality. Not only do they feel their bodies aging but they also begin to see people their own age dying. Some continue to live as if they were ageless—exercising, working hard, keeping up or even increasing the pace of their activities. Others become more reflective, retreating from the world. Some may turn outward, renewing their contacts with friends, relatives, and especially their children and grandchildren.

Later-Life Marriages

Later-life marriages represent the last two stages (Stages VII and VIII) of the family life cycle. In families with children, a later-life marriage is one in which the children have been launched and the partners are middle aged or older. Later-life families tend to be significantly more satisfied than families at earlier stages in the family life cycle (Mathis and Tanner, 1991). Compared with middle-aged couples, older couples showed less potential for conflict and greater potential for engaging in pleasurable activities together and separately, such as dancing, travel, or reading (Levenson, Cartensen, and Gottman, 1993). Research in the 1990s showed that older people without children experienced about the same level of psychological well being, instrumental support, and care as those who have children (Allen, Blieszner, and Roberto, 2000).

During this period, the three most important factors affecting middle-aged and older couples are health, retirement, and widowhood (Brubaker, 1991). In addition, these women and men must often assume roles as caretakers of their own aging parents or adjust to adult children who have returned home. Later-middle-aged men and women tend to enjoy good health, are firmly established in their work, and have their highest discretionary spending power because their children are gone (Voydanoff, 1987). As they age, however, they tend to cut back on their work commitments for both personal and health reasons.

As they enter old age, men and women are better off, on the average, than young Americans (Peterson, 1991). Beliefs that the elderly are neglected and isolated tend to reflect myth more than reality (Woodward, 1988). Over half of all people aged 65 and over live in either the same house or in the same neighborhood as one of their adult children (Troll, 1994). In addition, a national study of people over 65 found that 41 percent of those with children see or talk with them daily; 21 percent, twice a week; and 20 percent, weekly. Over half have children within 30 minutes' driving time (U.S. Census Bureau, 1988).

Beliefs that the elderly are a particularly poverty-stricken group are also misleading. Because of government programs initiated during the 1970s and 1980s benefiting the elderly, the poverty rate for the elderly declined from 25 percent in 1970 to 10.2 percent in 2000. In the meantime, the poverty level of young children soared to 15.6 percent; among young African-American and Latino children it skyrocketed to 30.4 percent and 27.3 percent respectively. From the mid- to late 1990s these rates declined slightly. We must not think, however, that all aged people benefited from the decline in poverty. The poverty rate in 2000 for aged African Americans was 22.4 percent, and 18.8 percent for Latinos (U.S. Census Bureau, 2002, Tables 669 and 671). Elderly African Americans tend to have lower levels of life satisfaction than do elderly whites (Krause, 1993). In part, this is because of greater financial strain and economic dependence on relatives among African Americans.

The health of the elderly also appears to be improving as they increase their longevity. Half of those between 75 and 84 years are free of health problems that require special care or limit their activities. Bernice Neugarten notes, "Even in the very oldest group, those above 85, more than one-third report no limitation due to health" (Toufexis, 1988). More married couples are living into old age, and there are fewer widows at younger ages.

THE INTERMITTENT EXTENDED FAMILY: SHARING AND CARING

Although many later-life families contract in size as children are launched, pushed, or cajoled out of the nest, other families may expand as they come to the assistance of family members in need. Families are most likely to become an intermittent extended family during their later-life stage (Beck and Beck, 1989). **Intermittent extended families** are families that take in other relatives during a time of need. These families "share and care" when younger or older relatives are in need or crisis: They help daughters who are single mothers; a sick parent, aunt, or uncle; or an unemployed cousin. When the crisis passes, the dependent adult leaves, and the family resumes its usual structure.

The incidence of intermittent extended families tends to be linked to ethnicity. Using national population studies, researchers estimate that the families of almost two-thirds of African-American women and one-third of white women were extended for at least some part of the time during their middle age (Beck and Beck, 1989; Minkler and Roe, 1993). Latina women are more likely than non-Latina women to form extended households (Tienda and Angel, 1982). Asian-American families are also more likely to live at some time in extended families. There are two reasons for the prevalence of extended families among certain ethnic groups. First, extended families are by cultural tradition more significant to African Americans, Latinos, and Asian Americans than to whites. Second, ethnic families are more likely to be economically disadvantaged. They share households and pool resources as a practical way to overcome short-term difficulties. In addition, there is a higher rate of single parenthood among African Americans, which makes mothers and their children economically vulnerable. These women often turn to their families of orientation for emotional and economic support until they are able to get on their own feet.

THE SANDWICH GENERATION

A relatively new phenomena, now referred to as the **sandwich generation,** are those middle-aged (or older) individuals who are sandwiched between the simultaneous responsibilities of caring for both their dependent children and their aging parents. Given the number of baby boomers now in their middle years, coupled with the increased longevity among their parents, we can anticipate that this type of dual care will become increasingly common. An estimated 20 to 30 percent of workers over age 30 are currently involved in caregiving to their parents, and this percentage is expected to grow (Field and Minkler, 1993). Daughters outnumber sons as caretakers by more than three to one (Allen, Blieszner, and Roberto, 2000; Cox, 1993), though among Asian Americans, the eldest son may be expected to be responsible for his elders (Kamo and Zhou, 1994). When sons are caretakers, whether in families with only sons or with sons and daughters, it is often actually daughters-in-law or grandaughters who provide the care (Allen, Blieszner, and Roberto, 2000).

As people live longer, their disabilities, dependency, and the number of their long-term chronic illnesses increases. Complicating this is the shrinking number of young workers, facilities, and resources to care for the old and frail. All of this puts additional pressure on families to provide support for their elders. Care that was traditionally handled by health-care professionals—injections, monitoring of medications, bathing, and physical therapy—is now often in the hands of family members.

The trend today, whenever possible, is for the dependent aged to be cared for in the home (Freedman, 1993). Placing added demands on family members' time, energy, and emotional commitment often results in exhaustion, anger, and in some cases, violence. Most people, however, are amazingly adept at meeting the needs of both their parents and their children. It is going to be an increasing challenge for society to acknowledge this phenomenon and provide services and support to both the elderly and those who care for them.

RETIREMENT

Retirement, like other life changes, has the potential for both satisfactions and problems. In a time of relative prosperity for the elderly, retirement is an event to which older couples generally look forward. The key to marital satisfaction in these later years is continued good health (Brubaker, 1991).

Retirement needs to be viewed as a process. People's experiences of retirement are often different and multilayered (Jensen-Scott, 1993). Usually as retirement approaches, men and women look forward to it; they slowly disengage themselves from their work roles. They experience less interest and greater dissatisfaction in their work. This disengagement prepares them for their actual retirement (Ekerdt and DeViney, 1993).

There is a growing trend toward early retirement among financially secure men and women. (Health limitations, however, increase both the likelihood of retirement as well as the chance that people will be unhappy during this time [Gradman, 1994; Solomon and Szwabo, 1994].) Changing social mores have led many to value leisure more than increased buying power and other job benefits. In fact, today, three-fourths of men and more than four-fifths of women choose to collect their Social Security checks before they reach the age of 65 years (Dentzer, 1990). Two scholars (Treas and Bengtson, 1987) note, "Together with greater prosperity, early retirement has probably reordered the preoccupations of later life toward greater concern with leisure activities—a development most compatible with the historic shift to companionate marriages." The bumper sticker "We're spending our children's inheritance" seen on campers and RVs captures this shift in sentiment.

This increased financial status is not true, however, for all older Americans (Gibson, 1993). Fewer African Americans than whites are financially able to retire. Within each racial group, older women are poorer than their male counterparts, with older African-American women the most likely to be poor.

When men reach their sixties, they usually retire from their jobs, losing a major activity through which they defined themselves. In spite of this, most men look favorably on retirement. Their role as husband becomes more important as they focus on leisure activities with their wives. In addition, many retirees are not really retired; many continue working part-time (Dentzer, 1990). Volunteer activities become more important. At home, the division of labor becomes somewhat more egalitarian; men participate in more household activities than they did when they were working (Rexroat and Shehan, 1987).

A study of older white and African-American women found that the psychological well-being of those who identified themselves as homemakers versus retirees did not differ significantly (Adelmann, 1993). A woman who retires may not find herself facing the same identity issues as a man, however, because women often have had (or continue to have) important family roles as wife, homemaker, grandmother, and possibly mother. In fact, women who identified themselves with dual roles as both homemakers and retirees had greater self-esteem and less depression than women who identified themselves in a single role. But retiring women also face greater financial insecurity because their traditionally lower wages result in lower retirement benefits (Perkins, 1992). Retired women who have adequate sources of income, high marital satisfaction, and overall good health have higher levels of satisfaction during retirement than do men (Slevin and Wingrove, 1995).

The marital relationship generally continues along the same track following retirement: Those who had vital, rewarding marriages will probably continue to have happy marriages, whereas those whose marriages were difficult will continue to have unsatisfying relationships (Brubaker, 1991). Various factors affect marital satisfaction during retirement (Higginbottom, Barling, and Kelloway, 1993). Good health is important, and financial security, contact with others, a sense of purpose, and the ability to structure time meaningfully contribute to marital satisfaction. Retired couples experience the highest degree of marital satisfaction since the first family stage, when they had no children (Johnson et al., 1986).

Although retirement affects marital interaction, changes in health are far more important over the long run. As long as both partners are healthy, the couple can continue their marital relationship unfettered. If one becomes ill or disabled, the other generally comes to his or her aid, providing care and nurturance. Ill spouses may also receive help from other family members, including their adult children, siblings, and other relatives. We will look more closely at caregiving in Chapter 11.

Chronic Illnesses

Most studies suggest that chronic illnesses have negative effects on the family. Family disruption and decreased marital satisfaction, for example, are common consequences (Hafstrom and Schram, 1984). Other studies find positive or inconsequential effects (Masters et al., 1983; Shapiro, 1983). Mark Peyrot and his colleagues (1988) found that the ultimate outcome of chronic illness is not necessarily bad. They found that after a period of disruption, some families resolved the crisis in a positive manner. Such families perceive the crisis as manageable and find the personal and social resources to create a favorable outcome. Several persons reported increased cohesiveness because they were compelled to spend more time with their families. One man with diabetes described the positive effect on his family: "I'm closer to my family than I was. Prior to this [the onset of diabetes] I was working 80 hours a week. I spent more hours at work than I did at home. For the first ten years my daughter never knew me because I was never at home" (Peyrot et al., 1988). Other families find hidden strengths and make their priorities more clear. Another respondent offered this comment:, "The traumatic experiences that we've been through pulled us closer together, made our marriage stronger. It's pointed out the more important things." (Peyrot et al., 1988)

Death and Dying in America

Families, like individuals, must face the final inevitability of death. Even though on a cognitive level we know that death comes to us all, when we actually confront death or dying, we are likely to be surprised, shocked, or at a loss about what to do.

ATTITUDES TOWARD DEATH

The nature and extent of our feelings and fears about death have to do with who we are—our age, sex, personality, spiritual beliefs, and so on. Our feelings also have to do with the person who has died—whether he or she was young or old, whether or not the death was expected, and what our relationship was.

REFLECTIONS

As your parents become dependent, will you become the primary caregiver? Why? Are you, your parents, or other close relatives involved in caregiving activities with an aging relative? If so, what effect has that had on you, your family, or your relative's family?

Cultural and Religious Context

Our cultural and religious context is important in determining our responses to death. Although Americans today seem somewhat more willing to talk and act realistically about death than they were a decade or two ago, as a society we remain ambivalent about the subject. Our responses to death and dying fall into three categories: denial, exploitation, and romanticization (Rando, 1987).

DENIAL. Although the existence of death is not specifically denied, there are practices in our culture that make it hard to view death as the real part of life it actually is. The removal of old or dying people from the home to the hospital has encouraged the belief that death is unnatural, frightening, or even disgusting. In a way, we treat death like sex: We try to shield children from it, and we use euphemisms when we speak of it ("Aunt Helen passed away," or "Fluffy was put to sleep").

EXPLOITATION. At the opposite extreme of denial is the exploitation of death. Therese Rando (1987) writes:

> While there has been a decline in the average individual's personal contact with death and dying, due to sociological changes and advances in technology, there has been an increase in exposure to violence and death through television, movies, and the print media. . . . Acts of murder, war, violence, terrorism, abuse, rape, crime, natural disasters, and so forth are exploited and sensationalized in the name of the public's right to know. Financial gains for the media and its sponsors are the obvious results.

The effects of this exploitation of death can be damaging. We may become overwhelmed by images of death and suffer from "annihilation anxiety." We may overcompensate and become aggressive, or we may withdraw and become desensitized to human pain. We may end up denying the realities of death altogether.

ROMANTICIZATION. Sometimes our images of death are glamorized and glorified. Although some deaths are in fact peaceful and "beautiful," many are not. Those who have been led to expect a beautiful death (their own or another's) may be shocked and feel betrayed to discover the messy, frustrating, or ugly aspects of dying.

Fear of Death and Dying

Many of us share a number of fears and anxieties regarding the dying process and death itself. When we think about dying, we may worry about being unable to care for ourselves and becoming a burden on those we love. We may worry about pain or physical impairment. We may fear isolation, loneliness, and separation from people who are dear to us. The finality of death is often fearsome; it implies the loss of relationships, the abandonment of goals, and an end to pleasure. For some, contemplating the possibilities of an unknown afterlife or the idea of eternal nothingness may be frightening. Others may be concerned about how their bodies will look after death and what will happen to them. One of the greatest fears aroused by the thought of death is the fear of losing control: The world will go on, but we will no longer have any effect on what happens in it.

Thanatologists—those who study death and dying—tell us that a certain amount of fear of death is a good thing. It certainly helps keep us alive. A certain amount of denial is healthy, too, for it prevents us from dwelling morbidly on the subject of death. What we need to develop is a realistic, honest view of death as part of life. Acknowledging that death exists can enrich our lives greatly. It can show us the importance of getting on with the "business of living," ordering our priorities, and appreciating what is around us.

THE PROCESS OF DYING

Unless death comes very quickly and unexpectedly, the person who is dying will undergo a number of emotional changes in addition to physical ones. These changes are often referred to as "stages," although they do not necessarily happen in a particular order. Indeed, some of them may not be experienced at all, or a person may return again and again to a particular part of the process.

Stages of Dying

Elizabeth Kübler-Ross constructed an influential model to represent the process that a person goes through when facing death (Kübler-Ross, 1969, 1982). Her model includes five stages:

1. *Denial and isolation* ("No, not me"). Coping with a diagnosis of a terminal illness often results in rejection of the news. Denial helps to overcome the shock, allows the person to begin to prepare himself or herself, and acts as a useful coping mechanism to allow the person to begin to gather together his or her resources.
2. *Anger* ("Why me?"). When the truth can no longer be denied, the dying person begins to feel

rage and resentment. This anger is often directed toward family, physicians, and God. Caretakers can do little but provide comfort and support.

3. *Bargaining* ("Yes, me but . . ."). In this stage, a person may try to bargain, usually with God, for a way to prolong life. In exchange, the patient often makes promises or amends. At this point, the person may also be his or her most vulnerable—grasping for anything that might help cure the condition.
4. *Depression* ("Yes, it's me"). Once people begin to accept their fate and face their impending death, they often become depressed about the unfinished business they are leaving behind. This anticipatory grieving is the most difficult time in the dying process, yet it is also an appropriate step in coming to terms with losses. It is eased when the patient is allowed to express sorrow among supportive and nonjudgmental friends.
5. *Acceptance* ("Yes, me; and I'm ready"). In this final stage, people facing their death come to some resolution and understanding. This time is neither frightening nor painful, neither sad nor happy—only inevitable. At the end, when death is near, patients may choose not to talk much, nor to visit with friends or family. This is part of the process of letting go.

Critics of the Kübler-Ross model raise a number of objections. These include the criticism that evidence is lacking to support the idea that there are these predictable stages in the dying process. Yet, by using "stages" as her construct, Kübler-Ross suggests that there is a set order or sequence of experiences. Additionally, people facing a terminal illness and approaching death do not necessarily experience all five stages. Because the model became so popular, the experiences of the dying person may have become somewhat artificially affected so as to "fit the model." The theory also doesn't mesh well with the unique and highly individualized experience of dying. Each of us, as we approach death, brings our individual personalities and identities to the experience, which is then further shaped by the environmental factors that surround us (Kastenbaum, 2000). There is no schedule to be anticipated or imposed; a dying person should not be expected to behave in a certain manner. Rather, each person facing death should be allowed to do so in his or her own way (DeSpelder and Strickland, 1996).

Needs of the Dying

Aside from basic physical care and perhaps relief from pain, what a dying person needs can be summed up very succinctly: to be treated like a human being. Robert Kavanaugh (1972) offers this observation:

> No matter how we measure [one's] worth, a dying human being deserves more than efficient care from strangers, more than machines and antiseptic hands, more than a mouthful of pills, arms full of tubes and a rump full of needles. . . . More than furtive eyes, reluctant hugs, medical jargon, ritual sacraments or tired Bible quotes, more than all the phony promises for a tomorrow that will never come.

Surgeon Sherwin Nuland (1994) writes that although "death with dignity" is the deepest wish of terminally ill patients, the realities of death for most people are far from dignified. Our society's own ambivalence and fear of death is all too often evident in the attitudes and actions of its doctors, nurses, and other medical personnel. Far too often, the dying are isolated from their loved ones during their final hours, until the machines that keep them in this world or measure their progress as they depart it emit their final beeps and blinks.

In response to the impersonality that has generally characterized death and dying in the hospital setting, the hospice movement has gained momentum in the last two decades. There are now more than a thousand hospice programs in the United States (DeSpelder and Strickland, 1996). A **hospice** may be an actual place where terminally ill people can be cared for with respect for their dignity, but it is also much more than that. It is a medical program that emphasizes both patient care (including management of pain and symptoms) and family support. The hospice provides education, grief counseling, financial counseling, and various types of practical assistance for home care of the dying person. It may also provide in-patient care in a noninstitutional setting.

BEREAVEMENT

For surviving family and friends, the loss of a loved one has profound social and emotional consequences. **Bereavement** is our response to the death of a loved one. It includes the customs and rituals that we practice within our culture or subculture. It also includes the emotional responses and expressions of feeling that we call the grieving process.

Mourning Rituals

Our culture, religion, and personal beliefs all influence the types of rituals we participate in after someone dies. By prescribing a specific set of formalized behaviors, bereavement rituals can give us security and comfort; we don't have to think, "What do I do now?" Social rituals, such as funerals and wakes, give us the opportunity to share our sorrow, to console and be consoled. A funeral also clearly marks the end of a life. Because we must face up to the fact that an important person in our life is gone, we can begin to move ahead to our "new" life. Religious rituals affirm a spiritual relationship for those who believe in them; for those who don't, they may be a source of tension or embarrassment.

For some Americans, rituals having to do with the dead consist basically of a funeral service followed by burial, entombment, or cremation. For others, there are important practices to be observed long after the burial (or cremation). Under Jewish law, for example, there are three successive periods of mourning. The first of these, **shiva**, is a seven-day period during which the immediate family undergoes certain austerities, such as refraining from haircutting, shaving, and using cosmetics; going to work; or engaging in sexual relations. During the second period, *shloshim,* the prohibitions become less strict, and the final period, *avelut,* applies only if one's mother or father has died. During this 11-month period, sons are to say *kaddish* (a form of prayer) for the parent daily. When the year of mourning is over, it is forbidden to continue practices that demonstrate grief (Kearl, 1989).

Among Latinos, *el Dia de los Muertos* (the Day of the Dead), an ancient ritual with Indian and Catholic roots that is observed on November 1, is making a comeback (Garcia, 1990). In Mexico and other Latin American countries, the dead are honored with prayers, gifts of food, and a nightlong graveside vigil; in the United States, parades and special exhibits not only commemorate the dead but also celebrate the cultural heritage of the participants. At home, altars may be set up with pictures of those who have died, religious figures or pictures, candles, orange marigolds (the Aztec flower of death), offerings of the honored one's favorite food or drink, and perhaps cartoons or darkly humorous verses that "laugh in the face of death."

Grief as a Healing Process

Like dying, grieving is a process. Thanatologists have variously described the stages of the grieving process (Kavanaugh, 1972; Kübler-Ross, 1982; Tallmer, 1987). There are also certain emotions or psychological states that may commonly be expected. Among these are shock, denial, depression, anger, loneliness, and feelings of relief. Guilt is also often experienced as part of the grieving process. If we felt resentment or anger toward the dead person, we feel guilty when he or she dies, as if we somehow caused the death. If we are spared and another dies—in an accident, for example—we feel guilty for surviving, and if the deceased was a burden to us while alive, we feel compelled to shoulder a load of guilt now that the burden has been lifted.

For some, most of the grieving process will be over in a matter of weeks or months. For others, it will occupy a year or two or maybe more. The first year will undoubtedly be the most difficult as holidays, birthdays, and anniversaries are experienced without the loved person. Grieving may occur sporadically for years to come, touched off by memories evoked by a particular date, a special piece of music, or a beautiful view that can no longer be shared. Healing, which is the goal of grieving, does not appear suddenly as the reward for all our suffering. Rather, it comes little by little as we work through grief (and around it and over it and under it and back through it again), until we look at ourselves one day and find we are whole.

Consoling the Bereaved

When a friend or relative is bereaved, we may feel awkward or embarrassed. We may want to avoid the family of the person who has died because we "just don't know what to say." That's okay. We really don't have to say much, except "I'm so sorry" and, perhaps, "How can I help?" Here are some suggestions for helping someone who is grieving:

- Listen, listen, listen. This may be the most helpful thing you can do.
- Express your own sadness about the death and your caring for the bereaved person, but don't

say "I know how you feel" unless you really do—that is, unless you've experienced a similar loss.

- Talk about the person who has died. Recall special qualities he or she possessed and the good times you may have experienced. If the bereaved person begins to speak of the one who has died, don't change the subject.
- Give practical support: Help with household tasks, do shopping or other errands, cook a meal, or help with child care.
- If there are children, involve them in remembering. Support their grieving process (it will be different from an adult's).
- Don't avoid the bereaved person because you are uncomfortable, worry about mentioning the dead person, or attempt to point out the "bright side."

Being able to share one's grief with others is a crucial part of healing. By "just" being around and "just" listening, we can be a positive part of the process.

Widowhood

Marriages are finite; they do not last forever. Eventually, every marriage is broken by divorce or death. Despite high divorce rates, most marriages end with death, not divorce. "Until death do us part" is a fact for most married people.

In 2000, 66.5 percent of those between ages 65 and 74 were married. Among those 75 years old and older, however, only 46 percent were married and 46 percent were widowed. Because women live about seven years longer on average than men, most widowed persons are women. Among women from 65 to 74, 56 percent were married, but only 31 percent over age 75 had a spouse. In contrast, among men 65 to 74 years old, 79.6 percent lived with their wives; among those over 75, 69.3 percent lived with a spouse (U.S. Census Bureau, 2001, Table 51). Three out of four wives will become widows.

Widowhood is often associated with a significant decline in income, plunging the grieving spouse into financial crisis and hardship in the year or so following death. This is especially true for poorer families (Smith and Zick, 1986). Feelings of well-being among both elderly men and elderly women are related to their financial situations. If the surviving spouse is financially secure, she or he does not have the added distress of a dramatic loss of income or wealth.

Recovering from the loss of a spouse is often difficult and prolonged. A woman may experience considerable disorientation and confusion from the loss of her role as a wife and companion. Having spent much of her life as part of a couple—having mutual friends, common interests, and shared goals—a widow suddenly finds herself alone. Whatever the nature of her marriage, she experiences grief, anger, distress, and loneliness. Physical health appears to be tied closely to the emotional stress of widowhood. Widowed men and women experience more health problems over the 14 months following their spouses' deaths than do those with spouses. Over time, however, widows appear to regain much of their physical and emotional health (Brubaker, 1991).

Eventually widows adjust to the loss. Some enjoy their new freedom. Others believe that they are too old to date or remarry; still others cannot imagine living with someone other than their former husband. (Those who had good marriages think of remarrying more often than those who had poor marriages.) A large number of elderly men and women live together without remarrying. For many widows, widowhood lasts the rest of their lives.

Enduring Marriages

Examining marriages and families in terms of the family life cycle is an important way of exploring the different tasks we must undertake at different times in our relationships. A number of those who have studied long-term marriages lasting 50 years or more have discovered several common patterns. Two researchers (Rowe and Lasswell, cited in Sweeney, 1982) have divided relationships into three categories: (1) couples who are happily in love, (2) unhappy couples who continue marriage out of habit and fear, and (3) couples in between who are neither happy nor unhappy and accept the situation. Lasswell and Row found that approximately 20 percent of long-term marriages to be very happy, while 20 percent were very unhappy.

Another way to look at marriage is according to stability rather than satisfaction. In other words, which marriages last? What researchers find is what many of us already know: Little correlation exists between happy marriages and stable ones. Many unhappily married couples stay together while some happily married couples undergo a crisis and breakup. In general, however, the quality of the marital relationship appears to show continuity over the years. Much of the discrepancy between happiness and stability results from the fact that happiness or satisfaction is an evaluative judgment of one's marriage relative to what one expected to get from marriage and what better alternatives are available. Stability results more from assessments of the costs and rewards of staying in or leaving a marriage. Unhappy marriages may be enduring ones because there are no better alternatives and/or because the costs of leaving exceed the costs of staying married.

Long-term marriages are not immune to conflict. As Figure 9.1 illustrates, as many as one-fourth of middle-aged couples, and between 12 and 20 percent of older couples acknowledge engaging in conflict over such issues as children, money, communication, recreation, sex, and in-laws. Surviving together does not require couples to eliminate or avoid conflict.

A study by Robert and Jeanette Lauer used a more modest definition of "long-term" to look at marriages that last. Their study of 351 couples married at least 15 years (in fact most were married a good deal longer) found the following to be the "most important ingredients" identified by men and women to explain their marital success: "my

FIGURE 9.1 ■ **Sources of Marital Conflict for Middle-Aged and Older Couples**

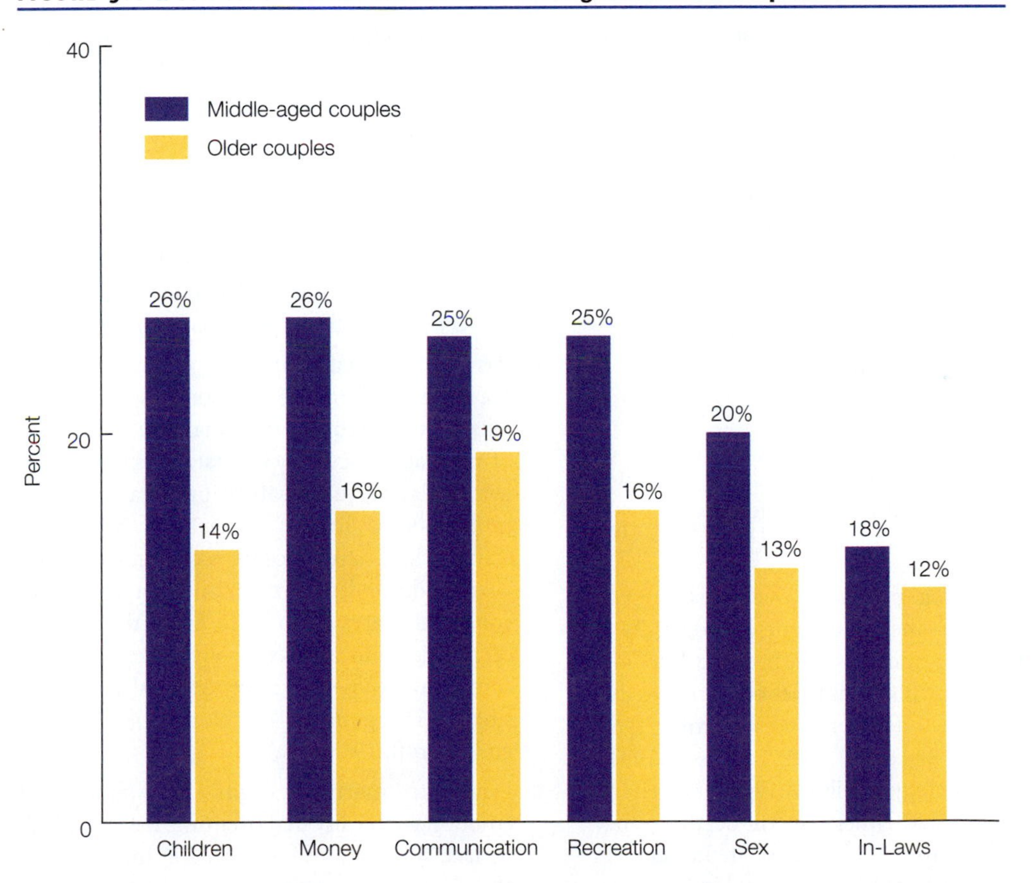

SOURCE: R. W. Levenson, L. L. Carstensen, and J. M. Gottman. "Long-Term Marriage: Age, Gender, and Satisfaction." *Psychology and Aging* 8 (1993): p. 307.

About 20 percent of long-term marriages can be classified as "very happy."

spouse is my best friend," "I like my spouse as a person," "marriage is a long-term commitment," "marriage is sacred," "we agree on aims and goals," and "my spouse has grown more interesting" (Lauer and Lauer, 1986). The correlation between husbands' and wives' lists was over .90, a remarkable consensus across gender lines. Summing up their results, the Lauers specify four keys to long-term satisfying marriages:

1. Having a spouse who is one's best friend and whom one likes as a person.
2. Believing in marriage as a long-term commitment and sacred institution.
3. Consensus on such fundamentals as aims and goals and one's philosophy of life.
4. Shared humor.

When assessing marriages one must keep in mind the fact that there is considerable diversity in married life. Thus, attempts have been made to document some of the different types of marriages that couples construct (Cuber and Harroff, 1965; Wallerstein and Blakeslee, 1995; Schwartz, 1994). One popular typology details five types of marriage, each of which could either last "till death do they part," or end in divorce. Thus these are not degrees of marital success but rather different kinds of marriage relationships (Cuber and Harroff, 1965).

- **Conflict-habituated marriages** are relationships in which tension, arguing and conflict "permeate the relationship" (Cuber and Harroff, 1965). One informant characterized his conflict-habituated marriage as a "long-running guerilla war," yet also acknowledged that neither he nor his wife had ever thought of ending the marriage. It may well be that conflict is what holds these couples together. It is at least understood to be a basic characteristic of this type of marriage.
- **Passive-congenial marriages** are relationships that begin without the emotional "spark" or intensity contained in our romantic idealizations of marriage. They may be marriages of convenience, that serve to satisfy practical needs in both spouses lives. Couples in which both spouses have strong career commitments and value independence may construct a passive-congenial marriage so as to enjoy the benefits of married life and especially parenthood. In some ways, these marriages are and have been since their beginning, "emotional voids" (Cuber and Harroff, 1965).
- **Devitalized marriages** begin with high levels of emotional intensity that over time has dwindled. In fact, from the outside looking in, they may closely resemble passive-congenial relationships. What sets them apart is that they have a history

of having been in a more intimate, sexually gratifying, emotional relationship that has become an "emotional void." Obligation and resignation may hold them together, along with the lifestyle they have built and the history they have shared.

- **Vital marriages** appeal more to our romantic notions of marriage because they begin and continue with high levels of emotional intensity. Such couples spend much of their time together and are "intensely bound together in important life matters" (Cuber and Harroff, 1965). The relationship is the most valued aspect of their lives and they allocate their time and attention based on such a priority. Conflict is not absent but it is managed in such a manner as to make quick resolution likely.
- **Total marriages** are relationships in which characteristics of vital relationships are present and multiplied. In some ways they may be seen as multifaceted vital relationships where the "points of vital meshing" extended across more aspects of daily coupled life.

Differentiating between these five types, Cuber and Harroff noted that the first three types were much more common than the last two. In fact, as many as 80 percent of the relationships among their sample were of one or another of these types. Both vital and total marriages (what they called *intrinsic marriages)* were relatively rare. Again, we must remember that the researchers were not sorting relationships into "successful" versus "unsuccessful" or "good" versus "bad." Marriages of all five types were enduring marriages, and any of the five types could end in divorce, though the reasons for divorce would differ.

At each stage in our individual and family life cycle we are presented the opportunity for growth and change as we enter our roles as husband/wife/partner, parent, stepparent, or grandparent. In all these stages, marriage requires a deep commitment. As David Mace and Vera Mace (1979) observe:

> Until two people, who are married, look into each other's eyes and make a solemn commitment to each other—that they will stop at nothing, that they will face any cost, any pain, any struggle, go out of their way so that they may learn and seek in order that they may make their marriage a continuously growing experience—until two people have done that they are not in my judgment married.

As we have seen, marriages and families never remain the same. They change as we change; as we learn to give and take; as children enter and exit our lives; as we create new goals and visions for ourselves and our relationships. In our intimate relationships, we are offered the opportunity to discover ourselves.

SUMMARY

- The eight developmental stages of the human life cycle described by Erik Erikson include (1) infancy: trust versus mistrust; (2) toddler: autonomy versus shame and doubt; (3) early childhood: initiative versus guilt; (4) school age: industry versus inferiority; (5) adolescence: identity versus role confusion; (6) young adulthood: intimacy versus isolation; (7) adulthood: generativity versus self-absorption; and (8) maturity: integrity versus despair. Each stage is intimately interconnected with family.
- The *family life cycle* perspective uses a developmental framework to look at how families change over time both in terms of the people who are members of the family and in terms of the roles they play. The family life cycle consists of eight stages: (1) beginning families, (2) childbearing families, (3) families with preschool children, (4) families with schoolchildren, (5) families with adolescents, (6) families as launching centers, (7) families in the middle years, and (8) aging families. Variations on the traditional

life cycle include disabled, adoptive, gay and lesbian, ethnic, immigrant, single-parent families, and stepfamilies.

- The relationships that precede marriage often predict marital success because marital patterns emerge during these times. Premarital factors correlated with marital success include (1) background factors (age at marriage, length of courtship, level of education, and childhood environment), (2) personality factors, and (3) relationship factors (communication, self-disclosure, and interdependence).
- Engagement is the culmination of the formal dating pattern. It prepares the couple for marriage by involving them in discussions about the realities of everyday life, it involves family members with the couple, and it strengthens the couple as a social unit. Individuals must deal with key psychological issues, such as anxiety, maturation and dependency needs, losses, partner choice, gender-role conflict, idealization and disillusionment, marital expectations, and self-knowledge. Cohabitation serves many of the same functions as engagement.
- A wedding is an ancient ritual that symbolizes a couple's commitment to each other. About two-thirds are formal church weddings. The wedding marks a major transition in life as the man and woman take on marital roles. Marriage involves many powerful traditional role expectations, including assumptions that the husband is head of the household and is expected to support the family and that the wife is responsible for housework and child rearing.
- The process of getting married and becoming spouses consists of six dimensions of experience that can be classified as the *stations of marriage:* (1) emotional, (2) psychic, (3) community, (4) economic, (5) legal, and (6) parental. One should also recognize the domestic responsibilities that marriage introduces as another part of becoming married.
- Gender-role attitudes and behaviors contribute to marital roles. Women are more egalitarian than men in marital-role expectations, but both genders expect men to earn more money. Marital tasks include establishing marital and family roles, providing emotional support for the partner, adjusting personal habits, negotiating gender roles, making sexual adjustments, establishing family and employment priorities, developing communication skills, managing budgetary and financial matters, establishing kin relationships, and participating in the larger community.
- Couples undergo *identity bargaining* in adjusting to marital roles. This is a three-step process: (1) person must identify with the role, (2) person must be treated by the other as if he or she fulfills that role, and (3) both people must negotiate changes in each other's roles.
- A critical task in early marriage is to establish boundaries separating the newly formed family from the couple's families of orientation. Ties to the families of orientation may include habits of subordination and economic dependency. In-law relationships tend to have little emotional intensity.
- In youthful marriages, about half of all working women leave the workforce to attend to child-rearing responsibilities. Motherhood more radically alters a woman's life than fatherhood changes a man's. Parental roles and child-care responsibilities need to be worked out.
- Middle-aged families must deal with issues of independence in regard to their adolescent children. Most women do not suffer from the *empty nest* syndrome. In fact, for many families, there is no empty nest because of the increasing presence of adult children in the home. As children leave home, parents reevaluate their relationship with each other and their life goals.
- In later-life marriages, usually no children are present. Marital satisfaction tends to be highest during this time. The most important factors affecting this life cycle stage are health, retirement, and widowhood. As a group, the aged have regular contact with their children, the lowest poverty level of any group, and good health through the early years of old age. Many families, especially among African Americans, Latinos, and Asian Americans, become *intermittent extended families* in which aging parents, adult children, or other relatives periodically live with them during times of need. This differs from the *sandwich generation* which finds itself caring for both their children and their aging parents at the same time.

- Cultural influences on our perception of death may cause us to respond with denial, exploitation, and romanticization. *Thanatologists*—people who study death and dying—tell us that the stages of dying are likely to include denial and isolation, anger, bargaining, depression, and acceptance. Apart from physical care and relief from pain, the most important need of a dying person is to be treated like a human being. The *hospice*—a place or program that provides care for terminally ill individuals—emphasizes both patient care and family support.
- *Bereavement*—the response to the death of a loved one—includes the customs and rituals of the grieving process (emotional responses and expressions of feeling). Mourning rituals include the funeral service and burial or cremation. The grieving process varies for different people; experiencing grief is a necessary part of healing.
- Long-term marriages may be divided into three categories: (1) couples who are happily in love, (2) unhappy couples who stay together out of habit or fear, and (3) couples who are neither happy nor unhappy. The percentage of couples who are happily in love is approximately 20 percent, the same percentage found for those who are unhappy.
- Some factors associated with long-term marriages are liking one's spouse as a person and thinking of one's spouse as one's best friend, believing in marriage as a commitment, spousal agreement on life's goals, and a sense of humor.
- Marriages differ from each other. One popular typology contrasts five types of marriage: (1) conflict-habituated, (2) devitalized, (3) passive-congenial, (4) vital and (5) total. These reflect different conceptualizations and experiences of marriage, not different degrees of marital success.

KEY TERMS

bereavement 308
boomerang generation 302
conflict-habituated marriage 311
devitalized marriage 311
duration-of-marriage effect 293
empty nest 301
family life cycle 282
honeymoon effect 288
hospice 308
identity bargaining 293
intermittent extended family 303
passive congenial marriage 311
sandwich generation 304
shiva 308
thanatologist 307
total marriage 312
vital marriage 312

SUGGESTED READINGS

Aldrous, John. *Family Careers; Rethinking the Developmental Perspective.* Newbury Park, CA: Sage, 1996. An examination of the expectable changes in today's families from the time it is first formed until it is dissolved.

Baber, Kristine, and Katherine Allen. *Women and Families: Feminist Reconstructions.* New York: Guilford Publications, 1992. A multidisciplinary examination of the diversity of women's experiences in family relationships, including intimacy, sexuality, reproduction, caregiving, and work.

DeSpelder, Lynne Ann, and Albert Lee Strickland. *The Last Dance: Encountering Death and Dying,* 4th ed. Mountain View, CA: Mayfield, 1996. A thorough and thoroughly readable examination of death and dying in America, including personal, sociocultural, medical, and spiritual aspects.

Hansson, Robert O., and B. N. Carpenter. *Relationships in Old Age: Coping with the Challenge of Transitions.* New York: Guilford, 1994. Discusses the common transitions people face as they age and the social network that is necessary to support aging.

Markides, Kyiados, and Charles Mindel. *Aging and Ethnicity; Perspectives on Gender, Race, Ethnicity and Class.* Newbury Park, CA: Sage, 1989. A good introduction to aging among African Americans, Latinos, Asian Americans, and other ethnic groups.

Whitbourne, Susan, and Joyce Ebmeyer. *Identity and Intimacy in Marriage: A Study of Couples.* New York: Springer-Verlag, 1990. An outstanding book describing how the individual identities of two people interact to create a relationship. Scholarly but readable.

RESOURCES ON THE INTERNET

Companion Web Site for This Book

http://sociology.wadsworth.com/strong/marriage9e
Gain an even better grasp on this chapter by going to the companion Web site to take one of the Tutorial Quizzes, use the Flash Cards to master key terms, or check out the many other study aids you'll find there. Visit the Marriage and Family Resource Center on the site. You'll also find special features such as GSS Data and Census 2000 information that will put data and resources at your fingertips to help you with that special project or to do some research on your own.

InfoTrac College Edition: Search Word Summary

single parents
sandwich generation
marital satisfaction
engagement
weddings
retirement

To learn more about these central topics in the study of the family, you can conduct an electronic search using InfoTrac College Edition. To aid in your search and to gain useful tips, see the Student Guide to InfoTrac College Edition that you can access through the companion Web site for this book.

Preview

To gain a sense of what you already know about the material covered in this chapter, answer "True" or "False" to the statements below.

1. The birthrate in the United States has risen steadily since 1990. True or False?
2. It is estimated that a third of women who marry will forgo having children. True or False?
3. Abortions have declined over the past decade. True or False?
4. It is usually unsafe for a woman to have sexual intercourse during the last two months of pregnancy. True or false?
5. Miscarriage and stillbirth are major life events for parents. True or false?
6. The rate of infant mortality in the United States is about what it is throughout the industrialized world. True or False?
7. Because of more effective contraceptives and an increase in the numbers of single mothers who choose to keep their babies, the number of adoptions in the United States has declined. True or False?
8. Men and women both can suffer from "postpartum blues." True or False?
9. There is often a decline in marital happiness following the transition to new parenthood. True or False?
10. Stress is common among both biological and adoptive new parents. True or False?

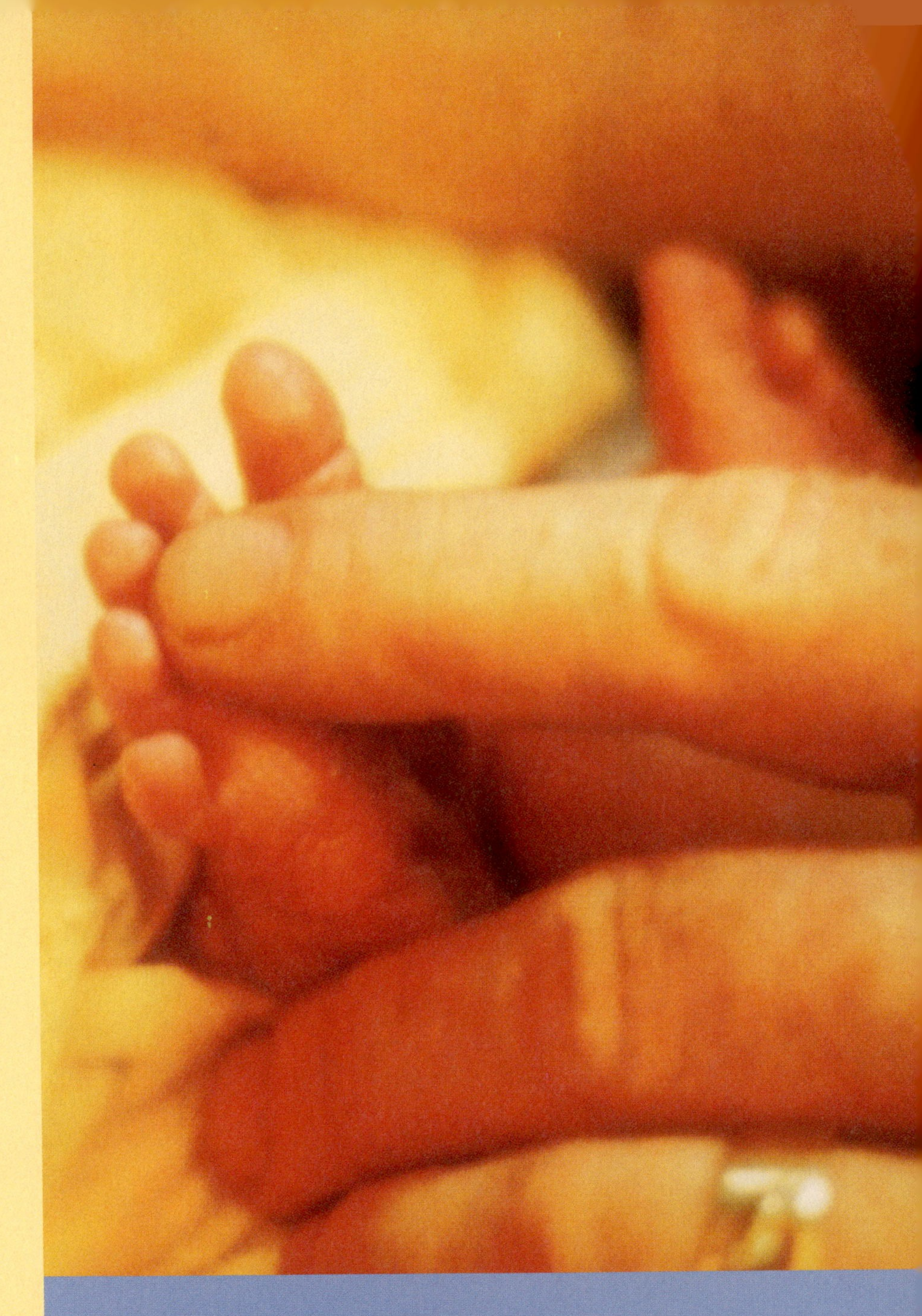

"You decide to have a baby and end up with a range of feelings that run from the rapturous to the murderous with four thousand stops in between. One minute you wish they had boarding school for two-year-olds and the next minute you dread the idea that they grow up and go away."

ELLEN GOODMAN

CHAPTER 10

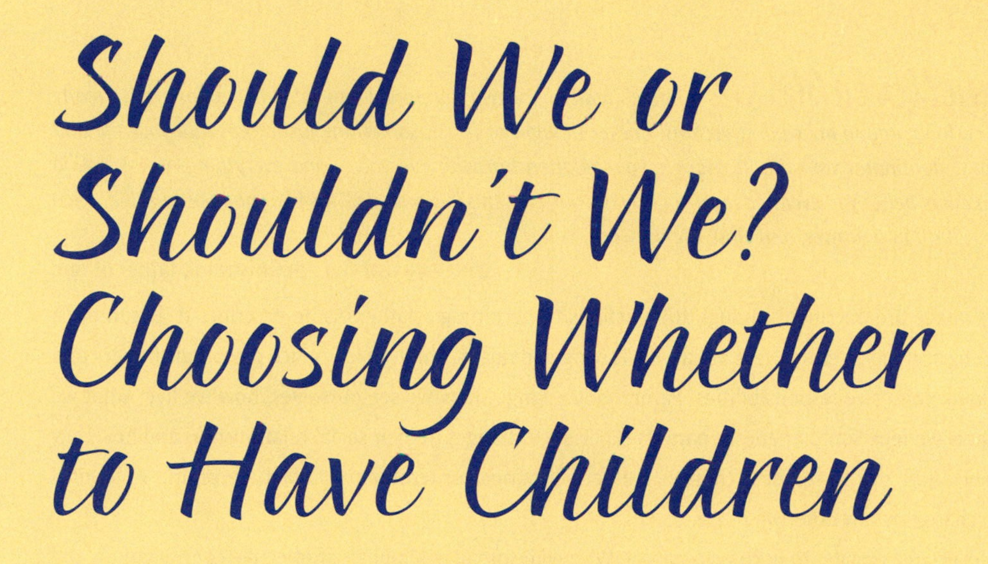

Should We or Shouldn't We? Choosing Whether to Have Children

Outline

It's unbelieveable . . . There's really no way a nonparent can think like a parent. It's really knocked me for a loop. And in my wildest dreams, I never thought of it. . . . Something just creeps into your life and all of a sudden it dominates your life. It changes your relationship with everybody and everything, you question every value and every belief you ever had. And you say to yourself, "this is a miracle." It's like you take your life, open up a drawer, put it all in a drawer, and close the drawer.

—33-year-old administrator, father of one

The comments above are one man's thoughtful reactions to becoming a father. As he describes it, becoming a parent is life-defining and life-altering. He is not alone. Having and raising children introduce profound changes and impose difficult, labor-intensive responsibilities. Parenthood changes how we see ourselves, how we live, what we think about, and how we feel. Simultaneously, parents experience changes in their social relationships and how they are viewed by others. And, of course, these changes are neither minor nor temporary. In fact, becoming a parent is as profound a life change as any other we make.

Of course, not everyone decides to become a parent. With widespread availability of effective contraception and access to legal abortion, women and men can decide whether and when to have children. The bulk of this chapter focuses on the process of deciding whether or not to have children and the range of factors that figure in to the decision process. The focus then shifts to pregnancy, especially the emotional and social changes that women and their partners encounter as they experience or witness the physiological changes the pregnant woman undergoes. Next, attention turns to the transition to parenthood and the changes it introduces into our lives. Chapter 11 will then explore the meaning and special challenges confronting contemporary mothers and fathers, particularly in the kind of society in which we currently live.

PREVIEW ANSWERS

1 False
2 False
3 True
4 False
5 True
6 False
7 True
8 True
9 True
10 True

Fertility Patterns in the United States

There were over 4 million births in the United States in 2002. As striking as this figure is, when we look at birth and fertility rates, we can see that the trend has been a downward one and continues to be toward smaller families. For example, the **crude birth rate,** a statistic reflecting the number of births per every thousand people in the population, was 13.9 in 2002. This is a slight decline from 2001, when the rate was 14.1, but represents a 17 percent reduction since 1990. It also represents the lowest such rate recorded for the United States since national data have been compiled (Hamilton, Martin, and Sutton, 2003). The United States has also experienced a decrease in the **fertility rate,** the number of births annually per 1,000 women 15 to 44 years old, from 118 in 1960 to 64.8 in 2002 (National Center for Health Statistics, 2003). This represents a more recent decline of 9 percent since 1990. Finally, the **total fertility rate,** a complicated statistic that estimates the number of births a hypothetical group of 1,000 women would have if they experience across their childbearing years the age-specific rates for a given year, indicates that there would be 2,012.5 births per 1,000 women, or—said another way—2 children per woman. This, too, reflects a 3 percent decline since 1990.

As seen below, fertility and birthrates vary considerably according to various social and demographic characteristics, such as race, ethnicity, education, and income. Figure 10.1 shows variation by ethnicity. Within the Latino population, rates vary from a high among Mexican Americans to a low among Cuban Americans. Cultural, social, and economic factors play a significant part in influencing the number of children a family has. Because of a combination of higher fertility rates and continuing immigration patterns, Hispanics have become our nation's largest minority group, thereby surpassing African Americans.

Fertility rates also vary by education. In 1995, for example, women with a high school education had the highest rate (67.4). College-educated women had the next highest rate (65.3), followed by women with graduate degrees (59.2) and women with less than high school (57.3).

Income effects on childbearing are a little more complex. For females between the ages of 15 and 29, birthrates are highest among women with the lowest incomes (<$10,000) and lowest among women with high incomes (>$75,000). Among women 30 to 44, we see almost the opposite pattern. The birthrate is highest among women with high incomes ($50,000–$74,999) followed closely by women with family incomes over $75,000. Meanwhile, the lowest birthrates among women between 30 and 44 years of age are found among women with family incomes between $10,000 and $29,999 (Legislative Commission on the Economic Status of Women Web site, 1999).

Approximately 18 percent of American women between the ages of 40 and 44 have not had children. Among women of that same age who had ever married, 12 percent were reported to be childless (U.S. Census Bureau, 2002). Among all women

FIGURE 10.1 ■ Fertility Rates by Race and Ethnicity: 2002

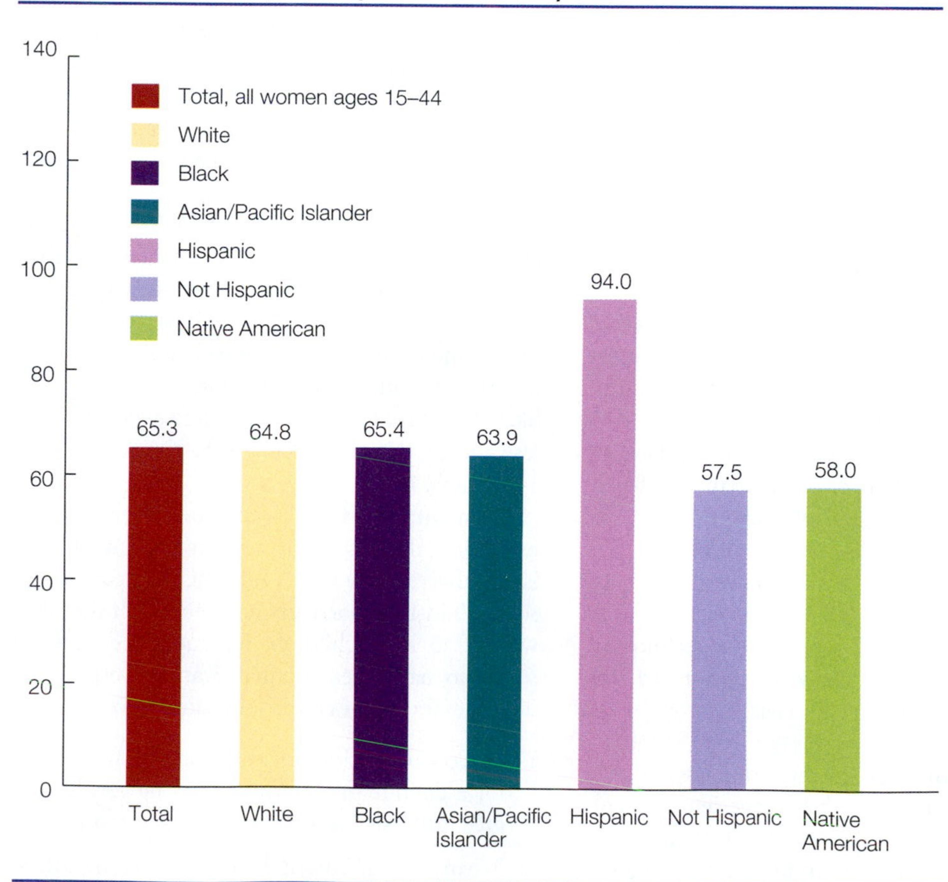

SOURCE: Hamilton, Martin, and Sutton, 2003.

Many couples today (especially those in middle- and upper-income brackets) defer having children until they have established their own relationships and built their careers. These parents are usually quite satisfied with their choice.

© Laurie DeVault Photography

without children, most expect to have at least one child. In fact, the percentage who intend to forgo parenting is estimated to be no more than 10 percent. However, it is important to consider the factors and dynamics associated with such a choice.

FORGOING PARENTHOOD

In the past, couples without children were referred to as *childless,* conveying the sense that such couples were missing something they wanted or were supposed to have. This certainly fits the experiences of the 8 to 9 percent of women 15 to 44 who have an "impaired ability" to have children, or the 7 to 8 percent of American couples who are involuntarily childless. Every year, more than 2 million couples seek help for infertility. In 2000, more than 25,000 births, and more than 35,000 infants were produced with the help of assisted reproductive technology. Nearly 100,000 procedures were performed, the most common being freshly fertilized embryos, using the patient's eggs.

In more recent years, the term *childless* has been joined by *child-free,* as we have experienced a cultural shift and demographic trend in the direction of increasing numbers of **child-free marriages**—couples who expect and intend to remain nonparents. The term *child-free* suggests that couples who do not choose to have children need no longer be seen as sympathetic figures, lacking something hitherto considered essential for personal and relationship fulfillment. In fact, the suffix *-free* suggests liberation from the bonds of a potentially oppressive condition (see Callan, 1985).

Women who choose to be child-free are generally well educated and career oriented (Ambry, 1992). Education and ethnicity are among the factors that influence their choices. Those women having the most education are more likely to anticipate a future without children. Hispanic women are less likely than black or white women to expect a childless future (Henslin, 2000; U.S. Census Bureau, 1998, Table 110).

Even with less familial and societal pressure to reproduce, the decision to remain without children is not always easily made. Although the partners in some child-free marriages have never felt that they wanted to have children, for most the decision seems to have been gradual. Jean Veevers (1980) identifies four stages of this decision process:

1. The couple decides to postpone having children for a definite time period (until he gets his degree, until she gets her promotion, and so on).
2. When the time period expires, they decide to postpone having children indefinitely (until they "feel like it").

3. They increasingly appreciate the positive advantages of being child-free (as opposed to the disadvantages of being childless).
4. The decision is made final, generally by the sterilization of one or both partners.

Studies of child-free marriages show eight basic categories of reasons given by those who have chosen this alternative (Houseknecht, 1987). The categories, in descending order of importance, are as follows:

- Freedom from child-care responsibility and greater opportunity for self-fulfillment
- More satisfactory marital relationship
- Wife's career considerations
- Monetary advantages
- Concern about population growth
- General dislike of children
- Early socialization experiences and doubts about parenting ability
- Concern about physical aspects of childbirth

Couples usually have some idea that they will or will not have children before they marry. If the intent isn't clear from the start or if one partner's mind changes, the couple may have serious problems ahead.

Many studies of child-free marriages indicate a higher degree of marital adjustment or satisfaction than is found among couples with children. These findings are not particularly surprising if we consider the great amount of time and energy that child rearing entails. It has also been observed that divorce is more probable in child-free marriages, perhaps because child-free couples do not stay together "for the sake of the children," as do some other unhappily married couples.

DEFERRED PARENTHOOD

Although most women still begin their families while in their twenties, demographers predict that the trend toward later parenthood will continue to grow, especially in middle- and upper-income groups (Edmondson et al., 1993; Price, 1982; Whitehead, 1990). A number of factors contribute to this. More career and lifestyle options are available to single women today than in the past. Marriage and reproduction are no longer economic or social necessities. People may take longer to search out the "right" mate (even if it takes more than one marriage to do it), and they may wait for the "right" time to have children. Increasingly effective birth control (including safe, legal abortion) has also been a significant factor in the planned deferral of parenthood.

Besides giving parents a chance to complete education, build careers, and firmly establish their own relationship, delaying parenthood can be advantageous for other reasons. Raising children is expensive. Maternity and medical expenses, food, furniture and equipment, clothing, toys, babysitters, lessons, and summer camp are cumulatively quite costly. Using the U.S. Department of Agriculture's estimate of what families in different income brackets spend raising children through age 17, we can estimate expenses of between $169,750 for low-income families (under $39,700 income), and $338,370 for high-income families (incomes of more than $66,900). In middle-income families (with incomes between $39,700-$66,900), parents who had a child in 2002 will spend approximately $231,680 to raise an only child to age 18 (Lino, 2003).

The USDA estimates actually underrepresent the costs incurred from having and raising children. Costs associated with childbirth and prenatal care are excluded. Childbirth costs alone can range from $7,000 to more than $11,000, depending on the kind of delivery (vaginal or cesarean) and type of health insurance coverage one has. Also, by calculating only *through age 17*, college costs are not included. Neither does the USDA take into account *wages lost* (or forgone) while parents take leave from the labor force. Finally, the projections to age 17 are based on an annual inflation rate of 3.2 percent. If this increases, so would the costs associated with raising a child (Lino, 2003).

Cost estimates that have included both college expenses and cost of wages lost project that raising a "typical" child amounts to a 22-year investment of between $761,871 (lower-third income bracket) and $2.78 million dollars (upper-third income bracket). For middle-income-bracket families the estimate is $1.45 million (Longman, 1998). Note that these expense estimates are based on two-parent families and pertain to raising one child. Subsequent children cost less than the first child, and daughters are more expensive to raise than are sons. Obviously, parents who have had a chance to establish themselves financially will be better able to bear the economic burdens of child rearing.

Ann Goetting's (1986a) review of parental satisfaction research revealed that "those who postpone parenthood until other components of their lives—especially their careers—are solidified" express enhanced degrees of satisfaction. Older parents may also be more emotionally mature and thus more capable of dealing with parenting stresses (although age isn't necessarily indicative of emotional maturity). In addition, as Jane Price (1982) writes, "Combating the aging process is something of a national preoccupation. . . . In our society, the power of children to revitalize and refresh is part of the host of forces encouraging men and women to become parents much later than they did in the past."

REFLECTIONS

If you don't have any children now, do you want to have them in the future? When? How many? What factors do you need to take into consideration when contemplating a family for yourself? Does your partner (if you have one) agree with you about having children?

Being Pregnant

Women and men who become parents enter a new phase of their lives. For those who bear their own children, this phase begins with pregnancy. Pregnancy is an important life event for both women and their partners. From the moment it is discovered, a pregnancy affects people's feelings about themselves, their relationship with their partner, and the interrelationships of other family members as well.

In 1999, there were 6.3 million pregnancies, down 7 percent from the 1990 peak of 6.8 million. These figures reflect 102 pregnancies per 1,000 women aged 15 to 44 years. Although the decline in pregnancy rates was across the board among all women under 30 years of age, the drop was steepest among teens, a decline of 15 percent from its record high in 1991. Pregnancy rates remain highest for women in their twenties.

Of the more than 6 million pregnancies in the United States, 62 percent (3.9 million) resulted in births, 22 percent (1.3 million) in abortions, and 16 percent (1 million) in stillbirths or miscarriages. Since 1990, trends in birth, abortion, and fetal loss have all declined: live births by 9 percent, abortions by 22 percent, and fetal losses by 4 percent. Women in the United States average 3.2 pregnancies each: 2.0 live births, 0.7 abortions, and 0.5 miscarriages and stillbirths. Only 1.8 of these 3.2 pregnancies result in wanted births.

Both marital status and race affect pregnancy outcomes. Three out of every four pregnancies among married women result in a live birth; 7 percent result in an abortion. Married women have a birthrate more than ten times greater than their abortion rate; among unmarried women, birth and abortion rates are more similar. In 1999, about half of pregnancies of unmarried women resulted in live

Both expectant parents may feel that the fetus is already a member of the family. They begin the attachment process well before birth.

© Scott Barrow

births; 40 percent ended in an abortion. Black women and white women report that they want about the same number of births, but black women experience more pregnancies. Among black women there is an average of 4.6 pregnancies per woman, compared with just 2.7 for white women. Only 39 percent of black women's pregnancies result in wanted births, compared with about 60 percent for Hispanic and non-Hispanic white women. Black women's pregnancies are twice as likely to end in abortions as pregnancies among white and Hispanic women (National Center for Health Statistics, 2000).

DID YOU KNOW?

It is fairly simple to figure out the date on which a baby's going to be born: add 7 days to the first day of the last menstrual period. Then subtract 3 months and add 1 year. For example, if a woman's last menstrual period began on July 17, 2004, add 7 days (July 24). Next subtract 3 months (April 24). Then add 1 year. This gives the expected date of birth as April 24, 2005. Few births actually occur on the date predicted, but 60 percent of babies are born within 5 days of the predicted time.

EXPERIENCING PREGNANCY: EMOTIONAL AND PSYCHOSOCIAL CHANGES

A woman's feelings during pregnancy will vary dramatically according to who she is, how she feels about pregnancy and motherhood, whether the pregnancy was planned, whether she has a secure home situation, and many other factors. Her feelings may be ambivalent; they will probably change over the course of the pregnancy.

A woman's first pregnancy is especially important because it has traditionally symbolized the transition to maturity. Even as social norms change and it becomes more common and acceptable for women to defer childbirth until they have established a career or to choose not to have children, the significance of first pregnancy should not be underestimated. It is a major developmental milestone in the lives of mothers—and fathers as well (Marsiglio, 1991; Notman and Lester, 1988; Snarey et al., 1987).

A couple's relationship is likely to undergo changes during pregnancy. It can be a stressful time, especially if the pregnancy was unanticipated. Communication is particularly important at this time because each partner may have preconceived ideas about what the other is feeling. Both partners may have fears about the baby's well-being, the approaching birth, their ability to parent, and the ways in which the baby will affect their own relationship. All of these concerns are normal. Sharing them, perhaps in the setting of a prenatal group, can deepen and strengthen the relationship (Kitzinger, 1989). If the pregnant woman's partner is not supportive or if she does not have a partner, it is important that she find other sources of support—family, friends, women's groups—and that she not be reluctant to ask for help.

A pregnant woman's relationship with her own mother may also undergo changes. In a certain sense, becoming a mother makes a woman the equal of her own mother. She can now lay claim to treatment as an adult. Women who have depended on their mothers tend to become more independent and assertive as their pregnancy progresses. Women who have been distant, hostile, or alienated from their own mothers may begin to identify with their mothers' experiences of pregnancy. Even women who have delayed childbearing until their thirties may be surprised to find their relationships with their mothers becoming more "adult." Working through the changing relationships is a kind of "psychological gestation" that accompanies the physiological gestation of the fetus (Silver and Campbell, 1988).

The first trimester (three months) of pregnancy may be difficult physically and emotionally for the expectant mother. She may experience nausea, fatigue, and painful swelling of the breasts. She may also fear that she will miscarry or that the child will not be normal. Her sexuality may undergo changes, resulting in unfamiliar needs (for more, less, or differently expressed sexual love), which may in turn cause anxiety. (Sexuality during pregnancy is discussed later in this chapter.) Education about the birth process and her own body's functioning and support from partner, friends, relatives, and health-care professionals are the best antidotes to her fear. Of course, as illustrated in Table 10.1 on the following page, not all pregnant women receive timely prenatal care, or care commencing during the first trimester.

TABLE 10.1 ■ Percent of Mothers Beginning Prenatal Care in First Trimester and Percent with Late or No Prenatal Care

	FIRST TRIMESTER	LATE OR NO CARE
All women	83.2	3.9
White (non Hispanic)	88.5	3.2
African American	74.3	6.7
Hispanic	74.4	6.3

SOURCE: Hamilton, B. E., Martin, J. A., Sutton, P. D. Births: Preliminary Data for 2002, *National Statistics Reports*, vol. 51, no. 11. Hyattsville, MD. National Center for Health Statistics, 2003.

During the second trimester, most of the nausea and fatigue disappear and the pregnant woman can feel the fetus move within her. Worries about miscarriage will probably begin to diminish, for the riskiest part of fetal development has passed. The pregnant woman may look and feel radiantly happy. She will very likely feel proud of her accomplishment and be delighted as her pregnancy begins to show. She may feel in harmony with life's natural rhythms. One mother writes (in C. Jones, 1988):

> I love my body when I'm pregnant. It seems round, full, complete somehow. I find that I am emotionally on an even keel throughout; no more premenstrual depression and upsets. I love the feeling that I am never alone, yet at the same time I am my own person. If I could always be five months pregnant, life would be bliss.

Some women, however, may be concerned about their increasing size; they may fear that they are becoming unattractive. A partner's attention and reassurance will ease this fear.

The third trimester may be the time of the greatest difficulties in daily living. The uterus, originally about the size of the woman's fist, has now enlarged to fill the pelvic cavity and is pushing up into the abdominal cavity, exerting increasing pressure on the other internal organs. Water retention (edema) is a fairly common problem during late pregnancy; it may cause swelling in the face, hands, ankles, and feet. It can often be controlled by cutting down on salt and refined carbohydrates (such as bleached flour and sugar) in the diet. If dietary changes do not help this condition, however, the woman should consult her physician. Another problem is that the woman's physical abilities are limited by her size. She may also be required by her employer to stop working at some point during her pregnancy. A family dependent on her income may suffer hardship.

The woman and her partner may become increasingly concerned about the upcoming birth. Some women experience periods of depression in the month preceding their delivery; they may feel physically awkward and sexually unattractive. Many, however, feel an exhilarating sense of excitement and anticipation marked by energetic bursts of industriousness. They feel that the fetus is a member of the family. Both parents may begin talking to the fetus and "playing" with it by patting and rubbing the expectant mother's belly.

The principal developmental tasks for the expectant mother and father may be summarized as follows (Valentine, 1982; also see Notman and Lester, 1988; Silver and Campbell, 1988; Snarey et al., 1987).

TASKS OF EXPECTANT MOTHER

- Development of an emotional attachment to the fetus
- Differentiation of the self from the fetus
- Acceptance and resolution of the relationship with own mother
- Resolution of dependency issues (generally involving parents or husband/partner)
- Evaluation of practical/financial responsibilities

TASKS OF EXPECTANT FATHER

- Acceptance of the pregnancy and attachment to the fetus
- Acceptance and resolution of the relationship with own father
- Resolution of dependency issues (involving wife/partner)
- Evaluation of practical/financial responsibilities

SEXUALITY DURING PREGNANCY

It is not unusual for a woman's sexual feelings and actions to change during pregnancy, although there is great variation among women in these expressions of sexuality. Some women feel beautiful, energetic, sensual, and interested in sex; others feel awkward and decidedly "unsexy." A woman's feelings may also fluctuate during this time. Some studies indicate a lessening of women's sexual interest during pregnancy and a corresponding decline in coital frequency. A study of 219 pregnant women found that although libido, intercourse, and orgasm declined, the frequency of oral and anal sex and masturbation remained at prepregnancy levels (Hart et al., 1991).

Men may feel confusion or conflicts about sexual activity during this time. They, like many women, may have been conditioned to find the pregnant body unerotic. Or they may feel deep sexual attraction to their pregnant partner, yet fear their feelings are "strange" or unusual. They may also worry about hurting their partner or the baby.

Although there are no "rules" governing sexual behavior during pregnancy, a few basic precautions should be observed:

- If the woman has had a prior miscarriage, she should check with her health practitioner before having intercourse, masturbating, or engaging in other activities that might lead to orgasm. Powerful uterine contractions could possibly induce a spontaneous abortion in some women, especially during the first trimester.
- If there is bleeding from the vagina, the woman should refrain from sexual activity and consult her physician or midwife at once.
- If the insertion of the penis into the vagina causes pain that is not easily remedied by a change of position, the couple should refrain from intercourse.
- Pressure on the woman's abdomen should be avoided, especially in the final months of pregnancy.
- During oral sex, care should be taken not to blow air into the vagina, as there is a possibility of causing an embolism (an air bubble in the bloodstream).
- Late in pregnancy, an orgasm is likely to induce uterine contractions. Generally this is not considered harmful, but the pregnant woman may want to discuss it with her practitioner. (Occasionally, labor is begun when the waters break as the result of orgasmic contractions.)

A couple, especially during their first pregnancy, may be uncertain as to how to express their sexual feelings. The following guidelines may be helpful (Strong and DeVault, 1997):

- Even during a normal pregnancy, sexual intercourse may be uncomfortable. The couple may want to try positions such as side by side or rear entry to avoid pressure on the woman's abdomen and to facilitate more shallow penetration.
- Even if intercourse is not comfortable for the woman, orgasm may still be intensely pleasurable. She may wish to consider masturbation (alone or with her partner) or cunnilingus.
- Both partners should remember that there are no rules about sexuality during pregnancy. This is a time for relaxing, enjoying the woman's changing body, talking a lot, touching each other, and experimenting with new ways—both sexual and nonsexual—of expressing affection.

MEN AND PREGNANCY

Obviously, pregnancy is something men do not experience directly. It is the woman's body that carries the fetus and undergoes profound change along the way. For men, pregnancy is only accessible vicariously. Still, how men navigate through the pregnancy process has consequence for their later conceptualization of and involvement in fathering (Marsiglio, 1998).

During pregnancy, men experience changes in their sexual relations with their partners, especially in the amount and nature of fantasies, and alterations in their patterns of dreams. In their sexual fantasies, they reported feeling as if they were fertilizing, nurturing, or "feeding" their fetuses or their wives, thus revealing the connection they draw between pregnancy and sexuality (Marsiglio, 1998). Early on, men's dreams occasionally take on qualities of mystery and awe, later on shifting to dreams of being neglected or rejected by their partners (Marsiglio, 1998).

Men's anxieties during pregnancy cover a number of different areas, including the health of both fetus and partner, whether one will be a good father, how fatherhood will affect their lives, and how well they will manage their economic responsibilities, especially given new expenses and reduced spousal income. Although a man's traditional role as father centered around providing, the concern over competence as a provider is not the major source of men's pregnancy anxieties. Of course, men whose employment is unstable or whose incomes are insufficient will experience more provider anxiety than will men who simply take for granted that they can meet their financial responsibilities (Cohen, 1993).

The roles men play in supporting their partners, participating in the preparation for parenthood, and at the birth also are significant. Not all men act in similar ways. Some may be relatively detached, others fully involved, and still others very practical in their participation in the pregnancy (Marsiglio, 1998). The way men act during pregnancy (reading material, attending prenatal classes, involving themselves in the birth process, and so on) may affect how they later relate with their newborns. Of particular note is the experience of witnessing the birth of their children, which reportedly opens men to a depth of emotional experience that is often otherwise absent from conventional cultural expressions of masculinity. Men are "feminized"; they speak poignantly, occasionally poetically about what that experience was like or meant to them (Cohen, 1987; Gerson, 1993).

PREGNANCY LOSS

The loss of a child through miscarriage, stillbirth, or death during early infancy is a devastating experience that has been largely ignored in our society. The statement "You can always have another one" may be meant as consolation, but it is particularly chilling to the ears of a grieving mother. In the past few years, however, the medical community has begun to respond to the emotional needs of parents who have lost a pregnancy or an infant.

Spontaneous Abortion

Spontaneous abortion (miscarriage) is a powerful natural selective force toward bringing healthy babies into the world. About one out of four women is aware she has miscarried at least once (Beck, 1988). Studies indicate that at least 60 percent of all miscarriages are due to chromosomal abnormalities in the fetus (Adler, 1986). Furthermore, as many as three-fourths of all fertilized eggs do not mature into viable fetuses (Beck, 1988). One study found that 32 percent of implanting embryos miscarried (Wilcox et al., 1988). The first sign that a pregnant woman may miscarry is vaginal bleeding ("spotting"). If a woman's symptoms of pregnancy disappear and she develops pelvic cramps, she may be miscarrying; the fetus is usually expelled by uterine contractions. Most miscarriages occur between the sixth and eighth weeks of pregnancy. Evidence is increasing that certain occupations involving exposure to chemicals or high levels of electromagnetism increase the likelihood of spontaneous abortions. Miscarriages may also occur because of uterine abnormalities or hormonal levels that are insufficient for maintaining the uterine lining.

Infant Mortality

The rate of **infant mortality** in the United States remains far higher than the rates in most of the developed world. The U.S. Public Health Service reported 6. 8 deaths for every 1,000 live births in 2001 (National Center for Health Statistics, 2002; U.S. Census Bureau, 1996). Nevertheless, among developed nations, the United States does not fare well in low infant mortality. In 1999, for example, the United States ranked twenty-eighth in a comparison of 37 countries with populations of at least 1 million, for which complete counts of live births and deaths were compiled (This means that 27 countries had *lower* infant mortality rates than the United States.) In the same comparison in 1990, the United States ranked eleventh (http://www.cdc.gov/nchs/data/hus/tables/2003/03hus025.pdf.). Of course, in comparison with developing countries, the rate in the United States is quite low (UNICEF State of the World's Children, 2001). The nation's capital has a higher infant mortality rate than any of the 50 states in the United States at 12.5 per 1,000 live births (U.S. Census Bureau, 2001, Table 104).

Of the more than 27,000 American babies less than 1 year old who die each year, most are victims of the poverty that often results from racial or ethnic discrimination. Up to a third of these deaths could be prevented if mothers were given adequate health care (Scott, 1990a). The infant mortality rate for African Americans is more than twice that for whites (14.0 versus 5.7 in 2001). Native Americans are also at high risk; for example, about 1 out of every 67 Navajo infants dies each year (Wilkerson, 1987). A study by the U.S. Centers for Disease Control found that mortality rates for many ethnic minorities have been severely underestimated because the infants were mistakenly classified as white ("Death Rates," 1992; Hahn, Mulinare, and Teutsch, 1992).

In 1981, the federal government began to cut pregnancy and infant-care programs such as WIC (Special Supplemental Nutrition Program for Women, Infants, and Children) dramatically to fund its record-breaking military budget. Medicare and Aid to Families with Dependent Children were also cut, leaving many families without prenatal or postnatal care. In recent years, the government has begun to restore funding to many of these programs, but many of them are currently threatened. The United States is far behind many other countries in providing health care for children and pregnant women. In France, Sweden, and Japan, for example, all pregnant women are entitled to free prenatal care. Free health care and immunizations are also provided for infants and young children. Working Swedish mothers are guaranteed one year of paid maternal leave, and French families in need are paid regular government allowances (Scott, 1990a). A recent analysis of 17 studies done between 1971 and 1988 found that WIC recipients were 25 percent less likely than comparable nonrecipients to have a low-birth-weight baby and 44 percent less likely to have a very low-birth-weight baby. The researchers calculated that one year of WIC expenditures may save society as much as $800 million in federal and state Medicaid expenditures (Avruch and Cackley, 1995).

Although many infants die of poverty-related conditions, others die from congenital problems (conditions appearing at birth) or from infectious diseases, accidents, or other causes. Sometimes, the causes of death are not apparent. Data from the CDC and the National Center for Health Statistics for 2001 attribute 2,234 infant deaths to **sudden infant death syndrome (SIDS)**, a perplexing phenomenon wherein an apparently healthy infant dies suddenly while sleeping (http://www.cdc.gov/nchs/fastats/pdf/mortality/nvsr52_03t32.pdf). A study from Australia identified four factors that appear to increase the chances of SIDS (Ponsonby et al., 1993): (1) a soft, fluffy mattress, (2) the baby being wrapped in a blanket, (3) the baby having a cold or other minor illness, and (4) allowing the baby to become too warm. Exposure to secondhand smoke also has been implicated (Klonoff-Cohen et al., 1995). It is also very important that an infant not be placed to sleep on its stomach until it is strong enough to turn over ("Sleeping on Back Saves 1,500 Babies," *San Mateo County Times*, June 25, 1996).

Coping with Loss

The depth of shock and grief felt by many who lose a child before or during birth is sometimes difficult to understand for those who have not had a similar experience. What they may not realize is that most women form a deep attachment to their children even before birth. At first, the attachment may be a "fantasy image of [the] future child" (Friedman and Gradstein, 1982). During the course of the pregnancy, the mother forms an actual acquaintance with her child through the physical sensations she feels within her. Thus, the death of the fetus can also represent the death of a dream and of a hope for the future. This loss must be acknowledged and felt before psychological healing can take place.

Women (and sometimes their partners) who lose a pregnancy or a young infant generally experience similar stages in their grieving process. Their feelings are influenced by many factors: the supportiveness of the partner and other family members, the reactions of social networks, life circumstances at the time of the loss, circumstances of the loss itself, whether other losses have been experienced, the prognosis for future childbearing, and the woman's unique personality. Physical exhaustion and, in the case of miscarriage, hormone imbalance often compound the emotional stress of the grieving mother.

The initial stage of grief is often one of shocked disbelief and numbness. This stage gives way to sadness, spells of crying, preoccupation with the loss, and perhaps loss of interest in the rest of the world. Emotional pain may be accompanied by physical sensations and symptoms, such as tightness in the chest or stomach, sleeplessness, and loss of appetite. It is not unusual for parents to feel guilty, as if they had somehow caused the loss, although this is rarely the case. Anger (toward the physician, perhaps, or God) is also a common emotion.

Experiencing the pain of loss is part of the healing process (Vredevelt, 1994). This process takes time—months, a year, perhaps more for some.

Support groups and counseling are often helpful, especially if healing does not seem to be progressing—if, for example, depression and physical symptoms do not appear to be diminishing. Keeping active is another way to deal with the pain of loss, as long as it is not a way to avoid facing feelings. Projects, temporary or part-time work, or travel (for those who can afford it) can be ways of renewing energy and interest in life. Planning the next pregnancy may be curative, too, though we must keep in mind that the body and spirit need some time to heal. It is important to have a physician's input before proceeding with another pregnancy; specific considerations may need to be discussed, such as a genetic condition that may be passed to the child or a physiological problem of the mother. If future pregnancies are ruled out, the parents need to take stock of their priorities and consider other options that may be open to them, such as adoption. Counselors and support groups can be invaluable at this stage.

MEDICALIZATION OF CHILDBIRTH

Women and couples planning the birth of a child have decisions to make in a variety of areas—birth place, birth attendants, medications, preparedness classes, circumcision, and breastfeeding, to name but a few. The "childbirth market" is beginning to respond to consumer concerns, so it's important for prospective parents to fully understand their options.

Hospital Birth

The impersonal, assembly-line quality of hospital birth (and hospital care in general) is increasingly being questioned. One woman described her initial feelings in the hospital (Leifer, 1990):

> When they put that tag around my wrist and put me into that hospital gown, I felt as if I had suddenly just become a number, a medical case. All of the excitement that I was feeling on the way over to the hospital began to fade away. It felt like I was waiting for an operation, not about to have my baby. I felt alone, totally alone, as if I had just become a body to be examined and not a real person.

Another woman was shocked at the impersonal treatment (Leifer, 1990):

> And then this resident gave me an internal [examination], and it was quite painful then. And I said: "Could you wait till the contraction is over?" And he said he had to do it now, and I was really upset because he didn't even say it nicely, he just said: "You'll have to get used to this, you'll have a lot of this before the baby comes."

Most hospitals have responded to the need for family-centered childbirth. Fathers and other relatives or close friends often participate today. Some hospitals permit rooming-in (the baby stays with the mother rather than in the nursery) or a modified form of rooming-in. Regulations vary as to

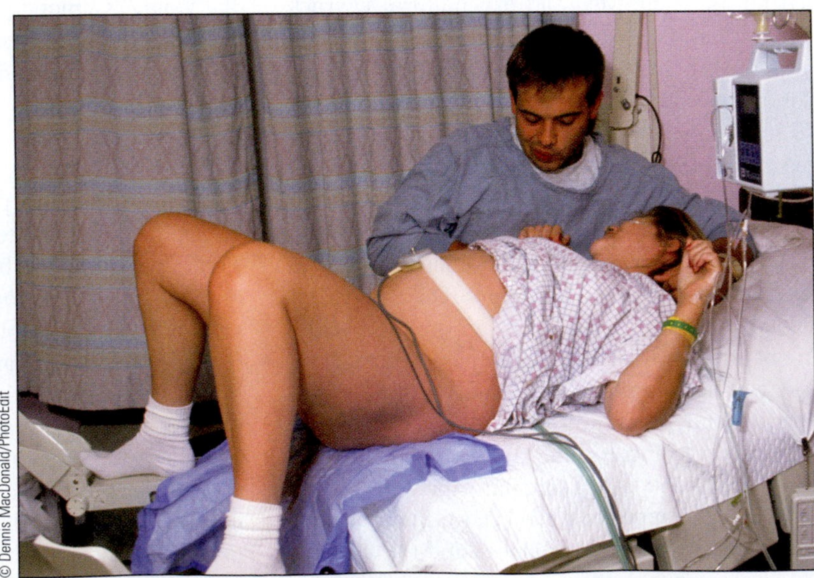

Family-centered childbirth allows fathers to participate alongside mothers in the birth process.

Exploring Diversity

COUVADE: HOW MEN GIVE BIRTH

Throughout the world, men envy and imitate both pregnancy and childbirth. In our own culture, there are sympathetic pregnancies in which a man develops physical characteristics similar to those of his pregnant partner. If she has morning sickness, so does he; if her belly begins to swell, so does his. As someone wrote, "Man is a rational animal, but only women can have babies." Thus, men often use images of pregnancy and childbirth to describe their creative work. A man "conceives" an idea. He "gives birth" to a new theory. He is in a "fertile" period in his artistic development. His book is "pregnant" with ideas. Less fortunate men find their careers "aborted," while others have "stillborn" ideas. It has even been suggested that men often discriminate against women in the arts and sciences precisely because, in these areas of highly creative work, they feel most jealous of women's biological creativity and thus attempt to keep them from entering this domain.

Other cultures have the ritual of **couvade.** The word comes from the French *couver,* meaning "to hatch or brood." The couvade involves a number of different activities in different cultures. The father must often follow certain rest patterns, work restrictions, and dietary practices. Among the Hopi, for example, a man is required to be careful not to hurt animals. If he does, his child may be born deformed. Among the Ifage in the Philippines, the husbands of pregnant women are prohibited from cutting or killing anything.

Perhaps the most startling to the Westerner are male imitations of childbirth. For instance, in many tribes throughout the world, a man wraps his arms around his belly and imitates his wife having labor contractions. After the birth, he may pretend he is entering a postpartum period. Among the Chaorti in South America, the man takes to his hammock for several days during labor and after the delivery.

The Huichol of Mexico traditionally practiced a ritual of couvade in which the husband squatted in the rafters of the house or the branches of a tree above his wife during labor. When the woman experienced a contraction, she pulled on ropes attached to his scrotum. In this way, the man shared the experience of childbirth.

The couvade is a dramatic symbol of the man's paternity and his "magical" relation to the child. By pretending he is pregnant, he distracts evil spirits from harming his baby. Describing the magical impact of the couvade, Arthur and Libby Colman (1971) write:

> The couvade phenomena have the important side effects of helping a husband play an important part in pregnancy and childbirth. . . . They help a man cope with the envy and competitiveness which he may feel at his wife's ability to perform such a fundamental and creative act. . . . In his activities to deceive the evil spirits, a man may also find a reasonable outlet for his own desire to take on something of the female role in life.

Some American men may experience what medical researchers call the "couvade syndrome." A study of the partners of 267 postpartum women found that 22.5 percent of the men experienced nausea, vomiting, anorexia, abdominal pain and bloating, and other symptoms of pregnancy that could not be objectively explained (Lipkin and Lamb, 1982). These men had four times more symptoms than they had had prior to their partners' pregnancies. Another study suggests that 11 to 65 percent of men with pregnant partners experience sympathetic pregnancies (White and Bulloch, 1980).

How the Husband Assists in the Birth of a Child by Guadalupe, widow of Ramón Medina Silva, after a yarn painting by Ramón. 23 1/2 x 23 1/2" (59.7 x 59.7 cm.). Yarn on plywood, beeswax. #74.21.12. Fine Arts Museums of San Francisco. Gift of Peter F. Young.

The Huichol people of Mexico traditionally practiced a form of couvade shown in this yarn painting by artist Guadalupe Medina.

when the father and other family members and friends are allowed to visit.

But the norm is still all too often the impersonal birth. During one of the most profound experiences of her life, a woman may have her baby among strangers to whom birth is merely business as usual. There are likely to be bright lights, loud noises, and people coming and going, moving her, poking at her, and asking questions when she's in the middle of a contraction. She may be given a routine enema, even though surgery is not anticipated. She and her unborn child are likely to be attached to various types of monitoring machines. Studies have shown that although fetal monitoring is helpful in high-risk cases, it is generally not helpful in normal (low-risk) situations. Generally, the baby's heart rate can be detected by stethoscope against the mother's abdomen.

Some form of anesthetic is administered during most hospital deliveries, as well as various hormones (to intensify the contractions and to shrink the uterus after delivery). The mother isn't the only recipient of the drugs, however; they go directly through the placenta to the baby, in whom they may reduce heart and respiration rates.

During delivery, the mother will probably be given an episiotomy—a surgical procedure to enlarge the vaginal opening by cutting through the perineum toward the anus. Although an episiotomy may be helpful if the infant is in distress, it is usually performed to reduce the risk of tearing, give the obstetrician more control over the birth, and speed up the delivery. Studies show that episiotomies are performed in about 80 percent of first vaginal births in hospitals (Hetherington, 1990; Klein et al., 1992); yet one midwife who has assisted at over 1200 births reports a rate of less than 1 percent (Armstrong and Feldman, 1990). A Canadian study of 703 uncomplicated births found no advantage to routine episiotomies; the disadvantages were pain and bleeding (Klein et al., 1992). The authors recommended that "liberal or routine use of episiotomy be abandoned."

The baby is usually delivered on a table, against the force of gravity. He or she may be pulled from the womb with a vacuum extractor (which has a small suction cup that fits onto the baby's head) or forceps. (In some cases of acute fetal distress, these instruments may be crucial in order to save the infant's life, but too often they are used by physicians as a substitute for patience and skill.) In most cultures, a woman gives birth while sitting in a birthing chair, kneeling, or squatting. Until the present century, most American women used birthing chairs; the delivery table was instituted for the convenience of the physician. A few hospitals use a motorized birthing chair that can be raised, lowered, or tilted according to the physician's and the woman's needs. In about one-fourth of births in the United States, the baby is not delivered vaginally but is surgically removed from the uterus.

Although only 5 to 10 percent of births actually require medical procedures, we seem to assume that childbirth is an inherently dangerous process. Of course, in high-risk cases, advances in technology can and do save lives. But the following questions remain: Why do women accept unnecessary, uncomfortable, demeaning, and even dangerous interventions in the birth process? Why do they tolerate a 24 percent cesarean rate, a 61 percent episiotomy rate, the almost universal administration of drugs, and the use of intrusive fetal monitoring—not to mention routine enemas? Why do they allow their infant boys to have their penises circumcised? Why do they accept, often without question, the physician's opinion over their own gut feelings?

Penny Armstrong and Sheryl Feldman (1990) suggest that society's increasing dependence on technology has hampered women's ability to view birth as a natural process for which they are naturally equipped. Women have allowed themselves to be persuaded that technology can do the job better than they can on their own.

Society expects birth technology to deliver a "product"—a "perfect" baby—without understanding that nature has already equipped women to deliver that product without much outside interference in the physical process (although encouragement and emotional support are paramount). Sheila Kitzinger (1989) writes:

> It is not advances in medicine but improved conditions, better food, and general health which have made childbirth much safer for mothers and babies today than it was 100 years ago. The rate of stillbirths and deaths in the first week of life is directly related to a country's gross national product and to the position of the mother in the social class.

The idea that the pain of childbirth is to be avoided at all costs is a relatively new one. When England's Queen Victoria accepted chloroform in 1853, she undoubtedly had little idea of the precedent she was setting. Today's advocates of prepared childbirth do not deny that there is pain involved; they argue, however, that it is a different pain from that of injury and that normally it is worth experiencing. This "pain with a purpose" is an intrinsic part of the birth process (Kitzinger, 1989). To obliterate it with drugs is to obliterate the mother's awareness and the baby's as well, depressing the child's breathing, heart rate, and general responsiveness in the process.

Another aspect of dependence on technology is that we get the feeling we are omnipotent and should be able to solve any problem. Thus, if something goes wrong with a birth—if a child is stillborn or has a disability, for example—we look around for something or someone to blame. We have become unwilling to accept that some aspects of life and death are beyond human control.

Prospective parents face a daunting array of decisions. The more informed they are, however, the better able they will be to decide what is right for them. According to Armstrong and Feldman (1990):

> Teaching women that they have a say—whether they consciously exercise it or not—is a major educational undertaking, one that requires breaking the hold obstetrical medicine has on the American imagination and helping women to rediscover their natural power at birth.

ADOPTIVE FAMILIES

It is difficult to say with certainty how frequent adoption is in the United States because of the absence of any dependable and comprehensive source from which data can be obtained (Grotevant and Kohler, in Lamb, 1999). The Center for Adoption Research and Policy estimated that more than a million children are currently in adoptive families, while more than five million adults and children have been adopted (Grotevant and Kohler, in Lamb, 1999).

Although adoption is being examined here as the traditionally acceptable alternative to pregnancy for infertile couples, it may also include the adoption of stepchildren in a remarriage, the adoption of a child by a relative, the adoption of adolescents, the adoption of two or more siblings, and the adoption of foster children who have been removed from their parental homes (Grotevant and Kohler, in Lamb, 1999). According to data collected in 1992 by the National Adoption Information Clearinghouse, 42 percent of adoptions were to stepparents or other relatives, and 16 percent were from foster care (http://naic.acf.hhs.gov/pubs).

In recent years, many Americans—married and single, child-free or with children—have been choosing to adopt for a variety of reasons beyond infertility. They may have concerns about overpopulation and the number of homeless children in the world. They may wish to provide families for older or disabled children. Although tens of thousands of parents and potential parents are currently waiting to adopt, there is a shortage of available healthy babies (especially healthy white babies) in this country. As a consequence of more effective birth control and an increase in the number of single mothers who choose to keep their children, there are fewer opportunities to adopt.

The costs of open adoption tend to run between $6,000 and $20,000, with some prospective parents paying up to $100,000 (Waldman and Caplan, 1994), depending on lawyer and agency fees and the birth mother's expenses (if the state allows these to be covered by the adoptive parents). Adoption laws vary widely from state to state; six states prohibit private adoption, whereas California and Texas have laws that are considered quite supportive of it.

With confidentiality no longer the norm, the trend is toward **open adoption** in which there is contact between the adoptive family and the birth parents (McRoy, Grotevant, and Ayers-Lopez, 1994). This involvement can be either mediated (through an adoption agency) or direct, where the birth mother and adoptive family have contact with each other. Many adoption experts agree that some form of open adoption is usually in the best interests of both the child and the birth parents.

One study of 720 adoptive families and birth mothers found that those participating in open adoptions reported more awareness of the adoption, an increased empathy toward the birth parents and child, a stronger feeling of permanence in the relationship with their child, and fewer fears that the birth mother might try to reclaim the child (Grotevant et al., 1994).

Grotevant and Kohler (in Lamb, 1999) suggest that, especially given the trend toward openness, there are five individual and family processes that are important elements in the experiences of adoptive families.

- *Acknowledgment of difference:* Unlike the past, when adoptive parents minimized or denied difference, sometimes even pretending to the child and outsiders that the child was biologically related to them, today there is a concern for *how much acknowledgment* is healthy. A curvilinear model, in which either too much acknowledgment of difference (e.g., introducing the child as "my adopted daughter") or too little can be problematic.

- *Compatibility:* Defining compatibility as a relationship in which parent and child communicate efficiently and accurately and are "meshed well," research suggests that *incompatibility* between parent and child (as perceived by parents) is a strong predictor of difficulties.
- *Control:* Adoption often results when adoptive parents as well as biological parents wrestle with having little or no control. Adoptive parents may have no control over their fertility, whereas biological parents may surrender their child because environmental conditions or circumstances were out of control or control was taken from them by courts. Although they may take measures to assert control (such as choosing the type of adoption or type of child, or choosing who ultimately adopts one's child), at various times they may have to negotiate around issues of the amount, timing, and frequency of contact between adoptive and biological parents.
- *Identity:* As articulated by Grotevant and Kohler, "Changes in adoption practice and policy have had significant bearing on the development of identity among all parties involved in adoption" (in Lamb, 1999). For the adopted person, identity issues translate into such questions as, How am I similar or different from my birth parents? How do we fit into each other's worlds? Further, there may be a need to integrate a pre-adoption history into one's identity, as well as confront differences (in such matters as appearance or ethnic or cultural background, for example) between oneself and one's adopted family.
- *Entitlement:* It is important for adoptive parents to claim the "emotional rights" that go along with the legal rights that the courts have bestowed upon them. In other words, they must come to see themselves as the child's full parent, and take responsibility for providing discipline and structure.

Several states have now enacted laws allowing adoptees to get copies of their original birth certificates. Other states have set up voluntary registries to assist adoptees and biological parents who wish to meet each other. Although many adoptees do not feel the need to search for their biological parents, others do. Those who succeed report a variety of outcomes. Many find "long-buried family problems that led to their being put up for adoption in the first place"; others need to figure out "what kind of relationship to have with a stranger who is also a parent" (Chira, 1993). Adoptees must also often deal with the bruised feelings of adoptive parents who wonder where they've "gone wrong."

Some professionals in the adoption field believe that it is not in the child's best interest to tell him or her of the adoption. Psychiatrist Dennis Donovan thinks that children should be given only the information they ask for (only if they ask) and no more. He believes that telling a child he or she was loved by the birth mother but was still given up blocks the child's ability to attach (cited in "At a Glance," 1990). Most adoption authorities, however, disagree with this view.

In addition to open adoptions, foreign adoptions are also increasingly favored. Approximately 15 percent of U.S. adoptions (about 8,000 per year) are of children born outside this country (Bogert, 1994). International adoption agencies are working to expand their programs in India, the Philippines, Romania, Colombia, and a number of other Asian and Latin American countries. For foreign adoptions, the waiting period is usually around a year, although it may be longer, depending on the political climate of the country in question. The costs vary widely, depending on the number and kind of agencies involved and whether or not the parents travel to pick up their child.

Families with children from other cultures face challenges in addition to those faced by other adoptive families. There is usually little information about the birth parents and no opportunity for continued contact. Older children from foreign countries must deal with the loss of their birth parents and other significant people. They must also adjust to different customs, strange food, and a baffling new language. To combat a sense of rootlessness, many parents of foreign-born children endeavor to give them an understanding of their birth country and its culture. They often participate in supportive networks with other adoptive families.

Adoptive families face unique problems and stresses. They may struggle with the physical and emotional strains of infertility; they endure uncertainty and disappointment as they wait for their child; and they may have spent all their savings and then some in the process. They often face insensitivity or prejudice. For example, an adopted child may be asked, "Who is your *real* mother?" or "Are you their *real* daughter?" Adoptive parents may be congratulated by well-meaning folks: "Oh, you're doing such a good thing!" as though they had made a sacrifice of some kind in choosing to build a family in this way. Even grandparents may reject adopted grandchildren (at least initially), especially if the adoption is interracial. The idea that adoption is not quite "natural" is all too common in our society.

Adopted children may feel uniquely loved. Suzanne Arms (1990) recounts, "When Joss was six, he was overheard explaining to a friend how special it was to be adopted. Apparently," she adds, "he made a good case for it, because when his friend got home, he told his mother he wanted to be adopted so he could be special too."

REFLECTIONS

Is the ability to create a child important to your sense of self-fulfillment? If you discovered that you were infertile, what do you think your responses would be? Would adoption be an option for you? Why or why not?

Becoming a Parent

The time immediately following birth is a critical period for family adjustment. No amount of reading, classes, and expert advice can prepare expectant parents for the real thing. The three months or so following childbirth (the "fourth trimester") constitute the **postpartum period.** This is a time of physical stabilization and emotional adjustment.

New mothers, who may well have lost most of their interest in sexual activity during the last weeks of pregnancy, will probably find themselves returning to prepregnancy levels of desire and coital frequency. Some women, however, may have difficulty reestablishing their sexual life because of fatigue, physiological problems such as continued vaginal bleeding, and worries about the infant (Reamy and White, 1987).

The postpartum period also may be a time of significant emotional upheaval. Even women who had easy and uneventful births may experience a period of "postpartum blues" characterized by alternating periods of crying, unpredictable mood changes, fatigue, irritability, and occasional mild confusion or lapses of memory. A woman may have irregular sleep patterns because of the needs of her newborn, the discomfort of childbirth, or the strangeness of the hospital environment. Some mothers may feel lonely, isolated from their familiar world. Many women blame themselves for their fluctuating moods. They may feel that they have lost control over their lives because of the dependency of their newborns.

Biological, psychological, and social factors are all involved in postpartum depression. Biologically, during the first several days following delivery, there is an abrupt fall in certain hormone levels. The physiological stress accompanying labor, dehydration, blood loss, and other physical factors contribute to lowering the woman's stamina. Psychologically, conflicts about her ability to mother, ambiguous feelings toward or rejection of her own mother, and communication problems with the infant or partner may contribute to the new mother's feelings of depression and helplessness. Finally, the social setting into which the child is born is important, especially if the infant represents a financial or emotional burden for the family. Postpartum counseling prior to discharge from the hospital can help couples gain perspective on their situation so they will know what to expect and can evaluate their resources.

Although the postpartum blues are felt by many women, they usually don't last more than a couple of weeks. Interestingly, men seem to get a form of postpartum blues as well. When infants arrive, many fathers do not feel prepared for their new parenting and financial responsibilities. Some men are overwhelmed by the changes that take place in their marital relationship. Fatherhood is a major transition for them, but their feelings are overlooked because most people turn their attention to the new mother.

The transition to parenthood can be made easier if the new parents understand in advance that a certain amount of fatigue and stress is inevitable. They need to ascertain what sources of support will be helpful to them, such as friends or family members who can help out with preparing meals or running errands. They also need to keep their lines of communication open—to let each other know when they are feeling overwhelmed or left out. It's also very important that they plan time to be together, alone or with the baby—even if it means telling a

Although becoming a parent is stressful, the role of mother or father is deeply fulfilling for many people.

© Laurie DeVault Photography

well-meaning relative or friend they need time to themselves.

For many women and men, the arrival of a child is one of life's most important events. It fills mothers and fathers alike with a deep sense of accomplishment. The experience itself is profound and totally involving. A father describes his wife (Kate) giving birth to their daughter (Colleen) (Armstrong and Feldman, 1990):

> Toward the end, Kate had her arms around my neck. I was soothing her, stroking her, and holding her. I felt so close. I even whispered to her that I wanted to make love to her—It wasn't that I would have or meant to—it's just that I felt that bound up with her.
>
> Colleen was born while Kate was hang-ing from my neck. . . . I looked down and saw Mimi's [the midwife's] hands appearing and then, it seemed like all at once, the baby was in them. I had tears streaming down my face. I was laughing and crying at the same time. . . . Mimi handed her to me with all the goop on her and I never even thought about it. She was so pink. She opened her eyes for the first time in her life right there in my arms. I thought she was the most beautiful thing I had ever seen. There was something about that, holding her just the way she was. . . . I never felt anything like that in my life.

TAKING ON PARENTAL ROLES AND RESPONSIBILITIES

Even more than marriage, parenthood signifies adulthood—the final irreversible end of youthful roles. A person can become an ex-spouse but never an ex-parent. The irrevocable nature of parenthood may make the first-time parent doubtful and apprehensive, especially during the pregnancy. Yet, for the most part, parenthood has to be learned experientially, although ideas can modify practices. A person may receive assistance from more experienced parents, but ultimately all new parents have to learn on their own.

The abrupt transition from a nonparent to a parent may create considerable stress. Parents take on parental roles literally overnight, and the job goes on without relief around the clock. Many parents express concern about their ability to meet all the responsibilities of child rearing.

There have been a number of important analyses of the transition to parenthood (Rossi, 1968; LaRossa and LaRossa, 1981; Cowan and Cowan, 1992). An early and influential analysis of what parents experience as they enter the new social reality of parenthood was offered by Alice Rossi (1968). According to Rossi, entering parenthood is stressful because of the nature of the role of parent and the characteristics of the parental role transition.

Perspective

A FATHER'S STORY

The following account describes one man's reactions to his entry into fatherhood. At the time, he was a 25 year-old food service worker, father of one son, 4 months old.

I guess I always knew I wanted to be a father someday, As far back as I can remember, from the time when I actually understood fathers and mothers, I always hoped I'd be. I wanted to be able to show a child the good points about people, the good virtues that they have, and offset a lot of what you see going on in the newspaper—people getting mugged, that sort of thing. I know I have ideals of the way people should act and should treat each other. I wanted those to be carried on. If I think about my own father—he didn't change us, or feed us, or anything like that. At that time most fathers didn't. But I change my son, feed him, bathe him. I don't know, I think I'm getting a lot more out of the relationship than most fathers are. I'm getting a lot out watching him grow up and develop little habits. I'm understanding a little more about why mothers feel about their children as strongly as they do, because of all the time and effort. Well, I was in the delivery room too. I really think that made a difference.

Before, there was no way you could feel as strongly, because you went to the hospital and waited for them to come out and tell you you had a baby. But I'd gone with Nancy, I had been trained in childbirth. Still, actually going there and being in the delivery room, in the labor room . . . . You'd never feel that something so small and so cute and so cuddly could be such great drama But it is. That's the only way I can describe it. It's almost like one of Shakespeare's plays—lots of pain, but it makes him a little more precious.

I guess when I thought about it beforehand, I thought that I would leave him with Nancy most of the time. If she had to go out, I could change a diaper once in a while, read him a book. But I never figured that I'd feel like I do about him and for his welfare. It's like an obsession in a way. I am constantly watching out for his head, making sure he doesn't fall, is he strapped in, did he eat enough, is the milk too hot, is the wind blowing on him, is this or that gonna wake him up . . . it's a good feeling. Probably crumbling the male macho image in a way, but I feel that's fine.

Sure, there is also a lot of frustration. There are a lot of times where you want to say, "Oh, geez, just go to sleep" or when you say, "Geez, hurry up and finish your bottle so I can watch the rest of my show or finish the rest of my book." But then I stop and think about how he's changing every day—he doesn't look now like he did when he came home. I just think, "Wow, that time is gone and I just have the memory, and after awhile that's gonna fade too." So I just try to enjoy him more day by day.

I think having him has calmed me down a lot more. When I see a sad movie on the t.v. I might get a lot closer to watering up than I ever would have. I'm surprised at the way I've totally changed on a few issues . . . I feel I've been completed in a way. It's almost like a little candle has been lit inside of me. It's there all the time. You can feel its warmth, you can use it for light if you want. It's very satisfying. It's like I have a focus in life. Before, everything I did really didn't matter much. But now, everything I do has a bearing on the future. It just feels like I'm filled with hope . . . I look much more at the better side of things right now. Where before I might have been a little lackadaisical, now I have a defined purpose. I used to stay up real late and watch the real old movies, maybe not go to bed until 3 or 4 A.M. I know I can't do that anymore. I might stay up until 1 A.M. but knowing in the back of my mind that he might get up as early as 4, I want to have the energy for him.

I'm getting used to saying to my son, "Where's Daddy?" It's becoming a part of my identity. I can almost see it happening, losing my identity as "Nancy's husband" to "the father of Nancy's child." I think especially once he gets to talking, we'll lose our identities and become "Daddy" and "Momma."

Rossi singled out the following features of entering parenthood:

- *Irreversibility:* Unlike nearly any other role you play, once you enter parenthood you cannot easily leave without incurring significant social or legal repercussions. Even "deadbeat parents," who have left their children and ceased to support them, are still considered responsible for their children's welfare.
- *Lack of preparation:* There is almost nowhere and no way to practice parenting. One can read books, attend classes, or baby-sit for children,

but these pale in comparison to the reality one faces upon having children. Furthermore, little systematic effort is made to equip people with more realistic understanding or even practical skills to more effectively parent.

- *Idealization and romanticization:* Related to the lack of preparation is the fact that the expectations we have about what parenthood will be like are often unrealistic and overly idealized. If and when reality turns out to be less than ideal, we become frustrated and disappointed.
- *Suddenness:* Despite what might be eight months of awareness of impending parenthood, the actual transition is sudden. There is no opportunity for expectant parents to ease into the role; one goes from nonparent to parent in the moment of childbirth, and assumes all of the role responsibilities with that same suddenness.
- *Role conflict:* The parental role affects all of the other roles a person plays, encroaching upon time spent with a spouse or partner, and complicating paid employment, especially for women.

Based on their research on new parents, Ralph and Maureen Mulligan LaRossa suggested that the major adjustment new parents face is *temporal.* Like a hospital or fire station, new parenthood is a **continuous coverage system;** infants must have someone available to care for them 24 hours a day, seven days a week. When direct care (*primary level of attention and accessibility*) is not needed (for example, during naps or nights when infants sleep), someone must at least be "on call" (*secondary level of attention and accessibility*), ready to move to more direct interaction and/or demanding attention. Finally, when at least two competent caregivers are present, one may move to a state of "downtime" (*tertiary level*), wherein one is free to pursue other activities without concern for the infant's needs. The LaRossas suggest that new parents (in a two-parent household) compete with each other and experience conflict over the ever-scarcer resource of downtime. Much of the LaRossas' analysis follows from that, looking at who gets more downtime (fathers), who does most of the primary parenting (mothers), and why. In answering the latter question, they note some of the wider cultural beliefs that value mother-care, as well as some of the relationship and individual-level factors that press mothers toward more of the work associated with children, and turn fathers into "helpers" and "playmates" (LaRossa and LaRossa, 1981).

More recently, Carolyn and Phillip Cowan identified five domains in which new parents experience change as a result of the arrival of children (1992, 2000):

- *Identity and inner-life changes:* New parents discover that they no longer think of themselves the same way they did before their children were born. Their priorities and personal values also change. Issues that previously seemed remote, unimportant, or abstract become very personal, meaningful, and real.
- *Shifts within the marital roles and relationship:* Parenthood alters how couples divide tasks or allocate responsibilities. Because they are also experiencing fatigue (from reduced sleep and more work), their relationship quality may diminish.
- *Shifts in intergenerational relationships:* Becoming parents alters—often improving and intensifying, sometimes straining—the relationship between new parents and *their* parents.
- *Changes in roles and relationships outside the family:* New parenthood, especially new motherhood, may force changes in other nonfamily roles and relationships, such as at work or in friendships. Although some of these changes may be temporary (for example, leaving work only for the length of a parental leave), they nonetheless compound other things to which new parents are adjusting.
- *New parenting roles and relationships:* New parenthood means that a couple must arrive at an agreeable division of child care. New parents learn how difficult it is to maintain equal and/or equitable divisions of child care. One parent may feel put upon or taken advantage of in the way the couple allocates their individual time and energy to child-care tasks.

The Cowans suggest that the difficulties associated with the parental transition are different and more difficult for contemporary parents because of some major features of the social climate in which they parent. First, contemporary parenthood is more discretionary or optional, making decisions about whether and when to have children subject to more discussion, negotiation, and potential dispute. Second, most new parents, especially middle-class parents, are relatively isolated, geographically, from their wider kin groups and other long-term social supports. Third, changes in women's roles have in-

© Myrleen Ferguson Cate/PhotoEdit

Becoming a parent introduces changes in intergenerational relationships.

troduced more role conflict for new mothers and have increased women's need and legitimate demand for more sharing by their partners. Fourth, the social policies that address the needs of parents are weak to nonexistent. Fifth, there are few enviable or attractive role models for effective parenting; *Leave-It-to-Beaver* families are unrealistic, and yet there is no equivalent cultural model of dual-earner parents to draw upon. If we cannot parent like our own parents did, who can we emulate? Sixth, today's families are supposed to fulfill all of our emotional needs. Parenting is stressful and requires mutual effort and sacrifice. But effort and sacrifice don't fit compatibly with individual emotional fulfillment. Thus, the difficulties may become sources of resentment and estrangement (Cowan and Cowan, 1992, 2000).

STRESSES OF NEW PARENTHOOD

Many of the stresses felt by new parents closely reflect gender roles. Overall, mothers seem to experience greater stress than fathers (Harriman, 1983). Although a couple may have an egalitarian marriage before the birth of the first child, the marriage usually becomes more traditional once a child is born. If the mother, in addition to the father, must continue to work, or if the woman is single, she will have a dual role as both homemaker and provider. She will also probably have the responsibility for finding adequate child care, and it will most likely be she who stays home to take care of a sick child. Multiple role demands are the greatest source of stress for mothers. In a *New York Times* survey, 83 percent of women acknowledged they (sometimes) felt torn between the demands of their job and wanting to spend more time with their family (Belkin, 1989).

There are various other sources of parental stress. In one study (Ventura, 1987), 64 percent of fathers described severe stress associated with their work. A number of mothers and fathers were concerned about not having enough money. Other sources of stress involve infant health and care, infant crying, interactions with the spouse (including sexual relations), interactions with other family members and friends, and general anxiety and depression (Harriman, 1983; McKim, 1987; Ventura, 1987; Wilkie and Ames, 1986).

Changes in marital quality and marital conflict were studied among a sample of white and African-American spouses as they transitioned to parenthood (Crehan, 1996). The results of this study showed a decline in marital happiness and more frequent conflicts among both white and African-American spouses. White parents also reported higher marital tension and a greater likelihood to become quiet and withdrawn after the birth of their child. This increase in avoidance behaviors may be due to the limited time and energy that new parents have to devote to conflict resolution.

DID YOU KNOW?

Most, though not all, mothers would choose parenthood again. Interviews with 2,000 mothers found that eight out of ten would have children if they had the chance to do it again (Genevie and Margolies, 1987).

Coping with Stress

Although the first year of child rearing is bound to be stressful, the couple experience less stress if they (1) have already developed a strong relationship, (2) are open in their communication, (3) have agreed on family planning, and (4) originally had a strong desire for the child. In spite of planning, the reality for most is that this is a very stressful time. Accepting this fact while developing time management skills, patience with oneself, and a sense of humor can be most beneficial. Psychiatrist and researcher Jerry M. Lewis and colleagues (Lewis, Owen, and Cox, 1989) stress the importance of "marital competence" in the "successful incorporation of the child into the family." Another important factor in maintaining marital quality during this time is the father's emotional and physical support (Tietjen and Bradley, 1985). The more involved the father is in the care of the baby, the easier it is on the recovering mother.

Ventura (1987) suggests that families can be assisted by improved health care and support in three areas. First, during the prenatal period, parents need help locating community support resources, such as La Leche League and child-care assistance, so that they can "restore their energy" during the postpartum period. Fathers need to be encouraged in their nurturing role. Employers need to restructure schedules, develop child-care programs, and institute parental leave policies.

Second, coordination of care is important. Parents need good communication and support from practitioners following the birth. They need "concrete explanations and demonstrations." Health practitioners, child-care providers, and community professionals need to coordinate their communication with each other.

Third, families need anticipatory care and help in problem solving. Teaching about development, childhood illnesses, and safety should be an ongoing part of clinic and child-care services. New parents should reach out to others with similar experiences and concerns for support and understanding.

REFLECTIONS

If you have children, did you plan to have them? What considerations led you to have them? What adjustments have you had to make? How did your relationship with your partner change?

Having a child is unlike any other experience we undertake. The changes in our lives are wide-ranging and irreversible, the potential rewards are great, and the sacrifices are many. Increasing numbers of women and couples are deciding to forgo parenthood, in large part to avoid its many and profound consequences, Most people, however, continue to decide to embark on the journey described in this chapter, and take on the challenging tasks that we look at in Chapter 11.

SUMMARY

- With increases in delayed parenthood and *child-free marriage*, parenthood may now be considered more a matter of choice.
- Women are guaranteed the constitutional right to abortion. About a fifth of all women of reproductive age have had abortions.

- The reasons for having an abortion are complex, including a womans developmental/life stage, relationships with others, educational goals, and economic circumstances. Most women suffer the greatest distress prior to the abortion. There is no scientific evidence supporting a so-called post-abortion trauma syndrome.
- A woman's feelings vary greatly during pregnancy. It is important for the woman to share her fears and to have support from her partner, relatives, friends, and possibly, women's groups. Her feelings about sexuality are likely to change during pregnancy. Men may also experience conflicting feelings. Sexual activity is generally safe during pregnancy unless there is a prior history of miscarriage, bleeding, or pain.
- About one out of four women is aware of having had a spontaneous abortion (miscarriage). Infant mortality rates in the United States are extremely high compared to those in other industrialized nations. Loss of pregnancy or death of a young infant is recognized as a serious life event.
- The medicalization of childbirth—making this natural process into a medical "problem"—has caused an overdependence on technology and an alienation of women from their bodies and feelings. Women can empower themselves by becoming informed about childbirth and their options.
- Adoptive families face unique problems and stresses; nevertheless, most report feeling greatly enriched. Issues faced by adoptive families include choosing open or closed adoption, dealing with feelings about the biological parents, and dealing with insensitivity and prejudice from society.
- Although their experiences are indirect and vicarious, men are affected by a partner's pregnancy. They experience new feelings about and during sex, find their dreams take on different qualities, and deal with a number of anxieties. Men's involvement in the pregnancy and birth process affect their later parenting.
- The transition to parenthood is unlike other role transitions. It is irreverisble and sudden, and it comes with little prior preparation.
- Parental roles are acquired virtually overnight and can create considerable stress. The main sources of stress for women involve traditional gender roles and multiple role demands (parent, spouse, and provider). Other sources of stress for mothers and fathers are associated with work; not having enough money; worries about infant care and health; and interactions with spouse, family, and friends

KEY TERMS

child-free marriage 320
continuous coverage system 336
couvade 329
crude birth rate 318
deferred parenthood 321
fertility rate 318
infant mortality 326
open adoption 331
postpartum period 333
spontaneous abortion 326
sudden infant death syndrome (SIDS) 327
total fertility rate 318

SUGGESTED READINGS

Armstrong, Penny, and Sheryl Feldman. *A Wise Birth.* New York: Morrow, 1990. Written with intelligence and warmth, a thought-provoking book exploring the effects of medical technology and technological thinking on modern childbirth.

Cowan, Carolyn Pape, and Philip Cowan. *When Partners Become Parents: The Big Life Change for Couples.* Mahwah, NJ: Lawrence Erlbaum, 2000. In this newly released edition of their book (originally published by Basic Books in 1992), Carolyn and Philip Cowan detail the life-altering impact couples face when they have children and become parents.

LaRossa, Ralph, and Maureen Mulligan LaRossa. *Transition to Parenthood: How Infants Change Families.* Beverly Hills, CA: Sage, 1981. Ralph and Maureen LaRossa use interview data with couples who are new parents to show how the birth of children changes couple relationships by reducing the amount of free time. Using conflict, interactionist, and exchange theories, they look at how new parents strategically negotiate and construct divisions of child care.

Marsiglio, William. *Procreative Man.* New York: New York University Press, 1998. Marsiglio reviews some of men's experiences around contraception, pregnancy, and reproduction issues. These materials nicely complement the growing literature on men as fathers.

McMahon, Martha. *Engineering Motherhood: Identity and Self-Transformation in Women's Lives.* New York: Guilford, 1996. An insightful book that examines the process by which women become mothers and the meaning that motherhood brings to them.

Nilsson, Lennart, and Lars Hamberger. *A Child Is Born.* New York: DPT/Seymour Lawrence, 1993. The story of birth, beginning with fertilization, told in stunning photographs and text.

RESOURCES ON THE INTERNET

Companion Web Site for This Book

http://sociology.wadsworth.com/strong/marriage9e
Gain an even better grasp on this chapter by going to the companion Web site to take one of the Tutorial Quizzes, use the Flash Cards to master key terms, or check out the many other study aids you'll find there. Visit the Marriage and Family Resource Center on the site. You'll also find special features such as GSS Data and Census 2000 information that will put data and resources at your fingertips to help you with that special project or to do some research on your own.

InfoTrac College Edition: Search Word Summary

ectopic pregnancy	low birthweight
infant mortality	labor
couvade	circumcision

To learn more about these central topics in the study of the family, you can conduct an electronic search using InfoTrac College Edition. To aid in your search and to gain useful tips, see the Student Guide to InfoTrac College Edition that you can access through the companion Web site for this book.

Preview

To gain a sense of what you already know about the material covered in this chapter, answer "True" or "False" to the statements below.

1. A maternal instinct has been proved to exist in humans. True or false?
2. There is a larger pay gap between mothers and women without children than between women and men. True or false?
3. Egalitarian marriages usually remain so after the birth of the first child. True or false?
4. The behavior of fathers has changed more than the cultural beliefs about fatherhood. True or false?
5. Studies consistently show that regular day care by nonfamily members is detrimental to intellectual and social development. True or false?
6. Children of higher-earning families are less likely to be cared for by parents only, or by other relatives, and are more likely to be cared for by nonrelatives. True or false?
7. Children who are raised by authoritarian parents tend to be less cheerful, more moody, and more vulnerable to stress. True or false?
8. Children of gay or lesbian parents are likely to be gay themselves. True or false?
9. In situations such as a parent's serious illness or death, or a parental divorce, children may become caregivers for their parents. True or false?
10. Many parents follow the advice of "experts" even though it conflicts with their own opinions, ideas, or beliefs. True or false?

"Baltimore Orioles shortstop Cal Ripken became a baseball immortal in 1995 just for being at the ballpark, day in and day out, attending more consecutive games (2131) than any other player in history. This feat prompted a note from a friend. 'To put Cal Ripken's recent achievement in perspective,' she wrote, 'we should all be congratulated for the number of days we have worked without a break, raising our children. I've been at it 3,316 days with Michael. You've been at it even longer with James. I think we also deserve a standing ovation in Camden Yards. . . . That will be the day.'"

ANN CRITTENDEN

CHAPTER 11

Experiencing Parenthood: Roles and Relationships of Parents and Children

Outline

What is it like to be a parent and to raise children? Consider these few observations. Journalist and novelist Anna Quindlen expresses how deep and broad her responsibility for her children is in the following way:

> *I am aghast to find myself in such a position of power over two other people (her sons). Their father and I have them in thrall simply by having produced them. We have the power to make them feel good or bad about themselves, which is the greatest power in the world. Ours will not be the only influence, but it is the earliest, the most ubiquitous, and potentially the most pernicious. Lovers and friends may make them blossom and bleed, but they move on to other lovers and friends. We are the only parents they will ever have.*

Economist, Sylvia Ann Hewlett, adds this:

> *Responsible parenthood involves the expenditure of a great deal of energy and effort. Done properly it is a noisy, exhausting, joyous business that uses up a chunk of one's best energy and taps into prime time. Well developing children dramatically limit personal freedom and seriously interfere with the pursuit of an ambitious career. When psychiatrist David Guttmann talks about the "'routine unexamined heroism of parenting,'" he is describing the manifold ways dedicated parents "'surrender their own claims to personal omnipotentiality'" in the wake of childbirth, conceding these instead to the new child" (Hewlett, 1992, p. 122).*

Finally, there is this comment from an interview one of the authors did with a 33-year-old father of one: "There's no way a nonparent can feel like a parent. It's unbelievable."

It is the objective of this chapter to examine some of the many aspects of being a parent and caring for children. Although the focus is largely on the relationships between parents and their pre-adult children, ultimately, we will also look at caring for adult children, children caring for parents, and the relationships between grandparents and grandchildren.

PREVIEW ANSWERS

1 False
2 True
3 False
4 False
5 False
6 True
7 True
8 False
9 True
10 True

Being Parents

Over the last four decades or so, major changes in society have profoundly influenced parental roles. Parents today cannot necessarily look to their own parents as models. Of course, the mothers and fathers of today's children have some things in common with mothers and fathers throughout history, such as the desire for their children's well-being. But in some areas they must chart a new course.

MOTHERHOOD

Many women see motherhood as their destiny. Given the choice of becoming mothers or not (made possible through birth control and abortion), most women would probably choose to become mothers at some point in their lives, and they would make this choice for very positive reasons (see Cook et al., 1982; Gallup and Newport, 1990; Genevie and Margolies, 1987). But many women make no conscious choice; they become mothers without weighing their decision or considering its effect on their own lives and the lives of their chil-

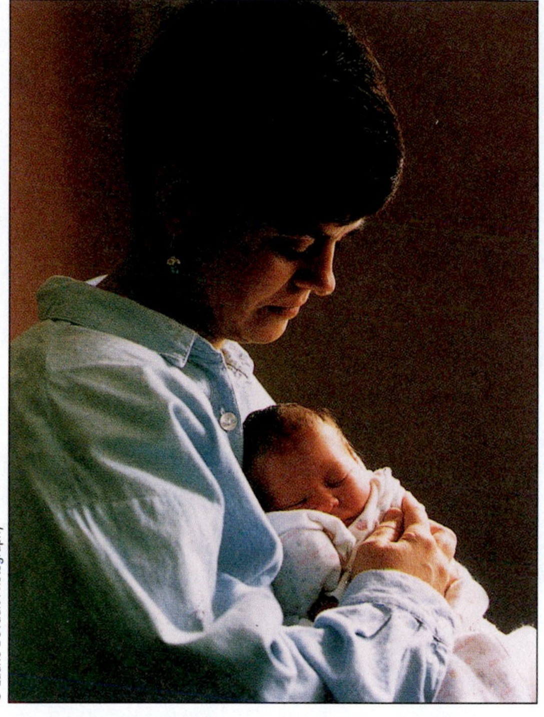

© Laurie DeVault Photography

Women become mothers for many reasons. Most mothers report overall satisfaction with their role.

dren and partners. The consequences of a nonreflective decision—bitterness, frustration, anger, or depression—may be great. Yet it is also possible that a woman's nonreflective decision will turn out to be "right" and that she will experience unique personal fulfillment as a result.

Although researchers are unable to find any instinctual motivation for having children among humans, they recognize many social motives impelling women to become mothers. When a woman becomes a mother, she may feel that her identity as an adult is confirmed. Having a child of her own proves her womanliness because from her earliest years, she has been trained to assume the role of mother. She has changed dolls' diapers and pretended to feed them, practicing infant care. She played house while her brother built forts. The stories a girl has heard, the games she has played, the textbooks she has read, the religion she has been taught, the television she has watched—all have socialized her for the mother role. Jessie Bernard (1982), a pioneer in family studies, writes, "An inbred desire is no less potent than an instinctive one. The pain and anguish resulting from deprivation of an acquired desire for children are as real as the pain and anguish resulting from an instinctive one." Whatever the reason, most women choose motherhood.

Still, many mothers feel ambivalence. Liz Koch (1987) writes:

> We fear we will lose ourselves if we stay with our infants. We resist surrendering even to our newborns for fear of being swallowed up. We hear and accept both the conflicting advice that bonding with our babies is vital, and the opposite undermining message that to be a good mother, we must get away as soon and as often as possible. We hear that if we mother our own babies full time, we will have nothing to offer society, our husbands, ourselves, even our children. We fear isolation, lack of self-esteem, feelings of entrapment, of emotional and financial dependency. We fear that we will be left behind—empty arms, empty home, empty women, when our children grow away. . . . The reality is that in many ways contemporary America does not honor Mothering.

Koch observes that the "job" of mother is not valued because it is associated with "menial tasks of housekeeper, cook, laundrymaid," and so on. Whether a mother is employed outside the home or works at home "full-time," her role deserves to be valued by society. Koch describes the "special state" of motherhood:

> Being mothers is truly immersing ourselves in a special state, a moment to moment state of being. It is difficult to look at our day and measure success quantitatively. The day is successful when we have shared moments, built special threads of communication, looked deeply into our children's eyes and felt our hearts open. . . . It is important that we see our job as vitally important to our own growth, to our community, to society, and to world peace. Building family ties, helping healthy, loved children grow to maturity is a worthwhile pursuit. . . . The transmission of values is a significant reason to raise our own children. We are there to answer their questions and to show children, through our example, what is truly important to us.

The expectation that mothering comes naturally can be frustrating and guilt-producing among many who are struggling with the new roles and

Perspective

PREPARING FOR PARENTHOOD, WITH HUMOR

As the previous chapter suggested and this chapter will show, becoming a parent and undertaking the roles involved in parenthood are difficult experiences to anticipate and prepare for. Unlike getting married—a process that unfolds gradually as two people form, maintain, and intensify a coupled relationship even before actually entering marriage—parenthood is more sudden. There is less opportunity to experience "being a parent" prior to actually having a child to care for. Most parents discover the extent to which they were either unprepared or incorrectly prepared for what awaits them as new parents. We present below an anonymously written preparation-for-parenthood test, circulated and posted widely on the Internet.

> Preparation for parenthood is not just a matter of reading books and decorating the nursery. Here are 12 simple tests for expectant parents to take to prepare themselves for the real-life experience of being a mother or father.

1. ***Women:*** To prepare for maternity, put on a dressing gown and stick a beanbag chair down the front. Leave it there for 9 months. After 9 months, take out 10 percent of the beans.
Men: To prepare for paternity, go to the local drug store, tip the contents of your wallet on the counter, and tell the pharmacist to help himself. Then go to the supermarket. Arrange to have your salary paid directly to their head office. Go home. Pick up the paper. Read it for the last time.
2. Before you finally go ahead and have children, find a couple who are already parents and berate them about their methods of discipline, lack of patience, appallingly low tolerance levels, and how they have allowed their children to run riot. Suggest ways in which they might improve their child's sleeping habits, toilet training, table manners and overall behavior. Enjoy it—it'll be the last time in your life that you will have all the answers.
3. To discover how the nights will feel, walk around the living room from 5 PM to 10 PM carrying a wet bag weighing approximately 8–12 lbs. At 10 PM put the bag down, set the alarm for midnight, and go to sleep. Get up at 12 and walk around the living room again, with the bag, till 1 AM. Put the alarm on for 3 AM. As you can't get back to sleep get up at 2 AM and make a drink. Go to bed at 2:45 AM. Get up again at 3 AM when the alarm goes off. Sing songs in the dark until 4 AM. Put the alarm on for 5 AM. Get up. Make breakfast. Keep this up for 5 years. Look cheerful.
4. Can you stand the mess children make? To find out, smear peanut butter onto the sofa and jam onto the curtains. Hide a fish stick behind

responsibilities that motherhood brings. Add to this socialization, the lack of confidence by both parties in a father's ability to parent, and the inherent ability of women to breastfeed and one can quickly see the enormous pressures that can face new mothers. Compounding the situation are our cultural expectations and standards for mothering, what Sharon Hays refers to as the *ideology of intensive mothering.* This ideology portrays mothers but not fathers as essential caregivers, and it pushes women to perceive child rearing as child centered, expert guided, emotionally absorbing, labor intensive, and financially expensive. As a result, mothers "see the child as innocent, pure, and beyond market pricing, they put the child's needs first, and they invest much of their time, labor, emotion, intellect, and money in their children" (Hays, 1996). In today's cultural climate, this version of motherhood stands in contrast to the ideology of the market, which is based on utilitarian emphases on such characteristics as efficiency, rationality, time saving, and profit.

The **intensive mothering ideology** confronts mothers as well as women who contemplate motherhood with cultural contradictions. Living up to its standards is difficult even for stay-at-home mothers. For women employed outside the home, the ideology can provoke self-doubt, guilt, and a sense of being judged by others. In fact, as Hays notes, there is almost no woman who can resolve the cultural no-win situation women face. Women who forgo childbearing may be perceived as "cold" and "unfulfilled as a woman." An employed woman with children may be told she is selfishly neglecting her children. If she scales back but stays in a job, she may be "mommy tracked," put in a less demanding but also less important and less upwardly mobile position. Finally, at-home mothers, in fulfilling many of the intensive mothering mandates, will be seen by some as "useless" or "unproductive" (Hays, 1996).

Because the strains of parenthood seem to fall more heavily upon women, guilt, depression, and conflict with other children are often the result. Pathe stereo and leave it there all summer. Stick your fingers in the flowerbeds then rub them on the clean walls. Cover the stains with crayons. How does that look?

5 Dressing small children is not as easy as it seems: first buy an octopus and a string bag. Attempt to put the octopus into the string bag so that none of the arms hang out. Time allowed for this: all morning.

6 Take an egg carton. Using a pair of scissors and a pot of paint, turn it into an alligator. Now take a toilet-paper tube. Using only scotch tape and a piece of foil, turn it into a Christmas candle. Last, take a milk container, a ping pong ball, and an empty package of Cocoa Pops and make an exact replica of the Eiffel Tower. Congratulations. You have just qualified for a place on the playgroup committee.

7 Forget the Miata and buy a Taurus. And don't think you can leave it out in the driveway spotless and shining. Family cars don't look like that. Buy a chocolate ice cream bar and put it in the glove compartment. Leave it there. Get a quarter. Stick it in the cassette player. Take a family-size packet of chocolate cookies. Mash them down the back seats. Run a garden rake along both sides of the car. There. Perfect.

8 Get ready to go out. Wait outside the toilet for half an hour. Go out the front door. Come in again. Go out. Come back in. Go out again. Walk down the front path. Walk back up it. Walk down it again. Walk very slowly down the road for 5 minutes. Stop to inspect minutely every cigarette butt, piece of used chewing gum, dirty tissue and dead insect along the way. Retrace your steps. Scream that you've had as much as you can stand, until the neighbors come out and stare at you. Give up and go back into the house. You are now just about ready to try taking a small child for a walk.

9 Always repeat everything you say at least five times.

10 Go to your local supermarket. Take with you the nearest thing you can find to a pre-school child—a fully grown goat is excellent. If you intend to have more than one child, take more than one goat. Buy your week's groceries without letting the goats out of your sight. Pay for everything the goats eat or destroy. Until you can easily accomplish this do not even contemplate having children.

11 Hollow out a melon. Make a small hole in the side. Suspend it from the ceiling and swing it from side to side. Now get a bowl of soggy Wheaties and attempt to spoon it into the swaying melon by pretending to be an airplane. Continue until half the Wheaties are gone. Tip the rest into your lap, making sure that a lot of it falls on the floor. You are now ready to feed a 12-month-old baby.

12 Learn the names of every character from *Barney and Friends*, *Sesame Street*, and *The Power Rangers*. When you find yourself singing Barney's theme song, "I love you. You love me . . ." at work, you finally qualify as a parent.

Now do you feel ready?

tience with oneself, support from family and friends, and a partner who takes equal responsibility in child rearing can go a long way toward alleviating the stresses and bringing greater joy to motherhood (Levy-Schiff, 1994).

FATHERHOOD

Beginning in the mid 1990s, a number of books appeared on fathers and fatherhood (Blankenhorn, 1995; Coltrane, 1996; Gerson, 1993; Popenoe, 1996; Hawkins and Dollahite, 1997). The depth of male involvement or absence in the lives of their children is a source of considerable, ongoing societal concern. In all the commentary and analysis, however, we find something short of a consensus about the state of fatherhood in America. This is even evident in the ambiguity of the idea of *fathering.*

When we speak of *mothering* a child, everyone knows what we mean: a process that involves nurturing, caring for, feeding, diapering, soothing, loving. Mothers generally "mother" their children almost every day of the year for at least 18 consecutive years. The meaning of *fathering* is quite different. Fathering a child need take no more than a few minutes if we understand the term in its traditional sense—that is, impregnating the child's mother. Nurturing behavior by a father toward his child has not typically been referred to as *fathering.* (*Mothering* doesn't seem appropriate, either, in this context.) *Parenting,* a relatively new word, more adequately describes the child-tending behaviors of both mothers and fathers (Atkinson and Blackwelder, 1993).

As we have seen, the father's traditional roles of provider and protector are instrumental; they satisfy the family's economic and physical needs. The mother's role in the traditional model is expressive; she gives emotional and psychological support to her family. However, the lines between these roles are becoming increasingly blurred because of economic

Perspective

THE COSTS OF MOTHERHOOD

Although the work associated with raising one's children may be the most meaningful and ultimately fulfilling work one can do, as writer and journalist Ann Crittenden (2001) notes, it is seriously *undervalued* in the United States. As a result, mothers pay a "price" that punishes them socially and economically, just for caring for their children.

Crittenden shows that women suffer a steep economic price for having invested themselves in raising their children. Among the more extreme aspects of the "price of motherhood" is her assessment that a typical, college-educated, mother in America loses around a million dollars of lifetime earnings as a result of having had and raising a child. How? There are a variety of interconnected issues; they may have to forgo, for at least a time, some income they could have earned. They receive no Social Security for time in which they are not "employed," but are, instead, caring for and raising their children. They also cannot count on any other pensions to assist them in their "retirement years," and—if they get divorced—cannot expect their contributions and sacrifices to count in their favor.

Crittenden makes the following additional points:

- America's 30-year-old childless women earn 90 percent of men's wages, but earnings for mothers of the same age and education level are only 70 percent of men's.
- The pay gap between mothers and women without children is actually larger than the gap between women and men.
- Some of this "parental pay gap" results from the fact that many more mothers than non-mothers are employed part-time, suffering not only lower wages but also a lack of benefits.
- The loss of income resulting from motherhood ("The Mommy Tax") is typically more than $1 million for college-educated American women.
- The United States is one of only six countries in the world that does not provide paid parental leave.
- More than one-third of all divorced mothers have to go on welfare because child-support formulas don't factor in the cost of being the primary caregiver.
- Fathers are statistically less likely than mothers to spend money on their children's health and education.
- Men pay too. Fathers who are in dual career families earn almost 20 percent less than fathers with at-home wives, even though they work only two fewer hours per week.
- Although many mothers feel as though employer policies force them onto shaky economic ground, only eight states have laws protecting them from discrimination in the workplace.
- Although these many aspects of "the mommy tax" are significant, Crittenden concludes that the "ultimate" price women pay is to not have children at all. A striking gender gap surfaced in a survey of 1,600 MBAs: although 70 percent of the males had children, only about 20 percent of the females did.

There is a potential danger in highlighting all of the above. One might conclude that, given the ways women are devalued and financially "punished" when they become mothers, perhaps women ought to rethink the desirability of becoming mothers. Instead of arriving at that interpretation, we should examine what changes might be made to lessen the "price of motherhood." Crittenden makes more than a dozen recommendations for needed changes. We present here a selection of just some of those:

- Extend parental leave to a full year with pay and implement more generous leaves for parents when their children are sick.
- Shorten the workweek, reversing the trend toward longer workweeks; at present Europeans work the equivalent of nine fewer 40-hour weeks per year than Americans.
- Pay part-time employees at the same rate as full-timers and offer them the same, but prorated, benefits (e.g., health insurance, pension plans, vacations, and so on).
- Redefine our ideas of who we count as being in the labor force to include primary caregivers, who should then be eligible for benefits such as unemployment, job training, and workers' compensation.
- Provide universal preschool for all three- and four-year-olds.
- Pay all primary caregivers, whether they are employed or not, a "child allowance."
- Enact divorce policies that would neither penalize mothers and children nor unduly reward either parent. Crittenden recommends that the federal government implement policies to achieve post-divorce "income sharing."
- Provide community support for parents and parental education.

Noting that it is unlikely that all of these recommendations would be implemented, Crittenden nonetheless believes that whatever might be put into practice would move us away from the punishing and unfair ways that mothers have been made to suffer.

pressures and new societal expectations and desires. From a developmental viewpoint, the father's importance to the family derives not only from his role as a representative of society, connecting his family and his culture, but also from his role as a developer of self-control and autonomy in his children. Research indicates that although mothers are inclined to view both sons and daughters as "simply children" and to apply similar standards to both sexes, fathers tend to be involved differently with their male and female children. Fathers tend to be more closely involved with their sons than their daughters (Morgan, Lye, and Condran, 1988; Smith and Morgan, 1994). This involvement generally involves sharing activities rather than sharing feelings or confidences (Cancian, 1989; Starrels, 1994). This may place a daughter at a disadvantage because she has less opportunity to develop instrumental attitudes and behaviors. It may also be disadvantageous to a son, as it can limit the development of his own expressive patterns and interests (Gilbert et al., 1982; Starrels, 1994).

In analyzing today's families, we find a diversity of opinion and a range of experiences of fathering. Some commentators point proudly to our embracement of the "new father" model, against which many men now measure themselves (Lamb, 1986, 1993). Feminist ideology is credited with being largely responsible for the shift in emphasis to a more expressive model of fathering, but many men pursue more involved versions of fatherhood as part of their own quest for deeper relationships with their children (Griswold, 1993; Daly, 1993). Most men today compare themselves favorably with their own fathers in both the quality and quantity of involvement they have with their children.

The new "nurturant father," as Michael Lamb (1997) refers to him, is able to participate in virtually all parenting practices (except, of course, gestation and lactation) and experience all of the emotional states that mothers experience. Psychologists Martin Greenberg and Norris offered the term **engrossment** to refer to a deep and broad bond many men form with their newborn and infant children, in which a father becomes totally absorbed and preoccupied—obsessed really—with his infant (Greenberg and Norris, 1974, cited in Doyle, 1995). Although researchers have questioned what triggers engrossment, and whether it has any innate aspect to it, it is clear that fathers can feel a connection to their infants that men were often thought to lack (Doyle, 1995). Furthermore, *father involvement* has been reconceptualized to include all of the many ways fathers are influential participants in their children's development. Men's more complete participation in the range of fathering activities (which include, for example, communicating, teaching, caregiving, protecting, sharing affection, and so on) is viewed as beneficial to the development and well-being of both children and adults (Palkovitz, 1997; Hawkins and Dollahite, 1997).

Although this new standard of fatherhood has been widely hailed, it is unclear how much it reflects actual fathering behavior (LaRossa, 1988; Gerson, 1993). As described by Ralph LaRossa, the **culture of fatherhood** has clearly changed in the direction described above; it is less clear and more debatable as to how much the **conduct of fatherhood** has kept pace. Mothers still do more of the hands-on, labor-intensive parenting. This distinction is important because it reminds us that reality may be quite different from rhetoric when it comes to what people do or believe they ought to do in their families.

Some observers are more critical of today's "new fathers." More traditional critics question the efficacy and desirability of a fatherhood that becomes

Fathers are increasingly involved in parenting roles—not just playing with their children, but changing their diapers, bathing, dressing, feeding, and comforting them.

THE RIGHTS OF BIOLOGICAL PARENTS VERSUS SOCIAL PARENTS: THE NUER OF THE SUDAN

Courtroom custody battles, replayed on the front pages of our newspapers and featured on the evening news, have brought the issue of parental rights into our living rooms. The very definition of parent has been brought into question. Which parents have a greater claim to a child? Is it the biological parents, who have contributed to the very genetic makeup of the child? Do they have superior claim to the child because they are its "natural" parents? What about the "social" parents, who through adoption, stepparenting, or fosterage have created a loving and enduring family for the child? Are they "unnatural" parents with only secondary rights because they do not share genes with the child? These are not only complex legal and social issues but also agonizing ones that pit "parents" by different definitions against one another.

A different light can be brought to these issues by looking at the solutions to these problems in other societies. For anthropologists, an especially illuminating case is provided by the Nuer of the Sudan in East Africa, a Nilotic (Nile basin) people whose livelihood has traditionally depended on pastoralism (Evans-Pritchard, 1951). The Nuer herd cattle, and the most important relationships in their society—as in other pastoralist societies—are created by the exchange of animals.

For the Nuer, even marriages are created by cattle. This has, as we will see, important implications for the rights of particular parents. In the arrangement of Nuer marriages, the elder males of the two different families engage in lengthy negotiations. They must decide how many cattle the bride's family will receive from the groom's family, as well as the timing of the cattle "installments." At marriage, the bride will leave her parents' home and move to the groom's family home. The marriage cattle move in precisely the other direction, from the groom's home to the bride's.

The transfer of cattle at marriage establishes a husband's inalienable rights to the children his wife will bear, regardless of who is the biological father. Take, for example, the case of adultery in Nuer society. Adultery is not socially condoned. It causes conflict between spouses and makes a wife's lover liable for expensive compensation payments. If it is established that a man has committed adultery with another man's wife, he becomes obligated to pay the wronged husband cattle as compensation. But there will be no contest over children born of an adulterous union. They belong unquestionably to the wife's husband. In essence, the Nuer distinguish genitor, or biological father, from pater, or legal and social father. The cattle paid at marriage establish a husband's rights in the children of his wife, regardless of who fathers them. Because she is a

too much like motherhood. They remain proponents of more traditional models of men as fathers (Blankenhorn, 1995). Still others focus more narrowly on the behaviors of irresponsible fathers, especially those who, after divorce, neither support nor maintain contact with their children. Instead, they simply disappear. Other negative expressions of fathering can be seen in data on child abuse. Although it is mothers who actually more often abuse their children, much of this phenomenon results from the fact that they spend so many more hours with them than fathers do. When one controls for mothers' and fathers' different levels of responsibility and involvement in child care, males are more often physically abusive to their children (Gelles, 1998). These negative aspects of fathering seem to contradict the cultural celebration of the new, more nurturant father.

One way to resolve the apparent contradiction between positive and negative depictions of fathers is to recognize the two sides or faces of contemporary fatherhood. This is what Frank Furstenberg Jr. (1988) had in mind when differentiating between "good dads" and "bad dads." This *bifurcation of fatherhood* results from the declining division of labor in the family, especially the decline of the male *good-provider role.* Rejecting this narrower notion of men's responsibilities "freed" some men from their sense of duty toward their spouses and children (especially children of ex-spouses) at the same time that it liberated others to construct expanded, more expressive versions of parenting.

Like both LaRossa and Furstenberg, sociologist Kathleen Gerson (1993) has looked at men's commitments to fatherhood and the models of fatherhood to which they feel committed. In her interview married woman, a woman's children are her husband's, inheriting membership in his clan and rights in his cattle at death. The contribution of the biological father is not unacknowledged, however. For example, upon the marriage of his biological daughter, the biological father receives one or two head of the many marriage cattle that are received by the bride's legal father. The legal father receives most of the cattle because he is pater.

Several interesting cultural practices further underscore the distinction the Nuer make between genitor and pater, as well as the superior claim of the pater over children in this society. Divorce does occur in Nuer society, and when it does take place, the marriage cattle must be returned to the husband's family. (This can be difficult if the couple have been married several years and many of the cattle have been parceled out to the bride's various kin.) Once the cattle are returned, there is no particular stigma attached to remarriage, and a divorced woman will usually marry again. If, however, two or more children have been born to a married couple, divorce will probably not occur. Instead, the unhappy couple will separate and the wife will set up house with a lover in a different village. The children born to a wife and her lover will belong not to the genitor-lover but to the separated husband, who is their pater. Although these children may be raised by their mother's lover—who then becomes their foster father—at adolescence they will return to their pater's home to herd his cattle and marry.

Because this is a patrilineal society, it is important that a man have a son to carry on his name and descent line. If a man dies before marriage and therefore dies without an heir, a spirit marriage will be arranged to remedy the situation. In this case, a brother will use some of the cattle belonging to his deceased brother in order to marry a wife for him. The brother will father "his brother's" children. The genitor will become foster father to his own biological children. These offspring are considered the legal children of the deceased brother by whose cattle the wife was acquired. In other words, the children of the spirit marriage belong to their deceased pater. If male, they inherit cattle from him; if female, they are married with his cattle. The foster father, who is actually the biological father, has no rights to his biological children by that wife. His own legal children and heirs will be children born to his own wife, who is acquired with his own cattle.

The unquestioned claim of a husband to all children born to his wife is further illustrated in the case where a man dies, leaving behind a widow. She may choose to live with one of her deceased husband's brothers or leave her husband's family completely and set up housekeeping with a lover of her choosing. In either event, the payment of cattle by the deceased husband created rights in her children—even those unborn at the time of his death. All children born to this wife will belong not to their genitor or biological father but to their legal father, who, it is believed in Nuer society, is the true and rightful father.

study of 138 fathers, Gerson identified an interesting diversity in men's perceptions of their family roles. Roughly a third of her sample was *traditional,* identifying themselves largely in terms of their jobs and allocating their time largely to paid work and career development. They saw their parental responsibilities largely in terms of providing financially for their families. Another near-third developed primary commitment to their family roles, especially their roles as fathers, which they conceptualized in deeply nurturant ways. The final third avoided involvement in child rearing because of how it would impose on their freedom and autonomy. They either had no children or were estranged from their children by virtue of having divorced or separated from their children's mothers.

As the above examples show, it is difficult to generalize about today's fathers. Although more of today's fathers may aim to be more broadly involved with their children than what they perceive fathers to have been in the past, and while most may recognize father involvement as beneficial, many are confused. They are unsure of what is expected of them. Women often can't understand why men don't automatically know what to do. Such stresses between mothers and fathers are common, according to a study by the Families and Work Institute (Levine, 1997; Martin, 1993). Although men are often willing to "help out" their wives, this can pose a problem (as we will discuss at greater length in Chapter 12). Although women generally appreciate any help they can get, they often wish their partners would take on an equal share of the work, rather than simply "helping." Fathers still see their roles as breadwinners as making important contributions because doing so provides financial resources for the family (Cohen, 1993). Though not as involved as their wives, most are also still very emotionally involved with their children.

Most mothers have acquired some of their parenting skills by modeling their own mothers, but many fathers have not had the advantage of such a parenting model. Because their own fathers were not highly involved in nurturing roles, today's fathers tend "to focus on being a model to their children to create for them a new set of standards for *who the father is*" (Daly, 1993). The creation of a new role is understandably accompanied by doubt and anxiety. Recently, a number of fathers have written books to help guide their peers through the joys and perils of involved fatherhood. Based on his intensive study, Kyle Pruett (1987) contends that "fatherhood is changing, with fastball speed, especially compared with the languid pace of social evolution." He concludes that fathers must be encouraged to develop the nurturing quality in themselves in order to experience the "unimagined rewards" of parenthood. Psychologist Jerrold Shapiro (1993), who writes about fatherhood, says, "Whether men have been enticed or cajoled, the fact is that we're around our kids a lot more. And when you're around your kids, you get to like it."

REFLECTIONS

How should child-rearing tasks be delegated between spouses (or partners)? Are there any particular tasks that you believe either men or women should not do? How are tasks delegated in your household? What was the role of your father in the care and nurturing of you and your siblings?

Who Actually Takes Care of the Children?

Child-care responsibility varies according to the marital status of parents and their employment roles and schedules. In a two-parent family, care for children is much more the responsibility of mothers than fathers. When one examines data on actual involvement in tasks associated with child care, mothers are much more involved in such tasks than fathers. In fact, mothers do more actual labor as well as more mental labor in the division of child care (Walzer, 1998). In other words, they take care of as well as "think about" their children more than fathers do.

ACTIVE CHILD CARE

Active, hands-on child care is much more in the hands of mothers than fathers. With the exception of households in which fathers are home full-time and mothers are employed full-time, in most two-parent households, mother's child-care responsibility and involvement greatly exceed fathers' involvement (Bird, 1997; Aldous and Mulligan, 2002). For every hour that fathers spend actively involved with their children, mothers spend between three and five (Bird, 1997). Additional research suggests that in terms of either "engagement" or "accessibility" fathers' involvement is less than half of mothers' levels of involvement. (Doherty, Kouneski, and Erickson, 1998). In two-parent households, mothers carry 90 percent of the responsibility for arranging child care (Bird, 1997).

There are circumstances that either increase or decrease the level of father involvement and participation in child care. The gender and age of the child, birth order, and number of children in the home all seem to make a difference. Specifically, fathers are more involved with sons than daughters, younger children than older children, with firstborn than with later born children, and when they have larger numbers of children. (Pleck, 1997, cited in Doherty, Kouneski, and Erickson, 1998). Additional research indicates that younger fathers, fathers with more egalitarian beliefs about gender roles, more child-centered beliefs about child rearing, more positive attitudes about their wives' employment, and more positive psychological characteristics such as higher self-esteem and lower levels of depression, are more likely to be involved and to be supportive and warm toward their children. Conversely, research suggests that fathers who work more hours and who have prestigious but time-demanding occupations tend to be less engaged in childrearing (NICHD Early Child Care Research Network, 2000).

MENTAL CHILD CARE

In her book, *Thinking about the Baby: Gender and Transitions into Parenthood,* sociologist Susan Walzer (1998), examines the division of responsibility for infants in 25 two-parent households. Her focus is less on "who does what" with their children than on "who thinks what, and how often" about their children. Walzer identifies this "invisible" parenting as **mental labor**—the process of worrying about the baby, seeking and processing information about infants and their needs, and managing the division of infant care in the household. Let's briefly consider each of these three categories:

- *Worrying:* Whether at home full-time, or in the paid labor force, full or part-time, mothers worry about their babies more than fathers do. To Walzer, this stems from cultural ideals of motherhood: mothers are supposed to worry about their children, which is part of "good mothering." Either extreme—not worrying or excessive worrying—might be seen as a problem, but worrying itself is expected. Worrying actually consists of two related but distinct phenomena: baby worry and mother worry. *Baby worry* refers to all the things that women as primary caregivers must concern themselves with, from babyproofing to making sure that the baby's nutritional needs are being met. *Mother worry* refers to whether one is being a good enough mother.
- *Processing information:* Mothers are the ones who seek out additional information or advice about children's development and needs. They decide when such outside input is needed or helpful, where to obtain it, whether it is helpful, how to share it with one's partner, and whether and how to implement it in their actual hands-on caregiving. The father's involvement is more as a passive recipient of the information the mother actively sought and processed. Thus, mothers spend more time and energy in the seeking, finding, and disseminating than men do in receiving advice or information.
- *Managing the division of labor:* Women bear the brunt of having to seek assistance with child care from their partners. Furthermore, they have to decide what type of help to ask for, when to seek it, and what to do if it is not forthcoming. Complicating these decisions is the fact that "help" may have its own costs. Needing and getting assistance from their partners may make some women feel as if they are not meeting the standards of mothering that they expect of themselves. It also may put them in a position of "owing" gratitude to their spouses or partners for whatever assistance they received. Thus, the thought process that goes into "managing" child care arrangements is far from simple.

Most studies of the division of child care do not include these kinds of responsibilities. One can say, as Walzer (1998) does, that even in instances where men do a fully equal share of caregiving tasks, mothers are still more heavily engaged in these invisible aspects of parenting .

NONPARENTAL CHILD CARE

Supplementary child care is a crucial issue for today's parents of young children. Given the prevalence of two-earner households (addressed more in Chapter 12) and single-parent households, many parents must look outside their homes for assistance in child rearing. In 2001, 63 percent of married women with children younger than 6 were in the labor force. Seventy percent of never-married mothers of preschool-age children and 76 percent of divorced widowed or separated women with preschool-age children were also in the paid labor force (U.S. Census Bureau, 2002, Table 570). In fact, in 2001, 58 percent of married mothers with husbands present and children under 1 year of age were employed outside the home (U.S. Census Bureau, 2002, Table 571). The combination of trends in employment status, marital status, and childbearing has increased the need for outside caregivers.

Based upon sample estimates from the National Center for Education Statistics, 77 percent of the more than 8 million 3- to 5-year-olds are in some form of nonparental child care (U.S. Census Bureau, 2002, Table 550). This varies by age of child, as 31 percent of 3-year-olds, compared with 18 percent of 4-year-olds, and 13.5 percent of 5-year-olds are in parental care only. Income also makes a difference, as children of higher-earning families are less likely to be cared for by parents only, or by other relatives, and are more likely to be cared for by nonrelatives. In fact, two out of three children of families earning between $50,000 and $75,000, and three out of four children from families earning over $75,000, spend some time in "center-based programs" (including day-care centers, nursery schools, preschools, and Head Start programs). Among children whose families earn less than $40,000, little more than half spend time in such center-based programs (U.S. Census Bureau 2002, Table 550). Finally, race and ethnicity also make a difference. A third of Hispanic families with children between the ages of 3 and 5, compared to nearly one-fourth of white families and 13 percent of black families rely only on parental child care (U.S. Census Bureau, 2002, Table 550).

Day-care homes and centers, nursery schools, and preschools can relieve parents of some of their child-rearing tasks and also furnish them with some valuable time of their own. Most experts agree that the ideal environment for raising a child is in the home with the parents and family. Intimate daily parental care of infants for the first several months to a year is particularly important. Because this ideal is often not possible, the role of day care needs to be considered.

What is the effect of child care on children? The results of research are mixed. In evaluating such data, it is important to keep in mind the family's education, the personalities involved, and the family interests—key factors that play a part in which parents choose to return to work or must return to work once a child is born (Crouter and McHale, 1993). Furthermore, a child's personality, age at which the custodial parent reentered the workforce, involvement of the other parent in the home, quantity of time, nature of work, along with the quality of care all contribute to how child care effects the child.

When mothers of infants enter the workforce, there is some evidence that these infants are at risk for insecure attachments between the ages of 12 to 18 months (Brooks, 1996). They are also at risk for being considered noncompliant and aggressive at ages 3 to 8 years (Howes, 1990). Other consequences, such as behavior problems, lowered cognitive performance, distractibility, and inability to focus attention have also been noted. These negative effects are not necessarily the consequence of being cared for by outside caregivers. Rather, they may be the result of *poor-quality* child care. It has been noted that high-quality care, that given by sensitive, responsive, and stimulating caregivers in a safe environment with low teacher-to-student ratio, can actually facilitate the development of positive social qualities, consideration, and independence (Field, 1991). In school-age and adolescent children, maternal employment is associated with self-confidence and independence, especially for girls whose mothers become role models of competence (Hoffman, 1979).

As more women return to the workforce, a critical issue is the quality of the day care for their children. High-quality day care can facilitate the development of positive social qualities.

National concern periodically is focused on day care by revelations of sexual abuse of children by their caregivers. Although these revelations have brought providers of child care under close public scrutiny and have alerted parents to potential dangers, they have also produced a backlash within the child-care profession. Some caregivers are now reluctant to have physical contact with the children; male child-care workers feel especially constrained and may find their jobs at risk (Chaze, 1984). A national study (Finkelhor, 1988), however, found that children have a far greater likelihood of being sexually abused by a father, stepfather, or other relative than by a day-care worker. In 1985, the U.S. Department of Health and Human Services announced day-care guidelines, calling for training of staff in the prevention and detection of child abuse, thorough checks on prospective employees, and allowances of parental visits at any time. Critics believe that the government should go further in establishing standards for day care itself.

What can parents do to ensure quality care for their children? In addition to the obvious requirements of cleanliness, comfort, nutritious food, and a safe environment, parents should be familiar with the state licensure regulations for child care. They should also check references and observe the caregivers with the child. Although the needs of young children differ from those of older ones, the American Academy of Child and Adolescent Psychiatry (1992) suggests that parents seek day-care services that meet specific standards:

- More adults per child than older children require.
- A lot of individual attention provided for each child.
- Trained, experienced teachers who understand, praise, and enjoy children.
- The same day-care staff for a long period of time.
- Opportunity for creative work, imaginative play, and physical activity.
- Space to move indoors and out.
- Enough teachers and assistants—ideally, at least one for every five (or fewer) children.
- An ample supply of drawing and coloring materials and toys, as well as equipment such as swings, wagons, and jungle gyms.
- Small rather than large groups if possible. (Studies have shown that five children with one caregiver is better than 20 children with four caregivers.)

Although parents may worry about how their children will adjust to the day-care setting, they should be enthusiastic about the prospect and encourage them to succeed. If the child shows persistent fear about leaving home, parents should discuss the problem with the child-care provider and their pediatrician.

As with a number of critical services in our society, those who most need supplementary child care are those who can least afford it. Child care done properly is a costly business (even though child-care work remains a relatively low-paying, low-status job). The United States is one of the few industrialized nations that do not have a comprehensive national day-care policy. In fact, beginning in 1981, the federal government dramatically cut federal contributions to day care; many state governments followed suit.

Theories of Child Socialization

Attitudes and beliefs about parenting flow from attitudes and beliefs about children and their development. Current attitudes about children have been influenced by a number of psychological theories concerning child socialization. Ultimately, as we will see, these theories have been influential in shaping some of the parenting advice offered by prominent authors in their child-rearing advice books.

PSYCHOLOGICAL THEORIES

Psychological theories of human development give prime importance to the role of the mind, particularly the subconscious mind, which, according to psychoanalytic theory, motivates much of our behavior without our being consciously aware of the process. According to these theories, many aspects of our psychological makeup are inborn; our minds grow and develop along with our bodies.

Psychoanalytic Theory

The emphasis by Sigmund Freud (1856–1939) on the importance of unconscious mental processes and on the stages of psychosexual development has greatly influenced modern psychology. Freud's **psychoanalytic theory** of personality development holds that we are driven by instinct to seek pleasure, especially sexual pleasure. This part of the personality, called the **id**, is kept in check by the **superego**—what we might call the conscience. The third component of personality, the rational **ego**, mediates between the demands of the id and the constraints of society. Freudian theory views the uninhibited id of the infant as gradually becoming controlled as the individual internalizes societal restraints. Too much restraint, however, leads to repression and the development of **neuroses**—psychological disorders characterized by anxiety, phobias, and so on.

Freud viewed the parents as the primary force responsible for the child's psychological development. He posited that between the ages of 4 and 6, the child identifies with the parent who is of the same sex. Not becoming like that parent was seen as a failure to reach maturity. Freud divided psychosexual development into five stages spanning the time from birth through adolescence: (1) oral, (2) anal, (3) phallic, (4) latency, and (5) genital (see Table 11.1). The scientific thought of Freud's time was greatly influenced by Charles Darwin's theories on evolution. As Jerome Kagan (1984) has pointed out, Freud viewed evolution as an apt metaphor for human behavior and constructed his theories in accordance with the scientific thought of his day.

Psychosocial Theory

Erik Erikson (1902–1994) based much of his work on psychoanalytic theory, but he emphasized the effects of society on the developing ego, creating a model that has come to be known as **psychosocial theory** (Erikson, 1963). Stressing parental and societal responsibilities in children's development, each of Erikson's life cycle stages (see Table 11.1) is centered around a specific emotional concern based on individual biological influences and external sociocultural expectations and actions.

LEARNING THEORIES

Learning theorists emphasize the aspects of behavior that are acquired rather than inborn or instinctual.

Behaviorism

Pioneering American psychologists such as John B. Watson (1878–1958) and B. F. Skinner (1904–1990) sought to explain human behaviors entirely on the basis of what could be observed; this model of human development is known as **behaviorism.** Behaviorists reject the concept of hidden "drives" that cannot be seen. Skinner developed the concept of **reinforcement** to explain how behaviors may be increased or decreased. He labeled the process of increasing the frequency of a behavior by adding a reinforcing stimulus **operant conditioning.** When we praise a child for picking up her blocks, we are posi-

TABLE 11.1 ■ Stages of Development: Freud, Piaget, and Erikson Compared

	FREUD	PIAGET	ERIKSON
Infancy	Oral	Sensorimotor	Trust vs. mistrust
	Anal		Autonomy vs. shame and doubt
Early childhood	Phallic	Preoperational	Initiative vs. guilt
Late-middle childhood	Latency	Concrete operational	Industry vs. inferiority
Adolescence	Genital	Formal operational	Identity vs. confusion
Early adulthood			Intimacy vs. isolation
Middle adulthood			Generativity vs. stagnation
Late adulthood			Ego integrity vs. despair

tively reinforcing her behavior, increasing the likelihood that she will pick them up in the future.

Social Learning Theory

Developed by psychologists such as Julian Rotter and Albert Bandura, **social learning theory** emphasizes the role of cognition, or thinking, in learning (Bandura, 1977; Rotter, Liverant, and Crowne, 1961). Human nature is formed by the interactions of culture, society, and the family with the inner qualities of the individual (Bandura, 1986, 1989). Children are first socialized through parental direction of their behavior. Parents teach children what is good, what is bad, what to eat, what not to eat, what to keep, what to share, how to talk, what to feel, and what to think. Parents influence their children by modeling (serving as examples for their children) and by defining (establishing expectations for them) (Cohen, 1987). Social learning theory accepts many of the tenets of behavioral psychology but adds to it the individual's innate ability to think and make choices to change his or her environment (Schickedanz et al., 1993).

COGNITIVE DEVELOPMENTAL THEORY

Beginning in the 1930s, Swiss psychologist Jean Piaget (1896–1980) began intensively observing and interviewing children, formulating what has become known as **cognitive developmental theory.** Piaget observed that cognitive development occurs in discrete stages through which all infants and children pass. Based on the development of the brain and the nervous system, these stages occur at about the same time in the development of all children (unless they are mentally impaired).The stages can be seen as building blocks, each of which must be completed before the next one can be put into place. In Piaget's view, children develop their cognitive abilities through interaction with the world and adaptation to their surroundings. Children adapt by **assimilation** (making new information compatible with their world understanding) and **accommodation** (adjusting their cognitive framework to incorporate new experiences) (Dworetsky, 1990). Piaget identified four stages of cognitive development: (1) sensorimotor, (2) preoperational, (3) concrete operational, and (4) formal operational (see again Table 11.1).

THE DEVELOPMENTAL SYSTEMS APPROACH

Parents do not simply give birth to children and then "bring them up." According to the **developmental systems approach,** the growth and development of children takes place within a complex and changing family system that both influences and is influenced by the child. The family system is part of a number of larger systems (extended family, friends, health care, education, and local and national government, to name a few), all of which mutually interact with one another. Models or theories that use a developmental systems approach include Bronfenbrenner's (1979) ecological model, Lerner's (1986) developmental contextual theory, Dewey and Bentley's (1949) transactional model, and Magnuson's (1988) interactive approach. (Family development theory was discussed in Chapter 2.)

Parent-Child Interactions

Not only are children socialized by their parents but they are also socializers in their own right. When an infant cries to be picked up and held, to have a diaper changed, or to be burped, or when he or she smiles when being played with, fed, or cuddled, the parents are being socialized. The child is creating strong bonds with the parents (see the discussion of attachment later in the chapter). Although the infant's actions are not at first consciously directed toward reinforcing parental behavior, they nevertheless have that effect. In this sense, even very young children can be viewed as participants in creating their own environment and in contributing to their further development (see Peterson and Rollins, 1987).

In the developmental systems model of family growth, social and psychological development are seen as lifelong processes, with each family member having a role in the development of the others. In terms of the eight developmental stages of the human life cycle described by Erikson, parents are generally at the seventh stage (generativity) during their children's growing years, and the children are probably anywhere from the first stage (trust) to the fifth (identity) or sixth (intimacy). The parents' need to establish their generativity is at least partly met by the child's need to be cared for and taught. The parents' approach to child rearing will inevitably be modified by the child's inherent nature or temperament.

Sibling Interactions

Over 80 percent of American children have one or more siblings. Siblings influence one another according to their particular needs and personalities. They are also significant agents for socialization. While rivalry and aggression may appear to be the foundation of such interactions, young siblings at home spend a large percentage of their time actually playing together. Sibling influence (or the lack of it, in the case of only children) is important in subtle yet powerful ways as the result of birth order and spacing (the number of years between sibling births). A study at Colorado State University, for example, found that a firstborn's self-esteem suffers if a sibling is born two or more years later but is not affected if the sibling is born less than two years later (Goleman, 1985). Furthermore, if the firstborn child is already five or six years old, the birth of a second sibling does not have the same impact. A study of African-American and white families indicated that race has no impact on the dynamics of birth order (Steelman and Powell, 1985). (See Sulloway [1996] for a study of the effects of birth order on personality and achievement.)

The quality of sibling interaction may have consequences for the child's later behavior (Newcombe, 1996). Close, affectionate sibling relationships contribute to the development of desirable characteristics such as social sensitivity, communication skills, cooperation, and understanding of social roles.

SYMBOLIC INTERACTION THEORY

Symbolic interaction theory is the sociological theory that most applies to the process of socialization. Symbolic interactionists such as Charles Horton Cooley and George Herbert Mead stressed the processes through which we develop a *social self,* the sense of who we are and how we are perceived by those around us. To interactionists, the self is not with us at birth but emerges out of interactions with others. In Cooley's formulation, three key components comprise the **looking-glass self,** the self-concept that develops from our sense of how others view us. First, we imagine how others perceive us. Second, we draw conclusions about how others judge us. And third, based on these, we develop our ideas about ourselves (Henslin, 2000).

George Herbert Mead emphasized that the self consists of both an active, spontaneous part (the "*I*") and a more passive, acted upon part (the "*me*"), in which we see ourselves as an object of other people's actions toward us (Henslin, 2000). The "me" is the reflective part of the self. This social self develops early in life and can be seen in the developing sophistication of children's play. Play forces children to see things from someone else's vantage point, what Mead called **taking the role of the other.** Initially, we can take the role of "*significant others*"; later we develop the capacity to take the perspective of the "*generalized other,*" the point of view of the group as a whole (Henslin, 2000). In looking at play activities, Mead noted that until about age 3 children really don't "play" but rather engage in imitative behavior. In this stage, they lack the capacity to see themselves as others might. In the **play stage** (3 to 6 years old), children play at being specific individuals, often by dressing up. By the **game stage,** they have developed sufficient self-awareness to be able to simultaneously take into account multiple perspectives and anticipate how other players might act in a given situation or outcome. In other words, they take the perspective of the *generalized other.* Thus, they can play more complex games (for example, baseball, a favorite of Mead's) (Henslin, 2000). More importantly, they have acquired a fully developed social self.

In symbolic interactionist terms, family members, especially parents, are among the more "significant" significant others in influencing the opinions we form of ourselves. They are perhaps the purest example of what Cooley called *primary groups,* characterized by intimate, face-to-face interaction, and crucial in the development of our social selves.

From the Theoretical to the Practical: Expert Advice on Child Rearing

About 150 years ago, Americans began turning to books to learn how to act and live rather than turning to one another. They began to lose confidence in their own abilities to make appropriate judgments. The vacuum that formed when traditional ways broke down under the impact of industrialization was filled by the so-called experts. The old values and ways had been handed down from parents to child in an unending cycle; men and women had learned how to be mothers and fathers from their own parents. But with increasing mobility, the continuity of generations ceased. A woman's mother was often not physically present to help her with her first child. New mothers were not able to turn to their more experienced kin for help. Instead, they enlisted the aid of new authorities—the experts who, through education and training, supposedly knew what to do. If your baby was colicky, for example, the experts recommended a drop of laudanum in his or her bottle. (The laudanum would put the baby to sleep, but it would also make him or her a heroin addict by the end of a year.)

Contemporary parents may still follow experts' advice even if it conflicts with their own beliefs. Yet if an expert's advice counters their own understanding, parents should carefully examine that advice, as well as their own beliefs. All parents should take an expert's advice with at least a grain of salt. It is, after all, the responsibility of the parents, not the experts, to raise their children.

Twentieth-century parenting was shaped by child-rearing advice from such notable authorities as Benjamin Spock, T. Berry Brazelton, and Penelope Leach. Cumulatively, these three authors sold well over 40 million copies of their books advising parents, especially mothers, as to the best ways to raise their children. Building on psychological theories of development including Freud, Piaget, and Erikson, their work stressed the importance of parents understanding their child's cognitive and emotional development. As sociologist Sharon Hays (1996) suggests, although the three offer their advice in different ways with somewhat different emphases, they share a "set of assumptions about the elements of good child rearing." Since few new parents actually read the psychological or sociological sources that inform Spock, Brazelton, and Leach, their more direct advice has had greater effect on parenting than the theories summarized above.

So what do these experts advocate as effective parenting? Hays suggests that the advice literature basically advocates the *ideology of intensive mothering,* which was discussed earlier in this chapter. Aside from the belief in the special nurturing capacities of mothers, this ideology contains the following assumptions about what children need from parents:

- Raising children is and should be an emotionally absorbing experience characterized by affectionate nurture. Emotional attachment is essential for healthy development; unconditional love and loving nurture are seen as the most important thing one can give one's child, no less essential, Spock asserts, than "vitamins and calories" (Spock, 1985, quoted in Hays, 1996).
- The child, more than the parent, guides the process of child rearing, because it is the mother's job to respond to the needs and wants of her child. Parents should follow the cues given off by their children, submit to the child's desires, and understand "what every baby knows" it needs from its parent (Brazelton, 1987, quoted in Hays, 1996). This requires both knowledge of children's needs and developmental phases as well as parental sensitivity since young children cannot articulate what it is they need from their parents.
- Parents must ultimately develop a sensitivity to the particular needs of their own children. This may entail learning to decode and recognize the different meanings of one's child's crying, or to understand the unique and individual developmental pattern of one's own child.
- Physical punishment is frowned upon. Instead, limit setting, providing a good example of what one expects from one's child, and giving the child lots of love are seen as preferred ways to get the child to internalize and act upon parents' standards. Punishment consists of "carefully managed temporary withdrawal of loving attention," a labor-intensive, emotionally absorbing method of discipline. Once a child can question, parents are urged to reason with the child, negotiate, and discuss motives and alternative ways of acting. This strategy involves much more time and effort than a threatened or actual spanking, so again effective parenting is presented as a labor-intensive process.

CONTEMPORARY CHILD-REARING STRATEGIES

One of the most challenging aspects of child rearing is knowing how to change, stop, encourage, or otherwise influence children's behavior. We can request, reason, explain, command, cajole, compromise, yell and scream, or threaten with physical punishment or the suspension of privileges; or we can just get down on our knees and beg. Some of these approaches may be appropriate at certain times; others clearly are never appropriate. Some may prove effective some of the time, some may never work very well, and no technique will work every time. The techniques of child rearing currently taught or endorsed by educators, psychologists and others involved with child development differ somewhat in their emphasis, but share most of the tenets that follow.

Respect

Mutual respect between children and parents must be fostered in order for growth and change to occur. One important way to teach respect is through modeling—treating the child and others respectfully. Child psychologist Rudolph Dreikurs (1964) stresses the importance of treating children with kindness and firmness simultaneously. Counselor Jane Nelsen (1987) writes, "Kindness is important in order to show respect for the child. Firmness is important in order to show respect for ourselves and the situation."

Consistency and Clarity

Consistency is crucial in child rearing. Without it, children become hopelessly confused and parents become hopelessly frustrated. Patience and teamwork (a united front) on the parents' part help ensure consistency. Because consistency means following through with whatever has been said, parents should beware of making promises or threats they won't be able to keep. Clarity is important for the same reason. A child needs to know the rules and the consequences for breaking them. This eliminates the possibility of the child's being unjustly disciplined or wiggling out through loopholes. ("But, Mom, I didn't know you meant not to walk on *this* clean carpet with my muddy shoes!")

Logical Consequences

One of the most effective ways to learn is by experiencing the logical consequences of our actions. Some of these consequences occur naturally—we forget our umbrella, and then we get wet. Sometimes parents need to devise consequences that are appropriate to their child's misbehavior. Dreikers and Soltz (1964) distinguish between logical consequences and punishment. The "Three R's" of logical consequences dictate that the solution must be *Related* to the problem behavior, *Respectful* (no humiliation), and *Reasonable* (designed to teach, not to induce suffering).

Open Communication

The lines of communication between parents and children must be kept open. Numerous techniques exist for fostering communication. Among these are active listening and the use of "I" messages, both important components of Thomas Gordon's (1978) Parent Effectiveness program. In *active listening,* the parent verbally feeds back the child's communications in order to understand the child and help him or her understand the nature of the problem. "I" messages are important because they impart facts without placing blame and are less likely to promote rebellion in children than are "you" messages.

Family meetings are another important way in which families can communicate. Regular weekly meetings provide an opportunity for being together and a forum for airing gripes, solving problems, and planning activities. Decisions are best reached by consensus rather than majority vote, as majority rule can lead to a "tyranny of the majority" in which the minority is consistently oppressed. If consensus can't be reached, the problem can be put on the next meeting's agenda, allowing time for family members to come up with alternative solutions (Nelsen, 1987).

No Physical Punishment

Many physicians, psychologists, and sociologists have become harsh and vocal critics of physical punishment. The American Psychological Association notes that "physical violence imprinted at an early age, and the modeling of violent behavior by punishing adults, induces habitual violence in children" (cited in Haferd, 1986). The American Medical Association also opposes physical punishment of children.

Many sociologists, most notably family violence authority, Murray Straus, are staunch opponents of corporal punishment, noting that it is related to later aggressive behavior from children, including later perpetration of spousal violence (Straus and Yodanis, 1996). Although such punishment is used widely (Straus and Yodanis estimate more than 90 percent of parents of toddlers use corporal punishment) and may "work" in the short run by stopping undesirable behavior, its long-range results—anger, resentment, fear, hatred, aggressiveness, family violence—may be extremely problematic (Dodson, 1987, Straus and Yodanis, 1996; McLoyd and Smith, 2002). Besides, it often makes parents feel confused, miserable, and degraded right along with their kids.

Behavior Modification

More effective types of discipline use some form of behavior modification. Rewards (hugs, stickers, or special activities) are given for good behavior, and privileges are taken away when misbehavior is involved. Good behavior can be kept track of on a simple chart listing one or several of the desired behaviors. Undesirable behavior may be met with the revocation of TV privileges or the curtailment of other activities. Time-outs—sending the child to his or her room or to a "boring" place for a short time or until the misbehavior stops—are useful for particularly disruptive behaviors. They also give the parent an opportunity to cool off (Dodson, 1987; see also Canter and Canter, 1985).

Styles and Strategies of Child Rearing

A parent's approach to training, teaching, nurturing, and helping a child will vary according to cultural influences, the parent's personality, the parent's basic attitude toward children and child rearing, and the role model that the parent presents to the child.

AUTHORITARIAN, PERMISSIVE, AND AUTHORITATIVE PARENTS

One popular formulation contrasts three basic styles of child rearing: (1) authoritarian, (2) permissive, and (3) authoritative (Baumrind, 1971, 1983).

Parents who practice **authoritarian child rearing** typically require absolute obedience. The parents' maintaining control is of primary importance. "Because I said so" is a typical response to a child's questioning of parental authority, and physical force may be used to ensure obedience. Working-class families tend to be more authoritarian than middle-class families. Diana Baumrind (1983) found that children of authoritarian parents tend to be less cheerful than other children and correspondingly more moody, passively hostile, and vulnerable to stress.

Permissive child rearing is a more popular style in middle-class families than in working-class families. The child's freedom of expression and autonomy are valued. Permissive parents rely on reasoning and explanations. Yet permissive parents may find themselves resorting to manipulation and justification. The child is free from external restraints but not from internal ones. The child is supposedly free because he or she conforms "willingly," but such freedom is not authentic. This form of socialization creates a bind: "Do what we tell you to do because you want to do it." Baumrind (1983) found that although children of permissive parents are generally cheerful, they exhibit low levels of self-reliance and self-control.

Parents who favor **authoritative child rearing** rely on positive reinforcement and infrequent use of punishment. They direct the child in a manner that shows awareness of his or her feelings and capabilities. Parents encourage the development of the child's autonomy within reasonable limits and foster an atmosphere of give-and-take in parent-child communication. Parental support is a crucial ingredient in child socialization. It is positively related to cognitive development, self-control, self-esteem, moral behavior, conformity to adult standards, and academic achievement (Gecas and Seff, 1991). Control is exercised in conjunction with support by authoritative parents. Children raised by authoritative parents tend to approach novel or stressful situations with curiosity and show high levels of self-reliance, self-control, cheerfulness, and friendliness (Baumrind, 1983).

REFLECTIONS

In your family, what child-rearing attitudes (authoritarian, permissive, or authoritative) predominated? Do you think these attitudes influenced your own development? If so, how? Which might (or do) you find useful in raising your own child?

Children's Needs, Parents' Needs

Although the relative effects of physiology and environment on human development are still often much debated by today's experts, it is clear that both nature and nurture play important roles. In addition to biological factors, important factors affecting early development include the formation of attachments (especially maternal) and individual temperamental differences.

BIOLOGICAL FACTORS

According to biological determinists, much of human behavior is guided by genetic makeup, physiological maturation, and neurological functioning. Jerome Kagan (1984) has presented a strong case for the role of biology in early development. He holds that the growth of the central nervous system in infants and young children ensures that motor and cognitive abilities such as walking, talking, using symbols, and becoming self-aware will occur "as long as children are growing in any reasonably varied environment where minimal nutritional needs are met and [they] can exercise emerging abilities." Furthermore, according to Kagan, children are biologically equipped for understanding the meaning of right and wrong by the age of 2, but although biology may be responsible for the development of conscience, social factors can encourage its decline.

ATTACHMENT

The strong bond forged between an infant and his or her primary caregiver or caregivers is called **attachment.** Babies appear to be equipped to build relationships with those who care for them. They signal their needs by methods such as gazing, crying, and smiling, and they bond with the people who are most responsive to them (in most cases, the mother). Based on observations of babies in "the strange situation"—that is, in the presence of a stranger both with and without the mother—researcher Mary Ainsworth and her colleagues (1978) discovered three patterns of infants' attachment to their mothers: (1) *secure,* (2) *anxious/ambivalent,*

TABLE 11.2 ■ Attachment Patterns in 12- to 18-Month-Olds in the "Strange Situation"

ATTACHMENT PATTERN	BEHAVIOR BEFORE SEPARATION	BEHAVIOR DURING SEPARATION	REUNION BEHAVIOR	BEHAVIOR WITH STRANGER
Secure	Separates from mother to explore toys; shares play with mother; is friendly toward stranger when mother is there	May cry; play is subdued for a while, usually recovers, plays	If distressed during separation, contact with mother ends distress; if not distressed, greets mother with affection	Somewhat friendly; may play with stranger
Anxious/ ambivalent	Has difficulty separating to explore toys; wary of new situations and people; stays close to mother	Is very distressed; cries hysterically	Seeks comfort but then rejects it; may continue to cry; may be passive	Wary of stranger; rejects offers to play
Anxious/ avoidant	Separates to explore toys, but does not share with parent; shows little preference for parent over stranger	Shows no distress; continues to play; interacts with stranger	Ignores or moves away from mother	Does not avoid stranger

SOURCE: Compiled from M. D. S. Ainsworth, and B. A. Wittig. "Attachment and Exploratory Behavior of One-Year-Olds in a Strange Situation." In B. M. Foss (ed.), *Determinants of Infant Behavior* (Vol. 4). London: Methuen, 1969.

and (3) *anxious/avoidant.* (See Table 11.2.) A person's style of attachment in infancy is thought to affect his or her style of relating in adulthood, as was discussed in Chapter 4.

INDIVIDUAL TEMPERAMENT

A child's unique temperament, such as "inhibited/restrained/watchful" or "uninhibited/energetic/spontaneous," also influences the way in which he or she develops (Kagan, 1984). Temperamental differences may be rooted in the biology of the brain (Kagan and Snidman, 1991), but temperament is also developed by interaction with the environment. For example, a baby who is vigorous, strong, and outgoing will probably encourage her parents to play with her vigorously. She and her parents will reinforce the lively, extroverted, and spontaneous aspects of her personality. An infant who is shy, fearful, and cries easily, however, will not encourage his parents to play energetically with him and may, in fact, inhibit them from interacting with him, thus causing him to become more shy and fearful. It is important for parents to understand "how they create the meaning of the child's individuality by their own temperaments, and their demands, attitudes, and evaluations," according to psychologists Richard Lerner and Jacqueline Lerner (Brooks, 1994). Lerner and Lerner stress the importance of what they call "goodness of fit" between the child's characteristics and those of the parents and other environmental influences. If parents are sensitive to a child's unique temperament, they are better able to understand the child and to seek appropriate ways to influence the child's behavior.

BASIC NEEDS

Parents often want to know what they can do to raise healthy children. Are there specific parental behaviors or amounts of behaviors (say, 12 hugs a day?) that all children need to grow up healthy? Apart from saying that basic physical needs must be met (adequate food, shelter, clothing, and so on), along with some basic psychological ones, experts cannot give parents detailed instructions.

Noted physician Melvin Konner (1991) lists the following needs for optimal child development—which, he writes, "parents, teachers, doctors, and child development experts with many different perspectives can fairly well agree on":

- Adequate prenatal nutrition and care
- Appropriate stimulation and care of newborns
- The formation of at least one close attachment during the first five years
- Support for the family "under pressure from an uncaring world," including child care when a parent or parents must work
- Protection from illness
- Freedom from physical and sexual abuse
- Supportive friends, both adults and children
- Respect for the child's individuality and the presentation of appropriate challenges leading to competence
- Safe, nurturing, and challenging schooling
- An adolescence "free of pressure to grow up too fast, yet respectful of natural biological transformations"
- Protection from premature parenthood

In today's society, especially in the absence of adequate health care and schools in so many communities, it is difficult to see how even these minimal needs can all be met. Even when the necessary social supports are present, parents may find themselves confused, discouraged, or guilty because they do not live up to their own expectations of perfection. Yet children have more strength, resiliency, and resourcefulness than we may ordinarily think. They can adapt to and overcome many difficult situations. A mother can lose her temper and scream at her child, and the child will most likely survive, especially if the mother later apologizes and shares her feelings with the child. A father can turn his child away with a grunt because he is too tired to listen, and the child will not necessarily grow up neurotic, especially if the father spends some "special time" with the child later on.

SELF-ESTEEM

High self-esteem—what Erik Erikson called "an optimal sense of identity"—is essential for growth in relationships, creativity, and productivity in the world at large. Low self-esteem is a disability that afflicts children (and the adults they grow up to be) with feelings of powerlessness, poor ability to cope,

FAMILIES OF MEXICAN ORIGIN

The Socialization of Children

The Mexican-origin home is usually child-centered when children are young; yet the role of children is based on the belief that they should "be seen and not heard." Although both parents tend to be permissive, boys and girls are raised very differently in Mexican-American families. Boys are granted far more liberty, and loud, aggressive behavior is generally tolerated. The male child is often overindulged and accorded greater status than the female. Young girls are expected to be demure and feminine, and girls are usually taught feminine roles, just as boys are taught masculine roles. Playmates are often segregated by gender, rather than by age, particularly as they grow older.

Adolescence marks differences in behavior patterns between boys and girls (Locke, 1992). The adolescent male is given much more freedom, and his decisions and actions are seldom questioned. His activities are seldom restricted and are not closely monitored. He is free to date and pursue intimate relationships, and he is expected to take on the role of the protector for his sisters.

Activities for girls are often restricted and closely monitored. They are expected to remain much closer to home. Regardless of age, but particularly during adolescence, girls are expected to be subservient to their older brothers. At this stage, a strong relationship between mother and daughter is encouraged and is viewed as a means of preparing a girl for the role of wife and mother (Mirandé, 1985). In families of Mexican origin, the *quinceanera,* or fifteenth birthday, marks the coming of age of a young girl. Celebrated by a mass or prayer service with a sermon, the event reminds the young girl of her future responsibilities.

Assimilation and acculturation have had an impact on socialization patterns. Whereas adolescent males still maintain greater degrees of independence, many adolescent females have challenged traditional gender roles and are exerting greater autonomy. However, the servile attitude of females remains present in many Mexican-origin families and is still encouraged and expected, regardless of the level of acculturation and assimilation.

As Mexican-American children make the transition to adolescence, just like all teenagers, they may encounter difficulties with issues related to autonomy. Adolescence is a period during which individuals are attempting to assert their independence, and for many Mexican-origin adolescents, this is compounded by conflicts that arise between first and second generations.

When demands and desires accumulate and are in conflict with traditional expectations, teenagers have difficulty dealing with the competing influences in their lives. Adolescence is particularly problematic for children who are raised in homes with traditional cultural values. They often encounter a different set of values at school than at home. External expectations may be in direct opposition to traditional values, which reinforce that assimilation should be avoided altogether and that *la cultura* (the culture) should be maintained (Klor de Alva, 1988). The extended family system has been a mainstay in many Mexican-origin families, and supportive institutions of *la familia*—including *parentesco* (the concept of family), *compadrazgo* (godparents), and *confianza* (trust)—have

low tolerance for differences and difficulties, inability to accept responsibility, and impaired emotional responsiveness. Self-esteem has been shown to be more significant than intelligence in predicting scholastic performance.

A study of 3,000 children found that adolescent girls had lower self-images, lower expectations from life, and less self-confidence than boys (Brown and Gilligan, 1992). At age 9, most of the girls felt positive and confident, but by the time they entered high school, only 29 percent said they felt "happy" the way they were. The boys also lost some sense of self-worth, but not nearly as much as the girls. Ethnicity was an important factor in this study. African-American girls reported a much higher rate of self-confidence in high school than did white or Latina girls. Two reasons were suggested for this discrepancy. First, African-American girls often have strong female role models at home and in their communities; African-American women are more likely than others to have a full-time job and run a household. Second, many African-American parents specifically teach their children that "there is nothing wrong with them, only with the way the world treats them" (Daley, 1991). According to researcher Carole Gilligan, this survey "makes it im-

helped set the standard for families. These ideologies reinforce the view that family is the central and most important institution in life. Although these concepts influence actual behavior, their ideals can never be fully realized; however, they provide a basis that helps establish norms and expectations (Kane, 1993).

Mexican-origin families have an extended family structure that serves as a vital link between family and community in Mexican-American society. While it is a very strong institution, it varies from one generation to the next. *Compadrazgo,* or godparents, who have a moral obligation to act as guardian, provide financial assistance in times of need, and substitute as parents in the event of death. The *compadrazgo* relationship is formed usually through baptism and confirmation ceremonies in the church. The parental relationship is maintained throughout life and extends beyond the child. Godparents often refer to each other as *comadre* (co-mother) or *compadre* (co-father) and are expected to maintain a reciprocal relationship of support and mutual assistance.

Parentesco is a kinship concept that extends family sentiment to kin and nonkin, ensuring that there is an automatic family network. This concept often helps build networks of support and reciprocity and also helps establish a sense of community support among individuals who share regional or geographic origins.

Confianza, commonly referred to as trust, is essential to the relationship between *compadrazgo* and *parentesco*. Yet it means more than trust and includes the notions of respect and intimacy. It builds relationships and provides the foundation for reciprocity. *Confianza* is seen as an institution that has facilitated adaptation after immigration.

Much of what the culture condemns is related to kinship relationships. The family is perceived as more important than the individual. Selfishness is condemned, and its absence is considered a virtue. The strength of the family in providing security to its members is sometimes expressed through the sharing of material things with other relatives, even when there is precious little to meet one's own immediate needs (Locke, 1992). The concept that one should sacrifice everything for family has its costs, however.

In traditional Mexican life, a set of family, religious, and community obligations was significant. Women had certain legal and property rights that acknowledged the importance of their work, their families of origin, and their children. However, the imposition of American law and custom ignored and ultimately undermined many aspects of the extended family in Mexican culture. Because of economic changes that occurred in the family as a result of these laws, many Mexican-American women were forced to participate in the economic support of their families by working outside the home. The preservation of traditional customs, such as language, celebrations, and healing practices, became an important element in maintaining and supporting familial ties (Thornton-Dill, 1994).

The culture and identity of Mexican Americans will continue to change as they are affected by inevitable generational fusion with Anglo society and the influence of immigrants.

SOURCE: Yolanda M. Sanchez. "Families of Mexican Origin." In Mary Kay DeGenova (ed.), *Families in Cultural Context.*

possible to say that what happens to girls is simply a matter of hormones. . . . [It] raises all kinds of issues about cultural contributions, and it raises questions about the role of the schools, both in the drop of self-esteem and in the potential for intervention" (quoted in Daley, 1991).

Parents can foster high self-esteem in their children by (1) having high self-esteem themselves, (2) accepting their children as they are, (3) enforcing clearly defined limits, (4) respecting individuality within the limits that have been set, and (5) responding to their child with sincere thoughts and feelings. It is also important to single out the child's behavior—not the whole child—for comment (Kutner, 1988). Children (and adults as well) can benefit from specific information about how well they've performed a task. "You did a lousy job" not only makes us feel bad, but it also gives us no useful information about what would constitute a good job.

Misusing the concept of self-esteem with superficial praise is probably the most common way parents have it backfire. Children notice when praise is insincere. If, for instance, Martha refuses to comb her hair, yet we continually tell her how shiny it looks, Martha quickly realizes that we either have very low expectations or do not have a clue about hair care. Instead, parents can accomplish more by giving kids timely, honest, specific feedback. For example, "I like the way you discussed Benjamin Franklin's inventions in your essay" is more effective than, "You're wonderful!" Each time you treat your child like an intelligent, capable person, you increase your child's self-esteem.

PSYCHOSEXUAL DEVELOPMENT IN THE FAMILY CONTEXT

It is within the context of our overall growth, and perhaps central to it, that our sexual selves develop. Within the family we learn how we "should" feel about our bodies—whether we should be ashamed, embarrassed, proud, or indifferent. Some families are comfortable with nudity in a variety of situations: swimming, bathing, sunbathing, dressing, or undressing. Others are comfortable with partial nudity from time to time: when sharing the bathroom, changing clothes, and so on. Still others are more modest and carefully guard their privacy. Most researchers and therapists would allow that all these styles can be compatible with the development of sexually well-adjusted children as long as some basic needs are met:

1. The child's body (and nudity) is accepted and respected.
2. The child is not punished or humiliated for seeing the parent naked, going to the toilet, or making love.
3. The child's needs for privacy are respected.

Families also vary in the amount and type of physical contact in which they participate. Some families hug and kiss, give back rubs, sit and lean on each other, and generally maintain a high degree of physical closeness. Some parents extend this closeness into their sleeping habits, allowing their infants and small children in their beds each night. (In many cultures, this is the rule rather than the exception.) Other families limit their contact to hugs and tickles. Variations of this kind are normal. Concerning children's needs for physical contact, we can make the following generalization. First, all children (and adults) need a certain amount of freely given physical affection from those they love. Although there is no prescription for the right amount or form of such expression, its quantity and quality both affect children's emotional well-being and the emotional and sexual health of the adults they will become.

Second, children should be told, in a nonthreatening way, what kind of touching by adults is "good" and what kind is "bad." They need to feel that they are in charge of their own bodies, that parts of their bodies are private property and that no adult has the right to touch them with sexual intent. It is not necessary to frighten a child by going into great detail about the kinds of things that might happen. A better strategy is to instill a sense of self-worth and confidence in children so that they will not allow themselves to be victimized (Pogrebin, 1983). We also should learn to listen to children and to trust them. They need to know that if they are sexually abused, it is not their fault. They need to feel that they can tell an adult about it and still be worthy of love.

Parents' Needs

Although some needs of parents are met by their children, parents have other needs as well. Important needs of parents during the child-rearing years are personal developmental needs (such as social contacts, privacy, and outside interests) and the need to maintain marital satisfaction. Yet so much is expected of parents that they often neglect these needs. Parents may feel a deep sense of guilt if their child is not happy or has some defect, an unpleasant personality, or even a runny nose. The burden is especially heavy for mothers because their success is often measured by how perfect their children are. Children have their own independent personalities, however, and many forces affect a child's development and behavior.

Accepting our limitations as parents (and as human beings) and accepting our lives as they are (even if they haven't turned out exactly as planned) can help us cope with the many stresses of child rearing in an already stressful world. Contemporary parents need to guard against the "burnout syndrome" of emotional and physical overload. Parents' careers and children's school activities, organized sports, Scouts, and music, art, or dance lessons compete for the parents' energy and rob them of the unstructured (and energizing) time that should be spent with others, with their children, or simply alone.

EMBATTLED PARENTS AND SOCIETAL INSENSITIVITY TO CHILD REARING

Even under ideal conditions, parenting is bound to be a difficult undertaking. However, as some social critics have suggested, contemporary American society is far from an ideal environment for children or the parents who try to raise them. Despite our cultural celebration of families and children, we do surprisingly little to ensure that families can function effectively or that children are raised by involved and dedicated parents. Sylvia Hewlett and Cornel West (1998) note that in recent decades "public policy and private decision making have tilted heavily against the altruistic nonmarket activities that comprise the essence of parenting. In recent years, big business, government, and the wider culture have waged an undeclared and silent war against parents." Hewlett and West point to a number of specific instances or examples of societal indifference or neglect regarding the needs of parents and children:

- *Economic issues:* Matters such as corporate downsizing, declining wages, and longer work weeks have led to more instability, impoverishment, and uncertainty as well as less time between parents and children.
- *Popular culture:* Television programs, popular music, and movies undermine the efforts of parents through the parent-bashing, violence, and sex to which they expose children.
- *Government insensitivity and neglect:* In such areas as housing and taxes, government policies have failed to support parents' efforts to raise their children.
- *The diminishment and devaluation of fathers:* Some social programs, especially in policies of poverty and divorce, have contributed to undermining the role of fathers in children's lives. Combining these with alterations in household structure and increased economic vulnerability spells disaster for many fathers in their efforts to function effectively.

A Parents' Bill of Rights

Hewlett and West (1998) detail a set of policies to respond to such concerns and thus improve the conditions under which contemporary parents attempt to raise children. Their **Parents' Bill of Rights** addresses six areas:

1. *Time-enhancing policies:* Paid parenting leaves and flexible, family friendly workplaces would reduce the time famine that many parents identify as their number one problem.
2. *Policies to provide economic security:* These include wage, tax, and housing policies that would improve the economic conditions that families face and that affect the quality of parent-child relationships.
3. *Pro-family electoral policies:* The most provocative of the recommendations is to give parents the right to vote on behalf of their children, effectively doubling the size of the parent voting bloc.
4. *Pro-family legal policies:* Policies that would strengthen marriage, reduce divorce, increase adoptions, and mandate parental leave for new fathers would all strengthen parent-child relationships.
5. *Policies that improve the external environment that affects parenting effectiveness:* Here Hewlett and West review a number of policies designed to create safer communities, better schools, more responsible media, and more accessible medical care.
6. *Policies that strengthen the social importance and authority of the parental role":* Hewlett and West suggest such things as a *National Parents' Day* or specially designed education credit or tuition benefits for parents in keeping with the educational benefits of the G.I. Bill.

Hewlett and West (1998) call for a parents' movement within which parents can unify and mobilize, pressing for the policies and insisting upon the support that will better enable them to do the job of raising their children.

Issues of Diverse Families

The diversity of family forms in our country creates a variety of experiences, needs, and possibilities. The problems and strengths of single-parent and stepfamilies are discussed in Chapter 15. Here we look at the influences of ethnicity and sexuality (i.e., lesbian and gay parenthood) on today's families.

ETHNICITY AND CHILD SOCIALIZATION

A person's ethnicity is not necessarily fixed and unchanging. Researchers generally agree that ethnicity has both objective and subjective components. The objective component refers to one's ancestry, cultural heritage, and to varying degrees physical appearance. The subjective component refers to whether one feels he or she is a member of a certain ethnic group, such as African American, Latino, Asian American, or Native American. If both parents are from the same ethnic group, the child will probably identify as a member of that group. But if a child has parents from different ethnic groups, ethnic identification becomes more complex. In such cases, the child may identify with both groups, only one group, or according to the situation—Latino when with Latino relatives and friends or Anglo when with Anglo friends and relatives, for example. However we choose to identify ourselves, our families are the key to the transmission of ethnic identification.

A child's ethnic background affects how he or she is socialized. Ethnic minority families socialize their children to more highly value obligation, cooperation and interdependence (Demo and Cox, 2000). Mexican American parents highly value cooperation and family unity more than individualism and competition. Asian Americans and Latinos further stress the authority of the father in the family. In both groups, parents command considerable respect from their children, even when the children become adults. Older siblings, especially brothers, have authority over younger siblings and are expected to set a good example (Becerra, 1988; Tran, 1988; Wong, 1988).

Asian Americans tend to discourage aggression in children and expect them to sacrifice their own personal desires or interests out of loyalty to their elders and to family authority more generally (Demo and Cox, 2000). In disciplining their children, Asian parents rely on compliance based on the desire for love and respect.

African Americans, too, may have group-specific emphases in the ways they socialize their children. As reported in Chapters 3 and 4, African American parents tend to socialize their children into less rigid, more flexible gender roles. They reinforce certain traits, such as assertiveness and independence, in both their sons and their daughters. They also seek to promote such values as pride, closeness to other African Americans, and racial awareness (Demo and Cox, 2000).

Groups with minority status in the United States may be different from one another in some key ways, but they also have much in common. Such groups often emphasize education as the means for the children to achieve success. Studies show that immigrant children tend to excel as students until they become acculturated and discover that it's not "cool." Minority groups are often dual-worker families, which means that the children may have considerable exposure to television while the parents are away from home. This may be viewed as a mixed blessing: On the one hand, television may help children who need to acquire English language skills; on the other, it can promote fear, violence, and negative stereotypes of women and minority-status groups. Television can discourage creativity and encourage passivity.

Some American children are raised with a strong sense of ethnic identification, whereas others are not. Identification with a particular group gives a sense of pride, security, and belonging. At the same time, it can also give a sense of separateness from the mainstream society. Often, however, that sense of separateness is imposed by the greater society. Discrimination and prejudice shape the lives of many American children. For minority status children, there may be a lack of congruence between society's assumptions and values and their own healthy psychological development (Spencer, 1985). Parents of ethnic minority children may try to prepare their children for the harsh realities of life beyond the family and immediate community (Peterson, 1985). According to Mary Kay Genova (1997), one of the first steps in changing racist attitudes is to admit to racist thinking. People must examine their racist attitudes and the reasons for them. DeGenova writes about the search for similarity:

> No matter how many differences there may be, beneath the surface there are even more similarities. It is important to try to identify the similarities among various cultures. Stripping away surface differences will uncover a multiplicity of similarities: people's hopes, aspirations, desire to survive, search for love, and need for family—to name just a few. While superficially we may be dissimilar, the essence of being human is very much the same for all of us.

GAY AND LESBIAN PARENTING

Although numbers are difficult to obtain, researchers believe the number of gay families number in the millions. There are between 2 million and 14 million children with at least one gay parent (Kantrowitz, 1996; Stacey and Biblarz, 2001). The breadth of this range results from a number of factors, including the age of children considered and the definition of parental sexual orientation. There are an estimated 800,000 to 7 million gay or lesbian parents, aged 18 to 59, with between 1.6 million to 14 million children. The high ends of these estimates include many parents with adult children no longer in the home and use generous definitions of sexual orientation (including anyone with homoerotic desires). If one restricts the estimates to families with children 19 or younger, there are anywhere from 1 million to 9 million children of lesbian or gay parents, representing between 1 percent and 12 percent of all children 19 and younger. (Stacey and Biblarz, 2001).

Families headed by lesbians or gay men generally experience the same joys and pains as those headed by heterosexuals—with one exception: They are likely to face insensitivity or discrimination from society.

Most of these parents are, or have been, married (Patterson and Chan, 1999). Heterosexual concerns about gay and lesbian parents center around parenting abilities, fear of sexual abuse, and worry that the children will become gay or lesbian. The research that has been conducted on gay and lesbian parents has failed to uncover significant negative outcomes for their children. In fact, much research has failed to identify any meaningful differences between children of gay and heterosexual parents. Sociologists Judith Stacey and Timothy Biblarz (2001) and psychologist Charlotte Patterson's (2000) reviews of more than 20 studies on the effect of parental sexual orientation on children finds that most research supports either a "no effects" or "beneficial effects" interpretation. In summarizing the research on children of gay and lesbian parents as they compare with children of heterosexual parents, Patterson notes that there are no significant differences in their gender identities, gender-role behaviors, self-concepts, moral judgment, intelligence, success with peer relations, behavioral problems, or successful relations with adults of both genders (Patterson, 2000). Stacey and Biblarz suggest that there may be some defensiveness on the part of researchers, especially from those who are sympathetic to gay and lesbian parents. Aware of the social stigma and lack of support gay and lesbian families face, there may be a tendency to minimize differences. In so doing, some differences that might actually be strengths of gay and lesbian families are underemphasized.

Some lesbians, especially those in committed relationships, are choosing to create families through artificial insemination. Nearly a third of lesbians have become mothers through some form of assisted reproductive technology (Salholz, 1990). The nonbiological (or nonadoptive) parent usually has no legal tie to the child, although "second-parent adoptions" by lesbians and gays have been approved in California, Oregon, Alaska, Washington, New York, New Jersey, Massachusetts, and Washington, D.C. (Schenden, 1993). Even so, society may not recognize the "second parent" as a "real" parent because children are expected to have only one real mother and one real father.

Fears about Gay and Lesbian Parenting

Heterosexual fears about the parenting abilities of lesbians and gay men are unwarranted. Fears of the sexual abuse of children by gay parents or their partners are completely unsubstantiated. A review of the literature on the children of gay men and lesbians found that there were virtually no documented cases of sexual abuse by gay parents or their lovers; such exploitation appears to be committed disproportionately by heterosexuals (Cramer, 1986).

Fears about gay parents' rejecting children of the other sex also seem unfounded. Such fears reflect the popular misconception that being gay or lesbian is a rejection of members of the other sex. Many gay and lesbian parents go out of their way to make sure that their children have role models of both sexes (Kantrowitz, 1996). Many also say that they hope their children will develop heterosexual identities in order to be spared the pain of growing up gay in a homophobic society. Research finds children of gay males and lesbians to be well adjusted, and no more likely to be gay as adults (Golemann, 1992; Flaks et al., 1995; Kantrowitz, 1996). Ultimately, it is the quality of parenting—not the sexuality of the parents—that matters most to kids.

Parenting and Caregiving in Later Life

PARENTING ADULT CHILDREN

Some years ago, a Miami Beach couple reported their son missing (Treas and Bengtson, 1987):

> Joseph Horowitz still doesn't understand why his mother got so upset. He wasn't "missing" from their home in Miami Beach: he had just decided to go north for the winter. Etta Horowitz, however, called authorities. Social worker Mike Weston finally located Joseph in Monticello, N.Y., where he was visiting friends. Etta, 102, and her husband, Solomon, 96, had feared harm had befallen their son Joseph, 75.

Parenting does not end when children grow up. Most elderly parents still feel themselves to be parents, but they are parents in different ways. Their parental role is considerably less important in their daily lives. They generally have some kind of regular contact with their adult children, usually by letters, phone calls, or e-mails; parents and adult children also visit each other fairly frequently and often celebrate holidays and birthdays together. Researcher Joan Aldous (1987b) found that middle-class parents typically assisted their children financially or provided services. Parents made loans, gave gifts, or paid bills for their children around six times a year; they also provided child care about the same number of times. They assisted in shopping, house care, and transportation and also helped in times of illness.

Parents tend to assist those whom they perceive to be in need, especially children who are single or divorced. Parents perceive their single children as being "needy" when they have not yet established themselves in occupational and family roles. These children may need financial assistance and may lack intimate ties; parents may provide both until the children are more firmly established. Parents often assist divorced children, especially if grandchildren are involved, by providing financial and emotional support. They may also provide child-care and housekeeping services. Parents generally provide the greatest assistance to daughters who are single mothers.

In providing assistance to their children, the parents in Aldous's study exercised considerable control over whom they would help and how involved they would become. Despite their continued parental concern, they tended to be maritally rather than parentally oriented. They valued their independence from their children. When one woman was asked how she would feel about an adult child's coming back home to live, she exclaimed, "Oh, God, I wouldn't like it. I'm tired of waiting on people" (Aldous, 1987b). In Aldous's study, adult children generally reciprocated in terms of physical energy. They helped their parents in household chores and yard work about six times a year; they also assisted to some extent with physical care during illness.

Parents tend to be deeply affected by the circumstances in which their adult children find themselves. Adult children who seem well-adjusted and who have fulfilled the expected life stages (such as becoming independent, starting a family, etc.) provide their aging parents with a vicarious gratification. On the other hand, adult children who have stress-related or chronic problems (e.g., with alcohol) cause higher levels of parental depression (Allen, Blieszner, and Roberto, 2000).

Some elderly parents never cease being parents because they provide home care for children who are severely limited either physically or mentally. Many elderly parents, like middle-aged parents, are taking on parental roles again as children return home for financial or emotional reasons. Although we don't know how elderly parents "parent," presumably they are less involved in traditional parenting roles.

REFLECTIONS

Think about your grandparents. How many are alive? What kind of relationship do you (or did you) have with them? What role do they (or did they) play in your life and your family's life?

© Spencer Grant/PhotoEdit

Grandparents are important to their grandchildren as caregivers, playmates, and mentors.

GRANDPARENTING

The image of the lonely, frail grandmother in a rocking chair needs to be discarded. Grandparents are often not very old, nor are they very lonely, and they are certainly not absent in contemporary American family life. Grandparents are "a very present aspect of family life, not only for young children but young adults as well," writes Gregory Kennedy (1990).

Grandparenting is expanding tremendously these days, creating new roles that relatively few Americans played a few generations back. Three-quarters of people aged 65 and older are grandparents (Aldous, 1995). Grandparents play important emotional roles in American families; the majority appear to establish strong bonds with their grandchildren (Kennedy, 1990; Strom et al., 1992–1993). They help to achieve family cohesiveness by conveying family history, stories, and customs. Grandparents influence grandchildren directly when they act as caretakers, playmates, and mentors. They influence indirectly when they provide psychological and material support to parents who may consequently have more resources for parenting (Brooks, 1996).

Grandparents seem to take on greater importance in single-parent and stepparent families and among certain ethnic groups (see Figures 11.1 and

FIGURE 11.1 ■ **Percentage of Population, Age 30 Years or Older Living with and Responsible for Grandchildren, 2000**

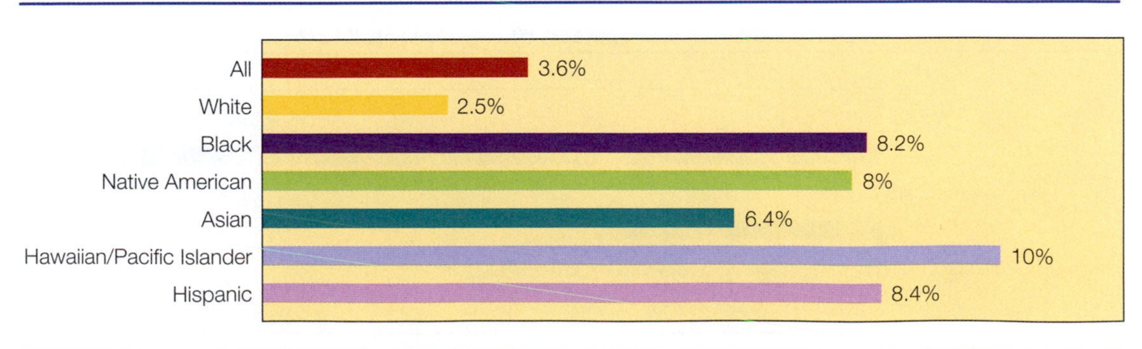

SOURCE: Tavia Simmons and Jane Lawler Dye. "Grandparents Living with Grandchildren, 2000." Table 1, *Census 2000 Brief,* U.S. Department of Commerce, October 2003.

11.2). They frequently act as a stabilizing force for their children and grandchildren when the families are divorcing and re-forming as single-parent families or stepfamilies. Kennedy and Kennedy (1993) found that the significance of grandparents varies by family form. When compared with children from intact families, children in single-parent families report greater closeness and active involvement with their grandparents; children in stepfamilies are even closer.

According to the 2000 Census, 5.8 million grandparents live in the same home as one of their grandchildren. In 42 percent of these 4.1 million households (some households have more than one grandparent), grandparents had primary caregiving responsibility for their grandchildren, age 18 or younger. Of these "grandparent caregivers" 39 percent had cared for their grandchildren for at least five years (Simmons and Dye, 2003).

Grandparents, especially grandmothers, are often involved in the daily care of their grandchildren (see Figure 11.2). A recent study found that African Americans had twice the odds of becoming caregiving grandparents, partly reflecting the long tradition of caregiving that goes back to West African cultures. In the crack cocaine epidemic, grandmothers and great-grandmothers play critical roles in rearing the children of addicted parents (Minkler and Roe, 1993).

Andrew Cherlin and Frank Furstenberg (1986) identified three distinct styles of grandparenting:

- *Companionate:* Most grandparents perceive their relationships with their grandchildren as companionate. The relationships are marked by affection, companionship, and play. Because these grandparents tend to live relatively close to their grandchildren, they can have regular interaction with them. Companionate grandparents do not perceive themselves as rule makers or enforcers; they rarely assume parentlike authority.
- *Remote:* Remote grandparents are not intimately involved in their grandchildren's lives. Their remoteness, however, is due to geographic rather than emotional remoteness. Geographic distance prevents the regular visits or interaction with their grandchildren that would bind the generations together more closely.
- *Involved:* Involved grandparents are actively involved in what have come to be regarded as parenting activities: making and enforcing rules and disciplining children. Involved grandparents (most often grandmothers) tend to emerge in times of crisis—for example, when the mother is an unmarried adolescent or enters the workforce following divorce. Some involved grandparents may become overinvolved, however. They may cause confusion as the family tries to determine who is the real head of the family.

What determines the amount of interaction between grandparents and grandchildren? Cherlin and Furstenberg (1986) succinctly sum up the three most important factors: "distance, distance, and distance." Other factors include age, health, employment, personality, and other responsibilities (Troll, 1985a). The middle generation is usually responsible for determining how much interaction takes place between grandparents and grandchildren. If rivalry rather than cooperation ensues, the middle generation may restrict grandparent-grandchildren interaction. But most families gain from such interactions (Barranti and Ramirez, 1985).

FIGURE 11.2 ■ **Percentage of Residential Grandparents Who Are Responsible for Grandchildren**

SOURCE: Tavia Simmons and Jane Lawler Dye. "Grandparents Living with Grandchildren, 2000." Table 1, *Census 2000 Brief,* U.S. Department of Commerce, October 2003.

Single parenting and remarriage have made grandparenthood more painful and problematic for many grandparents. Stepfamilies have created step-grandparents, who are often confused about their grandparenting role. Are they really grandparents? The grandparents whose sons or daughters do not have custody often express concern about their future grandparenting roles (Goetting, 1990). Although research indicates that children in stepfamilies tend to do better if they continue to have contact with both sets of grandparents, it is not uncommon for the parents of the noncustodial parent to lose contact with their grandchildren (Bray and Berger, 1990).

A variety of circumstances may lead to situations in which the grandparent role and the relationships with one's grandchildren are strained if not altogether disrupted. Divorce and single parenthood may be the most prominent of such circumstances, but death of a spouse, distance, or estrangement between parents and children can all impede grandparent-grandchild relationships (Keith and Wacker, 2002). Over the past 40 years, grandparent visitation statutes have been enacted in all 50 states and grandparents' visitation rights have been increased. Generally, courts have not wanted to expand grandparents' rights at the expense of parents' rights, especially parents' rights to control the custody of their children (Keith and Wacker, 2002).

CHILDREN CARING FOR PARENTS

A common experience faced by many American families is the need to provide care for aging or ill parents. The idea of the "sandwich generation," discussed in Chapter 9, captures the experience of many adults, sandwiched between raising their own children and caring for their parents. In fact, though, there are circumstances that create **parentified children**—children who are forced to become caregivers for their parents well before adulthood (Boszormenyi and Spark, 1973, quoted in Winton, 2003). In situations of "parentification," children may be pressed into taking care of parents who have become chronically ill, chemically dependent, mentally ill, incapacitated after a divorce or widowhood, or in any way socially isolated or incapacitated (Winton, 2003).

Much of the psychological and sociological literature depicts parentified children as pathological or deviant. Psychologists may focus on how taking on caregiving responsibilities for one's parent(s) while still a child or adolescent disrupts normal developmental processes. Sociologists tend to focus on the nonnormative nature of children being responsible for their parents. However, definitions of normative and nonnormative vary by culture. Among many populations other than white, middle-class, European Americans, parentification is expected and obligatory. Similarly, rather than pathological, parentification under certain circumstances may be beneficial for the development of certain personality traits, the maintenance of certain family relationships, and the acquisition of particular skills. In fact, Winton (2003) suggests that parentification may be a normative part of childhood in many contemporary American families, where children *temporarily* take care of a parent (e.g., after surgery, during an illness). This fits Gregory Jurkovic's continuum of caretaking roles, where parentification is normal and adaptive under certain conditions. "Destructive parentification" occurs when the circumstances become extreme and long-term, and the responsibilities children carry are age-inappropriate (Jurkovics, 1997, cited in Winton, 2003).

Winton suggests the following as possible consequences of parentification:

- *Delayed entry into marriage:* if children have had to care for parents (or siblings) over a number of years, they may decide to delay taking on the caretaking that comes with marriage, and choose—instead—to take time for themselves where they can concentrate on their own needs more than or instead of the needs of others.
- *Acquisition of certain personality characteristics:* Having played a parentified role over time might lead to the development of such traits or tendencies as the following:
    - masochistic or self-defeating behavior because of having had to meet others' needs and suppress their own compulsive behavior, such as perfectionism;
    - feelings of excessive responsibility for others that make it difficult to say "no" to people, to set limits, or to concentrate on their own needs
- *Relationship and intimacy problems:* Parentified children may seek as adult partners people who they can be caretakers for, in other words "dependent, needy people" who have emotional or physical disabilities or have been emotionally "wounded" by past experiences.

- *Career choices:* The "caretaker syndrome" associated with parentification may lead people to jobs where they can physically or emotionally take care of people, such as jobs in social work, medicine, nursing, teaching, or preschool child care.

CARING FOR AGING PARENTS

Most elder care is provided by women, generally daughters or daughters-in-law (Mancini and Blieszner, 1991). Psychologist Rita Ghatak estimates that "eighty percent of the time it's the female sibling who is taking most of the responsibility" (quoted in Rubin, 1994). Elder caregiving seems to affect husbands and wives differently. Women report greater distress, a greater decline in happiness, more hostility, less autonomy, and more depression from caregiving than do men (Fitting et al., 1986; Marks, Lambert, and Choi, 2002) when becoming a caregiver to a parent in the household. Men experienced less hostility and a smaller decline in happiness. In part, this may be because men approach their daily caregiving activities in a more detached, instrumental way. Another factor may be that women frequently are not only mothers but also workers; an infirm parent can sometimes be an overwhelming responsibility to an already burdened woman (Rubin, 1994). Interestingly, when caring for a parent out of the household, women get a "caregiver gain," a greater sense of purpose in life than that felt by noncaregiving women (Marks, Lambert, and Choi, 2002). Fortunately, most adult children in a given family participate in parental caregiving in some fashion, whether it involves doing routine caregiving, providing backup, or giving limited or occasional care (Mancini and Blieszner, 1991).

A study of 539 older participants found that although there are psychological benefits associated with intergenerational support, excessive support received from adult children may be harmful by virtue of eroding competence and imposing excessive demands (Silverstein, Chen, and Heller, 1996). The authors point out that "given the norm of intergenerational independence that prevails between adults in many American families, it is reasonable to expect that an overabundance of social support from adult children may sometimes do more harm than good." In the process of balancing personal needs with those of families, it is important to define the level of care that is both appropriate and necessary.

Caregiver Conflicts

Even though elder care is often done with love, it can be the source of profound stress. Caregivers often experience conflicting feelings about caring

Successful caregiving to elders necessitates a balance between our personal and family needs and those of our elders. For most of us, this is not an easy process.

for an elderly relative. The conflicts experienced by primary caregivers include the following (Springer and Brubaker, 1984):

- Earlier unresolved antagonisms and conflicts
- The caregiver's inability to accept the relative's increasing dependence
- Conflicting loyalties between spousal or child-rearing responsibilities and caring for the elderly relative
- Resentment toward the older relative for disrupting family routines and patterns
- Resentment by the primary caregiver for lack of involvement by other family members
- Anger or hostility toward an elderly relative who tries to manipulate others
- Conflicts over money or inheritance

Coping Strategies

Caregiver education and training programs, self-help groups, caregiver services, and family therapy can provide assistance in dealing with the problems encountered by caregivers. In addition, elders receiving Medicaid may be eligible for respite care and homemaker/housework assistance. Because elder care involves complex emotions raised by issues of dependency, adult children and their parents often postpone discussions until a crisis occurs.

Although no child-rearing technique or caregiving approach is guaranteed to be 100 percent successful all the time, it is important for families to keep seeking ways to improve their communication and satisfaction. It is also important for parents to develop and maintain confidence in their own parenting skills, their common sense, and especially their love for their children.

Finally, it is crucial that, as a society, we close the gap between pro-family/pro-child/pro-parent rhetoric and reality. We need to live up to our stated "family values" and create the kinds of social institutional environment that gives parents the opportunity and support to more effectively care for their children.

SUMMARY

- Many women find considerable satisfaction and fulfillment in motherhood. Although there is no concrete evidence of a biological maternal drive, it is clear that socialization for motherhood does exist. The role of the father in his children's development is currently being reexamined. The traditional instrumental roles are being supplemented, and perhaps supplanted, by expressive ones.
- There appear to be two extremes among contemporary fathers: Many men aspire for active, meaningful involvement with their children, while others, especially divorced fathers, maintain little actual contact with their children.
- Most child care is done by mothers. This is true whether one speaks of hands-on actual care of children or the "mental labor" that goes into having and raising children.
- Supplementary child care outside the home is a necessity for many families. The development and maintenance of quality day-care programs should be a national priority.
- Theories of child socialization include Sigmund Freud's *psychoanalytic theory,* in which the personality is seen to be composed of the pleasure-seeking *id,* the controlling *superego,* and the rational *ego;* Erik Erikson's *psychosocial theory,* in which society and family are seen as influencing the individual during specific life cycle stages; *behaviorism,* propounded by psychologists such as B. F. Skinner, who developed the concept of *reinforcement* and stressed the importance of observing actual behaviors; and *social learning theory,* which emphasizes the role of thinking in human development. Jean Piaget's *cognitive developmental theory* emphasizes the importance of specific stages of mental development. Children's abilities are developed through the processes of *assimilation* and *accommodation.* In the *developmental systems approach,* family members are viewed as being interdependent in their growth, which continues throughout their lives. Birth order, spacing between births, and sibling interactions are also important. The most relevant sociological theory of socialization is

symbolic interaction theory with its emphasis on how social selves develop out of interaction with meaningful others.

- Important factors in children's development include *attachment* (the bond between the infant and the primary caregiver or caregivers) and temperament (the child's basic personality and way of relating to the world).
- Children have a number of basic physical and psychological needs, including adequate prenatal care; formation of close attachments; protection from illness and abuse; and respect, education, and support from family, friends, and community. High self-esteem is essential for growth in relationships, creativity, and productivity. Adolescent girls, especially nonblacks, are likely to be low in self-esteem. Parents can foster high self-esteem in their children by encouraging the development of a sense of connectedness, uniqueness, and power, and by providing models.
- Psychosexual development begins in infancy. Infants and children learn from their parents how they should feel about themselves as sexual beings.
- American society is not particularly supportive of the needs of parents and children. Economic, cultural, and political institutions have neglected to adopt policies that would allow parents and children deeper and more frequent contact with each other.
- Ethnicity profoundly influences the way children are socialized. Identification with a group gives a sense of pride and belonging. Minority-status parents may try to give their children special skills for dealing with prejudice and discrimination.
- Most gay and lesbian parents are, or have been, married. Lesbian mothers share many similarities with heterosexual mothers. Studies indicate that children of both lesbians and gay men fare best when the parents are secure in their sexual orientation. Although the children of gay and lesbian parents may have difficulty accepting their parents' sexual orientation at first, they generally maintain close relationships with their parents, are well-adjusted, and develop the same sexual orientations and gender roles as children of heterosexuals.
- Basic attitudes toward child rearing can be classified as *authoritarian, permissive,* and *authoritative.* Today's parents often rely on expert advice. It needs to be tempered by parents' confidence in their own parenting abilities and in their children's strength and resourcefulness. Contemporary strategies for child rearing include the elements of mutual respect, consistency and clarity, logical consequences, open communication, and behavior modification in place of physical punishment.
- Parenting roles continue through old age. Older parents provide financial and emotional support to their children; they often take active roles in child care and housekeeping for their daughters who are single parents. Divorced children and those with physical or mental limitations may continue living at home.
- Grandparenting is an important role for the middle-aged and aged; it provides them and their grandchildren with a sense of continuity. Grandchildren report feeling close to grandparents; grandparents are important role models. They often provide extensive child care for grandchildren, and they take on greater importance in single-parent and stepparent families. Grandparenting can be divided into three styles: (1) companionate, (2) remote, and (3) involved.
- Family caregiving activities often begin when an aged parent becomes infirm or dependent. Conflicts that may arise involve previous unresolved problems, the caregiver's inability to accept the parent's dependence, conflicting loyalties, resentment, anger, and money or inheritance conflicts. Family therapy may facilitate coping; other coping strategies include planning ahead with siblings to manage expenses, arranging long-term care, assessing family capabilities, dividing responsibilities, and determining what services are available.

KEY TERMS

accommodation 357
assimilation 357
attachment 362
authoritarian child rearing 361
authoritative child rearing 361
behaviorism 356
cognitive developmental theory 357
conduct of fatherhood 349
culture of fatherhood 349

developmental systems approach 357
ego 356
engrossment 349
game stage 358
id 356
intensive mothering ideology 346
looking-glass self 358
mental labor 353
neuroses 356
operant conditioning 356
parentified children 373
Parents' Bill of Rights 367
permissive child rearing 361
play stage 358
psychoanalytic theory 356
psychosocial theory 356
reinforcement 356
social learning theory 357
superego 356
taking the role of the other 358

SUGGESTED READINGS

Crittenden, Ann. *The Price of Motherhood: Why the Most Important Job in the World Is Still the Least Valued.* New York: Metropolitan Books. 2001.

Gottfried, A. E. and Allen W. Gottfried. *Redefining Families: Implications for Children's Development.* New York: Plenum, 1994. A collection of articles on the diversity of families in the United States.

Hays, Sharon. *Cultural Contradictions of Motherhood.* New Haven: Yale University Press, 1996. Sociologist Hays draws upon advice literature and her own interviews with a sample of mothers to show how the *intensive mothering ideology* affects women's ideas about what children need and how mothers are needed.

Hewlett, Sylvia and Cornell West. *The War Against Parents: What We Can Do for America's Beleagured Moms and Dads.* Boston: Houghton-Mifflin, 1998. Hewlett and West collaborate to document a range of societal and cultural forces that make parenting difficult and which expose parents and children to undue stress.

Lamb, Michael (ed.). *Parenting and Child Development in "Nontraditional" Families.* Mahwah, NJ: Lawrence Erlbaum Associates. 1999. A broad ranging collection of parenting and child development in a variety of family circumstances.

Pollack, Jill. *Lesbian and Gay Families: Redefining Parenting in America.* Danbury, CT: Frank Watts Inc., 1995. A look at parenting in the past decade and a half, from a personal, social, and psychological viewpoint.

Walzer, Susan. *Thinking About the Baby: Gender and Transitions into Parenthood.* Philadelphia: Temple University Press. 1998. Walzer's interviews with new mothers and fathers highlight the division of mental labor (thinking or worrying about one's infants) in two-parent households.

RESOURCES ON THE INTERNET

Companion Web Site for This Book

http://sociology.wadsworth.com/strong/marriage9e
Gain an even better grasp on this chapter by going to the companion Web site to take one of the Tutorial Quizzes, use the Flash Cards to master key terms, or check out the many other study aids you'll find there. Visit the Marriage and Family Resource Center on the site. You'll also find special features such as GSS Data and Census 2000 information that will put data and resources at your fingertips to help you with that special project or to do some research on your own.

InfoTrac College Edition: Search Word Summary

child rearing
fatherhood
grandparents
caregiving
authoritative parenting
self-esteem

To learn more about these central topics in the study of the family, you can conduct an electronic search using InfoTrac College Edition. To aid in your search and to gain useful tips, see the Student Guide to InfoTrac College Edition that you can access through the companion Web site for this book.

Preview

To gain a sense of what you already know about the material covered in this chapter, answer "True" or "False" to the statements below.

1. In contrast to single-worker couples, dual-career couples tend to divide household work almost evenly. True or false?
2. Over 1 million American men are full-time homemakers with no outside employment. True or false?
3. It is generally agreed by economists that welfare encourages poverty. True or false?
4. Families make up about 25 percent of the homeless. True or false?
5. Women in the United States currently make 90 cents for every dollar that men earn. True or false?
6. Family economic well-being is a national priority. True or false?
7. The majority of female welfare recipients are on welfare as a result of a change in their marital or family status. True or false?
8. The majority of families are dual-earner families. True or false?
9. Women tend to interrupt their work careers for family reasons over 30 times as often as men. True or false?
10. Married women tend to earn more and have higher-status jobs than single women. True or false?

"Far and away the best prize that life offers is the chance to work hard at work worth doing."

THEODORE ROOSEVELT

CHAPTER 12

Marriage, Work, and Economics

Outline

Here are three brief exchanges overheard at a party. Identify what is wrong with each of them.

- *Exchange number 1:* "Do you work?" the man asks. "No, I'm just a mother," the woman replies.
- *Exchange number 2:* "What do you do?" the first woman inquires politely as she is introduced to another woman. "Nothing. I'm a housewife," the second responds. "Oh, that's nice," the first replies, losing interest.
- *Exchange number 3:* "What do you do?" the man asks. "I'm a doctor," the woman responds as she picks up her child, who is impatiently tugging on her. "And I'm an architect," her husband says while nursing their second child with a bottle.

In the first exchange, the woman ignores that as a mother, she works. In the second exchange, both women ignore the fact that as a homemaker, the second woman works. They also devalue such unpaid work in comparison with paid work. In the third exchange, the woman identifies herself as a physician without acknowledging that she is also a mother. Her husband makes the same mistake as he identifies himself as an architect without also noting that he is a father participating in the care of his infant. As husband and wife, father and mother, both the architect and the physician are unpaid family workers making important—but generally unrecognized—contributions to the family's economy.

Now, consider this story told by a 32-year-old woman about her husband's encounter at a party:

This girl came up to my husband at a party . . . (where) he was talking with a bunch of people. They were all talking about what they did, and they said, "Yeah, he (husband) used to work at . . . And then this ditz says to him, "Oh, what do you do now?" And he says—oh what is it he calls himself now? I think he said, "I'm a full-time househusband." And she went, "Oh . . . how nice!" And then, later on, she came back over to him and said, "So what do you do? What exactly do you do?" And he said, "I stay home full-time and I raise my kids." And she just didn't get it. (Cohen and Durst, 2001)

Like the previous exchanges, this reveals some of the ways in which we perceive work and family. In questioning what the husband "really does," the party guest reveals her belief that outside employment is what men "really do"; the home and children are for women. Cumulatively, these four examples point to some curious facts.

When we think of work, we tend to recognize only paid work. Family work, such as caregiving activities or household duties, is not regarded as "real" work. Because it is unpaid, family work is ignored and looked upon as being somehow inferior to paid work, regardless of how difficult, time consuming, creative, rewarding, and important it is for our lives and future as human beings. This is not surprising, for in the United States, employment takes precedence over family.

The role of work in families requires many of us to rethink the meaning of *family.* We ordinarily think of families in terms of relationships and feelings—the family as an emotional unit. But families are also economic units bound together by emotional ties (Ross, Mirowsky, and Goldsteen, 1991). Paid work and unpaid family work, as well as the economy itself, profoundly affect the way we live as families. Our most intimate relationships vary according to how husbands and wives participate in paid work and family responsibilities (Voydanoff, 1987).

PREVIEW ANSWERS

1 False
2 True
3 False
4 True
5 False
6 False
7 True
8 True
9 True
10 False

Family work is the work we perform in the home without pay: It includes housekeeping; household maintenance; and caring for children, the ill, and the aged. Women do the overwhelming majority of family work. Paid work is the work we do for salary or wages. Traditionally, men have been responsible for paid work, but currently, the vast majority of women—whether single or married, with children or without—are employed. Women either support themselves or make primary or significant contributions to family income.

Our paid work helps shape the quality of family life: It affects time, roles, incomes, spending, leisure, and even individual identities. Within our families, the time we have for one another, for fun, for our children, and even for sex is the time that is not taken up by paid work. The main characteristic of our paid work in relation to the family is its inflexibility. Work regulates the family, and for most families, as in the past, a woman's work molds itself to her family, whereas a man's family molds itself to his work (Ross, Mirowsky, and Goldsteen, 1991). We must constantly balance work roles and family roles. These facts are the focus of this chapter.

Workplace/Family Linkages

Outside of sleeping, probably the single activity to which the majority of employed men and women devote the most time is their jobs, and we are working more and more. According to Juliet Schor (1991), between 1970 and the 1990s, the average worker added the equivalent of an extra month of hours of work per year. As we work more and more, we have less time for our families and leisure. Most of us know from our own experiences that our work or studies affect our personal relationships.

It is unclear how much of an increase in work hours has occurred or how widespread it is. Some researchers suggest that instead of working *more*, Americans have experienced an increase in the pace of their lives, which has led to feeling overworked even when no real increase in hours at work has occurred. It has also been suggested that although some categories of workers (for example, professional and managerial) have experienced an increase in the time demands upon them, others are underemployed and would prefer to work more (see the review of work and family research by Perry-Jenkins, Repetti, and Crouters [2000]).

In addition to time availability, we carry workplace stress home with us. Work-family conflict, especially relating to work schedules and demands and to children at home, is often a painful source of stress (Menaghan and Parcel, 1991). However, most research has been devoted to descriptions of home life or work life, not to how the two interact with each other. Only more recently have scholars begun to look at how our jobs affect our family life, causing moodiness, fatigue, irritability, and so on (Crouter and Manke, 1994; Small and Riley, 1990). Furthermore, whether we love, loathe, or learn to live with them, our jobs structure the time we can spend as families (Hochschild, 1997).

WORK SPILLOVER

Common sense (as well as our own fatigue) suggests that our paid work affects other aspects of our lives. We can call this **work spillover**—the effect that work has on individuals and families, absorbing their time and energy and impinging on their

Work spillover and role strain affect many employed women.

psychological states. It links our home lives to our workplace (Small and Riley, 1990). Work is as much a part of marriage as love is. What happens at work—frustration or worry, a rude customer, an unreasonable boss—affects our moods, making us irritable or depressed. We take these moods home with us, affecting the emotional quality of our relationships. Workplace stress often causes us to focus on our problems at work rather than on our families. It can lead to fatigue, depression, stomach ailments, and increased drug and alcohol use (Crouter and Manke, 1994).

Work spillover affects men and women a bit differently. Whereas for men, excessive work time is the major cause of conflict between work and family, fatigue and irritability are the major causes of conflict for women.

As a result of work spillover, an employed woman's leisure is rarely spontaneous: It is fitted in or planned around her household work and paid work (Kinnunen, Gerris, and Vermulst, 1996). In addition, the stress a husband experiences at his work can diminish his wife's emotional health (Dooley and Catalano, 1991).

The relationship between paid work and family life cuts both ways. The demands of our home lives may impinge on our concentration, energy, or availability at work. Additionally, the emotional climate in our homes can affect our morale and performance in our jobs.

The family can help alleviate some workplace stress. The way family members respond helps shape the way in which the stressed worker expresses his or her frustration, anger, or fatigue. If, for example, your partner comes home feeling frustrated from a bad day at work, you can permit him or her to withdraw temporarily from family obligations, listen empathetically, and help him or her think of solutions or strategies for dealing with the work stress. Responding to your partner's stress with anger or a list of your own complaints is likely to add to the stress. If it's your partner's turn to make dinner, make dinner yourself, or order Chinese takeout.

ROLE CONFLICT, ROLE STRAIN, AND ROLE OVERLOAD

Two-parent families in which both partners are employed face more severe work-related problems than do nonparents. Being an employed parent usually means performing three demanding roles simultaneously: worker, parent, and spouse (Voydanoff and Donnelly, 1989). In juggling these roles one might experience role conflict, role strain, and/or role overload. When the multiple social statuses or positions that we occupy (for example, spouse, parent, and worker) present us with competing, contradictory, or simultaneous role expectations, we experience **role conflict.** When the role demands attached to any particular status (for example, mother, husband, or employee) are contradictory or incompatible we experience **role strain.** Finally, when the various roles we play require us to do more than we can comfortably or adequately handle, we experience **role overload.**

In the specific case of family and paid work roles, when we feel torn between spending time with our spouses or children and finishing up work-related tasks, we are experiencing role conflict. One cannot be in two places at once. Men who believe they are supposed to be traditional providers, but who are pressed into higher levels of housework or child care because of demands or needs of their spouses, are experiencing role strain. Similarly, employed wives who find themselves too exhausted by their housework and/or child-care responsibilities to engage in or enjoy sexual intimacy with their spouses are also experiencing role strain. Their housework responsibilities are incompatible with the companionship and sexual intimacy expected of them by their husbands. Finally, the sheer volume of the many things we are supposed to do as partners, parents, and employees may on occasion overwhelm us. With only 24 hours in a day, we may feel as if we cannot do all that is expected of us. We experience role overload.

Role overload and role conflict are fairly common among employed parents. In role overload we are expected to wear "too many hats." In the multiple responsibilities of parent/spouse/worker we must care for children, be intimate with our spouses, work eight hours a day, shop, clean house, take children to school, and so on. Role conflict confronts us with "too many balls to juggle." When our children are home sick but we need to work late for an important business meeting or pull a double shift because of production needs, the consequences include considerable distress. The emotional distress that may follow the job stress being experienced includes depression, which then may have an effect on marital or parent-child relations.

There is some evidence suggesting that job stress has a "crossover effect" on one's spouse or other family members. When a spouse feels a lot of pressure or overloads at work, the other spouse may begin to feel depressed or overloaded as well. This may be especially true regarding a cross-over effect of husbands' job stresses onto wives. It is unclear as to how much "crosses over" from parents to children However, parental "availability" to children is affected by the levels of stress parents experience. Particularly stressful days at work may be followed later by parents being withdrawn at home. This may even prove beneficial because, by withdrawing, less negative emotion is brought into the relationships (Perry-Jenkins, Repetti, and Crouter, 2000).

Some research indicates that individuals with high self-esteem feel less role conflict than those with low self-esteem (Long and Martinez, 1994). Women with high self-esteem, for example, accept lower housekeeping standards as necessary and realistic adjustments to their multiple roles rather than as signs of inadequacy. However, the more important sources of role conflict and overload are not within the person but rather within the person's role responsibilities.

Men experience role conflict when trying to balance their family and work roles. Because the workplace expects men to give priority to their jobs over their families, it is not easy for men to be as involved in their families as they may like. A recent study examined role conflict among men (O'Neil and Greenberger, 1994; see also Marks, 1994, and Greenberger, 1994). It found that men with the least role conflict fell into two groups. One group consisted of men who were highly committed to both work and family roles. They were determined to succeed at both. The other group consisted of men who put their family commitments above their job commitments. They were willing to work at less demanding or more flexible jobs, spend less time at work, and put their family needs first. In both instances, however, the men received strong encouragement and support from their spouses.

Married women employed full-time often prefer working fewer hours as a means of reducing role conflict (Warren and Johnson, 1995). Increasingly, women work some shift other than that of their spouses or partners. In fact, according to an AFL-CIO poll, almost half of employed women who are either married or cohabiting work different shifts than their partners. Among women with children under 18, the percentage climbs to 51 percent (Love, 2000). Not surprisingly, because they have less role conflict, single women (including those who are divorced) are often more advanced in their careers than married women (Houseknecht, Vaughan, and Statham, 1987). They are more likely to be employed full-time and have higher occupational status and incomes. They are also more highly represented in the professions and hold higher academic positions.

The various issues surrounding spillover, role conflict, overload, and role strain vary depending upon the household structure and division of labor. Single parent households with full-time working parents are easily susceptible to overload and role conflict. Two-parent dual-earner households also face different versions of work-family spillover than do households with one provider and a partner at home full-time.

REFLECTIONS

Much of the workplace-family linkage concept can be applied to the college environment. If you think of your student role as a work role and the college as the workplace, what types of work spillover do you experience in your personal or family life? If you are a homemaker or are employed (or both), what kinds of role strain do you experience?

The Familial Division of Labor: Traditional, Dual-Earning, and Nontraditional Patterns

Families divide their labor in a number of different ways. Some follow more traditional male-female patterns, most share wage earning, and a small but increasing number reverse roles. Even within a single family there will likely be a number of different divisions of labor over time, as they move through the various family life-cycle stages. How families allocate tasks and divide paid and unpaid work have tremendous impact on how a family functions.

THE TRADITIONAL PATTERN

In the traditional division of labor in the family, work roles are complementary: The husband is expected to work outside the home for wages, and the wife is expected to remain at home caring for children and maintaining the household. A man's family role is secondary to his provider role, whereas a woman's employment role is secondary to her family role (Blair, 1993). In 2002, among married couples with children under age 18, 30 percent had employed fathers and at-home mothers, whereas 61 percent had both parents employed. Among married couple families with children under 6, the percentage with employed fathers and at-home mothers was 38.7 percent. Fifty-four percent of such families had two employed parents (U.S. Census Bureau, 2003).

This difference in primary roles between men and women in traditional households profoundly affects such basic family tasks as who cleans the toilet, mops the floors, does the ironing, and washes the baby's diapers. Women—whether or not they are employed outside the home—remain primarily responsible for household tasks (Demo and Acock, 1993). This is the family form that most fits the **two-person career** model. Women become the domestic and child-rearing supports on whom families depend, freeing men to focus on wage earning and providing.

The division of family roles along stereotypical gender lines varies by race and class. It is more characteristic of white families than of African-American families. African-American women, for example, are less likely than white women to be exclusively responsible for household tasks. Latino and Asian families are more likely to be closer to the traditional than are African Americans or whites (Rubin, 1994). Class differences are somewhat ambiguous. Among middle-class couples, greater ideological weight is given to sharing and fairness. Working-class couples, though less ideologically traditional than in the past, are still not as openly enthusiastic about more egalitarian divisions of labor. However, in terms of *who does what,* working-class families are more likely than middle-class families to piece together work-shift arrangements that allow parents to take turns caring for the children and working outside the home. Such arrangements may force couples to depart from tradition, even if they neither believe they should nor boast that they do (Rubin, 1994).

MEN'S FAMILY WORK

Men's family work traditionally takes place outside the home, where men fulfill their primary economic role as provider. The husband's role as provider is probably his most fundamental role in marriage. The basic equation is that if the male is a good provider, he is a good husband and a good father (Bernard, 1981). This core concept seems to endure despite trends toward more egalitarian and androgynous gender roles. In fact, a woman's marital satisfaction is often related to how well she perceives her husband as fulfilling his provider role (Blair, 1993). It is not uncommon for women to complain of husbands who do not work to their full potential. They feel their husbands do not contribute their fair share to the family income.

Men are traditionally expected to contribute to family work by providing household maintenance. Such maintenance consists primarily of repairs, light construction, mowing the lawn, and other activities that are consistent with instrumental male norms. (But, as one woman asked, how often do you have to repair the toaster or paint the porch?) Men often contribute to housework and child care, although their contribution may not be notable in terms of the total amount of work to be done. Men tend to see their role in housekeeping or child care as "helping" their partner, not as assuming equal responsibility for such work. Husbands become more equal partners in the family work when they and/or their wives have egalitarian views of family work or when such a role is pressed upon them by either circumstantial necessity or ultimatum (Hochschild, 1989; Greenstein, 1996). Of course, men who believe they should act as traditional providers will resist performing more housework or do so only very reluctantly, whether their wives are employed outside the home or not. In marriages where both spouses share a traditional gender ideology (traditional beliefs about what each should contribute to paid and family work), men's low level of household participation will not be problematic.

WOMEN'S FAMILY WORK

A woman's family work may consist of unpaid work as homemaker and paid work outside the home as an employed or self-employed worker. In traditional divisions of labor, her work as homemaker is considered to be her primary role regardless of her employment status. Because housework is performed by women and is unpaid, however, it has been denigrated as "women's work," inconsequential and having no monetary value.

Women's family work is considerably more diverse than men's, permeating every aspect of the family. It ranges from housekeeping to child care, maintaining kin relationships to organizing recreation, socializing children to caring for aged parents and in-laws, and cooking to managing the family finances, to name but a few of the tasks. Ironically, family work is often invisible to the women who do most of it (Brayfield, 1992).

Although most women do earn salaries as paid employees, contributing more than 40 percent of family income in dual-earner households, neither traditional women nor their partners regard employment as a woman's fundamental role (Coontz, 1997). Women are not duty-bound to provide; they are duty-bound to perform household tasks (Thompson and Walker, 1991). No matter what kind of work the woman does outside the home or how nontraditional she and her husband may consider themselves to be, there is seldom equality when it comes to housework.

The greatest determinant of the amount of time a woman spends on housework is her employment status. Women spend less time on housework if they are employed (Coltrane, 2000; Greenstein, 1996). Men may help their partners with washing dishes, vacuuming, or doing the laundry, but as long as they merely "help," household responsibility continues to fall on women.

Sociologist Ann Oakley (1985) describes the **homemaker role,** which has four primary aspects:

- Exclusive allocation to women, rather than to adults of both sexes
- Association with economic dependence
- Status as nonwork, which is distinct from "real," economically productive paid employment
- Primacy to women—that is, having priority over other women's roles.

Housework consists primarily of household work and, if children are present, child-rearing tasks. It has the following characteristics (Bird and Ross, 1993; Hochschild, 1989; Oakley, 1985):

- Housework tends to isolate a woman at home. She cleans alone, cooks alone, launders alone, and cares for children alone. Loneliness is a common complaint.

© Laurie DeVault Photography

Researchers Linda Thompson and Alexis Walker (1991) observe, "Family work is unseen and unacknowledged because it is private, unpaid, commonplace, done by women, and mingled with love and leisure."

INDUSTRIALIZATION "CREATES" THE TRADITIONAL FAMILY

The American family has radically altered its economic functions since colonial times. During the colonial era, the family was basically an economic and social institution, the primary unit for producing most goods and caring for the needs of its members. The family planted and harvested food, made clothes, provided shelter, and cared for the necessities of life. Each member was expected to contribute economically to the welfare of the family. Husbands plowed, planted, and harvested crops. Wives supervised apprentices and servants, kept records, cultivated the family garden, assisted in the farming, and marketed surplus crops or goods, such as grain, chickens, candles, and soap. Older children helped their parents and, in doing so, learned the skills necessary for later life.

In the nineteenth century, industrialization transformed the face of America. It also transformed American families from self-sufficient farm families to wage-earning urban families. As factories began producing farm machinery such as harvesters, combines, and tractors, significantly fewer farm workers were needed. Looking for employment, workers migrated to the cities, where they found employment in the ever-expanding factories and businesses. Families no longer worked together in the fields or in the home producing food, clothing, and other necessities; instead these were produced by large-scale farms and factories. Food and other goods were purchased with wages earned in factories and stores. Because goods were now bought rather than made in the home, the family began to shift from being primarily a production unit to being a consumer and service-oriented unit.

With this shift, a radically new division of labor arose in the family. Men began working outside the home in factories or offices for wages to purchase the family's necessities and other goods. Men became identified as the family's sole providers or "breadwinners." Their work was given higher status than women's work because it was paid in wages. Men's work began to be identified as "real" work.

Industrialization also created the housewife, the woman who remained at home attending to household duties and caring for children. Previously, men and women were interdependent within the family unit because women produced many of the family's necessities. Because much of what the family needed had to be purchased with the husband's earnings, the wife's contribution in terms of unpaid work and services went unrecognized, much as it continues today.

Industrialization removed many of women's productive tasks from the home and placed them in the factory. Even then, although women made fewer things at home, they nevertheless continued their service and child-rearing roles: cooking, cleaning, sewing, raising children, and nurturing the family. With the rise of a money economy in which only paid work was recognized as real work, women's work in the home went unrecognized as necessary and important labor (Ferree, 1991).

Without its central importance as a work unit, the family became the focus and abode of feelings. In earlier times, the necessities of family-centered work gave marriage and family a strong center based on economic need. The emotional qualities of a marriage mattered little as long as the marriage produced an effective working partnership. Without its productive center, however, the family focused on the relationships between husband/wife and parent/child. Affection, love, and emotion became the defining qualities of a good marriage.

- Housework is unstructured, monotonous, and repetitive. A homemaker never feels that her work is done. There is always more dust, more dishes, more dirty laundry.
- The full-time homemaker role is restricted. For a woman who is not also employed outside the home, homemaking (which may also include child rearing) is essentially her only role. By contrast, a man's role is dual—employed worker and husband/father. If the man finds one role to be unsatisfactory, he may find satisfaction in the other. It is much more difficult for a woman to separate the satisfaction she may get from being a mother from the dissatisfaction of household work.
- Housework is autonomous. This is one of the most well-liked aspects of the homemaker role. Being her own boss allows a woman to direct a large part of her own life. In contrast to employment, this is a definite plus.

- Homemakers work long days and nights. This is especially true for employed mothers. For employed women, there is a "second shift" of work that awaits them at home. This adds a considerable burden onto women's paid work day that men don't experience.
- Homemaking can involve child rearing. For many homemakers, this is their most important and rewarding work. They enjoy the interactions with their infants and children. Although they can also feel overwhelmed and exhausted by the demands of child rearing, relatively few would choose not to have children.
- Homemaking often involves role strain and role conflict, especially for employed mothers as they try to perform their various obligations as parent, spouse, and worker.
- Housework is unpaid. The homemaker is "paid" in goods and services, primarily from her husband. Because she is unpaid, she is often dependent. She gets no increase in pay regardless of her skill. Her standard of living depends on her husband, not her own efforts. If the homemaker is a single woman, the husband-wife reciprocity is absent. There is no payment in goods and services from a spouse.

REFLECTIONS

List the different tasks that make up family work in your family. What family work is given to women? To men? On what basis is family work divided? Is it equitable?

Most full-time housekeepers feel the same about housework: It is routine, unpleasant, unpaid, and unstimulating, but it provides a degree of autonomy. Full-time male houseworkers, however, do not as often call themselves *housekeepers* or *homemakers.* Instead, they identify themselves as retired, unemployed, laid off, or disabled (Bird and Ross, 1993). Increasingly, they may call themselves "househusbands" but they are less likely to do so than full-time female homemakers are to call themselves "housewives."

Many women find satisfaction in the homemaker role, even in housework. Young women, for example, may find increasing pleasure as they experience a sense of mastery over cooking, entertaining, or rearing happy children. If homemakers have formed a network among other women—such as friends, neighbors, or relatives—they may share many of their responsibilities. They discuss ideas and feelings and give one another support. They may share tasks as well as problems.

Women in the Labor Force

Women have always worked outside the home. In fact, like many of today's families, early American families were **co-provider families**—families that were economic partnerships dependent on the efforts of both the husband and wife. Although women may have lacked the economic rights that men enjoyed, they worked with or alongside men in the tasks necessary for family survival (Coontz, 1997). Beginning in the early nineteenth century, "work" and "family work" were separated. Men were assigned the responsibility for the wage-earning labor that increasingly occurred away from the home in factories and other centralized workplaces. Women stayed within the home, tending to household tasks and child rearing. But this gendered division of labor was never total. Single women, who have no male partner on whom to depend for financial support, have traditionally been members of the paid labor force. There have also been large numbers of employed mothers, especially among lower-income and working-class families, African Americans, and many other ethnic minorities. By the late 1970s, the employment rate of white women began to converge with that of African-American women (Herring and Wilson-Sadberry, 1993).

The most dramatic changes in women's labor force participation have occurred since 1960, resulting in the emergence of a family form in which both husbands and wives/mothers are employed outside the home. Although many viewed that family type as abnormal, in the 1980s married women's employment came to be seen as the norm. Recent research indicates that women's employment has positive rather than negative effects on marriages and families (Crosby, 1991).

In 2002, more than 67 million women were employed in the civilian labor force. Women comprised

Understanding Yourself

THE DIVISION OF LABOR: A MARRIAGE CONTRACT

How do you expect to divide household and employment responsibilities in marriage? More often than not, couples live together or marry without ever discussing basic issues about the division of labor in the home. Some think that things will "just work out." Others believe that they have an understanding, although they may discover later that they do not. Still others expect to follow the traditional division of labor. Often, however, one person's expectations conflict with the other's.

The following questions cover important areas of understanding for a marriage contract. These issues should be worked out before marriage. Although marriage contracts dividing responsibilities are not legally binding, they make explicit the assumptions that couples have about their relationships.

Answer these questions for yourself. If you are involved in a relationship, live with someone, or are married, answer them with your partner. Consider putting your answers down in writing.

- Which has the highest priority for you: marriage or your job? What will you do if one comes into conflict with the other? How will you resolve the conflict? What will you do if your job requires you to work 60 hours a week? Would you consider that such hours conflict with your marriage goals and responsibilities? What would your partner think? Do you believe that a man who works 60 hours a week shows care for his family? Why? What about a woman who works 60 hours a week?
- Whose job or career is considered the most important—yours or your partner's? Why? What would happen if both you and your partner were employed and you were offered the "perfect" job 500 miles apart from your partner? How would the issue be decided? What effect do you think this would have on your marriage or relationship?
- How will household responsibilities be divided? Will one person be entirely, primarily, equally, secondarily, or not at all responsible for housework? How will this be decided? Does it matter whether a person is employed full-time as a salesclerk or a lawyer in deciding the amount of housework he or she should do? Who will take out the trash? Vacuum the floors? Clean the bathroom? How will it be decided who does these tasks?
- If you are both employed and then have a child, how will the birth of a child affect your employment? Will one person quit his or her job or career to care for the child? Who will that be? Why? If both of you are employed and a child is sick, who will remain home to care for the child? How will that be decided?

46.3 percent of the labor force, and 60 percent of adult women were employed. In comparison, 74 percent of adult men were employed (U.S. Census Bureau, 2003, Tables 592, 596). African-American women and white women had virtually the same rate of labor force participation (61.8 and 59.3 percent respectively); Hispanic women were slightly less likely to be employed (57.6 percent).

Between 1960 and 1997, the percentage of married women in the labor force almost doubled—from 32 percent to 62 percent. During that same period, the number of employed married women between 25 and 34 years of age (the ages during which women are most likely to bear children) rose from 29 to 72 percent. Over 70 percent of married women with children were in the labor force in 2000, including 76.8 percent of those with children 6 to 17 years of age, and 60.8 percent of those with children age 6 or younger (U.S. Census Bureau, 2003, Table 597).

In 2002 there were more than 3 million single mothers in the labor force; of these, almost 60 percent had preschool-aged children (U.S. Census Bureau, 2003, Table 598).

WHY WOMEN ENTER THE LABOR FORCE

Four factors influence a woman's decision to enter the labor force (Herring and Wilson-Sadberry, 1993):

- *Financial factors:* To what extent is income significant? For unmarried women and single mothers, employment may be their only source of income. The income of married women may be primary or secondary to their husbands' incomes.
- *Social norms:* How accepting is the social environment for married women and mothers work-

© Cindy Charles/PhotoEdit

Women seek the same gratifications from paid work that men seek. These include but go beyond wages.

ing at paid jobs? Does the woman's partner support her? If she has children, do her partner, friends, and family believe that working outside the home is acceptable? After the 1970s, social norms changed to make it more acceptable for white mothers to hold a job.

- *Self-fulfillment:* Does a job meet needs for autonomy, personal growth, and recognition? Is it challenging? Does it provide a change of pace?
- *Attitudes about employment and family:* Does the woman believe she can combine her family responsibilities with her job? Can she meet the demands of both? Does she believe that her partner and children can do well without her as a full-time homemaker?

Like men, women enter the labor force for largely financial reasons. Economic necessity is a driving force for many women in the labor force. According to Stephanie Coontz (1997), women's incomes are what keep approximately a third of dual-earner couples from falling into poverty. Economic pressures traditionally have been powerful influences on African-American women. Among many married women and mothers, entry into the labor force or increased working hours are attempts to compensate for their husbands' loss in earning power due to inflation. Additionally, the social status of the husband's employment often influences the level of employment chosen by the wife (Smits, Utee, and Lammers, 1996).

Among the psychological reasons for employment are an increase in a woman's self-esteem and sense of control. A comparison between African-American and white women found that personal preference was the primary employment motivation for about 42 percent of African-American women and 46 percent of white women (Herring and Wilson-Sadberry, 1993). Employed women are less depressed and anxious than nonemployed homemakers; they are also physically healthier (Gecas and Seff, 1989; Ross, Mirowsky, and Goldsteen, 1991). A 34-year-old Latina mother of three told social psychologist Lillian Rubin (1994) the following:

> I started to work because I had to. My husband got hurt on the job and the bills started piling up, so I had to do something. It starts as a necessity and it becomes something else.
>
> I didn't imagine how much I'd enjoy going to work in the morning. I mean, I love my kids and all that, but let's face it, being mom can get pretty stale. . . . Since I went to work I'm more interested in life, and life's more interested in me.
>
> I started as a part-time salesperson and now I'm assistant manager. One day I'll be manager. Sometimes I'm amazed at what I've accomplished; I had no idea I could do all this, be responsible for a whole business.

There are two reasons why employment improves women's emotional and physical well-being (Ross, Mirowsky, and Goldsteen, 1991). First, employment decreases economic hardship, alleviating stress and concern not only for the woman herself but for other family members as well. A single parent's earnings may constitute her entire family's income. Second, an employed woman receives greater domestic support from her partner. The more a woman earns relative to what her partner earns, the more likely he is to share housework and child care.

WOMEN'S EMPLOYMENT PATTERNS

The employment of women has generally followed a pattern that reflects their family and child-care responsibilities. Because of these demands, women must consider the number of hours they can work and what time of day to work. Their decision to work in the labor force is determined, in part, by the availability of timely work hours and adequate child care.

Traditionally, women's employment rates dropped during their prime child-bearing years, from ages 20 to 34 years. But this is no longer true. In fact, women's employment during these years has more than doubled. In 2002, about 61 percent of married women (and 77 percent of married men) were employed (U.S. Census Bureau, 2003). As seen above, a majority of women with children are in the labor force, regardless of age of child, marital status, and racial or ethnic affiliation. Women no longer automatically leave the job market when they become mothers. Either they need the income or they are more committed to work roles than in previous generations (Coontz, 1997).

Because of family responsibilities, many employed women work either part-time or shifts other than the 9-to-5 workday. Furthermore, when family demands increase, wives, not husbands, cut back on their job commitments (Folk and Beller, 1993). As a result of family commitments, women tend to interrupt their job and career lives much more frequently than do men.

Researchers have found that a woman's decision to remain in the workforce or to withdraw from it during her childbearing and early child-rearing years is critical for her later workforce activities. If a woman chooses to work at home caring for her children, she is less likely to be employed later. If she later returns to the workforce, she will probably earn substantially less than women who have remained in the workforce.

Dual-Earner Marriages

Since the 1970s, inflation, a dramatic decline in real wages, the flight of manufacturing, and the rise of a low-paying service economy have altered the economic landscape. These economic changes have reverberated through families, altering the division of household roles and responsibilities. Today, more than 60 percent of families with children under 18 are two-earner families. This includes two-thirds of two-parent families with children 6 to 18, and 54 percent of families with children under 6.

The sources of the dual-earner, or "co-provider," household are many. Over the past 30 years, wages have *declined* for male high school and college graduates. Since 1973, men aged 25 to 34 have had their wages decline 25 percent. Sylvia Hewlett and Cornell West (1998) note that even during the economic expansion of the mid-1990s, men's wages dropped. They point out that "32 percent of all men between 25 and 34 when working full-time now earn less than the amount necessary to keep a family of four above the poverty line."

Among women, wages of high school graduates have also declined, but the drop was less because women started at lower wages. College-educated women saw their wages increase, though they remained well behind the wages paid to male college graduates (Vobejda, 1994).

In 2001, the median income among families who depended on the wages of a male breadwinner was $50,926. Families in which both husbands and wives were employed had median incomes of $73,407. Families in which wives worked and husbands didn't had median incomes of $39, 566 (U.S. Census Bureau, 2003, Table 690).

Economic changes have led to a significant increase in dual-earner marriages. Even though 61 percent of all married women were employed in 2002, the majority of employed women are still segregated in low-paying, low-status, low-mobility jobs—secretaries, clerks, nurses, factory workers, waitresses, and so on. Rising prices and declining wages pushed most of them into the job market. Employed mothers generally do not seek personal fulfillment in their work as much as they do additional family income. Their families remain their top priorities.

Dual-career families are a subcategory of dual-earner families. They differ from other dual-earner families insofar as both husband and wife have high-achievement orientations, greater emphasis on gender equality, and a stronger desire to exercise their capabilities. Unfortunately, these couples may find it difficult to achieve both their professional and family goals. Often they have to compromise one goal to achieve the other because the work world generally is not structured to meet the family needs of its employees, as one study points out (Berardo et al., 1987):

> The traditional "male" model of career involvement makes it extremely difficult for both spouses to pursue careers to the fullest extent

possible, since men's success in careers has generally been made possible by their wives' assuming total responsibility for the family life, thus allowing them to experience the rewards of family life but exempting them from this competing set of responsibilities.

TYPICAL DUAL-EARNERS: HOUSEWORK, CHILD CARE, AND THE SECOND SHIFT

We are increasingly seeing that marital satisfaction is tied to fair division of household labor (Blair, 1993; Pina and Bengston, 1993; Suitor, 1991). A husband's wielding a vacuum cleaner or cooking dinner while his partner takes off her shoes to relax a few moments after returning home from work is sometimes better than his presenting her a bouquet of flowers—it may show better than any material gift that he cares. In a world where both spouses are employed, dividing household work fairly may be a key to marital success (Hochschild, 1989; Perry-Jenkins and Folk, 1994; Suitor, 1991).

Although we traditionally separate housework, such as mopping and cleaning, from child care, in reality the two are inseparable (Thompson, 1991). Although fathers have increased their participation in child care some, they have done less in terms of swinging a mop or scrubbing a toilet. If we continue to separate the two domains, men will take the more pleasant child-care tasks of playing with the baby or taking the kids to the playground, while women take on the more unpleasant duties of washing diapers, cleaning ovens, and ironing. Furthermore, someone must do behind-the-scenes dirty work in order for the more pleasant tasks to be performed. As Alan Hawkins and Tomi-Ann Roberts (1992) note:

> Bathing a young child and feeding him/her a bottle before bedtime is preceded by scrubbing the bathroom and sterilizing the bottle. If fathers want to romp with their children on the living room carpet, it is important that they be willing to vacuum regularly. . . . Along with dressing their babies in the morning and putting them to bed at night comes willingness to launder jumper suits and crib sheets.

If we are to develop a more equitable division of domestic labor, we need to see housework and child care as different aspects of the same thing: domestic labor that keeps the family running. (See Hawkins and Roberts [1992] and Hawkins [1994] for a description of a program to increase male involvement in household labor.)

Housework

Standards of housework have changed over the last few generations, as wryly noted by Barbara Ehrenreich (1993):

"My wife works, and I sit on the eggs. Want to make something of it?"

> Recall that not long ago, in our mother's day, the standards were cruel but clear: Every room should look like a motel room. The floors must be immaculate enough to double as plates, in case the guests prefer to eat doggie-style. The kitchen counters should be clean enough for emergency surgery, should the need at some time arise, and the walls should ideally be sterile. The alternative, we all learned in Home Economics, is the deadly scorn of the neighbors and probably the plague.

The engine of change was not the vacuum cleaner—which, in fact, seemed to increase hours spent in housework, as it promised the possibility of immaculateness if one worked hard enough. What changed was that working women could no longer hold up the standards of their mothers—or of household product advertisers. They now spend less time on housework. But Ehrenreich advises those who miss the good old days: "For any man or child who misses the pristine standards of yesteryear, there is a simple solution. Pitch in!"

Evidence indicates that although men do "pitch in" they are nowhere near sharing the burden of housework. Housework remains clearly unevenly divided between women and men. As Coltrane (2000) reports, the average married woman does more than three times the amount of routine housework as the average married man (32 hours versus 10 hours per week). This includes the most time-consuming chores such as cooking, cleaning, grocery shopping, laundry, and cleaning up after meals. Other less "routine" and more occasional tasks, such as repairs, yard care, and bill-paying were more the responsibilities of men, with the average married man doing 10 hours of such tasks per week versus six for the average married woman.

As a result of the division of household tasks, employed women have more to do, experience more stress, and have access to less leisure time than married men. They not only *do more*, they also "almost invariably" take on the responsibility of having to manage the housework that men do (Coltrane, 2000). Furthermore, whether or not married women are employed seems to have little impact on the division of housework. Studies suggest that employed women do three times as much housework as men (Coltrane, 2000). Cohabiting women, however, do significantly less housework than married women. Whether men are married or cohabiting does not affect their housework—it remains minimal (Shelton and John, 1993). It seems that marriage, rather than living with a man, transforms a woman into a homemaker. Marriage seems to change the house from a space to keep clean to a home with husband/family to care for.

In a study of over a thousand married couples with children at home, women devoted over 40 hours a week to household tasks, while men averaged less than 11 hours. These tasks included preparing meals, cleaning house, washing dishes, laundry, shopping, paying bills, and outdoor work (Demo and Acock, 1993). Most of the women in this study were employed outside the home over 30 hours per week. Whether or not she was employed outside the home, the woman contributed nearly the same major proportion of the total family time spent in housework. Employed wives' proportion of total family hours spent on household chores was still about 72 percent, compared with 81 percent for nonemployed wives.

Men tended to perceive that they do more housework than they actually do. Although husbands of employed women contributed slightly more hours to household work than did husbands of nonemployed women, these differences were relatively small.

The typical pattern in dual-earning families is nicely captured by Arlie Hochschild's notion of the **second shift**—the period at the end of a long, paid workday when women return home to a second job, and men just return home. Calculating the different levels of male and female participation in paid work, housework, and child care, Hochschild (1989) suggested that women in dual-earner households worked 15 hours a week more than men did, amounting to an *extra month* of 24 hour days each year.

Among Hochschild's own sample of 50 dual-earner couples, 20 percent of them shared equal responsibility for housework. In 70 percent of the couples, men did a "substantial amount" of housework, meaning more than third but less than half. Finally, in 10 percent of the couples men contributed less than a third of the domestic work (Hochschild, 1989). Other studies estimate that men do between 20 percent and a third of all housework. As we saw in Chapter 2, among dual-earner couples in the United States, wives and husbands estimated that men did between 19 and 25 percent of all housework (Baxter, 1997). Remember, as with "thinking about the baby" and childcare, women also do more of the arranging, monitoring, planning, and supervising of domestic work. Thus, even when at the office, women do not entirely escape the burdens of their domestic responsibilities.

One might assume that the stresses and inequalities undermine women's well-being, but research on marital and mental health consequences tells a different story. Women in dual-earner families are mentally healthier than are full-time housewives (Crosby, 1991). In juggling multiple roles, they suffer less depression, experience more variety, interact with a wider social circle, and have less dependency on their marital or familial roles to provide all of their needed gratification. These psychological benefits accrue despite the unequal division of labor. Undoubtedly, more equal sharing of housework would only increase the benefits by decreasing the stress and resentment that otherwise build.

Various factors seem to affect men's participation in housework. Men tend to contribute more to household tasks when they have fewer time demands from their jobs—that is, early in their employment careers and after retirement (Rexroad and Shehan, 1987) or when they have jobs that demand fewer hours of actual time at work (Coltrane, 2000). They also participate more in housework when their hours and their wives' hours at work do not overlap (see the discussion of "shift work" below).

As wives' income rises, they report more participation by husbands in household tasks; increased income and job status motivate women to secure their husbands' sharing of tasks. However, research by Brines reviewed by Coltrane (2000) suggests that men who are economically dependent on their wives do less housework.

Whether a couple has children or not is a factor affecting how much men participate in household labor. Even though the presence of young children increases women's and men's housework, it also skews the division of housework in more traditional directions. Men tend to work more hours in their paid jobs and women tend to work fewer hours at paid work and more in the home. Women then end up with a larger share of housework than prior to the arrival of children.

One factor that is not as strong a determinant as we might predict is the husband's **gender ideology**—what he believes he ought to do as a husband and how paid and unpaid work should be divided. As Hochschild's research showed, even traditional men can become more egalitarian if wives successfully utilize direct and indirect gender strategies. In some instances, repeated requests might be enough. In other cases, ultimatums may be necessary. Aside from these direct strategies, more indirect strategies—helplessness, withholding sexual intimacy, and so on—may work with husbands who otherwise would not do more (Hochschild, 1989). Furthermore, necessity may create more male involvement. Wives with particularly demanding jobs or who work unusual hours (see below) may force their husbands to share more, simply because they are not themselves available (Gerson, 1993; Rubin, 1994). Husbands in one study (Benin and Agostinelli, 1988) felt most satisfied in the division of household labor if tasks were divided equally, especially if their total time spent on chores was small. Women appeared to be more satisfied if their husbands shared traditional women's chores (such as laundry) rather than limiting their participation to traditional male tasks (such as mowing the lawn). But even men who contribute many hours to household labor tend, still, to do more traditional male tasks (Blair and Lichter, 1991). Children appear to be less gender segregated with their household work than their parents (Simons and Whitbeck, 1991). African Americans are less likely to divide household tasks along gender lines than whites.

Child-Rearing Activities

Men increasingly believe that they should be more involved as fathers than men have been in the past. Yet the shift of attitudes has not greatly altered men's behavior. One study (Darling-Fisher and Tiedje, 1990) found that the father's time involved in child care is greatest when the mother is employed full-time (fathers responsible for 30 percent of the care compared with mothers' 60 percent; the remaining 10 percent of care is presumably provided by other relatives, babysitters, or child-care providers). The father's involvement is less when the mother is employed part-time (fathers' 25 percent versus 75 percent for mothers) and least when she is a full-time homemaker (fathers' 20 percent versus 80 percent for mothers). At the other extreme, roughly 2 million fathers are the primary child-care providers while their wives are at work.

A review of studies (Lamb, 1987) on parental involvement in two-parent families concluded the following:

- Mothers spend from three to five hours of active involvement for every hour fathers spend, depending on whether the women are employed or not.
- Mothers' involvement is oriented toward practical daily activities, such as feeding, bathing, and dressing. Fathers' time is generally spent in play.
- Mothers are almost entirely responsible for child care: planning, organizing, scheduling, supervising, and delegating.
- Women are the primary caretakers; men are the secondary.

Although mothers are increasingly employed outside the home, fathers need to pick up more of the slack at home. Children especially suffer from the lack of parental time and energy when their fathers do not participate more. If children are to be given the emotional care and support they need to develop fully, their fathers must become significantly more involved (Hochschild, 1989; Hewlett and West, 1998).

MARITAL POWER

An important consequence of women's working is a shift in the decision-making patterns in a marriage. Although decision-making power in a family is not based solely on economic resources (personalities, for instance, also play a part), economics is a major factor. A number of studies suggest that employed wives exert greater power in the home than nonemployed wives (Blair, 1991; Schwartz, 1994). Marital decision-making power is greater among women who are employed full-time than among those who are employed part-time. Wives have the greatest power when they are employed in prestigious work, are committed to it, and have greater income than their husbands. Conversely, full-time housewives may find themselves taken for granted and, because of their economic dependency on their husbands, relatively powerless (Schwartz, 1994).

Some researchers are puzzled about why many employed wives, if they do have more power, do not demand greater participation in household work on the part of their husbands. Joseph Pleck (1985) suggests several reasons for women's apparent reluctance to insist on their husbands' equal participation in housework. These include (1) cultural norms that housework is the woman's responsibility, (2) fears that demands for increased participation will lead to conflict, and (3) the belief that husbands are not competent.

MARITAL SATISFACTION

How do patterns of employment and the division of family work affect marital satisfaction? Traditionally, this question was asked only of wives, not husbands; even then, it was rarely asked of African-American wives, who had a significantly higher employment rate. In the past, married women's employment, especially maternal employment, was viewed as a problem. It was seen as taking away from a woman's time, energy, and commitment for her children and family. In contrast, nonemployment or unemployment was seen as a major problem for men. But it is possible that the husband's work may increase marital and family problems by preventing him from adequately fulfilling his role as a husband or father: He may be too tired, too busy, or never there. It is also possible that a mother's not being employed may affect the family adversely: Her income may be needed to move the family out of poverty, and she may feel depressed from lack of stimulation (Menaghan and Parcel, 1991).

How does a woman's employment affect marital satisfaction? There does not seem to be any straightforward answer when comparing dual-earner and single-earner families (Piotrkowski et al., 1987). In part, this may be because there are trade-offs: a woman's income allows a family a higher standard of living, which compensates for the lack of status a man may feel for not being the "sole" provider. Whereas men may adjust (or have already adjusted) to giving up their sole-provider ideal, women find current arrangements less than satisfactory. After all, women are bringing home additional income but are still expected to do the overwhelming majority of household work. Role strain is a constant factor for women, and in general, women make greater adjustments than men in dual-earner marriages.

Studies of the effect of women's employment on the likelihood of divorce are not conclusive, but they do suggest a relationship (Spitze, 1991; White, 1991). Many studies suggest that employed women are more likely to divorce. Employed women are less likely to conform to traditional gender roles, which potentially causes tension and conflict in the marriage. They are also more likely to be economically independent and do not have to tolerate unsatisfactory marriages for economic reasons. Other studies suggest that the only significant factor in employment and divorce is the number of hours the wife works. Hours worked may be important because full-time work for both partners makes it more difficult for spouses to share time together. Numerous hours may also contribute to role overload on the part of the wife (Greenstein, 1990).

African-American women, however, are not more likely to divorce if they are employed. This may be because of their historically high employment levels and their husbands' traditional acceptance of such employment (Taylor et al., 1991).

Overall, despite an increased divorce rate, in recent years the overall effect of wives' employment on marital satisfaction has shifted from a negative impact to no impact or even a positive impact. If there are negative effects, they generally result from specific aspects of a woman's job, such as long hours or work stress (Spitze, 1991).

The effect of a wife's full-time employment on a couple's marital satisfaction is affected by such variables as social class, the presence of children, and the husband's and wife's attitudes and commitment to her working. Thus, the more the wife is satisfied with her employment, the higher their marital satisfaction will be. Also, the higher the husband's approval of his wife's employment, the higher the marital satisfaction.

Data on the effects of the division of domestic labor on marital satisfaction and divorce indicate a relationship. Couples who share report themselves as happier and are less at risk of divorce than couples where men do little of the family work. This appears to be true regardless of whether couples' gender ideologies are traditional, egalitarian, or transitional (somewhere between the other two) (Hochschild, 1989).

Atypical Dual-Earners: Shift Couples and Peer Marriages

Although less common than the typical dual-earning couple, there are some interesting lifestyle variations among dual-earner couples. These lifestyles differ from more common two-earner couples in either of two ways: They have constructed household arrangements in which the parents work opposite, mostly nonoverlapping shifts, and thus take turns working outside the home and caring for children, or they have quite consciously adopted a belief in equality and fairness into how they divide up domestic responsibilities. As a result of either of these differences, such atypical couples show much higher rates of male participation in child care and housework than among more typical dual-earners. Let's briefly consider each of these types.

SHIFT COUPLES

In 2001, nearly 15 million Americans worked hours other than the typical 9-to-5 or 8:30-to-4:30 daytime shifts. Couples in which one spouse works such a shift and the other remains in a more typical shift are **shift couples.** Shift couples structure their home and work lives into a turn-taking, alternating system of paid work and family work. When one is at work, the other is at home. When the at-work partner returns home, the at-home partner departs for work, giving them a kind of "Hello, goodbye," lifestyle. Although 1997 Labor Department data indicate one out of seven women and nearly one in five men work other than typical daytime shifts, this understates the percentages of couples who utilize this arrangement. According to a recent AFL-CIO survey of 765 employed women age 18 or older, nearly half of all married or cohabiting employed women with children work a different shift than do their spouses or partners. For roughly a third of these women, this is an arrangement imposed on them without their choice or decision (Love, 2000). But many select such schedules, particularly for the peace of mind and reduced costs associated with child care (Rubin, 1994).

When this lifestyle is the product of choice, shift couples see it as a reasonable trade-off. Through it they stress the importance of child rearing over the importance of marital relations. Spouses may not see each other very much, but they manage to communicate frequently via notes on refrigerators, calls during breaks, or even e-mail. Complicated color-coded calendars are often on display in kitchens, indicating the coordination and cooperation needed to make or keep such arrangements viable. Significantly, in order for the household to function, men are pressed to do a greater share of domestic work and especially child care than one finds among either traditional couples or more typical dual-earners (Rubin, 1994). If wives work second shift (late afternoon through midnight) or third shift (late night through morning) jobs, husbands must feed children dinner or breakfast, see that they do their homework, take baths, get to bed or get up for school, make lunches to take to school, and so on. With no one else there to do these child-care tasks, fathers do them.

PEER MARRIAGES

Among some dual-earner couples, it is agreed that household tasks will be divided along principles of fairness. Many couples believe their family's division of labor is fair (Spitze, 1991). Among couples who can afford household help, husbands may be excused from many household chores, such as cleaning and mopping. By virtue of their incomes, they are allowed to "hire" substitutes to do their share of housework (Perry-Jenkins and Folk, 1994).

It is important to note that an equitable division is not the same as an equal division. Relatively few couples, in fact, divide housework fifty-fifty. For women, a fair division of household work is more important than both spouses putting in an equal number of hours. There is no absolute standard of fairness, however (Thompson, 1991). What is fair is determined differently by different couples. Because most women work fewer hours than men in paid work, and wives tend to work more hours in the home, some women believe that the household labor should be divided proportionately to hours worked outside the home. Other women believe that it is equitable for higher-earning husbands to have fewer household responsibilities. Still others believe that the traditional division of labor is equitable, as household work is women's work by definition. Middle-class women are more likely to demand equity, whereas equity is less important for working-class women, who are more traditional in their gender-role expectations (Perry-Jenkins and Folk, 1994; Rubin, 1994).

Peer marriages (or *postgender marriages,* to use Risman's and Johnson-Sumerford's term) take concerns for fairness and sharing to heart in how they structure each facet of their relationships. Rarer than shift couples, they too depart from the model of typical dual-earners described above. Whereas shift arrangements may be the result of choice, necessity, or circumstance, peer relationships typically emerge from egalitarian values or conscious intent. Peer or postgender couples base their relationships on principles of deep friendship, fairness, and sharing. Hence, they monitor each other's level of commitment and involvement, maintain equally valued investments in their paid work, and share in household tasks and child care.

Research by Pepper Schwartz (1994) and Barbara Risman and Danette Johnson-Sumerford (1997) indicate that such relationships avoid many of the trappings that frequently accompany more traditional divisions of labor, including female powerlessness and resentment and male ingratitude or lack of respect. Furthermore, children receive attention and care from both parents, and men develop deeper relationships with their children than one commonly finds. Although such couples are rare, they show that the inequities in either the traditional or more typical dual-earner household are not inevitabilities. Couples can commit themselves to "doing it fairly" (Risman and Johnson-Sumerford, 1997).

COPING IN DUAL-EARNER MARRIAGES

Dual-earner marriages are here to stay. They are particularly stressful today because society has not pursued ways to alleviate the work-family conflict. The three greatest social needs in dual-earner marriages are (1) redefining gender roles to eliminate role overload for women, (2) providing adequate child-care facilities for working parents, and (3) restructuring the workplace to recognize the special needs of parents and families.

Coping strategies include reorganizing the family system and reevaluating household expectations. Husbands may do more housework. Children may take on more household tasks than before. Household standards—such as a meticulously clean house, elaborate meal preparation, and washing dishes after every meal—may be changed. Careful allocation of time and flexibility assist in coping.

Dual-earner couples often hire outside help, especially for child care, which is usually a major expense for most couples. One of the partners may reduce his or her hours of employment, or both partners may work different shifts to facilitate child care (but this usually reduces marital satisfaction as a result) (White and Keith, 1990).

The goal for most dual-earner families is to manage their family relationships and their paid work to achieve a reasonable balance that allows their families to thrive rather than merely survive. Achieving such balance will continue to be a struggle until society and the workplace adapt to the needs of dual-earner marriages and families.

REFLECTIONS

The chances are very good that if you cohabit or marry, you will be in a dual-earner relationship. How will you balance your employment and relationship or family needs?

AT-HOME FATHERS AND BREADWINNING MOTHERS

An additional departure from both the typical dual-earner and the traditional family is the family type in which spouses switch places or reverse roles. Although the term **role reversal** may be more familiar to us, it may be more accurate to suggest that what spouses do is switch traditional places; husbands move into the domestic realm and provide housework and child care while wives support the family financially with outside paid work. Calling them role reversed implies that men do and experience what women traditionally experienced and that wives approach work and wage earning as husbands traditionally did. This appears not to be the case (Russell, 1987; Cohen and Durst, 2000).

Although it is hard to estimate with precision, there are, *at minimum,* more than a million men who stay at home and care for their households and children while their wives work outside the home. Of the almost 25 million married couple families with children under 18, in 2002, in 5.3 percent of them (1.3 million) mothers were employed and fathers were not (U.S. Census Bureau, 2003, Table 599). A Bureau of Labor Statistics report indicated that of unemployed men 25 to 54 years of age, 8.4 percent elected not to look for work due to family responsibilities. In 1993, the Census Bureau estimated that 1.9 million unemployed men were the at-home caregivers to their children (Marin, 2000). In addition to the inexact nature of these sorts of estimates, it is likely that they undercount the numbers of households that ever take on this structure. With layoffs, more long-term unemployment, job changes, disabilities, and educational pursuits, in many families spouses may switch places for at least a period of time. As these examples illustrate, this arrangement may result from either choice or circumstances.

What happens to such couples? Based on research conducted in the 1980s by Graeme Russell and more recent research by Theodore Cohen and John Durst (2000), we can point to five areas in which couples experience some impact from having switched places:

1. *Economic impact:* Couples live on less money but spend less on child care. Hence, they may not suffer dramatic declines, especially if women's careers are enhanced and men's occupations were not high paying. Men gain an opportunity to take a "time-out," refocus, and try new career possibilities. They do, however, surrender the provider status and confront the reality of economic dependency. Interestingly, this dependency does not seem to have the same marital consequences for men as it does for at-home women.
2. *Social impact:* Socially, men experience some isolation, as they lose the primary source of social interaction—the workplace. Additionally, couples become the targets of curiosity, or even criticism, for their choices. Men, however, also receive supportive responses, especially from women. Women often receive envious reactions, especially from co-workers. In general, at-home fathers become visible in their domestic role in contrast to the invisibility that traditionally befell housewives.
3. *Marital impact:* This lifestyle leads to high levels of male involvement in housework and child care. While men don't take over everything to the same extent that housewives do, they are likely to share or do most of the domestic work. Additionally, couple relationships change. Whereas Russell (1987) found the changes to be negative, Cohen and Durst (2000) found high levels of communication, empathy, and appreciation among the couples they studied. Wives, in particular, know about the sacrifices and risks their husbands take by staying home and support them in ways that breadwinning husbands probably don't support housewives.
4. *Parental impact:* Perhaps the most noticeable area of impact is the enlarged relationship between fathers and children. Fathers get to know their children in ways that might not otherwise

LIFE AS AN AT-HOME FATHER

The following comments come from an interview with a 39-year-old, at-home father of three children: two daughters, ages 6 and 4, and a 1-year-old son. He describes how he and his wife became "role reversed," and then describes in great detail his daily routines and responsibilities.

On Becoming an At-Home Father

Let me see if I can remember. I think I was off Monday and Tuesday and I worked Wednesday, Thursday, and Friday. So we found a new day care called Tiny Town, and we had (our oldest daughter) there first. We had tried to put her in a woman's situation—that watches children—until she started telling me she was taking my daughter to the beach every day. And I said no. What really made me change my hours was when Tiny Town knocked down a wall and they put glass up. And I was the one who dropped my daughter off, before I used to go to work. That was the other thing that I did—I went to work late. 'Cause we still worked the opposite shift. My wife and I saw each other when we did, but I was more concerned . . . I really didn't believe in day care—to this day I don't believe in day care. I don't like it. So I used to go to work around 2:00, and I would drop her off before then. And what really broke my heart, I would say what created me—my daughter was in the room, on the other side of the glass and I was watching her play and I was watching her do things. And I watched her open her arms and go to this complete stranger, and I said, "Is that why I had children?", "Is that why my wife had children?", "All these people with children putting them in day care, are they missing something?", "Is there something wrong here?" And that's when I started going to work at 4:00, and then 5:00—and then I stopped (going to work).

On a Typical Day

Well I take care of my neighbor's daughter in the morning, so she's here at 7:30. So, Monday, Tuesday, and Wednesday is cereal day—so everyone gets cereal. Thursday is eggs. Friday is French toast or pancakes. We haven't gotten to the hot cereal yet because every time I give him hot cereal . . . it's just sitting there. Then (my son) gets up—I can honestly say that since I've had him, I can't tell you that I have a set schedule. I've never been so off balance in my life. I was saddened when my daughter went to school, I missed her. But now there's (my youngest daughter), "daddy, daddy"—I hear my name all day long. The best way that I can say it to you is that whether they're human, animal or bird—they all say my name constantly.

I've got my schedule where I start at that corner where the coffee pot is and I work my way around. If it gets dirty again, it's there until tomorrow. My dogs have a shedding problem, which only concerns me while there's one crawling around 'cause (my son) picks up every little bit of hair, he'll become like a dog himself. So vacuuming might be two times a day, alright? Then my routine in the morning is I feed the girls, he gets up . . . I change him, I get him down here, I feed him. After my son's done I usually drag him inside—we got cartoons on. The girls and I are all sittin' there watchin' cartoons or I'll be in here taking care of the kitchen. Then I'm on the stairway looking out that middle window watching for the bus for the girls 'cause I don't want them waiting at the bus stop 'cause it's too cold. When I see the bus, I kiss them good-bye and they run to the bus stop. They get on the bus, now I have (my other daughter) and I have (my son). My younger daughter sometimes doesn't like to eat with the rest of them because she's not hungry when she gets up, so then I feed her whatever she wants. I kind of run my house like a restaurant, meaning you don't have to have exactly what I'm making—I'll make something different for my wife, something different for myself . . . that's very rare but it's no problem for me. Then there's play time for him, there's play time for her, then there's her school time where she's learning how to trace her letters and doing that—or I put her on the computer; she does the computer. Then I go through my routine of cleaning and straightening out. Then there's my break time when he takes a nap. My

be possible, and children get to see fathers in nontraditional ways. Mothers maintain the same sorts of relationships as other employed mothers do with their children, but with greater peace of mind. Children are not in day care, at the sitter, or home alone. They are home with dad.

5. *Personal impact:* Being an at-home father changes the ways men look at their lives, resulting in a reshuffling of priorities and the construction of a new social identity. Breadwinning mothers may also enlarge their sense of themselves as providers, take advantage of the at-

break time will usually be when I go on the computer. I go up and lay on my bed and watch the weather channel or CNN—I'm not a soap opera person. Basically—then the dogs are crying, I've gotta walk the dog . . . then the birds have to be fed and cleaned. As much of the responsibility I can take off of my wife—she has nothing to do. She does the laundry. I take care of cleaning. There's the everyday cleaning and then there's the cleaning that you don't do everyday like taking the curtains down and cleaning the curtains. People look at me, when I say these things, and they're like, "Are you an ordinary guy?" I'm like, "Hey it's gotta be done—someone's gotta do it." And then Monday, Wednesday, and Friday (my other daughter) goes to nursery school so my day . . . I would say in the beginning when I started this, I was so frustrated but now I'm getting used to it because (my son) is getting older—his naps are changing. So I get him dressed, I get her ready—sometimes—I tell her for three hours sometimes, "Get dressed. Get your shoes and socks on." "Okay, okay . . ." —now I'm ready to leave she still don't have her shoes and socks on. Now I've got to get her shoes and socks on. And while I'm putting her shoes and socks on, he took his shoes and socks off. Then I have to put his shoes and socks on—then now that I have his shoes and socks on what does he do? He does this all the time—poop. And you can't put him in the car. And now you gotta run back upstairs, take off his coat, take everything off, change his diaper . . . "Okay, are we ready to go now?" Now we're ready to go, we get in the car. And then I pray, it's only a mile down the road, that he doesn't go to sleep 'cause I'd rather him come home and sleep so I can have my time. But lately, he gets into the car, she's in the car . . . and by the time I get to the school, he's sound asleep. I have to take him out of the car, I gotta hold him—the classroom is all the way down at the end of the hall. I gotta hold this 30-pound kid all the way down the end of the hall, drop her off and walk back. Lately he's been okay. I have to get him in the car, he's still sleeping. I come home, bring him up to his room and then it's my time to clean the bird cage. Out of my three children, my son is the nastiest. He cries, he has an attitude—he wants more attention, he wants to play with me. There are times everyday—actually I would say one to two times a day—I go up and I play with him. Then when my daughters are home from school, I'm making it their responsibility until dinner time—that they go up and that they play with their brother. But, basically that's it in a nut shell.

My wife thinks I'm crazy because I take on all of these responsibilities. I mean in one day you can hear, "Daddy, I'm thirsty," and my son is not talking yet, so he's crying. And then the dog starts barking wanting to go outside . . . she's my fetcher. She likes to play, she'll bring you the ball—so she'll bark all day long or she'll keep her nose in my hand telling me she wants to play. And if the birds are out of food all of a sudden they start screaming and I don't hear it as screaming—I hear it all as, "daddy." "Daddy, daddy," . . . I would say that as far as role reversal goes, I'm a woman and she's a man. I tell people—I have female friends that I tell, "I have husband problems." She comes home from work and everything's done—she's home for ten minutes, the kids get on her nerves and she's yelling at them . . . she's only with them ten minutes and she's yelling at them.

My wife leaves between 7:00 and 7:30 each morning and gets home between 5:30 and 6:00. Once she leaves . . . nobody else is here. I'm basically here from 7:30 to 6:00 every day. (And often she) has to go out of town. Because my wife is an executive, she has business meetings that go into the evening. She's got dinner dates. Sometimes she takes her employees out. (M)y wife thinks working is hard, and I've told her, "I don't care what you do, work is never as hard as what I'm doing or what we have done."

In a regular husband working - female home situation—you would say the female is in charge of cleaning the house, taking care of the laundry, and taking care of the children. Now with summertime, what are you responsible for? Or it snows, what are you responsible for? You're responsible for cutting the lawn, cutting the hedges, washing your cars, changing the oil and shoveling. I still do all of that—so where's my relief?

SOURCE: Cohen and Durst, 2000.

home resource, and make work a larger component of their own identities.

The increase in both actual involvement and social visibility of at-home fathers can be seen in a variety of ways and places. There are a variety of Web sites (such as Daddyshome.com or FullTimeDad.com), a number of newsletters (such as *At-Home Dad,* which also has a Web site), and an annual convention, which in 1999 drew more than 80 men from 20 different states, all catering to the needs and issues confronting at-home fathers (Marin, 2000).

Although only a small percentage of men are full-time caregivers, men are increasing their involvement in child care.

© Okoniwieski/The Image Works

There is good reason to think that this lifestyle will increase in coming years, but it is difficult to know by how much.

Employment and the Family Life Cycle

Throughout the life cycle, families try to balance work needs and family needs, which change as we pass through different stages of the family life cycle. Studies indicate that women spend considerably more total time in work and family roles than do men. This is because, in addition to out-of-home employment, they often spend more time with work in the home. Women have higher levels of role overload in every stage of the life cycle (Higgins, Duxbury, and Lee, 1994).

Much marital conflict in the contemporary family grows out of inequities between male and female work and family role experiences and expectations (Rogers, 1996). What it means to be male and to be female is not only influenced by biology, but even more by the way in which families define those roles in their work and home life. Role taking and role making are negotiated and renegotiated all through the family life cycle and are influenced by changing patterns in society (Zvonkovic et al., 1996).

Families may follow several patterns in combining family and employment responsibilities. Hypothetically, an individual family may be a typical dual-earner couple until they have children, moving to a traditional arrangement for a time when their first child is born. They may then switch back to being a dual-earner couple but working opposite shifts to handle their child-care needs. After a second child is born the husband may stay home for a while, allowing the wife to devote herself to occupational mobility and affording the husband a chance to switch to a new track. However, the rapid changes in society often leave couples without clear models for the allocation of roles and responsibilities. Such uncertainty sometimes leads to major differences in expectations and to feelings of exploitation and being misunderstood by both husbands and wives.

To reduce some of the complexity of the dual earner lifestyle, many couples display a pattern of sequential work/family role staging. This pattern reflects the adjustments women try to make in balancing work and family demands. Many of women's choices about employment and careers are based on their plans for a family and whether and when they will want to work. The key event is first pregnancy. Prior to pregnancy, most married women are employed. When they become pregnant, however, they begin leaving their jobs and careers to prepare for the transition to parenthood. By the last month of pregnancy, 80 percent have left the workforce. Within a year, more than half of these women have returned to employment. Most women who leave their paid work do so because of impending birth. Those who return to employment are strongly motivated by economic considerations or need.

There are four common forms of sequential work/family patterns:

- *Conventional:* A woman quits her job after marriage or the birth of her first child and does not return.
- *Early interrupted:* A woman stops working early in her career in order to have children and resumes working later.
- *Later interrupted:* A woman first establishes her career, quits to have children (usually in her thirties), and then returns.
- *Unstable:* A woman goes back and forth between full-time paid employment and homemaking, usually according to economic need.

A major decision for a woman who chooses sequential work/family role staging is at what stage in her life to have children. Should she have them early or defer them until later? As with most things in life, there are pros and cons. Early parenthood allows women to have children with others in their age group; they are able to share feelings and common problems with their peers. It also enables them to defer or formulate career decisions. At the same time, however, if they have children early, they may increase economic pressures on their beginning families. They also have greater difficulty in reestablishing their careers.

Women who defer parenthood until they reach their middle career stage often are able to reduce the role conflict and economic pressures that accompany the new parent/early career stage of the traditional pattern. Such women, however, may not easily find other new mothers of the same age with whom to share their experiences. They may find the physical demands to be greater than anticipated. Some may decide that they do not want children at all because motherhood would interfere with their careers.

In dual-earner families interrole conflict is often high as parents try to balance family and work obligations.

REFLECTIONS

Which work/family pattern will you adopt (or have you adopted)? What would its benefit be for you? Its drawbacks? Which pattern did your family of orientation adopt? What were its benefits for your parents? Its drawbacks? Does their experience influence your choice of patterns? How would single parenting affect the work/family pattern?

Family Issues in the Workplace

Many workplace issues, such as economic discrimination against women, occupational stratification, adequate child care, and an inflexible work environment, directly impact families. They are more than economic issues—they are also family issues.

DISCRIMINATION AGAINST WOMEN

A woman's earnings have a significant impact on family well-being, whether the woman is the primary or secondary contributor to a dual-earner family or the sole provider in a single-parent family. Thus, economic discrimination against women and sexual harassment are also important family issues.

Economic Discrimination

The effects of economic discrimination can be devastating for women. In 1997, women in the United States made 74 cents for every dollar that men earned. Median earnings of men who worked full-time, year-round were $38,884 in 2001. For women, the median was $29, 680,

Because of the great difference in women's and men's wages, many women are condemned to poverty and are forced to accept welfare and its accompanying stigma. Wage differentials are especially important to single women. Women face considerable barriers in their access to well-paying, higher-status jobs (Bergen, 1991). Although employment and pay discrimination are prohibited by Title VII of the 1964 Civil Rights Act, the law did not end the pay discrepancy between men and women. Much of the earnings gap is the result of occupational differences, gender segregation, and women's tendency to interrupt their employment for family reasons and to take jobs that do not interfere extensively with their family lives. Earnings are about 30 to 50 percent higher in traditionally male occupations, such as truck driver or corporate executive, than in predominantly female or sexually integrated occupations, such as secretary or schoolteacher. The more an occupation is dominated by women, the less it pays.

Sexual Harassment

Sexual harassment is a mixture of sex and power, with power often functioning as the dominant element. Such harassment may be a way to keep women in their place. **Sexual harassment** can be defined as two distinct types of harassment: (1) the abuse of power for sexual ends and (2) the creation of a hostile environment. In abuse of power, sexual harassment consists of unwelcome sexual advances, requests for sexual favors, or other verbal or physical conduct of a sexual nature as a condition of instruction or employment. Only a person with power over another can commit the first kind of harassment. In a **hostile environment,** someone acts in sexual ways so as to interfere with a person's performance by creating a hostile or offensive learning or work environment. Sexual harassment is illegal.

Issues of sexual harassment are complicated in the workplace because work, like college, is one of the most important places where adults meet potential partners. As a consequence, sexual undercurrents or interactions often are present. Flirtations, romances, and affairs are common in the work environment. Drawing the line, especially for men, between flirtation and harassment can be filled with ambiguity.

Some estimate that as many as half of employed women are harassed during their working years. Few harassed women report their harassment (Koss et al., 1994). There are significant gender differences that may contribute to sexual harassment. First, men are generally less likely to perceive activities as being harassing than are women (Jones and Remland, 1992; Popovich et al., 1992). Second, men misperceive women's friendliness as sexual interest (Johnson et al., 1991; Stockdale, 1993). Third, men are more likely to perceive male-female relationships as adversarial (Reilly et al., 1992). In addition to gender differences, power differences also affect perception. Behaviors exhibited by a supervisor, such as personal questions, are more likely to be perceived as sexual harassment than are the same behaviors of a coworker.

Sexual harassment can have a variety of consequences. One study of the workplace (Gutek, 1985) found that 9.1 percent of the women and 1 percent of the men quit their jobs because of harassment; almost 7 percent of the women and 2 percent of the men were dismissed from their jobs as part of their harassment. Victims often report depression, anxiety, shame, humiliation, and anger (Paludi, 1990).

LACK OF ADEQUATE CHILD CARE

As mothers enter the workforce in ever-increasing numbers, high-quality, affordable child care has become even more important. For many women, especially those with younger children and single mothers, the availability of child care is critical to their employment. In 1995, 60 percent of children under 6 were in some type of nonparental child-care arrangement. This represents almost 13 million children (U.S. Census Bureau, 1998, Table 634).

For a majority of employed mothers with children aged 5 to 14, school attendance is their primary day-care solution. Women with preschool children, however, do not have that option; in-home care by a relative is their most important resource. As more mothers with preschool children become employed, families are struggling to find suitable child-care arrangements. This may involve constantly switching arrangements, depending on who or what is available and the age of the child or children (Atkinson, 1994). One study found that 23 percent of employed mothers had to change child-care arrangements at least once in the previous year (Casper, Hawkins, and O'Connell, 1994). Relatives, especially fathers and grandparents, are important in caring for children, tending to the needs of almost 40 percent of children (Brayfield, 1995). Married women are more likely to use fathers, and single women to use other relatives (Folk and Beller, 1993). Twenty-eight percent of preschool children are cared for at home by fathers, grandparents, or other relatives. An additional 16 percent are cared for by a grandparent or other relative in the family member's home. Forty-four percent of preschool-aged children of employed mothers were cared for by fathers or other relatives. Twenty-one percent of children were cared for in their homes or in the caregiver's home by nonrelatives. Twenty-nine percent of preschoolers were cared for in organized child-care facilities (such as preschools and day-care centers). Six percent of children of employed mothers are cared for by their mothers at the mother's workplace (Grolier Web site, 1998).

Women often use multiple arrangements—the child's father, relatives living in or outside the household, day care, or a combination of these—before a child reaches school age. Thirty percent of employed mothers have two different child-care arrangements, 8 percent use three or more. Twenty percent of working mothers use two separate day-care centers (Gullo, 2000). For African-American and Latina single mothers, living in an extended family in which they are likely to have other adults to care for their children is an especially important factor allowing them to find jobs (Rexroat, 1990; Tienda and Glass, 1985).

Frustration is one of the most common experiences in finding or maintaining day care. Changing family situations, such as unemployed fathers' finding work or grandparents' becoming ill or overburdened, may lead to these relatives being unable to care for the children. Family day-care homes and child-care centers often close because of low wages or lack of funding. Furthermore, child care may be quite expensive. (Casper, Hawkins, and O'Connell, 1994). Costs may be as much as 10–35 percent of a family's budget (Children's Defense Fund, 1998). In fact, the high cost of child care is a major force that

About 10 percent of children are regularly cared for by grandparents.

in the past kept mothers on welfare from working (Joesch, 1991). As we shall see, the welfare reforms of the 1990s have left many low-income women struggling in jobs that don't pay much while having child care expenses eat up big chunks of their wages

In recent years, day care inside the child's home by relatives or babysitters has decreased. Over half the children of employed parents are cared for outside their own homes. This shift from parent care at home to day care outside the home may have profound effects on how children are socialized. It certainly disturbs many parents, who wonder whether they are, in fact, good parents. Many parents express concern about their children's being cared for outside their own family environment. Concern about their children's development tends to negatively affect parental well-being and feelings about work (Greenberger and O'Neil, 1990).

Parents who accept the home-as-haven belief—that the home provides love and nurturing—prefer placing their children in family day-care homes. They believe that a homelike atmosphere is more likely to exist in family day care than in preschools or children's centers, where greater emphasis is placed on education (Rapp and Lloyd, 1989). But we know relatively little about family day care: It is the most widely used and least researched form of American child care (Goelman, Shapiro, and Pence, 1990).

Impact on Employment and Educational Opportunities

The lack of child care or inadequate child care has the following consequences:

- It prevents many mothers from taking paid jobs.
- It keeps many women in part-time jobs, most often with low pay, few or no benefits, and little career mobility.
- It keeps many women in jobs for which they are overqualified and prevents them from seeking or taking job promotions or the training necessary for advancement.
- It conflicts sometimes with women's ability to perform their work.
- It restricts women from participating in education programs.

For women, lack of child care or inadequate child care is one of the major barriers to equal employment opportunity. Many women who want to work are unable to find adequate child care or to afford it. Child-care issues play a significant role in women's choices concerning work schedules, especially among women who work part-time. Eighteen percent of women with preschool children indicated that the need for better child-care arrangements dictated whether they worked day or night shifts. Among women working part-time, 47 percent said that the availability of child care was the prime consideration in choosing a work shift (Casper, Hawkins, and O'Connell, 1994). Studies of women in welfare-to-work programs report that the mothers' staying in the training program depends not only on the supply of child care but also on the quality and convenience of the care (Meyers, 1993).

Children's Self-Care

The lack of adequate after-school programs and prohibitive costs have resulted in **self-care**—a major form of child care in which children under the age of 14 care for themselves without supervision by an adult or older adolescent (Hochschild, 1997; Hewlett and West, 1998). Self-care increased through the 1980s and 1990s, and estimates of children in self-care range as high as 7 million, including a half-million preschoolers (Hewlett and West, 1998). It is a rapidly growing phenomenon, largely the result of inadequate or costly child care and the increasing numbers of married and single mothers who work outside the home (Folk and Yi, 1994). Self-care exists in families of all socioeconomic classes. In fact, there is little difference in the rates of self-care between the poor and the middle class (Casper, Hawkins, and O'Donnell, 1994).

The popular image of self-care arrangements is negative; such children are known pejoratively as "latchkey children." Research on self-care arrangements has been sketchy and often contradictory. Some studies find no differences in levels of achievement, fear, and self-esteem between children in day-care and self-care arrangements. Other studies do find differences. Some children are frightened, others get into mischief, and still others enjoy their independence. These differences may result from the child's own readiness to care for himself or herself, the family's readiness, and the community's suitability (Cole and Rodman, 1987). Rather than rejecting self-care altogether, researchers suggest that it may be appropriate for some children and not for others. Children who are mature physically, emotionally, and mentally, whose families are able to maintain contact, and who live in safe neighborhoods, for example, may have no problem in self-care. But if children are immature or live in unsafe neighborhoods, they may find themselves overwhelmed by anxiety and fear. Parents need to evaluate whether self-care is appropriate for their children (see Cole and Rodman [1987] for suggested guidelines). Because of the rise in self-care, educators are developing programs to teach children and their parents such self-care skills as basic safety, time management, and other self-reliance skills.

REFLECTIONS

Of the family economic issues discussed previously, have any affected you or your family? How? How were they handled?

INFLEXIBLE WORK ENVIRONMENTS AND THE TIME BIND

In dual-worker families, the effects of the work environment stem from not just one workplace but two. While some companies and unions are developing programs that are responsive to family situations (Crouter and Manke, 1994), the workplace in general has failed to recognize that the family has been radically altered during the last 50 years. Most businesses are run as if every worker were male with a full-time wife at home to attend to his needs and those of his children. But the reality is that women make up a significant part of the workforce, and they do not have wives at home. Allowances are not made in the American workplace for flexibility in work schedules, day care, emergency time off to look after sick children, and so on. Many parents would reduce their work schedules to minimize work-family conflict. Unfortunately, many do not have that option.

Carol Mertensmeyer and Marilyn Coleman (1987) note that our society provides little evidence that it esteems parenting. This seems to be especially true in the workplace, where corporate needs are placed high above family needs. Mertensmeyer and Coleman have offered some suggestions:

> Family policymakers should encourage employers to be more responsive in providing parents with alternatives that alleviate forced choices that are incongruent with parents' values. For example, corporate-sponsored child care may offset the conflict a mother feels because she is not at home with her child. Flextime and paid maternal and paternal leaves are additional benefits that employers could provide employees. These benefits would help parents fulfill self and family expectations and would give parents evidence that our nation views parenting as a valuable role.

Unfortunately, policies alone do not guarantee that employees will follow them. In her book, *The Time Bind: When Work Becomes Home and Home Becomes Work,* Arlie Hochschild (1997) describes the official policies and corporate culture at a large corporation that she calls Amerco to protect its anonymity. At Amerco, workers could utilize a number of "family friendly," time-enhancing policies, including job sharing, part-time work, parental leave, flextime, and "flexplace" (where workers could work from home). Despite the availability of such options, Hochschild notes that employees rarely made use of these opportunities. Only 3 percent of employees with children 13 or younger worked part-time. Nationally, less than 1 percent of Amerco employees shared a job. Parental leave was used by most eligible mothers at Amerco but by almost no fathers. Although working parents did make more use of flexible schedules, they also resisted cutting back their hours. In fact, employees who were parents of young children worked more hours than did those without children. More than half worked some weekends; almost three-fourths worked overtime. Among Amerco employees with children 12 and younger, only 13 percent of the women and 4 percent of the men worked less than 40 hours a week.

Hochschild notes that Amerco employees are typical of employees at other large corporations. Citing a 1990 study of 188 Fortune 500 manufacturing firms, Hochschild reports that although companies tended to offer family-friendly policies, few employees used them. Eighty-eight percent of the companies offered part-time work, but only between 3 and 5 percent of their employees chose to work part-time. Forty-five percent of the companies offered flextime, but only 10 percent of employees used it. Fewer companies offered job-sharing (6 percent) or work-at-home (3 percent) options, but among those who did few employees made use

Perspective

ARE FAMILY-FRIENDLY WORKPLACES UNFRIENDLY TO NONPARENTS?

According to journalist and historian Elinor Burkett (2000), the movement toward making the workplace more responsive to employees with children has come at the expense of employees without children. Two decades ago, social scientists Janet and Larry Hunt (1986) warned of the emergence of a two-tiered workplace environment that would discriminate against involved parents. Arguing that involved parents—mothers or fathers—would become "sociological women," potentially overlooked and undervalued in the workplace the same ways that women historically have been because of their commitments to being involved with their children. Women or men who either had no family commitments to compete with their work or had others (partners or paid help) who made the child-rearing sacrifices could function as "sociological men." They would be the candidates for key positions, quicker and higher promotions, and ultimately greater opportunities for success than their peers with family commitments.

Now, 15 years later, Burkett sees nonparents as being cheated by the spread of workplace family supports and flexibilities directed at employed parents. She notes both the informal culture that has emerged in the workplace and the unequal benefits that are available to parents and nonparents. More specifically, she points out the following:

In 1997, President Bill Clinton and Congress passed a middle-class tax break that totaled more than $5 billion a year. In the form of a tax credit for children, the taxes of those without children were left at a higher rate.

With a tax credit for college tuition, parents of college students could get a $1,500 tax credit to assist in paying for their children's schooling. Drawing a stark contrast, Burkett points out:

> A professional couple . . . with a six-digit income receives a $1,000 tax credit for their two younger children, a $960 tax credit for child care because the wife works two evenings a week, and an additional $1,500 tax credit because their daughter is in college, on top of five standard dependent exemptions. But a nonparent in poverty can receive only three months of food stamps every three years because he has no kids, and a nonparent earning as little as $10,000 a year receives a maximum Earned Income Tax Credit of $341—while an adult with a single child in that same bracket can claim up to $2,210 (Burkett, 2000, p. 8)

In 1999, President Clinton initiated policies that authorized states to extend unemployment insurance benefits to at-home parents of newborns.

Within the workplace, employees with children can take advantage of child-care centers (where available), parental leave packages, health insurance, and an "understanding" that they might need to scale back on certain time commitments due to the demands of child-rearing. Burkett cites a recent survey by the Families and Work Institute, which found that "88 percent of all companies with more than 100 employees allow parents time off for school functions."

Childless employees are expected to work late when necessary to cover for parents who leave early to be at kids' athletic or artistic involvements; to understand when workdays and meetings are disrupted by phone calls from children checking in after returning from school; to take vacations at less popular times; to take more business trips; to cover without additional pay for employees absent due to maternity or parental leaves; to be tolerant of coworkers who bring their children to the office, no matter how much disruption the children cause.

At academic institutions, employees are given tuition benefits for their children (and spouses). Assuming one has two children and both take advantage of the tuition benefit, employees with children are reaping hundreds of thousands of dollars in a benefit that childless employees have to forgo.

From examples like those above, Burkett builds a case for opening up many of the "family-friendly" policies to those without children, as some of the "most progressive companies in Corporate America" have done, and for evening out the current tax inequities that benefit parents without helping nonparents. To ignore the inequality created by our attempts to be more "family friendly" means to risk seeing a polarization of society as those without children will grow more dissatisfied and resentful of what they see as an unfair "baby boon."

of either opportunity. Less than 3 percent used flex-place options and 1 percent chose to share jobs.

In accounting for the lack of utilization of work-place policies, Hochschild considers and rejects a variety of explanations. Perhaps, employees can't afford to work fewer hours. Do they fear being laid off? Do employees even know about policies? Do they have insensitive and insincere supervisors? These explanations have partial validity. Some hourly employees do fear potential layoffs or reduced wages. There are some supervisors who seem reluctant to embrace and resentful at having to accommodate family-friendly policies. But the biggest reason employees do not use potential family time-enhancing initiatives is because they do not want to. They would rather be at work. In the 1980s and 1990s, with the dramatic changes in the division of labor and the growth of dual-earner families, home life has become more stressful and tightly scheduled. There is too much to do, too little time to do it, and not enough appreciation or recognition for what one does. On the other end of the work-family divide, many workplaces in the United States have implemented "humanistic management" policies designed to enhance worker morale and productivity and to reduce turnover. Thus, at work, people find social support, appreciation, and a sense of control and competence, which makes them feel better about themselves. In other words, work has become homelike, and home often feels like a job (Hochschild, 1997).

Because Hochschild studied only one company, it is hard to know how far we can generalize from her research. Other researchers have failed to support Hochschild's conclusions, at least to their fullest extent. For example, a study by Brown and Booth (2002), which uses the National Survey of Families and Households and is based on more than 1,500 dual-earner couples with children, indicates that Hochschild's findings may not be generalizable. Job status seems to be an important determinant of whether individuals see their jobs as more satisfying than their home lives. Brown and Booth claim that this is true only among workers in lower occupational status positions. Also, respondents who have high satisfaction with work and low satisfaction at home do not work significantly more hours at work. Only those who are satisfied with work, unsatisfied with home, and have adolescent children, work more hours (Brown and Booth, 2002).

Even if *The Time Bind* has somewhat more limited findings, it is important in that it shows that policies are not deterministic (see also Blair-Loy and Wharton, 2002). People must take advantage of policies. This suggests that people's values must be directed more toward home and family. Furthermore, cultural reinforcement for utilizing family-friendly policies must be more widespread and reflected in company "cultures." If "time equals commitment" to one's job, then one reduces one's work time at the risk of seeming undercommitted. But dual-earner family life must be made less stressful too. One way in which this can occur is by men doing more of the "second shift" work, thereby reducing the overload and time drain that their wives more consistently feel.

Employees who feel supported by their employer with respect to their family responsibilities are less likely to experience work-family conflict. It seems that having a family-friendly atmosphere is an integral part of how business organizations can help employees balance work and family obligations (Warren and Johnson, 1995). A model corporation would provide and support the use of family-oriented policies that would benefit both its employees and itself, such as flexible work schedules, job-sharing alternatives, extended maternity and/or paternity leaves and benefits, and child-care programs or subsidies. Such policies could increase employee satisfaction, morale, and commitment.

Living without Work: Unemployment and Families

Unemployment is a major source of stress for individuals, with its consequences spilling over into their families (Voydanoff, 1991). Even employed workers suffer anxiety about possible job loss due to economic restructuring and downsizing (Larson, Wilson, and Beley, 1994). Job insecurity leads to uncertainty that affects the well-being of both worker and spouse. They feel anxious, depressed, and unappreciated. For some, the uncertainty prior to

© Tony Freeman/PhotoEdit

Homeless families represent about a quarter of the homeless population; they are the fastest-growing segment of that population. Homeless children are likely to suffer depression, anxiety, and malnutrition.

losing one's job causes more emotional and physical upset than the actual job loss itself.

DID YOU KNOW?

In 1998, as much as 4.4 percent of the workforce over age 20 was unemployed. By ethnicity, 4 percent of whites, 9 percent of African Americans, and 7 percent of Latinos were unemployed (U.S. Bureau of Labor Statistics, 1998).

ECONOMIC DISTRESS

Those aspects of economic life that are potential sources of stress for individuals and families make up **economic distress** (Voydanoff, 1991). Major economic sources of stress include unemployment, poverty, and economic strain (such as financial concerns and worry, adjustments to changes in income, and feelings of economic insecurity).

In times of hardship, economic strain increases, and the rates of infant mortality, alcoholism, family abuse, homicide, suicide, and admissions to psychiatric institutions and prisons also sharply increase. Patricia Voydanoff (1991), one of the leading researchers in family-economy interactions, notes:

> A minimum level of income and employment stability is necessary for family stability and cohesion. Without it many are unable to form families through marriage and others find themselves subject to separation and divorce. In addition, those experiencing unemployment or income loss make other adjustments in family composition such as postponing childbearing, moving in with relatives, and having relatives or boarders join the household.

Further, economic strain is related to lower levels of marital satisfaction as a result of financial conflict, the husband's psychological instability, and marital tensions.

The emotional and financial cost of unemployment to workers and their families is high. A common public policy assumption, however, is that unemployment is primarily an economic problem. Joblessness also seriously affects health and the family's well-being.

The families of the unemployed experience considerably more stress than those of the employed. Reports by Ramsey Liem and his colleagues (cited in Gnezda, 1984) indicate that in the first few months of the male primary wage earner's unemployment, mood and behavior changes cause stress and strain in family relations. As families adapt to unemployment, family roles and routines change. The family spends more time together, but wives often complain of their husbands' "getting in the way" and not contributing to household tasks. Wives may assume a greater role in family finances by seeking employment if they are not already employed. After the first few months of their husbands' unemployment, wives of the unemployed begin to feel emotional strain, depression, anxiety, and sensitiveness in marital interactions. The children of the unemployed are more likely to avoid social interactions and tend to be more distrustful; they report more problems at home than do children in families with employed fathers. Families seem to achieve stable but sometimes dysfunctional patterns around new roles and responsibilities after six or seven months. If unemployment persists beyond a year, dysfunctional families become highly vulnerable to marital separation and divorce; family violence may begin or increase at this time (Teachman, Call, and Carver, 1994).

The types of families hardest hit by unemployment are single-parent families headed by women, African-American and Latino families, and young families. Wage earners in African-American, Latino, and female-headed single-parent families tend to remain unemployed longer than other types of families. Because of discrimination and the resultant poverty, they may not have important education and employment skills. Young families with preschool children often lack the seniority, experience, and skills to regain employment quickly. Therefore, the largest toll in an economic downturn is paid by families in the early years of childbearing and child rearing.

EMOTIONAL DISTRESS

Aside from the obvious economic impact of unemployment, job loss can have profound effects on how family members see each other and themselves. This in turn can alter the emotional climate of the family as much as lost wages alter the material conditions. Men are particularly affected by unemployment because wage earning is still a major way men satisfy their family responsibilities. Thus, when men fail as workers, they may feel as though they failed as husbands, fathers, and men (Rubin, 1994; Newman, 1988). As Rubin (1994) poignantly conveys in *Families on the Fault Line,* when men lose their jobs, "it's like you lose a part of yourself." Unemployed men may display a variety of psychological and relationship consequences, including emotional withdrawal, spousal abuse, marital distress, increased alcohol intake, and diminished self-identity (Rubin, 1994). Katherine Newman (1998) suggests that when families suffer downward mobility as a result of male unemployment, relations between spouses or between fathers and children are likely to be strained. Although children and spouses may be initially supportive, their support may wear thin or run out if joblessness lasts and other resources are unavailable, thus preventing families from maintaining their previous economic lifestyle.

Women, too, suffer nonmaterial losses when they lose their jobs, but those losses are different in degree and kind from those that men are likely to suffer. Men have more of their self-identities, and especially their gendered identities, tied up in working (success at work defines successful masculinity). Women have other acceptable ways of maintaining or achieving adult status (as mothers, for example). Thus, while both women and men will suffer from lost work relationships, lost gratification, even a loss of structure and purpose to one's day, women have not put as many of their "identity eggs" into the "work basket" as have most men.

COPING WITH UNEMPLOYMENT

Economic distress does not necessarily lead to family disruption. In the face of unemployment, some families experience increased closeness (Gnezda, 1984). Families with serious problems, however, may disintegrate. Individuals and families use a number of coping resources and behaviors to deal with economic distress. Coping resources include an individual's psychological disposition, such as optimism; a strong sense of self-esteem; and a feeling of mastery. Family coping resources include a family system that encourages adaptation and cohesion in the face of problems and flexible family roles that encourage problem solving. In addition, social networks of friends and family may provide important support, such as financial assistance, understanding, and a willingness to listen.

Several important coping behaviors assist families in economic distress caused by unemployment. These include the following:

- *Defining the meaning of the problem:* Unemployment means not only joblessness, but also diminished self-esteem if the person feels the job loss was his or her fault. If a worker is unemployed because of layoffs or plant closings, the individual and family need to define the unemployment in terms of market failure, not personal failure.
- *Problem solving:* An unemployed person needs to attack the problem by beginning the search for another job; dealing with the consequences of unemployment—for example, by seeking unemployment insurance and cutting expenses; or improving the situation—for example, by changing occupations or seeking job training or more schooling. Spouses and adolescents can assist by increasing their paid work efforts. Studies suggest that about a fifth of spouses or other family members find employment after a plant closing.

- *Managing emotions:* Individuals and families need to understand that stress may create roller-coaster emotions, anger, self-pity, and depression. Family members need to talk with one another about their feelings; they need to support and encourage one another. They also need to seek out individual or family counseling services to cope with problems before they get out of hand.

REFLECTIONS

Have you or your family experienced unemployment or job insecurity? How did it affect you? Your family? What coping mechanisms did you use?

Poverty

In 2002, the "poverty line" for a family of four with two children was drawn at $18,244. Although poverty and unemployment may appear to be largely economic issues, we know that they are also family issues. As we saw in Chapter 3, the family and economy are intimately connected to each other, and economic inequality directly affects the well-being of America's disadvantaged families. Poverty can drive families into homelessness. The poor have traditionally been isolated from the mainstream of American society (Goetz and Schmiege, 1996). Poverty is consistently associated with marital and family stress, increased divorce rates, low birth weight and infant deaths, poor health, depression, lowered life expectancy, and feelings of hopelessness and despair ("Poverty Helps Break Up Families," 1993). Poverty is a major contributing factor to family dissolution.

Catherine Ross and her colleagues (Ross, Mirowsky, and Goldsteen, 1991) suggest the connection between poverty and divorce:

> It is in the household that the larger social and economic order impinges on individuals, exposing them to varying degrees of hardship, frustration, and struggle. The struggle to pay the bills and to feed and clothe the family on an inadequate budget takes its toll in feeling run-down, tired, and having no energy, feeling that everything is an effort, that the future is hopeless, that you can't shake the blues, that nagging worries make for restless sleep, and that there isn't much to enjoy in life.

Despite stereotypes of the poor being African Americans and Latinos, the majority of the poor—and of welfare recipients—are white (Hacker, 1992). Although we tend to think of poverty as primarily an urban phenomenon, over 9 million poor live in America's rural areas. In some Iowa counties, poverty rates approach 30 percent. It is particularly ironic that hunger is a common problem in America's farming heartland (Davidson, 1990).

Poverty levels differ according to certain characteristics, such as ethnicity and family type. In Chapter 3 we saw how poverty rates vary by ethnicity and race. By family type 26.5 percent of single-mother families are below poverty (U.S. Census Bureau, 2003).

SPELLS OF POVERTY

The majority of those who fall below the poverty threshold tend to be there for spells of time rather than permanently (Rank and Cheng, 1995). About a quarter of the American population, in fact, requires welfare assistance at one time or another during their lives because of changes in families caused by divorce, unemployment, illness, disability, or death. About half of our children are vulnerable to poverty spells at least once during their childhood. Many families receiving assistance are in the early stages of recovery from an economic crisis caused by the death, separation, divorce, or disability of the family's major wage earner. Many who accept government assistance return to self-sufficiency within a year or two. Few depended heavily on welfare for more than seven out of ten years. Most of the children in these families do not receive welfare after they leave home.

Two major factors are related to the beginning and ending of spells of poverty: changes in income and changes in family composition. Thirty-eight percent of poverty spells begin with a decline in earnings of the head of the household, such as a job loss or a cut in work hours. Other causes include a decline in earnings of other family members (11 percent), the transition to single parenting (11 percent), the birth of a child to a single mother (9 percent), and the move of a youth to his or her own household (15 percent). One study of recipients under the old welfare system of Aid to Families with Dependent Children (AFDC) found that most women required assistance as a result of changes in their family situations—45 percent after separation or divorce, and 30 percent after becoming unmarried mothers. (AFDC was a government program designed to support poor families.) One-third of the women left the program within a year, half left at the end of two years, and two-thirds left within four years. About a third left the program because their income had increased, another third left when they remarried or reconciled with their mates, and 14 percent left when their children moved away from home or grew up.

Poverty spells are shorter if they begin with a decline in income than if they begin with transition to single parenthood or the birth of a child to a single mother. Half of poverty spells end with an increase in the earnings of the head of the household, and 23 percent end with an increase in earnings of other family members. Fifteen percent end when the family receives public assistance, and 10 percent end when a single mother marries.

THE WORKING POOR

Since 1979, the largest increase in the numbers of poor has been among the working poor because of low wages, occupational segregation, and the dramatic rise in single-parent families (Ellwood, 1988). Although their family members may be working or looking for work, these families cannot earn enough to raise themselves out of poverty .

An individual working full-time at minimum wage does not earn enough to support a family of three. Almost half of two-parent working poor families had at least one adult working full-time. Four out of five poor two-parent families are poor because of problems in the economic structure—low wages, job insecurity, or lack of available jobs (see Chilman [1991], for a literature review). The young are especially hard-hit. Families headed by men and women younger than 30 years of age are experiencing "a frightening cycle of plummeting earnings and family incomes, declining marriage rates, rising out-of-wedlock birth rates, increasing numbers of single-parent families, and skyrocketing poverty rates," writes Marian Wright Edelman (1988).

WOMEN, CHILDREN, AND POVERTY

The **feminization of poverty** is a painful fact that has resulted primarily from high rates of divorce, increasing numbers of unmarried women with children, and women's lack of economic resources (Starrels, Bould, and Nicholas, 1994). When women with children divorce, their income falls dramatically.

In 2002, 16.3 percent of children under 18 were poor. The rate was higher among younger children, as 18.6 percent of children living in families and under age 6 were poor. (Proctor and Dalaker, 2003). Like their parents, they move in and out of spells of poverty, depending on major changes in family structure, employment status of family members, or the disability status of the family head (Duncan and Rodgers, 1988). These variables affect ethnic groups differently and account for differences in poverty rates. African Americans, for example, have significantly higher unemployment rates and numbers of never-married single mothers than do other groups. As a result, their childhood poverty rates are markedly higher. Being poor puts the most ordinary needs—from health care to housing—out of reach, jeopardizes the children's schooling, and undermines their sense of self (Edelman, 1989).

THE GHETTO POOR

In the last two decades, the homeless and **ghetto poor**—inner-city residents, primarily African Americans and Latinos, who live in poverty—have become deeply disturbing features of American life, destroying cherished images of wealth and economic mobility. One commentator (DeParle, 1991) observes why many find the ghetto poor and homeless so disturbing:

> They suggest a second, separate America, a nation within a nation whose health, welfare, and social mobility evoke the Third World. Indeed visitors startled by rows of homeless lying in Grand Central Terminal in New York strain for analogies and produce the word "Calcutta."

It is not clear exactly who the ghetto poor are. They comprise not simply the poor, who have always existed in great numbers in the United States. They are primarily a phenomenon of the ghettos and barrios of decaying cities, where poor African Americans and Latinos are overrepresented. The ghetto poor feel excluded from society; indeed, they are often rejected by a society that neither understands nor empathizes with their plight (Applebome, 1991). Theirs is not a culture of poverty, however; the ghetto poor's behaviors, actions, and problems are often a response to lack of opportunity, urban neglect, and inadequate housing and schooling.

With the flight of manufacturing, few job opportunities exist in the inner cities; the jobs that do exist are usually service jobs that fail to pay their workers sufficient wages to allow them to rise above poverty. Schools are substandard. The infant death rate approaches that of Third World countries, and HIV infection and AIDS are epidemic. The housing projects are infested with crime and drug abuse, turning them into kingdoms of despair. Gunfire punctuates the night. A woman addicted to crack explained, "I feel like I'm a different person when I'm not here. I feel good. I feel I don't need drugs. But being in here, you just feel like you're drowning. It's like being in jail. I hate the projects. I hate this rat hole" (DePerle, 1991).

Within the inner city, residents struggle to maintain their dignity against surging hopelessness. They live day to day, fighting the forces that threaten to engulf them. A mother waiting for her child to return home from school said: "Mostly, you try to keep them away from the drugs and violence, but it's hard. I tell my oldest boy I don't want him hanging out with the boys who are getting in trouble, and he says, 'Aw, mama, ain't nobody else for me to be with'" (DeParle, 1991).

WELFARE REFORM AND POOR FAMILIES

Since the 1960s, when massive social programs known as the war on poverty cut the poverty rate almost in half, national priorities have shifted. In the last decade or so of the twentieth century, the war on poverty became a war on welfare—or, as some describe it, a war on the poor. Instead of viewing poverty as a structural feature of our society—caused by low wages, lack of opportunity, and discrimination—we increasingly blame the poor for their poverty (Aldous and Dumon, 1991; Katz, 1990). They are viewed as having become poor *because* they are "losers," "cheats," "lazy," "welfare queens," and "drug abusers"—people undeserving of assistance. Poverty is viewed as the result of individual character flaws—or even worse, as something inherently racial (Katz, 1990).

Nearly 13 million people received AFDC benefits in 1996 (U.S. Census Bureau, 1998, Table 605). Additionally, 27 million people received food stamps; their monthly value averaged $71. About 6.2 million children received free school breakfasts and 7.2 million pregnant women, infants, and children under 2 years of age participated in supplemental food programs known as the WIC (Women, Infants, and Children) Program (U.S. Census Bureau, 1998).

There has been considerable antagonism toward welfare and welfare recipients. Much of the antiwelfare sentiment is based on stereotypes of welfare recipients, especially young unmarried mothers. (Whereas women receiving welfare are often described as "welfare queens," there are no equivalent "welfare kings.")

Joel Handler, a longtime welfare researcher (quoted in Herbert, 1994), describes the stereotype of welfare recipients as "young women, without education, who are long-term dependents and whose dependency is passed on from generation to generation." He further notes: "The subtext is that these women are inner-city substance-abusing blacks spawning a criminal class." Furthermore, single mothers receiving welfare are stigmatized as incompetent and uncaring; some suggest that their children be placed in orphanages (Seeyle, 1994). Conservative thinker Charles Murray, for example, believes most adolescent girls "don't know how to be good mothers. A great many of them have no business being mothers and their feelings don't count as much as the welfare of the child" (quoted in Waldman and Shackelford, 1994).

Welfare became a central issue in 1990s politics, an emotional "hot-button" issue, what political commentator Mickey Kaus (1994) calls a "values issue." Many Americans who opposed welfare viewed it as violating the work ethic and destroying the traditional family. They believed that a person uses welfare as a way to avoid working and that welfare undermined the traditional family by "encouraging" women to become single mothers (Waldman and Shackelford, 1994). They accused unmarried adolescent mothers of getting pregnant in order to collect welfare benefits. But it is doubtful that adolescents are thinking of welfare benefits as they contemplate premarital sex. In fact, part of the problem is that adolescents often don't make the connection between sex and pregnancy. Finally, studies indicate that government welfare policies had little to do with the rise of divorce, single-parent families, and births to single mothers (Aldous and Dumon, 1991). Indeed, welfare benefits help stabilize families; those states with the most generous welfare benefits also have the lowest divorce rates (Zimmerman, 1991).

Numerous approaches to welfare reform were considered on both the federal and state level. On August 22, 1996, President Bill Clinton signed into law the Welfare and Medicaid Reform Act of 1996, also known as the Personal Responsibility and Work Opportunity Act of 1996. This legislation, which became Public Law 104-193, was proclaimed as an effort to "end welfare as we know it." Proponents in Congress believed that welfare had created a climate of irresponsibility and family pathology and saw the reform as a way to prevent or dramatically reduce out-of-wedlock pregnancy, out-of-wedlock births, and single-parent families. The legislation replaced AFDC with **Temporary Assistance for Needy Families (TANF)**, which sharply reduced the period during which one could receive governmental assistance and imposed more restrictive expectations on what recipients were compelled to do to remain eligible for assistance. TANF programs "include mandatory work- (public or private, subsidized or unsubsidized), education-, and job-related activities, including job training and job search, for the purpose of (1) providing such families with time-limited assistance in order to end their dependency on government benefits and achieve self-sufficiency; (2) preventing and reducing out-of-wedlock pregnancies, especially teenage ones; and (3) encouraging the formulation and maintenance of two-parent families" (Bill Summary, 104th Congress).

Beginning in October 1996, no family or individual was entitled to receive welfare help. Furthermore, recipients of TANF are limited to a maximum of five years, either consecutive or nonconsecutive, with exceptions allowed only for such misfortunes as battery or abuse victimization. The new legislation replaced AFDC entitlement with a block grant of federal funds given to states. States have the authority to decide how to provide assistance to eligible recipients, and the aid can be of some form other than money. Each state is required to operate a statewide welfare program and to provide certain social services (such as child care or health care for employed mothers) but the specifics may vary within and between the states. After a period of steady growth from the mid-1980s on, as a result of welfare reform, welfare rolls were sharply reduced (see Table 12.1). The figures for 1995 are "prereform," whereas the 2000 figures reflect the sharp reduction in welfare since the enactment of the 1996 reform act. As 2001 ended, the average number of monthly TANF cases was 57 percent lower than the number of AFDC cases prereform. The 5.4 million people receiving TANF was the lowest number to receive public assistance since 1961. In 2001, families on TANF received an average of $351 per month ($288 for one-child families, $362 for two-child families, $423 for three-child families, and $519 for families with four or more children). By September 2003, there had been still further reduction. There were 2,006,597 families and 4,880,037 individuals receiving TANF assistance (U.S. Department of Health and Human Services).

Moderates and liberals stress the importance of education and work training to prepare welfare recipients for employment. They believe that

TABLE 12.1 ■ Recipients of Aid to Families with Dependent Children (AFDC) and Temporary Assistance for Needy Families (TANF) 1975–2002

	1975	1980	1985	1990	1995	2000 (TANF)	2002
Total recipients (in thousands)	11,165	10,597	10,812	11,460	13,652	5,778	5,066
Percent of U.S. population	5.2	4.7	4.5	4.6	5.2	2.5	NA
Families receiving assistance (in thousands)	3,498	3,642	3,692	3,974	4,876	2,215	2,047

affordable child care should be made available in order for parents to work. Such solutions, however, entail spending public monies at a time many are demanding tax cuts and limits on spending. Moderates and liberals also criticize welfare programs that make children's welfare support dependent on their parents' reproductive or employment behavior (such as not having children if they are unmarried adolescents or finding employment, regardless of how low the pay). They point out that such programs penalize children if their parents "misbehave." Finally, they note that state bureaucracies may be as or more inefficient and unresponsive as the federal government. More important, states may not be equally willing to devote resources to helping welfare recipients out of poverty.

Other progressives argue that the problem was never welfare but poverty. People use welfare for the simple reason that they are poor. The best way to resolve welfare issues is by focusing on the poverty issues underlying it: low wages, unemployment, the high cost of housing, lack of affordable child care, economic discrimination against women and ethnic groups, and a deteriorating education system.

No doubt our welfare system was in trouble, but punitive approaches that blame the poor for their poverty do not resolve the problem. More imaginative approaches are needed. To deal with childhood poverty (see Figure 12.1), for example, we might use the approach used by all Western industrial nations (except ours): the provision of a minimum children's allowance. A children's allowance goes to all families and is based on the belief that a nation is responsible for the well-being of its children (Meyer, Phillips, and Maritato, 1991). By being universal, no poor child is missed nor is his or her family stigmatized as being "on welfare."

When we examine our attempt to reform and revamp the welfare system, we can't help but wonder what effect the interplay between politics and economics will have on children. As the state creates jobs for parents, it must also pave the way to providing available and affordable child care. But licensed day care is unlikely to meet the needs of the millions of welfare families and working poor who are mandated to work (Kilborn, 1997). Furthermore, in cities such as New York, Chicago, and Boston, the cost of care for even one child may be almost equal

FIGURE 12.1 ■ **Children under Eighteen Years Old below Poverty Level by Ethnicity, 1994**

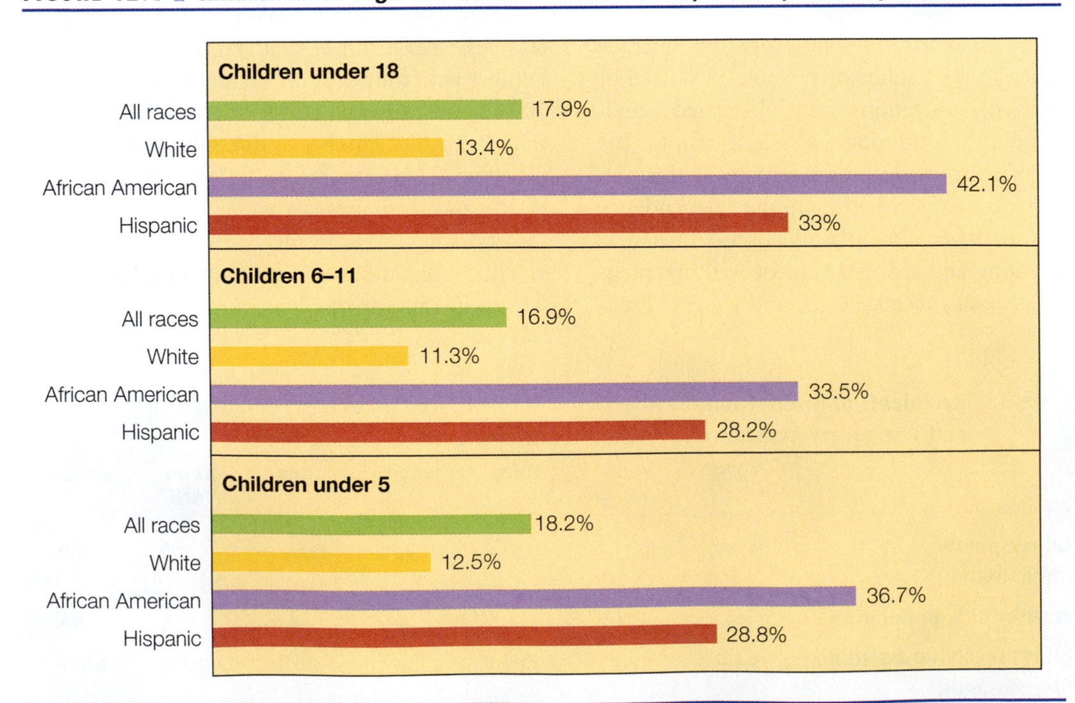

SOURCE: U.S. Bureau of the Census. *Statistical Abstract of the United States, 1996*. Washington, DC: U.S. Government Printing Office, 1996.

to the earnings of a minimum-wage worker. This situation could encourage wider use of unqualified child-care providers or a greater reliance on relatives. One consequence of welfare reforms has been the "reextension" of the family. As many single mothers enter the workforce, as is mandated by the new policies, it is often grandparents, especially grandmothers, who step into the child-care void they leave. The number of children in grandparental care has increased by 50 percent in the past decade. The new welfare policies may force it even higher.

The new law requires that recipients be working within two years and terminates their cash assistance in five years, so by then they must be able to support themselves. Critics of the reforms say the poor who have legitimate reasons for parental unemployment may be caught without a safety net, especially if the economy were to go into a recession (Livernois, 1997).

Welfare reform continues to be of acute concern. Evaluation of the legislative changes enacted in 1996 will continue for several years, along with various experimental programs. The ongoing challenge remains the same: We must find ways for people to have adequate food and shelter in an environment that facilitates the development of life skills and assists parents to succeed in the labor force. At the same time, we must provide for the safety, care, and guidance of our children.

REFLECTIONS

Do you believe that welfare helps or hinders families? Have you, your family, or your friends received welfare assistance? If so, were its effects positive, negative, or both? Why?

Workplace and Family Policy

Family policy is a set of objectives concerning family well-being and the specific government measures designed to achieve those objectives. As we examine America's priorities, it is clear that we have an implicit family policy that directs our national goals. Although it has never been articulated, it is very powerful in determining government and corporate policies. The policy is very simple: Families are not a national priority. Its corollary is equally simple: Neither are women and children.

Given the issues raised in this chapter, one might argue that if families were truly the cultural priorities we claim them to be, we would entertain and enact policy initiatives such as the following;

1. Paid parental leave for pregnancy and sick children; paid personal days for child and family responsibilities
2. Flexible work schedules for parents whenever possible; job-sharing alternatives
3. Increased minimum wage so that workers can support their families
4. Policies to ensure fair employment for all, regardless of ethnicity, gender, sexual orientation, or disability
5. Pay equity between men and women for the same or comparable jobs; affirmative action programs for women and ethnic groups
6. Corporate child-care programs or subsidies for families
7. Individual and family counseling services; provision of flexible benefit programs

Once enacted, policies such as these must be supplemented by sincere cultural support for families and children. People must believe that if they commit themselves to their families they will not suffer unfair economic consequences. This is harder to convey and carry out than are most of the specific workplace policies.

REFLECTIONS

If you were to construct a coherent family policy that meets your needs and reflects your values, what would it be like? How would it compare to the suggestions above?

Our marriages and families are not simply emotional relationships—they are also work relationships in which we divide or share many household and child-rearing tasks, ranging from changing diapers, washing dishes, cooking, and fixing leaking faucets to planning a budget and paying the monthly bills. These household tasks are critical to maintaining the well-being of our families. They are also unpaid and insufficiently honored. In addition to household work and child rearing, there is our employment, the work we do for pay. Our jobs usually take us out of our homes from 20 to 80 hours a week. They are not only a source of income; they also help our self-esteem and provide status. They may also be a source of work/family conflict.

As we enter the twenty-first century, we need to rethink the relationship between our work and our families. Too often, household work, child rearing, and employment are sources of conflict within our relationships. We need to rethink how we divide household and child-rearing tasks so that our relationships reflect greater mutuality. For many, poverty and chronic unemployment lead to distressed and unhappy families. We need to develop and support policies that help build strong families.

SUMMARY

- Families may be examined as economic units bound together by emotional ties. Families are involved in two types of work: paid work at the workplace and *family work*—unpaid work in the household.
- Employment affects family life. *Work spillover* is the effect that employment has on the time, energy, and psychological functioning of workers and their families at home. *Role strain* refers to difficulties that individuals have in carrying out the multiple responsibilities attached to a role. *Role overload* occurs when the total prescribed activities for one or more roles are greater than an individual can handle. *Role conflict* occurs when roles conflict with each other.
- Women enter the workforce for economic reasons and to raise their self-esteem. Employed women tend to have better physical and emotional health than do nonemployed women. In 1995, 59 percent of adult women and 75 percent of adult men were employed. Women's employment tends to be influenced by family needs; their labor-force participation is interrupted for family reasons over 30 times as often as is men's participation.
- The traditional division of labor in the family follows a complementary pattern: The husband works outside the home for wages and the wife works inside the home without wages. Men's participation in household work is traditionally limited to repairs, construction, and yard work. Women's primary responsibility for household work and child rearing is part of the traditional marriage contract.
- There are four characteristics that define the *homemaker role:* (1) its exclusive allocation to women; (2) its association with economic dependence; (3) its status as nonwork; and (4) its priority over other roles for women. Several characteristics may be noted about housework: (1) it isolates the person at home; (2) it is unstructured, monotonous, and repetitive; (3) it is often a restricted, full-time role; (4) it is autonomous; (5) it is "never done"; (6) it may involve child rearing; (7) it often involves role strain; and (8) it is unpaid.
- More than half of all married women are in dual-earner marriages. Husbands generally do not significantly increase their share of household duties when their wives are employed. Employed mothers remain primarily responsible for child rearing. Working wives are more independent than nonemployed women and have increased power in decision making. Women's employment has little or a slightly positive impact on marital satisfaction; there does seem to be a slightly greater likelihood of divorce when the woman is employed.
- Two more contemporary arrangements are (1) shift households, where spouses work opposite shifts and alternate domestic and caregiver responsibilities, and (2) households in which men stay home with children while women support the family financially.
- Families must balance family and work needs throughout the family life cycle. There are three basic work/family life cycle models: (1) the traditional-simultaneous work/family life cycle, (2)

sequential work/family role staging, and (3) symmetrical work/family role allocation. Major problems in this model are related to role strain. In the sequential pattern, women alternate work and mother roles rather than combine them. In the symmetrical pattern, men assume greater household and child-rearing responsibilities.

- Family issues in the workforce include economic discrimination against women; *sexual harassment;* lack of adequate child care; and an inflexible work environment.
- Economic distress refers to aspects of a family's economic life that may cause stress, including unemployment, poverty, and economic strain. Unemployment causes family roles to change; families spend more time together, but wives complain that unemployed husbands do not participate in housework. Unemployment most often affects female-headed single-parent families, African-American and Latino families, and young families. Coping resources for families in economic distress include individual family members' positive psychological characteristics, an adaptive family system, and flexible family roles. Coping behaviors consist of defining the problem in a positive manner, problem solving, and managing emotions.
- Almost 14 percent of the population of the United States lives in poverty. There are significant differences between whites and members of ethnic groups in terms of income and wealth. The disparity increased dramatically during the 1980s. As many as 25 percent of Americans go through spells of poverty, during which time they need welfare assistance. Poverty spells generally occur because of divorce; the birth of a child to an unmarried mother; or unemployment, illness, disability, or death of the head of the household. Young families are particularly vulnerable to poverty. The majority of poor people are women and children. The *ghetto poor* are inner-city poor, disproportionately African American and Latino.
- National priorities have shifted from the war on poverty to the war on welfare. Much of antiwelfare sentiment is based on stereotypes of welfare recipients, especially young unmarried mothers. Many Americans who oppose welfare view it as violating the work ethic and destroying the traditional family but research does not support these beliefs.
- Welfare reforms have been enacted by the U.S. government. Stricter limits now exist in determining and maintaining eligibility.
- Moderates and liberals have responded with various proposals that encourage self-sufficiency, such as education, work training, and affordable child care. Other progressives argue that the problem is not welfare but poverty. The best way to resolve welfare issues is by focusing on the poverty issues underlying them.
- Family policy is a set of objectives concerning family well-being and the specific government measures designed to achieve those objectives. Family policy would contain provisions affecting health care, social welfare, education, and the workplace.

KEY TERMS

co-provider families 387
economic distress 402
family policy 415
family work 381
feminization of poverty 411
gender ideology 393
ghetto poor 411
homemaker role 385
hostile environment 402
role conflict 382
role overload 382
role reversal 397
role strain 382
second shift 392
self-care 404
sexual harassment 402
shift couples 395
Temporary Assistance for Needy Families (TANF) 413
two-person career 384
work spillover 381

SUGGESTED READINGS

Galinsky, Ellen. *Ask the Children: The Breakthrough Study That Reveals How to Succeed at Work and Parenting.* New York: Harper-Collins, 1999. A comprehensive study that looks at how children and parents perceive and experience parents' jobs.

Gerson, Kathleen. *No Man's Land: Men's Changing Commitments to Work and Family.* Based on interviews with more than 100 men, Gerson explores the diversity among men in terms of their attachments to work and family. Furthermore, she illustrates the adult experiences that often redirect men from careers to family involvement and vice versa.

Gilbert, Lucia Albino. *Two Careers/One Family.* Newbury Park, CA: Sage, 1993. An excellent description of the two-career family, including research and theory, female and male perspectives, life in dual-career families, workplace policies, and future trends.

Hochschild, Arlie, with Anne Machung. *The Second Shift: Working Parents and the Revolution at Home.* New York: Viking, 1989. An influential study of the division of paid work and family work among two-earner couples. Hochschild examines how much housework women and men do, how couples arrive at their domestic arrangements, and what happens when domestic reality is incompatible with gender ideologies.

Hochschild, Arlie. *The Time Bind: When Work Becomes Home and Home Becomes Work.* New York: Metropolitan Books, 1997. A disturbing study of the work-family conflicts affecting workers at a large company Hochschild calls Amerco. Significantly, although Amerco offered workers a variety of options that would increase the flexibility of their jobs, most employees refused these initiatives. Hochschild ties this to the increasingly time-pressed home lives facing dual-earner couples.

Mahoney, Rhona. *Kidding Ourselves: Breadwinning, Babies, and Bargaining Power.* New York: Basic Books, 1995. A presentation of the thesis that women will not achieve economic equality until men do half the work of raising children.

RESOURCES ON THE INTERNET

Companion Web Site for This Book

http://sociology.wadsworth.com/strong/marriage9e
Gain an even better grasp on this chapter by going to the companion Web site to take one of the Tutorial Quizzes, use the Flash Cards to master key terms, or check out the many other study aids you'll find there. Visit the Marriage and Family Resource Center on the site. You'll also find special features such as GSS Data and Census 2000 information that will put data and resources at your fingertips to help you with that special project or to do some research on your own.

InfoTrac College Edition: Search Word Summary

interrole conflict	work and family
dual-career families	child care
unemployment	poverty

To learn more about these central topics in the study of the family, you can conduct an electronic search using InfoTrac College Edition. To aid in your search and to gain useful tips, see the Student Guide to InfoTrac College Edition that you can access through the companion Web site for this book.

Preview

To gain a sense of what you already know about the material covered in this chapter, answer "True" or "False" to the statements below.

1. Intimate relationships of any kind increase the likelihood of violence. True or false?
2. Rape by an acquaintance, date, or partner is less likely than rape by a stranger. True or false?
3. Male aggression is generally considered to be a desirable trait in our society. True or false?
4. Studies of family violence have helped strengthen policies for dealing with domestic offenders. True or false?
5. Physically abused children are often perceived by their parents as "different" from other children. True or false?
6. Sibling violence is the most widespread form of family violence. True or false?
7. Relatively few missing children have been kidnapped by strangers. True or false?
8. Deliberate fabrications of sexual abuse constitute nearly 25 percent of all reports. True or false?
9. Most people who were sexually abused as children at least partially remember the abuse. True or false?
10. Brother/sister incest is generally harmless. True or false?

"In the United States we spend more money on shelters for dogs and cats than for human beings. If we have effective animal rescue shelters for abused dogs, cats, and bunny rabbits, we should be able to spare something for people as well."

MURRAY STRAUS ET AL.

CHAPTER 13

Family Violence and Sexual Abuse

Outline

It seems a cruel irony that the relationships we most value are also the relationships in which we are most violent. The people we love and live with are the people most likely to hurt or assault us. It is an unhappy fact that intimacy or relatedness increases our likelihood of experiencing abuse, violence, or sexual abuse. Just think about who our society "permits" us to shove, hit, or kick. If we assault a stranger, fellow student or professor, coworker or employer, we would run great risk of being arrested. It is with our intimates that we are "allowed" to do such things.

Dating, loving, living with, or being related seems to give us permission to be violent when we are angry. Those closest to us are also those we are most likely to slap, punch, kick, bite, burn, stab, or shoot. And they are the most likely others to do these things to us (Gelles and Cornell, 1990; Gelles and Straus, 1988). For some partners, spouses, or parents, intimacy seems to confer a right to be physically or sexually abusive. Furthermore, living together provides people more opportunity to disagree, get angry at each other, and hurt each other. Although not as dangerous as war zones or urban riot scenes, families and households are dangerous places.

Consider the following points:

- Every 30 seconds, a woman is beaten by her boyfriend or husband.
- In various studies, 30 to 40 percent of college students report violence in dating relationships.
- At least a million American children are physically abused by their parents each year.
- Almost a million parents are physically assaulted by their adolescents or younger children every year.
- As many as 27 percent of American women and 16 percent of men have been the victims of childhood sexual abuse, much of it in their own families.

Until the 1970s, Americans believed that when they locked their homes at night, they locked out violence; the sad fact is that they also locked in violence. When the images of Hedda Nusbaum and Nicole Brown Simpson first appeared on our television screens, a permanent scar was marked in the minds of most Americans. The abuse inflicted by Hedda and the murder of Nicole and thousands like her was a wake-up call for policy makers, prosecutors, and individuals around the country to begin to listen and act on behalf of those who have been ignored too long. As their stories faded from our memory, others emerged to remind us of what often happens between spouses and partners, or parents and children. Consider these familiar cases: Scott Peterson, set to stand trial for the murder of his wife Laci and their unborn child; actor Robert Blake, set to stand trial soon for the murder of his wife, Bonnie Lee Bakley; Andrea Yates, convicted and sentenced to life for the murder of her five young children; Nikolay Soltys, who committed suicide in his jail cell after being convicted of murdering his pregnant wife, his 3-year-old son, two young cousins (aged 9 and 10) and his aunt and uncle. When Lyle and Erik Menendez were tried and convicted for murder for having shot and killed their parents, José and Kitty Menendez, after what they alleged was long-term abuse, many were riveted to the news and television coverage of their trial. These are just some of the more familiar cases over the past decade or so. Although we are more aware of the existence of violence and are better able to recognize and understand it, there is much work to be done toward reducing and eliminating it.

PREVIEW ANSWERS

1 True
2 False
3 True
4 True
5 True
6 True
7 True
8 False
9 True
10 False

In this chapter, we look at the models researchers use in studying family violence, and we discuss the dynamics that are present in battering relationships. We look at violence between husbands and wives (including marital rape), between gay and lesbian partners, between dating partners (including acquaintance rape), and between siblings, as well as violence committed against children by parents and against parents by grown children. We also discuss prevention and treatment strategies. The last section of the chapter is devoted to the discussion of child sexual abuse—its forms, participants, and effects as well as treatment and prevention strategies.

Family Violence and Abuse

Researchers typically differentiate between violence and abuse and with good reason. For the purpose of this book, we use the definition of **violence** offered by Richard Gelles and Claire Pedrick Cornell (1990): "an act carried out with the intention or perceived intention of causing physical pain or injury to another person." There are other prevalent forms of abuse, of course—such as neglect and emotional abuse, including verbal abuse—but the focus of this chapter is physical violence and sexual abuse. Thus, abuse is broader than family violence.

Interestingly, we need to differentiate in the opposite direction as well. Violence in families includes acts that are not defined by most members of society as abusive and are often seen as appropriate. In fact, violence may best be seen along a continuum, with "normal" and "routine" violence at one end and abusive, even lethal, violence at the other extreme (Gelles and Straus, 1988). Thus, family violence ranges from spanking to homicide. We must look at the continuum as a whole to be concerned with "families who shoot and stab each other as well as those who spank and shove, . . . [as] one cannot be understood without considering the other" (Straus, Gelles, and Steinmetz, 1980).

TYPES OF INTIMATE VIOLENCE

When we look specifically at violence in intimate couple relationships, we confront a wide range of behaviors that beg for differentiation. Michael Johnson and Kathleen Ferraro (2000) offer the following typology of partner violence.

- *Common couple violence:* This refers to violence that erupts in the course of an argument as one partner strikes at the other in the heat of the moment. Such violence is not part of a wider relationship pattern; it is as likely to come from a woman as a man or to be mutual, it rarely escalates, and it is less likely to lead to serious injury or fatality.

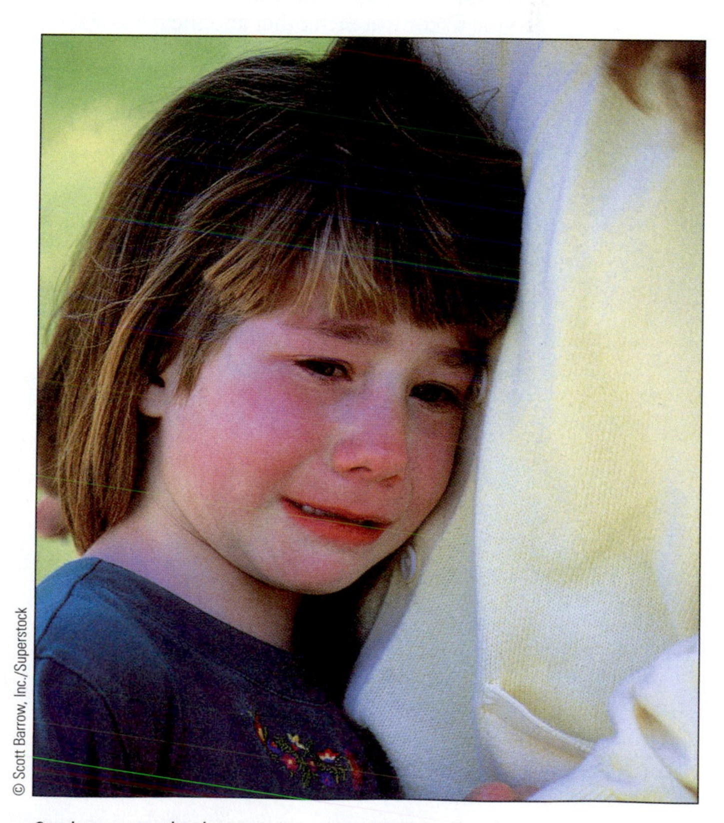

Our homes may be the most dangerous places for us to be, especially if we are young or female.

- *Intimate terrorism:* **Intimate terrorism** occurs in relationships characterized by the desire of one partner to dominate and control the other. Along with violent episodes that frequently escalate, there is often emotional abuse in intimate terrorism. As a result of the combination of physical and emotional or psychological abuse, the victims are left "demoralized and trapped" as their sense of self and their place in the world are greatly diminished by the subjugation they receive. The violence in intimate terrorism is more likely to recur, to escalate, and to lead to injury, and it is less likely to be mutual.
- *Violent resistance:* Johnson and Ferraro suggest that although it has been less often studied, violent resistance encompasses what is often meant by "self defensive" violence. It tends to be much more commonly perpetrated by women and may serve as a sign that women are moving toward leaving their abusive partners.
- *Mutual violent control:* Although likely rare and underresearched, this type encompasses the relationships in which both partners are violently trying to control each other and the relationship.

Distinctions such as these are important if we are to make sense of the data on who commits violence against a partner or spouse. Of the four types, common couple violence seems to be slightly more typical of men than of women, intimate terrorism is "essentially" perpetrated by men, and violent resistance is much more often committed by women (Johnson and Ferraro, 2000). When distinctions are not drawn and researchers look instead at overall rates of violent couple behavior, the genders may seem more alike than they really are as aggressors and victims.

WHY FAMILIES ARE VIOLENT: MODELS OF FAMILY VIOLENCE

To better understand violence within the family, we must look at its place in the larger sociocultural environment. Aggression is a trait that our society labels as generally desirable, especially for males. Getting ahead at work, being assertive in relationships, and winning at sports are all culturally approved actions. But does aggression necessarily lead to violence? All families have their ups and downs, and all family members at times experience anger toward one another. How do we explain the fact that violence erupts more frequently and with more severe consequences in some families than in others? The principal models used in understanding family violence are discussed in the following sections. Each of these models has valuable insight to offer concerning a very complex problem with no easy or single solution. They tend to emphasize such factors as gender, power, stress, and intimacy.

Individualistic Explanations and the Psychiatric Model

The psychiatric model finds an important source of family violence to be within the personality of the abuser (O'Leary, 1993). It assumes that he or she is violent as a result of a personality disorder, mental or emotional illness, or alcohol or drug misuse. Although research indicates that fewer than 10 percent of family violence cases are attributable to psychiatric causes, and about 25 percent of cases of wife abuse are associated with alcohol, the idea that people are violent because they are crazy or drunk is widely held (Gelles and Cornell, 1990). Gelles and Cornell also suggest that this model is compelling and appealing because "if we can persist in believing that violence and abuse are the products of aberrations or sickness, and, therefore, believe ourselves to be well, then our acts cannot be hurtful or abusive." But besides looking at the abuser, we must step back and look at the big picture—at the family and society that influence the abuser.

Ecological Model

The ecological model utilizes a systems perspective to look at the child's development within the family environment and the family's development within the community. Psychologist James Garbarino (1982) has suggested that cultural support for physical force against children combines with lack of family support in the community to increase the risk of intrafamily violence. Under this model, a child who doesn't "match" well with the parents (such as a child with emotional or developmental disabilities) and a family that is under stress (from, for example, unemployment or poor health) and that has little community support (such as child care or medical care) can be at increased risk for child abuse.

Feminist Model

Feminist explanations of domestic violence stress the role of male dominance in family violence. This approach draws its conclusions from a historical perspective. It holds that most social systems have traditionally placed women in a subordinate position to men, thus condoning or supporting the institution of male violence (Schechter and Gary, 1988; Yllo, 1993). There is no doubt that violence against women and children, and indeed violence in general, has had an integral place in most societies throughout history. Feminist theory must be credited for advancing our understanding of domestic violence by insisting that the patriarchal roots of domestic relations be taken into account. Taken alone, however, the patriarchy model does not adequately explain the variations in degrees of violence among families in the same society (Yllo, 1993). Furthermore, women are sometimes violent toward their husbands and partners too. More mothers are implicated in child abuse than fathers (although this has much to do with responsibility for and time with children). Finally, and most telling, rates of violence between lesbian partners may be as high as among heterosexual partners. Like heterosexual violence, when it does occur it is more likely to be a recurrent feature of the relationship than a one-time event.

Social Situational and Social Learning Models

The social models are related to the ecological and feminist models in that they view violence as originating in the social structure. The social situational model views family violence as arising from two main factors: (1) structural stress (such as low income or illness) and (2) cultural norms (such as the "spare the rod and spoil the child" ethic) (Gelles and Cornell, 1990). In this model, groups with few resources, such as the poor, are seen to be at greater risk for family violence than those who are well off. The social learning model holds that people learn to be violent from society at large and from their families (Ney, 1992). Although it is true that many perpetrators of family violence were themselves abused as children, it is also true that many victims of childhood violence do not become violent parents. These theories do not account for this discrepancy. (See Egeland [1993] and Kaufman and Zigler [1993] for conflicting views on the significance of the intergenerational transmission of abuse.)

Resource Model

William Goode's (1971) resource theory can be applied to family violence. This model assumes that social systems are based on force or the threat of force. A person acquires power by mustering personal, social, and economic resources. Thus, according to Goode, the person with the most resources is the least likely to resort to overt force. Gelles and Cornell (1990) describe the typical situation: "A husband who wants to be the dominant person in the family but has little education, has a job low in prestige and income, and lacks interpersonal skills may choose to use violence to maintain the dominant position."

Exchange/Social Control Model

Richard Gelles (Gelles, 1993b; Gelles and Cornell, 1990) posits the two-part exchange/social control theory of family violence. The first part, exchange theory, holds that in our interactions, we constantly weigh the perceived rewards against the costs. When Gelles says that "people hit and abuse family members because they can," he is applying exchange theory. The expectation is that "people will only use violence toward family members when the costs of being violent do not outweigh the rewards." (The possible rewards of violence might be getting one's own way, exerting superiority, working off anger or stress, or exacting revenge. Costs could include being hit back, being arrested, being jailed, losing social status, or dissolving the family.)

The second part of the theory, social control, raises the costs of violent behavior through such means as arrest, imprisonment, loss of status, or loss of income. Three characteristics of families that may reduce social control—and thus make violence more likely—are the following:

1. *Inequality:* Men are stronger than women and often have more economic power and social status. Adults are more powerful than children.
2. *Private nature of the family:* People are reluctant to look outside the family for help, and outsiders (the police or neighbors, for example) may hesitate to intervene in private matters. The likelihood of family violence goes down as the

number of nearby friends and relatives increases (Gelles and Cornell, 1990).

3. *"Real man" image:* In some American subcultures, aggressive male behavior brings approval. The violent man in these groups may actually gain status among his peers for asserting his "authority."

The exchange/social control model is useful for looking at treatment and prevention strategies for family violence, which we discuss later in this chapter.

DID YOU KNOW?

Female victims of violence were more likely to be injured when attacked by someone they knew than female victims who were attacked by a stranger (U.S. Department of Justice, 1995).

MULTIPLE CAUSES

Looking at these various theories, we see the following four factors surface repeatedly:

- *Gender:* Feminists stress the role of gender inequalities or masculinity as causes of violence. Although as we have seen there is female-on-male violence, and—as we shall see—female-on-female violence (see below), violence by males tends to be more extreme, has different causes (power and control versus self-defense), and results in different consequences (in terms of both physical injuries and domination). Thus, with regard to family violence, gender matters a lot, even though it may not define all perpetrators and victims.
- *Power and Control:* Central to both the feminist emphasis on patriarchy and the resource model emphasis on the objectives and outcomes of force and violence is the idea of power. Violence may be used as a tool to obtain and maintain power, in or outside of families. Within a family context, a power emphasis explains some male-on-female violence and adult-on-child violence, but domestic violence is more complicated. As we shall see, the most violent family relationship is between siblings. Children do strike and assault their parents, and wives do assault their husbands. Violence does not always flow in the direction of more powerful onto less powerful. Power is, however, a central motive in much intimate violence, especially the long-term and extreme forms of spousal violence that Michael Johnson calls intimate terrorism.
- *Stress:* Family relationships are stress-filled relationships, and households are stressful places. Social situational and ecological theories emphasize how stress can contribute to violence. As individuals are subjected to a variety of stresses (such as unemployment, underemployment, illness, pregnancy, and work-related relocations) tensions between family members may rise. Stress-based explanations work especially well to account for the greater prevalence of violence among lower-income families and households facing unemployment, but stress alone inadequately accounts for the breadth and depth of family violence (McCaghy, Capron, and Jamieson, 2000; Straus, Gelles, and Steinmetz, 1980). Stress may raise the likelihood of violence, but it is not the cause. Somewhere, one must also have learned that acting violently toward one's loved ones is appropriate and acceptable. (Gelles and Straus, 1988).
- *Intimacy:* The heightened emotions and long-term commitments that characterize family relationships are qualities we value about those relationships. Those same qualities lead to a greater likelihood that we will have disagreements, that those disagreements will be more emotional, and that escape from conflicts that arise and escalate will not be easy. Furthermore, the cultural beliefs about intimacy allow loved ones the right or responsibility to influence or affect each other's behavior. We risk "spoiling the child" by "sparing the rod." Some abusive men when asked why they assault their spouses say they do so "because they love them." We connect intimacy with a perceived right to be violent and treat intimate violence as different from other violent relationships.

Recall from Chapter 3 that two long-term historical trends in American families related to how we experience intimacy and what we expect of family relationships were the increasing *privatization* of family life and the heightened importance placed on *emotional fulfillment.* We grant and expect a kind of privacy and secrecy to family relationships that keeps them from public scrutiny. Even in public settings, we are often reluctant to intervene in "domestic disputes." In some ways, then, we legitimize violence and force within families, and then turn the other way when they occur. And since we expect our families to be the source of our greatest gratification (as they often are), we feel especially betrayed when things fall short of our expectations.

Prevalence of Family Violence

It is difficult to know exactly how much family violence there is in the United States. Part of the difficulty results from methodological limitations in the various data we gather. Depending on *how* one gathers the information, estimates of *how much* there is will vary. Can we use "official statistics," such as arrest records or emergency room visits? Can we trust survey data? What about shelter populations? Are they reflections of the extent of the problem?

As you can quickly see, any of the above is a potentially suspect source of data. Official reports require that some "officials" (police or medical) learn of incidents. Much of the family violence that occurs is unreported to the police (U.S. Bureau of Justice Statistics, 1998). This both understates and skews the direction of the data. Some people are better positioned to hide their abusive behavior from authorities. Some, especially upper- and middle-class individuals, may be given more credibility by police. If there are injuries that require medical attention, people who can afford to utilize nonhospital medical resources (such as family doctors rather than clinics) may not come under the same kind of suspicion. And they won't show up in hospital records.

Survey data are also prone to problems. In asking people to admit to experiencing or expressing family violence, researchers may well get underreports of overall rates. Even in anonymous surveys, individuals may downplay their involvement in socially undesirable behavior. Nevertheless, the estimates from large-scale, national surveys (Straus, 1993) give us our best estimate of the frequency and spread of family violence.

Based on survey data from large, representative samples of heterosexual couples in the United States, approximately 12 percent of adult intimates experience some form of physical abuse from their partners; out of every 1,000 couples, 122 wives and 124 husbands are assaulted by their spouses (Renzetti and Curran, 1999). Another national survey estimates nearly 9 million couples, one out of six marriages, experiencing some incident of violence every year (Gelles and Straus, 1988; Newman, 1999).

Tension and conflict are normal features of family life but can escalate into violence under certain conditions.

Using multiple sources of data, including responses from the annual National Crime Victimization Survey, the FBI's Uniform Crime Reports, the Bureau of Justice Statistics' Study of Injured Victims of Violence, and surveys of jail and prison inmates, the Bureau of Justice Statistics produced a report on violence between intimates (Bureau of Justice Statistics, 1998). Key findings are as follows:

- There are an estimated 1 million rapes, sexual assaults, robberies, or assaults (simple or aggravated) between intimates each year.
- Approximately 85 percent of these incidents had female victims.
- 150,000 men were victims of violent crimes committed by an intimate.
- In 2000, there were nearly 1700 murders attributed to spouses, ex-spouses, boyfriends, or girlfriends; one in every eleven homicides was a murder between intimate partners or ex-partners. Spousal homicides are down dramatically, however.
- Nearly 40 percent of violent incidents occur on weekends and most occur in or around the victim's home.
- A third of female murder victims and 4 percent of male murder victims in 2000 were killed by an intimate.

COMMON VERSUS EXTREME VIOLENCE

We are familiar with the phrase "battered women." We also recognize the phrase "men who batter." In fact, **battering,** as used in the literature on family violence, is a catchall term that includes, but is not limited to, slapping, punching, knocking down, choking, kicking, hitting with objects, threatening with weapons, stabbing, and shooting. By itself, the term *battering* does not specify the gender of the batterer or victim. In the survey research on domestic violence, a curious pattern results: The number of women who report expressing violence toward their male partners is the same as the number of men who report expressing violence toward their female partners. This is true of research on spousal, cohabiting, and dating relationships. Does that mean we should also talk of "battered men" and "women who batter"?

Although the term "battered husband syndrome" has been used (Steinmetz, 1978, as cited in Johnson, 1995), many feminist social scientists object to and reject this notion. Their opposition is not entirely inappropriate, as it appears that the majority of the violence perpetrated by women on men is of the more routine and relatively minor variety. Occasionally, though infrequently, this violence escalates, resulting in serious injury or even death. More often it does not. This is not only the form of violence that captures most of female-on-male violence but also most male-on-female violence. In other words, most partner violence is what Michael Johnson refers to as **common couple violence,** the violence that results from disputes and disagreements that go too far (Johnson, 1995). It is this kind of family violence for which survey estimates show something closer to gender symmetry. It is not the sort of violence that typically leads to hospitals or shelters. That less common and more extreme violence is more often committed by men against women (Johnson, 1995).

WOMEN AND MEN AS VICTIMS AND PERPETRATORS

In recent years the subject of intimate violence against women has gained public notoriety. It remains subordinate in the public mind to the physical and sexual abuse of children. In part, this may be because historically and culturally, women are considered "appropriate" victims of domestic violence (Gelles and Cornell, 1990). Many expect, understand, and accept the misogynistic idea that women sometimes need to be "put in their place" by men, thus providing a cultural basis for the physical and sexual abuse of women.

No one knows with certainty how many women are victims of partner violence each year, but as we saw above, the data we have paint a less than optimistic picture. Consider, too, these facts from the Bureau of Justice Statistics (1998, 2003):

- Twenty percent of all violent crime experienced by women is from an intimate (spouse, ex-spouse, or boyfriend). In 2001, intimates accounted for 3 percent of nonfatal violence against men.
- In 1996, at least a third of women who experienced violence reported having been assaulted more than once within the six months prior to the survey; 12 percent were assaulted at least six times.

- Half of victims report an injury; one in five injured women seeks medical treatment.
- More than half (56 percent) of female victims call the police. Police typically respond in ten minutes or less, though more than 40 percent of victims say police took an hour or more to arrive.
- The cumulative financial costs associated with intimate violence add up to more than $150 million. These include medical costs ($61 million), broken or stolen property ($35 million), and lost wages due to time out of work.

Women of all races, ages, and socioeconomic statuses are victimized. They are not victimized equally, however. Younger women, black women, lower-income women, and urban women are more frequent victims of partner violence. One out of every 50 women, ages 16 to 24, was a victim of intimate violence. This is the highest per capita rate of victimization. Black women suffered higher rates of nonlethal violence than did white women. As income increased, the rate of female victimization decreased (Bureau of Justice Statistics, 1998). Although no social class is immune to it, most studies find that marital violence is more likely to occur in low-income, low-status families (Gelles and Cornell, 1990). The "good news" on family violence is that it is declining. Between 1993 and 2001 intimate violence against women declined by nearly half. In that same time span, the rate against males dropped 42 percent (Bureau of Justice Statistics, 2003).

Personal Characteristics of Women and Men in Violent Relationships

Although early studies of battering relationships seemed to indicate a cluster of personality characteristics constituting a typical battered woman, more recent studies have not borne out this viewpoint. Factors such as self-esteem or childhood experiences of violence do not appear to be necessarily associated with a woman's being in an assaultive relationship (Hotaling and Sugarman, 1990). Two characteristics, however, do appear to be highly correlated with wife assault. First, a number of studies have found that wife abuse is both more common and more severe in families of lower socioeconomic status, a finding partly attributed to the fact that higher income adults have greater privacy and thus are better able to conceal domestic violence (Fineman and Mykitiuk, 1994). Second, marital conflict—and the apparent inability to resolve it through negotiation and compromise—is a factor in many battering relationships. Hotaling and Sugarman (1990) found that conflicts in these marriages often were associated with a difference in expectations about the division of labor in the family, frequent drinking by the husband, and the wife's having attained a higher educational level than the husband. These researchers concluded that it is not useful to focus "primarily on the victim in the assessment of risk to wife assault."

Characteristics of Perpetrators

A man who systematically inflicts violence on his wife or lover is likely to have some or all of the following traits (Edelson et al., 1985; Gelles and Cornell, 1990; Goldstein and Rosenbaum, 1985; Margolin, Sibner, and Gleberman, 1988; Vaselle-Augenstein and Erlich, 1992; Walker, 1979, 1984):

- He believes the common myths about battering (see "Understanding Yourself," page 432).
- He believes in the traditional home, family, and gender-role stereotypes.
- He has low self-esteem and may use violence as a means of demonstrating power or adequacy.
- He may be sadistic, pathologically jealous, or passive-aggressive.
- He may have a "Dr. Jekyll and Mr. Hyde" personality, being capable at times of great charm.
- He may use sex as an act of aggression.
- He believes in the moral rightness of his violent behavior (even though he may "accidentally" go too far).

Lenore Walker (1984) believes that a man's battering is not the result of his interactions with his partner or any kind of provocative personality traits of the partner. She writes:

> The best prediction of future violence was a history of past violent behavior. This included witnessing, receiving, and committing violent acts in [the] childhood home; violent acts toward pets, inanimate objects, or other people; previous criminal record; longer time in the military service; and previous expression of violent behavior toward women. If these items are added to a history of temper tantrums, insecurity, need to keep the environment stable, easily threatened by minor upsets, jealousy, possessiveness, and the

UPSCALE VIOLENCE

So much of the literature on intimate violence examines couples from lower income, financially distressed circumstances that it becomes easy to assume that this is where such violence occurs. Susan Weitzman's (2000) research on "hidden abuse in upscale marriages" shows that financial privilege does not protect women from victimization. Weitzman neither claims that marital violence and abuse happen with the same frequency among the upper and upper-middle classes as among those farther down the economic ladder, nor does she assert that the "upscale victims" have it worse than their lower-status counterparts. However, she convincingly shows that upscale victims have their own particular problems with which they must deal. Here, we look more closely at Weitzman's research.

Despite their predicaments, "upscale victims" are often excluded from or understudied in the research literature on domestic abuse. Weitzman found fewer than a dozen studies about highly educated, upper-income abused wives among the more than 500 books and articles published over the past 25 years. Thus, she set out to learn about how upper status wives experience marital abuse.

Who counts as "upscale"? Weitzman's definition included the following criteria: a combined income in excess of $100,000 a year, a minimum education of a bachelor's degree, a self-perception of being upper or upper-middle class, and residence in a neighborhood with a high ranking local reputation or ranked in the top 25 percent of its statewide area according to the U.S. Census Bureau.

Why don't we know more about such women? Weitzman considers upper-status battered women as "hidden" victims. They "do not report the tirades and tantrums, . . . refuse to press charges or even call the police despite the broken bones and blackened eyes inflicted on them by the men who purport to love them" (Weitzman, 2000, p. 5). Because of the embarrassment upscale victims feel, the resources at the disposal of affluent abusers (e.g., large, more insulated or protected living space, the means to retain more highly skilled legal representation) and insensitivity of "protective" services such as police or shelter personnel, upscale victims may suffer more in silence and be overlooked by researchers.

What sorts of abuse do upscale women suffer? All the women in Weitzman's research suffered from emotional abuse. This ranged from and included neglect, extreme selfishness, rage attacks and criticism—particularly about her suitability as a spouse and parent, bullying, public humiliation, threats, extramarital affairs, destruction of her property, and inducing fear (Weitzman, 2000, p. 84). More than half of the women Weitzman studied also suffered physical abuse that included pushing and shoving, choking and strangling, hair pulling, being pinned down, or punched, hit, or thrown against a wall or down stairs. For nearly three-fourths of the women, warning signs of such behavior were evident before marriage.

What makes "upscale abusers" abuse? Although she considers a range of psychological theories and approaches, ultimately Weitzman suggests that most of the upscale abusers she heard about or met appeared to suffer from narcissistic personality disorder, with characteristics such as an absence of empathy, a need for admiration, and a sense of grandiosity. In order to be diagnosed with narcissistic personality disorder, a person must demonstrate at least five of the following characteristics:

- Grandiose sense of self-importance
- Preoccupation with fantasies of unlimited power, success, brilliance, beauty, or ideal love

ability to be charming, manipulative, and seductive to get what he wants, and hostile, nasty and mean when he doesn't succeed—then the risk for battering becomes very high. If alcohol abuse problems are included, the pattern becomes classic.

Commonly, one reads or hears that a major, if not the major, factor in predicting partner violence is having grown up around violence in one's family of origin. This intergenerational transmission of violence is assumed despite evidence of very weak connections between childhood and adult experiences. According to research, parental violence accounts for 1 percent of dating violence and approximately the same proportion of violence in marriage or marriage-like relationships (see review by Johnson and Ferraro, 2000). Even in widely cited data indicating that sons of the most violent parents have *1,000 percent greater rate* of wife-beating than sons

- Belief that one is special and unique, and should only associate with other special people
- Need for excessive admiration
- Sense of entitlement
- Lack of empathy
- Envious or believes that others are envious of him or her
- Arrogance and haughtiness
- A sense of entitlement

With these sorts of traits, the upscale, narcissistic batterer gets gratification from remaining in a relationship with a spouse whom he consistently mistreats. Despite his abuse and mistreatment, she stays, thus validating his inflated sense of self-worth. As part of his psychological make-up, the narcissistic batterer fears abandonment, perceives slights that no one else will notice, and perceives that his wife's responsiveness is not swift or devoted enough. Compounding the problems created by the narcissistic personality disorder is the way such men are reinforced by others, which in turn enlarges their egos and insatiable expectations and causes them to believe that their demands are justified. Because they are financially successful and highly respected, they receive a kind of adoration that serves to increase their sense of entitlement with and over their wives. Because they control their ample financial resources, they also exploit their wives' fears of abandonment and impoverishment.

Where do upscale victims turn? Many of the typical resources and supports available to victims of domestic abuse fail the upscale abused wife. The various legal, medical, social, and mental health services to which they turn often "revictimize them" through disbelief and neglect. Their complaints may not be taken as seriously because it is assumed that a woman of her financial means would be able to act to protect herself or has the resources to remove herself from the abusive situation. In fact, her husband may control all financial resources to the point that she has nothing to use to extract herself from her victimization. Emergency rooms, domestic violence shelters, even the academic community (in sociology, psychology, and social work) tend to believe that the upscale abuse victim can take care of herself. Even friends and family may find it hard to fathom that a highly regarded man of means, a potential "pillar of the community" abuses and mistreats his spouse.

How can you help an upscale abuse victim? Weitzman offers some thoughtful recommendations if you suspect that someone you are close to is being victimized. Among her recommendations are the following:

If you observe injuries, ask how they were sustained. If her story seems implausible, tell her that you are worried and suspect that she is hiding something. Remind her that it is neither her fault nor something to feel ashamed of. You may need to be patient as she may be resistant to initial overtures. You may need to revisit the issue multiple times, incorporating anecdotal evidence of other women's situations to get her to self-disclose.

Help her see that she is neither alone nor responsible. Make her aware of her choices and reassure her of your support. Offer the following kinds of comments:

I'm afraid for your safety.

It will only get worse.

I'm here whenever you are ready to talk.

You don't deserve to be treated like this.

It happens everywhere, including to people like us.

The perks and privileges of your lifestyle are not worth this kind of pain and possible danger.

Help her to see her situation objectively by showing her profiles of abused women, abusive men, and early warning signs.

Help her create a plan for safe exit from her situation. This ought to include a safe but secret place (for example, a shelter, or a friend's house) that she can go to as she seeks help.

of nonviolent parents, the reality is that 80 percent of the sons of the most violent parents were nonviolent for at least the past 12 months (Johnson and Ferraro, 2000).

Male Victims

The incidence of "battered husbands" is unknown. We have, after all, the cartoon image of Blondie chasing Dagwood with a rolling pin. (We don't see Dagwood chasing Blondie with a gun or knife, however, which is a more realistic depiction of family violence.) Although it is undoubtedly true that some men are injured in attacks by wives or lovers, the overwhelming majority of victims of adult family violence are women. Ten times as many women as men are seriously victimized by an intimate partner or ex-partner (Campbell, 1995). In one study (Saunders, 1986) investigating "husband abuse," almost all the women reported that they acted in

Understanding Yourself

THE MYTHOLOGY OF FAMILY VIOLENCE AND SEXUAL ABUSE

The understanding of family violence and sexual abuse is often obscured by the different mythologies surrounding these issues. Below are twelve popular myths about family violence and sexual abuse in our society. Some of these myths may apply to some cases, but as generalizations they are not accurate. Many of these myths are accepted by the victims of family violence, as well as by the perpetrators themselves.

As you look at the following myths, which ones do you believe (or have you believed)? What was the basis or source of your beliefs? If you no longer believe a particular myth, what changed your mind?

- Family violence is extremely rare.
- Family violence is restricted to families with low levels of education and low socioeconomic status.
- Most family violence is caused by alcohol or drug abuse.
- Violent spouses or parents have psychopathic personalities.
- Violent families are not loving families.
- Battered women cause their own battering because they are masochistic or crazy.
- A battered woman can always leave home.
- Most child sexual abuse is perpetrated by strangers.
- Sexual abuse in families is a fairly rare occurrence.
- Abused children will grow up to abuse their own children.
- The police give adequate protection to battered women.
- Most of society does not condone domestic violence.

These myths hide the extent of physical and sexual abuse that actually takes place inside a painfully large number of American families. Belief in these myths makes it possible to avoid dealing with some of the unhappy realities—at least for a while.

self-defense; they did not initiate the violence. Their actions did not cause noticeable injury. Indeed, a woman may attempt to inflict damage on a man in self-defense or retaliation, but most women have no hope of prevailing in hand-to-hand combat with a man. A woman may be severely injured simply trying to defend herself. As Gelles and Cornell (1990) observe, although there may be similar rates of hitting, "when injury is considered, marital violence is primarily a problem of victimized women."

Although male violence dramatically overshadows female violence, by combining common couple violence and violent resistance, it may make female violence appear as though it occurs at about the same rate. Clearly, women tend to do less damage in assaulting a male partner than do men in assaulting female partners. Thus, we may not consider it as important as that committed by men (Straus, 1993). Suzanne Steinmetz (1987) suggests that some scholars "deemphasize the importance of women's use of violence." As such, there is a "conspiracy of silence [which] fails to recognize that family violence is never inconsequential." Murray Straus (1993) lists four reasons for taking the study of female violence seriously:

1. Assaulting a spouse—either a wife or a husband—is an "intrinsic moral wrong."
2. Not doing so unintentionally validates cultural norms that condone a certain amount of violence between spouses.
3. There is always the danger of escalation. A violent act—whether committed by a man or a woman—may well lead to increased violence.
4. Spousal assault is a model of violent behavior for children. Children are affected as strongly by viewing the violent behavior of their mothers as by viewing that of their fathers.

THE CYCLE OF VIOLENCE

Lenore Walker's (1979) research has revealed a three-phase wife-battering cycle. The duration of each phase may vary, but the cycle goes on and on:

- *Phase 1: tension building:* Tension is in the air. The woman tries to do her job well, to be conciliatory. Minor battering incidents may occur. She denies her own rising anger. Tension continues to build.
- *Phase 2: the explosion:* The man loses control. Sometimes the woman will precipitate the incident to "get it over with." He generally sets out to "teach her a lesson" and goes on from there. This

is the shortest phase, usually lasting several hours but sometimes continuing for two or three days or longer.

- *Phase 3: the "honeymoon":* Tension has now been released, and the batterer is contrite, begs forgiveness, and sincerely promises never to do it again. The woman chooses to believe him and forgives him. This "symbiotic bonding" (interdependence) makes intervention, help, or change unlikely during this phase. This behavioral process is not universal, however. Weitzman's research on upscale abuse and violence shows that "the man of means actually does little to seek his wife's forgiveness" for his violence (Weitzman, 2000).

Often the battered woman expresses surprise at what has touched off the battering incident. It may have been something outside the home—at the man's job, for example. He may have come home drunk, or he may have been drinking steadily at home. Alcohol is implicated in many battering incidents. Results of studies of alcohol and battering vary widely; alcohol problems were reported in 35 to 93 percent of cases of assaultive husbands, depending on the study (see Gelles [1993a] for discussions of conflicting studies about battering and alcohol use).

In a battering relationship, the woman may not only suffer physical damage but also be seriously harmed emotionally by a constant sense of danger and the expectation of violence that weaves a "web of terror" about her (Edelson et al., 1985). Walker (1993) suggests that women who are repeatedly abused may develop a set of psychological symptoms similar to those of post-traumatic stress disorder (PTSD). She labels these symptoms *battered woman syndrome.*

VIOLENCE IN GAY AND LESBIAN RELATIONSHIPS

Until recently, very little was known about violence in lesbian and gay relationships. One reason is that such relationships have not been given the same social status as those of heterosexuals. Recent research indicates that the rate of abuse in gay and lesbian relationships is comparable to that in heterosexual relationships: 11 percent to more than 45 percent according to various studies (Renzetti, 1995). Furthermore, Claire Renzetti found that violence in same-sex relationships is rarely a one-time event; once violence occurs it is likely to reoccur. It also appears to be as serious as violence in heterosexual relationships, including physical, psychological, and/or financial abuse. Johnson and Ferraro (2000) suggest that given Renzetti's research showing the prevalence of psychological abuse, jealousy, and struggles over power and control, it appears as if intimate terrorism can be found among lesbian couples.

One additional form of abuse, unique to same-sex couples, is the threat of "outing" (revealing another's gay orientation without consent). Threatening to out one's partner to co-workers, employers, or family may be used as a form of psychological abuse in same-sex relationships.

For battered partners in same-sex relationships, there is often nowhere to go for support. Services for gay men and lesbians are often nonexistent or uninformed about the multifaceted issues that face such victims. Renzetti (1995) points out several policy issues that must be addressed among service providers and domestic violence agencies:

- Consider how homophobia inhibits gay and lesbian victims of abuse from self-identifying as such.
- Recognize that battered gay men and lesbians of color experience a triple jeopardy: as victims of domestic violence, as homosexuals, and as racial/ethnic minorities.
- Address the issue of gay men and lesbians as both batterers and victims who may seek services at the same time from the same agency.

MARITAL RAPE

One of the most widespread and overlooked forms of family violence, marital rape is a form of battering. Most legal definitions of **rape** include "unwanted sexual penetration, perpetrated by force, threat of harm, or when the victim [is] intoxicated" (Koss and Cook, 1993). Rape may be perpetrated by males or females and against males or females; it may involve vaginal, oral, or anal penetration; and it may involve the insertion of objects other than the penis. Approximately 10 to 14 percent of wives have been forced by their husbands to have sex against their will (Yllo, 1995).

Historically, marriage has been regarded as giving husbands unlimited sexual access to their wives.

Beginning in the late 1970s, most states enacted legislation to make at least some forms of marital rape illegal. It was not until 1993, however, that laws were enacted to make marital rape a crime. Throughout the United States, a husband can be prosecuted for raping his wife, although many states limit the conditions, such as requiring extraordinary violence. Less than half of the states offer full legal protection for wives (Muehlenhard et al., 1992). The precise definition of **marital rape** differs from state to state, however. In several states, wife rape is illegal only if the couple has separated.

Because of the sexual nature of marriage, marital rape has not been regarded as a serious form of assault, as Kersti Yllo (1995) explains:

> A widely held assumption has been that an act of forced sex in the context of an ongoing relationship in which consensual sex occurs cannot be very significant or traumatic. This assumption is flawed because it overlooks the core violation of rape that is coercion, violence and in the case of wife rape, the violation of trust.

There still remains the problem of enforcing the law. Many people discount rape in marriage as a "marital tiff" that has little to do with "real" rape (Yllo, 1995). Many victims themselves have difficulty acknowledging that their husbands' sexual violence is indeed rape. White females are more likely than African-American females to identify sexual coercion in marriage as rape (Cahoon et al., 1995), and all too often judges seem more in sympathy with the perpetrator than with the victim, especially if he is very intelligent, successful, and well educated. There is also the "notion that the male breadwinner should be the beneficiary of some special immunity because of his family's dependence on him" (Russell, 1990). Diana Russell goes on to say:

> On the basis of such an argument, it follows that it would be a violation of the principle of equity to incarcerate men who beat up and/or rape women who are not their wives. Specifically, it would not be fair to the wives and children of employed stranger rapists, acquaintance rapists, date rapists, lover rapists, authority figure rapists, or rapists who rape their friends. Why should these families have to endure the loss of their breadwinners if the families of husbands who are rapists are spared this hardship?

Because these kinds of attitudes are so entrenched in the American psyche, it is estimated that two-thirds of sexual assault victims do not report the crime (U.S. Department of Justice, 1997).

Marital rape victims experience feelings of betrayal, anger, humiliation, and guilt. Following their rapes, many wives feel intense anger toward their husbands. One woman recounted, "'So,' he says, 'You're my wife and you're gonna . . .' I just laid there thinking 'I hate him, I hate him so much.'" Another expressed her humiliation and sense of "dirtiness" by taking a shower: "I tried to wash it away, but you can't. I felt like a sexual garbage can" (Finkelhor and Yllo, 1985). Some feel guilt and blame themselves for not being better wives. Some develop negative self-images and view their lack of sexual desire as a reflection of their own inadequacies rather than as a consequence of abuse.

DATING VIOLENCE AND RAPE

In the last two decades, researchers have become increasingly aware that violence and sexual assault can take place in all forms of intimate relationships. Violence between intimates is not restricted to family members. Even casual or dating relationships can be marred by violence or rape.

DID YOU KNOW?

Forty-four percent of rape victims are younger than 18 years old, and two-thirds of violent sex offenders serving time in state prisons said their victims were younger than 18 years (U.S. Department of Justice, 1997).

Dating Violence

The incidence of physical violence in dating relationships, including those of teenagers, is alarming. Evidence suggests that it even exceeds the level of marital violence (Lloyd, 1995). A survey on physical violence during courtship revealed that almost one-third of young adults (age 30 and under) had experienced or used physical violence in a dating relationship during the past 12 months (Lloyd, 1995). Sexual violence may be even more prevalent. Nearly two in five college women have experienced an actual rape or an attempted rape or have been coerced into intercourse against their will.

Although it may seem logical to assume that dating violence leads to marital violence, little actual research has been done in this area. It does appear, however, that the issues involved in dating violence are different than those generally involved in spousal violence. Whereas marital violence may erupt over domestic issues such as housekeeping and child rearing (Hotaling and Sugarman, 1990), dating violence is far more likely to be precipitated by jealousy or rejection (Lloyd and Emery, 1990; Makepeace, 1989). One young woman recounted the following incident (Lloyd and Emery, 1990):

> I was waiting for him to pick me up in front of school. I was befriended by some guys and we struck up a conversation. When my boyfriend picked me up he didn't say anything. When we got home, physical violence occurred for the first time in our relationship. I had no idea it was coming. He caught me on the jaw, and hit me up against the wall. I couldn't cry or scream or anything—all I could do was look at him. He picked me up and threw me against the wall and then started yelling and screaming at me that he didn't want me talking to other guys.

Lloyd and Emery found that dating violence might also involve the man's use of alcohol or drugs, "unpredictable" reasons, and intense anger.

Although many women leave a dating relationship after one violent incident, others stay through repeated episodes. Women who have "romantic" attitudes about jealousy and possessiveness and who have witnessed physical violence between their own parents may be more likely to stay in such relationships (Follingstad et al., 1992). Women with "modern" gender-role attitudes are more likely to leave than those with traditional attitudes (Flynn, 1990). Women who leave violent partners cite the following factors in making the decision to break up: a series of broken promises that the man will end the violence, an improved self-image ("I deserve better"), escalation of the violence, and physical and emotional help from family and friends (Lloyd and Emery, 1990). Apparently, counselors, physicians, and law enforcement agencies are not widely used by victims of dating violence (Pirog-Good and Stets [1989]; see Lloyd [1991] for implications for intervention in courtship violence; see Levy [1991] for perspectives on violence in adolescent relationships).

Date Rape

Sexual intercourse with a dating partner that occurs against his or her will with force or the threat of force—**date rape**—is the most common form of rape. Date rape is also known as **acquaintance rape.** One study found that women were more likely than men to define date rape as a crime. Disturbingly, date rape was considered less serious when the woman was African American (Foley et al., 1995).

Date rapes are usually not planned. Two researchers (Bechhofer and Parrot, 1991) describe a typical date rape:

> He plans the evening with the intent of sex, but if the date does not progress as planned and his date does not comply, he becomes angry and takes what he feels is his right—sex. Afterward, the victim feels raped while the assailant believes that he has done nothing wrong. He may even ask the woman out on another date.

Alcohol or drugs are often involved. When both people are drinking, they are viewed as more sexual. Men who believe in rape myths are more likely to see drinking as a sign that females are sexually available (Abbey and Harnish, 1995). In one study, 79 percent of women who were raped by their date had been drinking or taking drugs prior to the rape. Seventy-one percent said their assailant had been drinking or taking drugs (Copenhaver and Grauerholz, 1991). There are also high levels of alcohol and drug use among middle school and high school students who have unwanted sex (Rapkin and Rapkin, 1991).

Recently, certain "date-rape drugs," most often either Gamma hydroxybutyrate (GHB) or Rohypnol (flunitrazepam, popularly known as "roofies," "roofenol," "rochies," and other street names), have surfaced as major public safety concerns. Both drugs have sedative effects, especially when combined with alcohol. They may reduce inhibitions, and they affect memory. Both are used by some men to sedate and later victimize women, many of whom wake up unaware of where they are, how they got there, or what they have done. Fifteen-year-old Samantha Reid died as a result of drinking a soft drink that had been laced with GHB. Knowing only that the drink tasted funny, she died just hours later. Her friend, Melanie Sindone, recovered after having gone into a coma that lasted less than a day. According to a *New York Times* article, the Drug Enforcement Agency estimates that there have been 65 deaths since 1990 and 15 sexual assault cases involving 30 victims who had been given GHB. In Reid's death, three men were convicted of involuntary manslaughter, punishable by 15 years in prison (Bradsher, 2000). In 2000, former President Clinton signed into law the Hillory J. Farias and Samantha Reid Date-Rape Drug Prohibition Act of 2000, named for Reid and another teenage victim who died after unknowingly drinking a beverage mixed with GHB. It is a federal crime, punishable by up to 20 years in prison, to manufacture, distribute, or possess GHB (ABCNEWS.Go.com).

INCIDENCE OF DATE RAPE. Lifetime experience of date rape ranges from 15 to 28 percent for women, according to various studies. If the definition is expanded to include attempted intercourse as a result of verbal pressure or the misuse of authority, then women's lifetime incidence increases significantly. When all types of unwanted sexual activity are included, ranging from kissing to sexual intercourse, half to three-quarters of college women report sexual aggression in dating (Cate and Lloyd, 1992). There is also considerable sexual coercion in relationships between gay men. Coercion also exists in lesbian relationships, though less than in gay male and heterosexual ones.

In a large-scale study on sexual aggression, Mary Koss (1988) surveyed over 6,100 students in 32 colleges. Her findings indicated the following:

- Almost 54 percent of the women surveyed had been sexually victimized in some form; 15 percent had been raped.
- A quarter of the women surveyed had been the victims of rape or attempted rape; 84 percent knew their assailants.
- Forty-seven percent of the rapes were by first dates, casual dates, or romantic acquaintances.
- Twenty-five percent of the men had perpetrated sexual aggression, 3 percent had attempted rape, and 4 percent had actually raped.
- Almost three-quarters of the raped women did not identify their experiences as rape.

Physical violence often goes hand in hand with sexual aggression. One researcher found, in a study of acquaintance rape victims, that three-fourths of the women sustained bruises, cuts, black eyes, and internal injuries. Some were knocked unconscious (Belknap, 1989).

WHEN NO IS NO. There is considerable confusion and argument about sexual consent. Much sexual communication is done nonverbally and ambiguously,

Sexual assault peer educators at Brown University dramatize date rape to make students aware of its dynamics.

© Mark Peterson/Saba Press Photos

as Charlene Muehlenhard and her colleagues (1992) note:

> Most sexual scripts do not involve verbal consent. One such script involves two people who are overcome with passion. Another such script involves a male seducing a hesitant female, who, according to the sexual double standard, must not acknowledge her desire for sex lest she be labeled "loose" or "easy." Neither of these scripts involve explicit verbal consent from both persons.

That we don't usually give verbal consent for sex indicates the importance of nonverbal clues. Nonverbal communication is imprecise, however, as we saw in Chapter 6. It can be misinterpreted easily if it is not reinforced verbally. For example, men frequently mistake a woman's friendliness for sexual interest (Johnson, Stockdale, and Saal, 1991; Stockdale, 1993). They often misinterpret a woman's cuddling, kissing, and fondling as wishing to engage in sexual intercourse (Gillen and Muncher, 1995; Muehlenhard, 1988; Muehlenhard and Linton, 1987). A woman needs to make her boundaries clear verbally.

Our sexual scripts often assume yes unless a no is directly stated (Muehlenhard et al., 1992). This makes individuals "fair game" unless a person explicitly says no. The assumption of consent puts women at a disadvantage. First, because men traditionally initiate sex, men may feel it is legitimate to initiate sex whenever they desire without women's explicitly consenting. Second, women's withdrawal can be considered insincere because consent is always assumed. Such thinking reinforces a common sexual script in which men initiate and women refuse so as not to appear promiscuous. In this script, the man continues believing that her refusal is token. One study found that almost 40 percent of the women had offered a token no at least once (Muehlenhard and Hollabaugh, 1989). Some common reasons for offering token no's include not wanting to appear "loose," uncertainty about how the partner feels, inappropriate surroundings, and game playing (Muehlenhard and Hollabaugh, 1989; Muelhenhard and McCoy, 1991). Because some women sometimes say "no" when they mean "coax me," male-female communication may be especially unclear regarding consent (Muehlenhard and Cook, 1991). Furthermore, men are more likely than women to think of male-female relationships as a "battle of the sexes" (Reilly et al., 1992). Because relationships are conflictual, no's are to be expected as part of the battle. A man, however, "should" persist because it is his role to conquer, even if he is not interested in sex (Muehlenhard and Schrag, 1991; Muehlenhard et al., 1991).

AVOIDING DATE RAPE. To reduce the risk of date rape, women should consider the following points:

- When dating someone for the first time, go to a public place, such as a restaurant, movie, or sports event.
- Share expenses. A common scenario is a date expecting you to exchange sex for his paying for dinner, the movie, drinks, and so on (Muehlenhard and Schrag, 1991; Muehlenhard et al., 1991).
- Avoid using drugs or alcohol if you do not want to be sexual with your date. Their use is associated with date rape (Abbey, 1991).
- Avoid ambiguous verbal or nonverbal behavior. Examine your feelings about sex and decide early if you wish to have sex. Make sure your verbal and nonverbal messages are identical. If you only want to cuddle or kiss, tell your partner that those are your limits. Tell him that if you say no, you mean no. If necessary, reinforce your statement emphatically, both verbally ("No!") and physically (pushing him away) (Muehlenhard and Linton, 1987).
- Be forceful and firm. Don't worry about being polite. Often men interpret passivity as permission and ignore or misunderstand "nice" or "polite" approaches (Hughes and Sandler, 1987).
- If things get out of hand, be loud in protesting, leave, and go for help.
- Be careful about what you drink, who you accept drinks from, and where you place your unfinished drink if you put it down; be suspicious of any open drink that tastes funny (salty or flat). These strategies will help reduce the likelihood of having your drink laced with date-rape drugs.

WHEN AND WHY SOME WOMEN STAY IN VIOLENT RELATIONSHIPS

Violence in relationships generally develops a continuing pattern of abuse over time. We know from systems theory that all relationships have some degree of mutual dependence, and battering relationships are certainly no different. Despite the mistreatment they receive, some women stay in or return to violent situations for many reasons. However, we need to be careful not to overstate the tendency for abuse victims to stay with their abusers. Johnson and Ferraro (2000) note, for example, "We need to watch our language; there is no good reason why a study in which two thirds of the women have left the violent relationship is subtitled, 'How and why women stay' instead of 'How and why women leave.'

For the women who do stay in abusive situations, their reasons include the following:

- *Economic dependence:* Even if a woman is financially secure, she may not perceive herself as being able to cope with economic matters. For low-income or poor families, the threat of losing the man's support—if he is incarcerated, for example—may be a real barrier against change.
- *Religious pressure:* She may feel that the teachings of her religion require her to keep the family together at all costs, to submit to her husband's will, and to try harder.
- *Children's need for a father:* She may believe that even a father who beats the mother is better than no father at all. If the abusing husband also assaults the children, the woman may be motivated to seek help (but this is not always the case).
- *Fear of being alone:* She may have no meaningful relationships outside her marriage. Her husband may have systematically cut off her ties to other family members, friends, and potential support sources. She has no one to go to for any real perspective on her situation. (See Nielsen, Endo, and Ellington [1992] for the relationship between social isolation and abuse.)
- *Belief in the American dream:* The woman may have accepted without question the myth of the perfect woman and happy household. Even though her existence belies this, she continues to believe that it is how it should (and can) be.
- *Pity:* She feels sorry for her husband and puts his needs ahead of her own. If she doesn't love him, who will?
- *Guilt and shame:* She feels that it is her own fault if her marriage isn't working. If she leaves, she believes, everyone will know she is a failure, or her husband might kill himself.
- *Duty and responsibility:* She feels she must keep her marriage vows "till death us do part."
- *Fear for her life:* She believes she may be killed if she tries to escape.
- *Love:* She loves him; he loves her. On her husband's death, one elderly woman (a university professor) spoke of her 53 years in a battering relationship (Walker, 1979): "We did everything together. . . . I loved him; you know, even when he was brutal and mean. . . . I'm sorry he's dead, although there were days when I wished he would die. . . . He was my best friend. . . . He beat me right up to the end. . . . It was a good life and I really do miss him."
- *Cultural reasons:* Women from nonmainstream cultural backgrounds may face great obstacles to leaving a relationship. They may not speak English; they may not know where to go for help and may fear they will not be understood. They often fear that the husband will lose his job, retaliate against them, or take the children back to their country of origin (Donnelly, 1993). Recent immigrants from Latin America, Asia, and South Asia may be especially fearful that their revelations will reflect badly on the family and community.
- *Nowhere else to go:* Women may have no alternative place to live. Shelter space is limited and temporary. Relatives and friends may be unable or unwilling to house a woman who has left, especially if she brings children with her.

Learned Helplessness

Lenore Walker (1979, 1993) theorizes that women stay in battering relationships as a result of "learned helplessness." According to Walker, women who are repeatedly battered develop much lower self-concepts than women in nonbattering relationships. They begin to feel that they cannot control the battering or the events that surround them. Through a process of behavioral reinforcement, they "learn" to become helpless. As Walker notes:

> Women are systematically taught that their personal wants, survival, and anatomy do not depend on effective and creative responses to life situations, but rather on their physical beauty to men. They learn that they have no direct control over the circumstances of their lives.

If violence is used against them, women may even become desensitized to the accompanying pain and fear. The more it happens, the more helpless they feel and the less they are able to see alternative possibilities. Walker points out that these women are not totally helpless or passive, but that "they narrow their choice of responses, opting for those that have the highest predictability of creating successful outcomes" (Walker, 1993).

Women's Coping Strategies

Some women think they can stop their partners' violence, and some in fact do. Lee Bowker (1983) reported that women used a variety of strategies to stop their husbands' abusiveness. These ranged from passive defense techniques (covering their bodies with arms or hands) to seeking informal help (friends) or formal help (counseling through a social service agency). The particular strategy was not important in stopping the violence, however. What made the crucial difference appeared to be the woman's determination that the violence must cease (Bowker, 1983; Gondolf, 1987, 1988).

REFLECTIONS

In your family (including your extended family), has there been spousal violence? Have you experienced violence in a dating relationship? If so, what were the factors involved in causing it? In sustaining it? If you or your family have not been involved in such violence, what factors do you think have protected against it?

ALTERNATIVES: POLICE INTERVENTION, SHELTERS, AND ABUSER PROGRAMS

Professionals who deal with domestic violence have long debated the relative merits of control versus compassion as intervention strategies (Mederer and Gelles, 1989). Although more understanding of the dynamics of abusive relationships and the deterrence process is clearly needed, we can see that both approaches have their place. Both controlling measures (which raise the "costs" of violent behavior), such as arrest, prosecution, and imprisonment, and compassionate measures, such as shelters, education, counseling, and support groups, have been shown to be successful to varying degrees under varying conditions. Used together, these interventions may be quite effective. Mederer and Gelles (1989) suggest that controlling measures may be used to "motivate violent offenders to participate in treatment programs."

Battered Women and the Law

Family violence studies and feminist pressure have spurred a movement toward the implementation of stricter policies for dealing with domestic offenders. Long ignored, domestic violence has only recently become a top concern for legislators and law enforcement agencies throughout the country (Wilson, 1997). California has introduced measures to crack down on spousal abuse and increase funding for shelters and other related agencies. In 1995 alone, nearly 22 laws against domestic violence were passed. Other states have quickly followed. Obviously too late for many like Nicole Brown Simpson, our collective conscience is finally drawn to such issues as policy misconduct, racism, child custody, and spousal abuse.

Still there is resistance by some law enforcement and judicial branches to listen to the victims of abuse. Prevention and law enforcement are necessary measures this society must take in order to reduce the incidence of domestic violence. Today, at least a third of the largest U.S. police forces are instructed to arrest the assailants, although the jury is still out as to whether mandatory arrest is actually effective. (See Berk [1993], and Buzawa and Buzawa [1993], for conflicting views on the efficacy of arrest as a deterrent to domestic violence.)

The legal and criminal justice systems can do only so much. When courts issue and police enforce restraining orders requiring abusive partners to avoid contact with their targeted victims, there is no guarantee that offenders will comply. In fact, the 1998 Bureau of Justice Statistics reports, of those men incarcerated for domestic assault, 9 percent had restraining orders on them at the time they committed the assault for which they were imprisoned. One-fourth were under some other criminal justice supervision (parole or probation, for example). Thus, no policy is a fail-safe one.

Battered Women's Shelters

At the point where a woman finds that she can leave an abusive relationship, even temporarily, she may have any number of serious needs. If she is fleeing an attack, she may need immediate medical attention and physical protection. She will need accommodation for herself and possibly her children. She will certainly need access to support, counseling, and various types of assistance—money, food stamps or other basic survival items for herself and her children. She will need to deal with informed, compassionate professionals such as police officers, doctors, and social workers.

In the late 1970s, the shelter movement developed to meet the needs of many battered women. The shelter movement has grown slowly, hampered by lack of funding and mixed reaction from the public. There are an estimated one thousand shelters throughout the United States, a vast improvement over the estimated five or six shelters in existence in 1976 (Gelles and Cornell, 1990). Besides offering immediate safe shelter (the locations of safe houses are usually known only to the residents and shelter workers), these refuges let battered women realize that they are not alone in their misery and help them form supportive networks with one another. The shelters also provide many other services for battered women who call, such as information, advice, or referrals.

© Michael Newman/PhotoEdit

Battered women's shelters provide safe havens for women in abusive relationships. Shelters provide counseling and emotional support as well as temporary lodging, meals, and other necessities for women and their children.

Women who seem to benefit most from shelter stays are those who have decided to "take charge of their lives," according to one study (Berk, Newton, and Berk, 1986). Hampton, Gelles, and Harrop (1989) suggest that the dramatic drop in wife abuse among African Americans may be the result of African-American women's "increased status" coupled with their apparent willingness to make use of shelters and other programs.

DID YOU KNOW?

More than 1 million children are confirmed to be abused and neglected each year in the United States. More realistic perhaps is the estimated number of children who are abused and neglected: nearly 3 million ("Child Abuse and Neglect," 1997).

Abuser Programs

According to Richard Tolman (1995), "A comprehensive solution to violence against women in intimate relationships demands that perpetrators of abuse be held accountable for their behavior and that direct efforts be made with batterers to change their behavior." Treatment services for men who batter provide one important component of a coordinated response to domestic violence (see Gondolf [1993] for program and treatment issues). Psychotherapy, group discussion, stress management, or communication skills classes may be available through mental health agencies, women's crisis programs, or various self-help groups.

The extent to which attending batterers' groups actually changes abusing men's violent behavior is difficult to measure (Gelles and Conte, 1991). What has become apparent is the ineffectiveness of the "one size fits all" approach and the need to adopt a more sophisticated understanding of individual's violent behaviors (Tolman, 1995). Studies by Edward Gondolf (1987, 1988) of men who have completed voluntary programs showed that two-thirds to three-quarters of these men were subsequently nonviolent. Gondolf also confirmed the conclusion of a number of studies of women who are successful in stopping abuse: Battered women play a crucial role in stopping the violence against them. Women's insistence on their partners' getting help (the "woman factor") apparently can influence men to "learn more or try harder" to change.

A coordinated community response that includes proactive police and criminal justice strategies, advocacy and services for battered women and their children, and responses by other community institutions that promote safety for battered women and sanctions for men who batter are necessary interventions (Tolman, 1995).

CHILD ABUSE AND NEGLECT

The history of children has not always been a particularly happy one. At various times and places, children have been abandoned to die of exposure in deserts, in forests, and on mountainsides or have simply been murdered at birth if they were deemed too sickly, too ugly, of the wrong sex, or just impractical. Male children have been subjected to castration to make them fit for guarding harems or singing soprano in church choirs. Millions of female children in the Middle East and Africa undergo devastating sexual mutilation. These practices (and many others) have all been socially condoned in their time and place. In the societies in which they have existed, they have not been (or are not) recognized as abusive.

Of course, we feel we are more "civilized" today. In our society, some degree of physical force against children, such as spanking, is generally accepted as normal. In fact, more than 90 percent of toddlers in the United States are reportedly spanked (Straus and Field, 2003). Most child-rearing experts, however, currently suggest that parents use alternative disciplinary measures.

Child abuse was not recognized as a serious problem in the United States until the 1960s. At that time, C. H. Kempe and his colleagues (1962) coined the medical term *battered-baby syndrome* to describe the patterns of injuries commonly observed in physically abused children. The Children's Defense Fund (1996) reports:

- Every 10 seconds, a child is reported abused or neglected.
- Every 14 seconds, a child is arrested.
- Every 2 hours a child is killed by firearms.

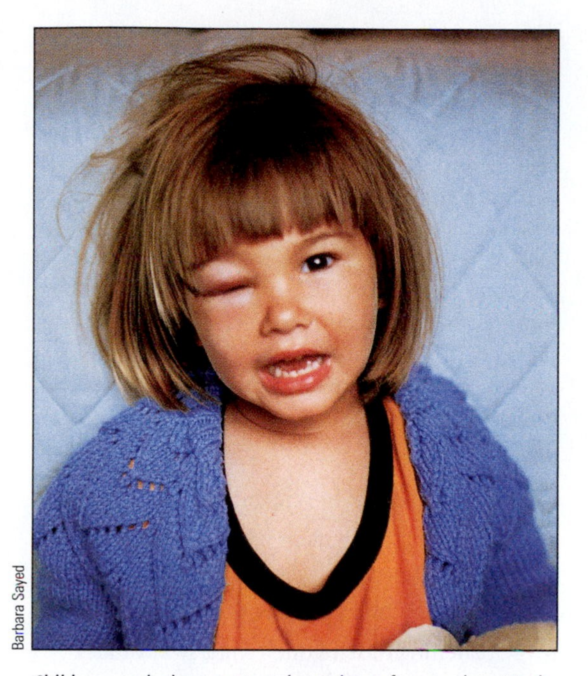

Barbara Sayed

Children are the least protected members of our society. Much physical abuse is camouflaged as discipline or as the parent "losing" his or her temper.

- Every 4 hours a child commits suicide.
- Every 5 hours a child dies from abuse or neglect.

When we look at violence among children from a global perspective, we see an even larger shadow cast over our nation. A study by the Centers for Disease Control and Prevention found that nearly three out of four child slayings in the industrialized world occur in the United States ("Violence Kills," 1997). The statistics show that the epidemic of violence in recent years that has hit younger and younger children is confined almost exclusively to the United States. The suicide rate alone for children 14 and under is double that of the rest of the industrialized world. No explanation for the huge gap between the rates of violent death for American children and those of other countries was given, though some experts speculate it is due to a growing faction of children who are unsupervised or otherwise at risk. The low level of funding for social programs, sexism, racism, and epidemic rates of poverty among our young are other factors that continue to embarrass our nation. Parental violence is among the five leading causes of death for children between the ages of one and 18. About 1,300 children are killed by their parents or other close relatives each year (McCormick, 1994).

As is true of partner relationships, children are subjected to other forms of mistreatment by parents. In examining the national prevalence of **psychological aggression** by parents, Straus and Field (2003) find that verbal attacks on children are so common as to be "just about universal." Based upon nearly a thousand interviews with a nationally representative sample of households with at least one child under 18 living at home, Straus and Field explore the prevalence of psychological aggression. They define psychological aggression as consisting of the following kinds of behaviors, with the latter three constituting "more severe" psychological aggression.

- Shouting, yelling, or screaming at one's child
- Threatening to spank or hit one's child but not actually doing it
- Swearing or cursing at one's child
- Threatening to send one's child away or kick him or her out of the house
- Calling one's child dumb or lazy, or making some other disparaging comment

Almost all (89 percent) of the sample parents reported having committed at least one of the five kinds of psychological aggression and 33 percent reported at least one instance of the more severe forms. The prevalence of the various forms of psychological aggression are illustrated in Table 13.1.

Use of psychological aggression varies along with the age of the child and the parent. "Only" 43 percent of parents of infants reported using psychological aggression, but that percentage increases quickly and dramatically. Nearly 90 percent of parents of two-year-olds use some form of psychological aggression. The percentage peaks at 98 percent at age 7, and as late as age 17 the rate remains very high, at 90 percent. Conversely, research on corporal punishment shows it declining with the age of the child; only 12 percent of parents of 17-year-olds report still using corporal punishment (Straus and Field, 2003).

Parents' ages matter too. Younger parents (ages 18 to 29) reported the most frequent use of psychological aggression (22 times in past 12 months) compared to parents 30 to 39 (19 times in past 12 months), and parents over 40 (15 times in past 12 months). Aside from age differences, there was "a lack of demographic differences in use of psychological aggression; this means that nearly all parents, regardless of sociodemographic characteristics, used at least some psychological aggression as a disciplinary tactic" (Straus and Filed, 2003, p. 805).

TABLE 13.1 ■ **Prevalence of Psychological Aggression**

MEASURE	PREVALENCE (% IN LAST YEAR)
Overall	88.6
Severe	33.4
Shouting, yelling, screaming	74.7
Threatening to spank	53.6
Swearing or cursing	24.3
Name-calling	17.5
Threatening to kick out of house	6.0

SOURCE: Straus and Field, 2003.

Families at Risk

Research suggests that three sets of factors put families at risk for child abuse and neglect: (1) parental characteristics, (2) child characteristics, and (3) the family ecosystem—that is, the family system's interaction with the larger environment (Burgess and Youngblood, 1987; Vasta, 1982). The characteristics described in the next sections are likely to be present in abusive families (Straus, Gelles, and Steinmetz, 1980; Turner and Avison, 1985).

DID YOU KNOW?

Approximately 80 percent of the perpetrators of child abuse and neglect were parents and other relatives ("Child Abuse and Neglect," 1997).

PARENTAL CHARACTERISTICS. Some or all of the following characteristics are likely to be present in parents who abuse their children:

- The abusing father was physically punished by his parents, and his father physically abused his mother.
- The parents believe in corporal discipline of children and wives.

- The marital relationship itself may not be valued by the parents. There may be interspousal violence.
- The parents believe that the father should be the dominant authority figure.
- The parents have low self-esteem.
- The parents have unrealistic expectations for the child.
- There is persistent role reversal in which the parents use the child to gratify their own needs, rather than vice versa.
- The parents appear unconcerned about the seriousness of a child's injury, responding, "Oh well, accidents happen."

CHILD CHARACTERISTICS. Who are the battered children? Are they any different from other children? Surprisingly, the answer is often yes; they are different in some way or at least are perceived to be so by their parents. Brandt Steele (1980) notes that children who are abused are often labeled by their parents as "unsatisfactory," a term that may describe any of the following:

- A "normal" child who is the product of a difficult or unplanned pregnancy, is of the "wrong" sex, or is born outside of marriage.
- An "abnormal" child—one who was premature or of low birth weight, possibly with congenital defects or illness.
- A "difficult" child—one who shows such traits as fussiness or hyperactivity.

Steele also notes that all too often, a child's perceived difficulties are a result (rather than a cause) of abuse and neglect.

FAMILY ECOSYSTEM. As discussed earlier in this chapter, the community and the family's relation to it may be relevant to the existence of domestic violence. The following characteristics may be found in families that experience child abuse:

- The family experiences unemployment.
- The family is socially isolated, with few or no close contacts with relatives, friends, or groups.
- The family has a low level of income, which creates economic stress.
- The family lives in an unsafe neighborhood, which is characterized by higher-than-average levels of violence.
- The home is crowded, hazardous, dirty, or unhealthy.
- The family is a single-parent family in which the parent works and is consequently overstressed and overburdened.
- One or more family members have health problems.

Notice the clustering of such socioeconomic characteristics as unemployment, low income, neighborhood, and housing. This combination tells an important story. Like spousal or partner violence, the mistreatment of children can be found across the socioeconomic spectrum. But like spousal violence, it happens more frequently at the lower levels.

The likelihood of child abuse increases with family size. Parents of two children have a 50 percent higher abuse rate than do parents of a single child. The rate of abuse peaks at five children and declines thereafter. The overall child abuse rate by mothers has been found to be 75 percent higher than that by fathers (Straus, Gelles, and Steinmetz, 1980). The responsibilities and tensions of mothering and the enforced closeness of mother and child may lead to situations in which women are likely to abuse their children. But, as David Finkelhor (1983) and others have pointed out, if we "calculate [child] vulnerability to abuse as a function of the amount of time spent in contact with a potential abuser, . . . we . . . see that men and fathers are more likely to abuse."

Single parents—both mothers and fathers—are at especially high risk of abusing their children (Gelles, 1989). According to Gelles, "the high rate of abusive violence among single mothers appears to be a function of the poverty that characterizes mother-only families." He states that programs must be developed that are "aimed at ameliorating the devastating consequences of poverty among single parents." Single fathers, who show a higher abuse rate than single mothers, "need more than economic support to avoid using abusive violence toward their children."

DID YOU KNOW?

American children are 12 times more likely to die by gunfire than their counterparts in the rest of the industrialized world (Meyer, 1997).

Intervention

The goals of intervention in domestic violence are principally to protect the victims and to assist and strengthen their families. In dealing with child abuse, professionals and government agencies may be called on to provide medical care, counseling, and services such as day care, child-care education, telephone crisis lines, and temporary foster care. Many of these services are costly, and many of those who require them cannot afford to pay. Our system does not currently provide the human and financial resources necessary to deal with these socially destructive problems.

The first step in treating child abuse is locating the children who are threatened. With heightened public awareness in recent years and mandatory reporting of suspected child abuse required of certain professionals (such as teachers, doctors, and counselors) in all 50 states, identifying these children is much easier now than it was two decades ago. Reported incidents of child abuse have increased greatly during this time, but the actual number of incidents appears to have decreased. This is good news as far as it goes. Still, levels of violence against children are unacceptably high, and not nearly enough resources are available to assist children. Child welfare workers are notoriously overburdened with cases, and adequate foster placement is often difficult to find (Gelles and Cornell, 1990).

Many of the interventions in child abuse appear to be the equivalent of putting a Band-Aid on a huge malignant tumor. We must address this societal cancer from a variety of levels:

- Parents must learn how to deal more positively and effectively with their children.
- Children need to be infused with self-esteem and taught skills in order to recognize and report abuse as soon as it occurs.
- Professionals working with children and families should be required to receive adequate training in child abuse and neglect and to be sensitive to cultural norms.
- Agencies should coordinate their efforts for preventing and investigating child abuse.
- Public awareness of child abuse needs to be created by methods such as posters and public service announcements.
- The workplace should promote educational programs to eliminate sexism, provide adequate child care, and help reduce stress among its workforce.
- Government should support sex education and family life programs in order to help reduce the number of unwanted pregnancies.
- Criminal statutes should be developed and enforced to impose felony sentences on those who perpetuate child maltreatment.
- Research efforts concerning family violence and child maltreatment should be supported.

REFLECTIONS

If you were (or are) a parent, would you consider it violent to spank your child with an open hand on the buttocks if the child was disobedient? To slap your child across the face? Is it acceptable to spank your small child to teach him or her not to run into a busy street? To spank because you are angry?

THE HIDDEN VICTIMS OF FAMILY VIOLENCE: SIBLINGS, PARENTS, AND THE ELDERLY

Most studies of family violence have focused on violence between spouses and on parental violence toward children. There is, however, considerable violence between siblings, between teenage children and their parents, and between adult children and their aging parents. These are the "hidden victims" of family violence (Gelles and Cornell, 1990).

Sibling Violence

Violence between siblings is by far the most common form of family violence (Straus, Gelles, and Steinmetz, 1980). Most of this type of sibling interaction is simply taken for granted by our culture—"You know how kids are!" Dating back more than 20 years, research has indicated that perhaps as many as three out of four children experience sibling violence every year. Straus and his colleagues report these additional findings:

- The rate of sibling violence goes down with the increasing age of the child.
- Boys of all ages are more violent than girls. The highest rates of sibling violence occur in families with only male children.
- Violence between children often reflects what they see their parents doing to each other and to the children themselves.

The full scope and implications of sibling violence have not been rigorously explored. However, Straus, Gelles, and Steinmetz (1980) have come to this conclusion:

> Conflicts and disputes between children in a family are an inevitable part of life. . . . But the use of physical force as a tactic for resolving their conflicts is by no means inevitable. . . . Human beings learn to be violent. It is possible to provide children with an environment in which nonviolent methods of solving conflicts can be learned. . . . If violence, like charity, begins at home, so does nonviolence.

Teenage Violence toward Parents

Most of us find it difficult to imagine children attacking their parents because it so profoundly violates our image of parent-child relations. Parents possess the authority and power in the family hierarchy. Furthermore, there is greater social disapproval of a child striking a parent than of a parent striking a child; it is the parent who has the "right" to hit. Finally, parents rarely discuss such incidents because they are ashamed of their own victimization; they fear that others will blame them for the children's violent behavior (Gelles and Cornell, 1985).

Although we know fairly little about adolescent violence against parents, scattered studies indicate that it is almost as prevalent as spousal violence (Gelles and Cornell, 1985; Straus, 1980; Straus, Gelles, and Steinmetz, 1980).

The majority of youthful children who attack parents are between the ages of 13 and 24. Sons are slightly more likely to be abusive than daughters; the rate of severe male violence tends to increase with age, whereas that of females decreases. Boys apparently take advantage of their increasing size and the cultural expectation of male aggression. Girls, in contrast, may become less violent because society views female aggression more negatively. Most researchers believe that mothers are the primary targets of violence and abuse because they may lack physical strength or social resources and because women are "acceptable" targets for abuse (Gelles and Cornell, 1985).

Abuse of the Elderly

Of all the forms of hidden family violence, only the abuse of elderly parents by their grown children (or, in some cases, by their grandchildren) has received considerable public attention. Elder mistreatment may be an act of commission (abuse) or omission (neglect) (Wolf, 1995). It is estimated that approximately 500,000 elderly people are physically abused annually. An additional 2 million are thought to be emotionally abused or neglected. Though mandatory reporting of suspected cases of elder abuse is the law in 42 states and the District of Columbia, much abuse of the elderly goes unnoticed, unrecognized, and unreported (Wolf, 1995). Older people generally don't get out much and are often confined to bed or a wheelchair. Many do not report their mistreatment out of fear of institutionalization or other reprisal. Although some research indicates that the abused in many cases were in fact abusing parents, more knowledge must be gained before we can draw firm conclusions about the causes of elder abuse (Egeland, 1993; Kaufman and Zigler, 1993; Ney, 1992).

The most likely victims of elder abuse are the very elderly—in the majority of cases, women—who are suffering from physical or mental impairments, especially those with Alzheimer's disease. Their advanced age renders them dependent on their caregivers for many, if not all, of their daily needs. It may be their dependency that increases their likelihood of being abused. Other research indicates that many abusers are financially dependent on their elderly parents; they may resort to violence out of feelings of powerlessness.

While researchers are sorting out the whys and wherefores of elder abuse, battered older people have a number of pressing needs. Karl Pillemer and Jill Suitor (1988) recommend the following services for elders and their caregiving families:

- Housing services, including temporary respite care to give caregivers a break and permanent housing (such as rest homes, group housing, and nursing homes)

- Health services, including home health care; adult day-care centers; and occupational, physical, and speech therapy
- Housekeeping services, including shopping and meal preparation
- Support services, such as visitor programs and recreation
- Guardianship and financial management

REDUCING FAMILY VIOLENCE

Based on the foregoing evidence, you may by now have concluded that the American family is well on its way to extinction as family members bash, thrash, cut, shoot, and otherwise wipe themselves out of existence. Statistically, the safest family homes are those with one or no children in which the husband and wife experience little life stress and in which decisions are made democratically (Straus, Gelles, and Steinmetz, 1980). By this definition, most of us probably do not live in homes that are particularly safe. What can we do to protect ourselves (and our posterity) from ourselves?

Prevention strategies usually take one of two paths: (1) eliminating social stress or (2) strengthening families (Swift, 1986). Family violence experts make the following general recommendations (Straus, Gelles, and Steinmetz, 1980):

- Reduce societal sources of stress, such as poverty, racism and inequality, unemployment, and inadequate health care.
- Eliminate sexism. Furnish adequate day care. Promote educational and employment opportunities equally for men and women. Promote sex education and family planning to prevent unplanned and unwanted pregnancies.
- Initiate prevention and early intervention efforts for young males before they become adult batterers.
- End social isolation. Explore means of establishing supportive networks that include relatives, friends, and community.
- Break the family cycle of violence. Eliminate corporal punishment and promote education about disciplinary alternatives. Support parent education classes to deal with inevitable parent-child conflict.
- Eliminate cultural norms that legitimize and glorify violence. Legislate gun control, eliminate capital punishment, and reduce media violence.

(For specific prevention and treatment strategies, see Hampton et al., 1993.)

Child Sexual Abuse

Whether it is committed by relatives or nonrelatives, **child sexual abuse** is defined as any sexual interaction (including fondling, erotic kissing, or oral sex, as well as genital penetration) between an adult or older adolescent and a prepubertal child. It does not matter whether the child is perceived by the adult as freely engaging in the sexual activity. Because of the child's age, he or she cannot legally give consent; the activity can only be considered as self-serving to the adult.

Estimates of the incidence of child sexual abuse vary considerably. A review of small-scale studies found estimates ranging from 6 to 62 percent for females and from 3 to 31 percent for males (Peters et al., 1986). The first national survey found that 27 percent of the women and 16 percent of the men surveyed had experienced sexual abuse as children (Finkelhor et al., 1990). Different definitions of abuse, methodologies, samples, and interviewing techniques account for the varied estimates (Gelles and Conte, 1991). Fabricated reports of sexual abuse do occur, but deliberate fabrications constitute only 4 to 8 percent of all reports (Finkelhor, 1995).

Child sexual abuse is generally categorized in terms of kin relationship. **Extrafamilial sexual abuse** is conducted by nonrelated individuals. **Intrafamilial abuse** is conducted by related individuals, including steprelatives. The abuse may be pedophilic or nonpedophilic. **Pedophilia** is an intense, recurring sexual attraction to prepubescent children. Nonpedophilic sexual interactions with children are not motivated as much by sexual desire as by nonsexual motives, such as power or affection (Groth, 1980). (For sexual abuse from an anthropological perspective, see Konker, 1992.)

The child's victimization may include force or the threat of force, pressure, or the taking advantage of trust or innocence. The most serious forms of sexual abuse include actual or attempted penile-vaginal penetration, fellatio, cunnilingus, and anilingus, with or without the use of force. Other serious forms range from forced digital penetration of the vagina to fondling of the breasts (unclothed) or simulated intercourse without force. The least traumatic sexual abuse ranges from kissing to intentional sexual touching of the clothed genitals, breasts, or other body parts with or without the use of force (Russell, 1984).

GENERAL PRECONDITIONS FOR SEXUAL ABUSE

Researchers have found that intrafamilial and extrafamilial sexual abuse share many common elements (Finkelhor, 1984). Because there are so many variables—such as the age and sex of the victims and perpetrators, their relationships, the type of acts involved, and whether there was force—we cannot automatically say that abuse within the family is more harmful than extrafamilial abuse, as we might assume.

David Finkelhor (1984) believes that four preconditions need to be met by the offender for sexual abuse to occur. These preconditions apply to pedophilic, nonpedophilic, intrafamilial, and extrafamilial abuse. According to Finkelhor, all four of these factors must come into play for sexual abuse to occur:

1. *Motivation to sexually abuse a child.* This motivation consists of three components: (1) emotional congruence, in which relating sexually to a child fulfills some important emotional need; (2) sexual arousal toward the child; and (3) blockage, in which alternative sources of sexual gratification are not available or are less satisfying.
2. *Overcoming internal inhibitions against acting on the motivation.* Internal inhibitions may be overcome through the use of alcohol, lack of impulse control, senility, social acceptance of sexual interest in children, and so on.
3. *Overcoming external obstacles to committing sexual abuse.* The most important obstacle appears to be the supervision and protection a child receives from others, such as family members, neighbors, and the child's peers. The mother is especially significant in protecting children. Growing evidence suggests that children are more vulnerable to abuse when the mother is absent, neglectful, or incapacitated in some way through illness, marital abuse, or emotional problems.
4. *Undermining or overcoming the child's potential resistance to the abuse.* The abuser may use outright force or select psychologically vulnerable targets. Certain children may be more vulnerable because they feel insecure, needy, or unsupported and will respond to the abuser's offers of attention, affection, or bribes. Children's ability to resist may be undercut because they are young, naïve, or have a special relationship with the abuser as friend, neighbor, or family member.

FORMS OF INTRAFAMILIAL CHILD SEXUAL ABUSE

The incest taboo, which is nearly universal in human societies, prohibits sexual activities between closely related individuals. There are only a few exceptions, and these concern brother-sister marriages in the royal families of ancient Egypt, Peru, and Hawaii. **Incest** is generally defined as sexual intercourse between persons too closely related to marry legally (usually interpreted to mean father-daughter, mother-son, or brother-sister intercourse). Sexual abuse in families can involve blood relatives (most commonly uncles and grandfathers) and steprelatives (most often stepfathers and stepbrothers). Grandfathers who abuse their granddaughters frequently sexually abused their children as well. Stepgranddaughters are at greater risk than are granddaughters (Margolin, 1992). (For a review of assessment and treatment of incest perpetrators, see Cole [1992].)

It is not clear what type of familial sexual abuse is the most frequent (Peters et al., 1986; Russell, 1986). Some researchers believe that father-daughter (including stepfather-stepdaughter) abuse is the most common; others think that brother-sister abuse is the most common. Still other researchers believe that abuse committed by uncles is the most common (Russell, 1986). Mother-son sexual relations are considered to be rare (or they are underreported).

Father-Daughter Sexual Abuse

There is general agreement that the most traumatic form of sexual victimization is father-daughter abuse, including that committed by stepfathers.

Over twice as many daughters abused by fathers reported serious long-term consequences compared to children who were victimized by other family members. Some factors contributing to the severity of reactions to father-daughter sexual relations include the following:

- Fathers were more likely to have engaged in penile-vaginal penetration than other relatives (18 percent versus 6 percent).
- Fathers sexually abused their daughters more frequently than other perpetrators abused their victims (38 percent of the fathers sexually abused their daughters 11 or more times, compared with a 12 percent abuse rate for other abusing relatives).
- Fathers were more likely to use force or violence than others (although the numbers for both fathers and others were extremely low).

In the past, many have discounted the seriousness of sexual abuse by a stepfather because incest is generally defined legally as sexual activity between two biologically related persons. The emotional consequences are just as serious, however. Sexual abuse by a stepfather still represents a violation of the basic parent-child relationship.

Brother-Sister Sexual Abuse

There are contrasting views concerning the consequences of brother-sister incest. Researchers generally have expressed little interest in it. Most have tended to view it as harmless sex play or sexual exploration between mutually involved siblings. The research, however, has generally failed to distinguish between exploitative and nonexploitative brother-sister sexual activity. Sibling incest needs to be taken seriously (Adler and Schultz, 1995). Diana Russell suggests that the idea that brother-sister incest is usually harmless and mutual may be a myth. In her study, the average age difference between the brother (age 17.9 years) and the sister (age 10.7 years) was so great that the siblings could hardly be considered peers (Russell, 1986). The age difference represents a significant power difference. Furthermore, not all brother-sister sexual activity is "consenting"; considerable physical force may be involved. Russell writes:

> So strong is the myth of mutuality that many victims themselves internalize the discounting of their experiences, particularly if their brothers did not use force, if they themselves did not forcefully resist the abuse at the time, if they still continued to care about their brothers, or if they did not consider it abuse when it occurred. And sisters are even more likely than daughters to be seen as responsible for their own abuse.

Two percent of the women in Russell's random sample had at least one sexually abusive experience with a brother.

Uncle-Niece Sexual Abuse

Both Alfred Kinsey (1953) and Diana Russell (1986) found the most common form of intrafamilial sexual abuse to involve uncles and nieces. Russell reported that almost 5 percent of the women in her study had been molested by their uncles, slightly more than the percentage abused by their fathers. The level of severity of the abuse was generally less in terms of the type of sexual acts and the use of force. Although such abuse does not take place within the nuclear family, many victims found it quite distressing. A quarter of the respondents indicated long-term emotional effects (Russell, 1986).

CHILDREN AT RISK

Not all children are equally at risk for sexual abuse. Although any child can be sexually abused, some groups of children are more likely to be victimized than others. A review of the literature (Finkelhor and Baron, 1986) indicates that children at higher risk for sexual abuse are the following: female children, preadolescent children, children with absent or unavailable parents, children whose relationships with parents are poor, children whose parents are in conflict, and children who live with a stepfather. A variety of studies have found little or no association between sexual abuse and race and socioeconomic status (Finkelhor, 1995).

The majority of sexually abused children are girls, but boys are also victims (Watkins and Bentovim, 1992). The ratio of girls to boys appears to be between 2.5 to 1 and 4 to 1 (Finkelhor and Baron, 1986). We have only recently recognized the sexual abuse of boys (Bera et al., 1991). Finkelhor (1979) speculates that men tend to underreport sexual abuse because they experience greater shame; they feel that their masculinity has been undermined.

Boys tend to be blamed more than girls for their victimization, especially if they did not forcibly resist: "A real boy would never let someone do that without fighting back" (Rogers and Terry, 1984).

Most sexually abused children are between 8 and 12 years of age when the abuse first takes place. At higher risk appear to be children who have poor relationships with their parents (especially mothers) or whose parents are absent or unavailable and have high levels of marital conflict. A child in such a family may be less well supervised and, as a result, more vulnerable to manipulation and exploitation by an adult. In this type of family, the child may be unhappy, deprived, or emotionally needy; the child may be more responsive to the offers of friendship, time, and material rewards promised by the abuser.

Finally, children with stepfathers are at greater risk for sexual abuse. Russell (1986) found that only 2.3 percent of the daughters studied were sexually abused by their biological fathers. In contrast, 17 percent were abused by their stepfathers. The higher risk may result from the incest taboo's not being as strong in stepfamily relationships and because stepfathers have not built up inhibitions resulting from parent-child bonding beginning from infancy. As a result, stepfathers may be more likely to view the stepdaughter sexually. In addition, stepparents may also bring into the family steprelatives—their own parents, siblings, or children—who may feel no incest-related prohibition about becoming sexually involved with stepchildren.

EFFECTS OF CHILD SEXUAL ABUSE

Until recently, much of the literature on child sexual abuse has been anecdotal or based on case studies or small-scale surveys of nonrepresentative groups. Numerous well-documented consequences of child sexual abuse exist for both intrafamilial and extrafamilial abuse (see Kendall-Tackett et al. [1993] for a review of the literature). These include both initial and long-term consequences. Many abused children experience symptoms of posttraumatic stress disorder (PTSD) (McLeer et al., 1992).

Initial Effects of Sexual Abuse

The initial consequences of sexual abuse—those occurring within the first two years—include these effects:

- *Emotional disturbances,* including fear, anger, hostility, guilt, and shame
- *Physical consequences,* including difficulty in sleeping, changes in eating patterns, and pregnancy
- *Sexual disturbances,* including significantly higher rates of open masturbation, sexual preoccupation, and exposure of the genitals (Hibbard and Hartman, 1992)
- *Social disturbances,* including difficulties at school, truancy, running away from home, and early marriages among abused adolescents

Ethnicity appears to influence how a child responds to sexual abuse. For example, one study compared sexually abused Asian-American children with a random sample of abused white, African-American, and Latino children (Rao, Diclemente, and Pouton, 1992). The researchers found that Asian-American children suffered less sexually invasive forms of abuse. They tended to be more suicidal and to receive less support from their parents than did non-Asians. They were also less likely to express anger or to act out sexually. These different responses point to the importance of understanding the cultural context when treating ethnic victims of sexual abuse. (For a discussion of child sexual abuse histories among African-American college students, see Priest [1992].)

Long-Term Effects of Sexual Abuse

Although the initial effects of child sexual abuse can subside to some extent, the abuse may leave lasting scars on the adult survivor (Beitchman et al., 1992).

This drawing was made by an adolescent who was impregnated by her father. According to psychologists, it expresses her inability to deal with body images, especially genitalia, and her rejection of her body's violation.

These adults often have significantly higher incidences of psychological, physical, and sexual problems than the general population. Abuse as a child may predispose some women to sexually abusive dating relationships (Cate and Lloyd, 1992).

Long-term problems include the following (Beitchman et al., 1992; Browne and Finkelhor, 1986; Elliott and Briere, 1992; Wyatt, Gutherie, and Notgrass, 1992):

- Depression, the most frequently reported symptom of adults sexually abused as children
- Self-destructive tendencies, including suicide attempts and thoughts of suicide (Jeffrey and Jeffrey, 1991)
- Somatic disturbances and dissociation, including anxiety and nervousness, eating disorders (anorexia and bulimia), feelings of "spaciness," out-of-body experiences, and feelings that things are "unreal" (DeGroot, Kennedy, Rodin, and McVey, 1992; Walker et al., 1992; Young, 1992)
- Negative self-concept, including feelings of low-self-esteem, isolation, and alienation
- Interpersonal relationship difficulties, including difficulties in relating to both sexes, parental conflict, problems in responding to their own children, and difficulty in trusting others
- Revictimization, in which women abused as children are more vulnerable to rape and marital violence (Wyatt, Gutherie, and Notgrass, 1992)
- Sexual problems, in which survivors find it difficult to relax and enjoy sexual activities, or they avoid sexual relations and experience hypoactive (inhibited) sexual desire and lack of orgasm

In recent years, some adults have been accusing family members or others of abusing them as children. They say that they repressed their childhood memories of abuse and only later, as adults, recalled them. These accusations have given rise to a fierce controversy about the nature of memories of abuse. A review of the research related to this topic was done by the American Psychological Association (1994) and the following conclusions were made:

- Most people who were sexually abused as children at least partially remember the abuse.
- Memories of sexual abuse that have been forgotten may later be remembered.
- False memories of events that never happened may occur.
- The process by which accurate or inaccurate recollections of childhood abuse are made is not well understood.

Because firm scientific conclusions cannot be made at this time, the debate is likely to continue.

Sexual Abuse Trauma

As we have seen, childhood sexual abuse has numerous initial and long-term consequences. Together, these consequences create a traumatic dynamic that affects the child's ability to deal with the world. David Finkelhor and Angela Browne (1986) suggest a model of sexual abuse that contains four components: (1) traumatic sexualization, (2) betrayal, (3) powerlessness, and (4) stigmatization. When these factors converge as a result of sexual abuse, they affect the child's cognitive and emotional orientation to the world. They create trauma by distorting a child's self-concept, worldview, and affective abilities. These consequences affect abuse survivors not only as children but also as adults.

TRAUMATIC SEXUALIZATION. The process in which a sexually abused child's sexuality develops inappropriately and the child becomes interpersonally dysfunctional is referred to as *traumatic sexualization.* Finkelhor and Browne note:

> It occurs through the exchange of affection, attention, privileges, and gifts for sexual behavior, so that the child learns sexual behavior as a strategy for manipulating others to get his or her other developmentally appropriate needs met. It occurs when certain parts of the child's anatomy are fetishized and given distorted importance and meaning. It occurs through the misconceptions and confusions about sexual behavior and morality that are transmitted to the child from the offender. And it occurs when very frightening memories and events become associated in the child's mind with sexual activity.

Sexually traumatized children learn inappropriate sexual behaviors (such as manipulating an adult's genitals for affection), are confused about their sexuality, and inappropriately associate certain emotions—such as loving and caring—with sexual activities.

As adults, sexual issues may become especially important. Survivors may suffer flashbacks, sexual dysfunctions, and negative feelings about their bodies. They may also be confused about sexual norms and standards. A fairly common confusion is the belief that sex may be traded for affection. Some women label themselves as "promiscuous," but this label may be more a result of their negative self-image than of their actual behavior. There seems to be a history of childhood sexual abuse among many prostitutes (Simons and Whitbeck, 1991).

BETRAYAL. Children feel betrayed when they discover that someone on whom they have been dependent has manipulated, used, or harmed them. Children may also feel betrayed by other family members, especially mothers, for not protecting them from abuse. As adults, survivors may experience depression as a manifestation, in part, of extended grief over the loss of trusted figures. Some may find it difficult to trust others. Other survivors may feel a deep need to regain a sense of trust and become extremely dependent. Distrust may manifest itself in hostility and anger or in social isolation and avoidance of intimate relationships. In adolescents, antisocial or delinquent behavior may be a means of protecting themselves from further betrayal. Anger may express a need for revenge or retaliation.

Children need to have someone, such as a teacher who they trust, in whom they can confide about their suffering.

POWERLESSNESS. Children experience a basic kind of powerlessness when their bodies and personal spaces are invaded against their will. A child's powerlessness is reinforced as the abuse is repeated. In adulthood, powerlessness may be experienced as fear or anxiety; a person feels unable to control events. Adult survivors often believe that they have impaired coping abilities. This feeling of ineffectiveness may be related to the high incidence of depression and despair among survivors. Powerlessness may also be related to increased vulnerability or revictimization through rape or marital violence; survivors may feel unable to prevent subsequent victimization. Other survivors, however, may attempt to cope with their earlier powerlessness by an excessive need to control or dominate others.

STIGMATIZATION. Ideas about being a bad person as well as feelings of guilt and shame about sexual abuse are transmitted to abused children and then internalized by them. Stigmatization is communicated in numerous ways. The abuser conveys it by blaming the child or, through secrecy, communicating a sense of shame. If the abuser pressures the child for secrecy, the child may also internalize feelings of shame and guilt. Children's prior knowledge that their families or communities consider such activities deviant may contribute to their feelings of stigmatization. As adults, survivors may feel extreme guilt or shame about having been sexually abused. They may have low self-esteem because they feel that the abuse made them "spoiled merchandise." They also feel different from others because they mistakenly believe that they alone have been abused.

TREATMENT PROGRAMS

Child sexual abuse, especially father-daughter incest, is increasingly being treated through therapy programs working in conjunction with the judicial system rather than through breaking up the family by removing the child or the offender (Nadelson and Sauzier, 1986). Because the offender is often also the breadwinner, incarcerating him may greatly increase the family's emotional distress. The district attorney's office may work with clinicians in evaluating the existing threat to the child and deciding whether to prosecute, refer the offender to therapy, or both. The goal is not simply to punish the offender but to try to assist the victim and the family in coming to terms with the abuse.

Many of these clinical programs work on several levels at once: they treat the individual, the father-daughter relationship, the mother-daughter relationship, and the family as a whole. They work on developing self-esteem and improving the family and marital relationships. If appropriate, they refer individuals to alcohol or drug abuse treatment programs.

A crucial ingredient in many treatment programs is individual and family attendance at self-help group meetings. These self-help groups are composed of incest survivors, offenders, mothers, and other family members. Self-help groups such as Parents United and Daughters and Sons United help the offender acknowledge his responsibility and understand the impact of the incest on everyone involved.

PREVENTING SEXUAL ABUSE

The idea of preventing sexual abuse is relatively new (Berrick and Barth, 1992). Prevention programs began about a decade ago, a few years after programs were started to identify and help child or adult survivors of sexual abuse. (For an evaluation of commercially available materials for preventing child abuse, see Roberts et al. [1990].) Such prevention programs have been hindered, however, by three factors (Finkelhor, 1986a, 1986b):

1. The issue of sexual abuse is complicated by differing concepts of appropriate sexual behavior and partners, which are not easily understood by children.
2. Sexual abuse, especially incest, is a difficult and scary topic for adults to discuss with children. Children who are frightened by what their parents tell them, however, may be less able to resist abuse than those who are given strategies of resistance.
3. Sex education is controversial. Even where it is taught, instruction often does not go beyond physiology and reproduction. The topic of incest is especially opposed.

In confronting these problems, child abuse prevention (CAP) programs have been very creative. These programs typically aim at three audiences: children, parents, and professionals (especially teachers). Children have the right to control their own bodies and genitals and to feel "safe," and they have the right not to be touched in ways that feel confusing or wrong. The CAP programs stress that the child is not at fault when such abuse does occur. They also try to give children possible courses of action if someone tries to sexually abuse them. In particular, children are taught that it's all right to say no, that they should get away from scary situations, and that it's very important to tell someone they trust about what has happened—and to keep telling until they are believed (Gelles and Conte, 1991).

REFLECTIONS

Assume for a moment that a young child disclosed to you the fact that she was hurt by her father. What would you say to her? How would you feel? Whom would you tell?

Other programs focus on educating parents who, it is hoped, will in turn educate their children. These programs aim at helping parents discover abuse or abusers by identifying warning signs. Such programs, however, need to be culturally sensitive, as Latinos and Asian Americans may be reticent about discussing these matters with their children (Ahn and Gilbert, 1992). Parents seem reluctant in general about dealing with sexual abuse issues with their children, according to David Finkelhor (1986a). First, many do not feel that their children are at risk. Second, parents are fearful of unnecessarily frightening their children. Third, parents feel uncomfortable talking with their children about sex in general, much less about such taboo subjects as incest. In addition, parents may not believe their own children or may feel uncomfortable confronting a suspected abuser, who may be a partner, uncle, friend, or neighbor.

CAP programs have also directed attention to professionals, especially teachers, physicians, mental health professionals, and police officers. Because of their close contact with children and their role in teaching children about the world, teachers are especially important. Professionals are encouraged to watch for signs of sexual abuse and to investigate children's reports of such abuse. A number of schools have instituted programs to educate both students and their parents. (For a research review of child sexual abuse prevention, see Berrick and Barth [1992].)

In recent years, both the American Medical Association (AMA) and the federal government have become more actively involved in fighting domestic violence. AMA guidelines advise doctors to question female patients routinely as to whether they have been attacked by their partners or forced to have sex. Physicians are also urged to investigate cases of injuries to women that are not well explained. (As we discussed earlier, there are already laws in place regarding the reporting of suspected child abuse.) At the federal level, a system is being set up by the Centers for Disease Control for tracking domestic violence cases and assessing ways to prevent abuse. Additionally, the Family Preservation and Support Act of 1993 provides $930 million for domestic abuse prevention. Donna Shalala, Secretary of Health and Human Services under President Bill Clinton, has emphasized that government alone cannot solve the problem of violence in the home (Marek, 1994). Professionals, such as physicians, law enforcement personnel, and social workers, also have important roles to fulfill, and within local communities, organizations and individuals need to promote awareness of abuse and reach out to affected families. Most important, we all need to look into ourselves to find nonviolent solutions to the problems we face in our own relationships.

Obviously, family violence is a complex phenomenon. It is the product of individual characteristics as well as social and cultural factors. Not every home becomes a boxing ring, and most families are not embattled. We need to realize that those families that are violent are products of a blend of qualities and are affected on multiple levels. This understanding is important if we hope to reduce the prevalence of violence and if we care to help those who are most at risk or already victimized.

SUMMARY

- Any form of intimacy or relatedness increases the likelihood of violence or abuse. *Violence* is defined as an act carried out with the intention or perceived intention of causing physical pain or injury to another person.
- Seven principal models are used to study sources of family violence: (1) the *psychiatric model*, which finds the source of violence within the personality of the abuser; (2) the *ecological model*, which looks at both the child's development in the family context and the family's development within the community; (3) the *feminist model*, which finds violence to be inherent in male-dominated societies; (4) the *social situational model*, which views family violence as arising from a combination of structural stress and cultural norms; (5) the *social learning model*, in which violence is seen as a behavior learned within the family and larger society; (6) the *resource model*, which assumes that force is used to compensate for the lack of personal, social, and economic resources; and (7) the *exchange/social control model*, which holds that people weigh the costs versus the rewards in all their actions and that they will use violence if the social controls (costs) are not strong enough. Three factors that may reduce social control are inequality of power in the family, the private nature of the family, and the "real man" image.
- Family violence is the result of multiple causes. Researchers have stressed the role played by gender, power and control, stress, and intimacy.
- High expectations for emotional fulfillment combine with a sense of family privacy to make conflict and violence more likely.
- It is difficult to know exactly how much violence there is in intimate relationships. The use of official records and/or survey data gives what are likely underestimates.
- There are at least four types of partner violence: (1) *common couple violence*, is more routine and

typically less severe; (2) *intimate terrorism,* which is a more severe, most often male-on-female form of violence and abuse in which power and domination are key motives; (3) *violent resistance,* which is often considered under the idea of "self defense" and is more often used by women; and (4) *mutual violent control,* which consists of couple violence where both partners are violently controlling.

- *Battering* is the use of physical force against another person. It includes slapping, punching, knocking down, choking, kicking, hitting with objects, threatening with weapons, stabbing, and shooting. Wife battering is one of the most common and most underreported crimes in the United States. Although there does not appear to be a "typical" battered woman, two characteristics correlate highly with wife assault: low socioeconomic status and a high degree of marital conflict. A man who batters his wife probably has some or all of the following characteristics: belief in common myths about battering, traditional beliefs about the family, low self-esteem, pathological personality characteristics, use of sex as aggression, and a belief in the moral rightness of his aggression.
- Age, race, and social class all factor into domestic violence. Younger women, black women, and lower-income women experience more intimate violence than do other women.
- The three-phase cycle of violence in battering relationships proposed by Lenore Walker (1979) includes (1) the tension-building phase, (2) the explosive phase, and (3) the resolution ("honeymoon") phase. This process may not characterize all violent partner relationships, as evidence indicates that "upscale" abusers may lack the remorse typically expected in phase 3.
- *Marital rape* is a form of battering. Many people, including victims themselves, have difficulty acknowledging that forced sex in marriage is rape, just as it is outside of marriage.
- The incidence of violence and sexual assault in dating relationships is alarming. Violence is often precipitated by jealousy or rejection. *Date rape* or *acquaintance rape* may not be recognized by either the assailant or the victim because they think that rape is something done by strangers.
- Recently, we have seen the appearance and use of date-rape drugs such as *Rohypnol (flunitrazepam)* and *gamma hydroxybutyrate (GHB)* to sedate and sexually victimize unsuspecting women. Deaths and sexual assaults have resulted, prompting the passage of date-rape drug prohibition laws.
- Women may stay in, or return to, battering relationships for a number of reasons, including economic dependency, religious pressure or beliefs, the perceived need for a father for the children, pity, guilt, a sense of duty, fear, love, and reasons pertaining to their particular culture. Women may also be paralyzed by "learned helplessness." When women try to stop their husband's violence, the most important factor appears to be the woman's own determination that the abuse must stop. Women may leave violent relationships when the level of violence is very high or when their children become threatened.
- Domestic violence intervention can be based on either control or compassion. Arrest, prosecution, and imprisonment are examples of control; shelters and support groups (including abuser programs) are examples of compassionate intervention.
- At least a million children are physically abused and neglected by their parents each year in the United States. The majority of abuse cases are unreported. Parental violence is one of the five leading causes of childhood death. Families at risk for child abuse often have specific parental, child, and family ecosystem characteristics. Parental characteristics include a father who was abused as a child, belief in corporal punishment, a devalued marital relationship and interspousal violence, father dominance, low self-esteem, unrealistic expectations for the child, parent-child role reversal, and lack of parental concern about the child's injury. Child characteristics include a "normal" child who is the product of a difficult or unplanned pregnancy, is the "wrong" sex, or is born outside of marriage; an "abnormal" child with physical or medical problems; or a "difficult" child. The family ecosystem includes the general social and economic environment in which the family lives. Characteristics in families at risk include such conditions as unemployment, social isolation, poverty, and unsafe neighborhoods.
- Psychological aggression by parents on children is fairly common. Nearly 90 percent of parents in

a representative sample acknowledged using some form of psychological aggression with at least one child during the prior 12-month period. Younger parents use such aggression more often, and as children move from infancy they are more often recipients of such behavior.

- Mandatory reporting of suspected child abuse may be helping to decrease the number of abused children in the United States. However, social workers are still overburdened, and services such as foster care are in short supply. Early intervention and education may be successful in reducing abuse, but there is a shortage of government funds for these and other programs to assist the victims of family violence.
- The hidden victims of family violence include siblings (who have the highest rate of violent interaction), parents assaulted by their adolescent or youthful children, and elders assaulted by their middle-aged children.
- Some recommendations for reducing family violence include (1) reducing sources of societal stress, such as poverty and racism; (2) eliminating sexism; (3) establishing supportive networks; (4) breaking the family cycle of violence; and (5) eliminating the legitimization and glorification of violence.
- *Incest* is defined as sexual intercourse between persons too closely related to marry. Sexual victimization of children may include incest, but it can also involve other family members and other sexual activities. The most traumatic form of child abuse is probably father-daughter (or stepfather-stepdaughter) abuse. Stepfathers abuse their stepdaughters at significantly higher rates than biological fathers abuse their daughters. Brother-sister abuse is often traumatic if it is exploitative or violent.
- Children most at risk for sexual abuse include females, preadolescents, children with absent or unavailable parents, children with poor parental relationships, children with parents in conflict, and children living with a stepfather.
- Child sexual abuse has both initial and long-term effects. The initial effects include emotional disturbances, physical consequences, and sexual and social disturbances. The long-term effects include depression, self-destructive tendencies, somatic disturbances and dissociation, negative self-concept, interpersonal relationship difficulties, revictimization, and sexual difficulties. The survivors of sexual abuse frequently suffer from sexual abuse trauma, which is characterized by traumatic sexualization, betrayal, powerlessness, and stigmatization.
- Child sexual abuse offenders are increasingly being sent into treatment programs in an attempt to assist the incest survivor and family in coping with the crisis that incest creates. Self-help groups are important for many survivors of sexual abuse.

KEY TERMS

acquaintance rape 435
battering 428
child sexual abuse 446
common couple violence 428
date rape 435
extrafamilial sexual abuse 446
incest 447
intrafamilial sexual abuse 446
marital rape 434
intimate terrorism 424
pedophilia 446
psychological aggression 442
rape 433
violence 423

SUGGESTED READINGS

Barnett, Ola, Cindy Miller-Penn, and Robin Perrin. *Family Violence across the Lifespan.* Newbury Park, CA: Sage, 1996. Coverage of all types of abuse and methodology, etiology, prevalence, treatment, and prevention of family violence.

Browne, Angela. *When Battered Women Kill.* New York: Free Press, 1989. A harrowing study exploring extreme abuse and its lethal outcomes. Although the book is about a minority of the most extreme situations, Browne also entertains issues of more general relevance, such as the intergenerational transmission of violence, the predictable patterns within violent relationships, and why many battered women stay in abusive situations.

Fontes, Lisa Aronson, ed. *Sexual Abuse in Nine North American Cultures: Treatment and Prevention.* Thousand Oaks, CA: Sage, 1995. An examination of the impact of culture on child sexual abuse, including ways in which cultural norms can be used to protect children and help them recover from abuse.

Johann, Sara Lee. *Domestic Abusers: Terrorists in Our Homes.* Springfield, IL: Charles C. Thomas, 1994. An examination of judicial policy concerning domestic violence by an attorney.

Jones, Ann. *Next Time, She'll Be Dead: Battering and How to Stop It.* Revised and Updated Edition. Boston: Beacon Press. 2001. An excellent book detailing a range of issues about domestic violence, from what it is to why it happens, to what can be done about it.

Renzetti, Claire. *Violent Betrayal: Partner Abuse in Lesbian Relationships.* Newbury Park, CA: Sage, 1992. An exploration of lesbian relationships and the factors leading to abuse. It includes a section on seeking help.

Sipe, Beth, and Evelyn Hall. *I Am Not Your Victim; Anatomy of Domestic Violence.* Thousand Oaks, CA: Sage, 1996. A moving firsthand account by a victim of domestic violence and a system that didn't believe her.

Stark, Evan, and Ann Flitcraft. *Women at Risk.* Thousand Oaks, CA: Sage, 1996. An exploration of the theoretical perspectives as well as health consequences of the abuse of women and clinical interventions to reduce the incidence of abuse.

Wolfe, David, Christine Wekerle, and Katreena Scott. *Alternatives to Violence: Empowering Youth to Develop Healthy Relationships.* Thousand Oaks, CA: Sage, 1996. A practical and broad-based book addressing the important topic of preventing youth violence. It works well with *The Youth Relationship Manual* (also published by Sage).

RESOURCES ON THE INTERNET

Companion Web Site for This Book

http://sociology.wadsworth.com/strong/marriage9e
Gain an even better grasp on this chapter by going to the companion Web site to take one of the Tutorial Quizzes, use the Flash Cards to master key terms, or check out the many other study aids you'll find there. Visit the Marriage and Family Resource Center on the site. You'll also find special features such as GSS Data and Census 2000 information that will put data and resources at your fingertips to help you with that special project or to do some research on your own.

InfoTrac College Edition: Search Word Summary

family violence	acquaintance rape
child abuse	sexual abuse
pedophilia	battering

To learn more about these central topics in the study of the family, you can conduct an electronic search using InfoTrac College Edition. To aid in your search and to gain useful tips, see the Student Guide to InfoTrac College Edition that you can access through the companion Web site for this book.

Preview

To gain a sense of what you already know about the material covered in this chapter, answer "True" or "False" to the statements below.

1. Half of all those currently marrying end up divorcing within seven years. True or false?
2. Divorce occurs as a single event in a person's life. True or false?
3. Americans have one of the highest marriage, divorce, and remarriage rates among industrialized nations. True or false?
4. The critical emotional event in a marital breakdown is the separation rather than the divorce. True or false?
5. Anglos have a higher divorce rate than Latinos. True or false?
6. Divorce is an important element of the contemporary American marriage system because it reinforces the significance of emotional fulfillment in marriage. True or false?
7. The higher an individual's employment status, income, and level of education, the greater the likelihood of divorce. True or false?
8. Many problems assumed to be due to divorce are actually present before marital disruption. True or false?
9. Those whose parents are divorced have a significantly greater likelihood of themselves divorcing. True or false?
10. Marital conflict in an intact two-parent family is generally more harmful to children than living in a tranquil single-parent family or stepfamily. True or false?

"Divorce should not be undertaken lightly. . . . Every effort should be made to sustain marriages with some strengths and satisfactions. . . . But divorce is a reasonable solution to an unhappy, acrimonious, destructive marital relationship. It can be a gateway to pathways associated with joy, satisfaction, and attainments, not just with loss, pain, and failure."

E. MAVIS HETHERINGTON

CHAPTER 14

Coming Apart: Separation and Divorce

Outline

Are Americans pro-marriage? Are we soft on divorce? Do we believe in the importance of marriage and the commitment we make when we exchange wedding vows, or when we say "I do" are we really adding under our breath or in our heads, "for now"? Americans' feelings about marriage and divorce are paradoxical. Consider the following generalizations (Ganong and Coleman, 1994; White, 1991):

- Americans like marriage: They have one of the highest marriage rates in the industrialized world.
- Americans don't like marriage: They have one of the highest divorce rates in the world.
- Americans like marriage: They have one of the highest remarriage rates in the world.

What sense can we make from the fact that we are one of the most marrying, divorcing, and remarrying nations in the world? What does our high divorce rate actually tell us about how we feel about marriage? In this chapter we hope to explain the paradox of high rates of marriage and divorce as we examine the divorce process, marital separation, divorce consequences, children and divorce, child custody, and divorce mediation. This exploration will help you better understand what parents, children, and families experience and how they cope with what increasingly has become part of our marriage system—divorce.

Some scholars suggest that divorce represents not a devaluation of marriage but an idealization of it. They reason that we would not divorce if we did not have so much hope about marriage fulfilling our various needs. In fact, according to Frank Furstenberg and Graham Spanier (1987), divorce may very well be a critical part of our contemporary marriage system, which emphasizes emotional fulfillment and satisfaction:

> Divorce can be seen as an intrinsic part of a cultural system that values individual discretion and emotional gratification. Divorce is a social invention for promoting these cultural ideals. Ironically, the more divorce is used, the more exacting the standards become for those who marry. . . . Divorce . . . serves not so much as an escape hatch from married life but as a recycling mechanism permitting individuals a second (and sometimes third and fourth chance) to upgrade their marital situation.

Our high divorce rate also tells us that we may no longer believe in the permanence of marriage. Norval Glenn (1991) suggests that there is a "decline in the ideal of marital permanence and . . . in the expectation that marriages will last until one of the spouses dies." Instead, marriages disintegrate when love goes or a potentially better partner comes along. Divorce is a persistent fact of American marital and family life and one of the most important forces affecting and changing American lives today (Furstenberg and Cherlin, 1991).

Before 1974, the view of marriage as lasting "till death us do part" reflected reality. However, a surge in divorce rates that began in the mid-1960s did not level off until the 1990s. In 1974, a watershed in American history was reached when more marriages ended by divorce than by death. Today approximately 50 percent of all new marriages are likely to end in divorce (U.S. Census Bureau, 1996).

Not only does divorce end marriages and break up families; it also creates new forms from the old ones. It creates remarriages (which are very different from first marriages). It gives birth to single-parent families and stepfamilies. Today about one out of every five American families is a single-parent family; more than half of all children

PREVIEW ANSWERS

1 True
2 False
3 True
4 True
5 True
6 True
7 False
8 True
9 True
10 True

will become stepchildren and nearly half of current marriages include at least one spouse who is remarried (U.S. Census Bureau, 1996). Within the singles subculture is an immense pool of divorced men and women (most of whom are on their way to remarriage). But divorce does not create these new forms easily. It gives birth to them in pain and travail.

© Ghislain & Marie David de Lossy/Getty Images/The Image Bank

Dating again after divorce announces to the world that one is available to choose a new partner. It can be a way of enhancing self-esteem.

Researchers traditionally looked on divorce from a deviance perspective (Coleman and Ganong, 1991). It was assumed that normal, healthy individuals married and remained married. Those who divorced were considered abnormal, immature, narcissistic, or unhealthy in some manner. Social scientists, however, are increasingly viewing divorce as one path in the normal family life cycle. Those who divorce are not necessarily different from those who remain married. If we begin to regard divorce in this light, social scientists reason, part of the pain accompanying divorce may be diminished, as those involved will no longer regard themselves as "abnormal" (Raschke, 1987).

The greatest concern that social scientists express about divorce is its effect on children (Aldous, 1987; "Study Reveals," 1997; Wallerstein and Blakeslee, 1989). But even in studies of the children of divorce, the research may be distorted by traditional assumptions about divorce being deviant (Amato, 1991). For example, problems that children experience may be attributed to divorce rather than to other causes, such as personality traits. Although some effects are caused by the disruption of the family itself, others may be linked to the new social environment—most notably poverty and parental stress—into which children are thrust by their parents' divorce (McLanahan and Booth, 1991; Raschke, 1987). Some therapists suggest that we begin looking at those factors that help parents and children successfully adjust to divorce rather than focusing on risks, dysfunctions, and disasters (Abelsohn, 1992).

Measuring Divorce: How Do We Know How Much Divorce There Is?

How common is divorce and how likely is it to happen to us? The U.S. Census Bureau (2000) shows that there are nearly 20 million divorced people aged 15 and older in the United States, representing over 9 percent of the population And many if not most of you have probably heard the gloomy news: that one out of two marriages ends in divorce. What exactly do those statistics mean and on what are they based? There are a variety of ways to measure and represent the prevalence of divorce in the United States. Let's look briefly at the most common measures.

RATIO MEASURE OF DIVORCES TO MARRIAGES. The **ratio measure of divorce** is calculated by taking the number of divorces and the number of marriages in a given year and producing a ratio to represent how often divorce occurs relative to marriage. In 1998, for example, there were 1,135,000 divorces and 2,244,000 marriages—a ratio of 1 divorce for every

1.98 marriages. But recognize the difference between that statistic and a statement indicating that one of every two marriages *will end* in divorce. What the ratio measure truly reflects is the relative popularity or commonality of marriage and divorce. Ratios can be misleading. If the number of divorces over five years remained relatively constant but the number of marriages dropped, the ratio of divorces to marriages would increase. In fact, if the number of divorces dropped but the number of marriages dropped more, the ratio of divorce to marriage would *increase* even though the actual frequency of divorce was decreasing.

CRUDE DIVORCE RATE. The **crude divorce rate** represents the number of divorces in a given year for every 1,000 people in the population. In 2002, there were 4.0 divorces for every 1,000 Americans. Obviously, counting every 1,000 people in the population means including many unmarried people, children, the elderly, the already divorced, and so on. It is therefore a statistic that is highly susceptible to the age distribution, proportions of married and single persons in the population, and to changes in such population characteristics. As sociologist J. Ross Eshelman (1997) notes, this statistic is affected as much by a sizable proportion of unmarried people as it is by the reality of divorce. In fact, under certain population characteristics "a sizable percentage of those married could get divorced, yet the country or city would have a seemingly low divorce rate."

REFINED DIVORCE RATE. Considered to be the most useful measure of divorce, the **refined divorce rate** measures the number of divorces that occur in a given year for every 1,000 marriages (as measured by married women aged 15 and older). In 1998, the refined rate was 19 to 20 divorces per 1,000 married women, meaning that 2 percent of marriages ended in divorce.

Note that the range of available statistics produces different impressions about the reality of divorce in the United States. The ratio measure gives the most alarming impression, the one most closely approximating "one out of two marriages" ending in divorce. When one uses the refined rate of 2 percent of marriages ending in divorce annually, the picture seems much less bleak. The reality represented by each statistic is the same, but the meanings we attach to each statistic, and therefore the understanding it creates, vary significantly.

Another divorce statistic worth mentioning is the **predictive divorce rate.** This calculation (too complicated for our purposes) allows researchers to estimate how many new marriages will likely end in divorce. The prevailing estimate is that somewhere between 40 and 50 percent of marriages entered into this year are likely to become divorces, but some put the estimate as high as 60 percent. Scott M. Stanley, author of *The Heart of Commitment,* describes the projected data as follows:

> The 40–50% number comes from detailed analyses of various population demographics, including ages, divorce rates by ages, lifespan projections, etc. It represents a sophisticated projection—much like the projected life span for babies being born today. As with any projection, the number could change if conditions in society change, but it is a very valid projection under current conditions.

Of course, estimating future trends is a tricky business. One makes predictions based on the experiences of previous birth cohorts (people born between specific years), taking into account such things as the ages at which they married, whether and how many children they had, their economic and educational resources, and their divorce experiences. One then estimates what younger cohorts will experience, based on these patterns. But because this estimate is based on experience of prior cohorts, we cannot be confident that current and future cohorts will make the same choices or face the same circumstances as their predecessors. The most frequent predictive estimate is that somewhere around half of new marriages will end in divorce. Hence, we are back where we began—at a rate of nearly one out of two.

Divorce Trends in the United States

If one looks at long-term divorce trends, the unmistakable conclusion is that the twentieth century saw dramatic increases in marital breakups. However, if one looks, instead, over the past 20 years, a different picture emerges. In more recent decades, the divorce rate has dropped (see Table 14.1).

TABLE 14.1 ■ Divorce and Marriage through the Twentieth Century and Beyond

	MARRIAGES		DIVORCES		
YEAR	NUMBER	RATE PER 1,000	NUMBER	RATE PER 1,000	RATE PER 1,000 MARRIED WOMEN
1900	709,000	9.3	55,751	0.7	3
1920	1,274,476	12.0	170,506	1.6	8
1940	1,595,879	12.1	264,000	2.0	9
1960	1,523,000	8.5	393,000	2.2	9.2
1970	2,158,802	10.6	708,000	3.5	14.9
1980	2,406,708	10.6	1,189,000	5.2	22.6
1985	2,413,000	10.2	1,178,000	5.0	21.7
1990	2,448,000	9.8	1,182,000	4.7	20.9
1995	2,336,000	8.9	1,169,000	4.4	19.8
1998	2,244,000	8.4	1,135,000	4.2	NA
2001	2,327,000	8.4	NA	4.0	NA

Both marriage and divorce rates have declined. The marriage rate is at its lowest point since the 1930s, and the 2.3 million marriages in 2001 decline from the 2,384,000 marriages performed in 1997. As to divorce, we can see that after three-quarters of a century worth of increases (minus, of course, the "time-out" of the 1950s), in more recent years the rate has declined. There were 2 percent fewer divorces in 1998 than in 1997 (when there were 1,163,000 divorces) and 7 percent fewer than the 1,215,000 divorces occurring in 1992, which represented the all-time high in numbers of divorce. In addition, the 2001 crude divorce rate of 4.0 is lower than it has been since the 1970s. There are multiple stories to tell about trends in divorce and causes of divorce.

Factors Affecting Divorce

Sometimes it is easy to point to the cause of a particular divorce. Perhaps one spouse was unfaithful or abusive and the marriage was brought to a quick end. In other instances, even the divorcing parties can't identify the cause or causes that *ultimately* led to divorce. Researchers have looked at factors affecting wider divorce rates as well as divorce decisions. In this section, we look at both the larger societal or demographic factors as well as the individual and couple characteristics as they relate to the marriage or family.

SOCIETAL FACTORS

As seen above, even the reduced divorce rates of the late 1990s were *six times* the rate at the beginning of the twentieth century. They were twice as high as the rates in 1960. Additionally, as Table 14.2 shows, divorce rates in the United States are higher than rates elsewhere in the industrialized world.

Within the United States, rates of divorce vary across the 50 states. Higher rates are found in Nevada, New Mexico, Oklahoma, Wyoming, California, and other western states. Lower rates are found in Connecticut, Illinois, Massachusetts, New Jersey, and Maryland (U.S. Census Bureau, 1998, Table 1346, 2001; Table 118).

These dramatic differences point to the need for societal rather than individual explanations. Although social factors are involved in divorce, it is often difficult for us to view divorce in sociological terms because the pain of divorce seems so uniquely personal. "Social and structural factors," notes Joseph Guttman (1993), "are largely invisible to

TABLE 14.2 ■ International Variation in Refined Divorce Rate

	DIVORCES PER 1,000 MARRIED WOMEN		
COUNTRY	1980	1990	1995
United States	23	21	20
Canada	10	11	11
Denmark	11	13	12
France	6	8	9
Germany	6	8	9
Italy	1	2	2
Japan	5	6	6
Sweden	11	12	14
United Kingdom	12	13	13

SOURCE: U.S. Census Bureau, 1998, Table No. 1346.

people who must cope with the personal consequences of these factors on a daily basis."

Changed Nature of the Family

The shift from an agricultural society to an industrial one undermined many of the family's traditional functions. Schools, the media, and peers are now important sources of child socialization and child care. Hospitals and nursing homes manage birth and care for the sick and aged. Because the family pays cash for goods and services rather than producing or providing them itself, its members are no longer interdependent.

As a result of losing many of its social and economic underpinnings, the family is less of a necessity. It is now simply one of many choices we have: We may choose singlehood, cohabitation, marriage—or divorce—and if we choose to divorce, we enter the cycle of choices again: singlehood, cohabitation, or marriage and possibly divorce for a second time. A second divorce leads to our entering the cycle for a third time, and so on.

Social Integration

Social integration—the degree of interaction between individuals and the larger community—is emerging as an important factor related to the incidence of divorce. The social integration approach regards such factors as urban residence, church membership, and population change as especially important in explaining divorce rates (Breault and Kposowa, 1987; Glenn and Shelton, 1985; Glenn and Supancic, 1984).

The greater likelihood of divorce in the West and Southwest may be caused by the higher rates of residential mobility and lower levels of social integration with extended families, ethnic neighborhoods, and church groups (Glenn and Supancic, 1984; Glenn and Shelton, 1985). Among African Americans, the lowest divorce rate is found among those born and raised in the South; African Americans born and raised in the North and West have the highest divorce rates. One study found that urban residence was the highest correlate of divorce (Breault and Kposowa, 1987). Those who live in urban areas, where the divorce rate is higher than in rural areas, for example, are less likely to be subject to the community's social or moral pressures. They are more independent and have greater freedom of personal choice.

Individualistic Cultural Values

American culture has traditionally been individualistic. We value individual rights; we cherish images of an individual battling nature; we believe in individual responsibility. It should not be surprising that many view the individual as having priority over the family when the two conflict. Since the 1950s, perhaps as a reaction to the alienation and stifling conformity of the time, we have increasingly valued self-fulfillment and personal growth (Guttman, 1993). As marriage and the family lost many of their earlier social and economic functions, their meaning shifted. Marriage and family are viewed as paths to individual fulfillment. We marry for love and expect marriage to bring us happiness. When individual needs conflict with family demands, however, we no longer automatically submerge our needs to those of the family. We often struggle to balance individual and family needs. But if we are unable to do so, divorce has emerged as an alternative to an unhappy or unfulfilling marriage and as an escape from a mean-spirited or violent marriage.

DEMOGRAPHIC FACTORS

There are a number of demographic factors that appear to have a correlation with divorce.

Employment Status

Among whites, a higher divorce rate is more characteristic of low-status occupations, such as factory worker, than of high-status occupations, such as executive (Greenstein, 1985; Martin and Bumpass, 1989). Unemployment, which contributes to marital stress, is also related to increased divorce rates.

Studies conflict as to whether employed wives are more likely than nonemployed wives to divorce; overall, though, the findings seem to suggest that female employment contributes to the likelihood of divorce because the employed wife is less dependent on her husband's earnings (White, 1991). Wives' employment may lead to conflict about the traditional division of household labor, child-care stress, and other work spillover problems that, in turn, create marital distress.

Income

The higher the family income, the lower the divorce rate for both whites and African Americans. It is interesting, however, that the higher a woman's individual income, the greater her chances of divorce perhaps because with greater incomes women are not economically dependent on their husbands or because conflict over inequitable work and family roles increases marital tension.

Educational Level

For whites, the higher the educational level, the lower the divorce rate. Divorce rates among African Americans are not as strongly affected by educational levels. Men and women with only a high school education are more likely to divorce than those with a college education (Glick, 1984b).

There is evidence that the effect of education is a little different for women than for men (Coontz, 1997). Women who fail to complete high school have higher divorce rates than women who do complete it, whereas women who have some college education have lower divorce rates than women who do not. But here the picture changes a little. Divorce rates rise again among women who pursue education beyond college. Stephanie Coontz (1997) also points out that a bigger "education effect" may be the **glick effect,** named for demographer Paul Glick, in which pursuing but failing to obtain a particular level of education exposes women to a greater risk of divorce than obtaining a certain level of education or degree.

Race and Ethnicity

African Americans are more likely than whites to divorce. In 1998, 10 percent of the white population, compared to 12 percent of African Americans, were of divorced status. The relation between ethnicity and divorce is not surprising because of the strong correlation between socioeconomic status and divorce: The lower the socioeconomic class, the more likely a person is to divorce. As income levels for African Americans increase, in fact, divorce rates decrease; they become similar to those of whites (Raschke, 1987; Garfinkel, McLanahan, and Robins, 1994). In 1995, 9.1 percent of whites, 10.7 percent of African Americans, and 7.9 percent of Latinos were divorced (U.S. Census Bureau, 1996).

The percentage of divorced Latinos is smaller than either divorced whites or African Americans. In 1998, 7.7 percent of Latinos were of divorced status. The difference between Latinos varies within the Hispanic ethnicities. Mexican Americans and Cuban Americans have lower divorce rates than do Anglos, whereas the divorce rate among Puerto Ricans approaches that of African Americans (Brisbie, 1986). These recent findings stand in contrast to the old belief that Latino families are more stable than Anglo families (Vega, 1991).

Religion

Frequency of attendance at religious services (not necessarily the depth of beliefs) tends to be associated with the divorce rate (Glenn and Supancic,

Exploring Diversity

DIVORCE IN THE CHINESE-AMERICAN FAMILY

Despite long periods of separation in split-household families and the stresses caused by overwork in small-producer families, most Chinese-American marriages remained intact. The low divorce rate, however, does not say much about the quality of marriages. It is clear that marriages were looked upon by early Chinese immigrants as unions whose purpose was to produce offspring; the quality of the marital relationship was immaterial. Women who later joined their husbands or who were "hasty brides" suffered from many adjustment problems in their new environment. They often felt isolated because they spoke little English and had no supportive network of friends and relatives. Husbands were often too busy to pay close attention to the problems their wives faced. Long separations made even longtime mates strangers to each other. Brides who came to the United States after arranged marriages hardly knew their husbands, and their relationships were sometimes strained also by a disparity in age. In many cases, men in their forties, who had worked in the United States for years, married women in their teens or early twenties.

The low divorce rate, then, reflects the lack of choices Chinese-American women have, rather than a high level of marital quality. In small-producer families, interdependence made it impossible for spouses to survive without each other. At the same time, divorced women were considered to be an embarrassment to the community; a stigma was attached to divorce. Some desperate Chinese-American women even took their lives, believing there was no other way out of a miserable marriage and living conditions. In one study, Sung (1967) found that the suicide rate among Chinese Americans in San Francisco was four times that in the city as a whole, and that victims were predominantly women.

The Chinese-American divorce rate continues to be low. In 1990, only 2.3 percent and 3.3 percent of Chinese-American men and women, respectively, were divorced (U.S. Census Bureau, 1990). These figures are considerably lower than those of white men (7.5 percent) and white women (9.4 percent), and slightly lower than those of other Asian-American groups (e.g., 4.2 percent and 6.5 percent for Japanese-American men and women, respectively, and 3.7 percent and 5.1 percent for Filipino-American men and women) (U.S. Census Bureau , 1990).

We know little about the prevalence of remarriage among Chinese Americans. However, because most white divorced people eventually remarry, there is no reason to think that this is not the case with Chinese Americans. A high remarriage rate among Chinese Americans can also be predicted on the basis of an increasing number of interracial marriages, the stigma attached to divorce, and the strong belief among Chinese Americans that children fare better with two parents.

There is little information about the frequency of single parenthood in the Chinese-American population. According to the U.S. Census Bureau (1990), a vast majority of Chinese-American children (87.6 percent) reside with two parents. This figure is slightly higher than that for whites (77.2 percent) and considerably higher than that for African Americans (35.6 percent).

SOURCE: Masako Ishii-Kuntz. "Chinese American Families." In Mary Kay DeGenova (ed.), *Families in Cultural Context* Copyright © 1997 Mayfield Publishing Company. Reprinted with permission.

1984). Among white males, the rate of divorce for those who never attend religious services is three times as high as for those who attend two or three times a month. By religion, the lowest divorce rate is for Jews, followed by Catholics and then Protestants. Fundamentalist Protestants have a higher divorce rate than those of more moderate Protestant denominations (Guttman, 1993). Because the Roman Catholic church now allows divorce through annulments and no longer excommunicates divorced people by refusing them the sacraments, the annulment rate has increased from 450 in 1968 to over 50,000 in 1994 (Woodward, Quade, and Kantrowitz, 1995).

The greater the involvement in religious activities, the less likelihood there is of divorce. Since the major religions discourage divorce, highly religious men and women are less likely to accept divorce because it violates their values. It may also be that a shared religion and participation in organized religious life affirms the couple relationship (Guttman, 1993; Wineberg, 1994).

LIFE COURSE FACTORS

Different aspects of the life course may affect the probability of divorce for some individuals.

Age at Time of Marriage

Adolescent marriages are more likely to end in divorce than are marriages that take place when people are in their twenties or older (Kurdek, 1993). This is true for both whites and African Americans. Younger partners are less likely to be emotionally mature. After age 26 for men and age 23 for women, however, age at marriage seems to make little difference (Glenn and Supancic, 1984).

Premarital Pregnancy and Childbirth

Premarital pregnancy by itself does not significantly increase the likelihood of divorce. But if the pregnant woman is an adolescent, drops out of high school, and faces economic problems following marriage, the divorce rate increases dramatically. If a woman gives birth prior to marriage, the likelihood for divorce in a subsequent marriage increases, especially in the early years. This negative effect on marriage is stronger for whites than for African Americans (White, 1991).

Remarriage

The divorce rate among those who remarried in the 1980s is so far about 25 percent higher than it is for those who entered first marriages in that decade (White, 1991). It is not clear why there is a higher divorce rate in remarriages. Some researchers suggest that the cause may lie in a "kinds-of-people" explanation. The probability factors associated with the kinds of people who divorced in first marriages—low levels of education, unwillingness to settle for unsatisfactory marriages, and membership in certain ethnic groups—are present in subsequent marriages, which increases the likelihood of divorce (Martin and Bumpass, 1989). Others argue that the dynamics of second marriages, especially the presence of stepchildren, increase the chances of divorce (White and Booth, 1985). Stepfamily research, however, does not provide much support for this hypothesis (see Ganong and Coleman, 1994).

Intergenerational Transmission

Those whose parents divorce are subject to **intergenerational transmission**—the increased likelihood that divorce will later occur to them (Raschke, 1987; Amato, 1996). It is now estimated that parental divorce increases the chance of their children's marriage ending within the first five years by as much as 70 percent. This holds for both whites and blacks but is more likely among whites (Amato, 1996).

How can we explain this intergenerational cycle? Research by Paul Amato points to a number of variables that are associated with parental divorce and that may then lead to subsequent divorce. He notes that compared with children from nondivorced homes, children whose parents divorce are more likely to marry younger, cohabit, and experience higher levels of economic hardship. Attitudinally, they become more pessimistic about the chance for lifelong marriage and develop more liberal attitudes toward divorce. Finally, they develop personal characteristics and lifestyles that interfere with stable, adult intimate relationships (Amato, 1996). Additionally, females whose parents divorce develop less traditional attitudes about women's family and paid work roles, value self-sufficiency, and possess stronger attachments to paid employment. Each of these could raise one's susceptibility to divorce. Using survey data from over 1,300 individuals from the Study of Marriage over the Life Course, Amato examined the relative role of these factors. He found that the major effects of parental divorce that led to later divorce were "the problematic behaviors" one acquired (such as anger, jealousy, infidelity) and life course variables (such as age at marriage). On the other hand, the intergenerational connection was not well explained by people's attitudes toward divorce. Amato (1996) draws four other interesting conclusions:

1. The increased risk of divorce holds in second marriages as well as first marriages.
2. The intergenerational risk is especially great if both spouses experienced parental divorce.
3. The effects are especially pronounced in "offspring marriages" (marriages by children of divorced parents) of short duration but are not present in marriages of long duration.

4. The effects are strongest when parents divorce early in their children's lives (age 12 or younger).

One must keep in mind that, as with intergenerational cycles of family violence, this relationship is neither automatic nor inevitable. It is, however, an important factor that can undermine marital success. Perhaps, children of divorce need to more consciously guard against behaviors that might undermine their marriages.

FAMILY PROCESSES

The actual day-to-day marital processes of communication—handling conflict, showing affection, and other marital interactions—may be the most important factors holding marriages together or dissolving them (Gottman, 1994).

Marital Happiness

Although it seems reasonable that there would be a strong link between marital happiness (or, rather, the lack of happiness) and divorce, this is true only during the earliest years of marriage. Those who have low marital-happiness scores in the first years of marriage are four or five times more likely to divorce within three years than those with high marital happiness (Booth et al., 1986). The strength of the relationship between low marital happiness and divorce decreases in later stages of marriage, however (White and Booth, 1991). In fact, alternatives to one's marriage and barriers to divorce appear to influence divorce decisions more strongly than does marital happiness. With nothing better to leave for, or if there are too many obstacles to overcome in leaving, a couple might stay married even if unhappy. Although the opposite is also true—even if happy one might leave for a more attractive alternative—it is probably less common.

Children

Although 60 percent of divorces involve children, couples with children divorce less often than couples without children. The birth of the first child reduces the chance of divorce to almost nil in the year following the birth (White, 1991). Furthermore, couples with two children divorce less often than couples with one child or no children. This does not mean that having children will spare you from a divorce or that troubled spouses should become parents so that their troubles will disappear. In fact, it may well be that troubled spouses hold off having children, or if they have a child resist having more because of their troubles. Thus, the quality of the marriage may lead to childbearing more than vice versa. One of the most significant findings indicates that parents of sons are less likely to divorce than parents of daughters. The researchers suggest that fathers participate more in the parenting of sons than daughters, thereby creating greater family involvement for the men (Morgan, Lye, and Condran, 1988; Katzen, Warner, and Acock, 1994). Additionally, because boys have more problems adjusting to divorce than do girls, parents of sons may resist divorce in recognition of that fact.

There are some situations in which the presence of children may be related to higher divorce rates. Premaritally conceived (during adolescence) children and physically or mentally limited children are associated with divorce. Children in general contribute to marital dissatisfaction and possibly divorce, according to one researcher (Raschke, 1987): "It could be expected that normal children at least contribute to strains in an already troubled marriage, given the consistent findings that children, especially in adolescent years, lower marital satisfaction." At the same time, however, women without children have considerably higher divorce rates than women with children.

Marital Problems

If you ask divorced people to give the reasons for their divorce, they are not likely to say, "I blame the changing nature of the family" or "It was demographics." They are more likely to respond, "She was on my case all the time" or "He just didn't understand me," or if they are charitable, they might say, "It wasn't right for us." Personal characteristics leading to conflicts are obviously very important factors in the dissolution of relationships.

Studies of divorced men and women cite such problems as alcoholism, drug abuse, marital infidelity, sexual incompatibility, and conflicts about gender roles as leading to their divorces. Kitson and Sussman (1982) found that the four most common reasons given were, in descending order of frequency, (1) personality problems, (2) home life, (3) authoritarianism, and (4) differing values. Extramarital affairs ranked seventh. Complaints associated with gender roles accounted for 35 percent of the men's responses and 41 percent of women's responses. We know from studying enduring marriages that marriages often continue in the face of such problems. More recent research (Amato and Rogers, 1997) on the connections between marital problems and divorce reveals that reports of marital problems in 1980 were associated with later divorce between 1980 and 1992. Based on interviews with almost 2,000 people, Amato and Rogers found the following:

- Although men's and women's reports differed in the particular problems they emphasized, both predicted divorce equally well.
- Certain problems such as jealousy, moodiness, anger, poor communication, and drinking increased the odds of later divorce; sexual infidelity was an especially strong predictor of divorce.
- People who later divorce report a higher number of problems as early as 9 to 12 years prior to their divorce. Thus, their assessments of problems are not after-the-fact justifications concocted to account for or justify their divorce.
- Marital problems are *proximal causes* of later divorce. They are features of the relationship that directly raise the probability of divorce. There are also background characteristics, such as age at marriage, prior cohabitation, education, income, church attendance, and parental divorce that operate as more *distal causes.* These are brought by each spouse to the relationship and raise the likelihood that marital problems will later arise.

NO-FAULT DIVORCE

Since 1970, beginning with California's Family Law Act, all 50 states have adopted **no-fault divorce**—the legal dissolution of a marriage in which guilt or fault by one or both spouses does not have to be established. Although no-fault divorce has had no effect on divorce rates, it has decreased the time involved in the legal process (Kitson and Morgan, 1991). No-fault divorce has changed four basic premises about divorce (Weitzman, 1985):

- It has eliminated the idea of fault-based grounds. Under no-fault divorce, no one is accused of desertion, cruelty, adultery, impotence, crime, insanity, or a host of other melodramatic acts or omissions. Neither party is found guilty of anything; rather, the marriage is declared unworkable and is dissolved. Husband and wife must agree that they have irreconcilable differences (which they need not describe) and that they believe it is impossible for their marriage to survive the differences.
- It has eliminated the legal adversary processas well as the stress and strain of the courtroom. There is little research, however, to document whether no-fault divorce has been successful in lowering the distress level or the conflict among divorcing couples (Kitson and Morgan, 1991).
- It has established the bases for no-fault divorce settlements as being equity, equality, and need rather than fault or gender. "Virtue" is no longer financially rewarded, nor is it assumed that women need to be supported by men. Community property is to be divided equally, reflecting the belief that marriage is a partnership with each partner contributing equally, if differently. The criteria for child custody are based on a sex-neutral standard of the "best interests of the child" rather than on a preference for the mother.
- It has promoted gender equality by redefining the responsibilities of husbands and wives. The husband is no longer considered head of the household but is an equal partner with his wife. The husband is no longer solely responsible for support, nor is the wife solely responsible for the care of the children. The limitations placed on alimony assume that a woman will work.

REFLECTIONS

As you examine the four premises of no-fault divorce, do you agree with each premise? Why? Would you change or delete any of the premises? Would you add new elements to no-fault?

The Stations of the Divorce Process

Divorce is not a single event. You don't wake up one morning and say, "I'm getting a divorce," and then leave. It's a far more complicated process (Kitson and Morgan, 1991). It may start with little things that at first you hardly notice—a rude remark, thoughtlessness, an unreasonable act, a closedness. Whatever these things are, they begin to add up. Other times, however, the sources of unhappiness are more blatant—yelling, threatening, or battering. For whatever reasons, the marriage eventually becomes unsatisfactory; one or both partners become unhappy.

Over half of those who marry undergo this distressing process. Yet we know very little about the process of marital breakdown and divorce. We understand more about falling in love and courtship than we do about falling out of love and divorce (Furstenberg and Cherlin, 1991).

Anthropologist Paul Bohannan (1970b) developed one of the most influential psychological models to describe the divorce process. (For a discussion of other models, see Guttman [1993].) Bohannan views divorce as consisting of six "stations" (processes) or "divorces": (1) emotional, (2) legal, (3) economic, (4) co-parental, (5) community, and (6) psychic. As people divorce, they undergo these divorces though not necessarily in a particular order or simultaneously. The level of intensity of these different divorces varies at different times.

The Emotional Divorce

The emotional divorce, when one spouse (or both) begins to disengage, begins well before the legal divorce. But even as divorce papers are filed, the couple may find themselves feeling ambivalent. Because the emotional divorce is not complete, they may try to reconcile.

The partners may undermine each other's self-esteem with indifference or destructive criticism. From the outside, the marriage may appear to be functioning adequately, but its heart is missing. In fact, the marriage may even appear to be functioning better, to be more relaxed, and to have less conflict. But in reality, tension is diminished because one partner has decided to ignore or overlook problems—it is no longer worth the effort. "Why try to get him to do his share of the housework? I'm leaving!" a woman may ask. A husband may quit battling his wife over smoking. "I'm outta here," he thinks. "She can puff herself sick, for all I care." At the same time, ignoring the problems often adds to them, reinforcing the decision to divorce.

The Legal Divorce

The legal divorce is the court-ordered termination of a marriage. It permits divorced spouses to remarry and conduct themselves in a way that is legally independent of each other. The legal divorce also sets the terms for the division of property and child custody, issues that may lead to bitterly contested divorce battles. Many of the unresolved issues of the emotional divorce, such as feelings of hurt and betrayal, may be acted out during the legal divorce. Of course, no-fault divorce was intended to minimize these issues.

The Economic Divorce

The economic basis of marriage often becomes most painfully apparent during the economic divorce. Most property acquired during a marriage is considered joint property and is divided between the divorcing spouses. The property settlement is based on the assumption that each spouse contributes to the estate. This contribution may be nonmonetary, as in the case of the homemakers whose "moral assistance and domestic services" permitted their husbands to work outside the home. Alimony and child support may be required. As the partners go their own ways, they often suffer dramatic decreases in their standards of living because they must set up separate households and no longer pool their resources. Women usually experience the greatest decline in their standards of living, as we shall see.

The Co-Parental Divorce

Marriages end, but parenthood does not. Spouses may divorce each other, but they do not divorce their children. (Even those parents who never see their children remain in some sense fathers and mothers.) This may be the most complicated aspect of divorce, for it also gives birth to single-parent families and, in the majority of cases, stepfamilies, considered in more detail in Chapter 15. As parents divorce, issues of child custody, visitation, and support must be dealt with. The impact of divorce on children must be understood, negative consequences must be minimized as much as possible, and new ways of relating to the children and former spouses must be developed, keeping the children's best interest foremost in mind.

The Community Divorce

When people divorce, their social context changes. In-laws become ex-laws; often they lose (or stop) contact. (This is particularly troublesome when in-laws are also grandparents.) Old friends may choose sides or drop out; they may not be as supportive as one wants. New friends may replace old ones as divorced men and women begin dating again. They may enter the singles subculture, where activities center around dating. Single parents may feel isolated from such activities because child rearing often leaves them no leisure, and diminished income leaves them no money.

REFLECTIONS

From what you know about divorce, either from your own experience as a child or partner, or from the experiences of friends or other family members, how well does Bohannon's six-station model describe the experience? Are some stages more difficult than others? Why?

The Psychic Divorce

The psychic divorce is accomplished when your former spouse becomes irrelevant to your sense of self and emotional well-being. You are psychically divorced when you learn that your ex-spouse has gotten a promotion; married someone smarter, funnier, more sensitive, more understanding, and better looking than you; bought a 4 x 4; received an honorary doctorate; and looks terrific—and you don't care. You have your own life to live. Bohannan regards the psychic divorce as the most important element in the divorce process. It is in this stage that each partner develops her or his own sense of independence, completeness, and stability. Neither misses the other or blames him or her for mistakes or misfortunes. Both are responsible for their own lives and are going forward with them. Navigating one's way through the psychic station may be more difficult and take a good deal longer than it does to experience the other stations of divorce. It may, in fact, stretch well beyond the legal divorce. Thus, although "technically" divorced, ex-spouses may remain "attached" or connected to their former partner, unable to completely redefine themselves as single, autonomous individuals.

The divorce process, as we can see, is complex. It takes place on many different levels. Those who go through divorce experience both pain and liberation, but eventually emerge as new women and men.

Marital Separation

The crucial event in a marital breakdown is the act of separation. Divorce is a legal consequence that follows the emotional fact of separation (Melichar and Chiriboga, 1988). Although separation generally precedes divorce, not all separations lead to divorce. As many as one couple out of every six that remains married is likely to have separated for at least two days (Kitson, 1985). The majority of separated women are between 15 and 45 years old, and a greater proportion of African Americans than whites are separated. Unfortunately, we do not know much about marital separation. We don't know, for example, which separations are likely to lead to reconciliations, divorce, or long-term separation (Morgan, 1988). Those who reconcile may have separated in order to dramatize their complaints, create emotional distance, or dissipate their anger (Kitson, 1985).

DID YOU KNOW?

A study of separation and reconciliation found that of those who separated, 40 percent reconciled at least once; 18 percent reconciled twice or more. Almost all who reconciled did so within a year; 45 percent reconciled within a month (Bumpass, Martin, and Sweet, 1991).

UNCOUPLING: THE PROCESS OF SEPARATION

The trends in divorce are fairly clear, but the causes are not. Sociologists can describe societal, demographic, and family factors that appear to be associated with divorce. Unfortunately, however, such

In the early phases of the process of separation estrangement can grow before both parties are fully aware of what has happened.

© Walter Hodges/Getty Images/Stone

variables tell us about groups rather than individuals. Similarly, divorced men and women can tell us what they believe were the causes of their own divorces. But human beings do not always know the reasons for their own actions. They can deceive themselves, blame others, or remain ignorant of the causes.

Sometimes marital complaints are culled from long-term marital problems as a justification for the split-up. Gay Kitson and Marvin Sussman (1982) observe that the study of marital complaints as causes of divorce is merely the study of people's perceptions: "It is perhaps an impossibility to determine what 'really' broke up the marriage." Robert Weiss (1975) uses the term *account* to describe the individual's personal perception of the breakup. These accounts focus on a few dramatic events or factors in the marriage. Because accounts are personal perceptions, each spouse's account is often very different from the other's; what is important to one partner may not be important to the other. Sometimes the accounts of ex-spouses seem to describe entirely different marriages.

People do not suddenly separate or divorce. Instead, they gradually move apart through a set of fairly predictable stages. Sociologist Diane Vaughan (1986) calls this process *uncoupling.* The process appears to be the same for married or unmarried couples and for gay or lesbian relationships. The length of time together does not seem to affect the process.

"Uncoupling begins," Vaughan observes, "as a quiet, unilateral process." Usually one person, the initiator, is unhappy or dissatisfied but keeps such feelings to himself or herself. The initiator often ponders fundamental questions about his or her identity: "Who am I? Who am I in this relationship? What do I want out of my life? Can I find it in this relationship?" The dissatisfied partner may attempt to make changes in the relationship, but these are often unsuccessful, as he or she may not really know what the problem is.

Because the dissatisfied partner is unable to find satisfaction within the relationship, he or she begins turning elsewhere. This is not a malicious or intentional turning away; it is done to find self-validation without leaving the relationship. In doing so, however, the dissatisfied partner "creates a small territory independent of the coupled identity" (Vaughan, 1986). This creates a division within the relationship. Gradually, the dissatisfied partner voices more and more complaints, which make the relationship and partner increasingly undesirable. The initiator begins thinking about alternatives to the relationship and comparing the costs and benefits of these alternatives. Meanwhile, both the initiator and his or her partner try to cover up the seriousness of the dissatisfaction, submerging it in the little problems of everyday living.

Eventually, the initiator decides that he or she can no longer go on. She or he may actually go through

"That's right, Phil. A separation will mean—among other things—watching your own cholesterol."

a process of mourning the demise of what is still an intact marriage (Emery, 1994, cited in Amato, 2000). There are several strategies for ending the relationship. One way is simply to tell the partner that the relationship is over. Another way is to break—consciously or unconsciously—a fundamental rule in the relationship, such as by having an extramarital affair and letting the partner know or discover it.

REFLECTIONS

From your own experience, how well does "uncoupling" describe the process of separating from someone you care about? Are there missing elements or elements that should be emphasized? What about separation distress? In your own experience, what was it like? What things were you able to do to alleviate it? What advice would you give others about it?

Uncoupling does not end when the end of a relationship is announced, or even when the couple physically separate. Acknowledging that the relationship cannot be saved represents the beginning of the last stage of uncoupling. Vaughan (1986) describes the process:

> Partners begin to put the relationship behind them. They acknowledge that the relationship is unsaveable. Through the process of mourning they, too, eventually arrive at an account that explains this unexpected denouement. "Getting over" a relationship does not mean relinquishing that part of our life that we shared with another; but rather coming to some conclusion that allows us to accept and understand its altered significance. Once we develop such an account, we can incorporate it into our lives and go on.

THE NEW SELF: SEPARATION DISTRESS AND POSTDIVORCE IDENTITY

Our married self becomes part of our deepest self. Therefore, when people separate or divorce, many feel as if they have "lost an arm or a leg." This analogy, as well as the traditional marriage rite in which a man and a woman are pronounced "one," reveals an important truth of marriage: The constant association of both partners makes each almost a physical part of the other. This dynamic is true even if two people are locked in conflict; they, too, are attached to each other (Masheter, 1991).

Separation Distress

Most newly separated people do not know what to expect. There are no divorce ceremonies or rituals to mark this major turning point. Yet people need to understand divorce in order to alleviate some of its pain and burden. Except for the death of a spouse, divorce is the greatest stress-producing event in life (Holmes and Rahe, 1967). The changes that take place during separation are crucial because at this point a person's emotions are at their rawest and most profound. Men and women react differently during this period. Many people experience **separation distress,** situational anxiety caused by separation from an attachment figure.

Researchers have considerable knowledge about the negative consequences accompanying marital separation, some of which we discuss here. In looking at this negative impact, however, we need to keep in mind a caution from Helen Raschke (1987): "The psychological and emotional consequences of separation and divorce have been more distorted than any of the other consequences as a result of the deviance perspective." The negative aspects of separation are balanced sooner or later, notes Raschke, by positive aspects, such as the possibility of finding a more compatible partner, constructing a better (or different) life, developing new dimensions of the self, enhancing self-esteem, and marrying a better parent for one's children. These positive consequences may follow, or be intertwined with, separation distress. In the pain of separation, we may forget that a new self is being born.

Almost everyone suffers separation distress when a marriage breaks up. The distress is real but, fortunately, does not last forever (although it may seem so). The distress is situational and is modified by numerous external factors. About the only men and women who do not experience distress are those whose marriages were riddled by high levels of conflict. In these cases, one or both partners may view the separation with relief (Raschke, 1987).

During separation distress, almost all attention is centered on the missing partner and is accompanied by apprehensiveness, anxiety, fear, and often panic. "What am I going to do?" "What is he or she doing?" "I need him . . . I need her . . . I hate him . . . I love him . . . I hate her . . . I love her. . . ." Sometimes, however, the immediate effect of separation is not distress but euphoria. This usually results from feeling that the former spouse is not necessary, that one can get along better without him or her, that the old fights and the spouse's criticism are gone forever, and that life will now be full of possibilities and excitement. That euphoria is soon gone. Almost everyone falls back into separation distress.

Whether a person had warning and time to prepare for a separation affects separation distress. An unexpected separation is probably most painful for the partner who is left. Separations that take place during the first two years of marriage, however, are less difficult for the husband and wife to weather. Those couples who separate after two years find separation more difficult because it seems to take about two years for people to become emotionally and socially integrated into marriage and their marital roles (Weiss, 1975). After that point, additional years of marriage seem to make little difference in the spouses' reaction to separation.

As the separation continues, separation distress slowly gives way to loneliness. Eventually, loneliness becomes the most prominent feature of the broken relationship. Old friends can sometimes help provide stability for a person experiencing a marital breakup, but those who give comfort need to be able to tolerate the other person's loneliness.

Establishing a Postdivorce Identity

A person goes through two distinct phases in establishing a new identity following marital separation: *transition* and *recovery* (Weiss, 1975). The transition period begins with the separation and is characterized by separation distress and then loneliness. In this period's later stages, most people begin functioning in an orderly way again, although they still may experience bouts of upset and turmoil. The transition period generally ends within the first year. During this time, individuals have already begun making decisions that provide the framework for new selves. They have entered the role of single parent or noncustodial parent, have found a new place to live, have made important career and financial decisions, and have begun to date. Their new lives are taking shape.

The recovery period usually begins in the second year and lasts between one and three years. By this time the separated or divorced individual has already created a reasonably stable pattern of life. The marriage is becoming more of a distant memory, and the former spouse does not arouse the intense passions she or he once did. Mood swings are not as extreme, and periods of depression are fewer. Yet the individual still has self-doubts that lie just beneath the surface. A sudden reversal, a bad time with the children, or doubts about a romantic involvement can suddenly destroy a divorced person's confi-

Perspective

GENDER AND DIVORCE-RELATED STRESSORS

Ask anyone who has experienced a divorce what one word summarizes the process and you probably won't be surprised when you hear "Stressful!" A short look at the issues faced by divorced couples gives us some idea of the distress they can experience:

- Severance of marital bonds
- Establishment of a new lifestyle
- Economic stressors
- Negotiation of custody arrangements
- Adjustments in parenting
- Changes in social support system

To some degree, gender influences how individuals respond to divorce. Research indicates that divorced men experience greater emotional distress and report more suicidal thoughts than do women (Riesman and Gerstel, 1985; Rosengren, Wedel, and Wilhelmensen, 1989; Wallerstein and Kelly, 1980). Because women are more likely to initiate divorce, research suggests that they experience fewer postdivorce psychological problems. This may be because they have begun the detachment process earlier than men (Lawson and Thompson, 1996). Furthermore, divorced men exhibit higher rates of auto accidents, alcohol abuse, diabetes, heart disease, and mental illness than do divorced women. Higher rates of mortality have been found to exist among divorced men and women, especially if they have remarried or are cohabiting (Hemstrom, 1996).

The immediate impact of divorce on women is economic. This is especially true if they become the primary custodial parent. Nearly half (48 percent) of women who are granted child support do not receive the full amount and 24 percent received nothing (U.S. Census Bureau, 1988, Table 632). A combination of lowered earning power, increased expenses, and lack of financial support results in a decreased standard of living for the divorced mother and her children.

The psychological responses experienced among partners are numerous, ranging from anger to depression to ambivalence. Though some men suffer little distress following divorce (Albrecht, 1980), generally men seem to experience the greater emotional distress, possibly because of their more frequent social isolation (Reismann, 1990). In addition, men report greater attachment to their former spouses and are more likely to desire to rekindle the marriage (Bloom, and Kindle, 1985).

Almost 60 percent of divorces involve children (Kitson and Morgan, 1991), and since 80 percent of these children end up living with their mothers (with an additional 14 percent living with other relatives), fathers must face new emotional territory regarding these issues and their relationships with their children. Single parenting for the mother involves added responsibility to an already overburdened workload. Noncustodial parenting raises new role expectations concerning the quality of the parent-child relationship, normative behaviors, and discipline.

The more resources a person has, the better he or she may handle the separation crisis. Social support is positively correlated with lower distress and positive adjustment. These resources may be emotional as well as social and financial. Parents become an important resource for their divorcing adult children (Johnson, 1988). They may provide economic assistance and emotional support; they are also important for their grandchildren in easing the divorce transition. One study found that reliance on family and friends, involvement in church-related activities, social participation in community activities, and establishment of intimate heterosexual relationships correlated positively to postdivorce adjustment among African-American men (Lawson and Thompson, 1996). Friends are especially important in overcoming the isolation and loneliness that generally accompanies a separation.

As with other stressors in a person's life, it is often the individual's perception of the event, not the stress itself, that influences how a person adjusts to change. If those experiencing separation and divorce can begin to view and accept their changing circumstances as presenting new challenges and opportunities, there is a greater likelihood that the physiological and psychological symptoms of stress that follow divorce can be reduced.

---

dence. By the end of the recovery period, the distress has passed.

It takes some people longer than others to recover because each person experiences the process in his or her own way. But most are surprised by how long the recovery takes—they forget that they are undergoing a major discontinuity in their lives.

DATING AGAIN

A new partner reduces much of the distress caused by separation. A new relationship prevents the loneliness caused by emotional isolation. It also reinforces a person's sense of self-worth. But it does not necessarily eliminate separation distress caused by the disruption of intimate personal relations with the former partner, children, friends, and relatives.

A first date after years of marriage and subsequent months of singlehood evokes some of the same emotions felt by inexperienced adolescents. Separated or divorced men and women who are beginning to date again may be excited and nervous; worry about how they look; and wonder whether or not it is okay to hold hands, kiss, or make love. They may feel that dating is incongruous with their former selves or be annoyed with themselves for feeling excited and awkward. Furthermore, they have little idea of the norms of postmarital dating (Spanier and Thompson, 1987).

For many divorced men and women, the greatest problem is how to meet other unmarried people. They believe that marriage has put them "out of circulation," and many are not sure how to get back in. Because of the marriage squeeze, separated and divorced men in their twenties and thirties are at a particular disadvantage: considerably fewer women are available than men. The squeeze reverses itself at age 40 when there are significantly fewer single men available. The problem of meeting others is most acute for single mothers who are full-time parents in the home because they lack opportunities to meet potential partners. Divorced men, having fewer child-care responsibilities and more income than divorced women, tend to have more active social lives.

Dating fulfills several important functions for separated and divorced people. First, it is a statement to both the former spouse and the world at large that the individual is available to become someone else's partner (Vaughan, 1986). Second, dating is an opportunity to enhance one's self-esteem (Spanier and Thompson, 1987). Free from the stress of an unhappy marriage, dating may lead people to discover, for example, that they are more interesting and charming than either they or (especially) their former spouses had imagined. Third, dating initiates individuals into the singles subculture, where they can experiment with the freedom about which they may have fantasized when they were married. Interestingly, Spanier and Thompson (1987) found no relationship between dating experience and well-being following separation.

Several features of dating following separation and divorce differ from premarital dating. First, dating does not seem to be a leisurely matter. Divorced people are often too pressed for time to waste time on a first date that might not go well. Second, dating may be less spontaneous if the divorced woman or man has primary responsibility for children. The presence of children thus inhibits divorced women's opportunities to date others more than it does divorced men's. The parent must make arrangements about child care; he or she may wish not to involve the children in dating. Third, finances may be strained; divorced mothers may have income only from low-paying or part-time jobs or AFDC benefits while having many child-care expenses. In some cases a father's finances may be strained by paying alimony or child support. Finally, separated and divorced men and women often have a changed sexual ethic based on the simple fact that there are few (if any) divorced virgins (Spanier and Thompson, 1987).

Sexual relationships are often an important component in the lives of separated and divorced men and women. Engaging in sexual relations for the first time following separation may help people accept their newly acquired single status. Because sexual fidelity is an important element in marriage, becoming sexually active with someone other than one's ex-spouse is a dramatic symbol that the old marriage vows are no longer valid. Men initially tend to enjoy their sexual freedom following divorce, but women generally do not find it as satisfying as do men. For men, sexual experience following separation is linked with their well-being. Sex seems to reassure men and bolster their self-confidence. Sexual activity is not as strongly connected to women's well-being (Spanier and Thompson, 1987).

DID YOU KNOW?

About one-third of divorced men and women remarry within a year of divorce (Ganong and Coleman, 1994).

Consequences of Divorce

Most divorces are not contested; between 85 and 90 percent are settled out of court through negotiations between spouses or their lawyers. But divorce, whether it is amicable or not, is a complex legal process involving highly charged feelings about custody, property, and children (who are sometimes treated by angry partners as property to be fought over).

ECONOMIC CONSEQUENCES OF DIVORCE

Probably the most damaging consequences of the no-fault divorce laws are that they systematically impoverish divorced women and their children. Following divorce, women are primarily responsible for both child rearing *and* economic support (Maccoby et al., 1993). As a result, women are at a greater risk for poverty than they were during their marriage. Even if a woman is not plunged into poverty, she often experiences a dramatic downward turn in her economic condition (Garrison, 1994; Morgan, 1991). A single mother's income shows about a 27 percent decline, whereas the income of a divorced man results in a 10 percent decline of his pre-divorce income (Peterson, 1996; Smock, 1993). Because over half of the children born today will live in a single-parent family at some point during their childhood, through no-fault divorce rules "we are sentencing a significant proportion of the next generation of American children to periods of financial hardship" (Weitzman, 1985).

Sociologist Lenore Weitzman (1985) explains how no-fault divorce's gender-neutral rules, designed to treat men and women equally, actually place older homemakers and mothers of young children at a great economic disadvantage:

> Since a woman's ability to support herself is likely to be impaired during marriage, especially if she is a full-time homemaker and mother, she may not be "equal" to her former husband at the point of divorce. Rules that treat her as if she is equal simply serve to deprive her of the financial support she needs.

One of the most striking differences between two-parent and single-parent families is poverty. More than a third of middle-income women and a quarter of upper-income women find themselves needing welfare following divorce. The majority of single mothers become poor as a result of their marital disruption. Contributing factors are (1) the mother's low earning capacity, (2) lack of child support, and (3) inadequate welfare benefits (discussed in Chapter 12).

Husbands typically enhance their earning capacity during marriage. In contrast, wives generally decrease their earning capacity because they either quit or limit their participation in the workforce to fulfill family roles. This withdrawal from full participation limits their earning capacity when they reenter the workforce. Divorced homemakers have outdated experience, few skills, and no seniority. Furthermore, they continue to have the major responsibility and burden of child rearing.

When marriage ends, many women must face the triple consequences of gender, ethnic, and age discrimination as they seek to support themselves and their children. Because the workplace favors men in terms of opportunity and income, separation and divorce does not affect them as adversely. (See Chapter 12 for a discussion of economic discrimination against women.)

About a quarter of divorced women enter a spell of poverty sometime during the first five years following divorce. Whereas the disparities in income between white and African-American women are significant during marriage, following a divorce white women suffer a relatively greater decline in their standards of living, and the income levels of white and African-American women converge (Morgan, 1991). Mexican-American women suffer relatively less decline in economic status than do Anglo-American women because Latinas are already more economically disadvantaged. But because their lives have prepared them for greater economic adversity, Latinas' emotional well-being appears to suffer less than does that of Anglo-American women following divorce (Wagner, 1993).

Employment

The economic impact of divorce on women with children is especially difficult, as their employment opportunities are often constrained by the necessity

Perspective

LESBIANS, GAY MEN, AND DIVORCE

Although there are no reliable studies, it is estimated that about one-fifth of gay men and one-third of lesbians have been married. Estimates of bisexual men and women who are married run into the millions (Hill, 1987; Gochoros, 1989). Relatively few gay men, lesbians, and bisexuals are consciously aware of their sexual orientation at the time they marry. Those who are aware rarely disclose their feelings to their prospective partners (Gochros, 1989). When married lesbians and gay men acknowledge their gayness to themselves, they often feel that they are "living a lie" in their marriage. While they may deeply love their spouses, the majority eventually divorce.

How is it that lesbians and gay men marry heterosexuals in the first place? As we saw in Chapter 7, the gay/lesbian identity process is difficult and complex. Because of fear and denial, some gay men and lesbians are unable to acknowledge their sexual feelings. They believe or hope they are heterosexual and do their utmost to suppress their same-sex fantasies or behaviors. They often believe that their homosexuality is just a "phase." Typically they hold negative stereotypes about homosexuality and cannot bring themselves to believe or accept that they might be "one of them." Marriage is one way of convincing themselves that they are heterosexual. In addition to "curing" or denying one's gayness, their motivations to marry are no different from heterosexuals (Bozett, 1987). Like heterosexuals, gay men and lesbians marry because of pressure from family, friends, and fiancé, genuine love for one's fiancé, the wish for companionship, and the desire to have children.

When husbands or wives discover their partner's homosexuality or bisexuality they may initially experience shock; others experience temporary relief. Mysteries get explained: why one's spouse disappears for periods of time, why mysterious phone calls occur, the spouse's lack of sexual interest. But whether one is shocked or relieved, inevitably the heterosexual spouse feels deceived or stupid. Many feel shame (Hays and Samuels, 1987). One woman, who felt ashamed to tell anyone of her distress, recalled, "His coming out of the closet in some ways put the family in the closet" (Hill, 1987). At the same time, the gay, lesbian, or bisexual spouse often feels deeply grieved (Voeller, 1980):

Many people date, marry, and become parents, only to realize too late the error they made. They then find themselves deeply pained, fearful of losing their children through court suits, of losing spouses they care for but are ill suited to, of depriving their spouses and themselves of more deeply appropriate and meaningful relationships, and of causing their friends and other relatives deep pain.

When gay men, lesbians, or bisexuals disclose their orientation to their spouses, separation and divorce is the usual outcome. Many gay men and lesbians are also parents at the time they separate from their spouses. It is generally important for them to affirm their identities both as gay or lesbian and as a parent (Bozett, 1989c). This is especially important as negative stereotypes portray gay men and lesbians as "antifamily." Men and women begin to fuse their identities as gay or lesbian with their parental role.

A study of gay fathers reported that gay men usually do not reveal their orientation to their children unless the parents are separating or the gay father develops a gay love relationship (Bozett, 1989c). As with divorced fathers in general, gay fathers usually do not have custody of the children, but lesbians, like other divorced women, are more likely to have custody (Bozett, 1989b).

of caring for children (Maccoby et al., 1993). Child-care costs may consume a third or more of a poor single mother's income. Women may work fewer hours because of the need to care for their children.

Separation and divorce dramatically change many mothers' employment patterns (Morgan, 1991). If a mother was not employed prior to separation, she is likely to seek a job following the split-up. The reason is simple: If she and her children relied on alimony and child support alone, they would soon find themselves on the street. Most employed single mothers are still on the verge of financial disaster, however. On the average, they earn only a third as much as married fathers. This is partly because women tend to earn less than men and partly because they work fewer hours, primarily because of child-care responsibilities (Garfinkel and McLanahan, 1986). The general problems of women's lower earning capacity and lack of adequate child care are particularly severe for single mothers. Gender discrimination in employment and lack of societal support for child care condemn millions of single mothers and their children to poverty.

Alimony and Child Support

Alimony is the money payment a former spouse makes to the other to meet his or her economic needs. Alimony is different from **child support**—the monetary payments made by the noncustodial spouse to the custodial spouse to assist in child-rearing expenses. For many women, their source of income changes upon divorce from primarily joint wages earned during marriage to their own wages, supplemented by child support payments, alimony, help from relatives, and welfare. The Child Support Enforcement Amendments, passed in 1984, together with the Family Support Act of 1988 requires states to deduct delinquent support from fathers' paychecks, authorizes judges to use their discretion when support agreements cannot be met, and mandates periodic reviews of award levels to keep up with the rate of inflation. In addition, all states implemented automatic wage withholding of child support in 1994. Recent research has shown that enforcement has had a beneficial impact on compliance with child support orders (Meyer and Bartfeld, 1996). Nevertheless, child support awards are historically small, usually amounting to 10 percent of the noncustodial father's income and less than half of a child's expenses.

In 1989, about 16 percent of divorcing white women and 11 percent of divorcing African-American and of divorcing Latina women received alimony (U.S. Census Bureau, 1994). Child support was awarded to 64 percent of white women, 36 percent of African-American women, and 35 percent of Latina women. Altogether about 56 percent of single mothers were awarded child support. Of this group, 76 percent of the mothers received full payment (U.S. Census Bureau, 1996). Even though many men have the financial resources to support their children and ex-wives, they are less inclined to pay it if they disagree with the child-custody arrangements (Finkel and Roberts, 1994).

Even when fathers pay support, however, the amount is generally low, averaging slightly less than $5,500 annually (U.S. Census Bureau, 1996). The amount paid generally depends more on the father's circumstances than on the needs of the mother and children (Teachman, 1991). Alimony and child support payments make up only about 10 percent ($1,246) of the income of white single mothers and 3.5 percent ($322) of the income of African-American single mothers (McLanahan and Booth, 1991). Although some argue that divorced and remarried fathers cannot pay additional support without pushing themselves and their new families into poverty, evidence indicates that this is not true in the majority of cases (Duncan and Hoffman, 1985).

People generally approve, at least in principle, of child support, but alimony is more controversial. In the past, alimony represented the continuation of the husband's responsibility to support his wife. Currently, laws suggest that alimony should be awarded on the basis of need to those women who would otherwise be indigent. Although the courts award alimony in about 15 percent of all divorce cases, a much smaller percent actually receive it. At the same time, there is a strong countermovement in which alimony represents the return of a woman's "investment" in marriage (Oster, 1987; Weitzman, 1985). Weitzman argues that a woman's homemaking and child-care activities must be considered important contributions to her husband's present and future earnings. If divorce rules do not give a wife a share of her husband's enhanced earning capacity, then the "investment" she made in her spouse's future earnings is discounted. According to Weitzman, alimony and child support awards should be made to divorced women in recognition of the wife's primary child-care responsibilities and her contribution to her ex-husband's work or career. Such awards will help raise divorced women and children above the level of poverty to which they have been cast as a result of no-fault divorce's specious equality. A landmark court decision, in fact, upheld the "investment" doctrine by ruling that a woman who supported her husband during his medical education was entitled to a portion of his potential lifetime earnings as a physician (Oster, 1987).

Noneconomic Consequences of Divorce

In comparison to married people, the picture one gets about divorced individuals is fairly bleak. Reviewing the research literature of the 1990s, Amato

(2000) notes the following. Compared with married people, divorced individuals experience more psychological distress, poorer self-concepts, lower levels of psychological well-being, lower levels of happiness, more social isolation, less satisfying sex lives, and more negative life events. They also have greater risks of mortality and report more health problems.

Linda Waite and Maggie Gallagher, in their book *The Case for Marriage* (2000), take on the question of whether being married makes people happier or whether it is happier people who get married and *stay* married. Citing research that compared the emotional health of a sample of people over time— some who married and stayed married, some who never married or remained divorced, and others who married and divorced—they report the following:

> When people married, their mental health improved—consistently and substantially. Meanwhile over the same period, when people separated and divorced, they suffered substantial deterioration in mental and emotional well-being, including increases in depression and declines in reported happiness. . . . Those who dissolved a marriage also reported less personal mastery, fewer positive relations with others, less purpose in life, and less self-acceptance than their married peers did.

Waite and Gallagher (2000) also note that compared to married people, divorced (and widowed) women and men were three times as likely to commit suicide. Among the divorced, as among the general population, more men than women commit suicide. However, divorced women are "the most likely to commit suicide, followed by widowed, never-married and married, in that order".

As parents, divorced individuals have more difficulty raising children. They display more role strain, whether they are custodial or noncustodial parents, and they display less authoritative parenting styles (Amato, 2000).

Despite the stark picture that surfaces above, it is true that for some people, divorce is associated with positive consequences. These include higher levels of personal growth, greater autonomy, and—for some women—improvements in self-confidence, career opportunities, social lives, and happiness as well as a stronger sense of control (Amato, 2000).

REFLECTIONS

Why are alimony and child support often such emotional issues in divorce? On what basis should alimony be awarded? Child support? Why do many noncustodial parents fail to pay child support? What could be done to improve their likelihood of supporting their children?

Children and Divorce

Slightly over half of all divorces involve children. Popular images of divorce depict "broken homes," but it is important to remember that an intact nuclear family, merely because it is intact, does not guarantee children an advantage over children in a single-parent family or a stepfamily. A traditional family wracked with spousal violence, sexual or physical abuse of children, alcoholism, neglect, severe conflict, or psychopathology creates a destructive environment that is likely to inhibit children's healthy development. Living in a two-parent family with marital conflict is often more harmful to children than living in a tranquil single-parent family or stepfamily. Children living in happy two-parent families appear to be the best adjusted, and those from conflict-ridden two-parent families appear to be the worst adjusted. Children from single-parent families are in the middle. The key to children's adjustment following divorce is a lack of conflict between divorced parents (Kline, Johnston, and Tschann, 1991).

Telling children that their parents are separating is one of the most difficult and unhappy events in life. Whether or not the parents are relieved about the separation, they often feel extremely guilty about their children. Many children may not even be aware of parental discord (Furstenberg and Cherlin, 1991). Even those that are may be upset by the separation, but their distress may not be immediately apparent.

As psychologist Judith Wallerstein suggested in her book, *Second Chances* (Wallerstein and Blakeslee, 1989), divorce is differently experienced within the family. For at least one of the divorcing

© Kindra Clineff/Index Stock

Children react differently to divorce, depending on their age. Most feel sad, but the eventual outcome for children depends on many factors, including having a competent and caring custodial parent, siblings, and friends, and their own resiliency. The postdivorce relationship between parents and the custodial parent's economic situation are also important factors.

spouses, divorce is welcomed as an escape from an unpleasant or unfulfilling relationship. Ultimately, both spouses may come to appreciate the "second chance" they get with divorce: the opportunity to make a better choice and build themselves a better relationship. Children rarely see the breakup of their parents' marriage as an "opportunity."

A meta-analysis (a research technique combining statistical data from previous studies and reanalyzing it) used earlier divorce studies on the impact of parental divorce on the well-being of children in adulthood (Amato and Keith, 1991). The study found very little difference in the well-being of children from divorced families and intact families. A study by psychologist Judith Wallerstein found that children from divorced families suffered both emotionally and developmentally ("Study Reveals," 1997). Young children fare worse than do older children. In the "crisis period" of the two years following separation, boys tend to do less well than girls. This may be due to having internalized different gendered styles of reacting to distress. It is also the case, though, that after separation most boys live with their mothers and not their fathers. This can exacerbate their suffering (Furstenberg and Cherlin, 1991).

THE THREE STAGES OF DIVORCE FOR CHILDREN

Growing numbers of studies have appeared on the impact of divorce on children, but these studies frequently contradict one another. Part of the problem is a failure to recognize divorce as a process for children as opposed to a single event. Divorce is a series of events and changes in life circumstances. Many studies focus on only one part of the process and identify that part with divorce itself. Yet at different points in the process, children are confronted with different tasks and adopt different coping strategies. Furthermore, the diversity of children's responses to divorce is the result, in part, of differences in temperament, gender, age, and past experiences.

Children experience divorce as a three-stage process, according to Judith Wallerstein and Joan Kelly (1980b). Studying 60 California families during a five-year period, these researchers found that for children, divorce consisted of the initial, transition, and restabilization stages:

- *Initial stage:* The initial stage, following the decision to separate, was extremely stressful; conflict escalated, and unhappiness was endemic. The children's aggressive responses were magnified by the parents' inability to cope because of the crisis in their own lives.

- *Transition stage:* The transition stage began about a year after the separation, when the extreme emotional responses of the children had diminished or disappeared. The period was characterized by restructuring of the family and by economic and social changes: living with only one parent and visiting the other, moving, making new friends and losing old ones, financial stress, and so on. The transition period lasted between two and three years for half the families in the study.
- *Restabilization stage:* Families had reached the restabilization stage by the end of five years. Economic and social changes had been incorporated into daily living. The postdivorce family, usually a single-parent family or stepfamily, had been formed.

CHILDREN'S RESPONSES TO DIVORCE

A decisive element in children's responses to divorce is their developmental stage (Guttman, 1993). A child's age affects how she or he responds to one parent's leaving home, changes (usually downward) in socioeconomic status, moving from one home to another, transferring schools, making new friends, and so on.

Developmental Tasks of Divorce

Children must undertake six developmental tasks when their parents divorce (Wallerstein, 1983). The first two tasks need to be resolved during the first year. The other tasks may be worked on later; often they may need to be reworked because the issues often recur. How children resolve these tasks differs by age and social development. The tasks are as follows:

1. *Acknowledging parental separation.* Children often feel overwhelmed by feelings of rejection, sadness, anger, and abandonment. They may try to cope with them by denying that their parents are "really" separating. They need to accept their parents' separating and to face their fears.
2. *Disengaging from parental conflicts.* Children need to psychologically distance themselves from their parents' conflicts and problems. They require such distance so that they can continue to function in their everyday activities without being overwhelmed by their parents' crisis.
3. *Resolution of loss.* Children lose not only their familiar parental relationship but also their everyday routines and structures. They need to accept these losses and focus on building new relationships, friends, and routines.
4. *Resolution of anger and self-blame.* Children, especially young ones, often blame themselves for the divorce. They are angry with their parents for disturbing their world. Many often "wish" their parents would divorce, and when their parents actually do, they feel responsible and guilty for "causing" it.
5. *Accepting the finality of divorce.* Children need to realize that their parents will probably not get back together. Younger children hold "fairy-tale" wishes that their parents will reunite and "live happily ever after." The older the child is, the easier it is for him or her to accept the divorce.
6. *Achieving realistic expectations for later relationship success.* Children need to understand that their parents' divorce does not condemn them to unsuccessful relationships as adults. They are not damaged by witnessing their parents' marriage; they can have fulfilling relationships themselves.

Younger Children

Younger children react to the initial news of a parental breakup in many different ways. Feelings range from guilt to anger and from sorrow to relief, often vacillating among all of these. The most significant factor affecting children's responses to the separation is their age. Preadolescent children, who seem to experience a deep sadness and anxiety about the future, are usually the most upset. Some may regress to immature behavior, wetting their beds or becoming excessively possessive. Most children, regardless of their age, are angry because of the separation. Very young children tend to have more temper tantrums. Slightly older children become aggressive in their play, games, and fantasies—for example, pretending to hit one of their parents.

A recent study using longitudinal data collected over a twelve-year period examines parent-child relationships before and after divorce. Researchers found that marital discord may exacerbate children's behavior problems, making them more difficult to manage (Amato and Booth, 1996). Because discord between parents often preoccupies and distracts them from the tasks of parenting, they appear unavailable and unable to deal with their children's needs. This study reinforced a growing body of evidence showing that many problems assumed to be due to divorce are actually present before marital disruption.

Children of school age may blame one parent and direct their anger toward him or her, believing the other one innocent. But even in these cases the reactions are varied. If the father moves out of the house, the children may blame the mother for making him go, or they may be angry at the father for abandoning them, regardless of reality. Younger schoolchildren who blame the mother often mix anger with placating behavior, fearing she will leave them. Preschool children often blame themselves, feeling that they drove their parents apart by being naughty or messy. They beg their parents to stay, promising to be better. It is heartbreaking to hear a child say, "Mommy, tell Daddy I'll be good. Tell him to come back. I'll be good. He won't be mad at me anymore."

A study of 121 white children between the ages of 6 and 12 found that about a third initially blamed themselves for their parents' divorce. After a year, the figure dropped to 20 percent (Healy, Stewart, and Copeland, 1993). The largest factor in self-blaming was being caught in the middle of parental conflict. Children who blamed themselves displayed more psychological symptoms and behavior problems than those who did not blame themselves.

When parents separate, children want to know with whom they are going to live. If they feel strong bonds with the parent who leaves, they want to know when they can see him or her. If they have brothers or sisters, they want to know if they will remain with their siblings. They especially want to know what will happen to them if the parent they are living with dies. Will they go to their grandparents, their other parent, an aunt or uncle, or a foster home? These are practical questions, and children have a right to answers. They need to know what lies ahead for them amid the turmoil of a family split-up so that they can prepare for the changes. Some parents report that their children seemed to do better psychologically than they themselves did after a split-up. Children often have more strength and inner resources than parents realize.

The outcome of separation for children, Weiss (1975) observes, depends on several factors related to the children's age. Young children need a competent and loving parent to take care of them; they tend to do poorly when a parenting adult becomes enmeshed in constant turmoil, depression, and worry. With older, preadolescent children, the presence of brothers and sisters helps because the children have others to play with and rely on in addition to the single parent. If they have good friends or do well in school, this contributes to their self-esteem. Regardless of the child's age, it is important that the absent parent continue to play a role in the child's life. The children need to know that they have not been abandoned and that the absent parent still cares (Wallerstein and Kelly, 1980b). They need continuity and security, even if the old parental relationship has radically changed.

Adolescents

Many adolescents find parental separation traumatic. Studies indicate that much of what appears to be negative results of divorce for children (personal changes, parental loss, economic hardships, psychological adjustments) are probably the result of parental conflict that precedes and surrounds the divorce (Amato and Keith, 1991; Morrison and Cherlin, 1995; Amato and Booth, 1996).

Adolescents tend to protect themselves from the conflict preceding separation by distancing themselves. Although they usually experience immense turmoil within, they may outwardly appear cool and detached. Unlike younger children, they rarely blame themselves for the conflict. Rather, they are likely to be angry with both parents, blaming them for upsetting their lives. Adolescents may be particularly bothered by their parents' beginning to date again. Some are shocked to realize that their parents are sexual beings, especially when they see a separated parent kiss someone or bring someone home for the night. The situation may add greater confusion to the adolescents' emerging sexual life. Some may take the attitude that if their mother or father sleeps with a date, why can't they? Others may condemn their parents for acting "immorally."

REFLECTIONS

As you look at the adjustments that children must make when their parents divorce, are there others you would add? Which ones do you believe are the most important? Most difficult? If you were a divorcing parent, what strategies would you use to help your children adjust to divorce? How would your strategies differ according to the age of the child or adolescent? What do you think the experience might be of adult children whose parents divorce?

Helping Children Adjust

Helen Raschke's (1987) review of the literature on children's adjustment after divorce found that the following factors were important:

- Prior to separation, open discussion with the children about the forthcoming separation and divorce and the problems associated with them.
- The child's continued involvement with the noncustodial parent, including frequent visits and unrestricted access.
- Lack of hostility between the divorced parents.
- Good emotional and psychological adjustment to the divorce on the part of the custodial parent.
- Good parenting skills and the maintenance of an orderly and stable living situation for the children.

Continued involvement with the children by both parents is important for the children's adjustment. The greatest danger is that children may be used as pawns by their parents after a divorce. The recently divorced often suffer from a lack of self-esteem and a sense of failure. One means of dealing with the feelings caused by divorce is to blame the other person. To prevent further hurt or to get revenge, divorced parents may try to control each other through their children. A recent study has shown that children are likely to suffer long-term psychological damage—well into adulthood—if the parents do not consider their emotional needs during the divorce process ("Study Reveals," 1997).

MULTIPLE PERSPECTIVES ON THE LONG-TERM EFFECTS OF DIVORCE ON CHILDREN

There is a variety of perspectives on how and why divorce affects children (Amato, 1993). Specified outcomes range from negative through neutral to positive (Whitehead, 1996; Coontz, 1997). There is enough divergent information that one could selectively cite research to make either a more pessimistic or more optimistic generalization. We review some of these mixed findings below.

A variety of studies reviewed by Barbara Dafoe Whitehead, in her antidivorce book, *The Divorce Culture* (1997), suggest multiple ways in which children suffer after their parents divorce. First, across racial lines, children of divorce suffer substantial reduction in family income as a direct result of divorce. Second, a majority of children experience a weakening of ties with their fathers, suffering damage when and after fathers leave. She suggests that separation and later divorce induce a "downward spiral" in father-child relationships, wherein distance between them grows, and children eventually lose their fathers' "love, support, and substantial involvement." Third, children suffer a loss of "residential stability," often having to move from the family home due to drops in their economic standing. Whitehead goes on to detail other measurable ways in which children suffer: reduced school performance, increased likelihood of dropping out, worsened and increased behavioral problems, a greater likelihood of becoming teen parents. Many of these same outcomes were identified as among the "risks and problems associated with stepfamily life" (Whitehead, 1996).

Stephanie Coontz, in her more optimistic book *The Way We Really Are: Coming to Terms with America's Changing Families* (1997), tempers some of this distressing news. While acknowledging the "agonizing process" that accompanies divorce and the ways in which children, especially, can be hurt by divorce, Coontz qualifies the more pessimistic interpretations. In a subtle but important comparison she notes that research shows "*not* that children in divorced families have more problems but that *more* children of divorced parents have problems" (Coontz, 1997, emphases in original). In other words, all children of divorce do not suffer the negative consequences identified by researchers and reported by people such as Whitehead.

Coontz reminds us that although more children in divorced homes do drop out or get pregnant than do children whose parents stay married, "divorce does not account for the majority of such social problems as high school dropout rates and unwed teen motherhood" (Coontz, 1997). Finally, Coontz goes even further in an optimistic direction, noting that there are some measures on which large proportions of children of divorced homes score higher than do average children from homes with two parents. She reports that children of single parents (usually single mothers) spend more time talking with their custodial parent, receive more praise for their academic successes, and face fewer pressures toward conventional gender roles. Thus, she argues, in some ways, single-parent households may be beneficial environments within which to be raised (Coontz, 1997).

As illustrated above, researchers find and commentators can selectively use mixed outcomes in their quest to document the effects of divorce on children. Sociologist Paul Amato (1993) suggests that there are competing theoretical emphases in the research on divorce and children. Contrasting four different perspectives, Amato notes that each differs in its underlying assumptions about when and why divorce hurts children.

1. The *economic hardship* perspective stresses the role of declining economic resources in creating the negative effects of divorce and single-parenthood on children. Because most custodial parents are mothers who have suffered some declining standard of living, they and their children suffer accordingly.
2. The *parental adjustment* perspective emphasizes that what hurts or helps children is the effectiveness of the parenting and supervision they receive after divorce. Because custodial parents are under considerable stress, their effectiveness may diminish, especially during the "crisis" phase of divorce. According to this approach, if parents continue to responsibly and effectively support and supervise their children, children's suffering can be minimized.
3. The *parental loss* perspective emphasizes the essential nature of fathers and mothers. Both parents are portrayed as necessary sources of healthy and successful child development. Furthermore, children come to depend on the presence of both parents in their lives, and they suffer emotionally and socially when one leaves because of divorce. Here, the lack of contact and meaningful involvement with the absent parent undermines children. This is the perspective closest to Whitehead's emphasis.
4. The *parental conflict* perspective suggests that what most hurts children is the anger, conflict, and fighting that goes on either before, during, or after divorce. From this viewpoint children from intact but conflict-ridden households will suffer more than will children from stable single-parent homes whose divorced parents maintain amicable relationships. This is the viewpoint that appears most strongly supported by research (Amato, 1993; Newman, 1999).

These perspectives not only offer different interpretations of what causes children's distress but also lead to different recommendations about how best to help children survive with minimal suffering. In fact, children of divorce may suffer from all of these things. For children, as for adults, divorce is complex and multidimensional.

HOW BAD ARE THE LONG-TERM CONSEQUENCES OF DIVORCE?

The message about the long-term consequences varies according to the research one examines. Very influential longitudinal research conducted by Judith Wallerstein highlights fairly extensive, long-term trauma and distress that stays with and affects children of divorce well into adulthood. Beginning with *Surviving the Breakup: How Children and Parents Cope With Divorce* (Wallerstein and Kelly, 1980), through *Second Chances: Men, Women, and Children a Decade After Divorce* (Wallerstein and Blakeslee, 1989), and culminating with *The Unexpected Legacy of Divorce: A 25-Year Landmark Study* (Wallerstein, Lewis, and Blakeslee, 2000), Wallerstein has followed a sample of (originally) 60 families with 131 children among them, as they divorced and went through the subsequent adjustment processes at 18 months, 5 years, 10 years, 15 years, and ultimately 25 years. Seventy-five percent of the original families, and 71 percent of the 131 children were studied for all three books.

Wallerstein found that at the five-year mark, more than a third of the children were struggling in school, experiencing depression, had difficulty with friendships, and continued to long for a parental reconciliation. At the ten-year follow-up, she indicated that almost half of the children carried lingering problems, and they had become worried, sometimes angry, underachieving young adults. Three-fifths of the children of divorce retained a lingering sense of rejection by one or both parents, and suffered especially poor relationships with their fathers. Finally, at the quarter-century point, Wallerstein asserted that the effects of divorce on children reached their peak in adulthood, where the ability to form and maintain committed intimate relationships was negatively affected (see Amato [2003]).

A more moderate view of the long-term effects of divorce emerges from other studies (for example, see Hetherinton and Kelly [2002] and Amato [2003]). Hetherinton undertook the Virginia Longitudinal Study of Divorce and Remarriage, which initially consisted of following a sample of 144 families with a four-year-old "target child." Half of the sample families were divorced, half were married. Initially they were to be followed and restudied at two years in order to compare how those who divorced fared in comparison to those who did not. Eventually, the sample was expanded, and subsequent research was conducted at 2, 6, 11, and 20 years postdivorce. As the "target children" (i.e., the initial 4-year-olds) married, had a child, or cohabited for more than six months, they were further studied (Hetherington and Kelly, 2002). Meanwhile, families were added to the sample at each wave, to reach a final sample of 450, evenly split between nondivorced, divorced, and remarried families. Throughout the research, a variety of qualitative and quantitative data were collected on personalities of parents and children, adjustment, and relationships within and outside the family (Hetherington, 2003).

The impression that Hetherington's research leaves is much more encouraging than what one gets from Wallerstein's studies. For example, most adults and children adapt to the divorce within two to three years. Although at the one-year mark, 70 percent of the divorced parents were wrestling with animosity, loneliness, persistent attachment, and doubts about the divorce, by six years, most were moving toward building new lives. More than 75 percent of the sample said that the divorce had been a good thing, more than half of the women and 70 percent of the men had remarried, and most had embarked on the postdivorce paths they would continue to take (Hetherington, 2003).

In considering the effects of divorce on children, Hetherington reports that 20 percent of her sample of youths from divorced and remarried families was troubled and displayed a range of problems, including depression and irresponsible, antisocial behavior. They had the highest dropout rate, highest divorce rate (as they themselves married) and were most likely to be struggling economically. But perhaps more important, "80 percent of children from divorced homes eventually are able to adapt to their new life and become reasonably well adjusted" (Hetherington and Kelly, 2002, p. 228). Given that 10 percent of youths from nondivorced homes also were struggling, the difference for children from divorced as opposed to nondivorced homes was fairly small (10 percent).

Looking at the outcomes for the women and men who divorce, Hetherington identified distinctive patterns that divorced men and women displayed at one year and ten years postdivorce. These, too, show that the long-term effects are neither all of one kind, nor do they necessarily stay as they start out. The patterns fall into six categories:

- *Enhancers:* Mostly females, enhancers grew more well adjusted over time. They compared quite favorably against women in unhappy intact marriages, and had moved on to success at work, in remarriages, as parents, and in social relationships. The "enhanced" pattern was more evident as an outcome than as a starting point.
- *Goodenoughs:* Involved in "the average coping with divorce and life's challenges," this group had a combination of strengths and problems. They scored in the middle on items such as health, self-esteem, and antisocial behavior. They were the largest group and equally divided between women and men. By ten years postdivorce, goodenoughs had constructed lives that looked a lot like their predivorce lives.
- *Seekers:* These were the women and men who wanted to remarry or find new partners as quickly as they could. Men sought new partners more actively than women, and they tended to be low in self-esteem, prone to anxiety and depression, and more needy for the validation that would come from a wife. They chose mates more quickly than other divorced people, seeking partners that would be undemanding and would look after their needs without high expectations in return. At one year postdivorce there were many seekers. As people either found new part-

ners or adjusted more to their single life, the numbers of seekers diminished.

- *Swingers:* This group consisted mostly of males, partly because women typically had custody of children, which inhibited their ability to assume this pattern of adjustment. Within two years the swinger group dropped in size. Typically, swingers spent more time in pursuit of other sexual relationships, were more likely to go to bars seeking to meet casual partners, and were more likely to use drugs or alcohol. For some swingers, there appeared to be a short-term "getting it out of your system" approach to the sexual opportunities that they were now free to pursue. But as Hetherington suggests, this wore thin, often within a year postdivorce, as they missed their families. By six years postdivorce, most swingers had moved on to committed relationships.
- *Competent loners:* This was a small, mostly female group. They were healthy, socially skilled, well adjusted, had careers and social lives that were gratifying, a wide range of hobbies, and often went through a series of intimate relationships. They were comfortable being outside of a committed relationship. Hetherington claims that competent loners, like enhancers, "are divorce winners."
- *The defeated:* As the group that most embodies the negative outcome of divorce that tends to be assumed as "the standard outcome" of a divorce, the defeated score high on depression and antisocial behavior, and often struggle with alcoholism, drug abuse, poverty, despair, and legal problems. Although initially there are more men among the defeated than women, over time the gender difference diminishes. Hetherington found that one year after the divorce about a third of the divorced women and men in her sample were among the defeated. Long after the divorce, only 10 percent remained as defeated.

As Hetherington points out, the optimal outcome for adults and their children is to be in a happily married household. But at the same time, her research indicates that we may overstate the risks and fail to recognize the resilience of men, women, and children of divorce.

Paul Amato (2003) suggests that much research supports Wallerstein's claims that divorce is "disruptive and disturbing" in the lives of children, but he fails to find the same strength and pervasiveness of the supposed effects. Using longitudinal data that were gathered as part of the Marital Instability Over the Life Course Study, Amato reports that 90 percent of children with divorced parents achieve the same level of adult well-being as children of "continuously married parents" (Amato, 2003). Amato further suggests that children who experience multiple family transitions (i.e., parental divorce, remarriages, subsequent divorces, and so on) are the ones who most suffer. In fact, he found that children who experienced only a single parental divorce (without any additional parental transitions), were no different in their psychological well-being than children of continuously married parents.

Child Custody

Of all the issues surrounding separation and divorce, custody issues are generating the "greatest attention and controversy" among researchers (Kitson and Morgan, 1991). When the court awards custody to one parent, the decision is generally based on one of two standards: the *best interests of the child* or the *least detrimental of the available alternatives.* In practice, however, custody of the children is awarded to the mother in about 90 percent of the cases. Three reasons can be given for this: (1) women usually prefer custody, and men do not; (2) custody of the mother is traditional; and (3) the law reflects a bias that assumes women are naturally better able to care for children.

Sexual orientation has also been a traditional basis for awarding custody (Baggett, 1992; Beck and Heinzerling, 1993). In the past, a parent's homosexuality per se has been sufficient grounds for denying custody, but increasingly, courts are determining custody on the basis of parenting ability rather than sexual orientation. Interviews with children whose parents are gay or lesbian testify to the children's acceptance of their parents' orientation without negative consequences (Bozett, 1987).

TYPES OF CUSTODY

The major types of custody are sole, joint, and split. In **sole custody,** the child lives with one parent, who has sole responsibility for physically raising the child and making all decisions regarding his or her upbringing. There are two forms of **joint custody:** legal and physical. In **joint legal custody,** the children live primarily with one parent, but both share jointly in decisions about their children's education, religious training, and general upbringing. In **joint physical custody,** the children live with both parents, dividing time more-or-less equally between the two households. Even though joint custody does not necessarily mean that the child's time is evenly divided between parents, it does give children the chance for a more normal and realistic relationship with each parent (Arnetti and Keith, 1993). Under **split custody,** the children are divided between the two parents; the mother usually takes the girls and the father, the boys. Split custody often has harmful effects on sibling bonds and should be entered into only cautiously (Kaplan, Hennon, and Ade-Ridder, 1993).

Parental satisfaction with custody arrangements depends on many factors (Arditti, 1992; Arditti and Allen, 1992). These include how hostile the divorce was, whether the noncustodial parent perceives visitation as lengthy and frequent enough, and how close the noncustodial parent feels to his or her children. In addition, the amount of support payments also affects satisfaction. If parents feel they are paying too much or were "cheated" in the property settlement, they are also likely to feel that the custody arrangements are unfair. Unfortunately, custodial satisfaction is not necessarily related to the best interests of the child.

The anger and conflict surrounding custody arrangements helped give rise to a fathers' rights movement (Coltrane and Hickman, 1992). The fathers' rights movement depicts its participants as caring fathers who want equal treatment regarding child custody, visitation, and support (Bertoia and Drakich, 1993). Given the nature of changing gender roles and the reality of economic hardships, more mothers are relinquishing their children to the fathers. This trend of fathers seeking and gaining custody of their children comes in spite of many judges' traditional attitudes about gender and established child-care patterns. Research concerning the effects of a father's custody on the psychological well-being of children reveals no conclusive evidence to preclude or prefer it (Rosenthal and Keshert, 1980). The chances of a father gaining custody are improved when the children are older at the time of the divorce, the oldest is male, and when the father is the plaintiff in the divorce (Fox and Kelly, 1995). Regardless of who gets custody, however, it is important for children, if possible, to maintain close ties with both parents following a divorce (Howell, Brown, and Eichenberger, 1992).

Sole Custody

Most children continue to live with their mothers after divorce. This occurs for several reasons. First, because women have traditionally been responsible for child rearing, sole custody by mothers has seemed the closest approximation to the traditional family, especially if the father is given free access. Second, many men have not had the day-to-day responsibilities of child rearing and do not feel (or are not perceived to be) competent in that role.

Sole custody does not mean that the noncustodial parent is prohibited from seeing his or her children. Wallerstein and Kelly (1980b) believe that if one parent is prohibited from sharing important aspects of the children's lives, he or she will withdraw from the children in frustration and grief. Children experience such withdrawal as a rejection and suffer as a result. Generally, it is considered in the best interests of the children for them to have easy access to the noncustodial parent. Changes in the noncustodial parent's relationship with his or her children may be related to the difficulties and psychological conflicts arising from visitation and divorce, the noncustodial parent's ability to deal with the limitations of the visiting relationship, and the age and gender of the child (Wallerstein and Kelly, 1980a).

Joint Custody

Joint custody, in which both parents continue to share legal rights and responsibilities, is becoming increasingly common. A number of advantages accrue to this type of arrangement. First, it allows both parents to continue their parenting roles. Second, it avoids a sudden termination of a child's relationship with one of his or her parents. Joint-custody fathers tend to be more involved with their children; they spend time with them and share responsibility and decision making (Bowman and Ahrons, 1985). Third, dividing the labor lessens many of the burdens of constant child care experienced by most single parents.

Joint custody, however, requires considerable energy from the parents in working out both the logistics of the arrangement and their feelings about each other. Many parents with joint custody find it difficult, but they nevertheless feel that it is satisfactory. The children do not always like joint custody as much as the parents do. In actual practice, children relatively rarely split their time evenly between parents (Little, 1992).

Any custody arrangement has both benefits and drawbacks, and joint custody is no exception. Sometimes, although it may be in the best interests of the parents for each of them to continue parenting roles, it may not necessarily be in the best interests of the child. For parents who choose joint custody, it appears to be a satisfactory arrangement. But when joint custody is mandated by the courts over the opposition of one or both parents, it may be problematic. Joint custody may force two parents to interact (*cooperate* is too benign a word) when they would rather never see each other again, and the resulting conflict and ill will may end up being detrimental to the children. Parental hostility may make joint custody the worst form of custody (Opie, 1993).

REFLECTIONS

What form of custody do you believe is the most advantageous to a child? What factors would you consider important in deciding which is the best type of custody for a particular child? If two parents constantly battled over their children, what are some of the consequences you might expect for the children? How do children cope in such circumstances?

NONCUSTODIAL PARENTS

Only recently is research emerging about noncustodial parents. Popular images of noncustodial parents depict them as absent and noncaring, but a more accurate picture depicts varying degrees of involvement (Bray and Depner, 1993; Depner and Bray, 1993). Noncustodial parent involvement exists on a continuum in terms of caregiving, decision making, and parent-child interaction. Involvement also changes depending on whether the custodial family is a single-parent family or a stepfamily (Bray and Berger, 1993).

Researchers still know relatively little about noncustodial and nonresidential fathers (Depner and Bray, 1990). They know even less about noncustodial mothers, who account for about 13 percent of noncustodial parents (Christensen, Dahl, and Rettig, 1990). What we do know about men, however, tells us that they often suffer grievously from the disruption or disappearance of their father roles following divorce. They feel depressed, anxious, and guilt ridden; they feel a lack of self-esteem (Arditti, 1990). The change in status from full-time father to noncustodial parent leaves fathers bewildered about how they are to act; there are no norms for an involved noncustodial parent. Men often act irresponsibly after a divorce, failing to pay child support and possibly becoming infrequent parts of their children's lives. This lack of norms makes it especially difficult if the relationship between the former spouses is bitter. Without adequate norms, fathers may become "Disneyland Dads," who interact with their children only during weekends, when they provide treats such as movies and pizza, or they may become "Disappearing Dads," absenting themselves from any contact at all with their children. For many concerned noncustodial fathers, the question is simple but painful: "How can I be a father if I'm not a father anymore?"

Noncustodial fathers often weigh the costs of continued involvement with their children, such as emotional pain and role confusion, against the benefits, such as emotional bonding (Braver et al., 1993a, 1993b). Those fathers who maintain their connections are generally older and remarried; they have little or no conflict with their ex-spouses and no significant problems with their children (Wall, 1992). For others, however, the costs outweigh the benefits. They are not successful in being noncustodial fathers and abandon the role altogether. A study of noncustodial parents in a support group found that common themes included children rejecting parents and parents rejecting children (Greif and Kristall, 1993).

Children tend to have little contact with the nonresidential parent. A national survey revealed that over 60 percent of fathers did not visit their children or have contact with them over a one-year period (Bianchi, 1990).

The reduced contact between nonresidential fathers and children seems to weaken the bonds of affection. A study of 18- to 22-year-olds whose parents were divorced found that almost two-thirds had

Exploring Diversity

CROSS-CULTURAL ISSUES IN PARENTING: WHO GETS THE CHILDREN?

Foster parent, biological parent, adoptive parent, genetic parent, stepparent: Who is the "real" parent? Who has the greatest rights in relation to a child, and who has the most responsibilities toward him or her? Our answers to these questions are deeply conditioned by our own kinship system, which is a product of our own cultural history. We tend to believe that real parents are established by "nature." We believe that if we can learn a child's genetic or biological makeup, we can identify his or her real parents. We also believe that parent-child emotional bonds follow naturally from the biological tie. A child's relationship to his or her biological or genetic mother—"the mother-child bond"—is believed to be the most nurturant, basic, and (some would say) even mystical parental bond. A biological father's bond with his child seems more distant, though protective, and this is somehow also established by nature.

Of course, these cultural beliefs—as anthropologists argue—leave all foster parents, adoptive parents, and stepparents in a secondary category, with lesser claims to their children. Their parental rights are wrestled out in courtrooms across the country.

Kinship Systems

Perhaps the best way to appreciate the cultural character of our assumptions about parents and their relationships with their children is to look at other societies in which these roles and relationships have other meanings. Anthropologists who study kinship systems tell us that families in other societies are shaped by very different understandings of who is related to whom—and which relationships are most important. Learned in families as children grow up, different kinship systems become the natural way of seeing relationships. In each society, the kinship system comes to exist in people's minds as a kind of blueprint or technical manual that instructs people on the different meanings of genealogical relationships. Even words as basic as *mother* and *father* do not have the same meanings the whole world over.

Anthropologists give kinship systems descriptive names. Some are called *patrilineal* because—as was discussed in Chapter 1—the most important relationships are those traced on the father's side of the family, not the mother's. For example, paternal grandparents in these societies play a greater role in the lives of their grandchildren than do maternal grandparents; they have greater say in their grandchildren's futures and a greater claim on their labor and loyalties. In patrilineal societies, property rights and titles are passed down to children from their male relatives on the father's side of the family. Children inherit membership in their father's clan or lineage. They take their family identity and name from their fathers and paternal grandfathers—who took theirs from their own paternal grandfathers and great-grandfathers. Although Bedouin, Chinese, Masai, and Nuer societies have their own distinctive cultures, all these societies are called patrilineal, which describes their most important kinship relations.

Patrilineal Kinship

The fact that a particular society has patrilineal kinship also gives strong clues about people's ideas on parenting. Cross-culturally, there is great diversity in who counts as the more significant parent—and which other relatives have important roles in the "parenting" of children. Just knowing that a particular society is patrilineal does not tell us everything about parenting. (It does not tell us what work mothers and fathers do in that society, for example.) But it does tell us something basic about a child's identity. At birth, a child is assigned to the father's group—the father's family, descent line, lineage, or clan. This in turn tells us which parent and group has greater rights in the child.

In patrilineal societies, one important repercussion of the primary identity with the father and his group can be seen when divorce occurs: Children remain with their father and his relatives. They may be raised by another "mother," who is often a female member of the father's family (a paternal aunt or grandmother), or they may be "mothered" by another of the father's wives. Children will not leave their father to go with their mother when she returns to live with her own family after divorce. (In some societies, very young children might accompany their mother until after weaning or later; then these children are expected to return to their rightful and natural home with their father.) Thus, in patrilineal societies, children's basic identity is established at birth, and it links them to their father and his group, descent line, family, and relatives. Divorce has serious implications for a woman as mother in patrilineal societies: She surrenders her children to their father, who has greater rights in them. A father in these societies is the more fundamental—and even the more "natural"—parent in the lives of children.

Matrilineal Kinship

In this book we have looked at several patrilineal societies. Let's now turn to consider a matrilineal one. Traditional Iroquois society in nineteenth-century New York has been described by anthropologists as matrilineal. This means that the female descent line was dominant in the kinship system. The mother's side of the family counted more, and her relatives (both male and female) were more significant in the lives of her children. Important aspects of identity—clan membership and chiefly titles—were inherited through the female descent line. Residence after marriage was the reverse of the situation in most patrilineal societies: In Iroquois society, husbands moved to live with the wife in her family's house, the Iroquois longhouse. That house included the wife's own extended family—her sisters and their husbands and children, the wife's mother and her husband, perhaps the wife's mother's sister and her husband, and the wife's maternal grandmother. Thus, in Iroquois society, the husband was the newcomer and the outsider. In his wife's longhouse, he came under the scrutiny of her family. They were concerned that he be a good worker. A husband helped clear land for his wife's family to farm, and he provided them with meat from his hunting ventures. In addition, a husband was expected to father children to continue their mother's descent line. Thus, in Iroquois society, a man's children continued his wife's family, not his own.

Couples divorced when a husband failed in one of these capacities or if the couple could not get along. A husband's belongings set outside the longhouse signaled that the marriage was over. When this happened, it was the husband who moved; he returned to the longhouse belonging to his own mother and sisters. His children remained behind with his wife, to whose family they belonged.

In matrilineal societies, fathers were important in the parenting of their children. However, maternal uncles (the mother's brothers) were often the more significant male relatives. Maternal uncles were members of the same lineage group as their sisters and their children. Whether a sister divorced or remained married to her husband, her brother assumed an active role in raising and overseeing her children, and he passed on family titles to her sons. Thus, children took their primary identity from the mother and her family. The mere fact that a society is matrilineal thus tells us that relatives on the mother's side of the family are more significant in the lives of children. (Other examples of matrilineal societies include the Pueblo Indians of the U.S. Southwest and Trobriand Islanders of the South Pacific.)

Patrilineal/Matrilineal Kinship

In traditional Samoan society, the newly married couple could choose to live in either the husband's or the wife's village, with relatives of either side, in a large extended family household. Each person traced kinship on both the mother's and father's side of the family, through blood relationship, marriage, and adoption. Residence was a matter of choice, and people could choose fairly easily to go and live with other relatives on the other side of the family, in another village. (The head of the household had to agree, and all members helped farm the family fields.) There were no exclusive clans or lineages that tied a person to a particular family, household, or locale. In fact, each person could trace kinship to at least several different households in different villages, and thus had several alternative "homes." People could choose to go and live with any of these relatives.

According to anthropologist Margaret Mead, Samoans changed homes with relative ease. However, it was not just married couples or single adults who enjoyed this flexibility; children did, too. A child could choose to leave the house where his or her parents lived to go and live with an aunt or uncle, even in another village. There, a child called female relatives by the same term used for "mother"; and male relatives were called by the term for "father." These surrogate parents exercised authority over the child, supervising and educating him or her. Hence, children in Samoan society were sometimes raised by multiple sets of parents. This, according to Mead, meant that children could avoid intense conflict with any one set of parents. Diversity in kinship systems in cultures around the world has thus shaped diverse experiences in parenting.

© Pictor Images/ImageState

It is usually important for a child's postdivorce adjustment that he or she have continuing contact with the noncustodial parent. Noncustodial parents are involved with their children in varying degrees.

poor relationships with their fathers, and one-third had poor relationships with their mothers—about twice the rate of a comparable group from nondivorced families (Zill, Morrison, and Coiro, 1993).

Divorced fathers are less likely to consider their children as sources of support in times of need (Amato, 1994; Cooney, 1994). Furstenberg and Nord (1982) conclude that "marital dissolution involves either a complete cessation of contact between the nonresidential parent and child or a relationship that is tantamount to a ritual form of parenthood."

CUSTODY DISPUTES AND CHILD ABDUCTION

As many as one-third of all postdivorce legal cases involve children. Vagueness of the "best interests" and "least detrimental alternative" standards by which parents are awarded custody may encourage custody fights by making the outcome of custody hearings uncertain and increasing hostility. Any derogatory evidence or suspicions, ranging from dirty faces to child abuse, may be considered relevant evidence. As a result, child custody disputes are fairly common in the courts. They are often quite nasty.

As discussed in Chapter 13, about 350,000 children are abducted each year, most by family members in child custody disputes. Most are returned in two days to a week, and generally the parent from whom the child was taken knows the child's whereabouts (Hotaling, Finkelhor, and Sedlak, 1990; Finkelhor, Hotaling, and Sedlak, 1991). Research on the consequences for children of custody abductions is not reliable because much of it relies on parental impressions and on criminal and clinical populations (Greif and Hegar, 1992). According to researcher David Finkelhor (quoted in Barden, 1990), the number of family abductions could be reduced significantly. Under the present system, courts do not respond to people's needs and fears in bitter custody disputes. "For many people, going into court for custody is too risky, too expensive, and too time-consuming," Finkelhor notes, "so they grab the child." Finkelhor suggests that much child stealing could be prevented by the court's assigning a mediator to whom the distressed parents could turn in times of crisis. Mediation can ease parental anxiety and offer an alternative to legal proceedings, which tend to inflame the situation.

Divorce Mediation

The courts are supposed to act in the best interests of the child, but they often victimize children by their emphasis on legal criteria rather than on the children's psychological well-being and emotional development (Schwartz, 1994). There is increasing support for the idea that children are better served by those with psychological training than by those with legal backgrounds (Miller, 1993). Growing concern about the impact of litigation on children's well-being has led to the development of divorce mediation as an alternative to legal proceedings (Walker, 1993).

REFLECTIONS

If you were divorcing, what would be the pros and cons of entering divorce mediation? What would you personally do? Why?

Divorce mediation is the process in which a mediator attempts to assist divorcing couples in resolving personal, legal, and parenting issues in a cooperative manner. Over two-thirds of the states offer or require mediation through the courts over such legal issues as custody and visitation. Mediators act as facilitators to help couples arrive at mutually agreed-upon solutions. Mediators can either be private or court ordered. Mediators generally come from marriage counseling, family therapy, and social work backgrounds, though increasing numbers are coming from other backgrounds and are seeking training in divorce mediation (DeWitt, 1994).

Mediation has many different goals. A primary goal is to encourage divorcing parents to see shared parenting as a viable alternative and to reduce anxiety about shared parenting (Kruk, 1993). Mediators try to help couples develop communication skills to negotiate with each other. They help them clarify their personal, relationship, and parental goals. They suggest ways to minimize conflict. Mediators try to help parents determine whether their demands are based on their anger or on the best interests of their children. They may have parents role-play how their children feel. Mediators can assist parents to develop strategies for helping their children in postdivorce adjustment (Bonney, 1993).

When mediation is court mandated, topics are generally limited to custody and visitation issues. If the mediator is unsuccessful in getting the couple to cooperate, he or she becomes an arbitrator who makes decisions for the couple. The mediator role shifts from facilitator to decision maker. If a couple is unable to negotiate reasonable visitations during the summer, for example, the mediator will decide on a solution for them. As arbitrator, the mediator's decisions are accepted by the court.

Divorcing parents often find mediation helpful for resolving visitation and custody issues. In contrast to court settings, mediation provides an informal setting to work out volatile issues. Men and women both report that mediation is more successful at validating their perceptions and feelings than is litigation. Furthermore, women, the poor, and those from ethnic groups are less likely to experience bias in mediation than in a courtroom setting (Rosenberg, 1992).

Some courts order parents to participate in seminars covering the children's experience of divorce as well as problem solving and building coparent relationships (Petersen and Steinman, 1994). Parents report that these seminars help them become more aware of their children's reactions and give them more options for resolving child-related disputes.

Divorcing parents also report that mediation helps decrease behavioral problems in their children (Slater, Shaw, and Duquesnel, 1992). If parents can work through their differences apart from their children, the children are less likely to react to the anger and fear they might otherwise observe.

A study of fathers a year after divorce found those who mediated more satisfied with custody than those who litigated. They were also more likely to comply with child support (Emery, Matthews, and Kitzmann, 1994).

It is important, however, not to replace unrealistic images of "conflict-ridden postdivorce parenting" with equally unrealistic pictures of "happy-ever-after-postdivorce-parenting-thanks-to-mediation" (Walker, 1993). The stresses and conflicts of divorce are real and painful. Sometimes coparenting cannot work because of the personalities of the divorcing parents. But mediation is an important step forward in involving parents with therapists rather than with lawyers and courts to resolve difficult family matters.

What to Do about Divorce

As the previous pages have illustrated, getting divorced is a painful process for those involved and leaves families and individuals changed forever. Most people will agree that we would be better off reducing the rate of divorce, but how can that goal be achieved? First, we must decide what is the most important cause of the high divorce rates in the United States.

If we believe that divorce rates rose in part because we made it easier and more acceptable to divorce, should we restigmatize divorce? Make getting out of marriage more difficult? If divorce rates rose

Perspective

COVENANT MARRIAGE AS A RESPONSE TO DIVORCE

In 1996, as a way of trying to strengthen marriage and reduce divorce rates, Louisiana became the first state in the United States to establish a two-tiered system of marriage. Marrying couples could choose either a "standard marriage" or a covenant marriage (Hewlett and West, 1998). Following Louisiana's lead, many states have enacted their own covenant marriage legislation. Regardless of the state in question, covenant marriage usually consists of something very close to the following, which is drawn from the Louisiana law:

> We do solemnly declare that marriage is a covenant between a man and a woman who agree to live together as husband and wife for so long as they both may live. We have chosen each other carefully and disclosed to one another everything which could adversely affect the decision to enter into this marriage. We have received premarital counseling on the nature, purposes, and responsibilities of marriage. We have read the Covenant Marriage Act, and we understand that a Covenant Marriage is for life. If we experience marital difficulties, we commit ourselves to take all reasonable efforts to preserve our marriage, including marital counseling.
>
> With full knowledge of what this commitment means, we do hereby declare that our marriage will be bound by Louisiana law on Covenant Marriages and we promise to love, honor, and care for one another as husband and wife for the rest of our lives.

This is supplemented by an affidavit by the parties that they have discussed their intent to designate their marriage as a covenant marriage with a priest, minister, rabbi, clerk of the Religious Society of Friends, any clergyman of any religious sect, or a marriage counselor, which included a discussion of the obligation to seek marital counseling in times of marital difficulties and the exclusive grounds for legally terminating a covenant marriage by divorce or by divorce after a judgment of separation from bed and board.

Minnesota law mirrors almost to the letter Louisiana law:

> We do solemnly declare that our marriage will be a covenant marriage under Minnesota law and we agree to live together as husband and wife as long as we both live. We have chosen each other carefully and told one another everything that could adversely affect the decision to enter into this marriage. We have received premarital counseling on the nature, purposes, and responsibilities of marriage. We have read the Covenant Marriage Act, and we understand that a covenant marriage is for life. If we experience marital difficulties, we will take all reasonable efforts to preserve our marriage, including marital counseling that emphasizes the principles of reconciliation.
>
> A declaration of intent to enter into a covenant marriage must include an affidavit by the parties that they have received premarital counseling from a licensed or ordained minister of any religious denomination, or a person authorized to solemnize marriages by section 517.18, or a person authorized to practice marriage and family therapy under section 148B.33. The counseling must include a discussion of the seriousness of covenant marriage, communication of the fact that a covenant marriage is a commitment for life, and a discussion of the obligation to seek marital counseling in times of marital difficulties. The affidavit must state that the parties were given an informational pamphlet developed by the office of the attorney general that provides a full explanation of the terms and conditions of a covenant marriage. The affidavit must include or have attached to it a signed statement from the counselor confirming that the parties received the counseling required by this paragraph.
>
> The declaration of intent to enter into a covenant marriage must include the notarized signature of both parties or, if one or both of the parties are minors, the notarized written consent of those persons required to consent to or authorize their marriage.

We cannot say whether covenant marriage will "work" to reduce the

along with the increasing economic independence of women, how can we reduce divorce? Do we need to encourage employed women to stay home? How then do their families survive without their incomes (see Chapter 12)? If part of the explanation for rising divorce rates is to be found in the increasing importance given to self-fulfillment and the decline of both familistic self-sacrifice and religious constraints, how can we reduce divorce? Can we change people's values? Finally, if increases in divorce result from the weakening of all but the emotional function of marriage and the reduction, especially, of the family's economic role, can *anything* be done about divorce?

Part of the dilemma has to do with how one perceives divorce. Is divorce the *problem* or is it a *solution* to other problems? Do we want to impose restrictions on divorce that require people to remain in unfulfilling, possibly dangerous relationships? The societal reactions to reducing divorce have been largely prevalence of divorce. In fact, it may have no effect, as the people who elect to enter such a marriage may be the types who would resist and reject divorce and who already perceive marriage as a relationship to keep "till death" they do part.

This certainly seems to be the case based on recent research by Sanchez, Nock, Wright, and Gager (2002). After interviews with three Louisiana focus groups of about a dozen participants each, representing different views on marriage and divorce, the researchers suggest that advocates and opponents of covenant marriage have different perceptions of marriage, marriage reform, divorce, and children's well-being. The six conservative Christian couples they interviewed, married 11 to 56 years, expressed great concern about the vitality and future of marriage. They also saw a dangerous decline of traditional two-parent families, a decline in the value placed on motherhood, a general unwillingness to sacrifice for one's spouse and children, and the emergence of a "culture of divorce." They themselves had converted their marriages to covenant marriages just months before they were interviewed.

The second focus group, a dozen feminist activists (11 females and one male, ages 20 to 50), saw traditional marriage as "inherently patriarchal," and detrimental to women's independence and rights. They also suggested that marriage (from courtship through weddings) is a commercialized competition for men, with "victory" (i.e., marriage) celebrated with indulgent and conspicuous consumption. Finally, they believed that contemporary patterns of marriage imposed unrealistic and gender-stereotyped expectations on women and men. They were strongly suspicious of and against covenant marriage, perceiving such legislation to shift responsibility for social and family problems onto individuals, thus ignoring the importance of external forces or the need for wider social supports.

The third focus group consisted of ten low-income women (nine black, 1 white, all residents of public housing). Two of the ten women were married (18 years and 26 years each), a few were divorced, a few cohabited, some never married. These women were chosen in order to explore issues related to poverty and welfare, and how attitudes about marriage might be affected by or affect their socioeconomic status. This group had more practical and less politically ideological views of marriage. They valued marriage and saw numerous disadvantages faced by unmarried women. They perceived no-fault divorce as a source of a reduced commitment to marital responsibility, allowing people easy opportunities to leave rather than fix marriages. They also felt that divorce and single-parenthood harmed children. Marriage was portrayed as an ideal worth aspiring toward but they also acknowledged the problems of "falling out of love, growing apart, and modern strains on women and men in marriage" (Sanchez, Nock, Wright, and Gager, 2002, p. 103).

On the specific question of *covenant marriage,* the public-housing residents had more of a range of viewpoints. For example, some thought that mandatory counseling provisions were beneficial, others thought they would do little good once one person decided she or he wanted to end their marriage. They applauded the objective of strengthening marriage but also expressed doubt or ambivalence that a "covenant" could make a failed marriage succeed or a law could restore the love necessary to make marriage work.

The values expressed by the three focus groups suggest that in the short run, covenant marriage will appeal to those who already endorse its assumptions about marriage. To those who have concerns about inequalities in traditional marriage or worry about women's rights in families, covenant marriage will be entirely unappealing. To do more than "preach to the choir" —appealing to those who already share the covenant marriage philosophy—will be more difficult for proponents of such reform.

of two kinds: cultural and legal. From a cultural perspective, some commentators bemoan the popular cultural denigration of marriage (Whitehead, 1993, 1997; Popenoe, 1993). They suggest that we "dismantle the divorce culture" we have constructed by more consistently championing and effectively demonstrating the benefits of stable, lifelong marriage. Instead of celebrating "family diversity" and glorifying single-parent households, they believe we should consistently reiterate the idea that marriage is a lifelong commitment involving considerable sacrifice. If that means to "restigmatize divorce," then that is what we should do (Whitehead, 1997).

The other emphasis has been a legal one. Believing that marriage was weakened and divorce increased by no-fault divorce legislation, some have argued that we make divorce *harder to get.* Some states have contemplated repealing no-fault divorce legislation or raising marriage ages. Many states have enacted a two-tiered system of marriage in which couples are allowed and encouraged to consider **covenant marriages**—marriage laws that require couples to undergo premarital counseling, swear to the lifelong commitment of marriage, and promise to divorce only under extraordinary circumstances and only after seeking marriage counseling (see the Perspective on covenant marriages). Too new to yet evaluate, the covenant marriage system has appealed to both those who wish to reduce divorce and those who wish to establish a more traditional, even religious, understanding of marriage commitments.

The difficulty behind both cultural and legal efforts is that in attempting to make divorce harder or less attractive, they do little to make staying married easier. This, too, could be done. It might entail enacting some work-family policy initiatives to ease the stress and strain facing two-earner households. Additionally, eliminating the marriage penalty in the tax code might help create stronger marriages by eliminating the financial penalties associated with marriage (Hewlett and West, 1998). On the subject of financial resources, since we know that divorce hits hardest at lower- and working-class levels, bolstering the economic stability and security of low-income families might also lead to less divorce.

If we can't reduce or eliminate divorce, we should at least do what we can to protect those who go through divorce, especially children (Coontz, 1997; Furstenberg and Cherlin, 1991). We should devote resources that will help custodial parents to parent more effectively. This means, among other things, ensuring their access to quality child care when they are at work, guaranteeing their receipt of financial obligations (such as child support and alimony) from their former spouses, and helping them avoid the devastating plunge into poverty. Additionally, ex-spouses must be instructed in how to display more amicable relationships with each other and should be expected to do so. Since at least some of the effects of divorce are tied to the level of postdivorce conflict and adjustment, taking steps to reduce conflict and ensure more effective adjustment will benefit children and their parents. Early and aggressive intervention into the postdivorce family (such as teaching anger management or instructing fathers about the vital roles they can still play) constitutes such intervention (Coontz, 1997; Furstenberg and Cherlin, 1991).

There is no denying that separation or divorce is filled with pain for everyone involved—husband, wife, and children. But as one family ends, new family forms emerge. These include new relationships and possibilities, new circumstances and responsibilities, and new families with unique relationships: the single-parent or the stepfamily. These are the families that we explore in the next chapter.

SUMMARY

- Divorce is an integral part of the contemporary American marriage system, which values individualism and emotional gratification. Divorce serves as a recycling mechanism, giving people a chance to improve their marital situations by marrying again. The divorce rate increased significantly in the 1960s but leveled off in the early 1990s. About half of all current marriages end in divorce.
- Researchers are increasingly viewing divorce as a normal part of the family life cycle rather than as a form of deviance. Divorce creates the single-parent family, remarriage, and the stepfamily.
- A variety of factors can affect the likelihood of divorce. Societal factors include the changed nature of the family, social integration, and individualistic cultural values. Demographic factors include socioeconomic status, employment status, income, educational level, ethnicity, and religion. Life course factors are intergenerational transmission, age at time of marriage, premarital pregnancy and childbearing, and remarriage. The most important factors may be family processes: marital happiness, presence of children (in some cases), and marital problems.
- Divorce can be viewed as a process involving six "stations" or processes: (1) emotional, (2) legal, (3) economic, (4) co-parental, (5) community, and (6) psychic. As people divorce, they undergo these "divorces" more or less simultaneously. The intensity level of these stages varies at different times.

- Uncoupling is the process by which couples drift apart in predictable stages. Initially, uncoupling is unilateral; the initiator begins to turn elsewhere for satisfactions, creating an identity independent of the couple. The initiator voices more complaints and begins to think of alternatives. Eventually the initiator ends the relationship. Uncoupling ends when both partners acknowledge that the relationship cannot be saved.
- In establishing a new identity, newly separated people go through transition and recovery. Transition begins with the separation and is characterized by *separation distress*, which is usually followed by loneliness. Separation distress is affected by (1) whether the person had any forewarning of the separation, (2) the length of time married, (3) who took the initiative in leaving, (4) whether someone new is found, and (5) available resources. The more personal, social, and financial resources a person has at the time of separation, the easier the separation generally will be.
- Dating is important for separated or divorced people. Their greatest social problem is meeting other unmarried people. Dating is a formal statement of the end of a marriage; it also permits individuals to enhance their self-esteem.
- *No-fault divorce* has revolutionized divorce by eliminating fault finding and the adversarial process and by treating husbands and wives as equals. The most damaging unintended consequence of no-fault divorce is the growing poverty of divorced women with children.
- Women generally experience dramatic downward mobility after divorce. The economic consequences of divorce include the impoverishment of women, changed female employment patterns, and very limited *child support* and *alimony.*
- Children in the divorce process go through three stages: (1) the initial stage, lasting about a year, when turmoil is greatest; (2) the transition stage, lasting up to several years, in which adjustments are being made to new family arrangements, living and economic conditions, friends, and social environment; and (3) the restabilization stage, when the changes have been integrated into the children's lives. Children must undertake six developmental tasks when their parents divorce: (1) acknowledging parental separation, (2) disengaging from parental conflicts, (3) resolving loss, (4) resolving anger and self-blame, (5) accepting the finality of divorce, and (6) achieving realistic expectations for later relationship success.
- A significant factor affecting the responses of children to divorce is their age. Young children tend to act out and blame themselves, whereas adolescents tend to remain aloof and angry at both parents for disrupting their lives. Adolescents may be bothered by their parents' dating again. Many problems assumed to be due to divorce are actually present before marital disruption.
- Factors affecting a child's adjustment to divorce include (1) open discussion prior to divorce, (2) continued involvement with noncustodial parent, (3) lack of hostility between divorced parents, (4) good psychological adjustment to divorce by custodial parent, and (5) stable living situation and good parenting skills. Continued involvement with the children by both parents is important for the children's adjustment.
- Custody is generally based on one of two standards: the best interests of the child or the least detrimental of the available alternatives. The major types of *custody* are *sole, joint,* and *split.* Custody is generally awarded to the mother. Joint custody has become more popular because men are becoming increasingly involved in parenting.
- Noncustodial parent involvement exists on a continuum from absent to intimately and regularly involved. Noncustodial parents often feel deeply grieved about the loss of their normal parenting role. Children tend to have little contact with nonresidential parents.
- As a result of custody disputes, as many as 350,000 children are stolen from custodial parents each year. Most are returned home within a week. *Divorce mediation* is the process in which a mediator attempts to assist divorcing couples in resolving personal, legal, and parenting issues in a cooperative manner. A primary goal of mediation is to encourage divorcing parents to see shared parenting as a viable alternative, to ease parental anxiety, and to reduce custody-related abductions.
- Recent legislative initiatives such as *covenant marriage* are attempts to reduce the divorce rate by strengthening the marriage commitment.

KEY TERMS

alimony 479
child support 479
covenant marriage 496
crude divorce rate 462
divorce mediation 493
glick effect 465
intergenerational transmission 467
joint custody 488
joint legal custody 488
joint physical custody 488
no-fault divorce 469
predictive divorce rate 462
ratio measure of divorce 461
refined divorce rate 462
separation distress 474
social integration 464
sole custody 487
split custody 488

SUGGESTED READINGS

Ahrons, Constance. *The Good Divorce: Raising Your Family Together When Your Marriage Comes Apart.* New York: HarperCollins, 1994. An easy read based on the author's "Binuclear Family Study" of family functions after divorce.

Arendell, Terry. *Men and Divorce.* Thousand Oaks, CA: Sage, 1995. A look at the legal, economic, and social consequences of divorce, based on interviews with 75 men.

Gottman, John M. *What Predicts Divorce? The Relationship between Marital Processes and Marital Outcomes.* Hillsdale, NJ: Lawrence Erlbaum, 1994. An examination by a leading marital scholar of why some marriages fail and others thrive.

Hetherington, E. Mavis, and John Kelly. *For Better or Worse: Divorce Reconsidered.* New York: W. W. Norton, 2002. A thoughtful longitudinal study of the long-term consequences of divorce.

Irving, Howard, and Michael Benjamin. *Family Mediation; Contemporary Issues.* Thousand Oaks, CA: Sage, 1995. A broad-based look at mediation that succinctly addresses such issues as diversity of culture, the scope of feminist thought, and gender issues.

Kayser, Karen. *When Love Dies: The Process of Marital Disaffection.* New York: Guilford Press, 1993. A social psychological description of the gradual process of emotional estrangement from one's partner—which may or may not end in divorce. In-depth interviews add poignancy to the work.

Simons, Ronald. *Understanding Differences Between Divorced and Intact Families: Stress, Interaction, and Child Outcome.* Thousand Oaks, CA: Sage, 1996. An illustration of the special stresses both divorced and intact families suffer and the impact of divorce on children, based on two large-scale studies of Midwest families

Wallerstein, Judith, J. Lewis and Sandra Blakeslee. *The Unexpected Legacy of Divorce: A 25-Year Landmark Study.* New York: Hyperion, 2000. A best-selling book describing the long-term negative impact of divorce on families based on three decades of research by Wallerstein, one of the leading psychologists in the field.

Whitehead, Barbara Dafoe. *The Divorce Culture: Rethinking Our Commitment to Marriage and Family.* New York: Vintage Books, 1998. A strong, critical analysis of the rise in divorce in twentieth-century America and of the culture that supports or condones it.

RESOURCES ON THE INTERNET

Companion Web Site for This Book

http://sociology.wadsworth.com/strong/marriage9e
Gain an even better grasp on this chapter by going to the companion Web site to take one of the Tutorial Quizzes, use the Flash Cards to master key terms, or check out the many other study aids you'll find there. Visit the Marriage and Family Resource Center on the site. You'll also find special features such as GSS Data and Census 2000 information that will put data and resources at your fingertips to help you with that special project or to do some research on your own.

InfoTrac College Edition: Search Word Summary

divorce
child support
no-fault divorce
divorce mediation
custody

To learn more about these central topics in the study of the family, you can conduct an electronic search using InfoTrac College Edition. To aid in your search and to gain useful tips, see the Student Guide to InfoTrac College Edition that you can access through the companion Web site for this book.

Preview

To gain a sense of what you already know about the material covered in this chapter, answer "True" or "False" to the statements below.

1. Researchers are increasingly viewing stepfamilies as normal families. True or false?
2. Divorce does not end families. True or false?
3. Shared parenting tends to be the strongest tie holding former spouses together. True or false?
4. Second marriages are significantly happier than first marriages. True or false?
5. Most stepfamilies feel that they have become true families. True or false?
6. Children tend to have greater power in single-parent families than in traditional nuclear families. True or false?
7. Becoming a stepfamily is a process. True or false?
8. Stepmothers generally experience less stress in stepfamilies than stepfathers because stepmothers are able to fulfill themselves by nurturing their stepchildren. True or false?
9. Researchers are increasingly finding that remarried families and intact nuclear families are similar to each other in many important ways. True or false?
10. People who remarry and those who marry for the first time tend to have similar expectations. True or false?

"Second marriage is the triumph of hope over experience."

SAMUEL JOHNSON

CHAPTER **15**

New Beginnings: Single-Parent Families, Remarriages, and Blended Families

Today's families mark a definitive shift from the traditional family system, based on lifetime marriage and the intact nuclear family, to a pluralistic family system, including families created by divorce, remarriage, and births to single women. This new pluralistic family system consists of three major types of families: (1) intact nuclear families, (2) single-parent families (either never married or formerly married), and (3) stepfamilies. **Single-parent families** are families consisting of one parent and one or more children; the parent can be either divorced, widowed, or never married. **Stepfamilies** are families in which one or both partners have children from a previous marriage or relationship. Stepfamilies are sometimes referred to as *blended families*.

The dominance of the new system is attested to by the following facts (Bumpass, Sweet, and Martin, 1990; Coleman and Ganong, 1991; Coleman, Ganang, and Fine, 2000; Demo and Acock, 1991; Dainton, 1993; Ganong and Coleman, 1994; U.S. Census Bureau, 1996):

- The chances are more than two out of three that an individual will divorce, remarry, or live in a single-parent family or stepfamily as a child or parent sometime during his or her life.
- **Remarriage**—a marriage in which one or both partners have been previously married—is as common as first marriage. Half of all recent marriages involve at least one previously married partner, and one out of ten marriages is a third marriage for one or both partners.
- About half of remarried women give birth to at least one child, typically within two years of remarrying.
- More than one-fourth of all families with children are currently single-parent families. Single-parent families are growing faster in number than any other family form. Approximately half of all children born in the 1990s will live in single-parent families sometime during their childhoods.
- Over 2.3 million households have stepchildren living in them; the number of noncustodial stepfamilies without children living with them is significantly higher. One-sixth of all children are currently members of stepfamilies. Over a third of all children can expect to live with a biological parent and a stepparent at some time during their childhoods.

To better understand this evolving pluralistic family system, in this chapter we examine single-parent families, binuclear families, remarriage, and stepfamilies. Because of this shift to a pluralistic family system, researchers are beginning to reevaluate single-parent families and stepfamilies and to view them as normal rather than deviant family forms (Coleman and Ganong, 1991; Pasley and Ihinger-Tallman, 1987). It is useful to see these families as different structures pursuing the same goals as traditional nuclear families: the provision of intimacy, economic cooperation, the socialization of children, and the assignment of social roles and status.

If we shift our perspective from structure to function, the important question is no longer whether a particular family form is deviant. (If the statistical prevalence of a family form determines deviance, logic tells us that the traditional nuclear family may soon become deviant.) The important question becomes whether a specific family—regardless of whether it is a traditional family, a single-parent family, or a stepfamily—succeeds in performing its functions. In a practical sense, as long as a family is fulfilling its functions, it is a normal family.

PREVIEW ANSWERS

1 True
2 True
3 True
4 False
5 True
6 True
7 True
8 False
9 True
10 False

The dramatic rise in single-parent families and stepfamilies over the last four decades is the result of shifting social values and trends rather than individual shortcomings or pathologies. Single-parent families and stepfamilies have become a natural part of the contemporary American family system. As such, they are not problems in themselves. Instead, to a great extent, many of their problems lie in the stigma attached to them and their lack of support by the larger society (Ahrons and Rodgers, 1987; Gongla and Thompson, 1987). If we are going to strengthen these families, note Constance Ahrons and Roy Rodgers (1987), "We must unambiguously acknowledge and support them as normal, prevalent family types that have resulted from major societal trends and changes."

REFLECTIONS

What effect does it have on your views of single-parent families and stepfamilies to think of them as "normal" families? As "abnormal" or "deviant" families? If you were reared in a single-parent family or stepfamily, did your friends, relatives, schools, and religious groups treat your family as normal? Why?

Single-Parent Families

Throughout the world, single-parent families are increasing in number (Burns and Scott, 1994). In the United States, they are the fastest-growing family form; no other family type has increased in number as rapidly. Yet single-parent families are still all too often treated negatively in the popular imagination. They may be negated as "broken homes" or as headed by "welfare queens" who "breed" children "out of wedlock" only to collect benefits. Occasionally, they are still stereotyped as the product of teenage promiscuity. None of these images is true. The "broken home" image is based on the myth of the "happy" traditional family, the "welfare queen" mythology is based on a mixture of racism and moralism condemning women for bearing children outside of marriage, and the "promiscuous teenage mother" stereotype ignores the reality that more than two-thirds of births to single mothers are to women older than 20. In fact, the teen birth rate has been declining.

Between 1970 and 2002, the percentage of children living in single-parent families more than doubled, increasing from 13 to 28 percent (Fields, 2003). The single-parent family is a more significant departure from the traditional nuclear family than is the dual-worker family or the stepfamily in three important ways. First, both the dual-worker family and the stepfamily are two-parent families, whereas the single-parent family is not. Second, single-parent families are generally headed by women and are thus more vulnerable to poverty. In 2002, 23 percent of children under age 18 were in households headed by their mothers; another 5 percent were in households headed by single fathers (4 percent lived with neither parent). Third, the mother may never have married.

Unmarried adolescent mothers are empowered to build successful families when they have emotional and financial support from their families, educational and employment opportunities, and child care.

In previous generations, the life pattern most women experienced was (1) marriage, (2) motherhood, and (3) widowhood. Single-parent families existed in the past, but they were formed by widowhood rather than divorce or births to unmarried women; significant numbers were headed by men. But a new marriage and family pattern has taken root. Its greatest impact has been on women and their children. Divorce and births to unmarried mothers are key factors creating today's single-parent family.

The life pattern many married women today experience is (1) marriage, (2) motherhood, (3) divorce, (4) single parenting, (5) remarriage, and (6) widowhood. For those who are not married at the time of their child's birth, the pattern may be (1) dating/cohabitation, (2) motherhood, (3) single parenting, (4) marriage, and (5) widowhood.

CHARACTERISTICS OF SINGLE-PARENT FAMILIES

Single-parent families share a number of characteristics, including the following: (1) creation by widowhood, divorce, or births to unmarried women; (2) usually female headed; (3) significance of ethnicity; (4) poverty; (5) diversity of living arrangements; and (6) transitional character. In addition, some single-parent families are created intentionally through planned pregnancy, artificial insemination, and adoption. Others are headed by lesbians and gay men (Miller, 1992). Finally, many single-parent households actually contain two cohabiting adults and are therefore not *single-adult* households (Fields, 2003).

Creation by Divorce or Births to Unmarried Women

Single-parent families today are usually created by marital separation, divorce, or births to unmarried women rather than by widowhood. Throughout the world, including the United States, single-parent families created through births to unmarried women are increasing at a higher rate than are single-parent families created through divorce (Burns and Scott, 1994). In 2002, 34 percent of all births were to unmarried women. The number of children living with an unmarried couple more than tripled between 1980 and 2000 Today, 19 million children under age 18 live in 9.4 million households with either their mothers only or fathers only (Fields, 2003; U.S. Census Bureau, 2002, Table 58).

In comparison to single parenting by widows, single parenting by divorced or never-married mothers receives considerably less social support. A divorced mother usually receives less assistance from her own kin and considerably less (or none) from her former partner's relatives. Widowed mothers, however, often receive social support from their husband's relatives. Our culture is still ambivalent about divorce and tends to consider single-parent families deviant (Kissman and Allen, 1993). It is even less supportive of families formed by never-married mothers. Conservatives have recently returned to earlier forms of stigmatization by characterizing children of never-married women as "illegitimate" and their mothers as "unwed mothers." Eighty-seven percent of single-parent families are headed by women. This has important economic ramifications because of gender discrimination in wages and job opportunities, as discussed in Chapter 12.

Significance of Ethnicity

Ethnicity remains an important demographic factor in single-parent families. In 2002, among white children, 20 percent lived in single-parent families; among African-American children, 53 percent lived in such families; among Hispanics, 30 percent lived in single-parent families, and among Asian and Pacific Islander children, 15 percent lived in a single-parent household (Fields, 2003). White single mothers were more likely to be divorced than their African-American or Latino counterparts, who were more likely to be unmarried at the time of the birth or widowed.

Poverty

Married women usually experience a sharp drop in their income when they separate or divorce (as discussed in Chapter 14). Among unmarried single mothers, poverty and motherhood go hand in hand. Because they are women, because they are often young, and because they are frequently from ethnic minorities, single mothers have few financial resources. They are under constant economic stress in trying to make ends meet (McLanahan and Booth, 1991). They work at low wages, endure welfare, or both. They are unable to plan because of their constant financial uncertainty. They move more frequently than two-parent families as economic and living situations change, uprooting themselves and their children. They accept material support from kin, but often at the price of receiving unsolicited "free advice," especially from their mothers.

DID YOU KNOW?

Among children in divorced single-parent families, 32.4 percent live in poverty (U.S. Bureau of Statistics, 1996).

Diversity of Living Arrangements

There are many different kinds of single-parent households. Children are nearly five times as likely to live with a single mother as with a single father (23 percent to 5 percent). Single-parent families also show great flexibility in managing child care and housing with limited resources. In doing so, they rely on more of a variety of household arrangements than is suggested by the umbrella heading, "single parent household." For example, many young African-American mothers live with their own mothers in a three-generation setting.

Of perhaps more interest is the fact that many "single parent households" actually contain the parent and her or his unmarried partner. In 2002, for example, 11 percent (1.8 million) of the 16.5 million children living with single mothers lived with their mothers and the mothers' unmarried partners. A third (1.1 million) of the 3.3 million children living with an unmarried father actually lived with their fathers and their fathers' unmarried partners (Fields, 2003)

Transitional Form

Single parenting is usually a transitional state. A single mother has strong motivation to marry or remarry because of cultural expectations, economic stress, role overload, and a need for emotional security and intimacy.

Intentional Single-Parent Families

For many, especially single women in their thirties and forties, single parenting has become a more accepted, intentional, and less transitional lifestyle (Seltzer, 2000; Gongla and Thompson, 1987; Miller, 1992). Some older women choose unmarried single parenting because they have not found a suitable partner and are concerned about declining fertility. They may plan their pregnancies or choose donor insemination or adoption. If their pregnancies are unplanned, they decide to bear and rear the child. Others choose single parenting because they do not want their lives and careers encumbered by the compromises necessary in marriage. Still others choose it because they don't want a husband but they do want a child. Sociologist Judith Seltzer offers this explanation:

> Individuals who prefer autonomy or intimate partnerships with greater role symmetry than is common in marriage, may also see childbearing outside of marriage and cohabitation as relatively more beneficial than formalizing these ties by marriage (Seltzer, 2000, p. 1258).

Lesbian and Gay Single Parents

There may be as many as 2.5 million to 3.5 million lesbian and gay single parents. The majority were married before they were aware of their sexual orientation or got married with hopes of "curing" it. They became single parents as a result of divorce. Others were always aware of being lesbian or gay; they chose adoption or donor insemination in order to have children. Said one gay adoptive father, "I always knew I wanted to be a father." A lesbian who was artificially inseminated said, "I started to get this baby hunger. I just needed to have a child" (quoted in Miller, 1992).

CHILDREN IN SINGLE-PARENT FAMILIES

One of the most important and controversial questions surrounding single parenthood is what effects it has on children. Children born outside of marriage tend to suffer economic disadvantages that may then lead to other educational, social, and behavioral outcomes. Their disadvantages tend to be worse than those experienced by children of divorced parents or by children in two-parent, married households (Seltzer, 2000).

The bulk of research on the effects divorced, single-parent households have on children points to some negative outcomes in areas such as behavioral problems, academic performance, social and psychological adjustment, and health. The gaps between children in such households and those whose parents remain continuously married are relatively small but consistent. As Amato (2000) reports, especially when exposed to associated negative life events such as having to move or change schools, the effects of living in a divorced, single-parent home can create particular adjustment difficulties.

Parental Stability and Loneliness

After a divorce, single parents are usually glad to have the children with them. Everything else seems to have fallen apart, but as long as divorced parents have their children, they retain their parental function. Their children's need for them reassures them of their own importance. A mother's success as a parent becomes even more important to counteract the feelings of low self-esteem that result from divorce.

Feeling depressed, the mother knows she must bounce back for the children. Yet after a short period, she comes to realize that her children do not fill the void left by her missing spouse. The children are a chore as well as a pleasure, and she may resent being constantly tied down by their needs. Thus, minor incidents with the children—a child's refusal to eat or a temper tantrum—may get blown out of proportion. A major disappointment for many new single parents is the discovery that they are still lonely. It seems almost paradoxical. How can a person be lonely amid all the noise and bustle that accompany children? However, children do not ordinarily function as attachment figures; they can never be potential partners. Any attempt to make them so is harmful to both parent and child. Yet children remain the central figures in the lives of single parents. This situation leads to a second paradox: Although children do not completely fulfill a person, they rank higher in most single mothers' priorities than anything else.

Changed Family Structure

A single-parent family is not the same as a two-parent family with one parent temporarily absent. The permanent absence of one parent dramatically changes the way in which the parenting adult relates to the children. Generally, the mother becomes closer and more responsive to her children. Her authority role changes, too. A greater distinction between parents and children exists in two-parent homes. Rules are developed by both mothers and fathers. Parents generally have an implicit understanding to back each other up in child-rearing matters and to enforce mutually agreed-on rules. In the single-parent family, no other partner is available to help maintain such agreements; as a result, the children may find themselves in a much more egalitarian situation. Consequently, they have more power to negotiate rules. They can badger a single parent into getting their way about staying up late, watching television, or going out. They can be more stubborn, cry more often and louder, whine, pout, and throw temper tantrums. Any parent who has tried to get children to do something they do not want to do knows how soon an adult can be worn down. So single parents are more willing to compromise: "Okay, you can have a small box of Cocoa Puffs. Put that large one back, and promise you won't fuss like this anymore." In this way, children acquire considerable decision-making power in single-parent homes. They gain it through default: The single parent finds it too difficult to argue with them all the time.

Additional "handicaps" faced by single-parent families include the following:

- With only one adult in the household, if that adult is distressed, overwhelmed, or angry, the tone of the whole house is affected (Coontz, 1997).
- Because they face more intense time pressures, single parents are less able to participate in their children's schooling. They are less likely to meet with teachers, attend extracurricular and school programs, and spend less time monitoring their children's homework (Coontz, 1997).
- Parental depression, especially among custodial mothers, affects their abilities to parent effectively and thus exposes their children to more "adjustment problems" (Amato, 2000).

Children in single-parent homes may also learn more responsibility, spend more time talking with their custodial parent, and face less pressure to conform to more traditional gender roles (Coontz, 1997). They may learn to help with kitchen chores, to clean up their messes, or to be more considerate. In the single-parent setting, the children are encouraged to recognize the work their mother does and the importance of cooperation. One single parent related how her husband had always washed the dishes when they were still living together. At that time it had been difficult to get the children to help around the house, particularly with the dishes. Now, she said, the children always do the dishes—and they do the vacuuming and keep their own rooms straightened up, too (see also Greif [1985]).

Although single parents continue to demonstrate love and creativity in the face of adversity, research on their children reveals some negative long-term consequences. In adolescence and young adulthood, children from single-parent families had fewer years of education and were more likely to drop out of high school. They had lower earnings and were more likely to be poor. They were more likely to initiate sex earlier, become pregnant in their teens, and cohabitate but not marry earlier (Furstenberg and Teitler, 1994). Furthermore, they were more likely to divorce. These conclusions are consistent for whites, African Americans, Latinos, and Asian Americans. The reviewers note that socioeconomic status accounts for some, but not all, of the effects. Some of the effects are attributed to family structure.

Harriette Pipes McAdoo (1988, 1996) traces the cause to poverty not to single parenthood. She notes that African-American families are able to meet their children's needs in a variety of structures. "The major problem arising from female-headed families is poverty," she writes. "The impoverishment of Black families has been more detrimental than the actual structural arrangement" (McAdoo, 1988).

SUCCESSFUL SINGLE PARENTING

Single parenting is difficult, but for many single parents, the problems are manageable. Almost two-thirds of divorced single parents found that single parenting grows easier over time (Richards and Schmeige, 1993). As we discuss single parenting it is important to note that many of the characteristics of successful single parents and their families are shared by all successful families, as we shall see in Chapter 16.

Characteristics of Successful Single Parents

In-depth interviews with successful single parents found certain themes running through their lives (Olsen and Haynes, 1993):

- *Acceptance of responsibilities and challenges of single parenthood:* Successful single parents saw themselves as primarily responsible for their families; they were determined to do the best they could under varying circumstances. Following divorce, they were determined to get on with their lives.
- *Parenting as first priority:* In balancing family and work roles, their parenting role ranked highest. Romantic relationships were balanced with family needs.
- *Consistent, nonpunitive discipline:* Single parents realized that their children's development required discipline. They adopted an authoritative style of discipline that respected their children and helped them develop autonomy. They rejected authoritarian discipline as ineffective and damaging to parent-child relationships.
- *Emphasis on open communication:* They valued and encouraged expression of their children's feelings and ideas. Parents similarly expressed their feelings.
- *Fostering individuality that was supported by the family:* Children were encouraged to develop their own interests and goals; differences were valued by the family.
- *Recognition of the need for self-nurturance:* Single parents realized that they needed time for themselves so they would not be submerged by family responsibilities and roles. They needed to maintain an independent self that they achieved through other activities, such as dating, music, dancing, reading, classes, and trips.
- *Dedication to rituals and traditions:* Single parents maintained or developed family rituals and traditions, such as bedtime stories; family prayer or meditation; sit-down family dinners at least once a week; picnics on Sundays, fireworks on the Fourth of July; a special birthday dinner; visits to Grandma's; or watching television or going for walks together.

Single-Parent Family Strengths

Although most studies emphasize the stress of single parenting, some studies view it as building strength and confidence, especially for women (Amato, 2000; Coontz, 1997). A study of 60 white single mothers and 11 white single fathers (most of whom were divorced) identified five family strengths associated with successful single parenting (Richards and Schmeige, 1993):

1. *Parenting skills:* Successful single parents develop the ability to assume some of the roles and attributes of the absent parent—the ability to take on both expressive and instrumental roles and traits. Single mothers may teach their children household repairs or car maintenance; single fathers may become more expressive and involved in their children's daily lives.
2. *Personal growth:* Developing a positive attitude toward the changes that have taken place in their lives helps single parents, as does feeling success and pride in overcoming obstacles.
3. *Communication:* Through good communication, single parents can develop trust and a sense of honesty with their children, as well as an ability to convey their ideas and feelings clearly to their children and friends.
4. *Family management:* Successful single parents develop the ability to coordinate family, school, and work activities and to schedule meals, appointments, family time, and alone time.
5. *Financial support:* Developing the ability to become financially self-supporting and independent is important to single parents.

Among the single parents in the study, over 60 percent identified parenting skills as one of their family strengths. Forty percent identified family management as a strength in their families (Richards and Schmeige, 1993). About a quarter identified personal growth and communication among their family strengths.

In another study, Jean Miller (1982) concluded the following about single mothers:

> A significant number of the single women studied have solved many extraordinary problems in the face of formidable obstacles. Their single parenthood has led to personal growth for many. In adulthood they have made major revisions in their roles in life and in their self- and object-representations. Many have become contributors to their community, and their children are often a source of strength rather than difficulty.

Barbara Risman's research on custodial single fathers showed their abilities to be attentive, nurturing caregivers to their children (Risman, 1986). Rather than relying on paid help or female social supports, men became the nurturers in their children's lives. They were involved in their personal, social, and academic lives and saw to it that their emotional and physical needs were met. To Risman, they affirmatively answer the question in her title, "Can Men Mother?"

REFLECTIONS

If you are or have been a member of a single-parent family, what were its strengths and problems? What do you know of the strengths and problems of friends and relatives in single-parent families?

Binuclear Families

One of the most complex and ambiguous relationships in contemporary America is what some researchers call the **binuclear family**—a postdivorce family system with children (Ahrons and Rodgers, 1987; Ganong and Coleman, 1994). It is the original nuclear family divided in two. The binuclear family consists of two nuclear families—the maternal nuclear family headed by the mother (the ex-wife) and the paternal one headed by the father (the ex-husband). Both single-parent families and stepfamilies are forms of binuclear families.

Divorce ends a marriage, but not a family. It dissolves the husband-wife relationship but not necessarily the father-mother, mother-child, or father-child relationship. The family reorganizes itself into a binuclear family. In this new family, ex-husbands and ex-wives may continue to relate to each other and to their children, although in substantially altered ways. The significance of the maternal and paternal components of the binuclear family varies. In families with joint physical custody, the maternal and paternal families may be equally important to their children. In single-parent families headed by women, the paternal family component may be minimal.

COMPLEXITY OF BINUCLEAR FAMILIES

As an illustration of the complexity of the binuclear family system, consider the family history of two children, whom we have named Paige and Daniel

Brickman. (See Figure 15.1 for a diagram of their binuclear family.)

When Paige was 6 and Daniel 8, their parents separated and divorced. The children continued to live with their mother, Sophia, in a single-parent household while spending weekends and holidays with their father, David. After a year, David began living with Jane, a single mother who had a 5-year-old daughter, Lisa.

Three years after the divorce, Sophia married John, who had joint physical custody of his two daughters, Sally and Mary, aged 7 and 9. Paige and Daniel continued living with their mother, and their stepfather's children lived with them every other week. After two years of marriage, Sophia and John had a son, Joshua. About the same time, David and Jane split up; they continued to maintain close ties because of the bonds formed between David and Lisa. A year later, David married Julie, who had physical custody of two children, Sally and Gabriel; the next year, they had a son, David, Jr. Lisa visits every few weeks.

Although Paige and Daniel's binuclear family is not at all unusual, don't be surprised if it's hard to figure out who's related to whom. As Ahrons and Rodgers (1987) point out, "The variations in family structure that result from remarriage in binuclear families almost defy categorization." Eight years

FIGURE 15.1 ■ **Diagram of Binuclear Family**

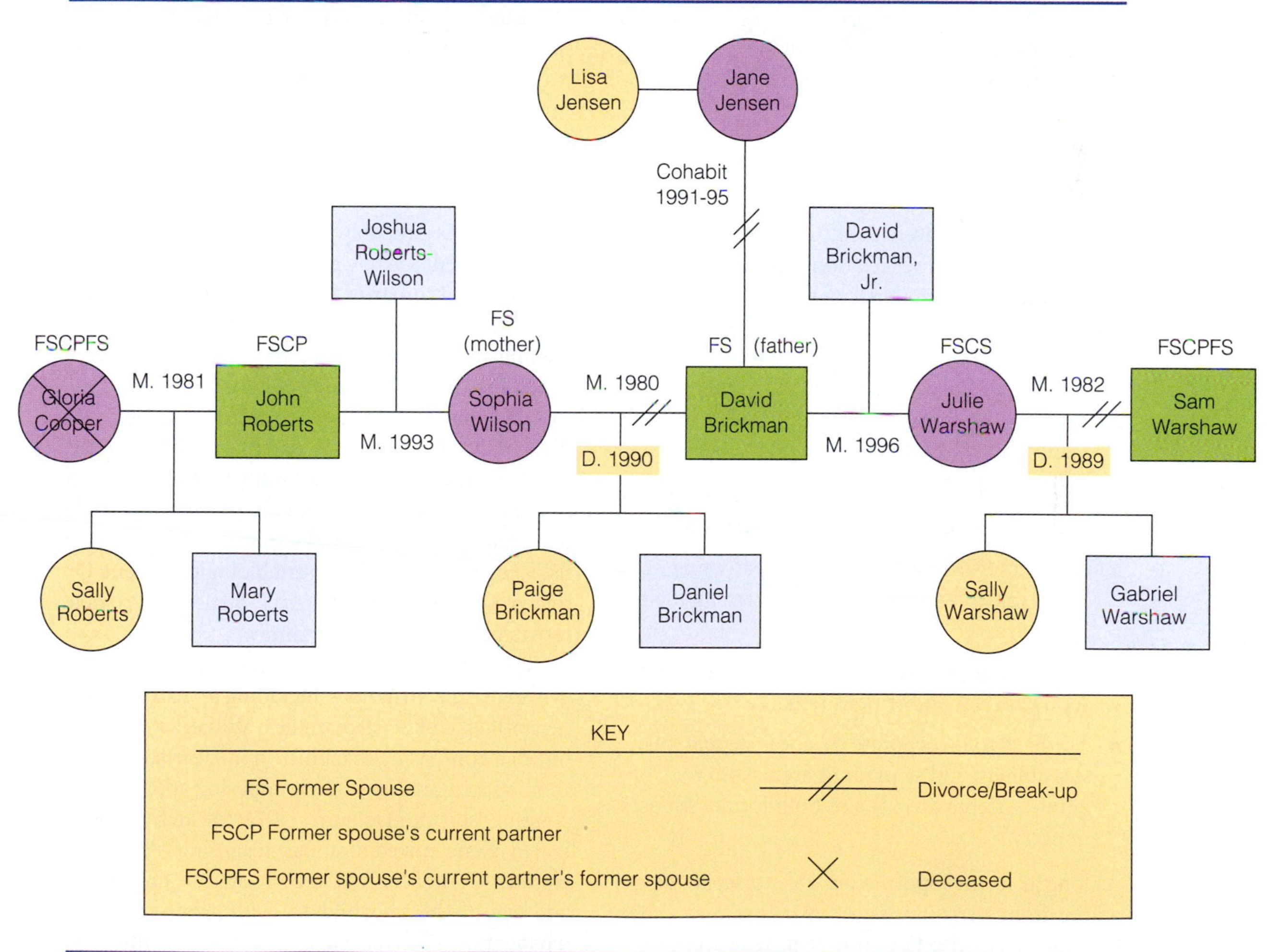

The binuclear family consists of five subsystems: former spouse, remarried couple, parent/child, sibling (biological, step, and half), and mother-stepmother/father-stepfather subsystems. Identify these subsystems in the diagram above.

after their parents divorced, Paige and Daniel's family consists of the following family members: two biological parents, two stepparents, three stepsisters, one stepbrother, and two half-brothers. In addition, they had a "cohabiting" stepmother and stepsister with whom they continued to have close ties. Their extended family includes two sets of biological grandparents and stepgrandparents, along with a large array of biological and step aunts, uncles, and cousins. Paige and Daniel continue to have two households to which they belong as children.

SUBSYSTEMS IN THE BINUCLEAR FAMILY

To clarify the different relationships, researchers Constance Ahrons and Roy Rodgers (1987) divide the binuclear family into five subsystems: (1) former spouse subsystem; (2) remarried couple subsystems; (3) parent-child subsystems; (4) sibling subsystems: stepsiblings and half-siblings; and (5) mother/stepmother-father/stepfather subsystems.

Former Spouse Subsystem

Although divorce severs the husband-wife relationship, the mother-father relationship endures as former spouses continue their parenting responsibilities. The degree of involvement for the noncustodial parent varies, but children generally benefit from the continued involvement of both parents. The former spouses, however, must deal with a number of issues. These include the following:

- Anger and hostility toward each other as a result of their previous marriage and separation
- Conflict regarding child custody, parenting styles, values, and aspirations concerning their children
- Shifting roles and relationships between former spouses when one or both remarry
- The need to incorporate others as stepparents, stepsiblings, and stepgrandparents into the family system when one or both former spouses remarry

As long as former spouses are able to separate parenting from personal issues involving each other, they may form effective co-parenting relationships. Constance Ahrons and Lynn Wallisch (1987) indicated that about half of the former spouses in their study are able to work well with each other; another quarter interact, but with substantial conflict. About a quarter are unable to co-parent because of the high degree of conflict.

Remarried Couple Subsystems

Remarried couples are generally unprepared for the complexities of remarried life. If they have physical custody of children from the first marriage, they must provide access between the children and the noncustodial parents. Both custodial and noncustodial parents must facilitate the exchange of children, money, decision-making power, and time. This may often be difficult. Typical marital issues such as power and intimacy may become magnified because they frequently involve not only the remarried couple but the former spouse as well. Because of custody arrangements, for example, the former spouse may exercise veto power over the remarried couple's plan to take a family vacation because it conflicts with visitation.

Parent-Child Subsystems

Remarriage is probably more difficult for children than for parents. While parents are caught up in the excitement of romance, their children may be reacting with anxiety and distress. Their parents are making choices that affect the children but over which the children themselves have little or no control. Furthermore, remarriage destroys the children's fantasies that their parents will reunite. In addition, children may not want a stepparent to function as a parent for various reasons. The children may feel, for example, that the stepparent is usurping the role of the absent biological parent. Or they may resent the stepparent's intrusion into their restructured single-parent family.

Researchers are still somewhat inconclusive regarding family dynamics, including levels and kinds of conflict within stepfamilies. While some point out that comparing dynamics in nuclear families to those in stepfamilies is like comparing apples to oranges (Olson and DeFrain, 1997), research has generated some mixed findings regarding stepfamily conflict. Conflict seems to be related to the gender and age of the stepchildren, with adolescent girls having more difficult relationships with stepfathers. In general, it appears as though adolescents have the greatest difficulty accepting the authority of their

© Christine Mendes/Buena Vista Photography

The binuclear family is the post-divorce family consisting of two nuclear families, the nuclear family that includes the mother (or ex-wife) and the one that includes the father (or ex-husband). The creation of the binuclear family presents both rewards and challenges for its members.

stepparents (Amato, 2000). Other research has failed to find higher levels of conflict in stepfamilies, and in fact, contrary to researchers' expectations, less frequent conflict has been found (MacDonald and DeMaris, 1995; Coleman, Ganong, and Fine, 2000).

What is certain is that both the biological parent and the stepparent must make adjustments. Former single parents, for example, must adjust to the presence of a second parent in decision-making and child-rearing practices. Stepparents, however, have the greatest adjustment to make. Because remarried families tend to model themselves after traditional nuclear families, stepparents often expect that the stepparent role will be similar to the parent role. Stepfathers who have their own children may feel conflict because they are "more" a father to their stepchildren than to their biological ones. Stepmothers, however, seem to experience greater stress than stepfathers. The stress may result from stepmothers' relatively high degree of involvement in child-care and nurturing activities, which are not adequately acknowledged or appreciated by stepchildren.

Sibling Subsystems: Stepsiblings and Half-Siblings

When parents remarry, their children may acquire "instant" brothers or sisters, who may differ considerably in age and temperament. The sibling relationships may be especially complex in binuclear families. Consider Paige and Daniel's stepsibling and half-sibling relationships. In their mother's family, the two children had to adjust to a half-brother and two stepsisters who lived with them half-time. In their father's family, they adjusted to a "cohabiting" sibling who was "like" a sister. When their father remarried, they gained two stepsiblings and, a little later, a half-brother. To make matters more complex, biological relationships do not guarantee emotional closeness. Paige feels closest, for example, to her "cohabitant" sibling, Lisa, and her stepsister Sally, neither of whom live with her. Daniel, however, feels closest to Mary, his stepsister, and Joshua, his half-brother.

All things being equal, stepsiblings are predisposed to bond with each other because family norms require affection between family members (Pasley, 1987). Bonding will occur most rapidly when siblings are of similar age and sex, have similar experiences and values, are interdependent, and perceive greater rewards than costs in their relationships.

Stepsiblings and siblings contend with one another for parental affection, toys, attention, physical space, and dominance. Sharing a parent with a new stepparent is often difficult enough, but to share the parent with a stepsibling can be overwhelming. Visiting biological children compete with stepchildren who are living with the visiting children's biological parent.

REFLECTIONS

Are you a part of a binuclear family? If so, in what role? Which subsystems are functional or dysfunctional in your binuclear family? How do you imagine that conflict within the former spouse subsystem would affect children in a binuclear family?

Mother/Stepmother-Father/Stepfather Subsystems

The relationship between new spouses and former spouses often influences the remarried family. The former spouse can be an intruder in the new marriage and a source of conflict between the remarried couple. Other times the former spouse is a handy scapegoat for displacing problems. Much of the current spouse–former spouse interaction depends on how the ex-spouses themselves feel about each other.

COURTSHIP

The norms governing courtship prior to first marriage are fairly well defined. As courtship progresses, individuals spend more time together; at the same time, their family and friends limit time and energy demands because "they're in love." Courtship norms for second marriages, however, are not clear (Ganong and Coleman, 1994; Rodgers and Conrad, 1986). For example, when is it acceptable for formerly married (and presumably sexually experienced) men and women to become sexually involved? What type of commitment validates "premarital" sex among postmarital men and women? How long should courtship last before a commitment to marriage is made? Should the couple cohabit? Without clear norms, courtship following divorce can be plagued by uncertainty about what to expect.

Remarriage courtships are short. As noted earlier, almost one-third of divorced individuals marry within a year of their divorces. This may indicate, however, that they knew their future partners before they were divorced. Furthermore, because many cohabit prior to remarriage, the courtship period may be even shorter than marriage dates indicate.

If neither partner has children, courtship for remarriage may resemble courtship before the first marriage, with one major exception: The memory of the earlier marriage exists as a model for the second marriage. Courtship may trigger old fears, regrets, habits of relating, wounds, or doubts. At the same time, having experienced the day-to-day living of marriage, the partners may have more realistic expectations. Their courtship may be complicated if one or both are noncustodial parents. In that event, visiting children present an additional element.

Cohabitation

Increases in the rates of cohabitation in the United States include many divorced women and men who cohabit prior to or instead of remarrying. Thus, although remarriage rates have declined some in recent years, "recoupling" via cohabitation remains common (Coleman, Ganong, and Fine, 2000).

Larry Ganong and Marilyn Coleman (1994) describe cohabitation as "the primary way people prepare for remarriage." In fact, cohabitation is a major difference between first-time marriages and remarriages. Although 15 to 25 percent of people marrying for the first time may be living together at the time of marriage, the majority of those remarrying are cohabiting (Coleman and Ganong, 1991; Ganong and Coleman, 1994). This larger percentage may reflect the desire to test compatibility in a "trial marriage" to prevent later marital regrets (Buunk and van Driel, 1989). At the same time, however, couples who lived together before remarriage did not discuss stepfamily issues any more than did those who did not cohabit (Ganong and Coleman, 1994).

We know very little about cohabitation prior to remarriage. A study of single mothers who cohabited, however, found that some of them partially cohabited prior to marriage (Montgomery et al., 1992); that is, the potential partner spent several days and nights a week in the mother's home before moving in on a full-time basis. Presumably, this gave both partners the chance to gauge the man's "fit" with his future stepfamily before marriage.

DID YOU KNOW?

Recent research has found that having children in the home has a strong positive effect on economic distress and a strong negative effect on income (Shapiro, 1996).

Courtship and Children

Courtship before remarriage differs considerably from that preceding a first marriage if one or both members in the dating relationship are custodial parents. Single parents are not often a part of the singles world because such participation requires leisure and money, which single parents generally lack. Children rapidly consume both of these resources.

Although single parents may wish to find a new partner, their children usually remain the central figures in their lives. This creates a number of new problems. First, the single parent's decision to go out at night may lead to guilt feelings about the children. If a single mother works and her children are in day care, for example, should she go out in the evening or stay at home with them? Second, a single parent must look at a potential partner as a potential parent as well. A person may be a good companion and listener and be fun to be with, but if he or she does not want to assume parental responsibilities, the relationship will often stagnate or be broken off. A single parent's new companion may be interested in assuming parental responsibilities, but the children may regard him or her as an intruder and try to sabotage the new relationship.

A single parent may also have to decide whether to permit a lover to spend the night when children are in the home. This is often an important symbolic act. For one thing, it brings the children into the parent's new relationship. If the couple have no commitment, the parent may fear the consequences of the children's emotional involvement with the lover; if the couple break up, the children may be adversely affected. Single parents are often hesitant to expose their children again to the distress of separation; the memory of the initial parental separation and divorce is often still painful. In addition, having a lover spend the night reveals to the children that their parent is a sexual being. This may make some single parents feel uncomfortable and may also make the parent vulnerable to moral judgment by his or her children. It may also raise questions about sex outside marriage for adolescents or younger children. Single parents often fear that their children will lose respect for them under such circumstances. Sometimes children do judge their parents harshly, especially their mothers. Parents are often deeply disturbed at being condemned by children who do not understand their need for love, companionship, and sexual intimacy. Finally, having someone sleep over may trigger the resentment and anger that the children feel toward their parents for splitting up. They may view the lover as a replacement for the absent parent and feel deeply threatened.

Remarriage

The eighteenth-century writer Samuel Johnson described remarriage as "the triumph of hope over experience." Americans are a hopeful people. Almost one-third remarry within a year of their divorces (Ganong and Coleman, 1994). About half of all divorced women remarry within five years of the dissolution of their marriage.

Many newly divorced men and women express great wariness about marrying again; yet at the same time they are actively searching for mates. Women often view their divorced time as important for their development as individuals, whereas men, who often complain that they were pressured into marriage before they were ready, become restless as "born-again bachelors" (Furstenberg, 1980).

REMARRIAGE RATES

Roughly half of all marriages in the United States are marriages in which at least one partner has been previously married (Coleman, Ganong, and Fine, 2000). Twenty percent remarry other divorced men and women, and approximately 22 percent marry never-married individuals (U.S. Census Bureau, 1996).

Remarriage is common among divorced persons, especially men, who have higher remarriage rates than women (Coleman, Ganong, and Fine, 2000). In recent years, the rate has slightly declined. The decline may be partly the result of the desire on the part of divorced men and women to avoid the legal responsibilities accompanying marriage. Instead of remarrying, many are choosing to cohabit. Still, more than half (54 percent) of divorced women remarry within five years, and 75 percent within ten years (Bramlett and Mosher, 2002).

Remarriage is more likely among younger women—women 25 or younger at the time of divorce—and among white divorced women. Eighty percent of women who were 25 or younger at the time of their first marriage ending remarry within ten years, compared to 68 percent of women older than 25 at the breakup of their marriage. Black women are less likely than white or Hispanic women to enter into a remarriage. Within five years after a divorce, approximately a third of black women, 44 percent of Hispanic women, and nearly 60 percent of white women had entered a remarriage (Bramlett and Mosher, 2002).

In addition to age and ethnicity, variables such as education or the presence of children may also affect remarriage rates. Education works differently for women's and men's likelihood of remarriage. Education raises a man's likelihood of remarriage but reduces a woman's (Coleman, Ganong, and Fine, 2000). Children lower the probability of remarriage for both women and men, but especially for women (Coleman, Ganong, and Fine, 2000).

Gender

Although a majority of both men and women remarry, there are a number of reasons that more men than women remarry. First, divorced women tend to be older than never-married women. Given the tendency for men to marry women younger than themselves and the fact that older women are seen as less attractive and therefore less desirable as spouses, women face more competition and possess fewer "resources" to bring to a remarriage. They are also more likely to have custody of children, which can reduce both the ease with which they socialize or date and their appeal as potential spouses.

Age

Divorced women and men marry when they are about 10 to 12 years older than those in first marriages. The average age for women remarrying is 34; for men, it is 37. A man's or woman's age at the time of separation is the greatest individual factor affecting remarriage. As with the "marriage market" in first marriages, as a divorced person ages, the number of potential partners declines; at the same time, one's own "marketability" decreases with increasing age. For women, the highest remarriage rate takes place in the twenties; it declines by a quarter in the thirties and by two-thirds in the forties. Because the pool of eligible partners is smaller for remarriage, men and women may be willing to "settle for less." They may choose someone they would not have chosen when they were younger (Ganong and Coleman, 1994).

DID YOU KNOW?

Research has concluded that remarriage indeed offers enhanced psychological well-being (Shapiro, 1996).

Presence of Children

Although children from earlier marriages traditionally have been thought to decrease the likelihood of remarriage, the evidence is mixed. A review of the literature found that the presence of children decreases the rate of remarriage by about one-quarter (Bumpass, Sweet, and Martin, 1990). The effects are most marked when a woman has three or more children. Most of the research, however, is 15 to 20 years old, and the increased incidence of single-parent families and stepfamilies may have decreased some of the negative impact of children. In fact, whereas researchers generally speculate that children are a "cost" in remarriage, others point out that some men may regard children as a "benefit" in the form of a ready-made family (Ganong and Coleman, 1994). Some research suggests that the stepparent with no biological children experiences the most negative effect (MacDonald and DeMaris, 1995).

REFLECTIONS

If you were seeking a marital partner, would you consider a previously married person? Why or why not? Would it make a difference if he or she already had children?

CHARACTERISTICS OF REMARRIAGE

Remarriage is different from first marriage in a number of ways. First, the new partners get to know each other during a time of significant changes in life relationships, confusion, guilt, stress, and mixed feelings about the past (Keshet, 1980). They have great hope that they will not repeat past mistakes, but there is usually some fear that the hurts of the previous marriage will recur (McGoldrick and Carter, 1989). The past is still part of the present. A Talmudic scholar once commented, "When a divorced man marries a divorced woman, four go to bed."

Remarriages occur later than first marriages. People are at different stages in their life cycles and may have different goals. A woman who already has had children may enter a second marriage with strong career goals. In her first marriage, raising children may have been more important (see Teachman and Heckert, 1985).

Divorced people have different expectations of their new marriages. Considering that divorce has been seen as psychologically detrimental, remarriage has often been regarded as the pathway to well-being. In a study of second marriages in Pennsylvania, Frank Furstenberg (1980) discovered that three-fourths of the couples had a different conception of love than couples in their first marriages. Two-thirds thought they were less likely to stay in an unhappy marriage; they had already survived one divorce and knew they could make it through another. Four out of five believed their ideas of marriage had changed. One woman (quoted in Furstenberg, 1980) explained remarriage this way:

> I think second marriages are less idealistic and a little more realistic. You realize that it's going to be tough sometimes but you also know that you have to work them out. You come into a second marriage with a whole new set of responsibilities. It's like coming into a ball game with the bases loaded. You've got to come through with a hit. Likewise, there's too much riding on the relationship; you've got to make it work and you realize it more after you've been divorced before. You just have to keep working out the rules of the game.

Finally, the majority of remarriages create stepfamilies. A single-parent family is generally a transition family leading to a stepfamily, which has its own unique structure, satisfactions, and problems.

MARITAL SATISFACTION AND STABILITY

According to various studies, remarried people are about as satisfied or happy in their second marriages as they were in their first marriages. As in first marriages, marital satisfaction appears to decline with the passage of time (Coleman and Ganong, 1991). Yet despite the fact that marital happiness and satisfaction are more or less the same in first and second marriages, remarried couples are more likely to divorce. In fact, as Coleman, Ganong, and Fine (2000) note, "serial remarriages are increasingly common." The divorce proneness of remarriages seems to lessen and become more like that of first marriages as people age. People who enter remarriage after turning 40 may face a lower divorce likelihood than that found among first marriages (Coleman, Ganong and Fine, 2000).

The vulnerability of remarriage to divorce is especially real if children from a prior relationship are in the home (Booth and Edwards, 1992). How do we account for this paradox? Researchers have suggested several reasons for the higher divorce rate in remarriage. (See Ganong and Coleman [1994] for a discussion of various models explaining the greater fragility of remarriage.)

First, persons who remarry after divorce often have a somewhat different outlook on marital stability and are more likely to use divorce as a way of resolving an unhappy marriage (Booth and Edwards, 1992). Furstenberg and Spanier (1987) note that they were continually struck by the willingness of remarried individuals to dissolve unhappy marriages: "Regardless of how unattractive they thought this eventuality, the great majority indicated that after having endured a first marriage to the breaking point they were unwilling to be miserable again simply for the sake of preserving the union."

Second, despite its prevalence, remarriage remains an "incomplete institution" (Cherlin, 1981). Society has not evolved norms, customs, and traditions to guide couples in their second marriages. There are no rules, for example, defining a stepfather's responsibility to a child: Is he a friend, a father, a sort of uncle, or what? Nor are there rules establishing the relationship between an individual's former spouse and his or her present partner: Are they friends, acquaintances, rivals, or strangers? Remarriages don't receive the same family and kin support as do first marriages (Goldenberg and Goldenberg, 1994).

Third, remarriages are subject to stresses that are not present in first marriages. Perhaps the most important stress is stepparenting. Children can make the formation of the husband-wife relationship more difficult because they compete for their parents' love, energy, and attention. In such families,

time together alone becomes a precious and all-too-rare commodity. Furthermore, although children have little influence in selecting their parent's new husband or wife, they have immense power in "de-selecting" them. Children have "incredible power" in maintaining or destroying a marriage, observe Marilyn Ihinger-Tallman and Kay Pasley (1987):

> Children can create divisiveness between spouses and siblings by acting in ways that accentuate differences between them. Children have the power to set parent against stepparent, siblings against parents, and stepsiblings against siblings.

Ganong and Coleman (1994) note that the presence of stepchildren is a major contributor to the higher divorce rate in stepfamilies compared with remarried families with no children.

Blended Families

Remarriages that include children are very different from those that do not. These **blended families,** that emerge from remarriage with children are traditionally known as stepfamilies. They are also sometimes called reconstituted, restructured, or remarried families by social scientists—names that emphasize their structural differences from other families. Satirist Art Buchwald, however, calls them "tangled families." He may be close to the truth in some cases. Nevertheless, there soon may be more stepfamilies in America than any other family form (Pill, 1990). If we care about families, we need to understand and support stepfamilies.

A DIFFERENT KIND OF FAMILY

When we enter a stepfamily, many of us expect to recreate a family identical to an intact family. The intact nuclear family becomes the model against which we judge our successes and failures. But researchers believe that blended families are significantly different from intact families (Ganong and Coleman, 1994; Papernow, 1993; Pill, 1990). If we try to make our feelings and relationships in a stepfamily identical to those of an intact family, we are bound to fail. But if we recognize that the stepfamily works differently and provides different satisfactions and challenges, we can appreciate the richness it brings us and have a successful stepfamily.

Structural Differences

Six structural characteristics make the stepfamily different from the traditional first-marriage family (Visher and Visher, 1979, 1991). Each one is laden with potential difficulties.

1. *Almost all the members in a stepfamily have lost an important primary relationship.* The children may mourn the loss of their parent or parents, and the spouses may mourn the loss of their former mates. Anger and hostility may be displaced onto the new stepparent.
2. *One biological parent typically lives outside the current family.* In stepfamilies that form after divorce, the absent former spouse may either support or interfere with the new family. Power struggles may occur between the absent parent and the custodial parent, and there may be jealousy between the absent parent and the stepparent.
3. *The relationship between a parent and his or her children predates the relationship between the new partners.* Children have often spent considerable time in a single-parent family structure. They have formed close and different bonds with the parent. A new husband or wife may seem to be an interloper in the children's special relationship with the parent. A new stepparent may find that he or she must compete with the children for the parent's attention. The stepparent may even be excluded from the parent-child system.
4. *Stepparent roles are ill defined.* No one knows quite what he or she is supposed to do as a stepparent. Some are reluctant to assume an active parenting role, some attempt to assume such a role too quickly. Children may resist the efforts made by stepparents to become involved in their lives. Most stepparents try role after role until they find one that fits.
5. *Many children in stepfamilies are also members of a noncustodial parent's household.* Each home may have differing rules and expectations. When conflict arises, children may try to play one household against the other. Furthermore, as Visher and Visher (1979) observe:

   > The lack of clear role definition, the conflict of loyalties that such children experience, the emotional reaction to the altered family pattern, and the loss of closeness with their parent who is now married to another person create inner turmoil and confused and unpredictable outward behavior in many children.

6. *Children in stepfamilies have at least one extra pair of grandparents.* Children get a new set of stepgrandparents, but the role these new grandparents are to play is usually not clear. A study by Spanier and Furstenberg (1980) found that stepgrandparents were usually quick to accept their "instant" grandchildren.

Numerous researchers have found that children in stepfamilies exhibit about the same number of adjustment problems as children in single-parent families and more problems than children in original, two-parent families (Furstenberg and Cherlin, 1991; McLenahan and Sandefor, 1994; Coleman, Ganong, and Fine, 2000).

As we examine the stepfamily further, we should keep in mind that when Ganong and Coleman (1984) reviewed the empirical literature comparing stepfamilies with traditional nuclear families, they found no significant differences in relationships with stepfathers as compared to fathers, in perceptions of parental happiness, in degree of family conflict, and in positive family relationships. Only the clinical literature upholds the image of stepfamilies beset by conflict and traumas (Ganong and Coleman, 1986). In fact, much research suggests that a "substantial proportion" of stable, long-term stepfamilies function in similar ways as first-marriage families (Coleman, Ganong, and Fine, 2000).

Although there is diversity among stepfamily forms, stepfamilies may well be very similar to traditional nuclear families. We perceive them as being different because we use a "deficit model" in examining them (Ganong and Coleman, 1994). Thus, we look for deficiencies and exaggerate problems or differences.

THE DEVELOPMENTAL STAGES OF STEPFAMILIES

Individuals and families become a stepfamily through a process—through a series of developmental stages. Each person—the biological parent, the stepparent, and the stepchild (or children)—experiences the process differently. For family members, it involves seven stages, according to a study of stepfamilies by Patricia Papernow (1993). The early stages are fantasy, immersion, and awareness; the middle stages are mobilization and action; and the later stages are contact and resolution.

Early Stages: Fantasy, Immersion, and Awareness

The early stages in becoming a stepfamily include the courtship and early period of remarriage, when each individual has his or her fantasy of their new family. It is a time when the adults (and sometimes the children) hope for an "instant" nuclear family that will fulfill their dreams of how families should be. They have not yet realized that stepfamilies are different from nuclear families.

FANTASY STAGE. During the fantasy stage, biological parents hope that the new partner will be a better spouse and parent than the previous partner. They want their children to be loved, adored, and cared for by their new partners. They expect their children to love the new parent as much as they do.

New stepparents fantasize that they will be loving parents who are accepted and loved by their new stepchildren. They believe that they can ease the load of the new spouse, who may have been a single parent for years. One stepmother recalled her fantasy: "I would meet the children and they would gradually get to know me and think I was wonderful. . . . I just knew they would love me to pieces. I mean, how could they not!" (quoted in Papernow, 1993). Of course, they did not.

The children, meanwhile, may have quite different fantasies. They may still feel the loss of their original families. Their fantasies are often that their parents will get back together. Others fear they may "lose" their parent to an interloper, the new stepparent. Some fear that their new family may "fail" again. Still others are concerned about upheavals in their lives, such as moving, going to new schools, and so on.

IMMERSION STAGE. The immersion stage is the "sink-or-swim" stage in a stepfamily. Reality replaces fantasy. "We thought we would just add the kids to this wonderful relationship we'd developed. Instead we spent three years in a sort of Cold War over them," recalled one stepparent (quoted in Papernow, 1993). Once the partners marry, everyday living collides with fantasies as family members begin to interact with one another. As one stepparent said, "The trouble with life is that it's daily" (Visher and Visher, 1991). The fantasy of recreating a traditional family gives way to the reality of creating a stepfamily. For adults, the challenge is to keep swimming through disappointment and doubt, to find out what will work.

For children, a man's transformation from "Mom's date" to stepfather may be the equivalent of the transformation from Dr. Jekyll to Mr. Hyde. Suddenly an outsider becomes an insider—with authority, as described by one 12-year-old (whose new stepmother also had children): "In the beginning it's fun. Then you realize that your whole life is going to change. Everything changes. We just had fun before, my Dad and me. And now there's all these new people and new rules" (quoted in Papernow, 1993). When children have a new stepparent, they may feel disloyal to their absent biological parent if they show affection. (Biological parents can make a difference: They can let their children know it's okay to love a stepparent.)

Mother's Day or Father's Day can become a stepchild's worst nightmare. A nine-year-old said: "On Mother's Day I didn't know what to do. . . . If I went with my stepmother, my mother would be furious. If I went with my mother, my stepmother would be upset. I couldn't even think about it. It's the worst situation." As the parent and stepparent get more involved with each other, children may feel new losses: The stepparent has replaced them as the parent's number one interest.

AWARENESS STAGE. The awareness stage in stepfamily development is reached when family members "map" the territory. They become aware that they are in new lands; they try to figure out where they are. They gather data about their new family members; they try to understand them—what they like and don't like about them. This stage involves individual and joint family tasks. The individual task is for each member to identify and name the feelings he or she experiences in being in the new stepfamily. A key feeling for stepparents to acknowledge is feeling like an outsider. They need to become aware of feelings of aloneness; they must discover their own needs; and they must set some distance between themselves and their stepchildren. They need to understand why their stepchildren are not warmly welcoming them, as they had expected.

Biological parents need to become aware of unresolved feelings from their earlier marriages and from being single parents. They may feel pulled from the multiple demands of their children and their partners. Biological parents may feel resentment toward their children, their partners, or both. They need to let go of residual feelings of creating the perfect family. They need to understand what it's like for their partners "to create an intimate relationship in the presence of a parent-child relationship that often feels more like ex-lovers" (Papernow, 1993).

Children in the awareness stage often feel "bumped" from their close relationship with the single parent. They miss cuddling in bed in the morning, the bedtime story, the wholehearted attention. When a new stepparent moves in, their feelings of loss over their parents' divorce are often rekindled. They resent the new stepparent for taking the place of their "real" mother or father. Loyalty issues resurface. If they are not pressured into feeling "wonderful" about their new family, however, they can slowly learn to appreciate the benefits of an added parent and friend who will play with them or take them places.

Middle Stages: Mobilization and Action

In the middle stages of stepfamily development, family members are more clear about their feelings and relationships with one another. They have given up many of their fantasies. They understand more of their own needs. They have mapped the new territory. The family, however, remains biologically oriented. Parent-child relationships are central. In this stage, changes involve the emotional structure of the family as a whole.

MOBILIZATION STAGE. In the mobilization stage, family members recognize differences. Conflict becomes more open. Members mobilize around their unmet needs. A stepmother described this change: "I started realizing that I'm different than Jim [her husband] is, and I'm going to be a different person than he is. I spent years trying to be just like him and be sweet and always gentle with his daughter. But I'm not always that way. I think I made a decision that what I was seeing was right" (quoted in Papernow, 1993). The challenge in this stage is to resolve differences while building the stepfamily's sense of family.

Stepparents begin to take a stand. They stop trying to be the ideal parent. They no longer are satisfied with being outsiders. Instead, they want their needs met. They begin to make demands on their stepchildren: to pick up their clothes, be polite, do the dishes. Similarly, they make demands on their partners to be consulted; they often take positions regarding their partners' former spouses. Because stepparents make their presence known in this stage, the family begins to change. The family begins to integrate the stepparent into its functioning. In doing so, the stepparent ceases being an outsider and the family increasingly becomes a real stepfamily. Some stepparents, however, become frustrated in their attempts to become an insider. They may metaphorically "move out" and withdraw from the family.

For biological parents, the mobilization stage can be frightening. The stepparents' desire for change leaves biological parents torn. Biological parents feel they must protect their own children and yet satisfy the needs of their partners. One father described the fighting between his wife and daughter (quoted in Papernow, 1993):

> My daughter is sulking upstairs, crying that nobody loves her. My wife is crying in the bedroom. If I comfort my wife, my daughter will accuse me of abandoning her. If I talk to my daughter, I'll catch hell from Gina for "giving in" to my daughter. What do I do?

At this stage, some parents step out of the middle. They no longer act as buffers between stepparents and children. The stepparent and child are allowed to establish their own relationship. Other parents begin to negotiate family roles with their partners, trying to determine what will work.

Children often attempt to resolve loyalty issues at this stage. They have been tugged and pulled in opposite directions by angry parents too long. Often the adults paid no attention to them. Finally, the children have had enough and can articulate their feelings. After hearing her parents squabble one time too many, one girl reflected: "I thought, this stinks. It's horrible. After the 50 millionth time I said, 'That's your problem. Talk to each other about it,' and they didn't do it again" (quoted in Papernow, 1993).

ACTION STAGE. In the action stage, the family begins to take major steps in reorganizing itself as a stepfamily. It creates new norms and family rituals. Although members have different feelings and needs, they begin to accept each other. The family finds a middle ground. They compromise: Some holidays will be spent with the nonresidential parent; bedrooms do not have to be immaculate; adults can play Harry Connick Jr., and kids can play alternative rock music. Most important, stepfamily members develop shared, realistic expectations and act on them.

Stepcouples begin to develop their own relationship independent from children. They also begin working together as a parental team. Stepparents begin to take on disciplinary and decision-making roles; they are supported by the biological parents. Stepparents begin to develop relationships with their stepchildren independent of the biological parents. Stepparent-stepchild bonds are strengthened.

Later Stages: Contact and Resolution

The later stages in stepfamily development involve solidifying the stepfamily. Much of the hard work has been accomplished in the middle stages.

CONTACT STAGE. In the contact stage, stepfamily members make intimate contact with each other. Their relationships become genuine. They communicate with a sense of ease and intimacy. The couple relationship becomes a sanctuary from everyday family life. The stepparent becomes an "intimate outsider" with whom stepchildren can talk about things "too hot" for their biological parents, such as sex, drugs, their feelings about the divorce, and religion.

For the stepparent, a clear role finally emerges—what is now called the **stepparent role.** The role is an individually created one because, as we saw earlier, it is undefined in our society. The role varies, however, from stepparent to stepparent and from stepfamily to stepfamily. It is mutually suitable to both the individual and the different family members.

RESOLUTION STAGE. The stepfamily is solid in its resolution stage. It no longer requires the close attention and work of the middle stages. Family members feel that earlier issues have been resolved. As one stepfather said (quoted in Papernow, 1993):

> I can feel that we've moved. Not easily, because it's been a pain in the ass. But I feel clear that our family works. . . . It's been proved over the years that we could do it, and we're doing it. We're happy for the most part. There's a lot of love.

Stepchildren can feel the benefits of having an "outsider" inside the family. Not all relationships in stepfamily are necessarily the same; they may differ according to the personalities of each individual. Some of the relationships develop more closely than others. But in any case, there is a sense of acceptance. The stepfamily has made it and has benefited from the effort.

It takes most stepfamilies about seven years to complete the developmental process. Some may complete it in four, and others take many, many years. Some only go through a few of the stages and get stuck. Others split up with divorce. But many are successful. Becoming a stepfamily is a slow process that moves in small ways to transform strangers into family members.

REFLECTIONS

If you are a member of a stepfamily, what were your experiences at the different stages? If you are not, ask friends or relatives who are members what their experiences were at the different stages. If you were to become a stepparent, how would you handle each stage?

PROBLEMS OF WOMEN AND MEN IN STEPFAMILIES

Most people go into stepfamily relationships expecting to recreate the traditional nuclear family: they are full of love, hope, and energy. Perhaps the hardest adjustment they have to make is realizing that stepfamilies are different from traditional nuclear families—and that being different does not make stepfamilies inferior. A nuclear family is neither morally superior to the stepfamily nor a guarantor of happiness.

Women in Stepfamilies

Stepmothers tend to experience more problematic family relationships than do stepfathers (Santrock and Sitterle, 1987; Kurdeck and Fine, 1993; Hetherington and Stanley-Hagan, 1999). To various degrees, women enter stepfamilies with certain feelings and hopes. Stepmothers generally expect to do the following (Visher and Visher, 1979, 1991):

- Make up to the children for the divorce or provide children whose mothers have died a maternal figure.

- Create a happy, close-knit family and a new nuclear family.
- Keep everyone happy.
- Prove that they are not wicked stepmothers.
- Love the stepchild instantly and as much as their own biological children.
- Receive instant love from their stepchildren.

Needless to say, most women are disappointed. Expectations of total love, happiness, and the like would be unrealistic in any kind of family, be it a traditional family or a stepfamily. The warmer a woman is to her stepchildren, the more hostile they may become to her because they feel she is trying to replace their "real" mother. If a stepmother tries to meet everyone's needs—especially her stepchildren's, which are often contradictory, excessive, and distancing—she is likely to exhaust herself emotionally and physically. It takes time for her and her children to become emotionally integrated as a family.

One of the things that makes stepmothering more difficult than stepfathering is the role women typically play in child rearing. Women are expected to and expect to become nurturing, primary caregivers. Stepmothers, more than stepfathers, involve themselves in caregiving and disciplining their stepchildren. This is particularly true if she plays a more active domestic role than her husband. Consequently, there are more opportunities for her to encounter stress and experience conflict with her stepchildren, and thus poorer relationships with her stepchildren may occur. This may be exacerbated by stepchildren, who tend to view relationships with stepmothers as more stressful than relationships with stepfathers. If their biological mothers are still living, they may feel as though their stepmothers threaten their relationships with their birth mothers (Hetherington and Stanley-Hagan, 1999).

Stepmothers married to men who have their children full-time often experience greater problems than stepmothers whose children are with them part-time or occasionally (Furstenberg and Nord, 1985). In part, it may be because children whose fathers have full-time custody may be more difficult for a number of reasons. Bitter custody fights may leave children emotionally troubled and hostile to stepmothers, whom they perceive as "forcibly" replacing their mothers. In other instances, children (especially adolescents) may have moved from their mother's home to their father's because their mother could no longer handle them. In either case, the stepmother may be required to parent children who have special needs or problems. Stepmothers may find these relationships especially difficult. Typically, stepmother-stepdaughter relationships are the most problematic (Clingempeel et al., 1984). Relationships become even more difficult when the stepmothers never intended to become full-time stepparents.

Men in Stepfamilies

Different expectations are placed on men in stepfamilies. Because men are generally less involved in child rearing, they usually have few "cruel stepparent" myths to counter. Nevertheless, men entering stepparenting roles may find certain areas particularly difficult at first (Visher and Visher, 1991). A critical factor in a man's stepparenting is whether he has children of his own. If he does, they are more likely to live with his ex-wife. In this case, the stepfather may experience guilt and confusion in his stepparenting because he feels he should be parenting his own children. When his children visit, he may try to be "Superdad," spending all his time with them and taking them to special places. His wife and stepchildren may feel excluded and angry.

A stepfather usually joins an already established single-parent family. He may find himself having to squeeze into it. The longer a single-parent family has been functioning, the more difficult it usually is to reorganize it. The children may resent his "interfering" with their relationship with their mother (Wallerstein and Kelly, 1980). His ways of handling the children may be different from his wife's, resulting in conflict.

Working out rules of family behavior is often the area in which a stepfamily encounters its first real difficulties. Although the mother usually wants help with discipline, she often feels protective if the stepfather's style is different from hers. To allow a stepparent to discipline a child requires trust from the biological parent and a willingness to let go. Disciplining often elicits a child's testing response: "You're not my real father. I don't have to do what you tell me." Homes are more positive when parents include children in decision making and are supportive (Barber and Lyons, 1994). Nevertheless, disciplining establishes legitimacy, because only a parent or parent figure is expected to discipline in our culture. Disciplining, however, may be the first step toward family integration, because it establishes the stepparent's presence and authority in the family.

In comparison to birth parents, stepfathers tend to have more limited and less positive relationships with their stepchildren. They communicate less, display less warmth and affection, and are less involved. Some research also indicates that among divorced, noncustodial fathers, remarriage and stepfathering may lead to development of closer relationships with stepchildren than with their biological children

The new stepfather's expectations are important. Though the motivations to stepparent are often quite different from those of biological parents, research from the 1987–1988 National Survey of Families and Households shows that 55 percent of stepfathers found it somewhat or definitely true that having stepchildren was just as satisfying as having their own children (Sweet, Bumpass, and Call, 1988). In spite of this, stepparents tend to view themselves as less effective than natural fathers view themselves (Beer, 1992). The complex role that the stepfather brings to his family often creates role ambiguity and confusion that takes time to work out.

CONFLICT IN STEPFAMILIES

Achieving family solidarity in the stepfamily is a complex task. When a new parent enters the former single-parent family, the family system is thrown off balance. Where equilibrium once existed, there is now disequilibrium. A period of tension and conflict usually marks the entry of new people into the family system. Questions arise about them: Who are they? What are their rights and their limits? Rules change. The mother may have relied on television as a baby-sitter, for example, permitting the children unrestricted viewing in the afternoon. The new stepfather, however, may want to limit the children's afternoon viewing, and this creates tension. To the children, everything seemed fine until this stepfather came along. He has disrupted their old pattern. Chaos and confusion will be the norm until a new pattern is established, but it takes time for people to adjust to new roles, demands, limits, and rules.

Conflict takes place in all families: traditional nuclear families, single-parent families, and stepfamilies. If some family members do not like each other, they will bicker, argue, tease, and fight. Sometimes they have no better reason for disruptive behavior than that they are bored or frustrated and want to take it out on someone. These are fundamentally personal conflicts. Other conflicts are about definite issues: dating, use of the car, manners, television, or friends, for example. These conflicts can be between partners, between parents and children, or among the children themselves. Certain types of stepfamily conflicts, however, are of a frequency, intensity, or nature that distinguishes them from conflicts in traditional nuclear families. Recent research on how conflict affects children in stepfather households found that parental conflict does not account for children's lower level of well-being (Hanson, McLanahan, and Thompson, 1996). These conflicts are about favoritism; divided loyalties; discipline; and money, goods, and services.

Favoritism

Favoritism exists in families of first marriages, as well as in stepfamilies. In stepfamilies, however, the favoritism often takes a very different form. Whereas a parent may favor a child in a biological family on the basis of age, sex, or personality, in stepfamilies favoritism tends to run along kinship lines. A child is favored by one or the other parent because he or she is the parent's biological child, or if a new child is born to the remarried couple, they may favor him or her as a child of their joint love. In American culture, where parents are expected to treat children equally, favoritism based on kinship seems particularly unfair.

Divided Loyalties

"How can you stand that lousy, low-down, sneaky, nasty mother (or father) of yours?" asks (or, more accurately, demands) a hostile parent. It is one of the most painful questions children can confront, for it forces them to take sides against someone they love. One study (Lutz, 1983) found that about half of the adolescents studied confronted situations in which one divorced parent talked negatively about the other. Almost half of the adolescents felt themselves "caught in the middle." Three-quarters found such talk stressful.

Divided loyalties put children in no-win situations, forcing them not only to choose between parents but to reject new stepparents. Children feel disloyal to one parent for loving the other parent or stepparent. But divided loyalties, like favoritism, can exist in traditional nuclear families as well. This is

Understanding Yourself

PARENTAL IMAGES: BIOLOGICAL PARENTS VERSUS STEPPARENTS

Parental Images Survey	Negative						Positive
1 Hateful/affectionate	1	2	3	4	5	6	7
2 Bad/good	1	2	3	4	5	6	7
3 Unfair/fair	1	2	3	4	5	6	7
4 Cruel/kind	1	2	3	4	5	6	7
5 Unloving/loving	1	2	3	4	5	6	7
6 Strict/not strict	1	2	3	4	5	6	7
7 Disagreeable/agreeable	1	2	3	4	5	6	7
8 Rude/friendly	1	2	3	4	5	6	7
9 Unlikable/likable	1	2	3	4	5	6	7

We seem to hold various images or stereotypes of parenting adults, depending on whether they are biological parents or stepparents. Such images affect how we feel about families and stepparents (Coleman and Ganong, 1987). The following instrument (modeled after one devised by Ganong and Coleman [1983]) will help give you a sense of how you perceive parents and stepparents.

The instrument consists of nine dimensions of feelings presented in a bipolar fashion—that is, as opposites, such as hateful/affectionate, bad/good, and so on. You can respond to these feelings on a 7-point scale, with 1 representing the negative pole and 7 representing the positive pole. For example, say you were using this instrument to determine your perceptions about aardvarks. You might feel that aardvarks are quite affectionate, so you would give them a 7 on the hateful/affectionate dimension. But you might also feel that aardvarks are not very fair, so you would rank them 2 on the unfair/fair continuum.

To use this instrument, take four separate sheets of paper. On the first sheet, write Stepmother; on the second, Stepfather; on the third, Biological Mother; and on the fourth, Biological Father. On each sheet, write the numbers 1 to 9 in a column, with each number representing a dimension. Number 1 would represent hateful/affectionate, and so on. Then, using the 7-point scale on each sheet, score your general impressions about biological parents and stepparents.

After you've completed these ratings, compare your responses for stepmother, stepfather, biological mother, and biological father. Do you find differences? If so, how do you account for them?

especially true of conflict-ridden families in which warring parents seek their children as allies.

REFLECTIONS

Think about conflicts involving favoritism, loyalty, discipline, and the distribution of resources. Do you experience them in your family of orientation? If so, how are they similar to, or different from, stepfamily conflicts? If you are in a stepfamily, do you experience them in your current family? How are these conflicts similar or different in your original family versus your current family? If you are a parent or stepparent, how are these issues played out in your current family?

Discipline

Researchers generally agree that discipline issues are among the most important causes of conflict among remarried families (Ihinger-Tallman and Pasley, 1987). Discipline is especially difficult to deal with if the child is not the person's biological child. Disciplining a stepchild often gives rise to conflicting feelings within the stepparent. Stepparents may feel that they are overreacting to the child's behavior, that their feelings are out of control, and that they are being censured by the child's biological parent. Compensating for fears of unfairness, the stepparent may become overly tolerant.

The specific discipline problems vary from family to family, but a common problem is interference by the biological parent with the stepparent (Mills, 1984). The biological parent may feel resentful or overreact to the stepparent's disciplining if he or she has been reluctant to give the stepparent authority. As one biological mother who believed she had a good remarriage stated (quoted in Ihinger-Tallman and Pasley, 1987):

> Sometimes I feel he is too harsh in disciplining, or he doesn't have the patience to explain why he is punishing and to carry through in a calm manner, which causes me to have to step into the matter (which I probably shouldn't do). . . . I do realize that it was probably hard for my husband to enter marriage and the responsibility of a family instantly . . . but this has remained a problem.

As a result of interference, the biological parent implies that the stepparent is wrong and undermines his or her status in the family. Over time, the stepparent may decrease his or her involvement in the family as a parent figure.

Money, Goods, and Services

Problems of allocating money, goods, and services exist in all families, but they can be especially difficult in stepfamilies. In first marriages, husbands and wives form an economic unit in which one or both may produce income for the family; husband and wife are interdependent. Following divorce, the binuclear family consists of two economic units: the custodial family and the noncustodial family. Both must provide separate housing, which dramatically increases their basic expenses. Despite their separation, the two households may nevertheless continue to be extremely interdependent. The mother in the custodial single-parent family, for example, probably has reduced income. She may be employed but still dependent on child support payments or TANF (Temporary Assistance for Needy Families). She may have to rely more extensively on child care, which may drain her resources dramatically. The father in the noncustodial family may make child support payments or contribute to medical or school expenses, which depletes his income. Both households have to deal with financial instability. Custodial parents can't count on always receiving their child support payments, which makes it difficult to undertake financial planning.

When one or both of the former partners remarry, their financial situation may be altered significantly. Upon remarriage, the mother receives less income from her former partner or lower welfare benefits. Instead, her new partner becomes an important contributor to the family income. At this point, a major problem in stepfamilies arises. What responsibility does the stepfather have in supporting his stepchildren? Should he or the biological father provide financial support? Because there are no norms, each family must work out its own solution.

Stepfamilies typically have resolved the problem of distributing their economic resources by using a *one-pot* or *two-pot pattern* (Fishman, 1983). In the one-pot pattern, families pool their resources and distribute them according to need rather than biological relationship. It doesn't matter whether the child is a biological child or a stepchild. One-pot families typically have relatively limited resources and consistently fail to receive child support from the noncustodial biological parent. By sharing their resources, one-pot families increase the likelihood of family cohesion.

In two-pot families, resources are distributed by biological relationship; need is secondary. These families tend to have a higher income, and one or both parents have former spouses who regularly contribute to the support of their biological children. Expenses relating to children are generally handled separately; usually there are no shared checking or savings accounts. Two-pot families maintain strong bonds between members of the first family. For these families, a major problem is achieving cohesion in the stepfamily while maintaining separate checking accounts.

Just as economic resources need to be redistributed following divorce and remarriage, so do goods and services (not to mention affection). Whereas a two-bedroom home or apartment may have provided plenty of space for a single-parent family with two children, a stepfamily with additional residing or visiting stepsiblings can experience instant overcrowding. Rooms, bicycles, and toys, for example, need to be shared; larger quarters may have to be found. Time becomes a precious commodity for harried parents and stepparents in a stepfamily. When visiting stepchildren arrive, duties are doubled. Stepchildren compete with parents and other children for time and affection.

It may appear that remarried families are confronted with many difficulties, but traditional nuclear families also face financial, loyalty, and discipline problems. We need to put these problems in perspective. (After all, half of all current marriages end in divorce, which suggests that first marriages are not problem free.) When all is said and done, the problems that remarried families face may not be any more overwhelming than those faced by traditional nuclear families (Ihinger-Tallman and Pasley, 1987).

STEPFAMILY STRENGTHS

Because we have traditionally viewed stepfamilies as deviant, we have often ignored their strengths. Instead, we have seen only their problems. Let us end this chapter, then, by focusing on the strengths of stepfamilies.

Family Functioning

Although traditional nuclear families may be structurally less complicated than stepfamilies, stepfamilies are nevertheless able to fulfill traditional family functions. A binuclear single-parent, custodial, or noncustodial family may provide more companionship, love, and security than the particular traditional nuclear family it replaces. If the nuclear family was ravaged by conflict, violence, sexual abuse, or alcoholism, for example, the single-parent family or stepfamily that replaces it may be considerably better, and because children now see happy parents, they have positive role models of marriage partners (Rutter, 1994). Second families may not have as much emotional closeness as first families, but they generally experience less trauma and crisis (Ihinger-Tallman and Pasley, 1987).

New partners may have greater objectivity regarding old problems or relationships. Opportunity presents itself for flexibility and patience. As family boundaries expand, individuals grow and adapt to new personalities and ways of being. In addition, new partners are sometimes able to intervene between former spouses to resolve longstanding disagreements, such as custody or child-care arrangements.

Impact on Children

Stepfamilies potentially offer children a number of benefits that can compensate for the negative consequences of divorce.

- Children gain multiple role models from which to choose. Instead of having only one mother or father after whom to model themselves, children may have two mothers or fathers: the biological parents and the stepparents.
- Children gain greater flexibility. They may be introduced to new ideas, different values, or alternative politics. For example, biological parents may be unable to encourage certain interests, such as music or model airplanes, because they lack training or interest; a stepparent may play the piano or be a die-hard modeler. In such cases, that stepparent can assist his or her stepchildren in pursuing their development. In addition, children often have alternative living arrangements that enlarge their perspectives.
- Stepparents may act as a sounding board for their children's concerns. They may be a source of support or information in areas in which the biological parents feel unknowledgeable or uncomfortable.
- Children may gain additional siblings, either as stepsiblings or half-siblings, and consequently gain more experience in interacting, cooperating, and learning to settle disputes among peers.
- Children gain an additional extended kin network, which may become at least as important and loving as their original kin network.
- A child's economic situation is often improved, especially if a single mother remarried.
- Children may gain parents who are happily married. Most research indicates that children are significantly better adjusted in happily remarried families than in conflict-ridden nuclear families.

It is clear that the American family is no longer what it was through most of the last century. The rise of the single-parent family and stepfamily, however, does not imply an end to the nuclear family. Rather, these forms provide different paths that contemporary families take as they strive to fulfill the hopes, needs, and desires of their members, and they are becoming as American as Beaver Cleaver's family and apple pie.

SUMMARY

- Single parenting is an increasingly significant family form in the United States. Single-parent families tend to be created by divorce or births to unmarried women, are generally headed by women, are predominantly African American or Latino, are usually poor, involve a wide variety of household types, and are usually a transitional stage.
- Relations between the parent and his or her children change after divorce: The single parent generally tends to be emotionally closer but to have less authority. Successful single parents have similar themes running through their lives: (1) acceptance of responsibilities and challenges of single parenthood; (2) parenting as first priority; (3) consistent, nonpunitive discipline; (4) emphasis on open communication; (5) fostering individuality; (6) recognition of the need for self-nurturance; and (7) dedication to rituals and traditions. Family strengths associated with successful single parenting include (1) parenting skills, (2) personal growth, (3) communication, (4) family management, and (5) financial support.
- The *binuclear family* is a postdivorce family system with children. It consists of two nuclear families: the mother-headed family and the father-headed family. Both single-parent families and stepfamilies are forms of binuclear families. The binuclear family consists of five subsystems: former spouse, remarried couple, parent-child, sibling, and mother/stepmother-father/stepfather subsystems.
- Courtship for second marriage does not have clear norms. If children are not involved from an earlier marriage, the only major difference between first and second marriage courtships is that the first marriage exists as a model in the second courtship. Courtship is complicated by the presence of children because remarriage involves the formation of a stepfamily. In addition, dating poses unique problems for single parents: They may feel guilty about going out, they must look at potential partners as potential parents, and they must deal with their children's judgments or hostility.
- Remarriage differs from first marriage in several ways: Partners get to know each other in the midst of major changes, they remarry later in life, they have different marital expectations, and their marriage often creates a stepfamily. Marital happiness appears to be about the same in first and second marriages. Remarried couples are more likely to divorce than couples in their first marriages. This may be accounted for either by their willingness to use divorce as a means of resolving an unhappy marriage or because remarriage is an "incomplete institution." Stresses accompanying stepfamily formation may also be a contributing factor.
- The stepfamily or blended family differs from the original family because (1) almost all members have lost an important primary relationship, (2) one biological parent lives outside the current family, (3) the relationship between a parent and his or her children predates the new marital relationship, (4) *stepparent roles* are ill defined, (5) children often are also members of the noncustodial parent's household, and (6) children have at least one extra pair of grandparents.
- *Stepfamilies* are families in which one or both partners have one or more children from a previous marriage or relationship. Traditionally, researchers have viewed stepfamilies from a "deficit" perspective. As a result, they have assumed that stepfamilies are very different from traditional nuclear families. More recently, the trend has been to view stepfamilies as normal families; researchers have found few significant differences in levels of satisfaction and functioning between stepfamilies and traditional nuclear families.
- Becoming a stepfamily is a process— a series of developmental stages. Each person—the biological parent, the stepparent, and the stepchild (or children)—experiences the process differently. For family members, it involves seven stages. The early stages are fantasy, immersion, and awareness; the middle stages are mobilization and action; the later stages are contact and resolution.
- Stepmothers tend to experience greater stress in stepfamilies than do stepfathers. In part this may be because families with stepmothers are more likely to have been subject to custody disputes or to include children with a troubled family history. They also may assume the role of the disciplinarian. Stepfathers tend not to be as involved in child rearing as stepmothers. Both often experience difficulty in being integrated into the family.

- A key issue for stepfamilies is family solidarity—the feeling of oneness with the family. Conflict in stepfamilies is often over favoritism; divided loyalties; discipline; and money, goods, and services.
- Stepfamily strengths may include improved family functioning and reduced conflict between former spouses. Children may gain multiple role models, more flexibility, concerned stepparents, additional siblings, additional kin, improved economic situation, and happily married parents.

KEY TERMS

binuclear family 508
blended family 516
remarriage 502
single-parent family 502
stepfamily 502
stepparent role 520

SUGGESTED READINGS

Bloomfield, Harold, and Robert B. Kory. *Making Peace in Your Stepfamily: Surviving and Thriving as Parents and Stepparents.* New York: Hyperion, 1993. A self-help book offering exercises, case studies, and techniques to help stepfamilies with conflicts and problems.

Burns, Alisa, and Cath Scott. *Mother-Headed Families and Why They Have Increased.* Hillsdale, NJ: Lawrence Erlbaum, 1994. An exploration of the worldwide rise of single-parent families and a discussion of the various cultural, economic, and social/psychological theories used to explain it.

Dickerson, Bette J., ed. *African-American Single Mothers: Understanding Their Lives and Families.* Newbury Park, CA: Sage, 1994. A collection of essays presenting an inside view, including the role of children, grandparents, religious values, government, and media stereotypes.

Gangon, Lawrence, and Marilyn Coleman. *Remarried Family Relationships.* Newbury Park, CA: Sage, 1994. A concise description of current scholarship on remarriage and stepfamilies.

Kelley, Patricia. *Developing Healthy Stepfamilies; Twenty Families Tell Their Stories.* Binghamton, NY: Hayworth Press, 1995. A look at what adults and children in stepfamilies say about such issues as discipline, money, family roles, relationships with ex-spouses, and the development of new traditions and rituals.

Papernow, Patricia. *Becoming a Stepfamily.* San Francisco: Jossey-Bass, 1993. An important book examining the normal development of stepfamilies and stepfamily relationships using a psychodynamic perspective.

Pasley, Kay, and Marilyn Ihinger-Tallman. *Remarriage and Stepparenting: Issues in Theory, Research, and Practice,* 2nd ed. Westport, CT: Greenwood Press, 1994. A superb collection of essays by leading researchers in the field.

RESOURCES ON THE INTERNET

Companion Web Site for This Book

http://sociology.wadsworth.com/strong/marriage9e
Gain an even better grasp on this chapter by going to the companion Web site to take one of the Tutorial Quizzes, use the Flash Cards to master key terms, or check out the many other study aids you'll find there. Visit the Marriage and Family Resource Center on the site. You'll also find special features such as GSS Data and Census 2000 information that will put data and resources at your fingertips to help you with that special project or to do some research on your own.

InfoTrac College Edition: Search Word Summary

single-parent families
family structures
stepfamilies
remarriage
courtship
kidnapping (parental)

To learn more about these central topics in the study of the family, you can conduct an electronic search using InfoTrac College Edition. To aid in your search and to gain useful tips, see the Student Guide to InfoTrac College Edition that you can access through the companion Web site for this book.

Preview

To gain a sense of what you already know about the material covered in this chapter, answer "True" or "False" to the statements below.

1. Researchers generally agree on the basic components of healthy families. True or false?
2. Among psychologically healthy families, family emotional health remains constant over the family life cycle. True or false?
3. A common way of measuring family strength is to see how a family responds to crisis. True or false?
4. The happiness of a couple before marriage is a good indicator of their happiness after they have children. True or false?
5. Although having a perfect family is difficult, it is possible. True or false?
6. Good communication is recognized by the overwhelming majority of researchers as the most important quality for family strength. True or false?
7. Children need to be taught respect and responsibility by their families. True or false?
8. Taking time to play is crucial to family health. True or false?
9. Seeking outside help for problems is a sign of family health. True or false?
10. Blood ties are more important than feelings in determining the strengths of our relationships with kin. True or false?

"It was an unspoken pleasure, that having come together so many years, ruined so much and repaired a little, we had endured."

LILLIAN HELLMAN

CHAPTER 16

Marriage and Family Strengths and Needs

Outline

This book has covered much ground. We have considered the history of families in the United States, the diversity of family forms, and the ways in which family life provides the bulwark for the stability and creativity of a society. We have also considered some of the pressures and problems facing twenty-first-century families and seen how those pressures and problems arise from both the internal and external environment. As a result, they require both individual and societal response and support. Finally, at times we have looked at the different approaches families, communities, and the wider society have used in an attempt to deal with and minimize family pressures and problems.

Especially over the past few decades, family life in the United States has become increasingly diverse and has endured a variety of changing conditions. However, the vital place of family in society has never been seriously questioned. There is wide agreement that *strong societies need strong families.* In fact, working with families is one of the first solutions suggested for virtually all social problems.

Multitudes of committees and agencies studying problems such as crime, violence, drugs, teenage pregnancy, and poverty invariably declare that the first place for prevention and remediation is the home. Assisting families, supporting families, and strengthening families require communities and governments to enact legal, economic, and social reforms to temper social attitudes and improve the treatment of individuals in different kinds of families who face different kinds of problems. Additionally, industry and government can work to see that families have more resources and opportunities to support and enjoy their lives together. Employment schedules, workplace expectations, and expenditures of resources should be directed toward ensuring that employment complements rather than competes with the work and maintenance of families.

If families were simply people-producing factories, they would need little and we would call all families that produce offspring "successful." Yet we know that families have important functions (and not all families produce offspring). As you will recall from Chapter 1, the basic tasks of families are (1) reproduction and socialization of children, (2) economic cooperation, (3) assignment of status and social roles, and (4) providing individuals with emotional support and intimacy. How well families accomplish these tasks affects how well a society functions and prospers. Success also depends on a number of characteristics and abilities of families and the communities and society that surround them. In this chapter, we look at the particular characteristics and qualities that make marriages and families strong. We explore how social and cultural characteristics can contribute to family success, and we examine how relationships with the extended family, with affiliative kin (friends who are like family), and with friends help sustain the family unit. Finally, we look at how the family interacts with the wider community and society. But first we take a look back.

PREVIEW ANSWERS

1 True
2 False
3 True
4 False
5 False
6 True
7 True
8 True
9 True
10 False

Recurring Themes of This Book

Throughout the many chapters and pages of this book you were introduced to a range of theories, exposed to much data, and looked at a number of family issues and relationships. It is worth reviewing the underlying themes that have been emphasized and illustrated repeatedly. There are five such themes, presented below.

FAMILIES ARE DYNAMIC

As we saw most profoundly in Chapter 3, the family is a dynamic social institution that has undergone considerable change in its structure and functions. Similarly, our values and beliefs about families have also changed over time. Major changes include a shift in the foundation: from more multifunctional to more specialized, from more public to more private, from more patriarchal to more egalitarian, and from more economic to more emotional. Our values have changed in the direction of individualistic self-fulfillment with less emphasis on familistic self-sacrifice.

We saw across the various topics and chapters how specific issues and relationships have changed over time. Compared with the past, there are more employed mothers, more cohabiting couples, higher rates of divorce, fewer children, more nurturing fathers, and more single-parent households. Changes such as these will only continue. In fact, the safest statement to make is that *families will continue to change.* In general, predictions are dangerous. One that can be counted on, though, is that *there will always be families.* Although traditional families may be on their way out, families are here to stay.

FAMILIES ARE DIVERSE

Beginning with Chapter 3, we have looked at a variety of factors that make families different from each other. Let's briefly review the major sources of patterned variation.

Race and Ethnicity

Since colonial days, American families have been diverse (Mintz and Kellogg, 1988). The rich ethnic diversity that makes up the melting pot of American society is not new. In fact, there were more than 240 different native cultures that lived in what is now the United States when the colonists first arrived (Mintz and Kellogg, 1988). Since then, American society has housed immigrant groups from the world over who bring with them some of the customs, beliefs, and traditions of their native lands, including traditions, beliefs, and customs about families.

Social Class

Ethnic diversity is joined by economic diversity. Different social classes have different experiences of family life. Due to the material and symbolic (including cultural and psychological) dimensions of social class, our chances of marrying, the marriages we form, the likelihood of those marriages lasting, the relationships with the children we have and raise, our tie with kin, our experience of juggling work and family, our divisions of domestic labor, our likelihood of experiencing violence or divorce, all vary. And this is but a partial and more general list. We have seen many of the specifics in more detail throughout this book.

Gender

Our family experiences are also distinctly shaped by our genders. Though themselves dynamic, gender differences surface in each area of marriage and family on which we have touched. Love and friendship, sexual freedom and expression, marriage responsibilities and gratifications, involvement with children, experience of abuse, consequences of divorce and single-parenting, and our chances for remarriage all differ between women and men.

Lifestyle Variation

The striking difference between twenty-first-century families and early American families is not in the presence of diversity but in the diversity of family lifestyles that people choose or experience. There is no *singular* family form that encompasses most people's aspirations or experiences. Statistically, the dual-earner household is the most common form of family household with children, but there is considerable variation among dual-earners, whose experiences may differ among themselves as much as they differ from traditional or single-parent families. Increasingly, people choose to cohabit, and same-sex cohabitants press increasingly closer to marriage rights. Increasing numbers of couples choose not to have children, while increasing numbers of others choose expensive procedures to assist their efforts and enable them to bear children. This diversity of family types and lifestyles will not soon abate. In fact, the joining of diversity to the messy business of future predictions leads to a third, very safe prediction: *Twenty-first-century families will continue in the direction of diversification,* creating a "salad bowl" of family types not unlike the "salad bowl" of ethnic groups. Rather than blending or melting into a single, dominating type, families will continue to maintain different structures and have different experiences.

FAMILIES SATISFY IMPORTANT SOCIETAL AND PERSONAL NEEDS

Family is the irreplaceable means by which most of the social skills, personality characteristics, and values of individual members of society are formed. Hope, purpose, and general attitudes of commitment, perseverance, and well-being are nurtured in the family.

Indeed, even the rudimentary maintenance and survival care provided by families is no small contribution to the well-being of a community. Some of the services provided by families are such a basic part of our existence that we tend to overlook them. These include such essentials as the provision of food and shelter—a place to sleep, rest, and play—as well as caretaking, including supervision of health and hygiene, transportation, and the accountability of family members involving their activities and whereabouts. We saw in the discussions of housework, paid work, child care, and family caregiving the many ways families provide for these basic needs. Without families, communities themselves would have to provide extensive dormitories and many personal-care workers with different levels of training and responsibility to perform the many activities in which families are engaged.

On a more emotional level, without families individuals must look elsewhere to satisfy basic needs for intimacy and support. We marry or form marriagelike cohabiting relationships, have children, and maintain contact with other kin (adult siblings, aging parents, extended kin) because such relationships retain importance as bases for our identities and sources of social and emotional sustenance. We bring to these relationships high affective expectations. When we don't get our intimacy needs met (in marriage or long-term cohabitation), we terminate those relationships and seek others that will provide them. We believe, however, that those needs are best met in families.

The health and stability of our society depend in large part on strong and stable families. When families fail, individuals must turn to society for assistance; social institutions must be designed to fill the voids left by failing families, and the pathologies created by weak family structures make society a less livable place. We might think of what economist Sylvia Hewlett (1991) casts in a kind of "pay me now or pay me later" warning: There are enormous costs that result from neglecting the needs of America's families and children.

FAMILIES NEED SOCIETAL SUPPORT

Many chapters of this book have pointed to places where families need outside assistance and support. Better child care, more flexible work environments, economic assistance for the neediest families, protection from violent or abusive partners or parents, and a more effective system for collecting child support are some examples of where families clearly have needs for greater societal or institutional support. For maximum chances of successful, fulfilling lives families need at least three things: time, resources, and commitment. Of these, only the last can be produced solely by individuals within their own families.

Time

Modern society deprives people of the opportunity to spend time with their families. Workplaces have been particular sources of *time deficits* (lack of family time, especially between parents and children) and *time binds* (the invasion of time pressures typical of workplaces into family life) (Hewlett, 1991; Hochschild, 1997). As a result, many families have neither the quantity nor quality of time needed to maintain effective parent-child relationships or spousal intimacy. Parents and children need more time together, as do spouses or partners. Without such time, children can drift more easily into activities that are dangerous or self-destructive and spouses can grow estranged. Thus, we need a variety of time-enhancing strategies and a time "movement" to reduce the drain currently experienced. We need family supports that allow people access to their families and which recognize and value the benefits of such access. We need corporate and governmental policies that make family time a priority, managers and supervisors who are sincerely committed to endorsing such policies, and more workers who use them. Of these needs, the latter two may require some reframing of our cultural values alongside our institutional priorities.

Resources

As we have seen, family life is easier to make work when we have adequate resources to meet our needs. Nearly every family problem or pathology we have examined (such as abuse, divorce, and teen pregnancy) is more common at lower economic levels. Furthermore, families who must worry about the ability to meet basic financial needs cannot fully enjoy their family life. Policies that provide greater economic security—from wage enhancement, access to and creation of decent and well-paying jobs, job training programs, and affordable housing—would enrich people's family lives as much as (or more than) their economic lives (Coontz, 1997; Hewlett and West, 1998).

Commitment

Families are not small or short-term investments. As we shall soon suggest, strong marriages and families require commitment, especially on the part of adult members, and a willingness to sacrifice. When we think of family commitment we often think of love, the emotional foundation and justification for making and honoring commitments. Even when people "fall out of love," they need to honor the commitments they made to their children if not to their partners. Society, through social policies, cannot make people love each other. Policies can, however, require people to comply with the responsibilities attached to their commitments.

Too often, we act as if the only thing that determines familial success is the strength of the love and commitment between individuals. This logic suggests that if only people were willing to love each other for life, stay together, and sacrifice individual self-interest for familial well-being we would have fewer family problems—or none. We cannot guarantee how people will feel or act toward each other. By emphasizing that love and commitment are what keep families together or make them successful, we implicitly justify neglecting the need for time and resources. In effect we say, "If you cared *enough* (or *more*), you wouldn't have these problems."

Commitment and perseverance are important. Individuals do need to be encouraged to live up to their responsibilities to each other, show respect for each other, and so on. But we can't stop there. In fact, more time together and more stability and security in one's economic resources make lifelong commitment more likely.

Additional Family Needs

The circumstances people face and the choices they make create other kinds of needs. Ethnic and racial minorities need to be freed from discrimination that inhibits their earnings or reduces their economic opportunities. Discrimination that cuts into one's access to dependable wages eats away at one's marriageability, much as prejudices that corrode one's self-esteem undermine one's abilities to be effective parents, partners, or providers. Gay and lesbian couples continue to press for recognition and support of the equivalent kind and quality given to heterosexual marriages. Given how similar same-sex and heterosexual couples are in such areas as relationship formation, relationship development, conflict, and reasons for success or failure, a strong argument emerges for allowing at least more marriage-like rights and protections to same-sex couples. If people enter relationships for the same reasons (love, intimacy, and companionship), seek the same things, face the same sorts of issues and conflicts (money, work, housework, sexual fidelity), does it make sense to privilege some such relationships over others? Families with children have particular needs for a safe environment within which to raise their children, better-quality schools, and a more responsible popular media, in addition to the workplace and economic issues raised above (Hewlett and West, 1998). Couples who depart from convention, reversing roles or working unconventional shifts, all need to be recognized and supported, not stigmatized.

This leads to our last recurring theme, which will be the subject of much of the remainder of this chapter: *Society needs strong marriages and families.*

Marital and Family Strengths

Virtually all marriages and families have strengths. All families feel more successful and more satisfied at some times than at other times. Some families go through periods of great stress and come out stronger; other families have great difficulty and nearly disintegrate. Research shows that strong, resilient families share some important patterns that enable them to survive and to grow, patterns that are labeled *family strengths.* These relationship patterns, interpersonal skills and competencies, and social and psychological characteristics create a sense of positive family identity. They promote satisfying and fulfilling interaction among family members, encourage the development of the potential of the family group and the individual family members, and contribute to the family's ability to deal effectively with stress and crisis (Stinnett and DeFrain, 1985).

© Michael Newman/PhotoEdit

Vows are an important part of most wedding ceremonies, but commitment to the marriage must be renewed every day.

MARITAL STRENGTHS

Marriage may be seen as a forum for negotiating the balance between the desire for intimacy and the need to maintain a separate identity. To negotiate this balance successfully, we need to develop what sociologist Nelson Foote (1955) called *interpersonal competence*, the ability to share and develop an intimate, growing relationship with another. (For practical insights, see Gottman, [1995].)

Marriage Strengths versus Family Strengths

Many of the traits of healthy marriages are also found in healthy families, as discussed later in this chapter. Indeed, marital competence can be viewed as a necessary basis for family success (Epstein, Bishop, and Baldwin, 1982). It might seem ideal if couples could perfect their interpersonal skills before the arrival of children. Child-free couples generally have more time for each other and substantially less psychological, economic, and physical stress. In reality, however, many of our marital skills probably develop alongside our family skills. Thus, as we improve communication with our spouses, for example, our communication with our children also improves.

There have been hundreds of studies of marital happiness and very few (and those in the last decade or so) on family happiness. Still, an argument can be made that certain kinds of strengths accrue only to families with children. The relationships of couples with children generally have greater stability than those of childless couples because the emotional cost of a breakup is much greater when children are present. Also, the relationships of parents to their children are rewarding in and of themselves and fulfill much of the need for intimacy. Sometimes the bonds between the individual parent and the children or between the children themselves can sustain a family during times of marital stress. Furthermore, the growing acceptance of singlehood and child-free marriages notwithstanding, society expects its adults to be parents and rewards the attainment of the parental role with approval and respect.

Essential Marital Strengths

Therapist and researcher David Mace (1980) cites the essential aspects of successful marriage as commitment, communication, and the creative use of conflict. Research indicates that the strongest predictor of marital success may be the effectiveness of communication experienced by the couple before marriage (Goleman, 1985). Numerous studies show that there is a strong correlation between a couple's communication patterns and marital satisfaction (Noller and Fitzpatrick, 1991). Classes and workshops are being developed in many communities to help teach these vital marital skills to couples before they commit themselves to marriage.

Mace defines *commitment* in terms of the relationship's potential for growth. He writes, "There must be a commitment on the part of the couple to ongoing growth in their relationship. . . . There is tremendous potential for loving, for caring, for warmth, for understanding, for support, for affirmation; yet, in so many marriages of today it never gets developed." Commitment to any endeavor, and marriage is certainly no exception, requires the willingness and ability to work. In their study of strong families, Nick Stinnett and John DeFrain (1985) quote one-half of a married couple:

> You know the stereotypical story of the couple who have the lavish wedding, expensive and exotic honeymoon and then settle down. The work is over. She gets dumpy and nags, he gets sloppy and never again brings flowers. We were like that until one day when we examined our life together and found it lacked something. Then we decided that the wedding, rings, and honeymoon marked the beginning, not the end. We had to renew the marriage all along.

Commitment to the sexual relationship within marriage appears to be an important aspect of marital strength. All the families in the Stinnett and DeFrain study recommended sexual fidelity. An extramarital relationship does not mean the end of a strong marriage, however. Sometimes an affair can be a "catalyst for growth" (Stinnett and DeFrain, 1985). A couple may not realize how far they have drifted apart, how much their communication has deteriorated, or how important the marital relationship is until the extramarital relationship brings things into clearer focus. When spouses who love each other have sexual liaisons outside their marriage, it is almost certainly a sign that their relationship is in trouble. Yet they can then take the painful lessons they have learned and use them to forge a stronger and better marriage.

Another aspect of commitment is compromise. Often people give up things—work demands, social activities, or material goods—for those they love without thinking twice about it. Sometimes difficult choices must be made. Within a committed relationship, there is give-and-take so that neither partner ends up in the martyr role (see Stinnett and DeFrain, 1985). When spouses give up something in order to be together, they nurture the marriage relationship, which in turn supports and strengthens the individual.

We can look at marriage as a kind of task to be performed (albeit willingly and joyfully) or as a growing thing to be nurtured. Whatever metaphor we choose, we see that the relationship requires time, energy, patience, thoughtfulness, planning, and perhaps a little more patience in order to grow and prosper. Commitment to this success (and the sacrifices that commitment may entail) is an essential component in the formation of strong marriages, and strong families as well.

FAMILY STRENGTHS

Family strengths are those characteristics that contribute to a family's satisfaction and its perceived success as a family. An important part of the work of being a family and building family strengths is the identification of family goals. Each family is unique. Therefore, the goals by which a family will identify its success and family satisfaction will be described differently by each family.

Family scholars integrating feminist principles state how important it is to view family in its full range of forms and compositions (Allen and Farnsworth, 1993; Dilworth-Anderson, Burton, and

Turner, 1993; Walker, 1993). In that regard, what is presented in this chapter are the strengths and qualities that are characteristic of successful, healthy families regardless of their structure. Some families may develop these strengths more easily than other families. Yet it is important to underscore that these strengths are available to all families. These are the strengths that are present in families describing themselves as satisfying and successful.

Family as Process

As the saying goes, "nobody's perfect," and neither is any family. Perfection in families, as in most other aspects of life, exists as an ideal, not as a reality. Family quality can be seen as a continuum, with a few very healthy or unhealthy families at either end and the rest somewhere in between. Also, family quality varies over the family life cycle. Generally, families experience increased stress during the childbearing and teenager phases of the cycle. The overall cohesiveness of the family is severely tested at these times, and although families often emerge stronger, they may have experienced periods of distrust, disorder, and unhappiness.

Families also are idiosyncratic: Each is different from all the others. Some may function optimally when the children are very young and not so well when they become teenagers, whereas the reverse may be true for other families. Some families may possess great strengths in certain areas and weaknesses in others. But all families constantly change and grow. The family is a process.

TEN CHARACTERISTICS OF STRONG FAMILIES

The area of research devoted to the study of the characteristics of successful families has developed descriptions and models that help families examine and strengthen their approaches to family living. As with success in any field, strong families benefit from knowledge and practice. One of the most well known of these research projects is the Family Strengths Research Project directed by Nick Stinnett and John DeFrain. In this project, approximately three thousand families were questioned about their family life. Some families lived in the United States; others were from Latin America, Germany, Switzerland, Austria, South Africa, and Iraq. They were from all socioeconomic levels, both rural and urban. Some were single-parent families. All were families who had been nominated as "successful families" by organizations or people in their communities.

When the mass of information from these families was analyzed, six qualities stood out repeatedly, even though the families presented considerable complexity and diversity: (1) appreciation, (2) spending time together, (3) commitment, (4) good communication patterns, (5) spiritual wellness, and (6) ability to deal with crises. Not all of these qualities appeared in every family, nor was the emphasis the same in all families. However, the patterns prevailed. Other researchers have discovered additional areas of family strength (Curran, 1983; McCubbin and McCubbin, 1988).

In surveying the rich collection of research and writing about life in successful families, we have identified nine areas of strength: (1) commitment; (2) affirmation, respect, and trust; (3) communication; (4) responsibility, morality, and spiritual orientation; (5) rituals and traditions; (6) crisis management; (7) ability to seek help; (8) spending time together; and (9) a family wellness orientation. These will be discussed in the following sections.

REFLECTIONS

What are the strengths of your family? How do these compare with those qualities found by Stinnett and DeFrain? Which of those cited in the literature do you value the most? Why?

Commitment

Interviews and surveys with strong families reveal that the members of those families feel an identification with their families. They are willing to work for the well-being, or defend the unity and continuity, of their families (Bichof, Stith, and Wilson, 1992). (Children and youth appearing in juvenile court for various acts of violence usually express a low sense of commitment to their families.) Most adults view marriage as a commitment for life. With the expectation of an enduring relationship, couples are motivated to develop satisfying styles of relating.

When an individual is committed to his or her family, motivation is high to solve problems and to deal with conditions that threaten the family. When families become busy and fragmented, those with high levels of commitment take the initiative to review their priorities and activities and make changes that will relieve this pressure. The members deliberately work together to set priorities for the use of family time.

Commitment involves the promotion of growth of other family members. Family members are concerned for one another's happiness and well-being. Commitment involves working on behalf of one's family—wanting the best for each person. Committed persons strive to help other family members actualize their potential and achieve in their areas of interest (Schvaneveldt and Young, 1992). Marital satisfaction studies report that the more the partners express a sense of personal responsibility for the success of the relationship, the greater their satisfaction with their marriage. A partner's giving of time or self is most effective when it is in terms of enhancing the quality of the relationship rather than in terms of returns expected to be received individually. From a negative perspective, the less an individual contributes to the well-being of the family, the less that person receives.

Commitment is a prevailing characteristic in strong families of all forms. Single-parent and remarried families show commitment to members of the family even though the makeup of the family constellation has changed due to divorce or remarriage.

A challenging aspect of commitment is finding the right balance between extending ourselves to the priorities and identity of family and maintaining respect for our own individuality. Successful handling of this challenge involves being aware that the family milieu and relationships contribute to our own identity and creativity. It also involves understanding that freedom to choose to be involved in service to the family is vital to the development of both the individual and the family (Carter and McGoldrick, 1989).

Commitment to the family involves the participation of family members in a world view that encompasses more than self-centered interest. The members feel they are part of something larger than themselves. Commitment differs from obligation and duty. Although vows and promises may be important as symbols of our intent to loyally support those we care about, true commitment is much more than words. Commitment is revealed and renewed in our actions. It is created every day in the choices we make to do things for the benefit of those we love.

Affirmation, Respect, and Trust

For most of us, the phrase that we like best to hear consists of three little words. "I love you" may be a short sentence, but it can go a long way toward soothing a hurt, drying a tear, restoring a crumbling sense of self-worth, and maintaining a feeling of satisfaction and well-being. Supporting others in our family—letting them know we're interested in their projects, problems, feelings, and opinions—and being supported and affirmed in return is essential to family health (Gecas and Seff, 1991).

Another way to show that we care about family members, friends, and fellow human beings is by according them respect for their uniqueness and differences, even if we may not necessarily understand or agree with them. Healthy families encourage the development of individuality in their members, though it may lead to difficulties when the children's views begin to diverge widely from the parents' views on such subjects as religion, premarital sex, consumerism, patriotism, and child rearing. Criticism, ridicule, and rejection undermine self-esteem and severely restrict individual growth. Families that hamper the expression of their children's attitudes and beliefs tend to send into society children who are unable to respect differences in others. In addition to exhibiting respect for others, a healthy family member also insists on being respected in return.

The establishment of trust, you will recall from Chapter 11, is a child's first developmental stage, according to Erik Erikson. Not only must an infant develop trust in his or her parents, but a growing child must also continue to feel that other family members can be relied on absolutely. In turn, children learn to act in ways that make their parents trust them. Children in healthy families are allowed to earn trust as deemed appropriate by their parents. Children who know they are trusted are then able to develop self-confidence and a sense of responsibility for themselves and others.

It is not only children who need to "earn" the right to be called trustworthy; parents do, too. It's important for parents to be realistic in their promises to children and honest about their own mistakes and shortcomings. It is also important for children to see that their parents trust each other.

REFLECTIONS

How are differences of opinion handled in your family? Do you feel loved and supported by your family? Do you feel trusted? How would you like to be treated differently? Do you let your family know that you love them, that you respect and trust them?

Many researchers agree that parental role modeling is a crucial factor in the development of qualities that ensure personal psychological health and growth. A loving relationship between parents, observes a family program administrator, "seems to breed security in the children, and, in turn, fosters the ability to take risks, to reach out to others, to search for their own answers, become independent, and develop a good self-image" (quoted in Curran, 1983). The more children observe their elders in situations that demonstrate mutual trust, respect, and care, the more they are encouraged to incorporate these successful and satisfying behaviors into their own lives. In divorced families, a continued positive relationship between the parents (if possible) is crucially important to the developing child (Goldsmith, 1982; Hanson, McLanahan, and Thompson, 1996).

Good Communication

We say "I love you" or "I'm angry with you" in many ways other than just in words. Our tone of voice, body language, eye contact, and silences, as well as a touch or a gift are all forms of communication (Satir, 1988). Many books and manuals are available to help people improve their communication as individuals, families, business associates, or in other groups. Chapter 6 highlights the dimensions of communication that are associated with effective problem solving.

In strong families, communication is direct. Strong families talk a lot. They have much to share, and they enjoy doing it. They trust one another. They are good listeners. Ultimately, when family members have truly been heard—not just the words, but the feelings, also—they know they are respected and appreciated because of the active attention and empathy of the listener.

In times of conflict, strong families seem to be able to keep their communication focused on the issues rather than on the personalities of those involved. In fact, recognizing conflict and actively attempting to resolve it can serve as a catalyst in deepening relationships (Rosenzweig, 1992). Strong families attempt to avoid accusation or labeling of others and to consider together the factors involved

© Esbin Anderson/The Image Works

© Bill Bachmann/Stock Boston, LLC.

In strong marriages and families, members communicate feelings of love and trust both verbally and nonverbally.

in the dispute. Studies of youth who are in trouble with the law or are in juvenile detention report that most youths with conduct disorders come from family situations characterized by poor communication among the members (Bichof, Stith, and Wilson, 1992).

Communication has been described as a huge umbrella that covers all that transpires among human beings. Virginia Satir (1988) estimates that by the time we reach age five, we have had a billion experiences in sharing communications. Communication is both the medium and the message of relationships among family members. Good communication is an art that is well developed in strong families. To some extent, they have developed this art through intentional learning and practice. Communication also facilitates the other family strengths presented in this chapter.

The aspects of healthy communication and the conflict-resolution skills discussed in Chapter 6 apply not only to the parents' relationship with each other but also to parent-child and child-child relationships. Among the most significant things a family can teach its children are the value of expressing their feelings effectively and the importance of truly listening to the expressions of others.

Responsibility, Morality, and Spiritual Orientation

One of the family's principal socializing tasks is to teach the individual responsibility for his or her own actions. Fostering morality—a code of ethics for dealing with our fellow human beings—is also essential. A spiritual orientation, whether defined as adherence to a particular religion or as reverence for life in general, gives family members a sense of being part of a larger whole.

The acquisition of responsibility is rooted in self-respect and an appreciation of the interdependence of people. When we as children develop a sense of our own self-worth, we begin to understand how much difference our own acts can make in the lives of others; this feeling of "making a difference" aids the further growth of self-esteem, as the following true story illustrates.

When Sally was ten years old, she went one very cold Saturday with a neighborhood youth group on an outing to the city shopping mall. There Sally noticed a girl about her own age. The girl had on a short, sleeveless sweater and was shivering in the cold. Impetuously, Sally went to the girl and asked if she were cold. The girl, with teeth chattering, nodded her head yes. Without really thinking of what she was doing, Sally took off her coat, put it on the girl, and ran to catch up with her group. The sponsor of the group told Sally to go back and get her coat. The thought horrified Sally; she said she could not do that. When the group returned to their neighborhood, the sponsor marched Sally home and declared to her mother that Sally had given her coat to a ragamuffin in the mall, implying that Sally should be severely punished. Sally's mother put her hand on Sally's shoulder and asked her if the little girl had been cold. Sally said yes, that she had been shivering. Sally's mother thanked the sponsor for taking care of the group. Then she took Sally to a clothing store nearby and told her to pick out any coat she wanted. When recalling this event in a college class, Sally recognized that her impulse to share with the shivering girl was an expression of her own personality and the values and attitudes of her family. From that time on, Sally had a new sense of her identity.

Successful families realize the importance of delegating responsibility (participating in household chores, sharing in family decisions, and so on) and of developing responsible behavior (completing homework, remembering appointments, keeping our word, and cleaning up our own messes). Sometimes children may have to accept more responsibility than they want, as when their dual-worker or single parents rely on them for assistance (Kissman and Allen, 1993). Although these situations may be detrimental for children in families already suffering severe economic (or other) hardships, in many families children have come to benefit from the added responsibility; they feel a sense of accomplishment and worthiness. Parental acknowledgment of a job well done goes a long way toward building responsibility in children.

Healthy families also know the importance of allowing children to make their own mistakes and face the consequences. These lessons can be very painful at times, but they are stepping stones to growth and success. Along with a sense of responsibility, healthy families help develop their children's ability to discriminate between right and wrong. This morality may be grounded in specific religious principles, but it need not be; it is, however, based on a firm conviction that the world and its people must be valued and respected. Harvard psychologists

Families can encourage the spiritual development of their members through formal religious education or by passing on their own values, beliefs, and traditions. Here, a family celebrates Hanukkah.

© Laurie DeVault Photography

Robert Coles and Jerome Kagan believe that children are naturally equipped with a basic moral sense; parents need to recognize and nurture that sense from early on. Besides helping children resolve their moral dilemmas, healthy parents take care to be responsible for their own behavior. Parental hypocrisy sets the stage for disrespect, disappointment, and rebellion.

REFLECTIONS

Do you consider yourself to be a responsible person? A moral person? How did you get that way? Have you had any noteworthy experiences with hypocrisy (yours or someone else's)? How did these experiences feel?

Families with a spiritual orientation see a larger purpose for their family than simply their own maintenance and self-satisfaction. They see their families as contributing to the well-being of their neighborhood, city, country, or world, and as being an avenue through which love, caring, and hospitality can be expressed. Many families find support and expression of spiritual strength and purpose in religious associations. These families find in religious activities a transcendent framework on which they formulate family values, behavior patterns, and goals, as well as a source of strength with which they attempt to live out those values.

Spirituality gives meaning, purpose, and hope. In the case of formal religions, it provides a sense of community and support (Abbott, Berry, and Meredith, 1990). In times of adversity, families tap their spiritual resources to gain the strength that they need to sustain themselves. Stinnett and DeFrain (1985) suggest seven ways to help spirituality work in families:

1. Set aside time each day for meditation, prayer, or contemplation. Try to get outdoors to enjoy the beauty of nature.
2. Join (or help form) a discussion group to consider religious or philosophical issues.
3. Examine your own values. Try keeping a journal of your thoughts.
4. Help your children to clarify their values.
5. Identify three of your strengths. Work on developing them more fully. Identify three of your weaknesses. Decide how you can improve in these areas.
6. Have regular family devotional times: "Read . . . inspirational material, pray, sing, count your blessings, reaffirm your love and commitment to each other."

7. Volunteer time, energy, and money to a cause that helps others.

Spiritual wellness, it should be noted, is not necessarily the same as religiosity. Though religion can provide a source of personal, family, and social support as well as an opportunity for a family to engage in religious services together (Robinson and Blanton, 1993), it can also create a wedge between couples, especially when they disagree about the type and place of religion in their lives.

Rituals and Traditions

Traditions, especially those rooted in our cultural background, help give us a sense of who we are as a family and as a community. Through tradition and ritual, families find a link to the past and, consequently, a hope for the future.

Speaking to the American Association for Counseling and Development, the renowned African-American writer Alex Haley said, "Americans should study their family histories, hold reunions, interview their elders and commemorate them with pictures" (Haley, 1986). Haley pointed to the far-reaching impact of his book *Roots* to illustrate the deep desire people have for family connectedness. More than two decades ago, *Roots* captured the interest of the nation. It tapped a process vital to all humankind, a process of telling and nurturing the family story, a process that is continually threatened by rapid social changes. "*Roots* was really born in the evenings of the summer when I was six years old," Haley said. He described how he would sit on his front porch, listening to his grandmother and her sisters talk about their parents, their grandparents, and the lives they had led. "I was greatly imprinted by my grandparents," he added.

REFLECTIONS

Make a list of your family traditions. Where did they come from? Which are your favorites? Your least favorites? Would you like to create any new ones? Make a list of your daily and weekly family rituals. Note your favorites. Are there others you'd like to incorporate into your family routine?

Most successful families have someone who maintains and transmits the family story (Gunn, 1980). They have rituals and traditions that help them keep alive their identity as a family. Those traditions and rituals help the family remember their

Family holiday gatherings are often occasions for both witnessing and transmitting family traditions.

bondedness of commitment—the goals for which the family endured difficulties and the purposes for which the family works. Young people's personal identity, their sense of style and distinctive way of doing things, their sense of who they are and who they might become have a foundation in the sense of continuity and rootedness that comes from their identification with their family history.

Traditions and rituals vary greatly, from elaborate holiday celebrations to daily or weekly routines. Children's bedtime rituals—with blankets, books, bears, or prayers—give security and comfort, as do Sunday morning rituals that find the family together in church or in bed with the funny papers. Large and small rituals, if not passed down from preceding generations, can be freshly created and passed forward. Therapist Mary Whiteside (1989) advises that stepfamilies may benefit particularly from the creation of new rituals, which can "generate a feeling of closeness" and "provide fuel for weathering the more difficult times." She adds, "As these experiments . . . succeed enough to recur and to emerge as customs, the stepfamily unit begins to feel as if it has a life of its own." Rituals, special practices, customs, and techniques unique to the family—whether ways of celebrating birthdays or the "correct" angle for mowing the lawn—imprint their family identity in the hearts and memories of its members.

Crisis Management

The growth and development of the family through the life cycle inevitably produces crises. Growth involves change and, with it, a certain amount of stress. Unexpected events and those not prepared for can cause great pressures. Some families are able to cope with these effectively, and some may become immobilized.

Research consistently identifies resilience, the capacity to deal effectively with family crises, as a characteristic of strong families (McCubbin and McCubbin, 1988; Burr and Klein, 1994). Members of strong families unite to face the challenges of a crisis. Each feels a part of the team and says, "What can I do?" No one carries the load alone.

Strong families are more able than other families to face challenges with confidence. The cumulative effect of other family strengths, such as commitment, cohesion, communication, and adaptability, enable strong families to evaluate the obstacle facing the family with resolve and to anticipate the resources and help needed. They are often able to recognize the task before the family as a normal developmental task, whereas less secure families might see the same situation as an unusual and overwhelming disaster.

Perception itself is an important factor in the approach and solution to a problem. Strong families are able to accept changes resulting from crises and to see possibilities for growth in them. Thus, they are able to approach with hope situations that might generate despair and paralysis in other families. Family strengths such as adaptability, communication skills, spiritual resources, and a sense of unity and trust based on family commitment contribute to effective crisis management.

Ability to Seek Help

Another characteristic of strong families is associated with effective crisis management. This is the family's ability to be open to resources outside itself. Strong families are able to acknowledge their vulnerabilities. They feel themselves to be accepted members of the community and are able to ask for and receive assistance in time of need. Strong families' experiences of interdependence within the family better equip them to recognize, in turn, the interdependence among families and individuals in the neighborhood and community. They feel less compelled to prove their independence and to "go it alone."

A vast range of experiences (borrowing an egg from a neighbor, cooperative work on the school board, serving as a block parent, giving financial assistance when an unexpected illness keeps a neighbor homebound and out of work, and the like) continually involve families in interdependent relationships with their communities. Some needs can be met by normal exchanges among friends. Some can be met by help from extended family, from church, or from other support groups. Some needs require professional assistance. For example, family counseling may provide the context and support a family needs to examine difficult relationships and behavior patterns (Browne, 1997). Many community programs are available, such as parent education, marriage enrichment, and support groups dealing with alcoholism or other specific problems.

REFLECTIONS

Does your family seek outside help for problems? Make a list of all the resources available to you that you might want to use sometime. Note the ones you've used in the past. The next time you have a family problem, take a look at your list.

Spending Time Together

When young couples look toward marriage, one of their anticipations is that they will be able to "be together." They anticipate the sharing of interests and activities that are mutually enjoyable. They look forward to companionship and shared psychological support. The expression of affection and meeting one another's emotional needs are sometimes romanticized as the central motivation for marriage. Certainly, these motivations are part of the vital glue that holds couples together.

As children arrive, the range of activities expands and relationships become more rich and complex. Having time to be together becomes increasingly crucial. The very process of making provisions for having time together becomes an important part of the building of family strengths. In some instances, it may be viewed as a "sacrifice" that is made by some members for the well-being of the whole family. As noted earlier, making time for family is an interlocking trait that brings together other characteristics found in strong families. It is an action that expresses commitment to the family.

The maturing of the love relationship that first brought the couple together requires their deliberately setting aside time to spend together. Competition of careers or other friendships may begin to erode the basic substance of their marriage without the couples' awareness unless they stop to listen to each other. New identities are being formed by both members of the marriage. Strong marriages are those in which couples create special occasions to be together. In strong families, "dating" does not stop after the wedding.

Spending time together is necessary to develop adequate communication and to build cohesion. Family traditions and rituals require family time. Family counselors indicate that one of the first signs of family difficulties is lack of family time together. Children and spouses learn they are appreciated, are valued, and have worth when other members spend time with them. The modeling and teaching that parents provide for their children require shared time, as does showing support to family members by attending school events and other occasions of special significance in the lives of family members.

Quality time together need not be spent on lengthy or expensive vacations: A relaxed meal or a game in which everyone participates can serve the same purpose a lot more often. Sometimes parents may have difficulty learning to leave a messy desk at work or a sticky kitchen floor at home in order to relax with each other or with their children. Yet time taken to play and relax with our loved ones pays off in ways that clean desks and shiny floors never can. When we are realistic, we see that the papers will never stop flowing onto the desk and the jam will never stop dripping onto the floor, but our children and our mates will never again be as they are now. If we don't take the time now to enjoy our families, we lose an opportunity forever. Healthy families know this and give play and leisure time high priority.

REFLECTIONS

Do you feel that your family has enough time for play and leisure? Can you give up anything to spend more time with your family? Can you reorder your priorities? Can they? Make a list of things you'd like to do with your family if you had the time. Take the time.

A Family Wellness Orientation

Having a *family wellness* orientation means making a conscious decision to live our lives in ways that move us toward optimal health in physical, emotional, intellectual, spiritual, and social dimensions. Wellness is positive and proactive; it focuses on being healthy and whole, as opposed to being fearful of disease and dysfunction.

Families with a wellness orientation identify their strengths. They recognize that strong, healthy family life does not just happen. It requires attention, preparation, and monitoring. Families oriented toward wellness take advantage of educational opportunities that help them gain perspective on family developmental processes. Marriage enrichment, parent education, and family retreats are occasions where families can exercise their family wellness orientation. There they can gain a perspective that helps them recognize, articulate, and prepare for upcoming changes in child development and family stages.

Different Families, Different Strengths

Each family is unique. However, as we saw in Chapter 3, families of different ethnic or economic backgrounds often have distinctive characteristics and special strengths. The same may be said of families of different structures or forms. However, an analysis of the strengths of diverse families requires first that we recognize the commonality of family processes among families of all types (Walker, 1993).

Studies of different types of families often reveal similarities beneath apparent differences. Bettie Sanderson and Lawrence Kurdek (1993), for example, in a comparison of African-American and white couples, found that despite the existence of ethnic differences, the processes that are linked to relationship satisfaction are similar between these two groups. Another study found relations between a variety of family processes and child well-being to be similar for both traditional (nuclear) families and nontraditional families (Bronstein et al., 1993).

FAMILY STRENGTHS AND ETHNIC IDENTITY

Ethnicity, writes Monica McGoldrick (1989), "is more than race, religion, and national or geographic origin. . . . It involves conscious and unconscious processes that fulfill a deep psychological need for identity and a sense of historical continuity. It is transmitted by an emotional language within the family and reinforced by the surrounding community."

In the melting pot of American society, ethnicity is a complicated and ever-changing phenomenon. Although some ethnic groups intermarry with, and adapt rapidly to, mainstream society, others, by choice or by pressure from without, have retained many of their traditions and values. In this sense, the United States today can be seen more as a cauldron of very complex stew or a stir-fry dish than as a homogenized melting pot (McAdoo, 1993). Appreciating and respecting diverse types of families may shed light on how families can adapt effectively to adverse circumstances (Fine, 1993).

African-American Families

An ethnic group may possess particular strengths that help its families to survive in a larger society that is often less than welcoming and sometimes overtly hostile. This was true of African-American families during slavery and its aftermath (and amid the racism and discrimination still present in U.S. society today). African Americans needed to rely on their own families and community strengths. Strong kinship networks characterized the West African families whose members were abducted to the Americas as slaves, and such networks have been a major resource of African-American families during times of trouble (Taylor, Chatters, and Jackson, 1993).

African-American families traditionally are centered around the children (creating a "pedi-focal" family system). Thus, the family unit can be defined as including all those involved in the nurturance and support of an identified child, regardless of household membership (Crosbie-Burnett and Lewis, 1993). Because of the high value placed on child rearing, the role of mother for African-American women is often more important than any other role, including that of wife, and because of the strong kinship bonds in the culture, the single mother is not left alone to raise her child (Ellison, 1990; Hampson, Beavers, and Huylgus, 1992). Black fathers also tend to be warm and loving toward their children.

Family values such as unconditional love for children, respect for self and others, and the "assumed natural goodness of the child" are other strengths of the African-American family (Nobles, 1988). Robert Taylor (1990) and his colleagues note that a distinctive task of African-American parents is to attempt to prepare their children for the "realities of being black in America." Family values also include a special consideration for the elderly.

The family strengths associated with African-American families include the following:

- *An extended kinship network:* The close relationship between family members provides economic and moral support in both day-to-day and crisis situations (Wilkinson, 1993).
- *Flexibility of roles:* Households have the ability to expand and contract in response to external and internal pressures.
- *Resilient children:* Youth are socialized to obtain the best from "both worlds," African American and white, through survival techniques that provide a wide repertoire from which they may choose.
- *Egalitarian parental relationships:* There is a reciprocal exchange of roles, duties, and rights by both parents.
- *Strong motivation to achieve:* African-American parents support the education of their children.

(See also Carter [1993] and Gary et al. [1986]; their research describes the strong religious orientation that has historically been an integral part of African-American family life.)

Characteristics of African-American extended kin systems include (1) a high degree of geographical closeness; (2) a strong sense of family and familial obligation; (3) fluidity of household boundaries, with a great willingness to absorb relatives—both real and affiliated, adult and minor—if the need arises; (4) frequent interaction with relatives; (5) frequent extended family get-togethers for special occasions and holidays; and (6) a system of mutual aid (Hatchett and Jackson, 1993).

Latino Families

Although there is great diversity among Spanish-speaking cultures, Latino families tend to live in nuclear families near others in the extended family network. Among Latinos, father-child relationships are often playful and companionable. Children's ties with mothers are primary and lifelong no matter what the geographic distance or age. Children are taught to carry family responsibilities, to prize family unity, and to respect their elders (Chilman, 1993). Close relationships with maternal and paternal grandparents are fundamental. Of special importance are the emotional ties with the mother's relatives. Maternal aunts often serve as "brokers," providing a link between parents and other adults in the family (Wilkinson, 1993). Latino culture emphasizes the family as a basic source of emotional support, especially for children. No sharp distinction is made between relatives and friends; in fact, friends are considered virtually kin if a close relationship has been formed. The term *compadrazgo* (co-parentship) is often used for this relationship (Chilman, 1993).

The major strengths associated with Latino families include the following (Vega, 1995):

- *Family focus:* The family is a major priority and as such receives strong commitment by all members.
- *Strong ethnic identity:* The importance of culture in the Latino family binds and gives families strength and identity.
- *High family flexibility:* In contrast to the family being male dominated, there is flexibility of family roles.
- *Supportive network of kin:* The extended family is both a tradition and a support of family.
- *Egalitarian decision making:* As traditions shift, input by all family members is valued.
- *Family cohesion:* Although cohesion decreases across generations living in the United States, Latino families still remain cohesive.

Asian-American Families

Resiliency marks the lives of many Asian-American families living in the United States. Even though they have faced prejudice and discrimination, Asian Americans have maintained both family values and ties and in many cases, economic stability. Responsibilities to aged parents and to close relatives are fundamental to the family institution among Japanese, Chinese, Filipino, and other Asian families living in the United States (Wilkinson, 1993). Most often, an elderly parent lives in the same household as the adult child.

In Chinese-American families, the family orientation is summarized in the concept of *hsiao,* or filial piety, which involves a series of obligations of child to parent. First, the child is to provide aid, comfort, affection, and contact with the parent in an attitude of loving warmth and reverence. Second, the child is to bring reflected glory to the parent by doing well in education and occupational activity (Lin and Liu,

Understanding Yourself

FAMILY STRENGTHS INVENTORY

The Family Strengths Inventory was developed by Nick Stinnett and John DeFrain to identify areas of strengths and weaknesses in families (Stinnett and DeFrain, 1985).

Complete the inventory, and then compute your score by adding the numbers you have circled (with 1 representing the strongest degree of agreement and 5 representing the least degree). You might also have other family members complete it independently to get a sense of their perceptions of your family. The score will fall between 13 and 65.

What does the score mean, according to the researchers? A score below 39 is below average. It indicates that there are areas in your family relationships that need improving. Look at the areas in which you scored low to help you target where you need work. Scores from 39 to 52 are average. Scores above 52 indicate strong families. Consider discussing the inventory results with your family.

Circle the number on the 5-point scale that best describes your family.

Family Strengths Inventory	Strongly agree	Slightly agree	Neither agree nor disagree	Slightly disagree	Strongly disagree
1 Spending time together and doing things with one another	1	2	3	4	5
2 Commitment to one another	1	2	3	4	5
3 Good communication (talking with one another often, listening well, sharing feelings with one another)	1	2	3	4	5
4 Dealing with crises in a positive manner	1	2	3	4	5
5 Expressing appreciation to one another	1	2	3	4	5
6 Spiritual wellness	1	2	3	4	5
7 Closeness of your relationship with your spouse	1	2	3	4	5
8 Closeness of your relationship with your children	1	2	3	4	5
9 Happiness of your relationship with your spouse	1	2	3	4	5
10 Happiness of your relationship with your children	1	2	3	4	5
11 Some people make us feel good about ourselves; that is, they make us feel self-confident, worthy, competent, and happy about ourselves. Your spouse makes you feel good about yourself.	1	2	3	4	5
12 You make your spouse feel good about himself or herself.	1	2	3	4	5
13 You make your children feel good about themselves.	1	2	3	4	5

1993). A child who misbehaves brings shame to the family name (Braun and Chao, 1978). Children are taught that everyone has to work for the welfare of the family. They are given a great deal of responsibility and are assigned specific chores. Adolescents are responsible for supervising young children and for work around the house or in the family business (McLeod, 1986). The concept of *hsiao* provides a stability, purpose, and structure for the activities and commitment of all the family members.

Japanese-American families are often cohesive units that act as agents of both socialization of their children and social control of their members. Typical features that characterize strengths of the Japanese-American family include: close family ties between generations indicated by strong feelings of loyalty to family; low divorce rates, which often are seen as an indicator of family stability; and a complex system of values and techniques of social control including guilt, shame, obligation, and duty, which are transmitted from parents to children. These characteristics reflect the continuities between traditional Japanese culture and contemporary Japanese Americans (Takagi, 1994).

Vietnamese-American families are the most recent Asian population to enter the United States in significant numbers. They bring a strong, traditional Vietnamese extended family structure with them.

This includes a group that stretches beyond immediate or nuclear family ties to include a wide range of kin. These extended households mesh in a large and active web of kinship relations in the neighborhood and general vicinity. These relations with kin often function as important sources of economic and social support.

Due to patterns of immigration and disruption of family relations, Vietnamese-American families have developed variations in the traditional extended family household and extended kin system. People are incorporated into the kin network who would not have been part of it in traditional Vietnam. These include more distant relatives and nonrelated friends. These reconstructed kin networks reflect the importance placed on familial relations by Vietnamese Americans. These extended kin networks also provide for mutual aid and for exchanging goods, services (like child care and cooking), and information (about government agencies, hospitals, and so on) (Kibria, 1994).

The major strengths of Asian-American families include the following:

- *Filial piety:* A series of obligations and great respect for elders exist among Asian-American families.
- *Family as a cohesive unit:* Both the nuclear family and the extended family play important roles in the lives of Asian-American families.
- *Value of education:* From preschool through college, parents support and encourage their children's education.
- *Feelings of loyalty:* Close family ties, low divorce rate, and maintenance of traditional values all reflect loyalty to family and culture.
- *Extended family support:* Emotional and financial support exists among nuclear and extended families.

Native-American Families

Native Americans are a diverse group. There are hundreds of Indian tribes, and although they share many traditions and beliefs, they also differ from one another in significant ways. Generally, however, Native-American families see human life as being in harmony with nature. They are usually group oriented and emphasize cooperation (Markstrom-Adams, 1990). Relations with kin are often characterized by residential closeness, obligatory mutual aid, active participation in life-cycle events, and the presence of central figures around whom family ceremonies revolve. Relatives tend to live near one another and become involved in the daily lives of the members of the kinship unit. Women play fundamental roles in these extended systems. One of the enduring features of the Native-American family has been its dynamic quality. Native-American families have a great capacity to adapt to a changing social environment.

Extended family networks may include several households of significant relatives that assume a village type of character. Transactions within and among these households occur within a community context. Despite a history of severe dislocation and disenfranchisement among Native Americans, the community-family configuration is a mark of an integrated social organization that serves to offset the ravages of perpetual displacement (Wilkinson, 1993). In a study of Native Americans' adjustments to urban life, anthropologist John Price (1981) found that kinship networks remained "strong and supportive." In Los Angeles, he found that Indians from over a hundred tribes shared pride in their common heritage through Native-American sports leagues and dance groups, powwows, and traditional crafts. Many urban families also commute frequently to their home reservations to renew their family and cultural ties.

REFLECTIONS

Do you consider yourself part of a particular ethnic group? Are there special strengths you see within your group? Would you consider your group to be assimilated into American society? Accepted by American society? If you don't consider yourself part of a particular group, do you nevertheless preserve certain aspects of your ancestors' ethnicity? If so, what are they?

A special role for the elderly has historically been recognized as a strength in Native-American families. Despite the tenuous economic conditions under which many Native American elderly live, they maintain social and cultural ties to their families. They perform a variety of important and beneficial roles, including instructing the young and helping care for children. For example, one-fourth of elderly

Native Americans take charge of caring for at least one grandchild and two-thirds reside within five miles of family, with whom they share socialization, chores, and routine obligations (Edwards, 1983). Elders are seen as a resource for young parents to assist them in understanding traditional roles of discipline and child rearing. They also maintain responsibility for remembering and relating tribal philosophies, myths, traditions, and stories peculiar to their tribal groups (Yellowbird and Snipp, 1994).

Contemporary research emphasizes the continuing importance of certain core family values that enable Native Americans to preserve important aspects of their culture (Yellowbird and Snipp, 1994). The concepts of *time, cooperation, leadership, sharing,* and *harmony with nature* are often viewed quite differently by Native Americans than by those in the dominant culture. For example, Native Americans tend to view the role of a leader as more sacred, more humanistic, more person-oriented, more honest, more intuitive, and less ambitious. The leader's role is one of servitude rather than one of assertiveness (Lewis and Gingerich, 1980).

The major strengths of Native-American families include the following:

- *Extended family network:* Strong ties exist among relatives as well as with extended kin and tribe.
- *Value placed on cooperation and groups:* Geographic proximity and solid family values give relatives the opportunity to support and become involved in each others' daily lives.
- *Respect for the elderly:* Social and cultural ties help to maintain respect for elders, who are seen as teachers and resources for the young.
- *Tribal support system:* There is reliance and trust in tribal support for advice and resolution of problems.
- *Preservation of culture:* Core family values are supported by means of the maintenance of native language, harmony with nature, respect for leadership, and high family cohesion.

FAMILY STRENGTHS AND FAMILY FORM

In single-parent, never-married families, a special form of loyalty may be found between the parent and child. This involves a conscious consideration of the tasks faced by the single parent and a strong determination to succeed in maintaining an effective family life. A significant contrast has been noted by researchers between single-parent, never-married families with a strong commitment to their families and single-parent, never-married families who do not affirm that commitment. In the committed families, the single parent accepts responsibility for maintaining a supportive family climate and often works with extended family members in developing the kind of psychological and social assistance needed for his or her family's success.

REFLECTIONS

What are the various strengths of your family? How do you contribute to them? What role do culture, religion, and extended family play in supporting your family?

Members of single-parent families often develop a sense of confidence and pride in maintaining and providing for their family in very difficult circumstances (Richards and Schmiege, 1993). Strengths of single-parent families include the following: (1) they have a more efficient decision-making system; (2) they have more direct communication, and parent and child share responsibilities in partnership fashion; (3) a greater sense of vitality is present in the work and contribution made by the child; and (4) children develop a more egalitarian view of the roles of men and women.

Resources of the extended family may come into play in unexpected ways as the family moves through life cycle changes or changes in family form. For example, research with college students (Kennedy and Kennedy, 1993) found that college students from stepfamilies tend to feel closer to a particular grandparent than do college students from intact or single-parent families. (Extended family strengths are discussed later in this chapter; stepfamily strengths are discussed in Chapter 15.)

Kin and Community

Strong, successful families neither emerge from nor exist in a vacuum. Nor do they depend on members alone to meet all needs or provide necessary strengths. In addition to the members of our immediate families, other networks of caring and identity provide strength and a sense of belonging to us as individuals and families. Such networks offer a larger context of purpose, continuity, and posterity. They also provide practical assistance and emotional support. Relationships with extended family members and family units (aunts, uncles, cousins), enduring familylike (affiliated) relationships, abiding friendships, and participation in social organizations and their attendant roles and responsibilities in the community create meaning and promote security.

INTIMACY NEEDS

Whether we are married or single, we need relationships in which we can be intimate. Neither couples nor individuals can function well in isolation. Robert Weiss (1969) notes that people have needs that can be met only in relationships with other people. These needs can be summarized as follows:

1. *Nurturing others:* This need is filled through caring for a partner, children, or other intimates, both physically and emotionally.
2. *Social integration:* We need to be actively involved in some form of community; if we are not, we feel isolated and bored. We meet this need through knowing others who share our interests and participating in community or school projects.
3. *Assistance:* We need to know that if something happens to us, there are people we can depend on for help. Without such relationships, we feel anxious and vulnerable.
4. *Intimacy:* We need people who will listen to us and care about us; if such people are not available, we feel emotionally isolated and lonely.
5. *Reassurance:* We need people to respect our skills as persons, workers, parents, and partners. Without such reassurance, we lose our self-esteem.

THE EXTENDED FAMILY: HELPING KIN

Few aspects of family life exist to which relatives (especially parents and siblings) do not make a significant contribution. Among African-American families, for example, grandparents are often a valuable resource for child care and child socialization (Flaherty, Facteau, and Garver, 1990; Kennedy, 1990). Adult siblings are also important sources of help in times of need (Chatters, Taylor, and Neighbors, 1989).

In many families, parents loan money to their adult children at low or no interest. They may give

Courtesy of Keegan Family/© Christine Mendes/Buena Vista Photography

Extended families are a source of both support and pleasure. This group of siblings and cousins is enjoying a reunion with their grandmother.

or loan them the down payment for a house, for example. The obligations that the loans and gifts entail differ according to a person's age and marital status. If the children are young and single, parents may still expect to exercise considerable control over their children's behavior in return for their support. But if the children are older or, more important, married, there are fewer obligations.

Even when extended families are separated geographically, they continue to provide emotional support. Contacts with kin are especially important in the lives of the aged (Kivett, 1993).

AFFILIATIVE KIN

Blood relationships do not define the type of feelings that a person will have. Instead, they provide a framework to encourage brotherly feelings toward a brother, motherly feelings toward a child, and so on. As we discussed earlier, the strength of kinship ties ultimately depends more on feeling than on biology. A brother or sister can seem like a stranger; a grandmother can be more of a mother than one's biological mother; a parent can be like a brother or sister. Feelings of kinship can extend beyond traditional kin. We form affiliative kin by transforming friends and neighbors into kin: We might say "He is like a brother to me" or "We are like cousins."

Among African Americans, affiliative kin are important in child rearing. Especially in single-parent African-American families, one or more people may be responsible for providing money for a child; others may contribute clothing or meals and still others, nurturance and guidance. Unrelated men and women take on family roles and often have the same rights and privileges as related family members (Crosbie-Burnett and Lewis, 1993). In Latino culture, *compadres* (godparents, literally coparents) are important figures in the family. The *compadre* (godfather) and *comadre* (godmother) are considered to be responsible for the child's spiritual development, and they are available as resources in time of need. Among lesbians and gay men, affiliative kin are important as a means of including one's partner and his or her relatives and friends as family.

Because divorce and geographic mobility are breaking down our abililties to interact with in-laws and biological kin, we are beginning to form new kinds of kin (Lindsey, 1982). Single persons may attempt to create families from friends by sharing time, problems, meals, and housing with one another. Single parents may form networks with other single parents for emotional support and exchange of child care. Family networks may be formed in which three or four families from the same neighborhood share problems, exchange services, and enjoy leisure time together. Sometimes involved neighbors may be called by a family name, such as "Auntie" or "Gramps."

FRIENDSHIP

Families today find close friends, casual acquaintances, and temporary support within a complex and changing social network. Traditional sources of community support, such as church groups, are augmented by other sources. Parents meet one another and form friendships in groups that center around their children: parent-teacher groups, Scouts and Campfire organizations, athletic groups, and so on. Peer support groups and self-help groups are available to fill a variety of short- or long-term needs; lasting friendships are often formed within such contexts. People may join political groups to work with others for change in society; they also join hobby groups, crafts clubs, or dance classes in order to share their special interests with others. Some networks of support provide specific information or services for dealing with specific needs, whereas others provide more general help (Cooke et al., 1988). Social networks enrich our day-to-day lives and provide vital assistance in times of stress and crisis.

FAMILY IN THE COMMUNITY

Urie Bronfenbrenner (1979) has proposed that we look at the family in an "ecological environment." He suggests that we think of this environment in terms of a set of nested Russian dolls. The developing child is the tiny innermost doll contained inside the various systems, such as home or school, that influence him or her. The interplay among the many systems profoundly affects the child in their midst. Furthermore, the influences of external factors, such as the flexibility of the parents' job schedules or the availability of good health care, also play a critical role.

The well-being of the family depends not only on its own resources, then, but also on the support it receives from the community in which it is embedded (Unger and Sussman, 1990). This community includes extended family, friends, schools, employers, health-care providers, and government agencies at local, state, and federal levels. As discussed in Chapter 12, the United States is far behind most industrialized nations in terms of the support it gives families in the areas of health, education, social welfare, and workplace policy. The need for creativity and energy in these areas is great and presents challenges and opportunities for those who wish to work with the families of today and tomorrow.

Strengthening Families through Family Policy

Since the late 1970s, interest has been rising in systematic family policy—family-oriented legislation that encompasses clearly defined objectives concerning family well-being and specific measures initiated and enforced by the government to achieve them. Interest has been increasing for several reasons related to issues and changes covered earlier in this book (Aldous and Dumon, 1991; Wisendale and Allison, 1989), including the following:

1. There has been an increase in female-headed households because of rising divorce rates and births to single mothers.
2. Women, especially mothers of young children, have entered the labor force in unprecedented numbers.
3. Problems confronting families—ranging from poverty to abuse, homelessness to inadequate child care—are both more visible and more widely recognized as social rather than individual problems.
4. "The family" as a symbol and "family values" have become ideological battlefields for both liberals and conservatives, with each proclaiming themselves "pro-family" as they question and criticize the viewpoints of the "other side."

If our marriages, families, and children are truly top national priorities, the implementation of family policies and the adoption of a "family perspective in policy making" must be considered by government and business (Bogenschneider, 2000). The distinction between "family policies" and "a family perspective in policy making" has to do with whether the policy is a direct attempt to affect families or has indirect familial consequences.

Family policy would include those explicit social policies designed to address issues such as teen pregnancy, child care, and child support. Policies with a "family perspective" could be any from among areas such as health care, social welfare, housing, poverty and education (Bogenschneider, 2000; Edelman, 1989; Macchiarola and Gartner, 1989; Hewlett and West, 1998). Tax policies might be either explicit family policies, as in the case of child care tax credits or a repeal of the "marriage tax," or they might be guided by a "family perspective" in recognizing the impact such reform might have on individuals' decisions to marry or stay married.

Bogenschneider (2000) identifies four areas that became major family policy issues in the 1990s: (1) family and work, (2) long-term caregiving, (3) family poverty, (4) marriage strengthening. All of these are likely to continue to require research, assessment, and policy design in the years ahead. The conflict between work and family, discussed in Chapter 12, commands continued thinking about how to restructure the workplace, how to teach employed women and men techniques to better manage instances of work-related stress on families, and how to get people to make and honor stronger personal commitments to their families. Families tend to supply the long-term care for elderly or disabled members. With advances in medical technology, people are living longer and forcing more families to cope with the caregiving needs that will result. Given the strong connections between poverty and a range of family issues (such as the stability of marriage, child development, and the well being of children), programs designed to fight poverty and support those most in need will continue to be critical issue areas. Policies designed to support marriage will continue to be debated and discussed, especially given the connection between marital status and the well-being of adults and children.

Regarding policies that have a family perspective, certain points come to mind. In the area of health care, we might implement the following:

1. Guaranteed adequate medical care for every citizen, with a national health-care policy.

2. Prenatal and infant care for all mothers; adequate nutrition, immunizations, and "well baby" clinics to monitor infant health.
3. Medical and physical care of the aged and the disabled, including support for family caregivers.
4. Education for young people about sexuality and pregnancy prevention; implementation of comprehensive and realistic programs that address the economic and social realities of youth.
5. Education for all Americans about the realities of STDs, HIV, and AIDS and their prevention through abstinence or condom use; guaranteed access to treatment.
6. Drug and alcohol rehabilitation programs available to all who want them.

REFLECTIONS

If you were to construct a coherent family policy that meets your needs and reflects your values, what would it be like? How would it compare with the author's suggestions?

Social welfare concerns would be enhanced by policies such as these:

1. Tax credits or income maintenance programs for families.
2. Child allowances to ensure a basic standard of living.
3. Child care for working or disabled parents; temporary respite child care when parents are ill or unable to care for their children; training of neighborhood day-care providers.
4. Advocacy for children, the aged, the disabled, and others who may not be able to speak for themselves.
5. Attention given to problems of the homeless, such as food, shelter, and medical and psychiatric care.
6. Regulation of children's television to promote literacy, good nutrition, humane values, and critical thinking.

In the area of education, the following policies should be implemented:

1. Implementation of preschool programs such as Head Start wherever needed.
2. Stress on the teaching of basic skills; use of innovative programs to reach all students, including outreach programs for adult literacy and English as a second language (ESL).
3. Work exposure for students who wish to work through programs such as the Job Corps.
4. Guarantees that all Americans receive the benefits of education through the implementation of bilingual and multicultural programs, special education for the developmentally disabled, and adult education.

Finally, as Hewlett and West (1998) recommend, we should aim to improve the external environment in which families live and restore honor and dignity to parental roles and responsibilities. For example:

1. Reduce violence in our neighborhoods and schools.
2. Make our communities drug-free by implementing drug education and prevention programs.
3. Make our popular media more responsible through regulation of children's programming and stricter rating systems for movies, television, and music CDs.
4. Establish a National Parents' Day to commemorate and celebrate the collaborative work involved in raising children—collaborations between parents raising their children and the partnership between communities and parents in the tasks of raising healthy children.
5. Create a "Family Movement." Building on and incorporating the wisdom of existing ideas about a "time movement" and "Parents' Movement" (Hochschild, 1997; Hewlett and West, 1998), we must collectively declare our commitment to our families and recognize that across the divides of age, race, class, ethnicity, sexual orientation, and household structure we are all trying to enjoy the benefits—intimacy, emotional support, and nurturance—that families uniquely yield. Collectively, we can press public and private institutions for greater commitments to families and children and for assistance in meeting our needs.

We can enjoy strong and successful families in the absence of these initiatives, as many people have found. But doing so is harder than it needs to be and often requires more than a little good fortune. This is a society that makes stable, loving, gratifying families too much the work of individuals. Rather than leave the quality of our family lives to chance, we need to implement a set of initiatives that make strong marriages and families a national priority.

This chapter has brought us right back to where we began. As stated in the Preface, there is one underlying and unifying assumption that captured our objectives in writing this book: our enduring belief that our families, whatever their forms, are the crucibles in which our humanity is born, nurtured, and fulfilled. They are what makes us human, and they need to be cherished, honored, and supported. Nothing less will do. The families that love, shelter, and teach us remain this country's greatest resource. They enrich our lives like nothing else. Nurturing, strengthening, and protecting them must become our greatest priority.

SUMMARY

- Marital success requires the development of interpersonal competence, the ability to share and develop an intimate, growing relationship with another. Commitment (including sexual fidelity and the willingness to sacrifice), communication, and the creative use of conflict are essential aspects of success in marriage.
- The family should be seen as a process. Family health may change over the course of the family cycle.
- The immense work of families includes socializing children and sustaining the care, supervision, and accountability of all family members. The "climate" of the family determines how these services are carried out. Strong families do not just happen. Developing and maintaining family strengths take work.
- *Family strengths* are the patterns that enable successful families to survive and grow. Strengths identified in this chapter include (1) sustaining a commitment to the family; (2) giving affirmation, respect, and trust; (3) developing communication skills; (4) promoting responsibility, morality, and a spiritual orientation; (5) preserving rituals and traditions; (6) managing crises creatively and effectively; (7) utilizing the ability to seek help; (8) spending time together; (9) processing a family wellness orientation; and (10) balancing levels of cohesion and adaptability.
- Specific family strengths may be associated with families that correspond to particular family structures. Single-parent families may benefit from efficient decision making, direct communication, a strong sense of contribution to the family by children, and more egalitarian views of the roles for women and men. Extended families, stepfamilies, and other types of families may be likely to profit from strengths specific to those forms.
- A sense of ethnicity or cultural heritage is an important source of strength for many American families. Often an ethnic group possesses particular strengths that help its families survive in an indifferent or unwelcoming larger society.
- We all have needs that can be met only in close relationships with others. These needs are to nurture others, to be socially integrated in some form of community, to have assistance we can rely on, to be intimate with others, and to be reassured about our skills. Kin, friends (including affiliative kin), neighbors, and a variety of social networks all contribute to our well-being. They provide assistance and emotional support that give our families added strength.
- The family exists in the environment of community and is influenced directly and indirectly by many systems existing outside of itself. The well-being of the family depends not only on its own resources but also on the support of the larger community. For this reason, policies supportive of the family are crucial to its survival.

KEY TERM

family strengths 535

SUGGESTED READINGS

Acock, Alan, and John Demo. *Family Diversity and Well-Being.* Thousand Oaks, CA: Sage, 1994. A comprehensive examination of families of varying structures by two leading family researchers.

Beavers, Robert W., and Robert Hampton. *Successful Families: Assessment and Intervention.* New York: Norton, 1990. A systems approach to building successful families.

Borenstein, M. H., ed. *Cultural Approaches to Parenting.* Hillsdale, NJ: Lawrence Erlbaum, 1991. A discussion of child-rearing practices in a wide variety of cultures in the United States and abroad.

Gottman, John M. *Why Marriages Succeed or Fail, and How You Can Make Yours Work.* New York: Simon and Schuster, 1995. A focus on the research on marriage conducted by this popular author, providing food for thought about what makes it strong.

Hogan, M. J., ed. *Initiatives for Families: Research, Policy, Practice, Education.* St. Paul, MN: National Council on Family Relations, 1995. An excellent and well-respected contribution to the United States International Year of the Family.

Kagan, Sharon L., and Bernice Weissbourd, eds. *Putting Families First: America's Support Movement and the Challenge of Change.* San Francisco: Jossey-Bass, 1994. Contributing authors discuss how families can be strengthened through policy in education, health care, social services, and religious organizations.

McAdoo, Harriette Pipes, ed. *Family Ethnicity: Strength in Diversity.* Newbury Park, CA: Sage, 1993. A collection of 21 essays on major American ethnic groups, stressing their strengths.

McCubbin, Hamilton I., ed. *Family Types and Strengths.* Edina, MN: Burgess International Group, 1988. An examination of family strengths, emerging family types, and recent developments in family stress therapy, coping, and support.

McGoldrick, Monica, John Pearce, and Joseph Giordano, eds. *Ethnicity and Family Therapy,* 2nd ed. New York: Guilford Press, 1996. Separate chapters examining over 25 ethnic groups. Each chapter includes a historical overview, cultural traits, and values. The book treats ethnicity as an important aspect of psychological functioning.

RESOURCES ON THE INTERNET

Companion Web Site for This Book

http://sociology.wadsworth.com/strong/marriage9e
Gain an even better grasp on this chapter by going to the companion Web site to take one of the Tutorial Quizzes, use the Flash Cards to master key terms, or check out the many other study aids you'll find there. Visit the Marriage and Family Resource Center on the site. You'll also find special features such as GSS Data and Census 2000 information that will put data and resources at your fingertips to help you with that special project or to do some research on your own.

InfoTrac College Edition: Search Word Summary

trust
filial piety
ethnicity
crisis management
friendship

To learn more about these central topics in the study of the family, you can conduct an electronic search using InfoTrac College Edition. To aid in your search and to gain useful tips, see the Student Guide to InfoTrac College Edition that you can access through the companion Web site for this book.

APPENDIX A

Sexual Structure and the Sexual Response Cycle

THE FEMALE REPRODUCTIVE SYSTEM

External Genitalia

The female external genitalia are known collectively as the *vulva,* which includes the mons veneris, labia, clitoris, urethra, and introitus. The *mons veneris* (literally, "mountain of Venus") is a protuberance formed by the pelvic bone and covered by fatty tissue. The *labia* are the vaginal lips surrounding the entrance to the vagina. The *labia majora* (outer lips) are two large folds of spongy flesh extending from the mons veneris along the midline between the legs. The outer edges of the labia majora are often darkly pigmented and are covered with pubic hair beginning in puberty. Usually the labia majora are close together, giving them a closed appearance. The *labia minora* (inner lips) lie within the fold of the labia majora. The upper portion folds over the clitoris and is called the *clitoral hood.* During sexual excitement, the labia minora become engorged with blood and double or triple in size. The labia minora contain numerous nerve endings that become increasingly sensitive during sexual excitement.

The *clitoris* is the center of erotic arousal in the female. It contains a high concentration of nerve endings and is highly sensitive to erotic stimulation. The clitoris becomes engorged with blood during sexual arousal and may increase greatly in size. Its tip, the *clitoral glans,* is especially responsive to touch.

Between the folds of the labia minora are the urethral opening and the *introitus.* The introitus is the opening to the vagina; it is often partially covered by a thin perforated membrane called the *hymen,* which may be torn accidentally or intentionally before or during first intercourse. On either side of the introitus is a tiny *Bartholin's gland* that secrets a small amount of moisture during sexual arousal.

Internal Genitalia

The *vagina* is an elastic canal extending from the vulva to the cervix. It envelops the penis during sexual intercourse and is the passage through which a baby is normally delivered. The vagina's first reaction to sexual arousal is "sweating," that is, producing lubrication through the vaginal walls.

A few centimeters from the vaginal entrance, on the vagina's anterior (front) wall, there is, according to some researchers, an erotically sensitive area that they have dubbed the "Grafenberg spot" or "G-spot." The spot is associated with female ejaculation, the

FIGURE A.1 ■ External Female Genitalia

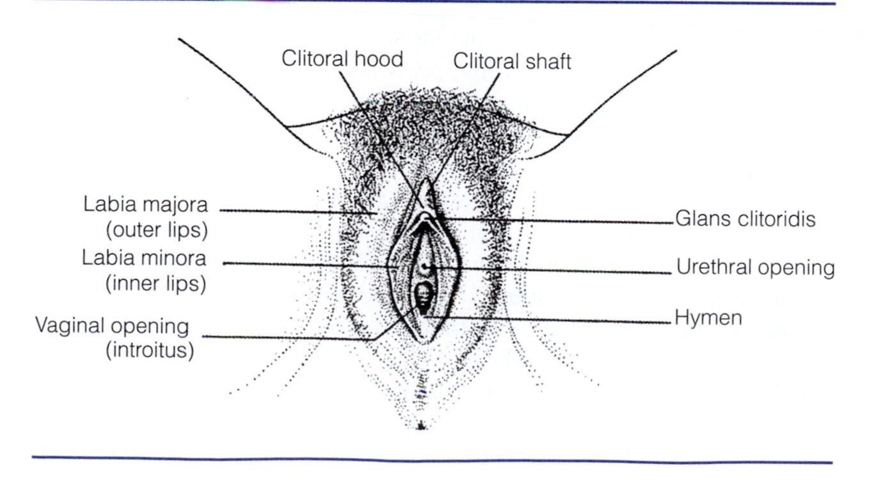

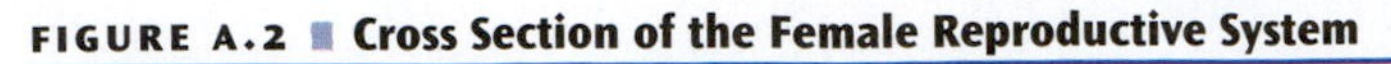
FIGURE A.2 ■ Cross Section of the Female Reproductive System

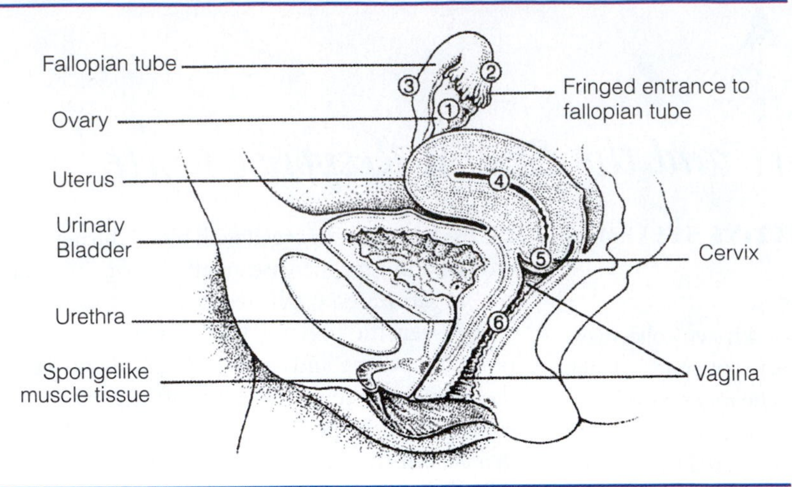

1. A follicle matures in the ovary and releases an ovum. 2. The fimbriae trap the ovum and move it into the fallopian tube. 3. The ovum travels through the fallopian tube to the uterus. 4. If the ovum is fertilized, the resulting blastocyst descends into the uterus. 5. If not fertilized, the ovum is discharged through the cervix into the vagina along with the shed uterine lining during the menstrual flow. 6. The vagina serves as a passageway to the body's exterior.

expulsion of clear fluid from the urethra, which is experienced by a small percentage of women.

A female has two *ovaries,* reproductive glands (gonads) that produce *ova* (eggs) and the female hormones *estrogen* and *progesterone.* At the time a female is born, she already has all the ova she will ever have—more than forty thousand of them. About four hundred will mature during her lifetime and be released during ovulation; ovulation begins in puberty and ends at menopause.

The Path of the Egg

The two *fallopian tubes* extend from the uterus up to, but not touching the ovaries. When an egg is released from an ovary during the monthly *ovulation,* it drifts into a fallopian tube, propelled by waving *fimbriae* (the fingerlike projections at the end of each tube). If it is fertilized by sperm, fertilization usually takes place within the fallopian tube. The fertilized egg will then move into the uterus.

The *uterus* is a hollow, muscular organ within the pelvic cavity. The pear-shape uterus is normally about 3 inches long, 3 inches wide at the top, and 1 inch at the bottom. The narrow, lower part of the uterus projects into the vagina and is called the *cervix.* If an egg is fertilized, it will attach itself to the inner lining of the uterus, the *endometrium.* Inside the uterus it will develop into an embryo and then into a fetus. If an egg is not fertilized, the endometrial tissue that developed in anticipation of fertilization will be shed during *menstruation.* Both the unfertilized egg and the inner lining of the uterus will be discharged in the menstrual flow.

THE MALE REPRODUCTIVE SYSTEM

The Penis

Both urine and semen pass through the penis. Ordinarily, the penis hangs limp and is used for the elimination of urine because it is connected to the bladder by the urinary duct (urethra). The penis is usually between 2.5 and 4 inches in length. When a man is sexually aroused, it swells to about 5 to 8 inches in length, is hard, and becomes erect (hence, the term *erection*). When the penis is erect, muscle contractions temporarily close off the urinary duct, allowing the ejaculation of semen.

The penis consists of three main parts: the root, the shaft, and the glans penis. The *root* connects the penis to the pelvis. The *shaft,* which is the spongy body of the penis, hangs free. At the end of the shaft

FIGURE A.3 ■ External Male Genitalia

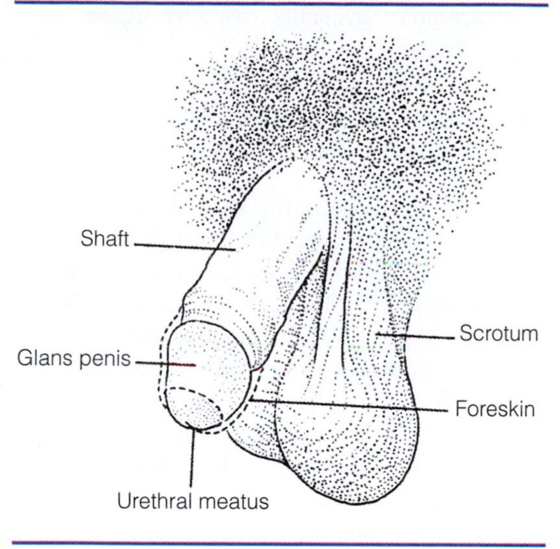

is the *glans penis,* the rounded tip of the penis. The opening at the tip of the glans is called the *urethral meatus.* The glans penis is especially important in sexual arousal because it contains a high concentration of nerve endings, making it erotically sensitive. The *frenulum,* a small area of skin on the underside of the penis where the glans and shaft meet, is especially sensitive. The glans is covered by a thin sleeve of skin called the *foreskin.* Circumcision, the surgical removal of the foreskin, may damage the frenulum.

When the penis is flaccid, blood circulates freely through its veins and arteries, but as it becomes erect, the circulation of blood changes dramatically. The arteries expand and increase the flow of blood into the penis. The spongelike tissue of the shaft becomes engorged and expands, compressing the veins within the penis so that the additional blood cannot leave it easily. As a result, the penis becomes larger, harder, and more erect.

The Testes

Hanging behind the male's penis is his *scrotum,* a pouch of skin holding his two *testes* (singular *testis;* also called *testicles*). The testes are the male reproductive glands (also called *gonads*), which produce both sperm and the male hormone *testosterone.* The testes produce sperm through a process called *spermatogenesis.* Each testis produces between 100 million and 500 million sperm daily. Once the sperm are produced, they move into the *epididymis,* where they are stored prior to ejaculation.

The Path of the Sperm

The epididymis merges into the tubular *vas deferens* (plural *vasa deferentia*). The vasa deferentia can be felt easily within the scrotal sac. Extending into the pelvic cavity, each vas deferens widens into a flasklike area called the *ampulla* (plural *ampullae*).

FIGURE A.4 ■ Cross Section of the Male Reproductive System

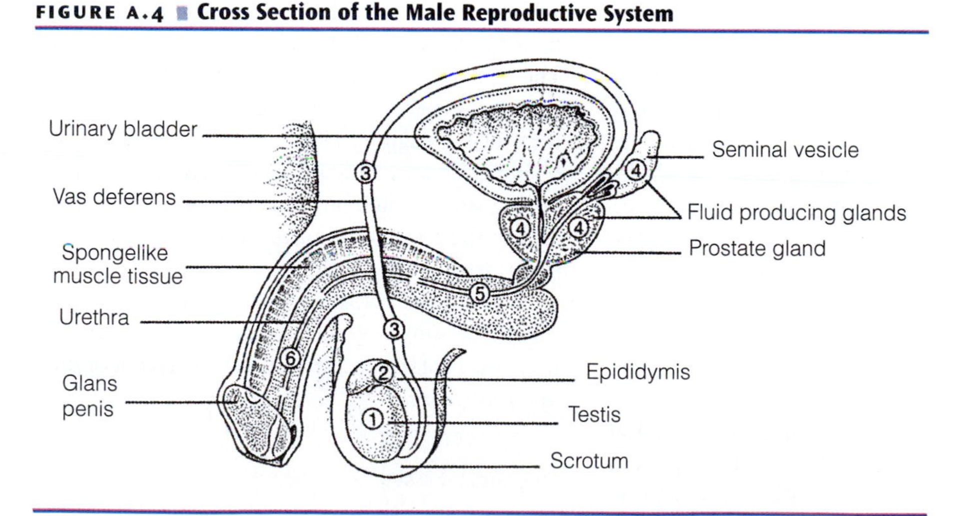

1. The testis produces sperm. 2. Sperm mature in the epididymis. 3. During ejaculation, sperm travel through the vas deferens. 4. The seminal vesicles and the prostate gland provide fluids. 5. Sperm mix with the fluids, making semen. 6. Semen leaves the penis by way of the urethra.

Within the ampullae, the sperm mix with an activating fluid from the *seminal vesicles.* The ampullae connect to the *prostate gland* through the *ejaculatory ducts.* Secretions from the prostate account for most of the milky, gelatinous liquid that makes up the *semen* in which the sperm are suspended. Inside the prostate, the ejaculatory ducts join to the urinary duct from the bladder to form the urethra, which extends to the tip of the penis. The two *Cowper's glands,* located below the prostate, secrete a clear, sticky fluid into the urethra that appears as small droplets on the meatus during sexual excitement.

If the erect penis is stimulated sufficiently through friction, an ejaculation usually occurs. *Ejaculation* is the forceful expulsion of semen. The process involves rhythmic contractions of the vasa deferentia, seminal vesicles, prostate, and penis. Altogether, the expulsion of semen may last from three to fifteen seconds. It is also possible to have an orgasm without the expulsion of semen.

The Sexual Response Cycle

PSYCHOLOGICAL AND PHYSIOLOGICAL ASPECTS

When we respond sexually, we begin what is known as the *sexual response cycle.* Helen Singer Kaplan (1979) developed a model to describe the sexual response cycle. According to this model, the cycle consists of three phases: the desire phase, the excitement phase, and the orgasmic phase. The desire phase represents the psychological element of the sexual response cycle; the excitement and orgasmic phases represent its physiological aspects.

SEXUAL DESIRE

Desire can exist separately from overtly physical sexual responses. It is the psychological component that motivates sexual behavior. We can feel desire but not be physically aroused. It can suffuse our bodies without producing explicit sexual stirrings. We experience sexual desire as erotic sensations or feelings that motivate us to seek sexual experiences. These sensations generally cease after orgasm.

PHYSIOLOGICAL RESPONSES: EXCITEMENT AND ORGASM

A person who is sexually excited experiences a number of bodily responses. Most of us are conscious of some of these responses: a rapidly beating heart, an erection or lubrication, and orgasm. Many other responses may take place below the threshold of awareness, such as curling of the toes, the ascent of the testes, the withdrawal of the clitoris beneath the hood, and a flush across the upper body.

The physiological changes that take place during sexual response cycle depend on two processes: vasocongestion and myotonia. *Vasocongestion* occurs when body tissues become engorged with blood. For example, blood fills the genital regions of both males and females, causing the penis and clitoris to enlarge. *Myotonia* refers to increased muscle tension as orgasm approaches. Upon orgasm, the body undergoes involuntary muscle contractions and then relaxes. (The word *orgasm* is derived from the ancient Sanskrit *urja,* meaning "vigor" or "sap.")

Excitement Phase

In women, the vagina becomes lubricated and the clitoris enlarges during the excitement phase. The vaginal barrel expands, and the cervix and uterus elevate, a process called "tenting." The labia majora flatten and rise; the labia minora begin to protrude. The breasts may increase in size, and the nipples may become erect. Vasocongestion causes the outer third of the vagina to swell, narrowing the vaginal opening. This swelling forms the *orgasmic platform;* during sexual intercourse, it increases the friction against the penis. The entire clitoris retracts but remains sensitive to touch.

In men, the penis becomes erect as a result of vasocongestion, and the testes begin to rise. The testes may enlarge to as much as 150 percent of their unaroused size.

Orgasmic Phase

Orgasm is the release of physical tensions after the buildup of sexual excitement; it is usually accompanied by ejaculation of semen in physically mature

FIGURE A.5 ■ Stages of Female Sexual Response (internal left; external right)

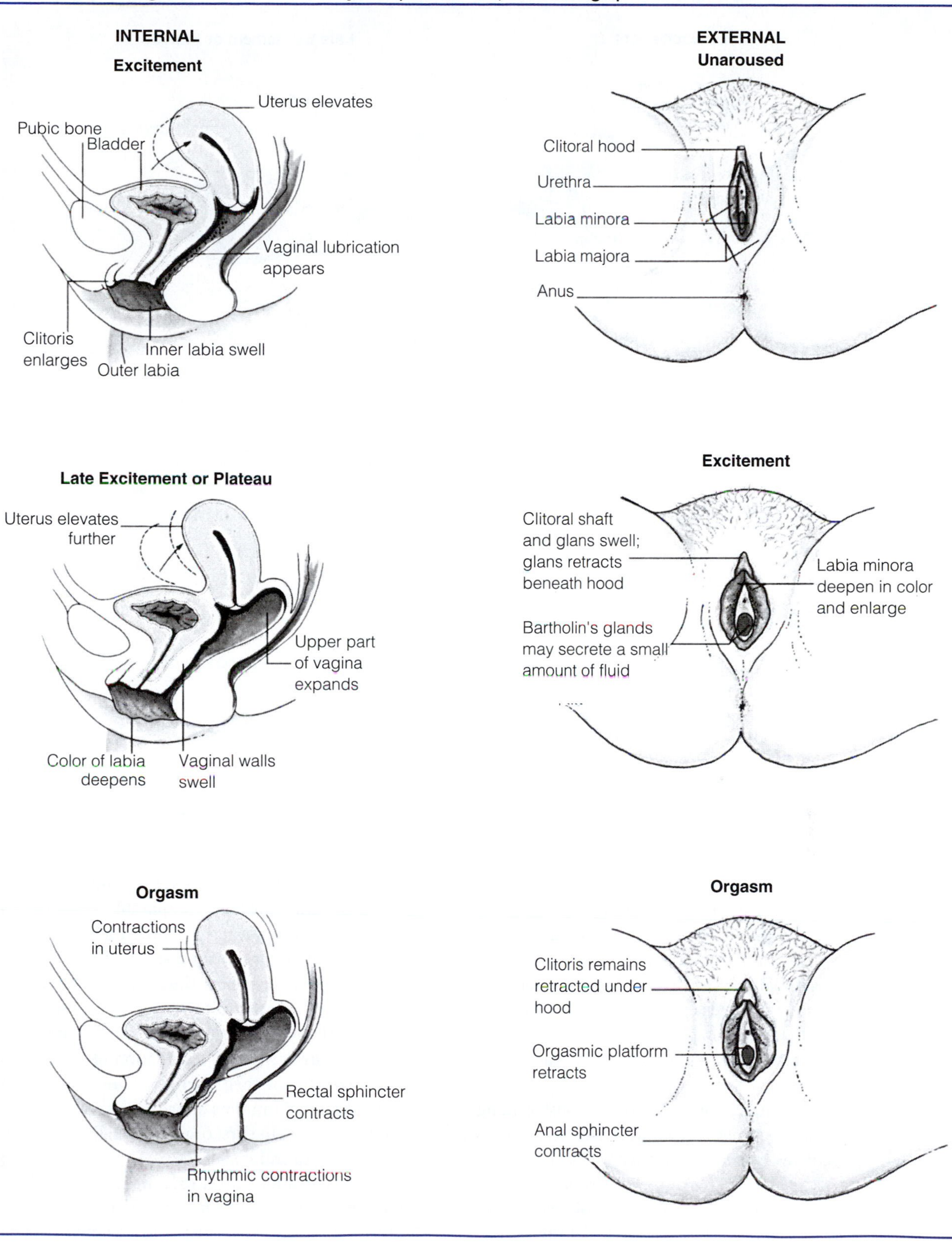

FIGURE A.6 ■ **Stages of Male Sexual Response**

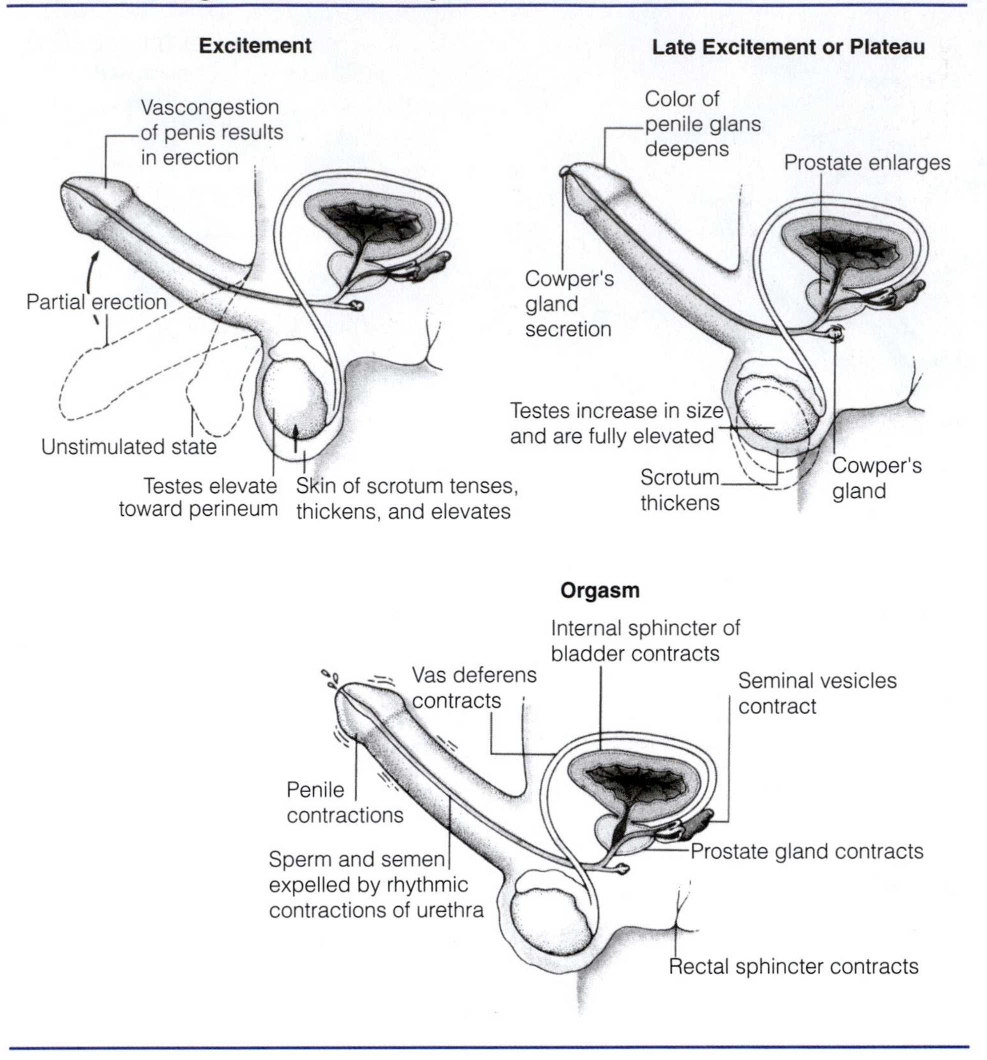

males. In women, the orgasmic phase is characterized by simultaneous rhythmic contractions of the uterus, orgasmic platform, and rectal sphincter. In men, muscle contractions occur in the vasa deferentia, seminal vesicles, prostate, and the urethral bulb, resulting in the ejaculation of semen; contractions of the rectal sphincter also occur. Ejaculation usually accompanies male orgasm, but ejaculation and orgasm are separate processes.

Following orgasm, one of the most striking differences between male and female sexual response occurs as males experience a *refractory period.* The refractory period denotes the time following orgasm during which male arousal levels return to prearousal or excitement levels. During the refractory period, additional orgasms are impossible. Females do not have any comparable period. As a result, they have greater potential for multiple orgasms—that is, for having a series of orgasms. Although most women have the potential for multiple orgasms, only about 13 to 16 percent regularly experience them. For multiple orgasms, women generally require continued stimulation of the clitoris. Most women (or their partners), however, do not seek additional orgasms after the first one because our culture uses the first orgasm (usually the male's) as a marker to end sexual activities.

In sexual intercourse, orgasm has many functions. For men, it serves a reproductive function by causing ejaculation of semen into a women's vagina.

For both men and women, it is a source of erotic pleasure, whether it is an autoerotic or relational context; it is intimately connected with our sense of well-being. We may measure both our sexuality and ourselves in terms of orgasm. Did I have one? Did my partner have one? When we measure our sexuality by orgasm, however, we discount activities that do not necessarily lead to orgasm, such as touching, caressing, and kissing. We discount erotic pleasure as an end in itself.

Men tend to be more consistently orgasmic than women, especially in sexual intercourse. If all women are potentially orgasmic, why do smaller proportion of women have orgasms than men? An answer may be found in our dominant cultural model that calls for female orgasm to occur as a result of penile thrusting during heterosexual intercourse in the face-to-face, male-above position. This traditional American model calls for a "no-hands" approach. The women is supposed to be orgasmic without manual or oral stimulation by her partner or herself. If she is orgasmic during masturbation or cunnilingus, such orgasms are usually discounted because they aren't considered "real" sex—that is, heterosexual intercourse.

The problem for women in sexual intercourse is that the clitoris frequently does not receive sufficient stimulation from penile thrusting alone to permit orgasm. In an influential study on female sexuality, Shere Hite (1976) found that only 30 percent of her three thousand respondents experienced orgasm regularly through sexual intercourse "without more direct manual clitoral stimulation being provided at one time of orgasm." Hite concludes that many women need manual stimulation during intercourse to be orgasmic. They also need to be assertive. There is no reason why a women cannot be manually stimulated by herself or her partner to orgasm before or after intercourse. But to do so, a woman has to assert her own sexual needs and move away from the idea that sex is centered around male orgasm.

APPENDIX B

Pregnancy, Conception, and Fetal Development

Once fertilization of the ovum by a sperm occurs, the birth will take place in approximately 266 days, if the pregnancy is not interrupted. Traditionally, physicians count the first day of the pregnancy as the day on which the woman began her last menstrual period; thus, they calculate the gestation (pregnancy) period to be 280 days, which is also 10 lunar months.

Following fertilization, which normally occurs within the fallopian tube, the fertilized ovum, or *zygote,* undergoes a series of divisions during which the cells replicate themselves. After four or five days, the zygote contains about a hundred cells and is called *blastocyst.* On about the fifth day, the blastocyst arrives in the uterine cavity, where it floats for a day or two before implanting itself in the soft, blood-rich uterine wall (endometrium), which has spent the past three weeks preparing for its arrival. This process of *implantation* takes about a week. The hormone human chorionic gonadotropin (HCG), which is secreted by the blastocyst, maintains the uterine environment in an "embryo-friendly" condition and prevents the shedding of the endometrium that would normally occur during menstruation.

The blastocyst, or pre-embryo, rapidly grows into an *embryo* (which will, in turn, be referred to as a *fetus* around the eighth week of development). During the first two or three weeks of development, the embryonic membranes, including the *amnion*—a membranous sac that will contain the embryo and *amniotic fluid*—and the *yolk sac* are formed.

During the third week, extensive cell migration occurs and the stage is set for the development of the organs. The first body segments and the brain begin to be formed. The digestive and circulatory systems begin to develop in the fourth week; the heart begins to pump blood. By the end of the first month, the spinal cord and nervous system have also begun to develop.

The fifth week sees the formation of arms and legs. In the sixth week, the eyes and ears form. At seven weeks, the reproductive organs begin to differentiate in the males; female reproductive organs continue to develop. At eight weeks, the fetus is about the size of a thumb, although the head is nearly as large as the body. The brain begins to function to coordinate the development of the internal organs. Facial features begin to form, and bones begin to develop.

Arms, hands, fingers, legs, feet, toes, and eyes are almost fully developed at twelve weeks. At fifteen weeks, the fetus has a strong heartbeat, fair digestion, and active muscles. Most bones are developed by then, and the eyebrows appear. At this stage, the fetus is covered with # a fine, downy hair called *lanugo.* (Figure B.1 and Figure B.2 shows the actual size of the developing embryo and fetus through its first sixteen weeks.)

Throughout its development, the fetus is nourished through the *placenta.* The placenta begins to develop from part of thee blastocyst following implantation. This organ grows larger as the fetus does, passing nutrients from the mother's bloodstream to the fetus, to which it is attached by the umbilical cord. The placenta blocks blood corpuscles and large molecules.

By five months, the fetus is 10 to 12 inches long and weighs between one-half and one pound. The internal organs are well developed, although the lungs cannot function well outside the uterus. At six months, the fetus is 11 to 14 inches long and weighs more than a pound. At seven months, it is 13 to 17 inches long, weighing about three pounds. At this point, most healthy fetuses are viable—capable of surviving outside the womb. (Although some fetuses are viable at five or six months, they require specialized care to survive.) The fetus spends the final two months of gestation growing rapidly. At term (nine months), it will be about 20 inches long and weigh about seven pounds.

FIGURE B.1 ■ **Embryonic and Fetal Development**

Fetal development, or gestation, takes approximately 266 days from fertilization of the ovum to birth. These photographs chronicle various stages of the process.

(a) After ejaculation, several million sperm move through the cervical mucus toward the fallopian tubes; an ovum has descended into one of the tubes. En route to the ovum, millions of sperm are destroyed in the vagina, uterus, or fallopian tubes. Some go the wrong direction in the vagina and others swim into the wrong tube.

(b) The ovum has divided for the first time following fertilization; the mother's and father's chromosomes have united. In subsequent cell divisions the genes will be identified. After about a week the blastocyst will implant itself into the uterine lining.

(c) The embryo is five weeks old and is two-fifths of an inch long. It floats in the embryonic sac. The major divisions of the brain can be seen as well as an eye, hands, arms, and a long tail.

(d) The embryo is now seven weeks old and is almost an inch long. Its outer and inner organs are developing. It has eyes, a nose, a mouth, lips, and a tongue.

(e) At twelve weeks, the fetus is over three inches long and weighs almost an ounce.

(f) At sixteen weeks, the fetus is more than six inches long and weighs about seven ounces. All organs have been formed. The time that follows is now one of simple growth.

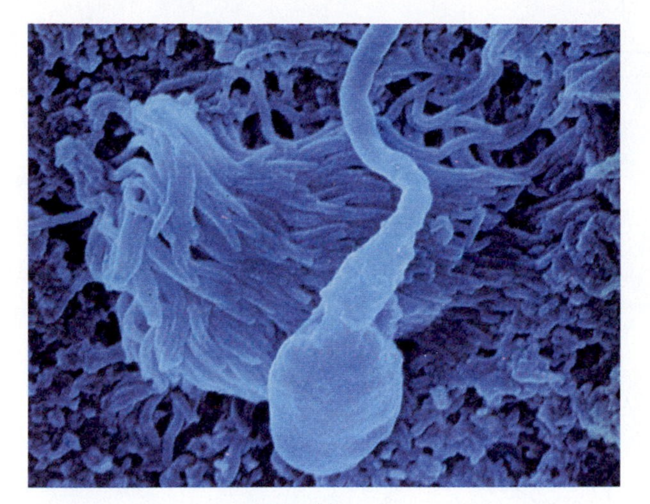

(a)

(b)

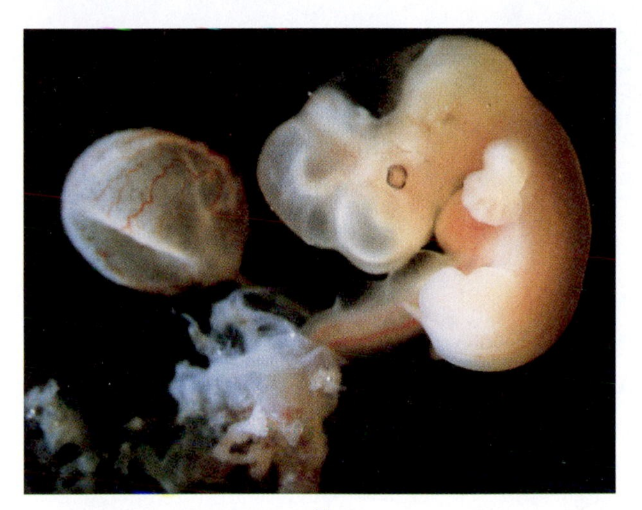

(c)

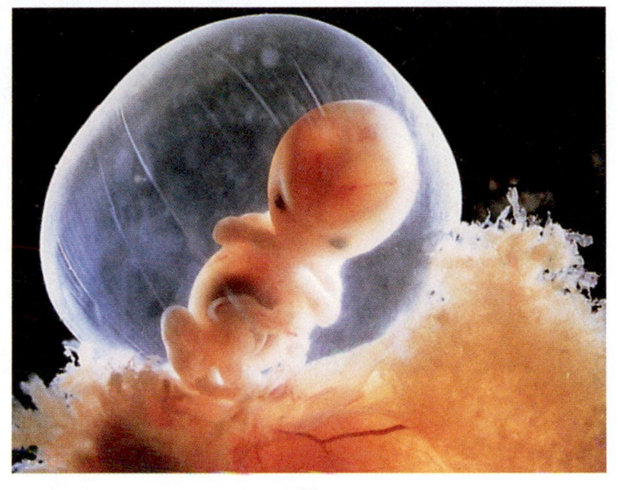

(d)

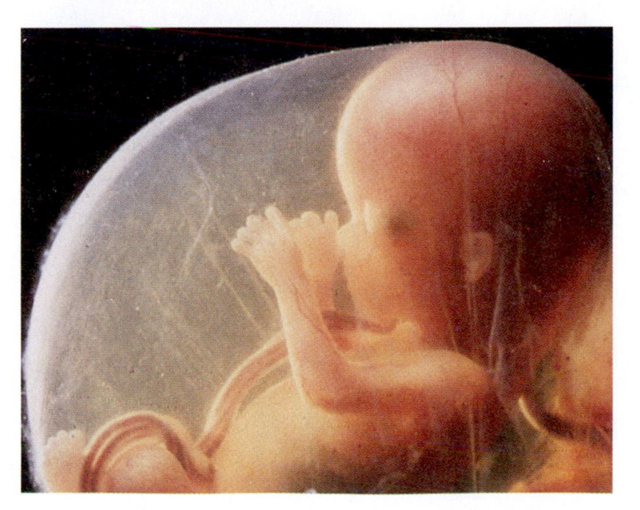

(e)

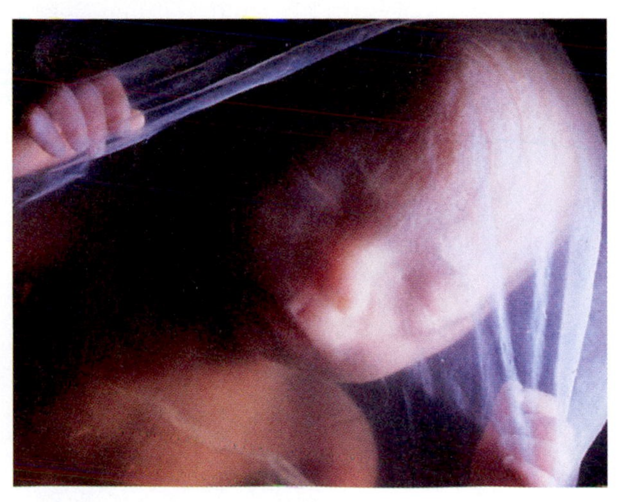

(f)

FIGURE B.2 ■ Embryo and Fetus Growth

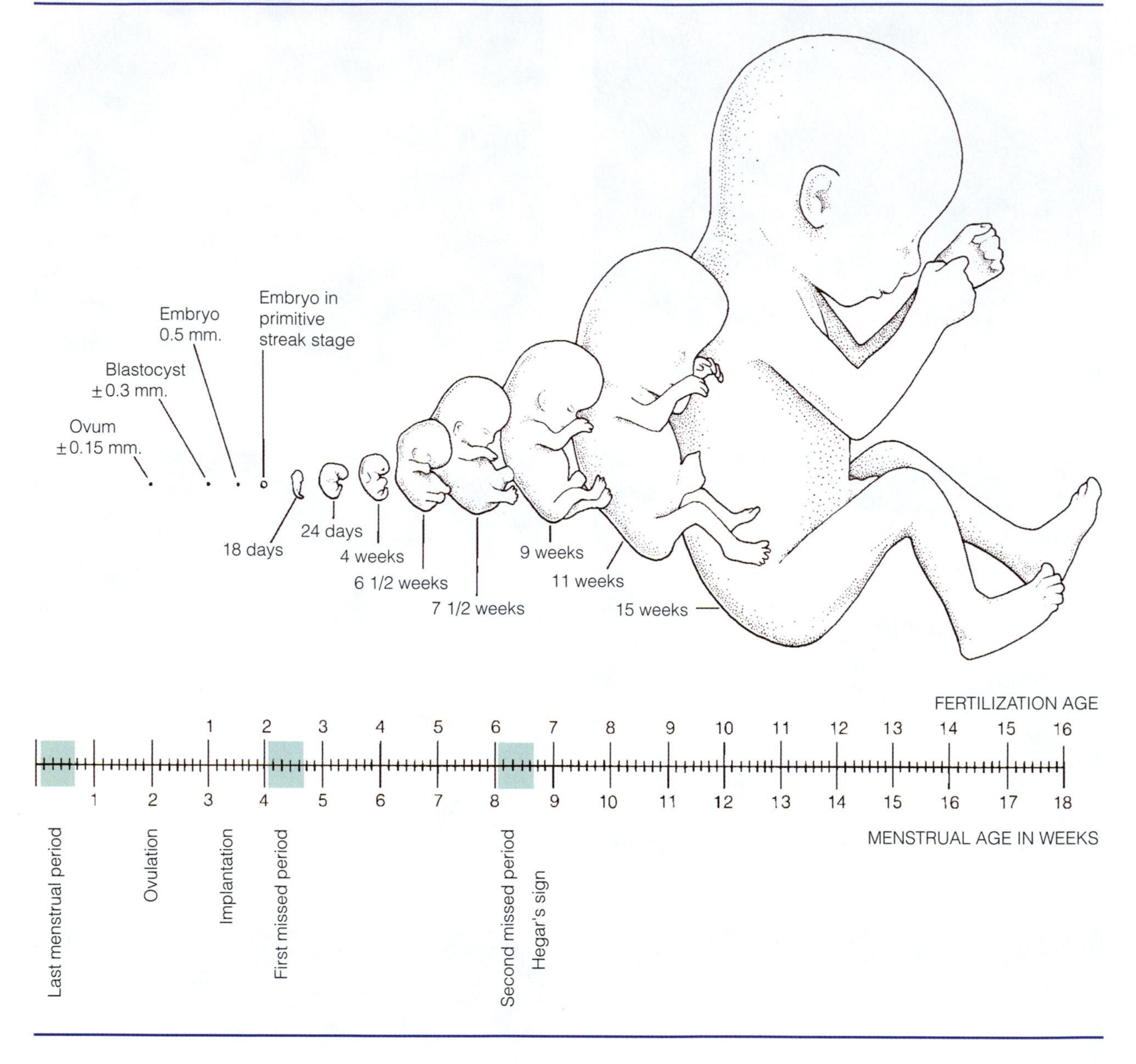

APPENDIX C

The Budget Process

A budget is a plan for spending and saving. It requires you to estimate your available income for a particular period of time and decide how to allocate this income toward your expenses. A working budget can help you implement your money management plan. A well-planned budget does several things for you and your household. It can help you do the following:

- Prevent impulse spending
- Decide what you can or cannot afford
- Know where your money goes
- Increase savings
- Decide how to protect against the financial consequences of unemployment, accidents, sickness, aging, and death

A working budget need not be complicated or rigid. However, preparing one takes planning, and following one takes determination. You must do several things to budget successfully.

First, communicate with other members of your household, including older children. Consider each person's needs and wants so that all family members feel they are a part of the plan. Everyone may work harder to make the budget a success and be less inclined to overspend if they realize the consequences. When families fail to communicate about money matters, it is unlikely that a budget will reflect a workable plan.

Second, be prepared to compromise. This is often difficult. Newlyweds, especially, may have problems. Each may have been living on an individual income and not be accustomed to sharing or may have been in school and dependent on parents. If, for example, one wants to save for things and the other prefers buying on credit, the two will need to discuss the pros and cons of both methods and decide on a middle ground that each can accept. A plan cannot succeed unless there is a financial partnership.

Third, exercise willpower. Try not to indulge in unnecessary spending. Once your budget plan is made, opportunities to overspend will occur daily. Each household member needs to encourage the others to stick to the plan

Fourth, develop a good record-keeping system. At first, all members of the household may need to keep records of what they spend. This will show how well they are following the plan and will allow intermediate adjustments in the level of spending. Record keeping is especially important during the first year of a spending plan, when you are trying to find a budget that works best for you. Remember: good budget is flexible, requires little clerical time, and, most important, works for you.

CHOOSING A BUDGET PERIOD

A budget may cover any convenient period of time—a month three months, or a year, for example. Make sure the period you us, is long enough to cover the bulk of household expenses and income. Remember: Not all bills come due monthly, and every household experiences some seasonal expenses. Most personal budgets are for twelve months. You can begin the twelve-month period at any time during the year. If this is your first budget, you may want to set up a trial plan for a shorter time to see how it works.

After setting up your plan, subdivide it into more manageable operating periods. For a yearly budget, divide income and expenses by 12, 24, 26, or 52, depending on your pay schedule or when your bills come due. Most paychecks are received weekly or every two weeks. Although most bills come due once a month, not all are due at the same time in the month. Try using each paycheck to pay your daily expenses and expenses that will be due within the next week or two. This way you will be able to pay your bills on time. You may also want to allocate something from each paycheck toward large expenses that will be coming due soon.

WORKSHEET 1 ■ Estimating Your Income

SOURCE	JANUARY	FEBRUARY	MARCH	APRIL	MAY	JUNE
Net salary:[a]						
Household member 1						
Household member 2						
Household member 3						
Household member 4						
Social Security payments						
Pension payments						
Annuity payments						
Veterans' benefits						
Assistance payments						
Unemployment compensation						
Allowances						
Alimony						
Child support						
Gifts						
Interest						
Dividends						
Rents from real estate						
Other						
Monthly Totals						

[a]Net salary is the amount that comes into the household for spending and saving after taxes, Social Security, and other deductions.

DEVELOPING A SUCCESSFUL BUDGET

Step 1: Estimate Your Income

Total the money you expect to receive during the budget period. Use Worksheet 1 as a guide in estimating your household income. Begin with regular income that you and your family receive—wages, salaries, income earned from a farm or other business, Social Security benefits, pension payments, alimony, child support, veterans' benefits, public assistance payments, unemployment compensation, allowances, and any other income. Include variable income, such as interest from bank accounts and investments, dividends from stock and insurance, rents from property you own, gifts, and money from any other sources.

If your earnings are irregular, it may be difficult to estimate your income. It is better to underestimate than to overestimate income when setting up a budget. Some households have sufficient income, but its receipt does not coincide with the arrival of bills. For these households, planning is very important.

Step 2: Estimate Your Expenses

After you have determined how much your income will be for the planning period, estimate your ex-

WORKSHEET 1 ■ ***(continued)***

JULY	AUGUST	SEPTEMBER	OCTOBER	NOVEMBER	DECEMBER	YEARLY TOTALS

penses. You may want to group expenses into one of three categories: fixed, flexible, or set-asides. Fixed expenses are payments that are basically the same amounts each month. Fixed regular expenses include such items as rent or mortgage payments, taxes, and credit installment payments. Fixed irregular expenses are large payments due once or twice a year, such as insurance premiums. Flexible expenses vary from one month to the next, such as amounts spent on food, clothing, utilities, and transportation. Set-asides are variable amounts of money accumulated for special purposes, such as for seasonal expenses, savings and emergency funds, and intermediate and long-term goals.

Use old records, receipts, bills, and canceled checks to estimate future expenses, if you are satisfied with what your dollars have done for you and your family in the past. If you are not satisfied, now is the time for change. Consider which expenses can be cut back and which expenses need to be increased. If you spent a large amount on entertainment, for example, your new budget may reallocate some of this money to a savings account to contribute to some of your future goals.

If you do not have past records of spending, or if this is your first budget, the most accurate way to find out how much Von will need to allow for each expense is to keep a record of your household spending. Carry a pocket notebook in which you jot down expenditures during a week or pay period, and total the amounts at the end of each week. You may prefer to keep an account book in a convenient place at home and make entries in it. Kept faithfully for a month or two, the record can help you find out what you spend for categories such as food, housing, utilities, household operation, clothing, transportation, entertainment, and personal items. Use this record to estimate expenses in your plan for future spending. You also need to plan for new situations and changing conditions that increase or decrease expenses. For example, the cost of your utilities may go up.

Total your expenses for a year and divide to determine the amounts that you will have to allocate toward each expense during the budgeting period. Record your estimate for each budgetary expense in the space provided on Worksheet 2. Begin with the regular fixed expenses that you expect to have. Next, enter those fixed expenses that come due once or twice a year. Many households allocate a definite amount each budget period toward these expenses to spread out the cost.

One way to meet major expenses is to set aside money regularly before you start to spend. Keep your set-aside funds separate from other funds so you will not be tempted to spend them impulsively. If possible, put them in an account where they will earn interest. You may also plan at this point to set aside a certain amount toward the long-term and intermediate goals you listed on Worksheet 1. Saving could be almost as enjoyable as spending, once you accept the idea that saving money is not punishment but instead a systematic way of reaching your goals. You do without some things now in anticipation of buying what will give you greater satisfaction later.

You may want to clear up debts now by doubling up on your installment payments or putting aside an extra amount in your savings fund to be used for this purpose. Also, when you start to budget, consider designating a small amount of money for emergencies. Extras always come up at the most inopportune times. Every household experiences occasional minor crises too small to be covered by insurance but too large to be absorbed into the day-to-day budget. Examples may be a blown-out tire or an appliance that needs replacing. Decide how large a cushion you want for meeting emergencies. As your fund reaches the figure you have allowed for emergencies, you can start saving for something else. Now, record money allocated for occasional major expenses, future goals, savings, emergencies, and any other set-asides in the space provided for them on Worksheet 2.

After you have entered your fixed expenses and your set-asides, you are ready to consider your flexible expenses. Consider including here a personal allowance, or "mad money," for each member of the household. A little spending money that does not have to be accounted for gives everyone a sense of freedom and takes some of the tedium out of budgeting.

Step 3: Balance

Now you are ready for the balancing act. Compare your total expected income with the total of your planned expenses for the budget period. If your planned budget equals your estimated future income, are you satisfied with this outcome? Have you left enough leeway for emergencies and errors? If your expenses add up to more than your income, look again at all parts of the plan. Where can you cut down? Where are you overspending? You may have to decide which things are most important to you and which ones can wait. You may be able to do some trimming on your flexible expenses.

Once you have cut back your flexible expenses, scan your fixed expenses. Maybe you can make some sizable reductions here, too. Rent is a big item in a budget. Some households may want to consider moving to a lower-priced apartment or making different living arrangements. Others may turn in a too-expensive car and seek less expensive transportation. Look back at Worksheet 1. You may need to reallocate some of this income to meet current expenses. Perhaps you may have to consider saving for some of your goals at a later date.

If you have cut back as much as you think you can or are willing to do and your plan still calls for more than you make, consider ways to increase your income. You may want to look for a better-paying job, or a part-time second job may be the answer. If only one spouse is employed, consider becoming a duel-earner family. The children may be able to earn their school lunch and extra spending money by doing odd jobs in your neighborhood, such as cutting grass or baby-sitting. Older children can work part-time on weekends to help out. Another possibility, especially for short-term problems, is to draw on savings. These are decisions each individual household has to make.

If your income exceeds your estimate of expenses—good! You may decide to satisfy more of your immediate wants or to increase the amount your family is setting aside for future goals.

CARRYING OUT YOUR BUDGET

After your plan is completed, put it to work. This is when your determination must really come into play. Can you and your family resist impulse spending?

Become a Good Consumer

A vital part of carrying out a budget is being a good consumer. Learn to get the most for your money, to recognize quality, to avoid waste, and to realize time costs as well as money costs in making consumer decisions.

Keep Accurate Records

Accurate financial records are necessary to keep track of your household's actual money inflow and outgo. A successful system requires cooperation from everyone in the household. Receipts can be kept and entered at the end of each budget period in a "Monthly Expense Record." It is sometimes a good idea to write on the back of each receipt what the purchase was for, who made it, and the date. Decide which family member will be responsible for paying bills or making purchases, and decide who will keep the record system up to date.

The household business record-keeping system does not need to be complex. The simpler it is, the more likely it will be kept current. Store your records in one spot—a set of folders in a file drawer or other fire-resistant box is a good place. You can assemble a folder for each of several categories, including budget, food, clothing, housing, insurance, investments, taxes, health, transportation, and credit. Use these folders for filing insurance policies, receipts, warranties, cancelled checks, bank statements, purchase contracts, and other important papers. Many households also rent a safe deposit box at the bank for storing deeds, stock certificates, and other valuable items.

EVALUATING YOUR BUDGET

The information on Worksheet 2 can help you determine whether your actual spending follows your plan. If your first plan did not work in all respects, do not be discouraged. A budget is not something you make once and never touch again. Keep revising until the results satisfy you.

Dealing with Unemployment

STEP 1: TAKE TIME TO TALK

Come right out and let your family know what's going on. Lay-off Plant closing? Depressed economy? Business down? Explain what happened. Break down the big words so that everyone understands, especially the kids.

Fill in everyone at a family meeting or on a one-to-one basis. The important thing is not to leave anyone in the dark. If a family meeting seems out of the question, take time to talk when cleaning up after meals, cutting or raking the lawn, or taking trips to the store. Don't sugar coat the facts or tell "fairy tales." Living with less money will force your family to make hard changes. Yet let your kids know that even though there's less money, they can still count on a loving family—maybe more loving than ever.

STEP 2: TAKE TIME TO LISTEN

Let everyone have a say about what these changes mean to him or her. Especially now, kids should be seen and heard.

Listen to words and actions. Is someone suddenly having a lot of crying spells, sleeping in late all the time, acting mean, drinking heavily, withdrawing, abusing drugs, complaining of stomach pains?

STEP 3: FIND OUT WHO'S HURTING

Let everyone say what he or she is really feeling from time to time.

WORKSHEET 2 ■ Expense Estimate and Budget Balancing Sheet, Fixed Expenses (Prepare for Each Month)

MONTH:	AMOUNT ESTIMATED	AMOUNT SPENT	DIFFERENCE
Rent			
Mortgage			
Installments:			
Credit card 1			
Credit card 2			
Credit card 3			
Automobile loan			
Personal loan			
Student loan			
Insurance:			
Life			
Health			
Property			
Automobile			
Disability			
Set-asides:			
Emergency fund			
Major expenses			
Goals			
Savings and investments			
Allowances			
Education:			
Tuition			
Books			
Transportation:			
Repairs			
Gas and oil			
Parking and tolls			
Bus and taxi			
Recreation			
Gifts			
Other			
Total Fixed Expenses for Month			

Just repeat whatever you hear, right when it's said. Then look for a nod to see if you heard it right. Is someone feeling helpless, sad, unloved, confused, worried, frightened, angry, like a burden to the family?

Try not to say "You shouldn't fell that way" because someone may be in real pain. The best you can do is let your loved ones have their say and get it off their chests.

STEP 4: LET YOUR FEELINGS OUT, TOGETHER OR ALONE

Give everyone in your family a space and time to let deep feelings out. Don't bottle them up or hide them from yourselves. If you're not comfortable showing others how you feel or fear you may strike someone who's dear to you, consider getting out of the house for a run or a brisk walk; having a good cry, alone; hitting a cushion or pillow; going to your room, shutting the door, and screaming; or all of the above.

STEP 5: SOLVE PROBLEMS TOGETHER

Every week, look at the changes taking place in your household, and work out ways to deal with them. Working together as a team, your family can do more than survive. It can grow together and come through stronger.

Decide together things like these: what we can't afford now; what things we can do for family fun that don't cost a lot of money; who will do what chores around the house; how we'll all get by with less. If your discussions break down, go back to Step 1.

If you have a lot of trouble going through these steps, professional help may be what you need. Call and make an appointment with the family service agency nearest you. Whether or not you have money to pay for the services, the agency will do its best to help your family. Remember: You're not alone.

GLOSSARY

A

aberration A departure from what is culturally defined as "normal" behavior.

abortifacient A substance that can induce abortion.

abortion The termination of a pregnancy either through miscarriage (spontaneous abortion) or through human intervention (induced abortion).

abstinence Refraining from sexual intercourse, often on religious or moral grounds.

abuse Mistreatment; wrong, bad, injurious, or excessive use.

accommodation According to Jean Piaget, the process by which a child makes adjustments in his or her cognitive framework in order to incorporate new experiences.

acquaintance rape Rape in which the assailant is personally known to the victim, usually in the context of a dating relationship. Also known as *date rape.*

acquired immunodeficiency syndrome (AIDS) An infection caused by the human immunodeficiency virus (HIV), which suppresses and weakens the immune system, leaving it unable to fight opportunistic infections.

adaptability Ability to adjust relationships, roles, and rules to changing circumstances.

adolescence The social and psychological state occurring during puberty.

adoption The process by which an individual or couple legally become the parents of a child not biologically their own.

advice/information genre Media such as self-help books, newspaper advice columns, radio and TV talk shows, and women's magazines that purport to offer factual and accurate information but are actually motivated by considerations such as the need to entertain and to make a profit.

AFDC See *Aid to Families with Dependent Children.*

affiliated kin Unrelated individuals who are treated as if they were related.

affinity 1. Relationship by marriage. 2. A close relationship.

affirmative action Programs that attempt to place qualified members of minorities in government, corporate, and educational institutions from which they have been historically excluded because of their minority status.

afterbirth The placenta and fetal membranes expelled from the uterus during the third stage of labor.

agape [AH-ga-pay] According to sociologist John Lee's styles of love, altruistic love.

agglutination test A urine analysis test used to determine the presence of human chorionic gonadotropin (HCG) secreted by the placenta, which is an indication of pregnancy.

AIDS See acquired immunodeficiency syndrome.

Aid to Families with Dependent Children (AFDC) A government program providing financial assistance to families with children during times of poverty. Cf. *Temporary Assistance to Needy Families.*

alimony Court-ordered monetary support to a spouse or former spouse following separation or divorce.

anal eroticism Sexual activities involving the anus.

anal intercourse Penetration of the anus by the penis.

androgyny The state of having flexible gender roles combining instrumental and expressive traits in accordance with unique individual differences.

annulment The legal invalidation of a marriage as if the marriage never occurred.

anonymity A state or condition requiring that no one, including the researcher can connect particular responses to the individuals who provided them.

antiabortion movement A social movement that takes a position against abortion. Also known as *pro-life movement.*

antigay prejudice Strong dislike, fear, or hatred of gay men and lesbians because of their homosexuality. See also *homophobia.*

Apgar score The rating given to a newborn immediately after birth, indicating the child's overall condition based on heart rate, respiration, coloring, reflexes, and muscle tone.

Asian American Collective term relating to Americans of Asian descent, such as Chinese American, Japanese American, Korean American, Vietnamese American, or Cambodian American.

assimilation In Jean Piaget's cognitive developmental theory, the process through which the developing child makes new information compatible with his or her world understanding.

asymptomatic Not showing symptoms.

attachment Close, enduring emotional bonds, especially those forged between an infant and his or her primary caregiver(s).

attachment theory of love A theory maintaining that the degree and quality of an infant's attachment to his or her primary caregiver is reflected in his or her love relationships as an adult.

authoritarian child rearing A parenting style characterized by the demand for absolute obedience.

authoritative child rearing A parenting style that recognizes the parent's legitimate power and also stresses the child's feelings, individuality, and need to develop autonomy.

autoeroticism Erotic behavior involving only the self; usually refers to masturbation but also includes erotic dreams and fantasies.

B

barrier method Any of a number of contraceptive methods that place a physical barrier between sperm and egg, such as the condom, diaphragm, and cervical cap.

basal body temperature (BBT) method A contraceptive method based on variations of the woman's resting body temperature, which rises 24 to 72 hours before ovulation.

basic conflict Pronounced disagreement about fundamental roles, tasks, and functions. Cf. *nonbasic conflict.*

battering A violent act directed against another, such as hitting, slapping, beating, stabbing, shooting, or threatening with weapons.

BBT method See *basal body temperature method.*

behaviorism A model of human development that explains behavior solely on the basis of that which can be observed.

bereavement The response to a loved one's death, including customs, rituals, and the grieving process.

bias A personal leaning or inclination.

Billings method See *mucus method.*

binuclear family A postdivorce family with children, consisting of the original nuclear family divided into two families, one headed by the mother, the other by the father; the two "new" families may be either single-parent or stepfamilies.

bipolar gender role model The traditional view of masculinity and femininity in which male and female gender roles are seen as polar opposites, with males possessing exclusively instrumental traits and females possessing exclusively expressive traits.

birth control Devices, drugs, techniques, or surgical procedures used to prevent conception or implantation or to terminate pregnancy.

birth plan A written plan made by an expectant mother or couple and shared with the birth attendant, detailing expectations regarding birth setting, medications, father's participation, visitors, circumcision, breastfeeding, and so on.

birth rate The number of births per year per thousand people in a given community or group. Cf. *fertility rate.*

bisexuality Sexual involvement with both sexes, usually sequentially rather than during the same time period.

blastocyst An early stage of the fertilized ovum, containing about 100 cells, that implants itself in the uterine wall; a pre-embryo.

blended family A family in which one or both partners have a child or children from an earlier marriage or relationship; a stepfamily. See also *binuclear family.*

bondedness The degree of emotional bonding or closeness within a family.

boomerang generation Individuals who, as adults, return to their family home and live with their parents.

boundary In systems theory, the emotional, psychological, or physical separation between subsystems or roles (such as between family members) required for adequate functioning.

Braxton Hicks contractions Uterine contractions that occur periodically throughout pregnancy and also initiate effacement and dilation of the cervix at the beginning of labor.

breech presentation A fetal position in which the baby enters the birth canal with the buttocks or feet appearing first.

bride price The goods, services, or money a family receives in exchange for giving their daughter in marriage.

brit milah In Judaism, the circumcision ceremony for an infant boy. Also known as *bris.*

bundling A colonial Puritan courtship custom in which a couple slept together with a board separating them.

C

calendar method A fertility awareness method based on calculating "safe" days according to the range of a women's longest and shortest menstrual cycles; also known as the *rhythm method.*

candidiasis A yeast infection caused by the *Candida albicans* organism; also called *moniliasis* and *yeast infection.*

caregiver role In family caregiving, the role of the person who provides the most ongoing physical work and decision making relating to the one who is being cared for.

case-study method In clinical research, the in-depth examination of an individual or small group in some form of psychological treatment in order to gather data and formulate hypotheses.

celibacy Abstinence from sexual intercourse.

cervical cap A thimble-shaped cap that fits snugly over the cervix (uterine opening) to prevent conception.

cervix The opening of the uterus within the vagina.

chancre A painless sore or ulcer that may be the first symptom of syphilis.

chastity The state of being morally or sexually pure.

Chicano [fem. **Chicana**] A Mexican American born in the United States.

child-free marriage A marriage in which the partners have chosen not to have children.

child neglect Failure to provide adequate or proper physical or emotional care for a child.

child sexual abuse Any sexual interaction, including fondling, erotic kissing, oral sex, or genital penetration, that occurs between an adult (or older adolescent) and a prepubertal child.

child snatching The kidnapping of one's own children, usually by the noncustodial parent.

child support Court-ordered financial support by the noncustodial parent to pay or assist in paying child-rearing expenses incurred by the custodial parent.

chlamydia A common sexually transmitted disease caused by *Chlamydia trachomatis;* affects the urinary tract or other organs.

Christmas A Christian holiday commemorating the birth of Jesus and celebrated on December 25, around the time of the winter solstice.

Circumplex Model See *Family Circumplex Model.*

clan A group of families related along matrilineal or patrilineal descent lines, regarded as the basic family unit in some cultures.

clinical research The in-depth examination of an individual or small group in clinical treatment in order to gather data and formulate hypotheses. See also *case-study method.*

closed field A setting in which potential partners may meet, characterized by a small number of people who are likely to interact, such as a class, dormitory, or party. Cf. *open field.*

coalition Within a family, an alliance between two or more family members.

cognition The mental processes, such as thought and reflection, that occur between the moment we receive a stimulus and the moment we respond to it.

cognitive developmental theory A theory of socialization associated with Swiss psychologist Jean Piaget in which the emphasis was placed on the child's developing abilities to understand and interpret their surroundings.

cohabitation The sharing of living quarters by two heterosexual, gay, or lesbian individuals who are involved in an ongoing emotional and sexual relationship. The couple may or may not be married.

cohesion Emotional closeness or connectedness among family members.

cohort A group of persons experiencing a specific event at the same time, such as a birth cohort consisting of persons born in the same year. See also *generation.*

coitus The insertion of the penis into the vagina and subsequent stimulation; sexual intercourse.

coitus interruptus The withdrawal of the penis from the vagina immediately before ejaculation; considered ineffective as a contraceptive measure.

coming out For gay, lesbian, and bisexual individuals, the process of publicly acknowledging one's sexual orientation.

common couple violence Sociologist Michael Johnson's term for the more routine forms of partner violence that results from disputes and disagreements, and for which there is a high degree of gender symmetry.

compadrazgo [Spanish] The Latino institution of godparentage (literally, coparentage).

compadre [Spanish] The godfather of one's child (literally, cofather).

commuter marriage A marriage in which couples who prefer living together live apart in pursuit of separate goals.

companionate love A form of love emphasizing intimacy and commitment.

companionate marriage A marriage characterized by shared decision making and emotional and sexual expressiveness.

comparable worth An economic model arguing that occupations traditionally employing women are compensated at a lower rate than those traditionally employing men as a result of gender discrimination and that to overcome income differences between men and women, pay should be based on experience, knowledge and skills, mental demands, accountability, and working conditions, not on specific occupations per se.

complementary marriage model A model in which male employment outside the home and female work within the home are viewed as separate but interdependent.

complementary needs theory A theory of mate selection suggesting that we select partners whose needs are different from and/or complement our own needs.

conception The union of sperm and ovum; impregnation.

conceptualization The specification and definition of concepts used by the researcher.

concubine In polygamous societies, a secondary wife.

condom A sheath made from latex rubber or animal intestine that fits over the erect penis to prevent the deposit of sperm in the vagina; used as a contraceptive device and also as protection against sexually transmitted diseases (latex condoms only).

conduct of fatherhood Men's actual participation in raising their children.

confidentiality An ethical rule according to which the researcher knows the identities of participants and can connect what was said to who said it but promises not to reveal such information publicly.

conflict theory A social theory that views individuals and groups as being basically in competition with each other. Power is seen as the decisive factor in interactions.

congenital Existing from birth.

conjugal extended family An extended family formed through marriages.

conjugal family A family consisting of husband, wife, and children. See also *nuclear family.*

conjugal relationship A relationship formed by marriage.

consanguineous extended family An extended family formed through blood ties.

consanguineous relationship A relationship formed by common blood ties.

continuous coverage system The responsibility facing new parents in which someone must be available to care for their infant around the clock or provide alternate caregiving arrangements. This new temporal reality introduces conflict over which parent will be most directly involved and how much free time each will retain.

contraception Devices, techniques, or drugs used to prevent conception (fertilization of the ovum by a sperm).

contraction During childbirth, the action of the uterine muscles that open the cervix and expel the fetus. See also *labor.*

contraindication A symptom, sign, or condition indicating that a particular drug or device should not be used.

conventionality In marriage and family research, the tendency of subjects to give conventional or conformist responses.

coping The process of utilizing resources in response to stress.

coprovider families Families that are dependent on economic activity from both men and women.

correlational study A study (clinical, survey, or observational) that measures, but does not manipulate, two or more naturally occurring variables.

courtship The period during which a commitment to marriage is developed.

couvade The psychological or ritualistic assumption of the symptoms of pregnancy and childbirth by the male.

covenant marriage A new antidivorce reform of legal marriage in which couples acknowledge the lifelong nature of their marital commitments. IThey are required to undergo premarital counseling, promise to seek marital counseling if they experience serious marital difficulties, and pledge to divorce only under extreme hardships via a fault-based divorce.

crisis A turning point; a crucial time, stage, or event. See also *predictable crisis* and *unpredictable crisis.*

crisis model A theoretical construct for studying family strengths, based on the family's ability to function during times of high stress.

crude birth rate A statistic reflecting the number of births per thousand people in the population

crude divorce rate A statistical measure of divorce calculated on the basis of the number of divorces per 1,000 people in the population.

cultivation theory In media research, a theory asserting that there are consistent images, themes, and stereotypes across all media genres that form a more or less consistent worldview.

cultural relativity The view that the practices of a particular culture should be evaluated in terms of how they fit within the culture as a whole and not judged in terms of another culture's standards or values.

culture of fatherhood Ralph LaRossa's term for the beliefs we have about the roles, responsibilities, and involvement of fathers in raising their children. LaRossa noted that these beliefs have changed more dramatically than has the conduct of fatherhood.

culture of poverty The view that the poor form a qualitatively different culture from the larger society and that their culture accounts for their poverty.

cunnilingus Oral stimulation of the female genitals.

custodial Having physical or legal custody of a child. Cf. *noncustodial.*

custody Legal responsibility for certain aspects of a child's well-being. See also *joint custody, joint legal custody, joint physical custody, sole custody,* and *split custody.*

cycle of violence According to Lenore Walker's research, the recurring three-phase battering cycle of (1) tension building, (2) explosion, and (3) reconciliation.

cystitis A urinary tract infection usually affecting women.

D

date rape Rape in which the assailant is personally known to the victim, usually in the context of a dating relationship. Also known as *acquaintance rape.*

dating A process during which two individuals meet to engage in activities together; dating may be either exclusive or nonexclusive.

Day of the Dead See *Dia de los Muertos.*

D&C Dilation and curettage.

D&E Dilation and evacuation.

death rate The number of deaths per year per thousand people in a given community or group.

Defense of Marriage Act Federal legislation signed into law by President Clinton denying recognition to same-sex couples, should any state legalize same-sex marriage.

deferred parenthood The intentional postponement of child-bearing until after certain goals have been fulfilled.

demographics The statistical characteristics of a population, such as family size, marriage and divorce rates, and ethnic and racial composition.

demography The study of population and population characteristics, such as family size, marriage and divorce rates, and ethnic and racial composition.

denial The conscious or unconscious refusal to recognize painful acts, situations, or ways of being.

dependent variable A variable that is observed or measured in an experiment and may be affected by another variable. See *independent variable.*

desire Erotic sensations or feelings that motivate a person to seek out or receive sexual experiences.

developmental systems approach An approach to human development that recognizes the importance of the individual's interactions within a complex and changing family system and within the numerous systems of the larger society.

developmental task Appropriate activities and responsibilities that individuals learn at different stages in the life cycle.

deviant Departing from social or cultural norms.

Dia de los Muertos, el [Spanish] The Day of the Dead, a Latino and Latin American holiday with Indian and Catholic roots honoring the dead, celebrated on November 1.

Diagnostic and Statistical Manual of Mental Disorders (DSM-IV) The fourth edition of a manual published periodically by the American Psychiatric Association establishing categories of psychiatric disorders and listing criteria for diagnosing such disorders.

diaphragm A flexible rubber cap placed in the vagina to block the passage of sperm, preventing conception.

dilation The opening up of the cervix during childbirth.

dilation and curettage (D&C) A first-trimester abortion technique in which the embryo is removed from the uterus with a sharp instrument (curette).

dilation and evacuation (D&E) Second-trimester abortion technique in which suction and forceps are used to remove the fetus.

direct sperm injection Fertilization technique involving the injection of a single sperm into an ovum in a laboratory dish and the subsequent implantation of the blastocyst in the mother's uterus.

disability A physical or developmental limitation. Preferred usage over "handicap."

discrimination The process of acting differently toward a person or group because the individual or group belongs to a minority.

disengagement An extremely low level of family cohesion occurring when families do not feel close to each other and are unable to communicate or to carry out minimum family tasks.

displaced homemaker A full-time homemaker who has lost economic support from her husband as a result of divorce or widowhood.

division of labor The interdependence of persons with specialized tasks and abilities. Within the family, labor is traditionally divided along gender lines. See also *complementary marriage model.*

divorce The legal dissolution of a marriage. Cf. *separation.*

divorce mediation The process in which a mediator (counselor) assists a divorcing couple in resolving personal, legal, and parenting concerns in a cooperative manner.

Domestic Partners Act Law granting certain legal rights similar to those of married couples to committed cohabitants, whether heterosexual, gay, or lesbian.

domestic partnership Cohabiting couples—lesbian, gay, or heterosexual—in committed relationships. Domestic partners are legally recognized in some cities and countries and have some of the protections enjoyed by married partners, such as shared insurance benefits.

double ABC-X model A model describing stress, in which Aa represents stressor pileup, B represents family coping resources, C represents perception of stressor pileup, and X represents outcome.

double standard of aging The devaluation of women in contrast to men in terms of attractiveness as they age.

douching Introducing water or liquid into the vagina for medical, hygienic, or contraceptive reasons.

Down syndrome A chromosomal error characterized by mental retardation.

dowry In many traditional cultures, the property a woman brings to her husband upon marriage.

DSM-IV See *Diagnostic and Statistical Manual of Mental Disorders.*

dual-career family A type of dual-earner family in which both husband and wife are committed to careers.

dual-earner family A family in which both husband and wife are employed. Also known as *dual-worker family.*

dual-worker family A dual-earner family.

duration-of-marriage effect The accumulation over time of various factors, such as poor communication, unresolved conflicts, role overload, heavy work schedules, and child-rearing responsibilities, that negatively affect marital satisfaction.

dyad A two-member group; a couple.

Dyadic Adjustment Scale A survey instrument developed by Graham Spanier that measures relationship satisfaction.

dysfunction Impaired or inadequate functioning, as in sexual dysfunction.

dyspareunia Painful sexual intercourse.

E

eclampsia A potentially life-threatening condition of late pregnancy brought on by untreated high blood pressure.

economic adequacy The psychological perception that one has sufficient income and economic resources.

economic distress The stressful aspects of the economic life of individuals or families, including unemployment, poverty, and worrying about money.

economies of scale Situations in which an economic unit, such as a family, is able to economize because the cost per individual goes down as the number of individuals increases.

edema Fluid retention and swelling.

effacement The thinning of the cervix during labor.

egalitarian gender roles Gender roles in which men and women are treated equally.

ego In psychoanalytic theory, the part of the personality that is rational and mediates between the demands of the id and the constraints imposed by society. See also *id* and *superego.*

egocentric fallacy The mistaken belief that one's own personal experience and values are those of others in general.

ejaculate [noun] Semen. [verb] To expel semen.

ejaculation The expulsion of semen during orgasm.

embryo In human beings, the early development of life between about one week and two months after conception.

embryo transplant The implantation of the blastocyst into the mother's (or surrogate's) uterus following a fertilization procedure such as in vitro fertilization or direct sperm injection.

emission The first stage of male orgasm during which the semen moves into the urethra.

empty nest The experience of parents when the last grown child has left home. The "empty nest syndrome," in which the mother becomes depressed after the children have gone, is believed to be more of a myth than a reality.

endogamy Marriage within a particular group. Cf. *exogamy.*

engagement A pledge to marry.

enmeshment An extremely high level of family cohesion resulting from overidentification with the family and resulting in a lack of individual independence.

environmental influences The wider context and external influences on families that are the focus of family ecological theory.

episiotomy An incision from the vagina toward the anus made during childbirth.

equity theory A theory emphasizing that social exchanges must be fair or equally beneficial over the long run.

erectile dysfunction Inability or difficulty in achieving erection.

erection An erect penis.

eros 1. From the Greek *eros* [love], the fusion of love and sexuality. 2. According to sociologist John Lee's styles of love, the passionate love of beauty.

erotic Pertaining to sexuality, sensuality, or sexual sensations.

ethical guidelines Standards agreed upon by professional researchers. These guidelines protect the privacy and safety of individuals who provide information in a research setting.

ethnic group A large group of people distinct from others because of cultural characteristics, such as language, religion, and customs, transmitted from one generation to another. See also *minority group* and *racial group.*

ethnicity Ethnic affiliation or identity.

ethnic stratification The hierarchical ranking of groups in superior and inferior positions according to ethnicity.

ethnocentric fallacy (also ethnocentrism) The belief that one's own ethnic group, nation, or culture is inherently superior to others. See also *racism.*

exchange theory See *social exchange theory.*

excitement phase The second stage in the sexual response cycle, denoting sexual arousal.

exogamy Marriage outside a particular group. Cf. *endogamy.*

exosystem In ecological theory, the settings in which the individual does not actively participate but that nonetheless affect his or her development.

experimental research A research method involving the isolation of specific factors (variables) under controlled circumstances to determine the effects of each factor.

expressive trait A supportive or emotional personality trait or characteristic.

extended family The family unit of parent(s), child(ren), and other kin, such as grandparents, uncles, aunts, and cousins. See also *conjugal extended family* and *consanguineous extended family.*

extended household A household composed of several different families.

extrafamilial sexual abuse Child sexual abuse that is perpetrated by nonrelated individuals. Cf. *intrafamilial sexual abuse.*

extramarital sex Sexual activities, especially sexual intercourse, occurring outside the marital relationship.

F

FAE See *fetal alcohol effect.*

fallacy A fundamental error in reasoning that affects our understanding of a subject.

familialism A pattern of social organization in which family loyalty and strong feelings for the family are important.

family A unit of two or more persons, of which one or more may be children who are related by blood, marriage, or affiliation and who cooperate economically and may share a common dwelling place.

Family Circumplex Model A model of family functioning in which cohesion, adaptability, and communication are the most important dimensions.

family ecosystem Family interactions and adaptations with the larger social environment, such as schools, neighborhoods, or the economy.

family hardship Difficulties specifically associated with a stressor, such as loss of income in the case of unemployment. See also *stressor.*

family life cycle A developmental approach to studying families, emphasizing the family's changing roles and relationships at various stages, beginning with marriage and ending when both spouses have died.

family of cohabitation The family formed by two people living together whether married or unmarried; may include children or stepchildren.

family of marriage The family formed through marriage. See also *family of procreation*; cf. *family of orientation.*

family of orientation The family in which a person is reared as a child. Cf. *family of procreation.*

family of origin See *family of orientation.*

family of procreation The family formed by a couple and their child or children. See also *family of cohabitation.*

family policy A set of objectives concerning family well-being and specific measures initiated by government to achieve them.

family power Power exercised by individuals in their family roles as mother, father, child, or sibling.

family role A social role within the family, such as husband or wife, father or mother. See also *kinship system.*

family rule A family's patterned or characteristic response to events, situations, or persons. See also *family systems theory, meta-rule,* and *hierarchy of rules.*

family strengths Those relationship patterns, interpersonal skills, and social and psychological characteristics that create fulfillment and satisfaction for a family as individuals and as a whole.

family stress An upset in the steady functioning of the family.

family systems theory A theory viewing family structure as created by the pattern of interactions between its various subsystems, and individual actions as being strongly influenced by the family context.

family tree Diagrammatic representation of family and ancestors.

family work The unpaid work that is undertaken by family members to sustain the family, such as housework, laundry, shopping, yard maintenance, budgeting and bill-paying, and care of children, the sick, and the elderly.

FAS See *fetal alcohol syndrome.*

fecundity A person's maximum biological capacity to reproduce.

feedback In communication, an ongoing process in which participants and their messages produce a result and are subsequently modified by the result.

fellatio Oral stimulation of the male genitals.

feminism 1. The principle that women should have equal political, social, and economic rights with men. 2. The social movement to obtain for women political, social, and economic equality with men.

feminization of poverty The shift of poverty to females, primarily as a result of high divorce rates and births to unmarried women.

feral child A child purportedly nursed and reared in the wild by animals such as wolves, bears, or lions.

fertility The ability to conceive; a person's actual reproductive performance.

fertility awareness methods Contraceptive method based on predicting a woman's fertile period and either avoiding intercourse or using an additional method of contraception during that interval.

fertility rate In a given year, the number of live births per 1,000 women aged 15–44 years. See also *birth rate.*

fertilization The union of an egg and sperm. Also known as *conception.*

fetal alcohol effect (FAE) Growth retardation caused by the mother's chronic ingestion of alcohol during pregnancy.

fetal alcohol syndrome (FAS) A syndrome caused by the mother's chronic ingestion of alcohol during pregnancy; may be characterized by unusual facial characteristics, small head and body size, poor mental capacities, and abnormal behavior patterns.

fetus The unborn young of a vertebrate; the human embryo becomes a fetus at about the eighth week following fertilization.

fictive kin ties The extension of kinshiplike attributes to non-blood relationships (such as friends or neighbors) to demonstrate their importance and to symbolize the mutual reciprocity found within them.

field of eligibles A group of individuals of the same general background and age who are culturally approved potential marital partners.

filial crisis Psychological conflict and stress experienced by adult children when aged parents become dependent on them.

flextime Flexible work schedules determined by employee/employer agreement.

foreplay Erotic activity prior to coitus, such as kissing, caressing, sex talk, and oral/genital contact; petting.

foreskin The sleeve of skin covering the tip of the penis; prepuce.

friendship An attachment between people; the foundation for a strong love relationship.

G

game stage According to symbolic interactionist theories of George Herbert Mead, the stage in self-development in which one can understand a variety of other perspectives simultaneously.

gamete A reproductive cell (sperm or ovum) that can unite with another gamete to form a zygote.

gamete intrafallopian transfer (GIFT) A fertilization technique in which sperm and ova are collected from the parents and deposited together in the mother's fallopian tube for fertilization.

gay Pertaining to same-sex relationships, especially among males.

gay male A male sexually oriented toward other males. Cf. *lesbian.*

gender The division into male and female, often in a social sense; sex.

gender differences The orienting focus in most feminist writing, research, and advocacy.

gender identity The psychological sense of whether one is male or female.

gender ideology Arlie Hochschild's term for what individuals believe they ought to do as husbands or wives, and how they believe paid and unpaid work should be divided.

gender-rebellion feminism Versions of feminism that emphasize the interconnectedness between multiple inequalities (race, class, sexual orientation, age, and gender), and see gender inequality as only one aspect of wider social inequality.

gender-reform feminism Versions of feminism that stress how similar women and men are and emphasize the need for equal rights and opportunities for both genders.

gender-resistant feminism Versions of feminism that advocate separatist strategies, wherein women establish women-only social institutions and settings.

gender role The culturally assigned role that a person is expected to perform based on male or female gender.

gender-role attitude A personal belief regarding appropriate male and female personality traits and behaviors.

gender-role behavior An actual activity or behavior in which males or females engage according to their gender role.

gender-role stereotype A rigidly held and oversimplified belief that all males and females possess distinctive psychological and behavioral traits as a result of their gender.

gender schema The cognitive organization of individuals, behaviors, traits, objects, and such by gender.

gender theory A theory in which gender is viewed as the basis of hierarchal social relations that justify greater power to males.

generation 1. A group of people born and living during the same general time period. 2. The approximately 30-year period between the birth of one generation and the next. See also *cohort.*

genital herpes A sexually transmitted disease caused by the herpes simplex virus type II, similar to cold sores or fever blisters but appearing on the genitals.

genitalia The external sex or reproductive organs; genitals.

genital warts Warts on the genitals caused by human papilloma virus (HPV), a sexually transmitted virus.

genogram A diagram of the emotional relationships of a family through several generations.

gestation The period of carrying young in the uterus from conception to birth.

gestational edema-proteinuria-hypertension complex A condition of pregnancy characterized by high blood pressure, toxemia, or preeclampsia.

getting together A courtship process in which men and women congregate ("get together") in groups to socialize or engage in common activities or projects. Cf. *dating.*

ghetto poor Inner-city residents, primarily African American and Latino, who live in poverty.

GIFT See *gamete intrafallopian transfer.*

gonorrhea A sexually transmitted disease caused by the *Neisseria gonorrhoeae* bacterium that initially infects the urethra in males and the cervix in females or the throat or anus in either sex, depending on the mode of sexual interaction.

grandparenting Performing the functions of a grandparent.

grieving process Emotional responses and expressions of feeling over the death of a loved one.

grounded theory A theory that is grounded by or emerges from observations of specific, concrete details.

H

halo effect The tendency to infer positive characteristics or traits based on a person's physical attractiveness.

Hana Matsuri The Japanese celebration of the Buddha's birthday; literally, Flower Festival.

handicap See *disability.*

health The state of physical and mental well-being.

hegemonic models of gender Dominant models of masculinity and femininity.

hepatitis A liver disease causing blood pigments to accumulate. Hepatitis B (serum hepatitis) may be sexually transmitted.

herpes simplex type II See *genital herpes.*

heterogamy Marriage between those with different social or personal characteristics. Cf. *homogamy.*

heterosexuality Sexual orientation toward members of the opposite sex.

heterosociality Close association with members of the opposite sex.

hierarchy of rules A ranking of family rules in order of significance and kind. Family rules are most important, followed by member (or role-based) rules and individual rules.

Hispanic Of Spanish or Latin American origin or background; may be of any race. See also *Latino.*

HIV See *human immunodeficiency virus.*

HIV-positive Infected with human immunodeficiency virus.

Holi A spring festival originating in Northern India.

homemaker role A family role usually allocated to women, in which they are primarily responsible for home management, child rearing, and the maintenance of kin relationships. Traditionally the role is associated with economic dependency and has primacy over other female roles.

homeostasis A social group's tendency to maintain internal stability or balance and to resist change.

homoeroticism Erotic attraction to members of the same sex.

homogamy Marriage between those with similar social or personal characteristics. Cf. *heterogamy.*

homophobia Irrational or phobic fear of gay men and lesbians.

homosexuality Sexual orientation toward members of the same sex. See also *gay male* and *lesbian.*

homosociality The tendency to associate mostly with members of the same sex.

honeymoon effect The tendency of newly married couples to overlook problems, including communication problems.

hormone A chemical substance, secreted by the endocrine glands into the bloodstream, that organizes and regulates physical development.

hospice A place or program caring for the terminally ill, emphasizing both patient care and family support.

hostile environment An environment created through sexual harassment in which the harassed person's ability to learn or work is negatively influenced by the harasser's actions.

housewife 1. A wife who manages the home. 2. A homemaker.

human chorionic gonadotropin (HCG) A hormone secreted by the placenta that helps sustain pregnancy.

human immunodeficiency virus (HIV) The virus causing AIDS.

human papilloma virus (HPV) The virus causing warts, including genital warts.

hypergamy A marriage in which one's spouse is of a higher social class or rank.

hypoactive sexual desire Inactive or limited sexual desire. Also known as inhibited sexual desire.

hypogamy A marriage in which one's spouse is from a lower social standing.

hypothesis An unproven theory or proposition tentatively accepted to explain a collection of facts.

hysterectomy The surgical removal of the uterus or part of the uterus.

hysterotomy A surgical method of abortion in which the fetus is removed through an abdominal incision.

I

id In psychoanalytic theory, the part of the personality that seeks to gratify pleasurable needs, especially sexual ones. See also *ego* and *superego.*

identity An individual's core sense of self.

identity bargaining The process of role adjustment in a relationship, involving identifying with a role, having the role validated by others, and negotiating with the partner to make changes in the role.

ideology of "intensive mothering" The term used by Sharon Hays to refer to beliefs about what mothers ought to provide their children. The key elements of intensive mothering are full-time attention, self-sacrificing devotion, and expert-guided, labor-intensive involvement with the child, whose needs are more pressing than those of mothers.

ie [EE-eh] The basic family unit in traditional Japanese society consisting of past, present, and future members of the extended family and their households.

illegitimate 1. Not based on law, right, or custom. 2. Born outside of marriage (sometimes used in a derogatory manner).

incest Sexual intercourse between individuals too closely related to marry, usually interpreted to mean father/daughter, mother/son, or brother/sister. See also *intrafamilial sexual abuse.*

independent variable A variable that may be changed or manipulated in an experiment.

induced abortion The termination of a pregnancy through human intervention.

induction The formation of arguments whose premises are intended to provide some (but not conclusive) support for their conclusions.

inductive research Research that is not hypothesis testing research but rather begins with a topical interest and perhaps some vague concepts.

infant mortality rate The number of deaths for every 1,000 live births.

institution An enduring social structure built around a significant and distinct cluster of social values. Institutions include the family, religion, education, and government.

instrument In social science, a research tool or device, such as a questionnaire, used to gather data about behaviors, attitudes, beliefs, or other such dimensions of an individual, group, or society.

instrumental trait A practical or task-oriented personality trait or characteristic.

interaction In communication, a reciprocal act that takes place between at least two people.

intermarriage Marriage between people of different ethnic or racial groups.

intermittent extended family The family that is formed when a family takes in other relatives in times of need.

interpersonal competence The ability to develop and share an intimate, growing relationship.

interrole conflict Conflict experienced when the role expectations of two or more roles are contradictory or incompatible. Also known as *role interference*; see also *role strain.*

intervening variable A variable that is affected by the independent variable and in turn affects the dependent variable.

intrafamilial sexual abuse Child sexual abuse that is perpetrated by related individuals, including steprelatives. See also *incest;* cf. *extrafamilial sexual abuse.*

intrauterine device (IUD) A device inserted into the uterus to prevent conception or implantation of the fertilized egg.

involution Following childbirth, the contracting process—over a period of about six weeks—by which the uterus returns to its prebirth state.

Issei First-generation Japanese (born in Japan).

I-statement In communication, a statement beginning with "I" that describes the speaker's feelings, such as "I feel upset when I see last week's dishes in the sink."

IUD See *intrauterine device.*

J

jealousy An aversive response occurring because of a partner's or other significant person's real, imagined, or likely involvement with or interest in another person.

joint custody Custody arrangement in which both parents are responsible for the care of the child. Joint custody takes two forms: *joint legal custody* and *joint physical custody.* See also *sole custody* and *split custody.*

joint legal custody Joint custody in which the child lives primarily with one parent but both parents jointly share in important decisions regarding the child's education, religious training, and general upbringing.

joint physical custody Joint custody in which the child lives with both parents in separate households and spends more or less equal time with each parent.

Juneteenth An African-American holiday commemorating freedom and celebrating black history and accomplishments.

K

kaddish In Judaism, a form of prayer.

kin Relatives.

kinship Family relationship.

kinship system The social organization of the family conferring rights and obligations based on an individual's status.

Kwanzaa An African-American harvest festival, observed between Christmas and New Year's Day and celebrating the culture and heritage of American blacks.

L

labor The physical efforts of childbirth.

Lamaze method A childbirth method in which the mother uses exercises and breathing techniques to assist her labor.

laparoscopy A tubal ligation technique using a laparoscope (viewing instrument) to locate the fallopian tubes, which are then closed or blocked.

latchkey children See *self-care.*

Latino [fem. **Latina**] A person of Latin American origin or ancestry; may be of any race. See also *Hispanic* and *Spanish-speaking.*

legitimacy The state or quality of being sanctioned by custom, rights, or law.

leisure gap Like the wage-gap, the leisure gap is a reflection of gender inequality. In their time away from their jobs, married men enjoy substantially more free time than do employed married women.

lesbian A female sexually oriented toward other females.

lesbian separatist A lesbian interested in creating a separate "womyn's" culture distinct from both heterosexual and gay culture. The lesbian separatist movement was strongest in the late 1960s through the early 1980s.

life chances Opportunities to enjoy a healthy and fulfilling life that are affected by one's social class standing.

life course A developmental perspective of individual change focusing on (1) individual time, an individual's own life span, (2) social time, social transition points, such as marriage, and (3) historical time, the times in which a person lives.

life cycle The developmental stages, transitions, and tasks individuals undergo from birth to death.

lochia A bloody vaginal discharge that appears for several weeks following childbirth.

looking-glass self In the symbolic interactionist theory of Charles Cooley, the looking-glass self refers to the influence of others' perceptions of us on how we come to perceive ourselves.

lower-middle class The socioeconomic class made up of white-collar service workers with incomes between $25,000 and $50,000, who own or rent more modest homes than the upper-middle class and purchase more affordable automobiles

ludus [LOO-dus] According to sociologist John Lee's styles of love, playful love.

M

macho [Spanish] In traditional Latin American usage, masculine, strong, or daring. In popular U.S. usage, excessively or stereotypically masculine.

macrosystem In ecological terms, the broadest level of environmental influences, encompassing the laws, customs, attitudes, and belief systems of the wider society, all of which influence individual development and experience.

madrina [Spanish] Godmother.

majority group A social category composed of people holding superordinate status and power and having the ability to impose their will on less powerful minority groups. Cf. *minority group.*

mania According to sociologist John Lee's styles of love, obsessive love.

marianismo [Spanish] In Latin American culture, the idealized mother role as represented by the Virgin Mary.

marital disruption Marital instability that includes marital separation as well as divorce.

marital exchange The process by which individuals trade resources with each other to secure the best marital partner. Traditionally, men exchanged their higher status and greater economic resources for women's physical attractiveness, expressive qualities, and childbearing and housekeeping abilities.

marital power The power exercised by individuals as husband or wife. Cf. *family power.*

marital rape Forced sexual contact by a husband with his wife; legal definitions of marital rape differ among states.

marriage The legally recognized union between a man and woman in which economic cooperation, legitimate sexual interactions, and the rearing of children may take place.

marriage contract 1. The legal and moral rights and responsibilities entailed by marriage. 2. An explicit contract delineating specific terms of marriage which, depending on the terms, may be legally binding. 3. A nonlegally binding agreement between partners, covering such areas as conflict resolution, division of household labor, employment, and child-rearing responsibilities.

marriage gradient The tendency for men to marry younger women of lower socioeconomic status and for women to marry older men of higher socioeconomic status. See also *marriage squeeze.*

marriage market An exchange process in which individuals bargain with each other using their resources in order to find the best available partner for marriage. See also *marital exchange.*

marriage squeeze The phenomenon in which there are greater numbers of marriageable women than marriageable men, particularly among older women and African-American women. See also *marriage gradient.*

masturbation Manual or mechanical stimulation of the genitals by self or partner; a form of autoeroticism.

matriarchal Pertaining to the mother as the head and ruler of a family. Cf. *patriarchal.*

matriarchy A form of social organization in which the mother or eldest female is recognized as the head of the family, kinship group, or tribe, and descent is traced through her. Cf. *patriarchy.*

matrilineal Descent or kinship traced through the mother. Cf. *patrilineal.*

mean world syndrome The belief, resulting from television viewing, that the world is more dangerous and violent than it is in actuality.

menarche [MEN-ar-kee] The first menstrual period, beginning in puberty.

menopause Cessation of menses for at least one year as a result of aging.

menses The monthly menstrual flow.

menstruation The discharge of blood and built-up uterine lining through the vagina that occurs approximately every four weeks among nonpregnant women between puberty and menopause.

mesosystem In ecological theory, the interconnections between microsystems

meta-analysis The reanalysis of combined statistical data from previous studies.

meta-rule An abstract, general, unarticulated rule at the apex of the hierarchy of rules upon which other rules are based.

Mexican American A U.S. citizen of Mexican ancestry.

midwife A person who attends and facilitates the birth of a child.

mifepristone See *RU-486.*

minority group A social category composed of people whose status places them at economic, social, and political disadvantage. Cf. *majority group*; see also *ethnic group.*

minority status Social rank having unequal access to economic and political power.

miscarriage A spontaneous abortion.

model 1. A person who demonstrates a behavior observed and imitated by others. 2. A prototype.

modeling The process of teaching or learning using imitation.

monogamy 1. The practice of having only one husband or wife at a time. 2. [colloq.] Sexual exclusiveness.

morality A set of social, cultural, or religious norms defining right and wrong.

morning-after pill See *postcoital birth control.*

morning sickness Nausea experienced by many women during the first trimester of pregnancy.

mucus method A contraceptive method that relies on predicting a woman's fertile period by observing changes in the appearance and character of her cervical mucus; also called *Billings* or *ovulation method.*

N

natural childbirth See *prepared childbirth.*

natural family planning A fertility awareness method of birth control that relies solely on predicting a woman's fertile period and avoiding intercourse on those days.

neonate A newborn infant.

neurosis A psychological disorder characterized by anxiety, phobias, and so on.

Nisei Second-generation Japanese, whose parents were born in Japan.

nocturnal orgasm Involuntary orgasm occurring in both females and males during sleep, usually accompanied by erotic dreaming. In males, it is usually accompanied by ejaculation ("wet dream").

no-fault divorce The dissolution of marriage because of irreconcilable differences for which neither party is held responsible.

nonbasic conflict Pronounced disagreement about nonfundamental or situational issues. Cf. *basic conflict.*

noncustodial Not having physical or legal custody of a child. Cf. *custodial.*

nonmarital sex Sexual activities, especially sexual intercourse, that take place among older single individuals. Cf. *premarital sex* and *extramarital sex.*

nonverbal communication Communication of emotion by means other than words, such as touch, body movement, and facial expression.

norm A cultural rule or standard.

normal Conforming to group or cultural norms.

normative Establishing or representative of a norm or standard.

nuclear family The basic family building block, consisting of a mother, father, and at least one child; in popular usage, used interchangeably with *traditional family.* Some anthropologists argue that the basic nuclear family is the mother and child dyad.

nursing Breastfeeding.

O

objective statement A factual statement presenting information based on scientifically measured findings, not on opinions or personal values.

objectivity Suspending the beliefs, biases, or prejudices we have about a subject until we have really understood what is being said.

observational research Research method using unobtrusive, direct observation.

obstetrician A physician specializing in pregnancy and childbirth.

occupational stratification The hierarchal ranking of jobs in superior and inferior positions based on pay and status.

open adoption A form of adoption in which the birth mother has an active part in choosing the adoptive parents; there is a certain amount of information exchanged between the birth mother and the adoptive parents, and there may be some form of continuing contact between the birth mother, the child, and the adoptive family following adoption.

open field A setting in which potential partners may not be likely to meet, characterized by large numbers of people who do not ordinarily interact, such as a beach, shopping mall, or large university campus. Cf. *closed field.*

open marriage A marriage in which the partners agree to allow one another to have openly acknowledged and independent sexual relationships outside the marriage.

operant conditioning A behavioral technique that uses a reinforcing stimulus to increase the frequency of a desired behavior.

operationalization The identification and/or development of research strategies to observe or measure concepts.

opinion An unsubstantiated belief or conclusion based on personal values or biases.

opportunistic disease An infection that is normally resisted by the healthy immune system, such as Kaposi's sarcoma and *Pneumocystis carinii* pneumonia, which are both associated with AIDS.

oral contraception Contraceptive taken orally; the pill.

oral-genital sex The erotic stimulation of the genitals by the tongue or mouth; fellatio, cunnilingus, or mutual oral stimulation.

orgasm The release of physical tensions after the build up of sexual excitement; usually accompanied by ejaculation in physically mature males.

orgasmic dysfunction The inability to have orgasm.

orgasmic phase The phase of the sexual-response cycle characterized by orgasm.

Outing The act of publicly disclosing the sexual orientation of gays, lesbians, or bisexuals.

ovulation method See *mucus method.*

ovum [plural **ova**] The egg produced by the ovary.

P

Pacific Islander Collective term referring to those of native Hawaiian, Fijian, Guamanian, Samoan, or other Melanesian, Micronesian, or Polynesian descent.

padrino [Spanish] Godfather.

Pap smear See *Pap test.*

Pap test The sampling of cervical cells to diagnose cancer or a precancerous condition.

paradigms Sets of concepts and assumptions about how families or other social phenomena work. These then guide the formation and interpretation of research.

parental image theory A theory of mate selection suggesting that we select partners similar to our opposite-sex parents.

parenting The rearing of children.

Parents' Bill of Rights Sylvia Hewlett and Cornel West's recommended policy initiatives and reforms to improve the conditions under which parents attempt to raise children.

parturition The process of childbirth.

passing [colloq.] Pretending to be heterosexual when actually gay or lesbian.

passionate love Intense, impassioned love. Cf. *companionate love.*

patriarchal Pertaining to the father as the head and ruler of a family. Cf. *matriarchal.*

patriarchal terrorism Michael Johnson's term for extreme, male-on-female partner violence in which men use violence, threats, isolation, and other tactics to subjugate and control their partners.

patriarchy A form of social organization in which the father or eldest male is recognized as the head of the family, kinship group, or tribe, and descent is traced through him. Cf. *matriarchy.*

patrilineal Descent or kinship traced through the father. Cf. *matrilineal.*

pedophilia Adult sexual attraction to prepubescent children that is intense and recurring; an adult's use of children for sexual purposes.

peer A person of equal status, as in age, class, position, or rank.

peer marriages Marriages built on principles of equity, equality, and "deep friendship," between spouses. Husbands and wives divide up housework and childcare more equally than is typical (between 50-50 and 60-40), and exhibit high levels of empathy and communication.

permissive child rearing A parenting style stressing the child's autonomy and freedom of expression, often over the needs of the parents.

permissiveness with affection Sexual norm permitting nonmarital sexual activity for both men and women in an affectionate relationship.

permissiveness without affection Sexual standard permitting nonmarital sexual activity without regard as to the nature of the relationship.

personality conflict Conflict based on personality characteristics; such conflicts are unlikely to be resolved. Cf. *situational conflict.*

Pesach A Jewish festival celebrated after the first full moon of spring, commemorating the flight of Moses and the Jews from Egypt; the Feast of Passover.

petting Foreplay; sexual contact usually referring to the manual or oral stimulation of the genitals or breasts.

phallus Penis.

phenotype A set of genetically determined anatomical and physical characteristics, such as skin and hair color and facial structure.

physical abuse Intentional violent mistreatment. See *violence.*

placenta The organ of exchange between the fetus and the pregnant female through which nutrients and waste pass. It also serves as an endocrine gland producing large amounts of the hormones progesterone and estrogen to maintain pregnancy; it is attached to the mother's uterine wall and connected to the fetus by the umbilical cord.

play stage In Mead's theory of self-development, the stage in which children "play at" being specific other people, taking on one role or viewpoint at a time.

pleasuring The giving and receiving of sensual pleasure through nongenital touching.

plural marriage The practice of having more than one husband or wife at the same time; polygamy.

polyandry The practice of having more than one husband at the same time. See also *polygamy*; cf. *polygyny.*

polygamy The practice of having more than one husband or wife at the same time; plural marriage. See also *polyandry, polygyny,* and *consanguineous extended family.*

polygyny The practice of having more than one wife at the same time. See also *polygamy;* cf. *polyandry.*

POSSLQ Person of opposite sex sharing living quarters, in U.S. Census Bureau terminology.

postcoital birth control Birth control that is administered after intercourse has taken place but before a diagnosis of pregnancy is possible; usually involves the administering of high-estrogen oral contraceptives. Also called "morning-after birth control."

postmarital sex Sexual intercourse among previously married individuals. See also *nonmarital sex.*

postpartum period A period of about three months following childbirth during which critical family adjustments are made.

power The ability to exert one's will, influence, or control over another person or group.

power conflict Pronounced disagreements concerning dominance.

powwow A Native American intertribal social gathering centering around drumming and traditional dances.

pragma Practical love, according to sociologist John Lee's styles of love.

predictable crisis Within the individual or family life cycle, normal but critical events, such as birth or death (of the elderly). Cf. *unpredictable crisis.*

predictive divorce rate A statistical calculation of the expected divorce rate of people who enter marriage in a given year.

preeclampsia Increasingly high blood pressure during late pregnancy; if untreated, it may lead to eclampsia. See also *gestational edema-proteinuria-hypertension complex.*

pre-embryo See *blastocyst.*

premarital sex Sexual activities, especially sexual intercourse, prior to marriage, especially among young, never-married individuals.

prematurity Birth before the normal gestation period has elapsed, often complicated by low birth weight.

prenatal Before birth.

prepared childbirth Birth philosophy stressing education and minimal use of anesthetics or other drugs; natural childbirth.

principle of least interest A theory of power in which the person less interested in sustaining a relationship has the greater power.

probability-based random samples Samples from which researchers can estimate the likelihood that their sample data can be safely inferred to the population in which they are interested.

pro-choice movement Social movement that advocates women's right to choose abortion.

profamily movement A social movement emphasizing conservative family values, such as traditional gender roles, authoritarian child rearing, premarital virginity, and opposition to abortion.

pro-life movement Social movement that advocates against abortion. Also known as *anti-abortion movement.*

prophylactic [adjective] Protecting against disease. [noun] Condom.

prostate gland A gland at the base of a man's bladder that produces most of the seminal fluid in the ejaculate.

prototype In psychology, concepts organized into a mental model.

proximity Nearness to another in terms of both physical space and time.

pseudokin See *affiliated kin.*

psychoanalytic theory The Freudian model of personality development, in which maturity is seen as the ability to gain control over one's unconscious impulses.

psychosexual Pertaining to the psychological aspects of sexuality.

psychosexual development The growth of the psychological aspects (such as attitudes and emotions) of sexuality that accompany physical growth.

psychosocial theory A theory of human psychological development that emphasizes the role of family and society in such development.

puberty The period during which the individual develops secondary sex characteristics and becomes capable of reproduction.

PWA Person with AIDS.

Q

qualitative research Small groups or individuals are studied in an in-depth fashion.

quantitative research Samples taken from a large number of subjects.

quickening The time of first movement of the fetus, which may be felt by the pregnant woman.

quinceañera, la [Spanish] The traditional Latin American celebration of a girl's fifteenth birthday, formally introducing her into society.

R

racial group A large group of people defined as distinct because of their phenotype (genetically transmitted anatomical and physical characteristics, especially facial structure and skin color). Cf. *ethnic group.*

racism The practice of discrimination and subordination based on the belief that race determines character and abilities. See also *ethnocentric fallacy* and *ethnocentrism.*

Ramadan A month-long Islamic holiday commemorating the prophet Mohammed's illumination; observed by extensive fasting and prayer, followed by feasting and celebration at the month's end.

rape Sexual act against a person's will or consent as defined by law, usually including sexual penetration by the penis or other object; it may not, however, necessarily include penile penetration of the vagina. Also known as *sexual assault.* See also *acquaintance rape* and *marital rape.*

rape trauma syndrome A group of symptoms experienced by a rape survivor, including fear, self-blame, anxiety, crying, sleeplessness, anger, or rage.

rating and dating game The process described by sociologist Willard Waller in which men and women rate potential dates on a scale of one to ten and then try to date the highest-rated individuals.

ratio measure of divorce A statistical calculation reflecting the ratio of the number of divorces in a given year to the number of marriages in that same year.

reactive jealousy Jealousy that occurs when a partner's past, present, or anticipated involvement with another is revealed. Cf. *suspicious jealousy.*

reconstituted family See *stepfamily.*

refined divorce rate A statistic reflecting the number of divorces in a given year for every thousand married couples.

refractory period Following orgasm, the period during which the penis cannot respond to additional stimulation.

reinforcement The process of influencing (increasing or decreasing) a behavior by adding or withholding a stimulus.

rejection sensitivity The tendency to anticipate and overreact to rejection.

relative love and need theory A theory of power in which the person gaining the most from a relationship is the most dependent.

relocation camps During World War II, camps in which Japanese Americans of all ages were imprisoned without cause by the U.S. government.

remarriage A marriage in which one or both partners have been previously married.

reproductive organs External and internal structures involved in reproduction.

residential propinquity A pattern in which the chances of two people marrying are greater the closer they live to each other.

resource Anything that can be called into use or used to advantage, such as love, money, or approval, to exert influence or power.

rhythm method See *calendar method.*

Roe v. Wade U.S. Supreme Court decision (1973) affirming a woman's constitutional right to abortion based on the right to privacy.

role The pattern of behavior expected of a person in a group or culture as a result of his or her social position, such as husband or wife in a family.

role conflict See *interrole conflict.*

role interference See *interrole conflict.*

role modeling A significant means by which children are taught role attitudes and behavior by learning to imitate adults whom they admire.

role overload The experience of having more prescribed activities in one or more roles than can be comfortably or adequately performed. See also *role strain.*

role reversed couples Couples in which women are the sole or dominant wage earners, while men stay home and assume domestic and child-care responsibilities traditionally performed by women.

role strain Difficulties, tensions, or contradictions experienced in performing a role, often because of multiple role demands. See also *interrole conflict* and *role overload.*

roleless role A role for which there are no clear guidelines for behavior, such as stepparent, widow, and ex-in-law roles.

romantic love Intense, passionate love. Cf. *companionate love.*

RU-486 An effective oral, postcoital birth control method containing mifepristone.

rule of thumb Prior to the nineteenth century, the legally sanctioned practice of disciplining one's wife with a rod, provided it was not wider than the husband's thumb.

S

safe sex Sexual practices, including the use of latex condoms, intended to prevent the transmission of bodily fluids, especially semen, that may contain HIV.

salpingitis Infection of a fallopian tube.

sample A group randomly and systematically selected from a larger group.

sandwich generation Individuals and families who care for both their own children and their aging parents at the same time.

Sansei Third-generation Japanese, whose grandparents were born in Japan.

scapegoating The conscious or unconscious singling out and blaming of an individual or group.

schema The cognitive organization of knowledge according to particular criteria.

scientific method A method of investigation in which a hypothesis is formed on the basis of impartially gathered data and is then tested empirically.

script A mental map, plan, or pattern of behavior. See also *role.*

secondary data analysis Use of research gathered by public sources of information.

secondary sex characteristic A physical characteristic other than the external genitals that distinguishes the sexes from each other, e.g., breasts and body hair.

second shift Arlie Hochschild's term for the domestic responsibilities awaiting employed women after their paid work hours are completed.

self-care Children under age 14 caring for themselves at home without supervision by an adult or older adolescent.

self-disclosure The revelation of deeply personal information about oneself to another.

self-esteem Feelings about the value of the self; high self-esteem includes feeling unique, having a sense of power, and feeling connected to others.

semen The fluid containing sperm, which is ejaculated, produced mostly by the prostate gland. Also known as *ejaculate.*

seminal fluid The fluid containing sperm; semen.

separation The state or condition of a married couple who have chosen to live together no longer. Cf. *divorce.*

separation distress A psychological state following separation that may be characterized by depression, anxiety, intense loneliness, or feelings of loss.

sequential work/family role staging A pattern of combining employment and family work in dual-earner families in which women leave employment during pregnancy and while their children are young and return to it at a later time.

sex 1. Biologically, the division into male and female. 2. Sexual activities.

sex hormones Hormones such as testosterone and estrogen that are responsible for the development of secondary sex characteristics and for activating sexual behavior.

sexism 1. The belief that biological differences between males and females provide legitimate bases for female subordination. 2. The economic and social domination of women by men.

sex organs Internal and external reproductive organs; commonly refers only to the penis, vulva, and vagina.

sex ratio The ratio of men to women in a group or society.

sex role See *gender role.*

sexual assault A legal term referring to rape. See also *acquaintance rape, marital rape,* and *rape.*

sexual behavior Behavior that is characterized by conscious psychological/erotic arousal (such as desire) and that may also be accompanied by physiological arousal (such as erection or lubrication) or activity (such as masturbation or coitus).

sexual coercion Nonconsensual sexual behavior such as rape, sexual assault, and sexual harassment

sexual desire The psychological component that motivates sexual behavior.

sexual dysfunction Recurring problems in sexual functioning that cause distress to the individual or partner; may have a physiological or psychological basis.

sexual enhancement Any means of improving a sexual relationship, including developing communication skills, fostering a positive attitude, giving a partner accurate and adequate information, and increasing self-awareness.

sexual harassment Deliberate or repeated unsolicited verbal comments, gestures, or physical contact that is sexual in nature and unwelcomed by the recipient. Two types of sexual harassment involve (1) the abuse of power and (2) the creation of a hostile environment. See also *hostile environment.*

sexual identity The individual's sense of his or her sexual self.

sexual intercourse Coitus; heterosexual penile/vaginal penetration and stimulation.

sexual orientation Sexual identity as heterosexual, gay, lesbian, or bisexual.

sexual script A culturally approved set of expectations as to how one should behave sexually as male or female and as heterosexual, gay, or lesbian.

sexuality The state of being sexual, which encompasses the biological, social, and cultural aspects of sex.

sexually transmitted disease (STD) An infection that can be transmitted through sexual activities, such as sexual intercourse, oral/genital sex, or anal sex.

sexual preference See *sexual orientation.*

sexual stratification The hierarchical ranking in superior and inferior positions according to gender.

sexual variation A departure from sexual norms; atypical sexual behavior.

shift couples Two-earner households in which spouses work different, often nonoverlapping shifts, so that one partner is home while the other is at work.

shiva In Judaism, a seven-day period of mourning for the dead during which certain austerities are practiced.

sibling A brother or sister.

SIDS See *sudden infant death syndrome.*

single-parent family A family with children, created by divorce or unmarried motherhood, in which only one parent is present. A family consisting of one parent and one or more children.

situational conflict Conflict arising as the result of specific acts, events, behaviors, or situations, which are amenable to resolution. Cf. *personality conflict.*

social classes Groupings of people who share a common economic position by virtue of their wealth, income, power, and prestige, and thus have similar social and familial experiences.

social construct An idea or concept created by society. As applied to gender, it refers to the ways in which we define gender and then act on our beliefs about it.

social exchange theory A theory that emphasizes the process of mutual giving and receiving of rewards, such as love or sexual intimacy, in social relationships, calculated by the equation Reward – Cost = Outcome.

social integration The degree of interaction between individuals and the larger community.

social learning theory A theory of human development that emphasizes the role of cognition (thought processes) in learning.

social mobility A term used to refer to movement up or down the socioeconomic ladder, which can occur within a person's lifetime or between generations.

social role A socially established pattern of behavior that exists independently of any particular person, such as the husband or wife role or the stepparent role.

social support Instrumental and emotional assistance, such as physical care and love.

socialization The shaping of individual behavior to conform to social or cultural norms.

socioeconomic status A term used to refer to the combined effects of income, occupational prestige, wealth, education, and income on a person's lifestyle and opportunities.

sole custody Child custody arrangement in which only one parent has both legal and physical custody of the child. See also *joint custody* and *split custody.*

Spanish-speaking Pertaining to Hispanic origin or ancestry.

spells of poverty The periodic movement in and out of poverty.

sperm The male gamete produced by the testis.

spermicide A substance toxic to sperm and used for contraception.

spinnbarkeit The elastic condition of cervical mucus just prior to ovulation.

spirit marriage In Canton, China, a marriage of two deceased persons, arranged by their families to provide family continuity.

split custody Custody arrangement when there are two or more children in which custody is divided between the parents, the mother generally receiving the girls and the father receiving the boys.

spontaneous abortion The natural but fatal expulsion of the embryo or fetus from the uterus; miscarriage.

status The position an individual occupies within a social hierarchy.

STD See *Sexually transmitted disease.*

stepfamily A family in which one or both partners have a child or children from an earlier marriage or relationship. Also known as a *blended family*; see also *binuclear family.*

stepparent role The role a stepparent forges for herself or himself within the stepfamily as there is no such role clearly defined by society.

stereotype A rigidly held, simplistic, and overgeneralized view of individuals, groups, or ideas that fails to allow for individual differences and is based on personal opinion and bias rather than critical judgment.

sterilization Intervention (usually surgical) making a person incapable of reproducing.

stigmatization The process of labeling and internalizing perceptions of self, other individuals, groups, behaviors, feelings, or ideas as deviant.

stimulus-value-role theory A three-stage theory of romantic development proposed by Bernard Murstein: (1) stimulus brings people together; (2) value refers to the compatibility of basic values; (3) role has to do with each person's expectations of how the other should fulfill his or her roles.

strain Tension lingering from previous stressors or tensions inherent in family roles. See also *stressor.*

storge [STOR-gay] According to sociologist John Lee's styles of love, companionate love.

stress Psychological or emotional distress or disruption.

stressor A stress-causing event.

structural functionalism A sociological theory that examines how society is organized and maintained by examining the functions performed by its different structures. In marriage and family studies, structural functionalism examines the functions the family performs for society, the functions the individual performs for the family, and the functions the family performs for its members.

subsystem A system that is part of a larger system, such as family, and religious and economic systems being subsystems of society and the parent/child system being a subsystem of the family.

sudden infant death syndrome (SIDS) The death of an apparently healthy infant during its sleep from unknown causes.

superego In psychoanalytic theory, the part of the personality that has internalized society's demands and acts as a sort of conscience to control the id. See also *ego* and *id*.

surrogate mother A woman who bears a child for another woman (often for money) and relinquishes custody upon birth; the pregnancy usually results from artificial insemination, in vitro fertilization, or embryo transplant.

survey research Research method using questionnaires or interviews to gather information from small, representative groups and to infer conclusions that are valid for larger populations.

suspicious jealousy Jealousy that occurs when there is either no reason for suspicion or only ambiguous evidence that a partner is involved with another. Cf. *reactive jealousy*.

symbiotic personality A personality characterized by excessive dependency and need for closeness.

symbol In communication, a word or gesture that represents something more than itself.

symbolic interaction A theory that focuses on the subjective meanings of acts and how these meanings are communicated through interactions and roles to give shared meaning.

symmetrical work/family role allocation The interface between family and employment in which family work is divided more equitably and females have greater commitment to work roles than in the traditional division of labor. Cf. *sequential work/family role staging* and *traditional-simultaneous work/family life cycle*.

syphilis A sexually transmitted disease whose first symptom is a painless chancre on the genitals, anus, or mouth; caused by the *Treponema pallidum* bacterium. Life-threatening if untreated.

T

TANF See *Temporary Assistance to Needy Families*.

taking the role of the other In symbolic interactionist theories of socialization, the ability to see things from someone else's vantage point.

Temporary Assistance to Needy Families (TANF) The current government program designed to financially assist families with children during times of poverty.

Tet The Vietnamese New Year, which coincides with the beginning of the lunar year.

thanatology The study of death and dying.

theoretical effectiveness The maximum effectiveness of a drug, device, or method if used consistently, correctly, and according to instructions.

theory A set of general principles or concepts used to explain a phenomenon and to make predictions that may be tested and verified experimentally.

total fertility rate A complicated statistic that estimates the number of births a hypothetical group of 1,000 women would have if they experience across their childbearing years the age-specific rates for a given year

toxemia A condition of pregnancy characterized by high blood pressure and edema. See also *gestational edema-proteinuria-hypertension complex*.

traditional family In popular usage, an intact, married two-parent family with at least one child, which adheres to conservative family values; an idealized family. Popularly used interchangeably with *nuclear family*.

traditional-simultaneous work/family life cycle The interface between family and employment in which the husband is regarded as the primary economic provider with little responsibility for family work, and the wife, regardless of her employment status, is primarily responsible for family work. Cf. *sequential work/family role staging* and *symmetrical work/family role allocation*.

trait A distinguishing personality characteristic or quality.

transactional pattern In communication, a habitual pattern of interaction.

transition 1. Passing from one stage or phase to another. 2. During childbirth, the process during which the fetus's head enters the birth canal, marking the end of the first stage of labor.

traumatic sexualization The process of developing inappropriate or dysfunctional sexual attitudes, behaviors, and feelings by a sexually abused child.

traveling time The time immediately following the Civil War when former slaves traveled throughout the South in search of relatives separated by sale.

trial marriage Cohabitation with the purpose of determining compatibility prior to marriage.

triangular theory of love A theory developed by Robert Sternberg emphasizing the dynamic quality of love as expressed by the interrelationship of three elements: intimacy, passion, and decision/commitment.

triangulation In social research, the use of multiple methods of data collection and/or analysis

trust Belief in the reliability and integrity of another.

tubal ligation A surgical method of female sterilization in which the fallopian tubes are tied off or closed, usually by laparoscopy.

two-person career An arrangement in which it takes the efforts of two spouses to ensure the career success of one. One spouse, typically the husband, can devote himself or herself fully to career pursuits because of the help and assistance received from his or her spouse. This help and assistance includes taking care of all family and domestic needs, but also often includes unpaid supportive roles (such as entertaining business colleagues).

typology A systematic categorization according to types, such as common traits or qualities.

U

umbilical cord A hollow cord that connects the circulation system of the embryo or fetus to the placenta.

underclass The socioeconomic class marked by persistent poverty and poor employability.

unpredictable crisis An unforeseen crisis, such as terminal illness in a child. Cf. *predictable crisis.*

unrequited love Love that is not returned.

upper-middle class A socioeconomic class consisting of college-educated, highly paid professionals (for example, lawyers, doctors, engineers) who have annual incomes that may reach into the hundreds of thousands of dollars.

urethritis An infection of the urethra.

user effectiveness The actual effectiveness of a drug, device, or method based on statistical information.

uterus A hollow, muscular organ within the pelvic cavity of a female in which the fertilized egg develops into the fetus; the womb.

V

vacuum aspiration First trimester abortion method in which the contents of the uterus are removed by suction.

vagina The passage leading from the vulva to the uterus that expands during intercourse to receive the erect penis or during childbirth to permit passage of the child; the birth canal.

vaginismus The involuntary constriction of the vaginal muscles which prohibits penetration.

vaginitis Vaginal infection, most commonly caused by *Trichomonas vaginalis, Candida albicans,* or *Gardnerella vaginalis,* which may be sexually transmitted. Men may also acquire these infections but often remain asymptomatic.

value judgment An evaluation based on ethics or morality rather than on objective observation.

values The social principles, goals, or standards held as acceptable by an individual, family, or group.

variable In experimental research, a factor, such as a situation or behavior, that may be manipulated. See also *independent variable* and *dependent variable.*

variation Departure from social or cultural norms.

vas deferens [plural **vasa deferentia**] One of two ducts that carry sperm from the testes to the seminal vesicles.

vasectomy A surgical form of male sterilization in which the vas deferens is severed.

VD (venereal disease) See *sexually transmitted disease.*

venereal warts Warts in the genital or anal area which may be sexually transmitted; caused by the human papilloma virus (HPV).

vernix A milky substance often covering infants at birth.

viability Ability to live and continue to grow outside the uterus.

Viagra An oral medication for erectile dysfunction that has restored many men's abilities to engage in sexual activity.

violence An act carried out with the intention of causing physical pain or injury to another.

virginity The state of not having engaged in sexual intercourse.

vulva The external female genitalia, including the mons veneris, labia majora and labia minora, clitoris, and the vaginal and urethral openings.

W

wealth Net worth, including income, savings, investments, property, and inheritances.

wedding 1. The act of marrying. 2. A marriage ceremony or celebration.

wellness Optimal health and well-being. Components of wellness include physical, emotional, intellectual, spiritual, social, and environmental health.

wheel theory of love A theory developed by Ira Reiss holding that love consists of four interdependent processes: rapport, self-revelation, mutual dependency, and intimacy fulfillment.

withdrawal See *coitus interruptus.*

womb Uterus.

work spillover The effect that employment has on time, energy, activities, and psychological functioning of workers and their families.

working class A socioeconomic class comprised of skilled laborers with high school or vocational educations. The working class lives somewhat precariously, with little savings and few liquid assets should illness or job loss occur.

workplace/family linkages Ways in which employment affects families.

Y

yeast infection See *candidiasis.*

Yonsei Fourth-generation Japanese whose great-grandparents were born in Japan.

Yuan Tan The Chinese New Year, beginning on the first day of the lunar year.

Z

ZIFT See *zygote intrafallopian transfer.*

zygote The fertilized ovum (egg).

zygote intrafallopian transfer (ZIFT) Fertilization technique in which ova and sperm are united in a laboratory dish and then transferred to the fallopian tube to begin cell division.

Bibliography

Abbey, Antonia. "Acquaintance Rape and Alcohol Consumption on College Campuses: How Are They Linked?" *Journal of American College Health* 39, 4 (January 1991): 165–169.

———. "Maternal Risk Factors and Fetal Alcohol Syndrome: Provocative and Permissive Influences." *Neurotoxicology and Teratology* 17, 4 (July 1995): 445.

Abbey, A., and R. J. Harnish. "Perception of Sexual Intent: The Role of Gender, Alcohol Consumption, and Rape Supportive Attitudes." *Sex Roles* 32, 5–6 (March 1995): 297–313.

Abbey, A., L. J. Halman, and F. M. Andrews. "Psychosocial, Treatment, and Demographic Predictors of the Stress Associated with Infertility." *Fertility and Sterility* 57, 1 (January 1992): 122–128.

Abelsohn, David. "A 'Good Enough' Separation: Some Characteristic Operations and Tasks." *Family Process* 31, 1 (1992): 61–83.

Absi-Semaan, Nada, Gail Crombie, and Corinne Freeman. "Masculinity and Femininity in Middle Childhood: Developmental and Factor Analyses." *Sex Roles: A Journal of Research* 28 (1993): 187–207.

Acitelli, Linda K. "Gender Differences in Relationship Awareness and Marital Satisfaction among Young Married Couples." *Personality and Social Psychology Bulletin* 18, 1 (1992): 102–110.

Acker, Joan. "From Sex Roles to Gendered Institutions." *Contemporary Sociology* 21, 5 (1992): 565–570.

Acker, Michele, and Mark H. Davis. "Intimacy, Passion, and Commitment in Adult Romantic Relationships: A Test of the Triangular Theory of Love." *Journal of Social and Personal Relationships* 9, 1 (1992): 21–50.

Ackerman, Diane. *A Natural History of the Senses.* New York: Random House, 1990.

———. *The Natural History of Love.* New York: Random House, 1994.

Acock, Alan C., and David H. Demo. *Family Diversity and Well-Being.* Thousand Oaks, CA: Sage, 1994.

Adams, Bert. "The Family Problems and Solutions." *Journal of Marriage and the Family* 47, 3 (August 1985): 525–529.

Adams, Candace B., et al. "Young Adults' Expectations about Sex-Roles in Midlife." *Psychological Reports* 69, 3 (1991): 823–830.

Adams, David. "Identifying the Assaultive Husband in Court: You Be the Judge." *Response to the Victimization of Women and Children* 13, 1 (1990): 13–16.

Adelmann, Pamela K. "Psychological Well-Being and Home-maker vs. Retiree Identity among Older Women." *Sex Roles* 29, 3–4 (1993): 195–212.

Adler, Alfred. "Individual Psychology Therapy." In *Psychotherapy and Counseling,* edited by W. S. Sahakian. Chicago: Rand McNally, 1976.

Adler, Jerry. "Learning from the Loss." *Newsweek* (March 24, 1986): 66–67.

Adler, Jerry et al. "The Joy of Gardening." *Newsweek* (July 26, 1982).

Adler, N. A., and J. Schultz. "Sibling Incest Offenders." *Child Abuse and Neglect.* 19, 7 (July 1995): 811–819.

Adler, Nancy, Susan Hendrick, and Clyde Hendrick. "Male Sexual Preference and Attitudes toward Love and Sexuality." *Journal of Sex Education and Therapy* 12, 2 (September 1989): 27–30.

Adler, Nancy E., H. P. David, B. N. Major, S. H. Roth, N. F. Russo, and G. E. Wyatt. "Psychological Responses after Abortion." *Science* 246 (April 1990): 41–44.

"Advance Report of Final Natality Statistics, 1991." *Monthly Vital Statistics Report* (Centers for Disease Control and Prevention) 42, 3 (Supplement) (September 9, 1993): 1–6.

"A Fatal, Unknowing Dose: With GHB, Line Between a High and Death is Narrow." ABC-News.Go.Com., March 2, 2000.

Ahlander, N., and K. Bahr, "Beyond Drudgery, Power, and Equity: Toward an Expanded Discourse on the Moral Dimensions of Housework in Families." *Journal of Marriage and the Family* 57,1 (February 1995): 54–68.

Ahn, Helen Noh, and Neil Gilbert. "Cultural Diversity and Sexual Abuse Prevention." *Social Service Review* 66, 3 (September 1992): 410–428.

Ahrons, Constance, and Roy Rodgers. *Divorced Families: A Multidisciplinary View.* New York: Norton, 1987.

Ahrons, Constance, and Jennifer Tanner. "Adult Children and Their Fathers: Relationship Changes 20 Years After Parental Divorce." *Family Relations* 52, 4 (340–351).

Ahrons, Constance, and Lynn Wallisch. "The Relationship between Former Spouses." In *Intimate Relationships: Development, Dynamics, and Deterioration,* edited by D. Perlman and S. Duck. Newbury Park, CA: Sage, 1987.

"AIDS and Children: A Family Disease." *World AIDS Magazine,* (November 1989): 12–14.

Ainsworth, Mary D., M. D. Blehar, E. Waters, and S. Wall. *Patterns of Attachment: A Psychological Study of the Strange Situation.* Hillsdale, NJ: Lawrence Erlbaum, 1978.

Alapack, Richard. "The Adolescent First Kiss." *Humanistic Psychologist* 19, 1 (March 1991): 48–67.

Aldous, Joan. *Family Careers: Developmental Change in Families.* New York: Wiley, 1978.

———ed. *Two Paychecks.* Beverly Hills, CA: Sage , 1982.

———. "American Families in the 1980s: Individualism Run Amok?" *Journal of Family Issues* 8, 4 (December 1987a): 422–425.

———. "New Views on the Family Life of the Elderly and Near-Elderly." *Journal of Marriage and the Family* 49, 2 (May 1987b): 227–234.

———. "Perspectives on Family Change." *Journal of Marriage and the Family* 52, 3 (August 1990): 571–583.

Aldous, Joan, and Wilfried Dumon. "Family Policy in the 1980s: Controversy and Consensus." In *Contemporary Families: Looking Forward, Looking Back,* edited by A. Booth. Minneapolis: National Council on Family Relations, 1991.

Aldous, Joan, and David Klein. "Sentiment and Services: Models of Intergenerational Relationships in Midlife." *Journal of Marriage and the Family* 53, 3 (August 1991): 595–608.

Aldous, Joan, and Gail M. Mulligan "Fathers' Child Care and Children's Behavior Problems: A Longitudinal Study." *Journal of Family Issues,* 23, 5 (July 2002): 624–647.

Allen, Katherine. *Single Women/Family Ties.* Newbury Park, CA: Sage , 1989.

Allen, Katherine, Rosemary Blieszner, and Karen Roberto. "Families in the Middle and Later Years: A Review and Critique of Research in the 1990's." *Journal of Marriage and the Family,* 62, 4 (November 2000): 911–926.

Allen, Katherine R., David H. Demo, and Mark A. Fine., *Handbook of Family Diversity.* New York: Oxford University Press, 2000.

Allen, Katherine R., and Elizabeth B. Farnsworth. "Reflexivity in Teaching about Families." *Family Relations* 42, 3 (July 1993): 351–356.

Allen, Katherine, and Robert Pickett. "Forgotten Streams in the Family Life Course." *Journal of Marriage and the Family* 49, 3 (August 1987): 517–528.

Allgeier, Elizabeth, and Naomi McCormick, eds. *Gender Roles and Sexual Behavior.* Palo Alto, CA: Mayfield, 1982.

Allport, Gordon. *The Nature of Prejudice.* Garden City, NY: Doubleday, 1958.

Altman, Dennis . *The Homosexualization of America, the Americanization of the Homosexual.* New York: St. Martin's Press, 1982.

———. *AIDS in the Mind of America.* Garden City, NY: Anchor/Doubleday, 1985.

Amato, Paul. "Parental Divorce and Attitudes toward Marriage and Family Life." *Journal of Marriage and the Family* 50 (May 1988): 453–461.

———. "Who Cares for Children in Public Places? Naturalistic Observation of Male and Female Caretakers." *Journal of Marriage and the Family* 51 (November 1989): 981–990.

———. "Parental Absence During Childhood and Depression in Later Life." *The Sociological Quarterly* 32 (1991): 543–556.

———. "Children's Adjustment to Divorce: Theories, Hypotheses, and Empirical Support." *Journal of Marriage and the Family,* 55 1 (February, 1993): 23–32.

———. "Explaining the Intergenerational Transmission of Divorce," *Journal of Marriage and the Family* 58 (August 1996): 628–640.

———. "Reconciling Divergent Perspectives: Judith Wallerstein, Quantitative Family Research, and Children of Divorce." *Family Relations* 52, 4 (October 2003) 332–339.

Amato, Paul, and Stacy Rogers. "A Longitudinal Study of Marital Problems and Subsequent Divorce." *Journal of Marriage and the Family* 59 (August 1997): 612–624.

Amato, Paul R. "Father-Child Relations, Mother-Child Relations, and Offspring Psychological Well-Being in Early Adulthood." *Journal of Marriage and the Family* 56, 4 (November 1994): 1031–1042.

———. "The Consequences of Divorce for Adults and Children," *Journal of Marriage and the Family,* 62, 4 (November 2000): 1269–1288.

Amato, Paul R., and Alan Booth. "A Prospective Study of Divorce and Parent-Child Relationships." *Journal of Marriage and the Family* 58, 2 (May 1996): 356–365.

Amato, Paul R., and Bruce Keith. "Parental Divorce and the Well-Being of Children: A Meta-Analysis." *Psychological Bulletin* 110 (1991): 26–46.

Ambry, Margaret K. "Childless Chances." *American Demographics* 14, 4 (April 1992): 55.

American Academy of Child and Adolescent Psychiatry. "Making Day Care a Good Experience." No. 20 (October 1992a).

______. "The Influence of Music and Rock Videos." No. 40 (October 1992b).

American College of Obstetricians and Gynecologists. "ACOG technical bulletin number 205: Preconception care." *International Journal of Gynaecology and Obstetrics* 50 (1995): 201–207.

American Fertility Society. "New Guidelines for the Use of Semen Donor Insemination: 1990." *Fertility and Sterility* 53, 3 (Supplement 1) (March 1990): 1S–13S.

American Psychological Association. *Interim Report of the APA Working Group on Investigation of Memories of Childhood Abuse.* Washington, DC: American Psychological Association, 1994.

American Psychiatric Association. *Diagnostic and Statistical Manual of Mental Disorders.* 4th ed. Washington, DC: American Psychiatric Association, 1994.

Andersen, D. A., M. W. Lustig, and J. F. Andersen. "Regional Patterns of Communication in the United States: A Theoretical Perspective." *Communication Monographs,* 54 (1987): 128–144.

Anderson, Kristin L. "Gender, Status and Domestic Violence: An Integration of Feminist and Family Violence Approaches." *Journal of Marriage and the Family* 59, 3 (August 1997) 655–669.

Anderson, Stephen. "Parental Stress and Coping during the Leaving Home Transition." *Family Relations* 37 (April 1988): 160–165.

Andreasen, Margaret S. "Patterns of Family Life and Television Consumption from 1945 to the 1990s." In *Media, Children and the Family: Social Scientific Psychodynamic and Clinical Perspectives,* edited by J. Bryant, D. Zillman, and A. C. Huston. Hillsdale, NJ: Lawrence Erlbaum, 1994.

Andrews, F. M., A. Abbey, and L. J. Halman. "Stress from Infertility, Marriage Factors, and Subjective Well-Being of Wives and Husbands." *Journal of Health and Social Behavior* 32, 3 (September 1991): 238–253.

Aneshensel, Carol, Eva Fielder, and Rosina Becerra. "Fertility and Fertility-Related Behavior among Mexican-American and Non-Hispanic White Females." *Journal of Health and Social Behavior* 30, 1 (March 1989): 56–78.

Aneshensel, C., and L. I. Pearlin. "Structural Contexts of Sex Differences in Stress." In *Gender and Stress,* edited by R. C. Barnett et al. New York: Macmillan, 1987.

Annon, Jack. *The Behavioral Treatment of Sexual Problems.* Honolulu, HI: Enabling Systems, 1974.

———. *Behavioral Treatment of Sexual Problems: Brief Therapy.* New York: Harper and Row, 1976.

Anson, Ofra. "Marital Status and Women's Health Revisited: The Importance of a Proximate Adult." *Journal of Marriage and the Family* 51, 1 (February 1989): 185–194.

"A.P.A. Says Television Has Potential to be Beneficial." *Brown University Child and Adolescent Behavior Letter* (March 1992).

Aponte, Robert, with Bruce Beal and Michelle Jiles. "Ethnic Variation in the Family: The Elusive Trend Toward Convergence." In *Handbook of Marriage and the Family,* 2nd ed., edited by M. Sussman, S. Steinmetz, and G. Peterson. New York: Plenum, 1999.

Applebome, Peter. "Although Urban Blight Worsens, Most People Don't Feel Its Impact." *New York Times* (January 26, 1991): 1, 12.

Arditti, Joyce A . "Noncustodial Fathers: An Overview of Policy and Resources." *Family Relations* 39, 4 (October 1990): 460–465.

———. "Factors Related to Custody, Visitation, and Child Support for Divorced Fathers: An Exploratory Analysis." *Journal of Divorce and Remarriage* 17, 3–4 (1992): 23–42.

Arditti, Joyce A., and Katherine R. Allen. "Understanding Distressed Fathers' Perceptions of Legal and Relational Inequities Postdivorce." *Family and Conciliation Courts Review* 31, 4 (1993): 461–476.

Arendell, Terry. *Mothers and Divorce: Legal, Economic, and Social Dilemmas.* Berkeley, CA: University of California, 1987.

———. "Conceiving and Investigating Motherhood: The Decade's Scholarship." *Journal of Marriage and the Family* 62, 4 (November 2000):1192–1207.

Aries, Phillipe. *Centuries of Childhood.* New York: Vintage, 1962.

Arms, Karen G., et al., eds. *Cultural Diversity and Families.* Dubuque, IA: William C. Brown, 1992.

Arms, Suzanne. *Adoption: A Handful of Hope.* Berkeley, CA: Celestial Arts, 1990.

Armstrong, Penny, and Sheryl Feldman. *A Wise Birth.* New York: Morrow, 1990.

Armsworth, Mary W. "Psychological Response to Abortion." *Journal of Counseling and Development,* 65, 4 (March 1991): 377–379.

Aron, Arthur. "Unrequited Love as Self-Expansion." Paper presented at Second Iowa Conference on Personal Relationships, Iowa City, Iowa, May 12, 1989.

Aron, Arthur, and Elaine Aron. "Love and Sexuality." In *Sexuality in Close Relationships,* edited by K. McKinney and S. Sprecher. Hillsdale, NJ: Lawrence Erlbaum, 1991.

Aron, Arthur, et al. "Experiences of Falling in Love." *Journal of Social and Personal Relationships* 6 (1989): 243–257.

Arond, M., and S. I. Parker. *The First Year of Marriage.* New York: Warner, 1987.

Aswad, Barbara. "Arab American Families." In *Families in Cultural Context,* edited by M. K. DeGenova. Mountain View, CA: Mayfield, 1997.

"At a Glance." *OURS: The Magazine of Adoptive Families* 23, 6 (November 1990): 63.

Athey, Jean L. "HIV Infection and Homeless Adolescents." *Child Welfare* 70, 5 (September 1991): 517–528.

Atkinson, Alice M. "Rural and Urban Families' Use of Child Care." *Family Relations* 43, 1 (January 1994): 16–22.

Atkinson, Alice, and Diedre James. "The Transition between Active and Adult Parenting: An End and a Beginning." *Family Perspective* 25, 1 (1991): 57–66.

Atkinson, Maxine P., and Stephen P. Blackwelder. "Fathering in the 20th Century." *Journal of Marriage and the Family* 55, 4 (November 1993): 975–986.

Atwood, J. D., and J. Gagnon. "Masturbatory Behavior in College Youth." *Journal of Sex Education and Therapy* 13 (1987): 35–42.

Avruch, S., and A. P. Cackley. "Savings Achieved by Giving WIC Benefits to Women Prenatally," *Public Health Reports* 110 (1995): 27–34.

Axelson, Leland, and Paula Dail. "The Changing Character of Homelessness in the United States." *Family Relations* 37, 4 (October 1988): 463–469.

Axelson, Marta, and Jennifer Glass. "Household Structure and Labor Force Participation of Black, Hispanic, and White Mothers." *Demography* 22 (1985): 381–394.

Axinn, William G., and Arland Thornton. "Mothers, Children, and Cohabitation: The Intergenerational Effects of Attitudes and Behavior." *American Sociological Review* 58, 2 (1993): 233–246.

Babbie, Earl. *Basics of Social Research— with SPSS.* Belmont, CA: Wadsworth. 2002

Baber, Kristen M., and Katherine R. Allen. *Women and Families: Feminist Reconstructions.* New York: Guilford Press, 1992.

Bachu, Amara. *Current Population Reports,* (Series P20–482), Washington, DC: U.S. Census Bureau, 1995.

Baggett, Courtney R. "Sexual Orientation: Should It Affect Child Custody Rulings?" *Law and Psychology Review* 16 (1992): 189–200.

Bahr, Kathleen. "Student Responses to Genogram and Family Chronology." *Family Relations* 39, 3 (July 1990): 243–249.

Baird, Donna D., and Allen J. Wilcox. "Cigarette Smoking Associated with Delayed Conception." *Journal of the American Medical Association* 253, 20 (May 1985): 2979–2983.

Baird, M. A., and W. J. Doherty. "Risks and Benefits of a Family Systems Approach to Health Care." *Family Medicine* 18 (1990): 5–17.

Baker, Sharon, Stanton Thalberg, and Diane Morrison. "Parents' Behavioral Norms as Predictors of Adolescent Sexual Activity and Contraceptive Use." *Adolescence* 23 (June 1988): 265–282.

Balay, Robert, ed. *Guide to Reference Books.* 11th ed. Chicago: American Library Association, 1996.

Baldwin, J. D., S. Whitely, and J. I. Baldwin. "The Effect of Ethnic Group on Sexual Activities Related to Contraception and STDs." *Journal of Sex Research* 29, 2 (May 1992): 189–206.

Bandura, Albert. *Social Learning Theory.* Englewood Cliffs, NJ: Prentice-Hall, 1977.

Bane, Mary Jo, and Paul A. Jargowsky. "The Links Between Public Policy and Family Structure: What Matters and What Doesn't." In *The Changing American Family and Public Policy,* edited by A. J. Cherlin. Washington, DC: Urban Institute Press, 1988.

Bankston, Carl, and Jacques Henry. "Endogamy Among Louisiana Cajuns: A Social Class Explanation." *Social Forces* 77 (4), 1999: 1317–1338.

Barbach, Lonnie. *For Each Other: Sharing Sexual Intimacy.* Garden City, NY: Doubleday, 1982.

Barber, B. L., and J. M. Lyons. "Family Processes and Adolescent Adjustment in Intact and Remarried Families. *Journal of Youth and Adolescence* 23, 4 (August 1994): 421–436.

Barden, J. C. "Many Parents in Divorces Abduct Their Own Children." *New York Times* (May 6, 1990): 10.

Barkas, J. L. *Single in America.* New York: Atheneum, 1980.

Barnes, Jessica and Claudette Bennett. "The Asian Population: 2000." *Census 2000 Brief.* (February 2002).

Barnett, R. C., and G. K. Baruch. "Determinants of Father's Participation in Family Work." *Journal of Marriage and Family* 49 (1987): 29–40.

Barret, R. L., and B. E. Robinson. *Gay Fathers.* Lexington, MA: Lexington Books, 1990.

Barrow, Georgia. *Aging, Ageism, and Society.* St. Paul, MN: West, 1989.

Bartholomew, Kim. "Avoidance of Intimacy: An Attachment Perspective." *Journal of Social and Personal Relationships* 7, 2 (1990): 147–178.

Baruch, Elaine Hoffman, et al., eds. *Embryos, Ethics, and Women's Rights: Exploring the New Reproductive Technologies.* New York: Harrington Park Press, 1988.

Basow, S. A. *Gender: Stereotyping and Roles.* Pacific Grove, CA: Brooks/Cole, 1992.

Bass, Ellen, and Louise Thornton, eds. *I Never Told Anyone: Stories and Poems by Survivors of Child Sexual Abuse.* New York: Harper and Row, 1983.

Bassuk, Ellen, and Lenore Rubin. "Homeless Children: A Neglected Population." *American Journal of Orthopsychiatry* 57, 2 (April 1987): 279–286.

Bassuk, Ellen, et al. "Characteristics of Sheltered Homeless Families." *American Journal of Public Health* 76 (1986): 1097–1101.

Baumeister, Roy F., Sara R. Wotman, and Arlene M. Stillwell. "Unrequited Love: On Heartbreak, Anger, Guilt, Scriptlessness, and Humiliation." *Journal of Personality and Social Psychology* 64, 3 (March 1993): 377–394.

Baumrind, Diana. "Current Patterns of Parental Authority." *Developmental Psychology Monographs* 4, 1 (1971): 1–102.

———. "Parental Disciplinary Patterns and Social Competence in Children." *Youth and Society* 9, 3 (March 1978): 239–276.

———. "Rejoinder to Lewis's Reinterpretation of Parental Firm Control Effects: Are Authoritative Families Really Harmonious?" *Psychological Bulletin* 94, 1 (July 1983): 132–142.

Baxter, Jean. "Gender Equality and Participation in Housework: A Cross-National Perspective," *Journal of Comparative Family Studies* 28 (Autumn 1997):220–248.

Baxter, L. A. "Cognition and Communication in Relationship Process." In *Accounting for Relationships: Explanation, Representation and Knowledge,* edited by R. Burnett, P. McGhee, and D. Clarke. London: Meuthen, 1987.

Baxter, Richard L., Cynthia De Riemer, Ann Landini, and Larry Leslie. "A Content Analysis of Music Videos." *Journal of Broadcasting and Electronic Media* 29, 3 (June 1985): 333–340.

Bean, Frank, and Marta Tienda. *The Hispanic Population of the United States.* New York: Russell Sage Foundation, 1987.

Beavers, Robert W. "Healthy, Midrange, and Severely Dysfunctional Families." In *Normal Family Processes,* edited by F. Walsh. New York: Guilford Press, 1982.

Beavers, Robert W., and Robert Hampson. *Successful Families.* New York: Norton, 1990.

Becerra, Rosina. "The Mexican American Family." In *Ethnic Families in America: Patterns and Variations,* 3rd ed., edited by C. Mindel et al. New York: Elsevier North Holland, 1988.

Bechhofer, L., and L. Parrot. "What Is Acquaintance Rape?" In *Acquaintance Rape: The Hidden Crime,* edited by A. Parrott and L. Bechhofer. New York: Wiley, 1991.

Beck, Joyce W., and Barbara M. Heinzerling. "Gay Clients Involved in Child Custody Cases: Legal and Counseling Issues." *Psychotherapy in Private Practice* 12, 1 (1993): 29–41.

Beck, Melinda. "Miscarriages." *Newsweek* (August 15, 1988): 46–49.

Beck, Melinda, and Geoffrey Cowley. "Mother Nature?" *Newsweek* (January 17, 1994): 54–58.

Beck, Rubye, and Scott Beck. "The Incidence of Extended Households among Middle-Aged Black and White Women." *Journal of Family Issues* 10, 2 (June 1989): 147–168.

Becker, Howard. *Outsiders.* New York: Free Press, 1963.

Becker, Penny Edgell, and Phyllis Moen. "Scaling Back: Dual Earner Couples' Work-Family Strategies," *Journal of Marriage and the Family* 61, 4 (November 1999) 995–1007.

Beer, William. *American Stepfamilies.* New Brunswick, NJ: Transaction, 1992.

Beggs, Joyce M., and Dorothy C. Doolittle. "Perceptions Now and Then of Occupational Sex Typing: A Replication of Shinar's 1975 Study." *Journal of Applied Social Psychology* 23, 17 (1993): 1435–1453.

Beitchman, Joseph H., et al. "A Review of the Short-Term Effects of Child Sexual Abuse." *Child Abuse and Neglect* 15, 4 (1991): 537–556.

———. "A Review of the Long-Term Effects of Child Sexual Abuse." *Child Abuse and Neglect* 16, 1 (January 1992): 101.

Belcastro, Philip. "Sexual Behavior Differences between Black and White Students." *Journal of Sex Research* 21, 1 (February 1985): 56–67.

Belkin, Lisa. "Bars to Equality of Women Seen as Eroding Slowly." *New York Times* (August 20, 1989).

Bell, Alan, and Martin Weinberg. *Homosexualities: A Study of Diversities among Men.* New York: Simon and Schuster, 1978.

Bell, Alan, et al. *Sexual Preference: Its Development in Men and Women.* Bloomington, IN: Indiana University Press, 1981.

Bell, Robert. *Worlds of Friendship.* Beverly Hills, CA: Sage, 1981.

Bellah, Robert, et al. *Habits of the Heart.* Berkeley, CA: University of California Press, 1985.

Belsky, Jay. "Stability and Change in Marriage across the Transition to Parenthood: A Second Study." *Journal of Marriage and the Family* 47, 4 (November 1985): 855–865.

———. "Infant Day Care, Child Development, and Family Policy." *Society* 27, 5 (July 1990): 10–12.

———. "Patterns of Marital Change and Parent-Child Interaction." *Journal of Marriage and the Family* 53, 2 (May 1991): 487–498.

Bem, Sandra. "The Measurement of Psychological Androgyny." *Journal of Consulting and Clinical Psychology* 42 (1974): 155–162.

———. "Androgyny versus the Tight Little Lives of Fluffy Women and Chesty Men." *Psychology Today* 9, 4 (September 1975a): 58–59 ff.

———. "Sex Role Adaptability: One Consequence of Psychological Androgyny." *Journal of Personality and Social Psychology* 31, 4 (1975b): 634–643.

———. *Bem Sex-Role Inventory: Professional Manual.* Palo Alto, CA: Consulting Psychologists Press, 1981a.

———. "Gender Schema Theory: A Cognitive Account of Sex Typing." *Psychological Review* 88 (1981b): 354–364.

———. "Gender Schema Theory and Its Implications for Child Development: Raising Gender-Aschematic Children in a Gender Schematic Society." *Signs* 8, 4 (June 1983): 598–616.

Bengston, Vern, and Joan Robertson, eds. *Grandparenthood.* Beverly Hills, CA: Sage, 1985.

Benin, Mary, and Joan Agostinelli. "Husbands' and Wives' Satisfaction with the Division of Labor." *Journal of Marriage and the Family* 50, 2 (May 1988): 349–361.

Benokraitis, Nijole. "How Family Wars Affect Us: Four Models of Family Change and Their Consequences." In *Feuds About Families: Conservative, Centrist, Liberal, and Feminist Perpectives,"* edited by N. Benokraitis. Upper Saddle River, NJ: Prentice Hall, 2000a: 14-24.

Benokraitis, Nijole V., ed. *Feuds About Families: Conservative, Centrist, Liberal, and Feminist Perpectives."* Upper Saddle River, NJ: Prentice Hall, 2000b.

Benson, D., C. Charlton, and F. Goodhart. "Acquaintance Rape on Campus: A Literature Review." *Journal of American College Health* 40 (1992): 157–165.

Bera, W., et al. *Male Adolescent Sexual Abuse.* Newbury Park, CA: Sage, 1991.

Berardo, Felix. "Trends and Directions in Family Research in the 1980s." In *Contemporary Families: Looking Forward, Looking Back,* edited by A. Booth. Minneapolis: National Council on Family Relations, 1991.

Bergen, David J., and John E. Williams. "Sex Stereotypes in the United States Revisited: 1972–1988." *Sex Roles* 24, 7/8 (1991): 413–423.

Berger, C. R. "Planning and Scheming: Strategies for Initiating Relationships." In *Accounting for Relationships: Explanations, Representation and Knowledge,* edited by R. Burnett, P. McChee, and D. Clarke. New York: Methuen, 1987.

Berger, Mark J., and Donald P. Goldstein. "Infertility Related to Exposure to DES in Utero: Reproductive Problems in the Female." In *Infertility: Medical, Emotional, and Social Considerations,* edited by M. Mazor and H. Simons. New York: Human Sciences Press, 1984.

Berk, Richard A. "What the Scientific Evidence Shows: On the Average, We Can Do No Better Than Arrest." In *Current Controversies in Family Violence,* edited by R. Gelles and D. Loseke. Newbury Park, CA: Sage, 1993.

Berk, Richard A., Sarah F. Berk, Phyllis J. Newton, and D. R. Loseke. "Cops on Call: Summoning the Police to the Scene of Spousal Violence." *Law & Society Review* 18, 3 (1984): 479–498.

Berk, Richard A., Phyllis J. Newton, and Sarah F. Berk. "What a Difference a Day Makes: An Empirical Study of the Impact of Shelters for Battered Women." *Journal of Marriage and the Family* 48 (August 1986): 481–490.

Berkowitz, Alan. "College Men as Perpetrators of Acquaintance Rape and Sexual Assault: A Review of Recent Research." *Journal of American College Health* 40, 4 (January 1992): 175–181.

Berman, William. "Continued Attachment After Legal Divorce." *Journal of Family Issues* 6, 3 (September 1985): 375–392.

Bernard, Jessie. *The Future of Marriage,* 2nd ed. New York: Columbia University Press, 1982.

Berrick, J. D., and R. P. Barth. "Child Sexual Abuse Prevention—Research Review and Recommendations." *Social Work Research and Abstracts* 28 (1992): 6–15.

Berscheid, Ellen. "Emotion." In *Close Relationships,* edited by H. H. Kelley et al. New York: Freeman, 1983.

———. "Interpersonal Attraction." In *Handbook of Social Psychology,* edited by G. Lindzey and E. Aronson. New York: Random House, 1985.

———. "Some Comments on Love's Anatomy: Or Whatever Happened to Old-fashioned Lust?" In *The Psychology of Love,* edited by Robert Sternberg and Michael Barnes. New Haven, CT: Yale University Press, 1988.

———. "Interpersonal Relationships." *Annual Review of Psychology* 45 (1994): 79–129.

Berscheid, Ellen, and J. Frei. "Romantic Love and Sexual Jealousy." In *Jealousy,* edited by G. Clanton and L. Smith. Englewood Cliffs, NJ: Prentice-Hall, 1977.

Berscheid, Ellen, and Elaine H. Walster. "A Little Bit About Love." In *Foundations of Interpersonal Attraction,* edited by T. L. Huston. New York: Academic Press, 1974.

———. *Interpersonal Attraction.* Reading, MA: Addison-Wesley, 1978.

Besharov, Douglas and Timothy Sullivan. "Welfare Reform and Marriage," *The Public Interest* 125 (1996): 81–94.

Betchen, Stephen. "Male Masturbation as a Vehicle for the Pursuer/Distancer Relationship in Marriage." *Journal of Sex and Marital Therapy* 17, 4 (December 1991): 269–278.

Bianchi, S. "America's Children." *Population Bulletin* 45, 1 (June 1990): 3–41.

"Bibulous America: Over Half of All Adults Are Drinkers." *Dialogue* 5, 1 (February 1995).

Bichoff, Gary P., Sandra M. Stith, and Stephan M. Wilson. "A Comparison of the Family Systems of Adolescent Sexual Offenders and Non-Sexual Offending Delinquents." *Family Relations* 41, 3 (July 1992): 318–323.

Bidwell, Lee Millar, and Brenda Vander Mey. *Sociology of the Family: Investigating Family Issues.* Needham Heights, MA: Allyn and Bacon, 2000.

Bielby, William, and James Baron. "Woman's Place Is with Other Women: Sex Segregation in the Workplace." National Research Council, Workshop on Job Segregation by Sex, 1982. Unpublished paper.

Bigler, Rebecca S., and Lynn S. Liben. "Cognitive Mechanisms in Children's Gender Stereotyping: Theoretical and Educational Implications of a Cognitive-Based Intervention." *Child Development* 63, 6 (1992): 1351–1364.

Billingsley, Andrew. "The Impact of Technology on Afro-American Families." *Family Relations* 37, 4 (October 1988): 420–425.

Billy, John, Nancy Landale, William Grady, and Denise Zimmerle. "Effects of Sexual Activity on Adolescent Social and Psychological Development." *Social Psychology Quarterly* 51, 3 (September 1988): 190–212.

Billy, John, Koray Tanfer, William R. Grady, and Daniel H. Klepinger. "The Sexual Behavior of Men in the United States." *Family Planning Perspectives* 25, 2 (March 1993): 52–60.

Binion, Victoria. "Psychological Androgyny: A Black Female Perspective." *Sex Roles* 22, 7–8 (April 1990): 487–507.

Bird, Chloe. "Gender Differences in the Social and Economic Burdens of Parenting and Psychological Distress." *Journal of Marriage and the Family* 59(4) 1997: 809–823.

Bird, Chloe E., and Catherine E. Ross. "Houseworkers and Paid Workers: Qualities of the Work and Effects on Personal Control." *Journal of Marriage and the Family* 55, 4 (November 1993): 913–925.

Bird, Gerald, and Gloria Bird. "The Determinants of Mobility in Two-Earner Families: Does the Wife's Income Count." *Journal of Marriage and the Family* 47, 3 (August 1985): 753–758.

Black, Rita D. "Women's Voices after Pregnancy Loss: Couples' Patterns." *Social Work in Health Care* 16, 2 (1991): 19–36.

Blackwell, Debra. "Marital Homogamy in the United States: The Influence of Individual and Paternal Education." *Social Science Research* 27 (2) 1998: 159–164.

Blair, Sampson Lee. "The Sex-Typing of Children's Household Labor: Parental Influence on Daughters' and Sons' Housework." *Youth and Society* 24, 2 (1992): 178–203.

———. "Employment, Family, and Perceptions of Marital Quality among Husbands and Wives." *Journal of Family Issues* 14, 2 (1993): 189–212.

Blair, Sampson, and Daniel Lichter. "Measuring the Division of Household Labor: Gender Segregation and Housework among American Couples." *Journal of Family Issues* 12, 1 (March 1991): 91–113.

Blair-Loy, Mary and Amy Wharton. "Employees Use of Work-Family Policies and the Workplace Social Context." *Social Forces* 80, 3 (March 2002): 813-845.

Blakely, Mary Kay. "Surrogate Mothers: For Whom Are They Working?" *Ms.* (March 1987): 18, 20.

Blankenhorn, David. *Fatherless America: Confronting Our Most Urgent Social Problem.* New York: Basic Books, 1995.

Blau, Elizabeth. "Study Finds Barrage of Sex on TV." *New York Times* (January 27, 1988).

Blieszner, Rosemary, and Janet Alley. "Family Caregiving for the Elderly: An Overview of Resources." *Family Relations* 39, 1 (January 1990): 97–102.

Block, Jeanne. "Differential Premises Arising from Differential Socialization of the Sexes: Some Conjectures." *Child Development* 54 (December 1983): 1335–1354.

Bloom, Bernard, et al. "Sources of Marital Dissatisfaction Among Newly Separated Persons." *Journal of Family Issues* 6, 3 (September 1985): 359–373.

Blum, R., et al. "American Indian—Alaska Native Youth Health." *Journal of American Medical Association* 267, 12 (March 25, 1992): 1637–1644.

Blumberg, Rae Lesser, ed. *Gender, Family, and the Economy.* Newbury Park, CA: Sage, 1990.

Blumenfeld, Warren, and Diane Raymond. *Looking at Gay and Lesbian Life.* Boston: Beacon Press, 1989.

Blumstein, Philip. "Identity Bargaining and Self-Conception." *Social Forces* 53, 3 (1975): 476–485.

Blumstein, Philip, and Pepper Schwartz. *American Couples.* New York: McGraw-Hill, 1983.

Bogenschneider, Karen. "Has Family Policy Come of Age? A Decade Review of the State of U.S. Family Policy in the 1990's." *Journal of Marriage and the Family* 62, 4 (November 2000): 1136–1159.

Bogert, Carroll. "Bringing Back Baby." *Newsweek* (November 21, 1994): 78–79.

Bohannan, Paul. "The Six Stations of Divorce." In *Divorce and After,* edited by P. Bohannan. New York: Doubleday, 1970a.

Bohannan, Paul, ed. *Divorce and After.* New York: Doubleday, 1970b.

Boken, Halcyone. "Gender Equality in Work and Family." *Journal of Family Issues* 5, 2 (June 1984): 254–272.

Boland, Joseph, and Diane Follingstad. "The Relationship between Communication and Marital Satisfaction: A Review." *Journal of Sex and Marital Therapy* 13, 4 (December 1987): 286–313.

Boles, Abner J., and Harriet Curtis-Boles. "Black Couples and the Transition to Parenthood." *American Journal of Social Psychiatry* 6, 1 (December 1991): 314–318.

Bollen, N., M. Camus, C. Staessen, H. Tournaye, P. Devroey, and A. C. Van Steirteghem. "The Incidence of Multiple Pregnancy after In Vitro Fertilization and Embryo Transfer, Gamete, or Zygote Intrafallopian Transfer." *Fertility and Sterility* 55, 2 (February 1991): 314–318.

Bonney, Lewis A. "Planning for Postdivorce Relationships: Factors to Consider in Drafting a Transition Plan." *Family and Conciliation Courts Review* 31, 3 (1993): 367–372.

Book, Cassandra L., et al. *Human Communication: Principles, Contexts, and Skills.* New York: St. Martin's Press, 1980.

Boone, Margaret S. *Capital Crime: Black Infant Mortality in America.* Newbury Park, CA: Sage, 1989.

Booth, A., and J. N. Edwards. "Starting Over: Why Remarriages Are More Unstable. *Journal of Family Issues* 13, 2 (June 1992): 179–194.

Booth, Alan. "Who Divorces and Why: A Review." *Journal of Family Issues* 6, 3 (September 1985): 255–293.

Booth, Alan, ed. *Contemporary Families: Looking Forward, Looking Back.* Minneapolis: National Council on Family Relations, 1991.

———. *Child Care in the 1990s: Trends and Consequences.* Hillsdale, NJ: Lawrence Erlbaum, 1992.

Booth, Alan, and John Edwards. "Age at Marriage and Marital Instability." *Journal of Marriage and the Family* 47, 2 (February 1985): 67–74.

Booth, Alan, and David R. Johnson. "Declining Health and Marital Quality." *Journal of Marriage and the Family* 56, 1 (February 1994): 218–223.

Booth, Alan, et al. "Predicting Divorce and Permanent Separation." *Journal of Family Issues* 6, 3 (September 1985): 331–346.

———. "Divorce and Marital Instability over the Life Course." *Journal of Family Issues* 7 (1986): 421–442.

Borhek, Mary. "Helping Gay and Lesbian Adolescents and Their Families: A Mother's Perspective." *Journal of Adolescent Health Care* 9, 2 (March 1988): 123–128.

Borisoff, Deborah, and Lisa Merrill. *The Power to Communicate: Gender Differences as Barriers.* Prospect Heights, IL: Waveland, 1985.

Borland, Dolores. "A Cohort Analysis Approach to the Empty-Nest Syndrome among Three Ethnic Groups of Women: A Theoretical Position." *Journal of Marriage and the Family* 44 (February 1982): 117–129.

Bosse, Raymond, et al. "Change in Social Support after Retirement: Longitudinal Findings from the Normative Aging Study." *Journal of Gerontology* 48, 4 (1993): 210–217.

Boston Women's Health Book Collective. *The New Our Bodies, Ourselves.* New York: Simon and Schuster, 1992.

Bostwick, Homer. *A Treatise on the Nature and Treatment of Seminal Disease, Impotency, and Other Kindred Affections,* 12th ed. New York: Burgess, Stringer, 1860.

Boswell, John. *Christianity, Social Tolerance, and Homosexuality.* Chicago: University of Chicago Press, 1980.

Bourne, Richard, and Eli Newberger, eds. *Critical Perspectives on Child Abuse.* Lexington, MA: Lexington Books, 1979.

Bowker, Lee. *Beating Wife Beating.* Lexington, MA: Lexington Books, 1983.

Bowlby, John. *Attachment and Loss.* 3 vols. New York: Basic Books, 1980.

Bowman, Madonna, and Constance Ahrons. "Impact of Legal Custody Status on Father's Parenting Post-Divorce." *Journal of Marriage and the Family* 47, 2 (May 1985): 481–485.

Bozett, Frederick. "Children of Gay Fathers." In *Gay and Lesbian Parents,* edited by F. W. Bozett. New York: Praeger, 1987a.

———. "Gay Fathers." In *Gay and Lesbian Parents,* edited by F. W. Bozett. New York: Praeger, 1987b.

———. "Gay Fathers: A Review of the Literature." *Journal of Homosexuality* 18 (1989): 137–62.

Bozett, Frederick W., ed. *Gay and Lesbian Parents.* New York: Praeger, 1987c.

Bozett, F. W., and M. B. Sussman, eds.. *Homosexuality and Family Relations.* New York: Harrington Park Press, 1990.

Bradbury, Thomas, Frank D. Fincham, and Steven R.H. Beach. "Research on the Nature and Determinants of Marital Satisfaction: A Decade in Review." *Journal of Marriage and the Family,* 62, 4 (November 2000): 964–980.

Bradsher, Keith. "3 Guilty of Manslaughter in Slipping Drug to Girl," *New York Times,* March 15, 2000.

Bramlett, M. D., and Mosher, W. D. "Cohabitation, Marriage, Divorce, and Remarriage in the United States." National Center for Health Statistics. Vital Health Stat 23(22). 2002.

Brand, H. J. "The Influence of Sex Differences on the Acceptance of Infertility." *Journal of Reproductive and Infant Psychology* 7, 2 (April 1989): 129–131.

Braun, J., and H. Chao. "Attitudes Toward Women: A Comparison of Asian-Born Chinese and American Caucasians." *Psychology of Women Quarterly* 2 (1978): 195–201.

Braver, Sanford, et al. "A Social Exchange Model of Nonresidential Parent Involvement." In *Nonresidential Parenting: New Vistas in Family Living,* edited by C. E. Depner, and J. H. Bray. Newbury Park, CA: Sage Publications, 1993a.

———. "A Longitudinal Study of Noncustodial Parents: Parents without Children." *Journal of Family Psychology* 7, 1 (June 1993b): 9–23.

Bray, James H., and Sandra H. Berger. "Noncustodial Father and Paternal Grandparent Relationship in Stepfamilies." *Family Relations* 39, 4 (October 1990): 414–419.

———. "Nonresidential Parent-Child Relationships Following Divorce and Remarriage: A Longitudinal Perspective." In *Nonresidential Parenting: New Vistas in Family Living,* edited by C. E. Depner and J. H. Bray. Newbury Park, CA: Sage, 1993.

Bray, James H., and Charlene Depner. "Nonresidential Parents: Who Are They?" In *Nonresidential Parenting: New Vistas in Family Living,* edited by C. E. Depner and J. H. Bray. Newbury Park, CA: Sage, 1993.

Bray, James H., and E. Mavis Hetherington. "Families in Transition: Introduction and Overview" [Special Section: "Families in Transition"]. *Journal of Family Psychology* 7, 1 (1993): 3–8.

Brayfield, April A. "Employment Resources and Housework in Canada." *Journal of Marriage and the Family* 54, 1 (February 1992): 19–30.

———. "Juggling Jobs and Kids: The Impact of Employment Schedules on Fathers' Caring for Children." *Journal of Marriage and the Family* 57, 2 (May 1995): 321–332.

Breault, K. D., and Augustine Kposowa. "Explaining Divorce in the United States: A Study of 3,111 Counties, 1980." *Journal of Marriage and the Family* 49, 3 (August 1987): 549–558.

Brenner, Harvey. *Mental Illness and the Economy.* Cambridge, MA: Harvard University Press, 1973.

———. "Influence of the Social Environment on Psychopathology: The Historic Perspective." In *Stress and Mental Disorder,* edited by J. Barrett et al. New York: Raven, 1979.

Bretschneider, Judy, and Norma McCoy. "Sexual Interest and Behavior in Healthy 80– to 102–Year-Olds." *Archives of Sexual Behavior* 17, 2 (April 1988): 109–128.

Brewster, Karin and Irene Padavic. "Change in Gender-Ideology, 1977–1996: The Contributions of Intracohort Change and Population Turnover. *Journal of Marriage and the Family,* 62, 2 (May 2000): 477–488.

Bridges, Judith S. "Pink or Blue: Gender Congratulations Cards." *Psychology of Women* 17, 2 (1993): 193–205.

Bringle, Robert G., and Glenda J. Bagby. "Self-Esteem and Perceived Quality of Romantic and Family Relationships in Young Adults." *Journal of Research in Personality* 26, 4 (1992): 340–356.

Bringle, Robert, and Bram Buunk. "Jealousy and Social Behavior: A Review of Person, Relationship, and Situational Determinants." In *Review of Personality and Social Psychology, Vol.* 6, *Self, Situation, and Social Behavior,* edited by P. Shaver. Newbury Park, CA: Sage, 1985.

———. "Extradyadic Relationships and Sexual Jealousy." In *Sexuality in Close Relationships,* edited by K. McKinney and S. Sprecher. Hillsdale, NJ: Lawrence Erlbaum, 1991.

Britton, D. M. "Homophobia and Homosociality: An Analysis of Boundary Maintenance." *Sociological Quarterly* 31, 3 (September 1990): 423–439.

Broderick, Carlfred, and James Smith. "The General Systems Approach to the Family." In *Contemporary Theories About the Family,* edited by W. Burr et al. New York: Free Press, 1979.

Brody, Jane. "Estrogen is Found to Improve Mood, Not Just Menopause Symptoms." *New York Times* (January 1, 1992a): 141.

———. "Personal Health: Maintaining Friendships for the Sake of Good Health." *New York Times* (February 5, 1992b): B8.

———. "Despite All the Benefits, Many Mothers Decide Against Breastfeeding." *New York Times* (April 6, 1994): B9.

Brodzinsky, David M., and Marshall D. Schecter, eds. *The Psychology of Adoption.* New York: Oxford University Press, 1990.

Broman, Clifford. "Satisfaction among Blacks: The Significance of Marriage and Parenthood." *Journal of Marriage and the Family* 50, 1 (February 1988): 45–51.

Bronfenbrenner, Urie. *The Ecology of Human Development.* Cambridge, MA: Harvard University Press, 1979.

Bronstein, Phyllis, JoAnn Clauson, Miriam F. Stoll, and Craig L. Abrams. "Parenting Behavior and Children's Social, Psychological, and Academic Adjustment in Diverse Family Structures." *Family Relations* 42, 3 (July 1993): 268–276.

Brooks, Jane B. *Parenting in the 90s.* Mountain View, CA: Mayfield, 1994.

Brown, Elizabeth, and William R. Hendee. "Adolescents and Their Music." *Journal of the American Medical Association* 262, 12 (September 22, 1989): 1659–1663.

Brown, Jane D., and Kenneth Campbell. "Race and Gender in Music Videos: The Same Beat but a Different Drummer." *Journal of Communications* 36, 1 (December 1986): 94–106.

Brown, Jane D., and Susan F. Newcomer. "Television Viewing and Adolescents' Sexual Behavior." *Journal of Homosexuality* 21, 1/2 (1991): 77–91.

Brown, J. D., and L. Schulze. "The Effects of Race, Gender, and Fandom on Audience Interpretations of Madonna's Music Videos." *Journal of Communication* 40 (1990): 88–102.

Brown, Lyn Mikel, and Carol Gilligan. *Meeting at the Crossroads: Women's Psychology and Girl's Development.* Cambridge, MA: Harvard University Press, 1992.

Brown, Patricia Leigh. "Where to Put the TV Set?" *New York Times* (October 4, 1990): B4.

Brown, Susan , and Alan Booth. "Cohabitation Versus Marriage: A Comparison of Relationship Quality," *Journal of Marriage and the Family* 58 (August) 1996: 668–678.

———. "Stress at Home, Peace at Work: A Test of the Time Bind Hypothesis." *Social Science Quarterly, 83,* 4 (December 2002a): 905–919.

______. "Bending the Time Bind: Rejoinder to Hochschild and Goodman." *Social Science Quarterly,* 83, 4, (December 2002b): 941–946.

Browne, Angela, and David Finkelhor. "Initial and Long-Term Effects: A Review of the Research." In *Sourcebook on Child Sexual Abuse,* edited by D. Finkelhor. Beverly Hills, CA: Sage, 1986.

Browne, Jane. "Terrific Tips from the Marriage Doctor." *New Woman.* (July 1997): 76–79ff.

Brownmiller, Susan. *Femininity.* New York: Fawcett Columbine, 1983.

Bryant, Lois, et al. "Race and Family Structure Stereotyping: Perceptions of Black and White Nuclear Families and Stepfamilies." *Journal of Black Psychology* 15 (1988): 1–16.

Bryant, Z. Lois, and Marilyn Coleman. "The Black Family as Portrayed in Introductory Marriage and Family Textbooks." *Family Relations* 37, 3 (July 1988): 255–259.

Bryon, K. "Family Composition Changing." Washington, DC: U.S. Census Bureau, 1996.

Buchholz, Ester, and Barbara Gol. "More than Playing House: A Developmental Perspective on the Strengths in Teenage Motherhood." *American Journal of Orthopsychology* 56, 3 (July 1986).

Buchta, Richard. "Attitudes of Adolescents and Parents of Adolescents Concerning Condom Advertisements on Television." *Journal of Adolescent Health Care* 10, 3 (May 1989): 220–223.

Budiansky, Stephen. "The New Rules of Reproduction." *U.S. News and World Report* (April 18, 1988): 66–69.

Buehler, Cheryl, and Bobbie H. Legg. "Selected Aspects of Parenting and Children's Social Competence Postseparation: The Moderating Effects of Child's Sex, Age, and Family Economic Hardship" [Special Issue: "Divorce and the Next . . ."]. *Journal of Divorce and Remarriage* 18, 3–4 (1992): 177–195.

Bumpass, Larry L., Teresa C. Martin, and James A. Sweet. "The Impact of Family Background and Early Marital Factors on Marital Disruption." *Journal of Family Issues* 12, 1 (1991): 22–42.

Bumpass, Larry, and James A. Sweet. "Children's Experience in Single Parent Families: Implications of Cohabitation and Marital Transition." *Family Planning Perspectives* 21, 6 (November 1989): 256–260.

Bumpass, Larry, James Sweet, and Teresa Castro Martin. "Changing Patterns of Remarriage." *Journal of Marriage and the Family* 52, 3 (August 1990): 747–756.

Burckly, William, Nicholas Reuterman, and Sondra Kopsky. "Dating Violence Among High School Students." *School Counselor,* 35, 5 (May 1988): 353–358.

Burgess, Ann W., ed. *Rape and Sexual Assault II.* New York: Garland Press, 1988.

Burgess, Ernest. "The Family as a Unity of Interacting Personalities." In *Family Roles and Interaction,* edited by J. Heiss. Chicago: Rand McNally, 1968.

———. "The Family as a Unity of Interacting Personalities." *The Family* 7, 1 (March 1926): 3–9.

Burkett, Elinor. *The Baby Boon: How Family Friendly America Cheats the Childless.* New York: Free Press, 2000.

Burleson, Brant, and Wayne Denton. "The Relationship Between Communication Skill and Marital Satisfaction: Some Moderating Effects." *Journal of Marriage and the Family* 59, 4 (November 1997):884–902.

Burns, Ailsa. "Perceived Causes of Marriage Breakdown and Conditions of Life." *Journal of Marriage and the Family* 46, 3 (August 1984): 551–562.

Burns, Ailsa, and Cath Scott. *Mother-Headed Families and Why They Have Increased.* Hillsdale, NJ: Erlbaum, 1994.

Burns, Scott. *The Household Economy.* New York: Harper and Row, 1972.

Burr, W. R., and S. R. Klein. *Reexamining Family Stress.* Thousand Oaks, CA: Sage, 1994.

Burr, Wesley, et al. *Contemporary Theories About the Family.* 2 vols. New York: Free Press, 1979.

Bush, Catherine R., Joseph P. Bush, and Joyce Jennings. "Effects of Jealousy Threats on Relationship Perceptions and Emotions." *Journal of Social and Personal Relationships* 5, 3 (August 1988): 285–303.

Buss, David M., Randy J. Larsen, Drew Westen, and Jennifer Semmelroth. "Sex Differences in Jealousy: Evolution, Physiology, and Psychology." *Psychological Science* 3, 4 (1992): 251–255.

Butts, June Dobbs. "Adolescent Sexuality and Teenage Pregnancy from a Black Perspective." In *Teenage Pregnancy in a Family Context,* edited by T. Ooms. Philadelphia: Temple University Press, 1981.

Buunk, Bram, and Ralph Hupka. "Cross-Cultural Differences in the Elicitation of Sexual Jealousy." *Journal of Sex Research* 23, 1 (February 1987): 12–22.

Buunk, Bram, and Barry van Driel. *Variant Lifestyles and Relationships.* Newbury Park, CA: Sage , 1989.

Buzawa, Eve S., and Karl G. Buzawa. "The Scientific Evidence Is Not Conclusive: Arrest Is No Panacea." In *Current Controversies in Family Violence,* edited by R. Gelles and D. Loseke. Newbury Park, CA: Sage, 1993.

Byers, E. S., and L. Heinlein. "Predicting Initiations and Refusals of Sexual Activities in Married and Cohabiting Heterosexual Couples." *Journal of Sex Research* 26 (1989): 210–231.

Byrd, W., et al. "A Prospective Randomized Study of Pregnancy Rates Following Intrauterine and Intracervical Insemination Using Frozen Donor Sperm." *Fertility and Sterility* 53, 3 (March 1990): 521–527.

Byrne, Donn, and Karen Murnen. "Maintaining Love Relationships." In *The Psychology of Love,* edited by R. Sternberg and M. Barnes. New Haven, CT: Yale University Press, 1988.

Cabai, Robert. "Gay and Lesbian Couples: Lessons on Human Intimacy." *Psychiatric Annals* 18, 1 (January 1988): 21–25.

Cado, Suzana, and Harold Leitenberg. "Guilt Reactions to Sexual Fantasies during Intercourse." *Archives of Sexual Behavior* 19, 1 (1990): 49–63.

Cahoon, D. E. M., Edmonds, R. M. Spaulding, and J. C. Dickens. "A Comparison of the Opinions of Black and White Males and Females Concerning the Occurrence of Rape." *Journal of Social Behavior and Personality* 10, 1 (March 1995): 91–100.

Calfin, Matthew S., James L. Carroll, and Jerry Schmidt. "Viewing Music-Videotapes Before Taking a Test of Premarital Sexual Attitudes." *Psychological Reports* 72, 2 (April 1993): 475–481.

Callan, Victor. "The Personal and Marital Adjustment of Mothers and of Voluntarily and Involuntarily Childless Wives." *Journal of Marriage and the Family* 47, 4 (November 1985): 1045–1050.

Camarillo, Albert. *Chicanos in a Changing Society.* Cambridge, MA: Harvard University Press, 1979.

———. *Latinos in the United States.* Santa Barbara, CA: ABCClio, 1986a.

———. *Mexican-Americans in Urban Society: A Selected Bibliography.* Berkeley, CA: Floricanto Press, 1986a.

Campbell, Jacquelyn. "Violence Toward Women: Homicide and Battering." In *Vision 2010: Families and Violence, Abuse and Neglect,* edited by R. J. Gelles. Minneapolis: National Council on Family Relations, 1995.

Canadian Paediatric Society, Fetus and Newborn Committee. "Neonatal Circumcision Revisited," *Canadian Medical Association Journal* 154, 6 (March 15, 1996): 769–780.

Cancian, F. M. "Gender Politics: Love and Power in the Private and Public Spheres." In *Family in Transition,* edited by A. S. Skolnick and J. H. Skolnick. Glenview, IL: Scott, Foresman, 1989.

Cancian, Francesca. "Gender Politics: Love and Power in the Private and Public Spheres." In *Gender and the Life Course,* edited by A. Rossi. Hawthorne, NY: Aldine, 1985: 253–262.

———. *Love In America: Gender and Self Development.* New York: Oxford University Press, 1987.

Canter, Lee, and Marlene Canter. *Assertive Discipline for Parents.* Santa Monica, CA: Canter and Associates, 1985.

Cantor, Muriel. "Popular Culture and the Portrayal of Women: Content and Control." In *Analyzing Gender,* edited by B. Hess and M. Marx Ferree. Newbury Park, CA: Sage, 1987.

Cantor, Muriel. "The American Family on Television: From Molly Goldberg to Bill Cosby." *Journal of Comparative Family Studies* 22, 2 (June 1991): 205–216.

Cappell, Charles, and Robert B. Heiner. "The Intergenerational Transmission of Family Aggression." *Journal of Family Violence* 5, 2 (June 1990): 121–134.

Cargan, Leonard, and Matthew Melko. *Singles: Myths and Realities.* Beverly Hills, CA: Sage , 1982.

Carl, Douglas. "Acquired Immune Deficiency Syndrome: A Preliminary Examination of the Effects on Gay Couples and Coupling." *Journal of Marital and Family Therapy* 12, 3 (July 1986): 241–247.

Carrasquillo, Hector. "Puerto Rican Families in America." In *Families in Cultural Context,* edited by M. K. DeGenova. Mountain View, CA: Mayfield, 1997.

Carroll, Jerry. "Tracing the Causes of Infertility." *San Francisco Chronicle* (March 5, 1990): B3 ff.

Carroll, Jerry, K. D. Volk, and J. J. Hyde. "Differences in Males and Females in Motives for Engaging in Sexual Intercourse." *Archives of Sexual Behavior* 14 (1985): 131–139.

Carson, David K., et al. "Family of Origin Characteristics and Current Family Relationships of Female Adult Incest Victims." *Journal of Family Violence* 5, 2 (June 1990): 153–172.

Carter, Betty, and Monica McGoldrick, eds. *The Changing Family Life Cycle,* 2nd ed. Boston: Allyn and Bacon, 1989.

Carter, D. Bruce. *Current Conceptions of Sex Roles and Sex Typing.* New York: Praeger, 1987a.

———. "Sex Role Research and the Future New Directions for Research." In *Current Conceptions of Sex Roles and Sex Typing,* edited by D. Bruce Carter. New York: Praeger, 1987b.

———. "Societal Implications of AIDS and HIV Infections, HIV Antibody Testing, Health Care, and AIDS Education." *Marriage and Family Review* 13, 1 (1989): 129–188.

Cary, Alice. "Big Fans on Campus." *TV Guide* (April 18, 1992): 26–31.

Cassell, Carole. *Swept Away.* New York: Simon and Schuster, 1984.

Cate, Rodney M., and Sally A. Lloyd. *Courtship.* Newbury Park, CA: Sage, 1992.

Cates, Jim A., Linda Graham, Donna Boeglin, and Steven Tielker. "The Effect of AIDS on the Family System." *Families in Society* 71, 4 (April 1990): 195–201.

CDC. *See* Centers for Disease Control and Prevention.

Centers for Disease Control and Prevention. "Statewide Prevalence of Illicit Drug Use by Pregnant Women—Rhode Island." *Morbidity and Mortality Weekly Report* 39, 14 (April 3, 1990): 225–227.

———. "HIV Survey in Childbearing Women." *National AIDS Hotline Training Bulletin* 103 (June 15, 1994a): 2.

———. "Surveillance Report: U.S. AIDS Cases Reported through June 1994." *HIV/AIDS Surveillance Report,* 1994b.

———. "Surveillance Report: U.S. AIDS Cases Reported through December 1996." *HIV/AIDS Surveillance Report,* 1996.

———. "Births, Marriages, Divorces and Deaths: Provisional Data for 2001." *National Vital Statistics Reports.* 50, 14 (September 2002).

Chasnoff, Ira J. "Drug Use in Pregnancy: Parameters of Risk." *Pediatric Clinics of North America* 35, 6 (December 1988): 1403–1412.

Chatters, Linda M., Robert J. Taylor, and Harold W. Neighbors. "Size of Informal Helper Network Mobilized during a Serious Personal Problem among Black Americans." *Journal of Marriage and the Family* 51, 3 (August 1989): 667–676.

Cherlin, Andrew. *Marriage, Divorce, Remarriage.* Cambridge, MA: Harvard University Press, 1981.

———. *The Changing American Family and Public Policy.* Washington, DC: Urban Institute Press, 1988.

———. *Marriage, Divorce, and Remarriage.* Rev. ed. Cambridge, MA: Harvard University Press, 1992.

———. *Public and Private Families: An Introduction,* 2nd ed. New York: McGraw-Hill College, 1997.

Cherlin, Andrew, and Frank Furstenberg, Jr. *The New American Grandparent.* New York: Basic Books, 1986.

Cheseboro, J. "Communication, Values, and Popular Television Series—A Four Year Assessment." In *Television: The Critical View,* 4th ed., edited by H. Newcomb. New York: Oxford University Press, 1987.

Chesser, Barbara Jo. "Analysis of Wedding Rituals: An Attempt to Make Weddings More Meaningful." *Family Relations* 29, 2 (April 1980).

Chiasson, M. A., R. L. Stoneburner, and S. C. Joseph. "Human Immunodeficiency Virus Transmission through Artificial Insemination." *Journal of Acquired Immune Deficiency Syndromes* 3, 1 (1990): 69–72.

"Child Abuse and Neglect Still a Widespread Problem in America." *Nation's Health* (May/June 1997): 9.

Children's Defense Fund. "Moments in America for Children." Washington, DC, 1996.

"Children's Health." *Nation's Health* (May/June 1997): 1.

Chilman, Catherine. "Working Poor Families: Trends, Causes, Effects, and Suggested Policies." *Family Relations* 40, 2 (April 1991): 191–198.

Chilman, Catherine Street. "Hispanic Families in the United States: Research Perspectives." In *Family Ethnicity: Strength in Diversity,* edited by H. Pipes McAdoo. Newbury Park, CA: Sage, 1993.

Chilman, Catherine, et al., eds. *Variant Family Forms.* Beverly Hills, CA: Sage, 1988.

Chira, Susan. "Years after Adoption, Adults Find Past, and New Hurdles." *New York Times* (August 30, 1993): B1, B6.

Chodorow, Nancy. *The Reproduction of Mothering: Psychoanalysis and the Sociology of Gender.* Berkeley: University of California Press, 1978.

Chojnacki, Joseph T., and W. Bruce Walsh. "Reliability and Concurrent Validity of the Sternberg Triangular Love Scale." *Psychological Reports* 67, 1 (August 1990): 219–224.

Christensen, Andrew. "Dysfunctional Interaction Patterns in Couples." In *Perspectives on Marital Interaction,* edited by P. Noller and M. A. Fitzpatrick. Philadelphia: Multilingual Matters, 1988.

Christensen, Donna, et al. "Noncustodial Mothers and Child Support: Examining the Larger Context." *Family Relations* 39, 4 (October 1990): 388–394.

Christian-Smith, Linda K. *Becoming a Woman through Romance.* New York: Routledge, 1990.

Christopher, F., and R. Cate. "Factors Involved in Premarital Decision-Making." *Journal of Sex Research* 20 (1984): 363–376.

———. "Anticipated Influences on Sexual Decision-Making for First Intercourse." *Family Relations* 34 (1985): 265–270.

Christopher, F. Scott, and Susan Sprecher. "Sexuality in Marriage, Dating, and Other Relationships: A Decade Review." *Journal of Marriage and the Family* 62, 4. (November 2000): 999–1017.

Christopher, F. S., and M. M. Frandsen. "Strategies of Influence in Sex and Dating." *Journal of Social and Personal Relationships* 7 (1990): 89–105.

Christopher, F. Scott, Richard A. Fabes, and Patricia M. Wilson. "Family Television Viewing: Implications for Family Life Education." *Family Relations* 38, 2 (April 1989): 210–214.

Ciancannelli, Penelope, and Bettina Berch. "Gender and the GNP." In *Analyzing Gender,* edited by B. Hess and M. Marx Ferree. Newbury Park, CA: Sage, 1987.

Cimons, Marlene. "American Infertility Rate Not Growing, Study Finds." *New York Times* (December 7, 1990): A3.

Claes, Jacalyn A., and David M. Rosenthal. "Men Who Batter Women: A Study in Power." *Journal of Family Violence* 5, 3 (September 1990): 215–224.

Clanton, Gordon, and Lynn Smith. *Jealousy.* Englewood Cliffs, NJ: PrenticeHall, 1977.

Clark, Danae. "Cagney and Lacey: Feminist Strategies of Detection." In *Television and Women's Culture: The Politics of the Popular,* edited by M. E. Brown. Newbury Park, CA.: Sage 1992.

Clark-Nicolas, Patricia, and Bernadette Gray-Little. "Effect of Economic Resources on Marital Quality in Black Married Couples." *Journal of Marriage and the Family* 53, 3 (August 1991): 645–656.

Clatterbaugh, Kenneth. *Contemporary Perspectives on Masculinity: Men, Women, and Politics in Modern Society,* 2nd ed. Boulder, CO: Westview, 1997.

Cleek, Margaret, and T. Allan Pearson. "Perceived Causes of Divorce: An Analysis of Interrelationships." *Journal of Marriage and the Family* 47, 2 (February 1985): 179–191.

Clemes, Harris, and Reynold Bean. *How to Raise Children's Self- Esteem.* San Jose, CA: Enrich, 1983.

Cleveland, Peggy, et al. "If Your Child Has AIDS . . . : Response of Parents with Homosexual Children." *Family Relations* 37, 2 (April 1988): 150–153.

Clingempeel, W. Glenn, and Eulalee Brand. "Quasi-kin Relationships, Structural Complexity, and Marital Quality in Stepfamilies: A Replication, Extension, and Clinical Implications." *Family Relations* 34, 3 (July 1985): 401–409.

Clingempeel, W. Glenn, et al. "Stepparent-Stepchild Relationships in Stepmother and Stepfather Families: A Multimethod Study." *Family Relations* 33 (1984): 465–473.

Cochran, Susan D., Vickie M. Mays, and Laurie Leung. "Sexual Practices of Heterosexual Asian-American Young Adults: Implications for Risk of HIV Infection." *Archives of Sexual Behavior* 20, 4 (August 1991): 381–394.

Cochran, Susan, et al. "Sexually Transmitted Diseases and Acquired Immunodeficiency Syndrome (AIDS): Changes in Risk Reduction Behaviors among Young Adults." *Sexually Transmitted Diseases* 16, 1 (January 1989): 80–86.

Coggle, Frances, and Grace Tasker. "Children and Housework." *Family Relations* 31 (July 1982): 395–399.

Cohan, Catherine, and Stacey Kleinbaum. "Toward a Greater Understanding of the Cohabitation Effect: Premarital Cohabitation and Marital Communication." *Journal of Marriage and the Family* 64, 1 (February 2002): 163–179.

Cohen, T .F. "What Do Fathers Provide? Reconsidering the Economic and Nurturant Dimensions of Men as Parents." In *Men, Work and Family,* edited by J. C. Hood. Newbury Park, CA: Sage , 1993.

Cohen, Theodore. *Men's Family Roles: Becoming and Being Husbands and Fathers.* Doctoral Dissertation, Boston University. (University Microfilms No. 86–09272) 1986.

———. "Remaking Men: Men's Experiences Becoming and Being Husbands and Fathers and Their Implications for Reconceptualizing Men's Lives," *Journal of Family Issues* 8 (1987): 57–77.

Cohen, Theodore, ed. *Men and Masculinity: A Text-Reader.* Belmont, CA: Wadsworth, 2001.

Cohen, Theodore, and John C. Durst. "Leaving Work and Staying Home: The Impact on Men of Terminating the Male Economic Provider Role." In *Men and Masculinity: A Text-Reader, edited by* T. Cohen. Belmont, CA: Wadsworth, 2001.

Cohler, Bertram, and Scott Geyer. "Psychological Autonomy and Interdependence within the Family." In *Normal Family Processes,* edited by F. Walsh. New York: Guilford Press, 1982.

Cole, R., and D. Reiss. *How Do Families Cope with Chronic Illness?* Hillsdale, NJ: Lawrence Erlbaum, 1993.

Cole, Robert. "Mental Illness and the Family." In *Vision 2010: Families and Health Care,* edited by B. A. Elliott. Minneapolis: National Council on Family Relations, 1993: 18–19.

Cole, W. "Incest Perpetrators: Their Assessment and Treatment." *Psychiatric Clinics of North America* 15, 3 (September 1992): 689–701.

Coleman, Marilyn, and Lawrence Ganong. "The Cultural Stereotyping of Stepfamilies." In *Remarriage and Stepparenting: Current Research and Theory,* edited by K. Pasley and M. Ihinger-Tallman. New York: Guilford Press, 1987.

———. "Remarriage and Stepfamily Research in the 1980s: Increased Interest in an Old Form." In *Contemporary Families: Looking Forward, Looking Back,* edited by A. Booth. Minneapolis: National Council on Family Relations, 1991.

Coleman, Marilyn, Lawrence Ganong, and Mark Fine. "Reinvestigating Remarriage: Another Decade of Progress." *Journal of Marriage and the Family* 62, 4 (November 2000): 1288–1307.

Coles, Robert. *The Spiritual Life of Children.* Boston: Houghton Mifflin, 1990.

Collier, J., M. Z. Rosaldo, and S. Yanagisako. "Is There a Family? New Anthropological Views." In *Rethinking the Family: Some Feminist Questions,* edited by B. Thorne and M. Yalom. New York: Longman, 1982.

Collins, Glenn. "U.S. Day-Care Guidelines Rekindle Controversy." *New York Times* (February 4, 1985): 20.

Collins, Patricia Hill. "The Meaning of Motherhood in Black Culture." In *The Black Family,* 4th ed., edited by R. Staples. Belmont, CA: Wadsworth, 1991.

Coltrane, Scott. *Family Man: Fatherhood, Housework, and Gender Equity.* New York: Oxford University Press, 1996.

———. "Research on Household Labor: Modeling and Measuring the Social Embeddedness of Routine Family Work." *Journal of Marriage and the Family,* 62, 4 (November 2000): 1208–1233.

Coltrane, Scott, and Michelle Adams. "The Social Construction of the Divorce 'Problem': Morality, Child Victims, and the Politics of Gender." *Family Relations* 52, 4 (October 2003): 363–372.

Coltrane, Scott, and Neal Hickman. "The Rhetoric of Rights and Needs: Moral Discourse in the Reform of Child Custody and Child Support Laws." *Social Problems* 39, 4 (1992): 400–420.

"A Comparative Survey of Minority Health." The Commonwealth Fund. New York, 1996.

Comstock, Jamie, and Krystyna Strzyzewski. "Interpersonal Interaction on Television: Family Conflict and Jealousy on Primetime." *Journal of Broadcasting and Electronic Media* 34, 3 (1990): 263–282.

Condron, John, and Jerry Bode. "Rashomon, Working Wives, and Family Division of Labor: Middletown." *Journal of Marriage and the Family* 44, 2 (May 1982): 421–426.

Condry, J., and S. Condry. "The Development of Sex Differences: A Study of the Eye of the Beholder." *Child Development* 47, 4 (1976): 812–819.

Conger, Rand, and Katherine Conger. "Resilience in Midwestern Families: Selected Findings from the First Decade of a Prospective Longitudinal Study." *Journal of Marriage and the Family* 64, 2 (May 2002): 361–373

Connell, Robert. *Gender and Power: Society, the Person, and Sexual Politics.* Stanford, CA: Stanford University Press, 1987.

———. *Masculinities.* Berkeley: University of California Press, 1995.

"Contract with America." *New York Times* (November 11, 1994): A10.

Cook, Alicia, et al. "Changes in Attitudes toward Parenting among College Women: 1972 and 1979 Samples." *Family Relations* 31, 1 (January 1982): 109–113.

Cook, Mark, ed. *The Bases of Human Sexual Attraction.* New York: Academic Press, 1981.

Cooney, Teresa M. "Young Adults' Relations with Parents: The Influence of Recent Parental Divorce." *Journal of Marriage and the Family* 56, 1 (February 1994): 45–56.

Coontz, Stephanie. *The Way We Never Were: American Families and the Nostalgia Trap.* New York: Basic Books, 1992.

———. *The Way We Really Are: Coming to Terms with America's Changing Families.* New York: Basic Books, 1997.

———. "Divorcing Reality: Other Researchers Question Wallerstein's Conclusions," *Children's Advocate,* Action Alliance for Children, January-February, 1998.

Cooper, A., and C. D. Stoltenberg. "Comparison of a Sexual Enhancement and a Communication Training Program on Sexual and Marital Satisfaction." *Journal of Counseling Psychology* 34 (July 1987): 309–314.

Cooper, Sheila McIssac. "Historical Analysis of the Family." In *Handbook of Marriage and the Family,* 2nd ed., edited by M. Sussman, S. Steinmetz, and G. Peterson. New York: Plenum. 1999.

Copenhaver, Stacey, and Elizabeth Grauerholz. "Sexual Victimization Among Sorority Women: Exploring the Link Between Sexual Violence and Institutional Practices." *Sex Roles* 24, (1/2) 1991:31–41.

Corby, Nan, and Judy Zarit. "Old and Alone: The Unmarried in Later Life." In *Sexuality in the Later Years: Roles and Behavior,* edited by R. Weg. New York: Academic Press, 1983.

Corea, Gena. *The Mother Machine: Reproductive Technology from Artificial Insemination to Artificial Wombs.* New York: Harper and Row, 1985.

Cortese, Anthony. "Subcultural Differences in Human Sexuality: Race, Ethnicity, and Social Class." In *Human Sexuality: The Societal and Interpersonal Context,* edited by K. McKinney and S. Sprecher. Norwood, NJ: Ablex, 1989.

Cosby, Bill. "The Regular Way." *Playboy* (December 1968): 288–289.

———. "Someone at the Top Has to Say: 'Enough of This.'" *Newsweek* (December 6, 1993): 60.

———. "We Are Losing." *Newsweek* (March 17, 1997): 58.

Coulanges, Fustel de. *The Ancient City.* 1867. Reprint, New York: Anchor Books, 1960.

Courtright, Joseph, and Stanley Baran. "The Acquisition of Sexual Information by Young People." *Journalism Quarterly* 57, 1 (March 1980): 107–114.

Cowan, Alison Leigh. "Can a Baby-Making Venture Deliver?" *New York Times* (June 1, 1992): C1, C4.

Cowan, Carolyn Pope, and Philip Cowan. *When Partners Become Parents: The Big Life Change for Couples.* New York: Basic Books, 1992.

Cowan, Philip, and Carolyn Cowan. "Becoming a Family: Research and Intervention." In *Methods of Family Research: Biographies of Research Projects,* edited by I. Sigel and G. Brody. Hillsdale, NJ.: Lawrence Erlbaum, 1990.

Cowley, Geoffrey. "AIDS: The Next Ten Years." *Newsweek* (June 25, 1990): 20–28.

Craig, R. Stephen. "The Effect of Television Day Part on Gender Portrayals in Television Commercials: A Content Analysis." *Sex Roles: A Journal of Research* 26, 5–6 (1992): 197–211.

Cramer, D. "Gay Parents and Their Children: A Review of Research and Practical Implications." *Journal of Counseling and Development* 64 (1986): 504–507.

Cramer, David, and Arthur Roach. "Coming Out to Mom and Dad: A Study of Gay Males and Their Relationships with Their Parents." *Journal of Homosexuality* 14, 1–2 (1987): 77–88.

Cramer, Robert Ervin, M. Dragna, R. G. Cupp, and P. Stewart. "Contrast Effects in the Evaluation of the Male Sex Role." *Sex Roles* 24, 3/4 (1991): 181–193.

Creti, L., and E. Libman. "Cognition and Sexual Expression in the Aging." *Journal of Sex and Marital Therapy* 15, 2 (June 1989): 83–101.

Crittenden, Ann. *The Price of Motherhood: Why the Most Important Job in the World Is Still the Least Valued.* New York: Metropolitan Books. 2001.

Crohan, Susan E. "Marital Quality and Conflict Across the Transition to Parenthood in African American and White Couples." *Journal of Marriage and Family* 58, 4 (November 1996): 933–944.

Crosbie-Burnett, Margaret, and Edith Lewis. "Use of African-American Family Structures and Functioning to Address the Challenges of European-American Postdivorce Families." *Family Relations* 42, 3 (July 1993): 245–248.

Crosby, Faye. *Juggling: The Unexpected Advantages of Balancing Career and Home for Women and Their Families.* New York: Free Press, 1991.

Crosby, John, ed. *Reply to Myth: Perspectives on Intimacy.* New York: Wiley, 1985.

Crouter, Ann C., and Beth Manke. "The Changing American Workplace: Implications for Individuals and Families." *Family Relations* 43, 2 (April 1994): 117–124.

Crouter, Ann C., and Susan M. McHale. "The Long Arm of the Job: Influences of Parental Work on Child Rearing." In *Parenting: An Ecological Perspective,* edited by T. Luster and L. Okagaki. Hillsdale, NJ: Lawrence Erlbaum, 1993.

Culp, Rex E., Alicia S. Cook, and Pat C. Housley. "A Comparison of Observed and Reported Adult-Infant Interactions: Effects of Perceived Sex." *Sex Roles* 9 (April 1983): 475–479.

Cupach, William, and J. Comstock. "Satisfaction with Sexual Communication in Marriage." *Journal of Social and Personal Relationships* 7 (1990): 179–186.

Cupach, William, and Sandra Metts, "Sexuality and Communication in Close Relationships." In *Sexuality in Close Relationships,* edited by K. McKinney and S. Sprecher. Hillsdale, NJ: Lawrence Erlbaum, 1991.

Curran, Dolores. *Traits of a Healthy Family.* New York: Ballantine, 1983.

Curry, Tim, Robert Jiobu, and Kent Schwirian. *Sociology for the 21st Century.* Prentice-Hall, 2002.

Dail, Paula W. "Prime-time Television Portrayals of Older Adults in the Context of Family Life." *Gerontologist* 28, 5 (1988): 700–706.

"The Daily Record." *National Law Journal.* (March 12, 1997).

Dainton, M. "The Myths and Misconceptions of the Stepmother Identity: Descriptions and Prescriptions for Identity Management. *Family Relations* 42, 1 (January 1993): 93–98.

Daley, Suzanne. "Girl's Self-Esteem Is Lost on Way to Adolescence, New Study Finds." *New York Times* (January 9, 1991): B1.

Daly, Kerry. "Reshaping Fatherhood: Finding the Models." *Journal of Family Issues* 14, 4 (December 1993): 510–530.

Danziger, S. "Antipoverty Policies and Child Poverty." *Social Work Research and Abstracts* 26, 4 (December 1990): 17–24.

Danziger, S. K., and S. Danziger. "Child Poverty and Public Policy—Toward a Comprehensive Antipoverty Agenda." *Daedalus* 122, 1 (December 1993): 57–84.

Darling, Carol A., and J. Kenneth Davidson. "Enhancing Relationships: Understanding the Feminine Mystique of Pretending Orgasm." *Journal of Sex and Marital Therapy* 12 (1986): 182–196.

Darling, Carol A., J. Kenneth Davidson, and Ruth P. Cox. "Female Sexual Response and the Timing of Partner Orgasm." *Journal of Sex and Marital Therapy,* 17, 1 (March 1991): 3–21.

Darling, Carol A., J. Kenneth Davidson, and D. A. Jennings. "The Female Sexual Response Revisited: Understanding the Multiorgasmic Experience in Women." *Archives of Sexual Behavior* 20, 6 (December 21, 1991): 527–540.

Darling-Fisher, Cynthia, and Linda Tiedje. "The Impact of Maternal Employment Characteristics on Fathers' Participation in Child Care." *Family Relations* 39, 1 (January 1990): 20–26.

Davidson, J. Kenneth, Carol Darling, and Colleen Conway-Welch. "The Role of the Grafenberg Spot and Female Ejaculation in the Female Orgasmic Response: An Empirical Analysis." *Journal of Sex and Marital Therapy* 15 (June 1989): 102–120.

Davidson, Kenneth, and Carol Darling. "Changing Autoerotic Attitudes and Practices among College Females: A Two-Year Follow-up Study." *Adolescence* 23 (December 1988a): 773–792.

———. "The Stereotype of Single Women Revisited." *Health Care for Women International* 9, 4 (October 1988b): 317–336.

Davidson, Kenneth, and Linda Hoffman. "Sexual Fantasies and Sexual Satisfaction: An Empirical Investigation of Erotic Thought." *Journal of Sex Research* 22 (May 1986): 184–205.

Davis, L. *The Black Family in the United States.*Westport, CT: Greenwood Press, 1986.

Davis, S. "Men as Success Objects and Women as Sex Objects: A Study of Personal Advertisements." *Sex Roles* 23 (July 1990): 43–50.

Davis, Sally M., and Mary B. Harris. "Sexual Knowledge, Sexual Interest, and Sources of Sexual Information of Rural and Urban Adolescents from Three Cultures." *Adolescence* 17, 66 (June 1982): 471–492.

Dawson, Deborah. "The Effects of Sex Education on Adolescent Behavior." *Family Planning Perspectives* 18, 4 (July 1986): 162 ff.

"A Day in the Life of China." *Time* (October 2, 1989).

Deal, James E., Karen S. Wampler, and Charles F. Halverson. "The Importance of Similarity in the Marital Relationship." *Family Process* 31, 4 (1992): 369–382.

De Armand, Charlotte. "Let's Listen to What the Kids Are Saying." *SIECUS Reports* (March 1983): 3–4.

"Death Rates for Minority Infants Were Underestimated, Study Says." *New York Times* (January 8, 1992): A10.

Deaux, K. "From Individual Differences to Social Categories: Analysis of a Decade's Research on Gender." *American Psychologist* 39, 2 (1984): 105–116.

DeBuono, Barbara, et al. "Sexual Behavior in College Women in 1975, 1986, and 1989." *New England Journal of Medicine* 322, 12 (March 22, 1990): 821–825.

DeCecco, John, ed. *Gay Relationships.* New York: Haworth Press, 1988.

DeCecco, John, and Michael Shively. "From Sexual Identity to Sexual Relationships: A Conceptual Shift." *Journal of Homosexuality* 9, 2–3 (December 1983): 1–26.

DeCecco, John P., and J. P. Elia. "A Critique and Synthesis of Biological Essentialism and Social Constructionist Views of Sexuality and Gender. Introduction." *Journal of Homosexuality* 24, 3–4 (1993): 1–26.

DeFleur, M., and S. Ball-Rokeach. *Theories of Mass Communication.* New York: Longman, 1989.

Degler, Carl. *At Odds.* New York: Oxford University Press, 1980.

DeGroot, J. M., et al. "Correlates of Sexual Abuse in Women with Anorexia Nervosa and Bulimia Nervosa." *Canadian Journal of Psychiatry* 37, 7 (September 1992): 516–518.

Delamater, J. D., and P. MacCorquodale. *Premarital Sexuality: Attitudes, Relationships, Behavior.* Madison: University of Wisconsin Press, 1979.

DelCarmen, Rebecca. "Assessment of Asian-Americans for Family Therapy." In *Mental Health of Ethnic Minorities,* edited by F. Serafica et al. New York: Praeger, 1990.

De Leon, Brunilda, "Sex Role Identity among College Students: A Cross-Cultural Analysis." *Hispanic Journal of the Behavioral Sciences* 15, 4 (1993): 476–489.

Demarest, Jack, and Jeanette Garner. "The Representation of Women's Roles in Women's Magazines over the Past 30 Years." *Journal of Psychology* 126, 4 (July 1992): 357–369.

DeMaris, Alfred, and K. Vaninadha Rao. "Premarital Cohabitation and Subsequent Marital Stability in the United States: A Reassessment." *Journal of Marriage and the Family* 54, 1 (February 1992): 178–190.

D'Emilio, John, and Estelle Freedman. *Intimate Matters: A History of Sexuality in America.* New York: Harper and Row, 1988.

Deming, Robert. "The Return of the Unrepressed: Male Desire, Gender, and Genre (New Directions in Television Studies: Essays in Honor of Beverle Ann Houston)." *Quarterly Review of Film and Video* 14, 1–2 (1992): 125–148.

Demo, David, and Alan Acock. "The Impact of Divorce on Children." In *Contemporary Families: Looking Forward, Looking Back,* edited by A. Booth. Minneapolis: National Council on Family Relations, 1991.

Demo, David H., and Alan C. Acock. "Family Diversity and the Division of Domestic Labor: How Much Have Things Really Changed?" *Family Relations* 42, 3 (July 1993): 323–331.

Demo, David, and Martha Cox. "Families with Young Children: A Review of Research in the 1990's." *Journal of Marriage and the Family,* 62, 4 (November 2000): 876–895.

Demos, John. *A Little Commonwealth.* New York: Oxford University Press. 1970.

Demos, Vasilikie. "Black Family Studies in the *Journal of Marriage and the Family* and the Issue of Distortion: A Trend Analysis." *Journal of Marriage and the Family* 52, 3 (August 1990): 603–612.

Dentzer, Susan. "Do the Elderly Want to Work?" *U.S. News and World Report* (May 14, 1990): 48–50.

Denzin, Norman. "Toward a Phenomenology of Domestic Family Violence." *American Journal of Sociology* 90, 30 (1984): 483–513.

DeParle, Jason. "Suffering in the Cities Persists as U.S. Fights Other Battles." *New York Times* (January 27, 1991): 15.

———. "The New Majority's Agenda: Welfare." *New York Times* (November 11, 1994): A14.

"Depiction of Women in Mainstream TV Shows Still Stereotypical, Sexual." *Media Report to Women* 21, 1 (1993): 4.

Depner, Charlene, and James Bray, eds. "Modes of Participation for Non-custodial Parents: The Challenge for Research, Policy, Practice, and Education." *Family Relations* 39, 4 (October 1990): 378–381.

———. *Nonresidential Parenting: New Vistas in Family Living.* Newbury Park, CA: Sage, 1993.

Derdeyn, A., and E. Scott. "Joint Custody: A Critical Analysis and Appraisal." *American Journal of Orthopsychiatry* 54 (April 1984): 199–209.

Derlega, Valerian J., Sandra Metts, Sandra Petronio, and S. Margulis. *Self-Disclosure.* Newbury Park, CA: Sage, 1993.

DeSpelder, Lynne Ann, and Albert Strickland. *The Last Dance: Encountering Death and Dying.* 5th ed. Mountain View, CA: Mayfield, 1996.

DeWitt, P. M. "Breaking Up Is Hard To Do." *American Demographics,* reprint package (1994): 14–16.

DiBlasio, Frederick A., and Brent B. Benda. "Gender Differences in Theories of Adolescent Sexual Activity." *Sex Roles* 27, 5/6 (1992): 221–236.

Dick-Read, Grantly. *Childbirth without Fear.* 4th ed. New York: Harper and Row, 1972.

Dietz, Tracy. "An Examination of Violence and Gender Role Portrayals in Video Games: Implications for Gender Socialization and Aggressive Behavior," *Sex Roles: A Journal of Research* 38 (March 1998): 5–6.

Dilworth-Anderson, Peggye, Linda M. Burton, and William L. Turner. "The Importance of Values in the Study of Culturally Diverse Families." *Family Relations* 42, 3 (July, 1993): 238–242.

Dilworth-Anderson, Peggye, and Harriette Pipes McAdoo. "The Study of Ethnic Minority Families: Implications for Practitioners and Policymakers." *Family Relations* 37, 3 (July 1988): 265–267.

Dinnerstein, Leonard, and David Reimers, eds. *Ethnic Americans: A History of Immigration,* 3rd ed. New York: Harper and Row, 1988.

Dion, Karen. "Physical Attractiveness, Sex Roles, and Heterosexual Attraction." In *The Bases of Human Sexual Attraction,* edited by M. Cook. New York: Academic Press, 1981.

Dion, Karen, et al. "What Is Beautiful Is Good." *Journal of Personality and Social Psychology* 24 (1972): 285–290.

Dodson, Fitzhugh. "How to Discipline Effectively." In *Experts Advise Parents,* edited by E. Shiff. New York: Dell, 1987.

Dodson, Jualynne. "Conceptualizations of Black Families." In *Black Families,* edited by H. Pipes McAdoo. 2nd ed. Newbury Park, CA: Sage, 1988.

Doherty, William J. "Research Update for Practitioners: Families and Health." Paper presented at the annual conference of the National Council on Family Relations, Baltimore, November 12, 1993a.

———. "Training Health Professionals about Families." In *Vision 2010: Families and Health Care,* edited by B. A. Elliott. Minneapolis: National Council on Family Relations, 1993b.

Doherty, William J. Edward Kouneski, Martha Erickson, "Responsible Fathering: An Overview and Conceptual Framework." *Journal of Marriage and the Family* 60, 2 (1998): 277–292.

Dolgin, Kim. "Men's Friendships: Mismeasured, Demeaned, and Misunderstood?" In *Men and Masculinity: A Text Reader,* edited by T. Cohen. Belmont, CA: Wadsworth, 2001.

Dolgin, Kim, and F. P. Rice, *The Adolescent: Development, Relationships, and Culture,* 10th ed, Boston: Allyn and Bacon, 2002.

Donnelly, Kathleen. "Breaking the Barriers." *San Jose Mercury News* (September 27, 1993): 1C, 8C.

Donovan, P. "New Reproductive Technologies: Some Legal Dilemmas." *Family Planning Perspectives* 18 (1986): 57 ff.

Dooley, Karen, and Ralph Catalano. "Stress Transmission: The Effects of Husbands' Job Stressors on the Emotional Health of Their Wives." *Journal of Marriage and the Family* 53, 1 (February 1991): 165–177.

Dorr, Aimee, Peter Kovaric, and Catherine Doubleday. "Age and Content Influences on Children's Perceptions of the Realism of Television Families." *Journal of Broadcasting and Electronic Media* 34, 4 (1990): 377–397.

Dorr, Aimee, et al. "Parent-Child Coviewing of Television." *Journal of Broadcasting and Electronic Media* 33, 1 (December 1989): 35–51.

Douglass, Frederick. *The Life and Times of Frederick Douglass.* New York: Collier Books, 1962. Originally published 1845.

Dreikurs, Rudolph, and V. Soltz. *Children: The Challenge.* New York: Hawthorne Books, 1964.

Drugger, Karen. "Social Location and Gender-Role Attitudes: A Comparison of Black and White Women." *Gender and Society* 2, 4 (December 1988): 425–448.

Duck, Steve, ed. *Dynamics of Relationships.* Thousand Oaks, CA: Sage, 1994.

Duncan, G., and S. Hoffman. "Economic Consequences of Marital Instability" In *Horizontal Equity, Uncertainty, and Economic Well-Being,* edited by M. David and T. Smeeding. Chicago: University of Chicago Press, 1985.

Duncan, Greg, and Willard Rodgers. "Longitudinal Aspects of Childhood Poverty." *Journal of Marriage and the Family* 50, 4 (November 1988): 1007–1022.

Duncombe, Jean, and Dennis Marsden. "Love and Intimacy: The Gender Division of Emotion and 'Emotion Work': A Neglected Aspect of Sociological Discussion of Heterosexual Relationships." *Sociology* 27, 2 (May 1993): 221–242.

Dunlop, Rosemary, and Ailsa Burns. "The Sleeper Effect—Myth or Reality?. *Journal of Marriage and the Family* 57, 2 (May 1995): 375–386.

Durant, Robert, R. Pendergast, and C. Seymore. "Contraceptive Behavior among Sexually Active Hispanic Adolescents." *Journal of Adolescent Health* 11, 6 (November 1990): 490–496.

Duvall, Evelyn. "Family Development's First Forty Years." *Family Relations* 37, 2 (April 1988): 127–134.

Duvall, Evelyn, and Brent Miller. *Marriage and Family Development,* 6th ed. New York: Harper and Row, 1985.

Dworetsky, John P. *Introduction to Child Development.* 4th ed. St. Paul, MN: West, 1990.

Eagly, Alice H. *Sex Differences in Social Behavior: A Social-Role Interpretation.* Hillsdale, NJ: Lawrence Erlbaum, 1987.

Edelman, Marian Wright. "Forward." In *Vanishing Dreams: The Growing Economic Plight of America's Young Families,* edited by C. Johnson et al. Washington, DC: Children's Defense Fund, 1988.

———. "Children at Risk." In *Caring for America's Children,* edited by F. J. Macchiarola and A. Gartner. New York: Academy of Political Science, 1989a.

———. "Children at Risk." *Proceedings of the Academy of Political Science* 37, 2 (1989b): 20–30.

Edelman, Marian, and Lisa Mihaly. "Homeless Families and the Housing Crisis in the United States." *Children and Youth Services Review* 11, 1 (1989): 91–108.

Edleson, Jeffrey, et al. "Men Who Batter Women." *Journal of Family Issues* 6, 2 (June 1985): 229–247.

Edmondson, Brad, Judith Waldrop, Diane Crispell, and Linda Jacobsen. "The Big Picture." *American Demographics* 15, 12 (December 1993): 28–30.

Edwards, Daniel. "Native American Elders: Current Issues and Social Policy Implications." In *Aging in Minority Groups,* edited by R. L. McNeely and J. N. Colen. Beverly Hills, CA: Sage, 1983.

Edwards, Gwenyth H. "The Structure and Content of the Male Gender Role Stereotype: An Exploration of Subtypes." *Sex Roles: A Journal of Research* 27, 9–10 (November 1992): 533–553.

Egan, Timothy. "Old, Ailing and Finally a Burden Abandoned." *New York Times* (March 26, 1992): A1, A9.

Egeland, Byron. "A History of Abuse Is a Major Risk Factor for Abusing the Next Generation." In *Current Controversies in Family Violence,* edited by R. Gelles and D. Loseke. Newbury Park, CA: Sage, 1993.

Ehrenkranz, Joel R., and Wylie C. Hembree. "Effects of Marijuana on Male Reproductive Function." *Psychiatric Annals* 16, 4 (April 1986): 243–248.

Ehrenreich, Barbara. *The Hearts of Men.* Garden City, NY: Anchor/Doubleday, 1984.

———. "Housework Is Obsolescent." *Time* (October 25, 1993): 92.

Eichler, Margrit. "Reflections on Motherhood, Apple Pie, the New Reproductive Technologies and the Role of Sociologists in Society." *Society-Société* 13, 1 (February 1989): 1–5.

Ekerdt, David J., and Stanley DeViney. "Evidence for a Preretirement Process among Older Male Workers." *Journal of Gerontology* 48, 2 (March 1993): S35–S43.

Elliott, Barbara, ed. *Vision 2010: Families & Health Care.* Minneapolis: National Council on Family Relations, 1993.

Elliott, Diana M., and John Briere. "Sexual Abuse Trauma Among Professional Women: Validating the Trauma Symptom Checklist (TSC-40)." *Child Abuse and Neglect* 16, 3 (May 1992): 391 ff.

Ellis, Albert. "The Justification of Sex Without Love." In *Reply to Myth: Perpectives on Intimacy,* edited by J. Crosby. New York: Wiley, 1985.

Ellison, Carol. "Intimacy-Based Sex Therapy." In *Sexology,* edited by W. Eicher and G. Kockott. New York: Springer-Verlag, 1985.

Ellison, Christopher. "Family Ties, Friendships, and Subjective Well-Being among Black Americans." *Journal of Marriage and the Family* 52, 2 (May 1990): 298–310.

Ellwood, David. *Poor Support: Poverty in the American Family.* New York: Basic Books, 1988.

Emanuele, M. A., J. Tentler, N. V. Emanuele, and M. R. Kelley. "Invivo Effects of Acute Etoh on Rat Alpha-Luteinizing and Beta-Luteinizing Hormone Gene Expression." *Alcohol* 8, 5 (September 1991): 345–348.

Emery, Robert E., Sheila G. Matthews, and Katherine M. Kitzmann. "Child Custody Mediation and Litigation: Parents' Satisfaction and Functioning One Year after Settlement." *Journal of Consulting and Clinical Psychology* 62, 1 (1994): 124–129.

Emery, Robert, and Melissa Wyer. "Child Custody Mediation and Litigation: An Experimental Evaluation of the Experience of Parents." *Journal of Consulting and Clinical Psychology* 55, 2 (1987): 179–186.

"Employment Characteristics of Families Summary." Labor Force Statistics from the *Current Population Survey.* http://stats.bls.gov.

Erikson, Erik. *Childhood and Society.* New York: Norton, 1963.

———. *Vital Involvements in Old Age: The Experience of Old Age in Our Time.* Boston: Norton, 1986.

Erikson, Erik H. *Identity and the Life Cycle.* New York: International Universities Press, 1959.

———. *Identity, Youth, and Crisis.* New York: Norton, 1968.

Eshelman, J. Ross. *The Family* (8th ed.). Needham Heights, MA: Allyn and Bacon, 1997.

Esp'n, Olivia M. "Cultural and Historical Influences on Sexuality in Hispanic/Latin Women: Implications for Psychotherapy." In *Pleasure and Danger: Exploring Female Sexuality,* edited by C. Vance. Boston: Routledge and Kegan Paul, 1984.

Evans, H. L., et al. "Sperm Abnormalities and Cigarette Smoking." *Lancet* 1, 8221 (March 21, 1981): 627–629.

Evans-Pritchard, E. E. *Kinship and Marriage among the Nuer.* Oxford: Oxford University Press, 1951.

Fabes, Richard, and Jeremiah Strouse. "Perceptions of Responsible and Irresponsible Models of Sexuality." *Journal of Sex Research* 23, 1 (February 1987): 70–84.

Fabricus, William. "Listening to Children of Divorce: New Findings That Diverge from Wallerstein, Lewis, and Blakeslee." *Family Relations* 52, 4 (October 2003): 385–396.

Faderman, Lillian. *Odd Girls and Twilight Lovers.* New York: Columbia University Press, 1991.

Fagot, Beverly, and Mary Leinbach. "Socialization of Sex Roles within the Family." In *Current Conceptions of Sex Roles and Sex Typing,* edited by D. Bruce Carter. New York: Praeger, 1987.

Fagot, Beverly I., Mary D. Leinbach, and Cherie O'Boyle. "Gender Labeling, Gender Stereotyping and Parenting Behaviors." *Developmental Psychology* 28, 2 (1992): 225–231.

Falik, Marilyn, and Karen Scott Collins, eds. *Women's Health.* Baltimore, MD: Johns Hopkins University Press, 1996.

Faludi, Susan. *Backlash: The Undeclared War Against American Women.* New York: Crown, 1991.

Farber, Bernard, Charles H. Mindel, and Bernard Lazerwitz. "The Jewish American Family." In *Ethnic Families in America: Patterns and Variations,* 3rd ed., edited by C. H. Mindel et al. New York: Elsevier North Holland, 1988.

Faust, Kimberly, and Jerome McKibben. "Marital Dissolution: Divorce, Separation, Annulment, and Widowhood. *Handbook of Marriage and the Family,* 2nd ed., edited by M. Sussman, S. Steinmetz, and G. Peterson. New York: Plenum. 1999.

Fay, Robert, Charles Turner, Albert Klassen, and John Gagnon. "Prevalence and Patterns of Same-Gender Sexual Contact among Men." *Science* 243, 4889 (January 20, 1989): 338–348.

Feeney, Judith A., and Patricia Noller. "Attachment Style as a Predictor of Adult Romantic Relationships." *Journal of Personality and Social Psychology* 58, 2 (February 1990): 281–291.

———. "Attachment Style and Verbal Descriptions of Romantic Partners." *Journal of Social and Personal Relationships* 8, 2 (1991): 187–215.

Feeney, Judith A., and Beverley Raphael. "Adult Attachments and Sexuality: Implications for Understanding Risk Behaviors for HIV Infection." *Australian and New Zealand Journal of Psychiatry* 26, 3 (1992): 399–407.

Fehr, Beverly. "Prototype Analysis of the Concepts of Love and Commitment." *Journal of Personality and Social Psychology* 55, 4 (1988): 557–579.

Fein, Robert. "Research on Fathering." In *The Family in Transition,* edited by A. Skolnick and J. Skolnick. Boston: Little, Brown, 1980.

Feirstein, Bruce. *Real Men Don't Eat Quiche.* New York: Pocket Books, 1982.

Feldman, Shirley, and Sharon Churnin. "The Transition from Expectancy to Parenthood." *Sex Roles* 11, 1/2 (1984): 61–78.

"Female Role Models Busy with Romance, Hair." *San Francisco Chronicle* (May 2, 1997).

Ferman, Lawrence. "After the Shutdown: The Social and Psychological Costs of Job Displacement." *Industrial and Labor Relations Report* 18, 2 (1981): 22–26.

Ferree, Myra Marx. "Beyond Separate Spheres: Feminism and Family Research." In *Contemporary Families: Looking Forward, Looking Back,* edited by A. Booth. Minneapolis: National Council on Family Relations, 1991.

Field, Tiffany. "Quality Infant Day Care and Grade School Behavior and Performance." *Child Development* 62 (1991): 863–870.

Fields, Jason. "Children's Living Arrangements and Characteristics: March 2002." *Current Population Reports* (Series P20, No. 547). Washington, DC: U.S. Census Bureau, June 2003.

Fields, Jason, and Lynne Casper. "America's Families and Living Arrangements: March 2000." *Current Population Reports* (Series P20, No. 537). Washington, DC: U.S. Census Bureau, 2001.

Figley, Charles, ed. *Treating Families Under Stress.* New York: Brunner/Mazel, 1989.

Filene, Peter. *Him/Her/Self: Sex Roles in Modern America* (2nd ed.). Baltimore: Johns Hopkins University Press, 1986.

Fine, Mark A. "Current Approaches to Understanding Family Diversity: An Overview of the Special Issue." *Family Relations* 42, 3 (July 1993): 235–237.

Fineman, Martha Albertson, and Roxanne Mykitiuk. *The Public Nature of Private Violence: The Discovery of Domestic Abuse.* New York: Routledge, 1994.

Finkel, J., and P. Roberts. *The Incomes of Noncustodial Fathers.* Washington, DC: Center for Law and Social Policy, 1994.

Finkel, Judith A., and Finy J. Hansen. "Correlates of Retrospective Marital Satisfaction in Long-Lived Marriages: A Social Constructivist Perspective." *Family Therapy* 19, 1 (1992): 1–16.

Finkelhor, David. *Sexually Victimized Children.* New York: Free Press, 1979.

———. "Common Features of Family Abuse." In *The Dark Side of Families,* edited by D. Finkelhor et al. Beverly Hills, CA: Sage, 1983.

———. *Child Sexual Abuse: New Theory and Research.* New York: Free Press, 1984.

———. "Prevention: A Review of Programs and Research." In *Sourcebook on Child Sexual Abuse,* edited by D. Finkelhor. Beverly Hills, CA: Sage Publications, 1986a.

———. "Prevention Approaches to Child Sexual Abuse." In *Violence in the Home: Interdisciplinary Perspectives,* edited by M. Lystad. New York: Brunner/Mazel, 1986b.

———. "Sexual Abuse in a National Survey of Adult Men and Women." *Child Abuse and Neglect* 14, 1 (1990): 19–28.

———. "Sexual Abuse of Children." In *Vision 2010: Families and Violence, Abuse and Neglect,* edited by R. J. Gelles. Minneapolis: National Council on Family Relations, 1995.

Finkelhor, David, and Larry Baron. "High Risk Children." In *Sourcebook on Child Sexual Abuse,* edited by D. Finkelhor. Beverly Hills, CA: Sage,, 1986.

Finkelhor, David, and Angela Browne. "Initial and Long-Term Effects: A Conceptual Framework." In *Sourcebook on Child Sexual Abuse,* edited by D. Finkelhor. Beverly Hills, CA: Sage, 1986.

Finkelhor, David, G. Hotaling, I. A. Lewis, and C. Smith. *Missing, Abducted, Runaway, and Throwaway Children in America.* Washington, DC: U.S. Department of Justice, 1990.

Finkelhor, David, Gerald Hotaling, and Andrea Sedlak. "Abduction of Children by Family Members." *Journal of Marriage and the Family* 53, 3 (August 1991): 805–817.

Finkelhor, David, and Karl Pillemer. "Elder Abuse: Its Relationship to Other Forms of Domestic Violence." In *Family Abuse and Its Consequences,* edited by G. T. Hotaling et al. Newbury Park, CA: Sage, 1988.

Finkelhor, David, and Kersti Ÿllo. *Forced Sex in Marriage: A Preliminary Research Report.* National Institute of Mental Health. Washington, DC: U.S. Government Printing Office, 1980.

———. *License to Rape: The Sexual Abuse of Wives.* New York: Free Press, 1987.

———. "Rape in Marriage." In *Abuse and Victimization across the Life Span,* edited by M. Straus. Baltimore, MD: Johns Hopkins University Press, 1988.

Finkelhor, David, et al. *The Dark Side of Families.* Beverly Hills, CA: Sage, 1983.

———. "Sexual Abuse in Day Care: A National Study." *National Center on Child Abuse and Neglect.* University of New Hampshire: Family Research Laboratory, 1988.

Finlay, B., and Scheltema, K. E. "The Relation of Gender and Sexual Orientation to Measures of Masculinity, Femininity, and Androgyny: A Further Analysis." *Journal of Homosexuality* 21, 3 (1991): 71–85.

Fireman, www.newsday.com, 2003.

Fischer, Lucy R. "Mothers and Mothers-in-Law." *Journal of Marriage and the Family* 45, 1 (February 1983): 187–192.

Fischl, Margaret, et al. "Evaluation of Heterosexual Partners, Children and Household Contacts of Adults with AIDS." *Journal of the American Medical Association* 257, 5 (February, 1987): 640–647.

Fisher, W. A., et al. "Erotophobia-Erotophilia as a Dimension of Personality." *Journal of Sex Research* 25, 1 (1988): 123–151.

Fisher, William, and Donn Byrne. "Social Background, Attitudes, and Sexual Attraction." In *The Bases of Human Sexual Attraction,* edited by M. Cook. New York: Academic Press, 1981.

Fishman, Barbara. "The Economic Behavior of Stepfamilies." *Family Relations* 32 (July 1983): 356–366.

Fitting, Melinda, Peter Rabins, M. Jane Lucas, and James Eastham. "Caregivers for Demented Patients: A Comparison of Husbands and Wives." *Gerontologist* 26 (1986): 248–252.

Fitzpatrick, Joseph. "The Puerto Rican Family." In *Ethnic Families in America,* 2nd ed., edited by C. Mindel and R. Habenstein. New York: Elsevier-North Holland, 1981.

Fitzpatrick, Joseph, and Lourdes Parker. "Hispanic-Americans in the Eastern United States." *Annals of the American Academy of Political and Social Science* 454 (March 1981): 98–110.

Flaherty, Mary Jean, Lorna Facteau, and Patricia Garver. "Grandmother Functions in Multigenerational Families." In *The Black Family: Essays and Studies,* 4th ed., edited by R. Staples. Belmont, CA: Wadsworth, 1990.

Flaks, David K., Ilda Ficher, Frank Masterpasqua, and G. Joseph. "Lesbians Choosing Motherhood: A Comparative Study of Lesbians and Heterosexual Parents and Their Children." *Developmental Psychology* 31, 1 (January 1995): 105–114.

Floge, Liliane. "The Dynamics of Child-Care Use and Some Implications for Women's Employment." *Journal of Marriage and the Family* 47, 1 (February 1985): 143–154.

Flowers, Blaine, Marla Veingrad, and Carmentxu Dominicis, "The Unbearable Lightness of Positive Illusions: Engaged Individuals Explanations of Unrealistically Positive Relationship Perceptions." *Journal of Marriage and the Family* 64, 2 (May 2002):450–460.

Floyd, Frank J., Stephen N. Haynes, Elizabeth R. Doll, David Winemiller et al. "Assessing Retirement Satisfaction and Perceptions of Retirement Experiences." *Psychology and Aging* 7, 4 (1992): 609–621.

Flynn, Clifton P. "Sex Roles and Women's Response to Courtship Violence." *Journal of Family Violence* 5 1 (March 1990): 83–94.

———. "Exploring the Link Between Corporal Punishment and Children's Cruelty to Animals." *Journal of Marriage and the Family* 61, 4 (November 1999): 971–981

Foa, Uriel G., Barbara Anderson, J. Converse, and W. A. Urbansky, "Gender-Related Sexual Attitudes: Some Cross-Cultural Similarities and Differences." *Sex Roles* 16, 19–20 (May 1987): 511–519.

Fogel, Daniel. *Junipero Serra, the Vatican, and Enslavement Theology.* San Francisco: ISM Press, 1988.

Foley, L., et al. "Date Rape: Effects of Race of Assailant and Victim and Gender of Subjects." *Journal of Black Psychology* 21, 1 (February 1995): 6–18.

Folk, Karen F., and Andrea H. Beller. "Part-Time Work and Child Care Choices for Mothers of Preschool Children." *Journal of Marriage and the Family* 55, 1 (February 1993): 147–157.

Folk, Karen Fox, and Yunae Yi. "Piecing Together Child Care with Multiple Arrangements: Crazy Quilt or Preferred Pattern for Employed Parents of Preschool Children." *Journal of Marriage and the Family* 56, 3 (August 1994): 669–680.

Follingstad, Diane R., L. L. Rutledge, B. J. Berg, and E. S. Haure. "The Role of Emotional Abuse in Physically Abusive Relationships." *Journal of Family Violence* 5, 2 (June 1990): 107–120.

Foote, Nelson, and Leonard Cottrell. *Identity and Interpersonal Competence.* Chicago: University of Chicago Press, 1955.

Ford, Donna. "An Exploration of Alternative Family Structures among University Students." *Journal of Marriage and the Family* 43, 1 (January 1994): 68–73.

Forehand, Rex L., Page B. Walley, and William M. Furey. "Prevention in the Home: Parent and Family." In *Prevention of Problems in Childhood: Psychological Research and Application,* edited by M. C. Roberts and L. Peterson. New York: Wiley, 1984.

Forste, Renata, and Tim Heaton. "Initiation of Sexual Activity among Female Adolescents." *Youth and Society* 19, 3 (March 1988): 250–268.

Foucault, Michel. *The History of Sexuality: An Introduction.* Vol.1. New York: Pantheon Books, 1980.

Fowler, Orison S. *Amativeness: or, Evils and the Remedies of Excessive Perverted Sexuality.* New York: Fowler and Wells, 1878.

Fox, Greer, and Velma McBride Murry. "Gender and Families: Feminist Perspectives and Family Research." *Journal of Marriage and the Family,* 62, 4 (November 2000): 1160-1172.

Fox, Greer L., and Robert F. Kelly. "Determinants of Child Custody Arrangements at Divorce." *Journal of Marriage and the Family* 57, 3 (August 1995): 693–708.

"Fractional Families (Dear Dr. Demo)." *American Demographics* 14, 12 (December 1992): 6.

Francke, Linda Bird. "Childless by Choice." *Newsweek* (January 14, 1980): 96.

Franklin, D. L. "The Impact of Early Childbearing on Development Outcomes: The Case of Black Adolescent Parenting." *Family Relations* 37 (1988): 268–274.

Franks, Lucinda J., James P. Hughes, Linda H. Phelps, and D. G. Williams. "Intergenerational Influences on Midwest College Students by Their Grandparents and Significant Elders." *Educational Gerontology* 19, 3 (1993): 265–271.

Franks, P., C. M. Clancy, M. R. Gold, and P. A. Nutting. "Health Insurance and Subjective Health Status." *American Journal of Public Health* 83, 9 (1993): 1295–1299.

Franz, Wanda, and David Readon. "Differential Impact of Abortion on Adolescents and Adults." *Adolescence* 27, 105 (March 1992): 161–172.

Freed, Doris, and Timothy Walker. "Family Law in the Fifty States." *Family Law Quarterly* 21 (1988): 417–573.

Freeman, Edith M., ed. "Substance Abuse Treatment: A Family Systems Perspective." Sage Sourcebooks for the Human Services Series, No. 25. Newbury Park, CA.: Sage, 1993.

French, J. P., and Bertram Raven. "The Bases of Social Power." In *Studies in Social Power,* edited by L. Cartwright. Ann Arbor: University of Michigan Press, 1959.

Friday, Nancy. *Men in Love.* New York: Delacorte, 1980.

Friedan, Betty. *The Feminine Mystique.* New York: Dell, 1963.

Friedman, Rochelle, and Bonnie Gradstein. *Surviving Pregnancy Loss.* Boston: Little, Brown, 1982.

Frisbie, W. Parker. "Variation in Patterns of Marital Instability among Hispanics." *Journal of Marriage and the Family* 48, 1 (February 1986): 99–106.

Fromm, Erich. *The Art of Loving.* New York: Perennial Library, 1974.

Fu, Haishan, and Noreen Goldman. "Incorporating Health into Models of Marriage Choice: Demographic and Sociological Perspectives." *Journal of Marriage and the Family* 58, 3 (August 1996): 740–758.

Fuller, Mary Lou. "Help Your Family Understand 'It's a Small, Small World.'" *PTA Today* (December 1989): 9–10.

Furstenberg, F. F., and A. J. Cherlin. *Divided Families: What Happens to Children When Parents Part.* Cambridge, MA: Harvard University Press, 1991.

Furstenberg, F. F., and J. O. Teitler. "Reconsidering the Effects of Marital Disruption: What Happens to Children of Divorce in Early Adulthood?" *Journal of Family Issues,* 15, 2 (June 1994): 173–190.

Furstenburg, Frank F. "The Sociology of Adolescence and Youth in the 1990's: A Critical Commentary." *Journal of Marriage and the Family* 62, 4 (November 2000): 896–910.

Furstenberg, Frank Jr. "Good Dads-Bad Dads: Two Faces of Fatherhood" In *The Changing Family,* edited by A. Cherlin. New York: Urban Institute Press, 1988.

Furstenberg, Frank K., Jr. "The New Extended Family: The Experience of Parents and Children after Remarriage." In *Remarriage and Stepparenting: Current Research and Theory,* edited by K. Pasley and M. Ihinger-Tallman. New York: Guilford Press, 1987.

———. "Reflections on Remarriage." *Journal of Family Issues* 1, 4 (1980): 443–453.

Furstenberg, Frank K., Jr., and Andrew Cherlin. *Divided Families.* Cambridge, MA: Harvard University Press, 1991.

Furstenberg, Frank K., Jr., and Christine Nord. "Parenting Apart: Patterns in Childrearing after Marital Disruption." *Journal of Marriage and the Family* 47, 4 (November 1985): 893–904.

Furstenberg, Frank K., Jr., and Graham Spanier, eds. *Recycling the Family—Remarriage after Divorce.* Rev ed. Newbury Park, CA: Sage, 1987.

Furstenberg, Frank K., Jr., R. Lincoln, and J. Menken, eds. *Teenage Sexuality, Pregnancy, and Childbearing.* Philadelphia: University of Pennsylvania Press, 1981.

Furukawa, S. "The Diverse Living Arrangements of Children." *Current Popultion Reports* (Series P70, No. 38). Washington, DC: U.S. Census Bureau, 1993.

Gagnon, John. "Attitudes and Responses of Parents to Pre-Adolescent Masturbation." *Archives of Sexual Behavior* 14, 5 (1985): 451–466.

———. "Sexual Scripts: Permanence and Change." *Archives of Sexual Behavior* 15, 2 (April 1986): 97–120.

Gagnon, John, and William Simon. *Sexual Conduct: The Social Sources of Human Sexuality.* Chicago: Aldine Publishing, 1973.

———. "The Sexual Scripting of Oral Genital Contacts." *Archives of Sexual Behavior* 16, 1 (February 1987): 1–25.

Gaines, Judith. "A Scandal of Artificial Insemination." *Good Health Magazine/New York Times Magazine* (October 7, 1990): 23 ff.

Gallup, George H., Jr. *The Gallup Poll: Public Opinion 1986.* Wilmington, DE: Scholarly Resources, 1987.

Gallup, George H., Jr., and Frank Newport. "Parenthood—A Nearly Universal Desire." *San Francisco Chronicle* (June 4, 1990), B3.

Ganong, Lawrence, and Marilyn Coleman. "The Effects of Remarriage on Children: A Review of the Empirical Literature." *Family Relations* 33 (1984): 389–405.

———. "A Comparison of Clinical and Empirical Literature on Children in Stepfamilies." *Journal of Marriage and the Family* 48 (May 1986): 309–318.

———. "Sex, Sex Roles, and Family Love." *Journal of Genetic Psychology* 148 (March 1987): 45–52.

———. "Gender Differences in Expectations of Self and Future Partner." *Journal of Family Issues* 13, 1 (March 1992): 55–64.

———. *Remarried Family Relationships.* Newbury Park, CA: Sage Publications, 1994.

———. "Family resilience in Multiple Contexts." *Journal of Marriage and the Family* 64, 2 (May 2002): 346-348.

Ganong, Lawrence, et al. "Stepparent: A Pejorative Term?" *Psychological Reports* 53, 3 (June 1983): 919–922.

———. "A Meta-Analytic Review of Family Structure Stereotypes." *Journal of Marriage and the Family* 52, 2 (May 1990): 287–289.

Gans, Herbert. "Symbolic Ethnicity: The Future of Ethnic Groups and Cultures in America." In *On the Making of Americans,* edited by H. Gans. Philadelphia: University of Pennsylvania, 1979.

Gao, Ge. "Stability of Romantic Relationships in China and the United States." In *Cross-Cultural Interpersonal Communication,* edited by S. Ting-Toomey and F. Korzenny. Newbury Park, CA: Sage, 1991.

Garbarino, James. *Children and Families in the Social Environment.* Hawthorne, NY: Aldine De Gruyter, 1982.

Garfinkel, I., S. S. McLanahan, and P. R. Robins, eds. *Child Support and Child Well-Being.*Washington, DC: Urban Institute Press, 1994.

Garfinkel, Irwin, and Sara McLanahan. *Single Mothers and Their Children: A New American Dilemma.*Washington, DC: Urban Institute Press, 1986.

Garfinkel, Irwin, Donald Oellerich, and Philip K. Robins. "Child Support Guidelines: Will They Make a Difference?" *Journal of Family Issues* 12, 4 (December 1991): 404–429.

Garnets, L.., et al. "Violence and Victimization of Lesbians and Gay Men: Mental Health Consequences." *Journal of Interpersonal Violence* 5 (1990): 366–383.

Garrison, James E. "Sexual Dysfunction in the Elderly: Causes and Effects." *Journal of Psychotherapy and the Family* 5, 1–2 (1989): 149–162.

Gary, Lawrence. "Strong Black Families: Models of Program Development for Black Families." In *Family Strengths: Positive Models for Family Life,* edited by S. Van Sandt et al. Lincoln, NE: University of Nebraska Press, 1980.

Gay, Peter. *The Bourgeois Experience: The Tender Passion.* New York: Oxford University Press, 1986.

Gecas, Viktor, and Monica Seff. "Social Class, Occupational Conditions, and Self-Esteem." *Sociological Perspectives* 32 (1989): 353–364.

———. "Families and Adolescents." In *Contemporary Families: Looking Forward, Looking Back,* edited by A. Booth. Minneapolis: National Council on Family Relations, 1991.

Geis, Frances, and Joseph Geis. *Marriage and the Family in the Middle Ages.* New York: Harper and Row, 1987.

Geist, Christopher. "Violence, Passion, and Sexual Racism: The Plantation Novel." *Southern Quarterly* 18, 2 (December 1980): 60–72.

Gelfand, Donald, and Charles Barresi, eds. *Ethnic Dimensions of Aging.* New York: Springer Publishing, 1987.

Gelles, Richard. "Through a Sociological Lens: Social Structure and Family Violence." In *Sociology of Families: Readings,* edited by C. Albers. Thousand Oaks, CA: Pine Forge, 1999: 299–308.

Gelles, Richard, and C. Cornell. *Intimate Violence in Families,* 2nd ed. Newbury Park, CA: Sage, 1990.

Gelles, Richard, and Murray Straus. *Intimate Violence: The Definitive Study of the Causes and Consequences of Abuse in the American Family.* New York: Simon and Schuster, 1988.

Gelles, Richard J. "Child Abuse and Violence in Single-Parent Families: Parent Absence and Economic Deprivation." *American Journal of Orthopsychiatry* 59, 4 (October 1989): 492–501.

———. "Alcohol and Other Drugs Are Associated with Violence—They Are Not Its Cause." In *Current Controversies in Family Violence,* edited by R. Gelles and D. Loseke. Newbury Park, CA: 1993a.

———. "Through a Sociological Lens: Social Structure and Family Violence." In *Current Controversies in Family Violence,* edited by R. Gelles and D. Loseke. Newbury Park, CA: Sage, 1993b.

———. *Intimate Violence in Families.* Thousand Oaks, CA: Sage 1997.

Gelles, Richard J., and Jon R. Conte. "Domestic Violence and Sexual Abuse of Children: A Review of Research in the Eighties." In *Contemporary Families: Looking Forward, Looking Back,* edited by A. Booth. Minneapolis: National Council on Family Relations, 1991.

Gelles, Richard J., and Claire Pedrick Cornell. *Intimate Violence in Families,* 2nd ed. Newbury Park, CA: Sage, 1987.

Gelles, Richard, and Donileen Loseke, eds. *Current Controversies in Family Violence.* Newbury Park, CA: Sage, 1993.

Gelles, Richard J., and Murray A. Straus. "Determinants of Violence in the Family: Toward a Theoretical Integration." In *Contemporary Theories About the Family,* edited by W. Burr. New York: Free Press, 1979.

Genevie, Lou, and Eva Margolies. *The Motherhood Report: How Women Feel about Being Mothers.* New York: Macmillan, 1987.

Genovese, Eugene. *Roll, Jordan, Roll.* New York: Harper and Row, 1976.

George, Kenneth, and Andrew Behrendt. "Therapy for Male Couples Experiencing Relationship Problems and Sexual Problems." *Journal of Homosexuality* 14, 1–2 (1987): 77–88.

Geraghty, Christine. *Women and Soap Opera: A Study of Prime Time Soaps.* London: Polity Press, 1990.

Gerbner, G., L. Gross, M. Morgan, and N. Signorielli. "Living with Television: The Dynamics of the Cultivation Process." In *Perspectives in Media Effects,* edited by J. Bryant and D. Zillman. Hillsdale, NJ: Lawrence Erlbaum, 1986.

Gerris, Jan, Maja Dekocic, and Jan Janssens. "The Relationship between Social Class and Childrearing Behaviors: Parents' Perspective Taking and Value Orientations."*Journal of Marriage and the Family* 59, 4(November 1997): 834–847.

Gerson, Kathleen. *Hard Choices: How Women Decide About Work, Career, and Motherhood.* Berkeley: University of California Press, 1985.

———. *No Man's Land: Men's Changing Commitments to Family and Work.* New York: Basic Books, 1993.

Gerstel, Naomi. "Divorce and Stigma." *Social Forces* 34 (April 1987a): 172–186.

———. *Families and Work.* Philadelphia: Temple University Press, 1987b.

Gerstel, Naomi, and Harriet Gross. *Commuter Marriage: A Study of Work and Family.* New York: Guilford Press, 1984.

Getlin, Josh. "Legacy of a Mother's Drinking." *Los Angeles Times* (July 24, 1989): V1 ff.

Giarretto, Henry. "Humanistic Treatment of Father-Daughter Incest." In *Child Abuse and Neglect: The Family and the*

Community, edited by R. Helfer and H. C. Kempe. Cambridge, MA: Ballinger, 1976.

Gibson, Rose. "Blacks at Middle and Late Life: Resources and Coping." *Annals of the American Academy* 464 (November 1982): 79–90.

Gilbert, Lucia, et al. "Perceptions of Parental Role Responsibilities: Differences between Mothers and Fathers." *Family Relations* 31 (April 1982): 261–269.

Gillen, K., and S. J. Muncher. "Sex Differences in the Perceived Casual Structure of Date Rape: A Preliminary Report." *Aggressive Behavior* 21, 2 (1995): 101–112.

Gilligan, Carol. *In a Different Voice: Psychological Theory and Women's Development.* Cambridge, MA: Harvard University Press, 1982.

Gilman, Lois. *The Adoption Resource Book.* Rev. ed. New York: Harper and Row, 1987.

Gilmore, David. *Manhood in the Making.* New Haven, CT: Yale University Press, 1990.

Givens, Ron. "A New Prohibition." *Newsweek on Campus* (April 1985), 7–13.

Glaser, Ronald, and Janice Kiecolt-Glaser. *Handbook of Human Stress and Immunity.* San Diego: Academic Press, 1994.

Glass, Robert H., and Ronald J. Ericsson. *Getting Pregnant in the 1980s.* Berkeley: University of California Press, 1982.

Glazer-Malbin, Nona, ed. *Old Family/New Family.* New York: Van Nostrand, 1975.

Glenn, Evelyn N., and Stacey G. H. Yap. "Chinese American Families." In *Minority Families in the United States: A Multicultural Perspective,* edited by R. L. Taylor. Englewood Cliffs, NJ: PrenticeHall, 1994.

Glenn, Norval. "Duration of Marriage, Family Composition, and Marital Happiness." *National Journal of Sociology* 3 (1989): 3–24.

———. "The Recent Trend in Marital Success in the United States," *Journal of Marriage and the Family* 53 (2) May 1991: 261–270.

———. "A Reconsideration of the Effect of No-Fault Divorce Legislation on Divorce Rates." *Journal of Marriage and the Family* 59, 4 (November 1997): 1023–1025.

"Who's Who in the Family Wars: A Characterization of the Major Ideological Factions," In *Feuds About Families: Conservative, Centrist, Liberal, and Feminist Perpectives,* edited by N. Benokraitis. Upper Saddle River, NJ: Prentice Hall, 2000: 2–13.

Glenn, Norval, and Kathryn Kramer. "The Marriages and Divorces of the Children of Divorce." *Journal of Marriage and the Family* 49, 4 (November 1987): 811–825.

Glenn, Norval, and Beth Ann Shelton. "Regional Differences in Divorce in the United States." *Journal of Marriage and the Family* 47, 3 (August 1985): 641–652.

Glenn, Norval, and Michael Supancic. "The Social and Demographic Correlates of Divorce and Separation in the United States: An Update and Reconsideration." *Journal of Marriage and the Family* 46, 3 (August 1984): 563–575.

Glick, Jennifer E., Frank D. Bean, and Jennifer V. W. Van Hook. "Immigration and Changing Patterns of Extended Family Household Structure in the United States: 1970–1990." *Journal of Marriage and the Family* 59, 1 (February 1997): 177–191.

Glick, Paul. "American Household Structure in Transition." *Family Planning Perspectives* 16, 5 (September/October 1984a): 205–211.

———. "Marriage, Divorce, and Living Arrangements: Prospective Changes." *Journal of Family Issues* 4, 1 (March 1984b): 7–26.

———. "The Family Life Cycle and Social Change." *Family Relations* 38, 2 (April 1989): 123–129.

———. "Fifty Years of Family Demography." *Journal of Marriage and the Family* 50, 4 (November 1988): 861–873.

———. "Remarried Families, Stepfamilies and Stepchildren: A Brief Demographic Analysis." *Family Relations* 38 (1989): 24–27.

Glick, Paul, and Sung Ling Lin. "Recent Changes in Divorce and Remarriage." *Journal of Marriage and the Family* 49, 4 (November 1986): 737–747.

Glick, Paul, and Graham Spanier. "Married and Unmarried Cohabitation in the United States." *Journal of Marriage and the Family* 42, 1 (February 1980): 19–30.

Gnezda, Therese. *The Effects of Unemployment on Family Functioning.* Prepared Statement to the Select Committee on Children, Youth and Families, House of Representatives, at Hearings on the New Unemployed, Detroit, March 4, 1984. Washington, DC: Government Printing Office, 1984.

Go, K. J. "Recent Advances in the Treatment of Male Infertility." *Naacogs Clinical Issues in Perinatal and Women's Health Nursing* 3, 2 (1992): 320–327.

Gochoros, J. S. *When Husbands Come Out of the Closet.* New York: Harrington Park Press, 1989.

Goelman, Hillel, et al. "Family Environment and Family Day Care." *Family Relations* 39, 1 (January 1990): 14–19.

Goetting, Ann. "Divorce Outcome Research." *Journal of Family Issues* 2, 3 (September 1981): 20–25.

———. "The Six Stages of Remarriage: Developmental Tasks of Remarriage after Divorce." *Family Relations* 31 (April 1982): 213–222.

______."The Developmental Tasks of Siblingship over the Life Cycle." *Journal of Marriage and the Family* 48, 4 (November 1986): 703–714.

———. "Patterns of Support among In-Laws in the United States: A Review of Research." *Journal of Family Issues* 11, 1 (1990): 67–90.

———. *Getting Out: Life Stories of Women Who Left Abusive Men.* New York: Columbia University, 1999.

Goetz, KathrynW. and Cynthia J. Schmiege. "From Marginalized to Mainstreamed: The HEART Project Empowers the Homeless." *Family Relations* 45, 4 (October 1996).

Goldenberg, H., and I. Goldenberg. *Counseling Today's Families,* 2nd ed. Pacific Grove, CA: Brooks/Cole, 1994.

Goldsmith, Jean. "The Postdivorce Family." In *Normal Family Process,* edited by F. Walsh. New York: Guilford Press, 1982.

Goldstein, Barry. *Scared to Leave, Afraid to Stay: Paths from Family Violence to Safety.* San Francisco, CA: Robert Reed Publishers, 2002.

Goldstein, David, and Alan Rosenbaum. "An Evaluation of the Self-Esteem of Maritally Violent Men." *Family Relations* 34, 3 (July 1985): 425–428.

Goleman, Daniel. "Marriage: Research Reveals Ingredients of Happiness." *New York Times* (April 16, 1985a): 19–20.

———. "Spacing of Siblings Strongly Linked to Success in Life." *New York Times* (May 28, 1985b): 17–18.

———. "How Viewers Grow Addicted to Television." *New York Times* (October 16, 1990): C1, 8.

———. "A Modern Tradeoff: Longevity for Health." *New York Times* (May 16, 1991): B10.

———. "Gay Parents Called No Disadvantage." *New York Times* (March 11, 1992).

Golombok, S. "Psychological Functioning of Infertility Patients." *Human Reproduction* 7, 2 (February 1992): 208–212.

Gondolf, Edward. "Male Batterers." In *Family Violence: Prevention and Treatment,* edited by R. L. Hampton et al. Newbury Park, CA: Sage, 1993.

Gondolf, Edward W. "Evaluating Progress for Men Who Batter." *Journal of Family Violence* 2 (1987): 95–108.

———. "The Effect of Batterer Counseling on Shelter Outcome." *Journal of Interpersonal Violence* 3, 3 (September 1988): 275–289.

Gongla, Patricia, and Edward Thompson, Jr., "Single Parent Families." In *Handbook of Marriage and the Family,* edited by M. Sussman and S. Steinmetz. New York: Plenum Press, 1987.

Goode, William. *World Revolution and Family Patterns.* New York: Free Press, 1963.

———. "Force and Violence in the Family." *Journal of Marriage and the Family* 33 (November 1971): 624–636.

Goode, William, ed. *The Family.* 2nd ed. Englewood Cliffs, NJ: Prentice-Hall, 1982.

Goodenow, Carol, and Oliva M. Esp'n. "Identity Choices in Immigrant Adolescent Females." *Adolescence* 28, 109 (1993): 173–185.

Goodman, Ellen. "Where Family Values Begin—And End." *Washington Post* (September 24, 1994): A27.

———. "Welfare Reform Plank Won't Work." *San Jose Mercury News* (December 9, 1994): A1.

Goodman, Leo. "How to Analyze Survey Data Pertaining to the *Time Bind,* and How *Not* to Analyze Such Data." *Social Science Quarterly,* 83, 4,(December 2002): 925–939.

Goodman, Walter. "TV's Sexual Circus Has a Purpose." *New York Times* (August 4, 1992): B2.

Gordon, Sol. "Parents as Sexuality Educators." *SIECUS Report* (March 1984): 10–11.

Gordon, Thomas. *P.E.T. in Action.* New York: Bantam Books, 1978.

Gottman, John, James Coan, Sybil Carrere, and Catherine Swanson. "Predicting Marital Happiness and Stability from Newlywed Interactions" *Journal of Marriage and the Family* 60, 1 (February 1998): 5-22.

Gottman, John, and Clifford Notarius. "Decade Review: Observing Marital Interaction." *Journal of Marriage and the Family,* 62, 4 (November 2000): 927–947.

Gottman, John M. *Why Marriages Succeed or Fail and How You Can Make Yours Work.* New York: Simon and Schuster, 1995.

Gottman, John M., and Lynn F. Katz. "Effects of Marital Discord on Young Children's Peer Interaction and Health." *Developmental Psychology* 25, 3 (May 1989): 373–381.

Gough, Kathleen. "Is the Family Universal: The Nayer Case." In *A Modern Introduction to the Family,* edited by N. Bell and E. Vogel. New York: Free Press, 1968.

Gove, Walter, et al. "Does Marriage Have Positive Effects on the Psychological Well-Being of the Individual?" *Journal of Health and Social Behavior* 24 (1983): 122–131.

Gray, John. *Men Are from Mars, Women Are from Venus.* New York: Harper Collins. 1993.

Greeley, Andrew. "Sit-Coms as Modern Morality Plays." *New York Times* (May 17, 1987).

Greenberg, B. S. "Content Trends in Media Sex." In *Media, Children, and the Family: Social Scientific, Psychodynamic, and Clinical Perspectives,* edited by D. Zillman, J. Bryant and A. C. Huston. Hillsdale, NJ: Lawrence Erlbaum, 1994.

Greenberg, Bradley S. *Life on Television: Content Analysis of U.S. TV Drama.* Norwood, NJ: Ablex Publishing , 1980.

Greenberg, Bradley S., and D. D. D'Alessio. "Quantity and Quality of Sex in the Soaps." *Broadcasting and Electronic Media* 29 (1985): 309–321.

Greenberg, Brigitte. "Barbie, GI Joe Get New Lines from a Doll-Liberation Group." *San Jose Mercury News* (December 29, 1993): 8A.

Greenberg, Dan, and Marsha Jacobs. *How to Make Yourself Miserable.* New York: Random House, 1966.

Greenberger, Ellen. "Explaining Role Strain: Intrapersona." *Journal of Marriage and the Family* 52, 1 (February 1994): 115–118.

Greenberger, Ellen, and Robin O'Neil. "Parents' Concern about Their Child's Development: Implications for Father's and Mother's Well-Being and Attitudes toward Work." *Journal of Marriage and the Family* 52, 3 (August 1990): 621–635.

Greenfield, Patricia M., Lisa Buzzone, Kristi Koyamatsu, and Wendy Satuloff. "What Is Rock Music Doing to the Minds of Our Youth? A First Experimental Look at the Effects of Rock Music Lyrics and Music Videos." *Journal of Early Adolescence* 7, 3 (1987): 315–329.

Greenstein, Theodore. "Marital Disruption and the Employment of Married Women." *Journal of Marriage and the Family* 52 (1990): 657–676.

Greenstein, Theodore N. "Husbands' Participation in Domestic Labor: Interactive Effects of Wives' and Husbands' Gender Ideologies." *Journal of Marriage and the Family* 58, 3 (August 1996): 585–595.

Greeson, Larry E. "Recognition and Ratings of Television Music Videos: Age, Gender, and Sociocultural Effects." *Journal of Applied Social Psychology* 21, 23 (1991): 1908–1920.

Greeson, Larry, and Rose Ann Williams. "Social Implications of Music Videos for Youth: An Analysis of the Content and Effects of MTV." *Youth and Society* 18, 2 (December 1986): 177–189.

Gregor, Thomas. *Anxious Pleasures: The Sexual Lives of an Amazonian People.* Chicago: University of Chicago Press, 1985.

Greif, Geoffrey. "Children and Housework in the Single Father Family." *Family Relations* 34, 3 (July 1985a): 353–357.

———. "Single Fathers Rearing Children." *Journal of Marriage and the Family* 47, 1 (February 1985b): 185–191.

Greif, Geoffrey L., and Rebecca L. Hegar. "Impact on Children of Abduction by a Parent: A Review of the Literature." *American Journal of Orthopsychiatry* 62, 4 (1992): 599–604.

Greif, Geoffrey L., and Joan Kristall. "Common Themes in a Group for Noncustodial Parents." *Families in Society* 74, 4 (1993): 240–245.

Gringlas, Marcy, and Marsha Weinraub. "The More Things Change . . . Single Parenting Revisited." *Journal of Family Issues* 16, 1 (January 1995): 29–52.

Griswold, Robert. *Fatherhood in America: A History.* New York: Basic Books, 1993.

Griswold del Castillo, Richard. *La Familia.* Notre Dame, IN: University of Notre Dame Press, 1984.

Grolier Inc. Web site.

Grossman, Michele, and Wendy Wood. "Sex Differences in Intensity of Emotional Experience: A Social Role Interpretation." *Journal of Personality and Social Psychology* 65, 5 (1993): 1010–1023.

Grotevant, Harold, and Julie Kohler. "Adoptive Families." In *Parenting and Child Development in "Nontraditional" Families,* edited by M. E. Lamb. Mahwah, NJ: Lawrence Erlbaum, 1999.

Grotevant, H. D., R. G. McCoy, C. Elde, and D. L. Frave. "Adoptive Family System Dynamics: Variations by Level of Openness in the Adoption." *Family Process* 33 (1994): 125–146.

Groth, Nicholas. *Men Who Rape: The Psychology of the Offender.* New York: Plenum Press, 1980.

Gruson, Lindsey. "Groups Play Matchmaker to Preserve Judaism." *New York Times* (April 1, 1985).

Grzywacz, Joseph, and Nadine Marks. "Family, Work, Work-Family Spillover, and Problem Drinking During Midlife." *Journal of Marriage and the Family* 62, 2 (May 2000): 336–348.

Gubman, Gayle, and Richard Tessler. "The Impact of Mental Illness on Families." *Journal of Family Issues* 8, 2 (June 1987): 226–245.

Guelzow, Maureen, et al. "Analysis of the Stress Process for Dual-Career Men and Women." *Journal of Marriage and the Family* 53, 1 (February 1991): 151–164.

Guerrero Pavich, Emma. "A Chicana Perspective on Mexican Culture and Sexuality." In *Human Sexuality, Ethnoculture, and Social Work,* edited by L. Lister. New York: Haworth, 1986.

Guidubaldi, John. "The Status Report Extended: Further Elaborations on the American Family." *School Psychology Review* 9, 4 (September 1980): 374–379.

Guldner, Gregory. "Long-Distance Romantic Relationships: Prevalence and Separation-Related Symptoms in College Students." *Journal of College Student Development* 37, 3 (May-June 1996): 289–296

Guldner, Gregory T., and Clifford H. Swensen. "Time Spent Together and Relationship Quality: Long-Distance Relationships as a Test Case." *Journal of Social and Personal Relationships* 12, 2 (1995): 313–320.

Gullo, Karen. "Parents Juggling Multiple Care Arrangements for Kids, Study Shows." Associated Press: March 8, 2000.

Gump, J. "Reality and Myth: Employment and Sex Role Ideology in Black Women." In *The Psychology of Women,* edited by F. Denmark and J. Sherman. New York: Psychological Dimensions, 1980.

Gunn, C. Douglas. "Family Identity Creation." In *Family Strengths: Positive Models for Family Life,* edited by N. Stinnett. Lincoln, NE: University of Nebraska Press, 1980.

Gutek, Barbara. *Sex and the Workplace.* San Francisco: Jossey-Bass, 1985.

Gutek, Barbara, et al. "Sexuality and the Workplace." *Basic and Applied Social Psychology* 1, 3 (1980): 255–265.

Guttentag, M., and P. Secord. *Too Many Women.* Newbury Park, CA: Sage, 1983.

Guttman, Herbert. *The Black Family: From Slavery to Freedom.* New York: Pantheon, 1976.

Guttman, Joseph. *Divorce in Psychosocial Perspective: Theory and Research.* Hillsdale, NJ: Lawrence Erlbaum, 1993.

Guzman, Betsy. "The Hispanic Population," Census 2000 Brief.

Gwartney-Gibbs, Patricia. "The Institutionalization of Premarital Cohabitation: Estimates from Marriage License Applications." *Journal of Marriage and the Family* 48, 2 (May 1986): 423–434.

Haaga, D. A. "Homophobia?" *Journal of Behavior and Personality* 6 (1991): 171–174.

Haas, Linda. "Families and Work." In *Handbook of Marriage and the Family,* 2nd ed., edited by M. Sussman, S. Steinmetz, and G. Peterson. New York: Plenum. 1999.

Haferd, Laura. "Paddling Returns to Child Rearing." *San Jose Mercury News* (December 20, 1986): D12.

Hafstrom, Jeanne, and Vicki Schram. "Chronic Illness in Couples: Selected Characteristics, Including Wife's Satisfaction with and Perception of Marital Relationships." *Family Relations* 33 (1984): 195–203.

Haine, Rachel, Irwin Sandler, Sharlene Wolchik, Jenn-Yun Tein, and Spring Dawson-McClure. "Changing the Legacy of Divorce: Evidence From Prevention Programs and Future Directions." *Family Relations* 52, 4 (October 2003): 397–405.

Hajal, Fady, and Elinor B. Rosenberg. "The Family Life Cycle in Adoptive Families." *American Journal of Orthopsychiatry* 61, 1 (January 1991): 78–85.

Haley, Alex. *Roots.* New York: Dell, 1976.

———. "Counselors Can Improve Society through Families." Paper presented at the annual meeting of the American Association for Counseling and Development, Los Angeles, 1986.

Hall, D. R., and J. Z. Zhao. "Cohabitation and Divorce in Canada: Testing the Selectivity Hypothesis." *Journal of Marriage and the Family* 57 (May 1995): 421–427.

Hallstrom, Tore, and Sverker Samuelsson. "Changes in Women's Sexual Desire in Middle Life: The Longitudinal Study of Women in Gothenburg [Sweden]." *Archives of Sexual Behavior* 19, 3 (1990): 259–267.

Hamburg, David A. *Today's Children.* New York: Times Books, 1992.

Hamby, Sherry and David Sugarman. "Acts of Psychological Aggression Against a Partner and Their Relation to Physical Assault and Gender." *Journal of Marriage and the Family* 61, 4 (November 1999):959–970.

Hamilton D. L. "A Cognitive-Attributional Analysis of Stereotyping." *Advances in Experimental Psychology* 12 (1979): 53–81.

Hampson, Robert, Robert Beavers, and Yosaf Hulgus. "Cross-Ethnic Family Differences: Interaction Assessments of White, Black, and Mexican-American Families." In *Cultural Diversity and Families,* edited by K. G. Arms et al. Dubuque, IA: William C. Brown, 1992.

Hampton, Richard. "Violence in Families of Color." In *Vision 2010: Families and Violence, Abuse and Neglect,* edited by R. J. Gelles. Minneapolis: National Council on Family Relations, 1995.

Hampton, Robert L., ed. *Black Family Violence: Current Research and Theory.* Lexington, MA: Lexington, 1991.

Hampton, Robert L., Richard J. Gelles, and John W. Harrop. "Is Violence in Black Families Increasing? A Comparison of 1975 and 1985 National Survey Rates." *Journal of Marriage and the Family* 51, 4 (November 1989): 969–980.

Hampton, Robert L., Thomas P. Gullotta, Gerald R. Adams, Earl H. Potter III, and Roger P. Weissberg, eds. *Family Violence: Prevention and Treatment.* Newbury Park, CA: Sage , 1993.

Handy, Bruce. "Roll Over, Ward Cleaver." *Time* (April 14, 1997): 78–85.

Hanley, Robert. "Surrogate Deals for Mothers Held Illegal in New Jersey." *New York Times* (February 4, 1988): 1 ff.

Hanline, Mary F. "Transitions and Critical Events in the Family Life Cycle: Implications for Providing Support to Families of Children with Disabilities." *Psychology in the Schools* 28, 1 (1991): 53–59.

Hansen, Christine H., and Ranald D. Hansen. "The Influence of Sex and Violence on the Appeal of Rock Music Videos." *Communication Research* 17, 2 (1990): 212–234.

Hansen, Gary. "Dating Jealousy among College Students." *Sex Roles* 12, 7–8 (April 1985): 713–721.

———. "Extradyadic Relations during Courtship." *Journal of Sex Research* 23, 3 (August 1987): 383–390.

———. "Balancing Work and Family: A Literature and Resource Review." *Family Relations* 40, 3 (July 1991): 348–353.

Hanson, B., and C. Knopes. "Prime Time Tuning Out Varied Cultures." *USA Today* (July 6, 1993).

Hanson, Sandra L., David Myers, Alan L. Ginsburg. "The Role of Responsibility and Knowledge in Reducing Teenage Out-of-Wedlock Childbearing." *Journal of Marriage and the Family* 49, 2 (May 1987): 241–256.

Hanson, Sandra, and Theodora Oooms. "The Economic Costs and Rewards of Two-Earner, Two-Parent Families." *Journal of Marriage and the Family* 53, 3 (August 1991): 622–644.

Hanson, Thomas L., Sara S. McLanahan, and Elizabeth Thomson. "Double Jeopardy: Parental Conflict and Stepfamily Outcomes for Children." *Journal of Marriage and the Family,* 58, 1 (February 1996): 141–154.

Hare-Mustin, Rachel T., and Jeanne Marecek. "Beyond Difference." In *Making a Difference: Psychology and the Construction of Gender,* edited by R. T. Hare-Mustin and J. Marecek. New Haven, CT: Yale University Press, 1990a.

———. "On Making a Difference." In *Making a Difference: Psychology and the Construction of Gender,* edited by R. T. Hare-Mustin and J. Maracek. New Haven, CT: Yale University Press, 1990b.

Hare-Mustin, R. T., and J. Marecek, eds. *Making a Difference: Psychology and the Construction of Gender,* New Haven, CT: Yale University Press, 1990.

Haring-Hidore, Marilyn, et al. "Marital Status and Subjective Well-Being: A Research Synthesis." *Journal of Marriage and the Family* 47, 4 (November 1985): 947–953.

Harriman, Lynda. "Personal and Marital Changes Accompanying Parenthood." *Family Relations* 32, 3 (July 1983): 387–394.

———. "Marital Adjustment as Related to Personal and Marital Changes Accompanying Parenthood." *Family Relations* 35, 2 (April 1986): 233–239.

Harrington, Michael. *The Other America: Poverty in the United States.* New York: Macmillan, 1962.

Harris, Kathleen, and S. Philip Morgan. "Fathers, Sons, and Daughters: Differential Paternal Involvement in Parenting." *Journal of Marriage and the Family* 53, 3 (August 1991): 531–544.

Harris, Sandra. "The Family and the Autistic Child." *Family Relations* 33, 1 (January 1984): 67–77.

Harris, Shanette M. "The Influence of Personal and Family Factors on Achievement Needs and Concerns of African-American and Euro-American College Women." *Sex Roles* 29: 9–10 (November 1993): 671–689.

Harrison, Margaret. "The Reformed Australian Child Support Scheme." *Journal of Family Issues* 12, 4 (December 1991): 430–449.

Harry, Joseph. "Decision Making and Age Differences among Gay Male Couples." In *Gay Relationships,* edited by J. DeCecco. New York: Haworth, 1988.

Hart, J., E. Cohen, A. Gingold, and R. Homburg. "Sexual Behavior in Pregnancy: A Study of 219 Women." *Journal of Sex Education and Therapy* 17, 2 (June 1991): 88–90.

Haskell, Molly. "2000–Year-Old Misunderstanding: Rape Fantasy." *Ms.* (November 1976): 84–86 ff.

———. *From Reverence to Rape.* 2nd ed. Chicago: University of Chicago Press, 1987.

Hatcher, Robert, et al. *Contraceptive Technology.* New York: Irvington, 1986.

Hatchett, Shirley J. "Women and Men." In *Life in Black America,* edited by J. S. Jackson. Newbury Park, CA: Sage, 1991.

Hatchett, Shirley, and James S. Jackson. "African-American Extended Kin System: An Assessment." In *Family Ethnicity: Strength in Diversity,* edited by H. Pipes McAdoo. Newbury Park, CA: Sage, 1993.

Hatfield, Elaine. "Passionate and Companionate Love." *The Psychology of Love,* edited by R. Sternberg and M. Barnes. New Haven, CT: Yale University Press, 1988.

Hatfield, Elaine, and Richard Rapson. "Passionate Love/Sexual Desire: Can the Same Paradigm Explain Both?" *Archives of Sexual Behavior* 16, 3 (June 1987): 259–278.

Hatfield, Elaine, and Susan Sprecher. *Mirror, Mirror: The Importance of Looks in Everyday Life.* New York: State University of New York, 1986.

Hatfield, Elaine, and G. William Walster. *A New Look at Love.* Reading, MA: Addison-Wesley, 1981.

Havemann, Joel. "Diagnosis: Healthier in Europe." *Los Angeles Times* (December 30, 1992): A1, A9.

Hawkins, Alan J., and Tomi-Ann Roberts. "Designing a Primary Intervention to Help Dual-Earner Couples Share Housework and Childcare." *Family Relations* 41, 2 (April 1992): 169–177.

Hawkins, Alan J., Tomi-Ann Roberts, Shawn Christiansen, and C. M. Marshall. "An Evaluation of a Program to Help Dual-Earner Couples Share the Second Shift." *Family Relations* 43, 2 (April 1994): 213–220.

Hawkins, Alan, and David Dollahite, eds. *Generative Fathering: Beyond Deficit Perspectives.* Vol. 3, *Current Issues in the Family.* Thousand Oaks, CA: Sage, 1997.

Hays, Sharon. *Cultural Contradictions of Motherhood.* New Haven, CT: Yale University Press, 1996.

Hayes, Robert M. "Homeless Children." In *Caring for America's Children,* edited by F. J. Macchiarola and A. Gartner. New York: Academy of Political Science, 1989.

Hays, Dorothea, and Aurele Samuels. "Heterosexual Women's Perceptions of their Marriages to Bisexual or Homosexual Men." *Journal of Homosexuality* 18, 2 (1989): 81–100.

Hazan, Cindy, and Philip Shaver. "Romantic Love Conceptualized as an Attachment Process." *Journal of Personality and Social Psychology* 52 (March 1987): 511–524.

Healy, Joseph M., Abigail J. Stewart, and Anne P. Copeland. "The Role of Self-Blame in Children's Adjustment to Parental Separation." *Personality and Social Psychology Bulletin* 19, 3 (1993): 279–289.

Heaton, Tim, and Stan Albrecht. "Stable Unhappy Marriages." *Journal of Marriage and the Family* 53, 3 (August 1991): 747–758.

Heaton, Tim B., and E. L. Pratt. "The Effects of Religious Homogamy on Marital Satisfaction and Stability." *Journal of Family Issues* 7, 2 (June 1990): 191–207.

Heaton, Tim, et al. "The Timing of Divorce." *Journal of Marriage and the Family* 47, 3 (August 1985): 631–639.

Hecht, Michael L., Peter J. Marston, and Linda Kathryn Larkey. "Love Ways and Relationship Quality in Heterosexual Relationships." *Journal of Social & Personal Relationships* 11, 1 (February 1994): 25–43.

Heer, D. M. "The Prevalence of Black-White Marriage in the United States, 1960 and 1970." *Journal of Marriage and the Family* 36 (1974): 246–259.

Hefner, R., et al. "Development of Sex-Role Transcendence." *Human Development* 18 (1975): 143–158.

Heilbrun, Alfred, et al. "Parent Identification and Gender Schema Development." *Journal of Genetic Psychology* 150, 3 (September 1989): 293–300.

Heilbrun, Carolyn. *Toward a Recognition of Androgyny.* New York: Norton, 1982.

Heiman, Julia, et al. *Becoming Orgasmic: A Sexual Growth Program for Women.* Englewood Cliffs, NJ: PrenticeHall, 1976.

Hemstrom, Orjan. "Is Marriage Dissolution Linked to Differences in Morbidity Risks for Men and Women?" *Journal of Marriage and the Family* 58, 2 (May 1996): 366–378.

Hendrick, Clyde, and Susan Hendrick. "A Theory and Method of Love." *Journal of Personality and Social Psychology* 50 (February 1986): 392–402.

———. "Research on Love: Does It Measure Up?" *Journal of Personality and Social Psychology* 56, 5 (May 1989): 784–794.

———. "Attachment Theory and Close Adult Relationships." *Psychological Inquiry* 5, 1 (1994): 38–41.

Hendrick, Clyde, et al. "Do Men and Women Love Differently?" *Journal of Personality and Social Psychology* 48 (1984): 177–195.

Hendrick, Susan. "Self-Disclosure and Marital Satisfaction." *Journal of Personality and Social Psychology* 40 (1981): 1150–1159.

Hendrick, Susan, and Clyde Hendrick. "Multidimensionality of Sexual Attitudes." *Journal of Sex Research* 23, 4 (November 1987): 502–526.

Hendricks, Glenn, et al., eds. *The Hmong in Transition.* Staten Island, NY: Center for Migration Studies of New York, 1986.

Henley, Nancy. *Body Politics: Power, Sex, and Nonverbal Communication.* Englewood Cliffs, NJ: PrenticeHall, 1977.

Henslin, James. *Essentials of Sociology: A Down to Earth Approach,* 3rd ed. Needham Heights, MA: Allyn and Bacon, 2000.

Henton, June, et al. "Romance and Violence in Dating Relationships." *Journal of Family Issues* 4, 3 (September 1983): 467–482.

Hepworth, J., S. McDaniel, and W. Doherty. "Medical Family Therapy with Families Coping with AIDS." In *Counseling Families with Chronic Illness.* American Counseling Association, 1993.

Herbert, Bob. "Scapegoat Time." *New York Times* (December 16, 1994): A19.

Herek, Gregory. "Beyond Homophobia: A Social Psychological Perspective on Attitudes toward Lesbians and Gay Men." *Journal of Homosexuality* 10, 1–2 (September 1984): 1–21.

———. "On Heterosexual Masculinity: Some Psychical Consequences of the Social Construction of Gender." *American Behavioral Scientist* 29, 5 (May 1986a): 563–567.

———. "The Social Psychology of Homophobia: Toward a Practical Theory." *Review of Law and Social Change* 14, 4 (1986b): 923–934.

———. "Hate Crimes against Lesbians and Gay Men: Issues for Research and Policy." *American Psychologist* 44 (June 1989): 948–955.

Herold, Edward, and Leslie Way. "Oral-Genital Sexual Behavior in a Sample of University Females." *Journal of Sex Research* 19, 4 (November 1983): 327–338.

Herring, Cedric, and Karen R. Wilson-Sadberry. "Preference or Necessity? Changing Work Roles of Black and White Women, 1973–1990." *Journal of Marriage and the Family* 55, 2 (May 1993): 314–325.

Hetherington, E. Mavis. "Divorce: A Child's Perspective." *American Psychologist* 34, 10 (October 1979): 851–858.

———. "Intimate Pathways: Changing Patterns in Close Personal Relationships Across Time." *Family Relations* 52, 4 (October 2003): 318–331.

Hetherington, E. Mavis and Margaret Stanley-Hagan. "Stepfamilies." In *Parenting and Child Development in 'Nontraditional' Families* edited by M. Lamb. Mahwah, New Jersey: Lawrence Erlbaum: 137–160.

Hetherington, E. Mavis and John Kelly. *For Better or Worse: Divorce Reconsidered.* New York: W.W. Norton, 2002.

Hetherington, E. Mavis, et al. "Family Interactions and the Social, Emotional, and Cognitive Development of Children Following Divorce." In *The Family Setting Priorities,* edited by V. C. Vaugh and T. B. Brazelton. New York: Science and Medicine Publishers, 1979.

Hevesi, Dennis. "Homeless in New York City: A Day on the Streets." *New York Times* (November 17, 1986): 13.

———. "TV News: Children's Scary Window on New York." *New York Times* (September 11, 1990): A 21.

Hewitt, J. "Preconceptual Sex Selection." *British Journal of Hospital Medicine* 37, 2 (February 1987): 149 ff.

Hewlett, Sylvia. *A Lesser Life: The Myth of Women's Liberation in America.* New York: Morrow, 1986.

———. *When the Bough Breaks: The Cost of Neglecting Our Children.* New York: Basic Books, 1991.

Hewlett, Sylvia, and Cornel West. *The War Against Parents: What We Can Do for America's Beleaguered Moms and Dads.* New York: Houghton Mifflin, 1998.

Heyl, Barbara. "Homosexuality: A Social Phenomenon." In *Human Sexuality: The Societal and Interpersonal Context,* edited by K. McKinney and S. Sprecher. Norwood, NJ: Ablex, 1989.

Hicks, Mary. "Dual Career/Dual Worker Families: A Systems Approach." In *Contemporary Families and Alternative Lifestyles,* edited by E. Macklin and R. Rubin. Beverly Hills, CA: Sage, 1983.

Higginbottom, Susan F., Julian Barling, and E. Kevin Kelloway. "Linking Retirement Experiences and Marital Satisfaction: A Mediational Model." *Psychology and Aging* 8, 4 (1993): 508–516.

Hill, Charles, et al. "Breakups before Marriage: The End of 103 Affairs." *Journal of Social Issues* 32 (1976): 147–168.

Hill, Ivan. *The Bisexual Spouse.* McLean, VA: Barlina Books, 1987.

Hill, Martha. "The Changing Nature of Poverty." *Annals of the American Academy of Political Science* 479 (May 1985): 31–37.

Hill, Reuben. "Generic Features of Families Under Stress." *Social Casework* 49 (February 1958): 139–150.

Hilton, Jeanne, and Virginia Haldeman. "Gender Differences in the Performance of Household Tasks by Adults and Children in Single-Parent and Two-Parent, Two-Earner Families." *Journal of Family Issues* 12, 1 (March 1991): 114–130.

Hilton, N. Zoe. "When Is an Assault Not an Assault? The Canadian Public's Attitudes Toward Wife and Stranger Assault." *Journal of Family Violence* 4, 4 (December 1989): 323–337.

———. "Life Expectancy for Blacks in U.S. Shows Sharp Drop." *New York Times* (November 20, 1990): A1, B7.

Hiltz, Roxanne. "Widowhood: A Roleless Role." *Marriage and Family Review* 1, 6 (November 1978).

Hite, Shere. *The Hite Report.* New York: Macmillan, 1976.

Hitt, Jack. "The Second Sexual Revolution," *The New York Times Magazine,* February 21, 2000.

Hobart, Charles. "Interest in Marriage among Canadian Students at the End of the Eighties." *Journal of Comparative Family Studies* 24, 1 (1993): 45–61.

Hobart, Charles, and Frank Griegel. "Cohabitation among Canadian Students at the End of the Eighties." *Journal of Comparative Family Studies* 23, 3 (1992): 311–338.

Hochschild, Arlie. *The Second Shift: Working Parents and the Revolution at Home.* New York: Viking Press, 1989.

Hochschild, Arlie Russell. *The Time Bind: When Work Becomes Home and Home Becomes Work.* New York: Holt, 1997.

———. "Reply: A Dream Test of the Time Bind." *Social Science Quarterly* 83, 4 (December 2002): 921–924.

Hodgson, Lynne G. "Adult Grandchildren and Their Grandparents: Their Enduring Bond." *International Journal of Aging and Human Development* 34, 3 (1992): 209–225.

Hoegerman, G., et al. "Drug Exposed Neonates." *Western Journal of Medicine* 152, 1 (May 1990): 559 ff.

Hoelter, Jon W. "Factoral Invariance and Self-Esteem—Reassessing Race and Sex Differences." *Social Forces* 61, 3 (March 1983): 834–846.

Hoelter, Jon, and Lynn Harper. "Structural and Interpersonal Family Influences on Adolescent Self-Conception." *Journal of Marriage and the Family* 49, 1 (February 1987): 129–139.

Hofferth, Sandra. "Updating Children's Life Course." *Journal of Marriage and the Family* 47, 1 (February 1985): 93–115.

Hoffman, Lois Wladis. "Maternal Employment: 1979." *American Psychologist* 34 (1979): 859–865.

Holmes, Steven. "On the Edge of Despair When Jobless Benefits End." *New York Times* (January 28, 1991): A11.

Holmes, T., and R. Rahe. "The Social Readjustment Rating Scale." *Journal of Psychosomatic Medicine* 11 (1967): 213–218.

Holtzen, D. W., and A. A. Agresti. "Parental Responses to Gay and Lesbian Children." *Journal of Social and Clinical Psychology* 9, 3 (September 1990): 390–399.

Honeycutt, James. *Memory, Gender, and Relationships.* New York: Guilford Press, 1994a.

Honeycutt, James, Lynn B. Wellman, and Mary S. Larson. "Social Learning Theory and the Calculation of Televised Family Communication Influence: A Time-Series Analysis of Turn-at-Talk on a Popular Family Program." Paper presented at the Annual Speech Communication Association Conference, New Orleans, November 19, 1992.

Hong, George K., and MaryAnna D. Ham. "Impact of Immigration of the Family Life Cycle: Clinical Implications for Chinese Americans." *Journal of Family Psychotherapy* 3, 3 (1992): 27–40.

Hopkins, Ellen. "Childhood's End." *Rolling Stone* (October 18, 1990): 66–72 ff.

Horowitz, R. *Honor and the American Dream.* New Brunswick, NJ: Rutgers University Press, 1983.

Hort, Barbara, et al. "Are People's Notions of Maleness More Stereotypically Framed Than Their Notions of Femaleness?" *Sex Roles* 23, 3 (February 1990): 197–212.

Hort, Barbara E., M. D. Leinbach, and B. I. Fagot. "Is There Coherence among the Cognitive Components of Gender Acquisition?" *Sex Roles* 24, 3–4 (1991): 195–207.

Horwitz, Alan, and Helene White, "The Relationship of Cohabitation and Mental Health: A Study of a Young Adult Cohort," *Journal of Marriage and the Family* 60 (May 1998): 505–514.

Hotaling, Gerald T., and David Finkelhor. "Estimating the Number of Stranger-Abduction Homicides of Children: A Review of Available Evidence." *Journal of Criminal Justice* 18, 5 (1990): 385–399.

Hotaling, Gerald T., and David B. Sugarman. "An Analysis of Risk Markers in Husband to Wife Violence: The Current State of Knowledge." *Violence and Victims* 1, 2 (June 1986): 101–124.

———. "A Risk Marker Analysis of Assaulted Wives." *Journal of Family Violence* 5, 1 (March 1990): 1–14.

Hotaling, Gerald T., et al., eds. *Coping with Family Violence: Research and Perspectives.* Newbury Park, CA: Sage, 1988.

———. *Family Abuse and Its Consequences.* Newbury Park, CA: Sage, 1988.

Houseknecht, Sharon K. "Voluntary Childlessness." In *Handbook of Marriage and the Family,* edited by M. B. Sussman and S. K. Steinmetz. New York: Plenum Press, 1987.

Houseknecht, Sharon K., Suzanne Vaughan, and Ann Statham. "The Impact of Singlehood on the Career Patterns of Professional Women." *Journal of Marriage and the Family* 49, 2 (May 1987): 353–366.

Howard, Judith. "A Structural Approach to Interracial Patterns in Adolescent Judgments about Sexual Intimacy." *Sociological Perspectives* 31, 1 (January 1988): 88–121.

Howes, Carollee. "Can the Age of Entry into Child Care and the Quality of Child Care Predict Adjustment in Kindergarten?" *Developmental Psychology* 26 (1990): 292–303.

Huang, Lucy. "The Chinese American Family." In *Ethnic Families in America: Patterns and Variations,*3rd ed., edited by C. Mindel et al. New York: Elsevier-North Holland, 1988.

Hudak, Mary A. "Gender Schema Theory Revisited: Men's Stereotypes of American Women." *Sex Roles: A Journal of Research* 28, 5–6 (1993): 279–293.

Hughes, Jean O'Gorman, and Bernice R. Sandler. "Friends Raping Friends—Could It Happen to You?" Project on the Status and Education of Women, Association of American Colleges, 1987.

Hunter, Andrea G., and James Earl Davis. "Constructing Gender: An Exploration of Afro-American Men's Conceptualization of Manhood." *Gender and Society* 6, 3 (September 1992): 464–479.

Hunter, James D. "The Family and the Culture War." In *Family in Transition,* 8th ed., edited by A. Skolnick and J. Skolnick. New York: HarperCollins, 1994: 537–547.

Hupka, Ralph B. "Cultural Determinants of Jealousy." *Alternative Lifestyles* 4, 3 (August 1981): 310–356.

Huston, Ted. "The Social Ecology of Marriage and Other Intimate Unions." *Journal of Marriage and the Family* 62, 2 (May 2000): 298–321.

Huston, Ted, S. M. McHale, and A. C. Crouter. "When the Honeymoon's Over: Changes in the Marriage Relationship over the First Year." In *The Emerging Field of Personal Relationships,* edited by S. Duck and R. Gilmour. Hillsdale, NJ: Lawrence Erlbaum, 1986.

Huston, Ted, et al. "From Courtship to Marriage: Mate Selection as an Interpersonal Process." In *Personal Relationships 2: Developing Personal Relationships,* edited by S. Duck and R. Gilmour. London: Academic Press, 1981.

Huston, Ted L., and Gilbert Geis. "In What Ways Do Gender-Related Attributes and Beliefs Affect Marriage?" *Journal of Social Issues* 49, 3 (1993): 87–106.

Hutchins, Loraine, and Lani Kaahumanu, eds. *Bi Any Other Name: Bisexual People Speak Out.* Boston: Alyson Publications, 1991a.

Hutchins, Loraine, and Lani Kaahuman. "Overview." In *Bi Any Other Name: Bisexual People Speak Out,* edited by L. Hutchins and L. Kaahuman. Boston: Alyson Publications, 1991b.

Hynie, Michaela, John E. Lydon, Sylvana Cote, Seth Wiener, "Relational Sexual Scripts and Women's Condom Use: The Importance of Internalized Norms." *Journal of Sex Research* 35, 4 (November 1998): 370–380.

Ickes, William. "Traditional Gender Roles: Do They Make, and Then Break, Our Relationships?" *Journal of Social Issues* 49, 3 (1993): 71–77.

Ihinger-Tallman, Marilyn. "Sibling and Stepsibling Bonding in Stepfamilies." In *Remarriage and Stepparenting Current Research and Theory,* edited by K. Pasley and M. Ihinger-Tallman. New York: Guilford Press, 1987.

Ihinger-Tallman, Marilyn, and Kay Pasley. "Divorce and Remarriage in the American Family: A Historical Review." In *Remarriage and Stepparenting: Current Research and Theory,* edited by K. Pasley and M. Ihinger-Tallmann. New York: Guilford Press, 1987a.

———. *Remarriage.* Newbury Park, CA: Sage, 1987b.

"Immunization Information." Washington, DC: Centers For Disease Control and Prevention (March 9, 1995).

Indivik, Julie, and Mary Fitzpatrick. "'If You Could Read My Mind, Love . . . ,' Understanding and Misunderstanding in the Marital Dyad." *Family Relations* 44, 4 (November 1982): 43–51.

Irving, Howard, et al. "Shared Parenting: An Empirical Analysis Utilizing a Large Data Base." *Family Process* 23 (1984): 561–569.

Isensee, Rik. *Love Between Men: Enhancing Intimacy and Keeping Your Relationship Alive.* New York: Prentice Hall, 1990.

Ishii-Kuntz, Masako. "Japanese American Families." In *Families in Cultural Context,* edited by M. K. DeGenova. Mountain View, CA: Mayfield, 1997.

Itard, Jean. "The Wild Boy of Aveyron." In *Wolf Children and the Problem of Human Nature,* edited by L. Malson. New York: Monthly Review Press, 1972. Originally published 1801.

Jackson, Robert L. "Panel Calls for U.S. to Curb Infant Deaths." *Los Angeles Times* (December 16, 1993): A37.

Jacoby, Arthur, and John Williams. "Effects of Premarital Sexual Standards and Behavior on Dating and Marriage Desirability." *Journal of Marriage and the Family* 47, 4 (November 1985): 1059–1065.

Jankowiak, W., and E. Fischer. "A Cross-Cultural Perspective on Romantic Love," *Ethnology,* 31, 1992: 149–155.

Janus, Samuel and Cynthia Janus. *The Janus Report.* New York: Wiley, 1993.

Jeffrey, T. B., and L. K. Jeffrey. "Psychologic Aspects of Sexual Abuse in Adolescence." *Current Opinion in Obstetrics and Gynecology* 3, 6 (December 1991): 825–831.

Jencks, Christopher. "The Homeless." *New York Review of Books* 41, 8 (April 1994a): 20–27.

———. *The Homeless.* Cambridge, MA: Harvard University Press, 1994b.

———. "Housing the Homeless." *The New York Review of Books* 41, 9 (May 1994c): 39–46.

Jenks, Richard. "Swinging: A Replication and Test of a Theory." *Journal of Sex Research* 21, 2 (May 1985): 199–210.

Jensen, Larry C., and Janet Jensen. "Family Values, Religiosity, and Gender." *Psychological Reports* 73, 2 (October 1993): 429–430.

Jensen, M. A. *Love's Sweet Return: The Harlequin Story.* Toronto: Women's Press, 1984.

Jensen-Scott, Rhonda L. "Counseling to Promote Retirement Adjustment." *Career Development Quarterly* 1, 3 (1993): 257–267.

Joe, Tom, and Douglas W. Nelson. "New Future for America's Children." In *Caring for America's Children,* edited by F. J. Macchiarola and A. Gartner. New York: Academy of Political Science, 1989.

Joesch, Jutta. "The Effects of the Price of Child Care on AFDC Mothers' Paid Work Behavior." *Family Relations* 40, 2 (April 1991): 161–166.

John, Robert. "The Native American Family." In *Ethnic Families in America: Patterns and Variations,* 3rd ed., edited by C. Mindel et al. New York: Elsevier North Holland, 1988.

Johnson, Beverly. "Single Parent Families." *Family Economics Review* (June 1980): 22–27.

Johnson, Catherine B., Margaret S. Stockdale, and Frank E. Saal. "Persistence of Men's Misperceptions of Friendly Cues across a Variety of Interpersonal Encounters." *Psychology of Women Quarterly* 15, 3 (September 1991): 463–475.

Johnson, Clifford, et al. *Vanishing Dreams: The Growing Economic Plight of America's Young Families.* Washington, DC: Children's Defense Fund, 1988.

Johnson, David, Lynn White, John Edwards, and Alan Booth. "Dimensions of Marital Quality: Toward Methodological and Conceptual Refinement." *Journal of Family Issues* 7 (1986): 31–49.

Johnson, Dirk. "At Colleges, AIDS Alarms Muffle Older Dangers." *New York Times* (March 8, 1990): B8.

Johnson, Leanor Boulin. "Perspectives on Black Family Empirical Research: 1965–1978. In *Black Families,* edited by H. Pipes McAdoo. Newbury Park, CA: Sage, 1988.

Johnson, Michael. "Patriarchal Terrorism and Common Couple Violence: Two Forms of Violence Against Women." *Journal of Marriage and the Family* 57 (2) (May 1995): 283–294.

Johnson, Michael, and Kathleen Ferraro. "Research on Domestic Violence in the 1990's: Making Distinctions," *Journal of Marriage and the Family,* 62, 4 (November 2000): 948–963.

Johnson, Michael P., Ted L. Huston, Stanley O. Gaines, and George Levinger. "Patterns of Married Life among Young Couples." *Journal of Social and Personal Relationships* 9, 3 (1992): 343–364.

Johnston, Thomas F. "Alaskan Native Social Adjustment and the Role of Eskimo and Indian Music." *Journal of Ethnic Studies* 3/4 (December 1976): 21–36.

Jones, Carl. *Mind Over Labor.* New York: Penguin, 1988.

Jones, Jennifer, and David Barlow. "Self-Reported Frequencies of Sexual Urges, Fantasies, and Masturbatory Fantasies in Heterosexual Males and Females." *Archives of Sexual Behavior* 19, 3 (1990): 269–279.

Jones, Maggie. *A Child by Any Means.* London: Piatkus, 1989.

Jones, T. S., and M. S. Remland. "Sources of Variability in Perceptions of and Responses to Sexual Harassment." *Sex Roles* 27, 3–4 (August 1992): 121–142.

Jorgensen, Stephen, and Russell Adams. "Predicting Mexican-American Family Planning Intentions: An Application and Test of a Social Psychological Model." *Journal of Marriage and the Family* 50, 1 (February 1988): 107–119.

Jorgensen, Stephen R., and A. C. Johnson. "Correlates of Divorce Liberality." *Journal of Marriage and the Family* 42 (1980): 617–622.

Julian, T. W., P. McKenry, and L. McKelvey. "Cultural Variations in Parenting: Perceptions of Caucasian, African American, Hispanic, and Asian American Parents." *Family Relations* (43), 1994: 30–37.

Juni, Samuel, and Donald W. Grimm. "Marital Satisfaction and Sex-Roles in a New York Metropolitan Sample." *Psychological Reports* 73, 1 (1993): 307–314.

Jurich, Anthony, and Cheryl Polson. "Nonverbal Assessment of Anxiety as a Function of Intimacy of Sexual Attitude Questions." *Psychological Reports* 57 (3, Pt. 2), (December 1985): 1247–1243.

Justice, Blair, and Rita Justice. *The Broken Taboo: Incest.* New York: Human Sciences Press, 1979.
———. *The Abusing Family.* Rev. ed. New York: Insight Books, 1990.
Kach, Julie, and Paul McGee. "Adjustment to Early Parenthood." *Journal of Family Issues* 3, 3 (September 1982): 375–388.
Kagan, Jerome, and N. Snidman. "Temperamental Factors in Human Development." *American Psychologist* 46, 8 (1991): 856–862.
Kain, Edward. *The Myth of Family Decline: Understanding Families in a World of Rapid Change.* Lexington, MA: Lexington Books, 1990.
Kalin, Tom. "Gays in Film: No Way Out." *US* [Special Issue: "The Sexual Revolution in Movie, Music and TV"], 175 (August 1992): 68–70.
Kalis, Pamela, and Kimberly Neuendorf. "Aggressive Cue-Prominence and Gender Participation in MTV." *Journalism Quarterly* 66, 1 (March 1989): 148–154, 229.
Kalmuss, Debra. "The Intergenerational Transmission of Marital Aggression." *Journal of Marriage and the Family* 46, 1 (February 1984): 11–20.
Kalof, Linda. "Dilemmas of Femininity: Gender and the Social Construction of Sexual Imagery." *Sociological Quarterly* 34, 4 (November 1994): 639–651.
Kamerman, Sheila, and C. D. Hayes. *Families That Work: Children in a Changing World.* Washington, DC: National Academy Press, 1982.
Kane, N., ed. *The Hispanic American Almanac: A Reference Work on Hispanics in the United States.* Detroit: MI: Gale Research, 1993.
Kantor, David, and William Lehr. *Inside Families.* San Francisco, CA: Jossey-Bass, 1975.
Kantrowitz, Barbara. "Who Keeps 'Baby M'?" *Newsweek* (January 19, 1987): 44–49.
———. "Gay Families Come Out." *Newsweek* (November 4, 1996): 51–57.
Kantrowitz, Barbara, and David A. Kaplan. "Not the Right Family." *Newsweek* (March 19, 1990): 50–51.
Kaplan, A., and J. P. Bean. "From Sex Stereotypes to Androgyny: Considerations of Societal and Individual Change." In *Beyond Sex-Role Stereotypes,* edited by A. Kaplan and J. P. Bean. Boston: Little, Brown, 1976.
Kaplan, Helen Singer. *Disorders of Desire.* New York: Simon and Schuster, 1979.
Kassop, Mark. "Salvador Minuchin: A Sociological Analysis of His Family Therapy Theory." *Clinical Sociological Review* 5 (1987): 158–167.
Katchadourian, Herant. *Midlife in Perspective.* San Francisco: Freeman, 1987.
Katz, Michael B. *The Undeserving Poor: From War on Poverty to War on Welfare.* New York: Pantheon, 1990.
Katzev, A. R., R. L. Warner, and A. C. Acock. "Girls or Boys? Relationship of Child Gender to Marital Instability." *Journal of Marriage and the Family* 56 (February 1994): 89–100.
Kaufman, Joan, and Edward Zigler. "The Intergenerational Transmission of Abuse Is Overstated." In *Current Controversies in Family Violence,* edited by R. Gelles and D. Loseke. Newbury Park, CA: Sage, 1993.
Kavanaugh, Robert. *Facing Death.* Baltimore: Penguin, 1972.
Kawamoto, Walter T., and Tamara C. Chesire. "American Indian Families." In *Families in Cultural Context,* edited by M. K. DeGenova. Mountain View, CA: Mayfield, 1997.
Kayal, P. M. "Healing Homophobia: Volunteerism and Sacredness in AIDS." *Journal of Religion and Health* 31, 2 (June 1982): 113–128.
Kaye, K., et al. "Birth Outcomes for Infants of Drug Abusing Mothers." *New York State Journal of Medicine* 144, 7 (May 1989): 256–261.
Kearl, Michael C. *Endings: A Sociology of Death and Dying.* New York: Oxford University Press, 1989.
Kehoe, Monika "Lesbians Over 60 Speak for Themselves." *The Gerontologist* 32 (April 1992): 280–282.
Keilor, Garrison. "It's Good Old Monogamy That's Really Sexy." *Time* (October 17, 1994): 71.
Keith, Pat, and Robbyn Wacker. "Grandparent Visitation Rights: An Inappropriate Intrusion or Appropriate Protection?" *International Journal of Aging and Human Development* 54, 3 (2002): 191–204.
Keith, Pat, et al. "Older Men in Employed and Retired Families." *Alternative Lifestyles* 4, 2 (May 1981): 228–241.
Keller, David, and Hugh Rosen. "Treating the Gay Couple within the Context of Their Families of Origin." *Family Therapy Collections* 25 (1988): 105–119.
Kellett, J. M. "Sexuality of the Elderly." *Sexual and Marital Therapy* 6, 2 (1991): 147–155.
Kelley, Douglas L., and Judee K. Burgoon. "Understanding Marital Satisfaction and Couple Type as Functions of Relational Expectations." *Human Communication Research* 18, 1 (1991): 40–69.
Kelley, Harold. "Love and Commitment." In *Close Relationships,* edited by H. Kelley et al. San Francisco: Freeman, 1983.
Kelley, Robert, and Patricia Voydanoff. "Work/Family Role Strain Among Employed Parents." *Family Relations* 34, 3 (July 1985): 367–374.
Kelly, Joan B. "Current Research on Children's Postdivorce Adjustment: No Simple Answers." *Family and Conciliation Courts Review* 31, 1 (1993): 29–49.
Kelly, Mary P., D. S. Strassberg, and J. R. Kircher. "Attitudinal and Experiential Correlates of Anorgasmia." *Archives of Sexual Behavior,* 19, 2 (April 1990): 165–167.
Kempe, C. Henry, and Ray Helfer, eds. *The Battered Child.* Rev. ed. Chicago: University of Chicago Press, 1980.
Kendall-Tackett, Kathleen, L. M. Williams, and D. Finkelhor. "Impact of Sexual Abuse on Children: A Review and Synthesis of Recent Empirical Studies." *Psychological Bulletin,* 113, 1 (January 1993): 164 ff.
Kennedy, Gregory E. "College Students' Expectations of Grandparent and Grandchild Role Behaviors." *Gerontologist* 30, 1 (1990): 43–48.

———. "Grandchildren's Reasons for Closeness with Grandparents." *Journal of Social Behavior and Personality* 6, 4 (1991): 697–712.

———. "Shared Activities of Grandparents and Grandchildren." *Psychological Reports* 70, 1 (1992a): 211–227.

———. "Quality in Grandparent/Grandchild Relationships." *International Journal of Aging and Human Development* 35, 2 (1992b): 83–98.

Kennedy, Gregory E., and C. E. Kennedy. "Grandparents: A Special Resource for Children in Stepfamilies." *Journal of Divorce and Remarriage* 19, 3–4 (1993): 45–68.

Keshet, Jamie. "From Separation to Stepfamily." *Journal of Family Issues* 1, 4 (December 1980): 517–532.

Kessler, R. C., et al. "Lifetime and 12–Month Prevalence of DSM-III-R Psychiatric Disorders in the United States. Results From the National Comorbidity Survey." *Archives of General Psychiatry* 51, 1 (January 1994): 8–19.

Kessler-Harris, Alice. *Women Have Always Worked: A Historical Overview.* New York: McGraw-Hill, 1981.

———. *A History of Wage-Earning Women in America.* New York: Oxford University Press, 1982.

Kett, Joseph. *Rites of Passage: Adolescence in America, 1970 to the Present.* New York: Basic Books, 1977.

Kibria, Nazli. "Vietnamese Families in the United States." In *Minority Families in the United States: A Multicultural Perspective,* edited by R. L. Taylor. Englewood Cliffs, NJ: Prentice-Hall, 1994.

Kikumura, Akemi, and Harry Kitano. "The Japanese American Family." In *Ethnic Families in America: Patterns and Variations,* 3rd ed., edited by C. Mindel et al. New York: Elsevier North Holland, 1988.

Kilborn, Peter T. "Job News Grim for High School Grads." *New York Times* (May 20, 1994): 1A.

———. "Day Care: Key to Welfare Reform." *San Francisco Chronicle* (June 1, 1997): A-4.

Kilmartin, Christopher. *The Masculine Self.* New York: MacMillan, 1994.

Kilzer, Louis. "Kid Fingerprinting a Sham, Foes Claim." *Denver Post* (May 13, 1985a): A1, 14.

———. "Public Often Not Told Facts in Missing Children Cases." *Denver Post* (September 22, 1985b): A1, 14.

Kilzer, Louis, and Diana Griego. "Missing-Child Reports Bring Out Best, Worst." *The Denver Post* (May 13, 1985): A1, 14.

Kimmel, Michael. "Masculinity as Homophobia: Fear, Shame, and Silence in the Construction of Gender Identity." In *Theorizing Masculinities,* edited by H. Brod and M. Kaufman. Thousand Oaks, CA: Sage, 1995.

———. *Manhood in America: A Cultural History.* Berkeley: University of California Press, 1996.

———. *The Gendered Society,* 2nd ed. New York: Oxford University Press, 2004.

Kimmel, Michael, and Michael Messner, eds. *Men's Lives,* 4th ed. Needham Heights, MA: Allyn and Bacon, 1998.

King, Laura A. "Emotional Expression, Ambivalence over Expression, and Marital Satisfaction." *Journal of Social and Personal Relationships* 10, 4 (1993): 601–607.

King, Patricia. "Not So Different After All: DomesticViolence within the Gay Community." *Newsweek* (October 4, 1993): 75.

Kingston, Paul, and Stephen Nock. "Consequences of the Family Work Day." *Journal of Marriage and the Family* 47, 3 (August 1985): 619–629.

Kinnunen, Ulla, Jan Gerris, and Ad Vermulst. "Work Experiences and Family Functioning Among Employed Fathers with Children of School Age." *Family Relations* 45, 4 (October 1996), 449–455.

Kinsey, Alfred, Wardell Pomeroy, and Clyde Martin. *Sexual Behavior in the Human Male.* Philadelphia: Saunders, 1948.

Kinsey, Alfred, Wardell Pomeroy, Clyde Martin, and P. Gebhard. *Sexual Behavior in the Human Female.* Philadelphia: Saunders, 1953.

Kirkpatrick, Lee A., and Phillip R. Shaver. "An Attachment—Theoretical Approach to Romantic Love and Religious Belief." *Personality and Social Psychology Bulletin* 18, 3 (1992): 266–275.

Kirkpatrick, Martha, et al. "Lesbian Mothers and Their Children: A Comparative Study." *American Journal of Orthopsychiatry* 51, 3 (July 1981): 545–551.

Kissman, Kris, and JoAnn Allen. *Single Parent Families.* Newbury Park, CA: Sage, 1993.

Kitano, K., and H. Kitano. The Japanese-American Family" In *Ethnic Families in America: Patterns and Variations,* 4th ed., edited by C. Mindel, R. Habenstein, and R. Wright, Jr. Upper Saddle River, NJ: PrenticeHall, 1998: 311–330.

Kitson, G. C., and L. A. Morgan. "The Multiple Consequences of Divorce: A Decade Review. In *Contemporary Families: Looking Forward, Looking Back,* edited by A. Booth. Minneapolis: National Council on Family Relations, 1991.

Kitson, Gay. "Marital Discord and Marital Separation: A County Survey." *Journal of Marriage and the Family* 47 (August 1985): 693–700.

Kitson, Gay C., Richard D. Clark, Norman B. Rushforth, Paul M. Brinich, Howard S. Sudak, and Stephen J. Zyranski. "Research on Difficult Family Topics: Helping New and Experienced Researchers Cope with Research on Loss," *Family Relations* 45, 2 (1996): 183–188.

Kitson, Gay, and Leslie Morgan. "Consequences of Divorce." In *Contemporary Families: Looking Forward, Looking Back* edited by A. Booth. Minneapolis: National Council on Family Relations, 1991.

Kitson, Gay, and Marvin Sussman. "Marital Complaints, Demographic Characteristics, and Symptoms of Mental Distress in Divorce." *Journal of Marriage and the Family* 44, 1 (February 1982): 87–101.

Kitzinger, Sheila. *Woman's Experience of Sex.* New York: Penguin, 1985.

Kitzinger, Sheila. *The Complete Book of Pregnancy and Childbirth.* New York: Knopf, 1989.

Kitzmann, Katherine, Robert Cohen, and Rebecca Lockwood. "Are Only Children Missing Out? Comparison of the Peer-Related Social Competence of Only Children and Siblings." *Journal of Social and Personal Relationships* 19 (2002): 299–316.

Kivett, Vira R. "The Grandparent-Grandchild Connection." *Marriage and Family Review* 16, 3–4 (1991): 267–290.

———. "Racial Comparisons of the Grandmother Role: Implications for Strengthening the Family Support System of Older Black Women." *Family Relations* 42, 2 (April 1993): 165–172.

Klein, Alan M. "Of Muscles and Men: Anthropological Study of the Culture of Bodybuilding." *Science* 33, 6 (1993): 32–38.

Klein, David M., and James M. White. *Family Theories: An Introduction.* Thousand Oaks, CA: Sage, 1996.

Kline, A., E. Kline, and E. Oken. "Minority Women and Sexual Choice in the Age of AIDS." *Social Science and Medicine* 34, 4 (February 1992): 447–57.

Kline, Marsha, Janet Johnson, and Jeanne Tschann. "The Long Shadow of Marital Conflict: A Model of Children's Postdivorce Adjustment." *Journal of Marriage and the Family* 53 (2) May 1991: 297–309.

Klinetob, Nadya, and David Smith. "Demand-Withdraw Communication in Marital Interaction: Tests of Interspousal Contingency and Gender Role Hypothesis." *Journal of Marriage and the Family*, 58, 4 (November 1996): 945–957.

Klinman, Deborah, et al. *Fatherhood, USA: The First National Guide to Programs, Services and Resources for and about Fathers.* New York: Garland Publishing, 1984.

Klor de Alva, J. "Telling Hispanics Apart: Latino Sociocultural Diversity." In *The Hispanic Experience in the United States: Contemporary Issues and Perspectives,* edited by E. Acosta-Belen and B. Sjostrom. New York: Praeger, 1988.

Kluwer, Esther, Jose Heesink, and Evert Van de Vliert. "Marital Conflict About the Division of Household Labor and Paid Work." *Journal of Marriage and the Family,* 58, 4. (November 1996): 958–969.

———. "The Marital Dynamics of Conflict Over the Division of Labor." *Journal of Marriage and the Family* 59, 3 (August 1997):635–654.

Knafo, D., and Y. Jaffe. "Sexual Fantasizing in Males and Females." *Journal of Research in Personality* 18 (1984): 451–462.

Knapp, J., and R. Whitehurst. "Sexually Open Marriage and Relationships: Issues and Prospects." In *Marriage and Alternatives: Exploring Intimate Relationships,* edited by R. Libby and R. Whitehurst. Glenview, IL: Scott, Foresman, 1977.

Knapp, Mark, et al. "Compliments: A Descriptive Taxonomy." *Journal of Communication* 34, 4 (1984): 12–31.

Knaub, P. K., et al. "Strengths of Remarriage." *Journal of Divorce* 7 (1984): 41–55.

Knox, David, Marty E. Zusman, Vivian Daniels, and Angel Brantley. "Absence Makes the Heart Grow Fonder? Long Distance Dating Relationships Among College Students." *College Student Journal* 36, 3 (2002): 364–366.

Knox, David, and Caroline Schacht. "Sexual Behaviors of University Students Enrolled in a Human Sexuality Course." *College Student Journal* 26, 1 (March 1992): 38–40.

Knox, David, and K. Wilson. "Dating Behaviors of University Students." *Family Relations* 30 (1981): 83–86.

Koblinsky, Sally, and Christine Todd. "Teaching Self-Care Skills to Latchkey Children: A Review of the Research." *Family Relations* 38, 4 (October 1989): 431–435.

Koch, Liz. "Mothering: An Honorable Profession." *The Doula* 2, 2 (September 1987): 4–6.

Kohlberg, Lawrence. "The Cognitive-Development Approach to Socialization." In *Handbook of Socialization Theory and Research,* edited by A. Goslin. Chicago: Rand McNally, 1969.

Kohn, M. L. "Social Class and Parental Values." *American Journal of Sociology* 64 (1959): 337–351.

Kolata, Gina. "Early Warnings and Latent Cures for Infertility." *Ms.* (May 1979): 86–89.

———. *The Baby Doctors.* New York: Delacorte Press, 1989.

———. "Racial Bias Seen in Prosecuting Pregnant Addicts." *New York Times* (July 20, 1990b): A10.

———. "New Pregnancy Hope: A Single Sperm Injected." *New York Times* (August 11, 1993): B7.

Kolker, A. "Advances in Prenatal Diagnosis: Social-Psychological and Policy Issues." *International Journal of Technology Assessment in Health Care* 5, 4 (1989): 601–617.

Komarovsky, Mirra. *Blue Collar Marriage.* New York: Vintage, 1962.

———. *Women in College.* New York: Basic Books, 1985.

———. *Blue-Collar Marriage.* 2nd ed. New Haven, CT: Yale University Press, 1987.

———. *Dilemmas of Masculinity.*

Konker, C. "Rethinking Child Sexual Abuse: An Anthropological Perspective." *American Journal of Orthopsychiatry* 62, 1 (January 1992): 147–153.

Konner, Melvin. *Childhood.* Boston: Little, Brown, 1991.

Kortenhaus, Carole M., and Jack Demarest. "Gender Role Stereotyping in Children's Literature: An Update." *Sex Roles: A Journal of Research* 28, 3–4 (1993): 219–323.

Koss, M., L. Goodman, L. Fitzgerald, N. Russo, G. Keita, and A. Browne. *No Safe Haven: Male Violence Against Women at Home, at Work, and in the Community.* Washington, D.C.: American Psychological Association, 1994.

Koss, Mary. "Hidden Rape: Sexual Aggression and Victimization in a National Sample of Students in Higher Education." In *Rape and Sexual Assault,* vol. 2, edited by A. W. Burgess. New York: Garland, 1988.

Koss, Mary, Thomas Dinero, Cynthia Seibel, and Susan Cox. "Stranger and Acquaintance Rape: Are There Differences in the Victim's Experience?" *Psychology of Women* 12, 1 (March 1988): 1–24.

Koss, Mary P., and Sarah L. Cook. "Facing the Facts: Date and Acquaintance Rape Are Significant Problems for Women." In *Current Controversies in Family Violence,* edited by R. Gelles and D. Loseke. Newbury Park, CA: Sage, 1993.

Koss, Mary P., and T. E. Dinero. "Discriminant Analysis of Risk Factors for Sexual Victimization among a National Sample of College Women." *Journal of Consulting and Clinical Psychology* 57 (April 1989): 242–250.

Kozol, Jonathan. *Rachel and Her Children: Homeless Families in America.* New York: Crown, 1988.

Kranichfeld, Marion. "Rethinking Family Power." *Journal of Family Issues* 8, 1 (March 1987): 42–56.

Krause, Neal. "Race Differences in Life Satisfaction among Aged Men and Women." *Journals of Gerontology* 48, 5 (1993): S235–S244.

Kruk, Edward. "Promoting Co-operative Parenting after Separation: A Therapeutic/Interventionist Model of Family Mediation." *Journal of Family Therapy* 15, 3 (1993): 235–261.

Kubey, Robert. "Media Implications for the Quality of Family Life" in *Media, Children and the Family: Social Scientific, Psychodynamic and Clinical Perspectives,* edited by D. Zillman, J. Bryant, and A. C. Huston. Hillsdale, NJ: Lawrence Erlbaum, 1994.

Kübler-Ross, Elisabeth. *On Death and Dying.* New York: Macmillan, 1969.

———. *Working It Through.* New York: Macmillan, 1982.

———. *AIDS: The Ultimate Challenge.* New York: Macmillan, 1987.

Kudson-Martin, C. and A. Mahoney. "Language Processes in the Construction of Equality in Marriages. *Family Relations,* 47 (1998): 81–91.

Kurdek, L. A. "Predicting Marital Dissolution: A 5-Year Prospective Longitudinal Study of Newlywed Couples." *Journal of Personality and Social Psychology* 64, 2 (1993): 221–242.

Kurdek, L. A., and M. A. Fine. "The Relation Between Family Structure and Young Adolescents' Appraisals of Family Climate and Parenting Behavior." *Journal of Family Issues* 14 (June 1993): 279–290.

Kurdek, Lawrence, and Albert Siesky. "Children's Perceptions of Their Parents' Divorce." *Journal of Divorce* 3, 4 (June 1980): 339–378.

Kurdek, Lawrence, et al. "Correlates of Children's Long-Term Adjustment to Their Parents' Divorce." *Developmental Psychology* 17, 5 (September 1981): 565–579.

Kutscher, Austin H., Arthur C. Carr, and Lillian G. Kutscher, eds. *Principles of Thanatology.* New York: Columbia University Press, 1987.

Ladas, Alice, et al. *The G Spot.* New York: Holt, Rinehart and Winston, 1982.

Laing, R. D. *The Politics of Experience.* New York: Random House, 1967.

———. *The Politics of the Family and Other Essays.* New York: Random House, 1972.

Lainson, Suzanne. "Breast-Feeding: The Erotic Factor." *Ms.* 11, 8 (February 1983): 66 ff.

Lamanna, Mary Ann, and Agnes Riedmann. *Marriages and Families; Making Choices in a Diverse Society.* Belmont, CA: Wadsworth, 1997.

Lamaze, Fernand. *Painless Childbirth,* 1st ed. 1956. Chicago: Regnery, 1970.

Lamb, Michael. "Book Review." *Journal of Marriage and the Family* 55, 4 (November 1993): 1047–1049.

Lamb, Michael, Kathleen Sternberg, and Ross Thompson. "The Effects of Divorce and Custody Arrangements on Children's Behavior, Development, and Adjustment." In *Parenting and Child Development in "Nontraditional" Families,* edited by M. Lamb. Mahwah, NJ: Lawrence Erlbaum, 1999.

Lamb, Michael E. "Nonparental Childcare." *Parenting and Child Development in "Nontraditional" Families,* edited by M. Lamb. Mahwah, NJ: Lawrence Erlbaum, 1999.

Landau, Rivka. "Affect and Attachment: Kissing, Hugging, and Patting as Attachment Behaviors." *Infant Mental Health Journal* 10, 1 (March 1989): 59–69.

Laner, Mary R. "Violence or Its Precipitators: Which Is More Likely to Be Identified as a Dating Problem." *Deviant Behavior* 11, 4 (October 1990): 319–329.

Lantz, Herman. "Family and Kin as Revealed in the Narratives of Ex-Slaves." *Social Science Quarterly* 60, 4 (March 1980): 667–674.

LaRossa, Ralph. "Fatherhood and Social Change." *Family Relations* 37 (1988): 451–458.

LaRossa, Ralph, and Maureen Mulligan LaRossa. *The Transition to Parenthood: How Infants Change Families.* Beverly Hills, CA: Sage, 1981.

Larsen, Andrea S., and David H. Olson. "Predicting Marital Satisfaction Using PREPARE: A Replication Study." *Journal of Marital and Family Therapy* 15, 3 (1989): 311–322.

Larson, Jeffry H., Stephan M. Wilson, and Rochelle Beley. "The Impact of Job Insecurity on Marital and Family Relationships." *Family Relations* 43, 2 (April 1994): 138–143.

Larson, Mary S. "Interaction between Siblings in Primetime Television Families." *Journal of Broadcasting and Electronic Media* 33, 3 (1989): 305–315.

Larson, Reed, Robert W. Kubey, and Joseph Colletti. "Changing Channels: Early Adolescent Media Choices and Shifting Investments in Family and Friends" [Special Issue: "The Changing Life Space of Early Adolescence"]. *Journal of Youth and Adolescence* 18, 6 (1989): 583–599.

Lasch, Christopher. *Haven in a Heartless World.* New York: Basic Books, 1977.

Lauer, Jeanette, and Robert Lauer. *'Til Death Do Us Part: How Couples Stay Together.* New York: Haworth Press, 1986.

Lauer, Robert, and Jeanette Lauer. "The Long-Term Relational Consequences of Problematic Family Backgrounds." *Family Relations* 40, 3 (July 1991): 286–291.

———. *Marriage and Family: The Quest for Intimacy,* 4th ed. Boston: McGraw-Hill, 2000.

Laurent, S. L., S. J. Thompson, C. Addy, C. Z. Garrison, and E. E. Moore. "An Epidemiologic Study of Smoking and Primary Infertility in Women." *Fertility and Sterility* 57, 3 (March 1992): 565–572.

Lavee, Yoav, and David Olson. "Family Types and Response to Stress." *Journal of Marriage and the Family* 53, 3 (August 1991): 786–798.

Laviola, Marisa. "Effects of Older-Brother Younger-Sister Incest: A Review of Four Cases." *Journal of Family Violence* 4, 3 (September 1989): 259–274.

Lawson, Erma Jean, and Aaron Thompson. "Black Men's Perceptions of Divorce-Related Stressors and Strategies for Coping with Divorce." *Journal of Family Relations* 17, 2 (March 1996): 249–273.

Leboyer, Frederick. *Birth Without Violence.* New York: Knopf, 1975.

Lederer, William, and Don Jackson. *Mirages of Marriage.* New York: Norton, 1968.

Lee, John A. *The Color of Love.* Toronto: New Press, 1973.

———. "Love Styles." In *The Psychology of Love,* edited by R. Sternberg and M. Barnes. New Haven, CT: Yale University Press, 1988.

Lee, Thomas, Jay Mancini, and Joseph Maxwell. "Contact Patterns and Motivations for Sibling Relations in Adulthood." *Journal of Marriage and the Family* 52, 2 (May 1990): 431–440.

Legislative Commission on the Economic Status of Women, "Birth Rates by Marital Status and Family Income (1995)." www.commissions.leg.state.

Lehrer, E., and C. Chiswick. "The Religious Composition of Unions." *Demography* 30 (1993): 385–404.

Leiblum, Sandra R. "Sexuality and the Midlife Woman" [Special Issue: "Women at Midlife and Beyond"]. *Psychology of Women Quarterly* 14, 4 (December 1990): 495–508.

Leifer, Myra. *Psychological Effects of Motherhood: A Study of First Pregnancy.* New York: Praeger, 1990.

Leigh, Barbara C. "Alcohol Expectancies and Reasons for Drinking: Comments from a Study of Sexuality." *Psychology of Addictive Behaviors* 4, 2 (1990): 91–96.

Leitenberg, H., M. J. Detzer, and D. Srebnik. "Gender Differences in Masturbation and the Relation of Masturbation Experience in Preadolescence and or Early Adolescence to Sexual Behavior and Sexual Adjustment in Young Adulthood." *Archives of Sexual Behavior* 22, 2 (April 1993): 87–98.

Lemkau, Jeanne Parr. "Emotional Sequelae of Abortion: Implications for Clinical Practice" [Special Issue: "Women's Health: Our Minds, Our Bodies." *Psychology of Women Quarterly* 12 (December 1988): 461–472.

Leonard, Kenneth E., and Theodore Jacob. "Alcohol, Alcoholism, and Family Violence." In *Handbook of Family Violence,* edited by V. B. Van Hasselt et al. New York: Plenum Press, 1988.

Leslie, Leigh, and Katherine Grady. "Changes in Mothers' Social Networks and Social Support Following Divorce." *Journal of Marriage and the Family* 47, 3 (August 1985): 663–673.

Levenson, Robert W., Laura L. Carstensen, and John M. Gottman. "Long-Term Marriage: Age, Gender, and Satisfaction." *Psychology and Aging* 8, 2 (1993): 301–313.

Levin, Irene. "The Model Monopoly of the Nuclear Family." Paper presented at the National Conference on Family Relations, Baltimore, November 1993.

Levin, Nora J. *How to Care for Your Parents.* Friday Harbor, WA: Storm King Press, 1987.

Levine, James. *Working Fathers: Strategies for Balancing Work and Family.* Reading, MA: Addison Wesley Longman, 1997.

Levine, Linda, and Lonnie Barbach. *The Intimate Male.* New York: Signet Books, 1983.

Levine, Martin. "The Life and Death of Gay Clones." In *Gay Culture in America: Essays from the Field,* edited by G. Herdt. Boston: Beacon, 1992.

Levinger, George. "Marital Cohesiveness and Dissolution: An Integrative Review." *Journal of Marriage and the Family* 27, 1 (February 1965): 19–28.

———. "A Social Psychological Perspective on Marital Dissolution." In *Divorce and Separation,* edited by G. Levinger and O. C. Moles. New York: Basic Books, 1979.

Levinger, George, and O. C. Moles, eds. *Divorce and Separation: Context, Causes, and Consequences.* New York: Basic Books, 1979.

Levinson, Daniel J. *The Seasons of a Man's Life.* New York: Ballantine, 1977.

Levy, Barrie, ed. *Dating Violence: Young Women in Danger.* Seattle: Seal Press, 1991.

Levy, Gary D. "High and Low Gender Schematic Children's Release from Proactive Inference." *Sex Roles* 30, 1–2 (January 1994): 93–108.

Levy-Shiff, R. "Individual and Contextual Correlates of Marital Change Across the Transition to Parenthood." *Developmental Psychology,* 30, 4 (1994): 591–601.

Levy-Shiff, Rachel, Ilana Goldshmidt, and Dov Har-Even. "Transition to Parenthood in Adoptive Families." *Developmental Psychology* 27, 1 (1991): 131–140.

Lewes, K. "Homophobia and the Heterosexual Fear of AIDS." *American Image* 49, 3 (September 1992): 343–356.

Lewin, Bo. "Unmarried Cohabitation: A Marriage Form in a Changing Society." *Journal of Marriage and the Family* 44, 3 (August 1982): 763–773.

Lewin, M. "Unwanted Intercourse: The Difficulty of Saying No." *Psychology of Women Quarterly* 9 (1985): 184–192.

———. "Drug Use During Pregnancy: New Issue before the Courts." *New York Times* (February 5, 1990): A1, 12.

Lewis, Jerry M., ed. *The Birth of the Family: An Empirical Inquiry.* New York: Brunner/Mazel, 1989.

Lewis, Karen. "Children of Lesbians: Their Point of View." *Social Work* 25, 3 (May 1980): 198–203.

Lewis, Lisa. "Consumer Girl Culture: How Music Video Appeals to Girls." In *Television and Women's Culture: The Politics of the Popular,* edited by M. E. Brown. Newbury Park, CA: Sage, 1992.

Lewis, Oscar. "The Culture of Poverty." *Scientific American,* 215, 4 (1966): 19–25.

Lewis, Robert, and Joseph Pleck, eds. "Men's Roles in the Family." *Family Coordinator* 28 (October 1979).

Lewis, Ronald, and Wallace Gingerich. "Leadership Characteristics: Views of Indian and Non-Indian Students." *Social Casework* 61, 10 (1980).

Libman, E. "Sociocultural and Cognitive Factors in Aging and Sexual Expression: Conceptual and Research Issues." *Canadian Psychology* 30, 3 (July 1989): 560–567.

Lieberman, B. "Extrapremarital Intercourse: Attitudes toward a Neglected Sexual Behavior." *Journal of Sex Research* 24 (1988): 291–299.

Lieberson, Stanley, and Mary Waters. *From Many Strands: Ethnic and Racial Groups in Contemporary America.* New York: Russell Sage Foundation, 1988.

Liem, Ramsay, and J. Liem, "Social Support and Stress: Some General Issues and Their Application to the Problem of Unemployment." In *Mental Health and the Economy,* edited by L. Ferman and J. Gordus. Kalamazoo, MI: Upjohn Institute, 1979.

Lin, Chien, and William T. Liu. "Intergenerational Relationships among Chinese Immigrant Families from Taiwan." In *Family Ethnicity: Strength in Diversity,* edited by H. Pipes McAdoo. Newbury Park, CA: Sage, 1993, 109–119.

Lindsey, Karen. *Friends as Family.* Boston: Beacon Press, 1982.

Lindsey, Linda. *Gender Roles: A Sociological Perspective,* 3rd ed. Upper Saddle River, NJ: Prentice Hall, 1997.

Lingren, Herbert, et al. "Enhancing Marriage and Family Competencies Through Adult Life Development." In *Family Strengths 4: Positive Support Systems,* edited by N. Stinnet et al. Lincoln: University of Nebraska Press, 1982.

Lino, Mark. "Expenditures on a Child by Husband-Wife Families." *Family Economics Review* 3, 3 (1990): 2–12.

———. *Expenditures on Children by Families: 2002.* Washington, DC: U.S. Department of Agriculture, Center for Nutrition Policy and Promotion, 2003.

Lips, Hilary. "Expenditures on a Child by Two-Parent Families." *Family Economics Review* 4, 1 (1991): 2–38.

———. *Sex and Gender.* 3rd ed. Mountain View, CA: Mayfield, 1997.

Little, Margaret A. "The Impact of the Custody Plan on the Family: A Five-Year Follow-up: Executive Summary." *Family and Conciliation Courts Review* 30, 2 (1992): 243–251.

Litwack, Leon. *Been in the Storm So Long.* New York: Alfred A. Knopf, 1979.

Liu, W. T. "Family Interactions Among Local and Refugee Chinese Families in Hong King." *Journal of Marriage and the Family,* 28 (1966): 314–323.

Livernois, Joe. "County Gears Up for Huge Welfare Shift." *Monterey County Herald* (June 1, 1997): A-1, A-8.

Lloyd, Sally A. "Conflict Types and Strategies in Violent Marriages." *Journal of Family Violence* 5, 4 (December 1990): 269–284.

———. "The Darkside of Courtship: Violence and Sexual Exploitation." *Family Relations* 40, 1 (January 1991): 14–20.

———. "Physical and Sexual Violence During Dating and Courtship." In *Vision 2010: Families and Violence, Abuse and Neglect,* edited by R. J. Gelles. Minneapolis: National Council on Family Relations, 1995.

Lloyd, Sally A., and Beth C. Emery. "The Dynamics of Courtship Violence." Paper presented at the annual meeting of the National Council on Family Relations, Seattle, WA, November 1990.

Locke, D. *Increasing Multicultural Understanding.* Newbury Park, CA: Sage, 1992.

Lockhart, Lettie. "A Reexamination of the Effects of Race and Social Class on the Incidence of Marital Violence: A Search for Reliable Differences." *Journal of Marriage and the Family* 49, 3 (August 1987): 603–610.

Lombardo, John P., and T. R. Kemper. "Sex Role and Parental Behaviors." *Journal of Genetic Psychology* 153, 1 (1992): 103–114.

Lombardo, W. K., et al. "Fer Cryin Out Loud—There Is a Sex Difference." *Sex Roles* 9 (1983): 987–995.

London, Richard, James Wakefield, and Richard Lewak. "Similarity of Personality Variables as Predictors of Marital Satisfaction." *Personality and Individual Differences* 11, 1 (1990): 39–43.

Long, Vonda O., and Estella A. Martinez. "Masculinity, Femininity, and Hispanic Professional Women's Self-Esteem and Self-Acceptance." *Journal of Counseling & Development* 73, 3 (November/December 1994): 183–186.

Longman, L. "Social Stratification." In *Handbook of Marriage and the Family,* edited by M. Sussman and S. Steinmetz. New York: Plenum, 1987.

Longman, Phillip. "The Cost of Children." *U.S. News and World Report* (March 30): 51–58.

LoPresto, C., M. Sherman, and N. Sherman. "The Effects of a Masturbation Seminar on High School Males' Attitudes, False Beliefs, Guilt, and Behavior." *Journal of Sex Research* 21 (1985): 142–156.

Lorber, Judith. *Paradoxes of Gender.* New Haven, CT: Yale University Press, 1994.

———. *Gender Inequality: Feminist Theories and Politics.* Los Angeles: Roxbury, 1998.

Lord, Lewis. "Desperately Seeking Baby." *U.S. News and World Report* 5 (October 1987): 58–65.

Losh-Hesselbart, S. "Development of Gender Roles." In *Handbook of Marriage and the Family,* edited by M.B. Sussman and S. Steinmetz. New York: Plenum, 1987: 535– 563.

Lott, B. *Women's Lives: Themes and Variations in Gender Learning.* 3rd ed. Pacific Grove, CA: Brooks/Cole, 1994.

Loulan, JoAnn. *Lesbian Sex.* San Francisco: Spinsters Books, 1984.

Love, Alice. "Poll Finds Many Women are Working Different Shifts from Spouses," Associated Press, March 9, 2000.

Lowry, Dennis, and David Towles. "Prime Time TV Portrayals of Sex, Contraception, and Venereal Disease." *Journalism Quarterly* 66, 2 (June 1989a): 347–352.

———. "Soap Opera Portrayals of Sex, Contraception, and Venereal Disease." *Journal of Communication* 39, 2 (March 1989b): 76–83.

Luker, Kristin. *Taking Chances.* Berkeley: University of California Press, 1975.

———. *Abortion and the Politics of Motherhood.* Berkeley: University of California Press, 1984.

Lunneborg, Patricia. *Abortion: The Positive Decision.* New York: Bergin & Garvy, 1992.

Lutz, Patricia. "The Stepfamily: An Adolescent Perspective." *Family Relations* 32, 3 (July 1983): 367–375.
Lyon, Jeff. *Playing God in the Nursery.* New York: Norton, 1985.
Macciarola, Frank J., and Alan Gartner, eds. *Caring for Americas Children.* New York: The Academy of Political Science, 1989.
Maccoby, Eleanor, and Carol Jacklin. *The Psychology of Sex Differences.* Stanford, CA: Stanford University Press, 1974.
Maccoby, Eleanor E., Christy M. Buchanan, Robert H. Mnookin, and Sanford M. Dornbusch. "Postdivorce Roles of Mothers and Fathers in the Lives of Their Children." *Journal of Family Psychology* 7, 1 (1993): 24–38.
MacCorquodale, Patricia. "Gender and Sexuality." In *Human Sexuality: The Societal and Interpersonal Context,* edited by K. McKinney and S. Sprecher. Norwood, NJ: Ablex, 1989.
Macdonald, Patrick T., Dan Waldorf, Craig Reinarman, and Sheigla Murphy. "Heavy Cocaine Use and Sexual Behavior." *Journal of Drug Issues* 18, 3 (June 1988): 437–455.
MacDonald, W. L., and A. DeMaris. "Remarriage, Stepchildren, and Marital Conflict. Challenges to the Incomplete Institutionalization Hypothesis. *Journal of Marriage and the Family* 57, 2 (May 1995): 387–398.
Mace, David, and Vera Mace. "Enriching Marriage." In *Family Strengths,* edited by N. Stinnet et al. Lincoln: University of Nebraska Press, 1979.
———. "Enriching Marriages: The Foundation Stone of Family Strength." In *Family Strengths: Positive Models for Family Life,* edited by N. Stinnet et al. Lincoln: University of Nebraska Press, 1980.
MacFarlane, Robin. "Summary of Adolescent Pregnancy Research: Implications for Prevention." In *The Prevention Researcher.* Eugene, OR: Integrated Research Services, 1997.
Macklin, Eleanor.. "Nonmarital Heterosexual Cohabitation." *Marriage and Family Review* 1 (March 1978): 1–12.
———. "Nontraditional Family Forms." In *Handbook of Marriage and the Family,* edited by M. Sussman and S. Steinmetz. New York: Plenum Press, 1987.
———. "AIDS: Implications for Families." *Family Relations* 37, 2 (April 1988): 141–149.
Macklin, Eleanor, ed. *AIDS and Families.* New York: Harrington Park, 1989.
Macmillan, Ross, and Rosemary Garner. "When She Brings Home the Bacon: Labor Force Participation and the Risk of Spousal Violence Against Women." *Journal of Marriage and the Family,* 61, 4 (November 1999): 947–958.
Madsen, William, ed. *Mexican-American Youth of South Texas.* 2nd ed. New York: Holt, Rinehart, and Winston, 1973.
Magno, Josefina. "The Hospice Concept of Care: Facing the 1990s." *Death Studies* 14, 2 (1990): 109–119.
Major, Brenda, and Catherine Cozzarelli. "Psychosocial Predictors of Adjustment to Abortion." *Journal of Social Issues* 48, 3 (1992): 121–142.
Makepeace, James. "Courtship Violence Among College Students." *Family Relations* 30, 1 (January 1981): 97–102.
———. "Gender Differences in Courtship Violence Victimization." *Family Relations* 35, 3 (July 1986): 383–388.
———. "Dating, Living Together, and Courtship Violence." In *Violence in Dating Relationships: Emerging Social Issues,* edited by M. Pirog-Good and J. Stets. New York: Praeger, 1989.
Malatesta, Victor, Dianne Chambless, Martha Pollack, and Alan Cantor. "Widowhood, Sexuality, and Aging: A Life Span Analysis." *Journal of Sex and Marital Therapy* 14, 1 (March 1989): 49–62.
Maloney, Lawrence. "Behind Rise in Mixed Marriages." *U.S. News and World Report* (February 10, 1986): 68–69.
Malson, Lucien. *Wolf Children and the Problem of Human Nature.* New York: Monthly Review Press, 1972.
Malveaux, Julianne. "The Economic Status of Black Families." In *Black Families,* 2nd ed., edited by H. Pipes McAdoo. Newbury Park, CA: Sage, 1988.
Mancini, Jay, and Rosemary Bliezner. "Research on Aging Parents and Adult Children." *Journal of Marriage and the Family* 51, 2 (May 1989): 275–290.
———. "Aging Parents and Adult Children: Research Themes in Intergenerational Relations." In *Contemporary Families: Looking Forward, Looking Back,* edited by A. Booth. Minneapolis: National Council on Family Relations, 1991.
———. "Social Provisions in Adulthood: Concept and Measurement in Close Relationships." *Journal of Gerontology* 47, 1 (1992): 14–20.
Marecek, Jeanne, et al. "Gender Roles in the Relationships of Lesbians and Gay Men." In *Gay Relationships,* edited by J. DeCecco. New York: Haworth Press, 1988.
Marek, Lynne. "U.S. Vows War on Domestic Violence." *Chicago Tribune* (March 12, 1994).
Margolin, Gayla, Linda Gorin Sibner, and Lisa Gleberman. "Wife Battering." In *Handbook of Family Violence,* edited by V. B. Van Hasselt et al. New York: Plenum Press, 1988.
Margolin, Leslie. "Sexual Abuse by Grandparents." *Child Abuse and Neglect* 16, 5 (September 1992): 735.
Margolin, Leslie, and Lynn White. "The Continuing Role of Physical Attractiveness in Marriage." *Journal of Marriage and the Family* 49, 1 (February 1987): 21–27.
Margolin, Malcolm. *The Ohlone Way.* Berkeley, CA: Heydey Books, 1978.
Marin, Peter. "Helping and Hating the Homeless." *Harper's* (January 1987): 39–49.
Marin, Rick. "At-Home Fathers Step Out to Find They Are Not Alone." *New York Times* (January 2, 2000): 1, 16.
Marker, Nadine F. "Flying Solo at Midlife: Gender, Marital Status, and Psychological Well-Being." *Journal of Marriage and the Family* 58 (November 1996): 917–932.
Markman, Howard. "Application of a Behavioral Model of Marriage in Predicting Relationship Satisfaction of Couples Planning Marriage." *Journal of Consulting and Clinical Psychology* 47 (1979): 743–749.
———. "Prediction of Marital Distress: A Five-Year Followup." *Journal of Consulting and Clinical Psychology* 49 (1981): 760–761.
———. "The Longitudinal Study of Couples' Interactions: Implications for Understanding and Predicting the

Development of Marital Distress." In *Marital Interaction,* edited by K. Hahlweg and N. S. Jacobsen. New York: Guilford Press, 1984.

Markman, Howard, et al. "The Prediction and Prevention of Marital Distress: A Longitudinal Investigation." In *Understanding Major Mental Disorders: The Contribution of Family Interaction Research,* edited by K. Hahlweg and M. Goldstein. New York: Family Process Press, 1987.

Marks, Nadine, James Lambert, and Heejeong Choi. "Transitions to Caregiving, Gender and Psychological Well-Being: A Prospective U.S. National Study," *Journal of Marriage and the Family* 64, 3, (August 2002): 657–667.

Marks, Stephen. "What Is a Pattern of Commitment?" *Journal of Marriage and the Family* 52, 1 (February 1994): 112–115.

Markstrom-Adams, C. "Coming of Age Among Contemporary American Indians as Portrayed in Adolescent Fiction." *Adolescence* 25 (1990): 225–237.

Marmor, Judd. "Homosexuality and the Issue of Mental Illness." In *Homosexual Behavior,* edited by J. Marmor. New York: Basic Books, 1980a.

———. "The Multiple Roots of Homosexual Behavior." In *Homosexual Behavior,* edited by J. Marmor. New York: Basic Books, 1980b.

———, ed. *Homosexual Behavior.* New York: Basic Books, 1980c.

Marrero, M. A., and S. J. Ory. "Unexplained Infertility." *Current Opinion in Obstetrics and Gynecology* 3, 2 (April 1991): 211–218.

"Married without Children." *American Demographics* 14, 7 (July 1992): A10.

Marsiglio, William. "Male Procreative Consciousness an Responsibility: A Conceptual Analysis and Research Agenda." *Journal of Family Issues* 12, 3 (September 1992): 268–290.

———. *Procreative Man.* New York: New York University, 1998.

Marsiglio, William, Paul Amato, Randall Day, and Michael Lamb. "Scholarship on Fatherhood in the 1990's and Beyond." *Journal of Marriage and the Family,* 62, 4 (November 2000):1173-1191.

Marsiglio, William, and Denise Donnelly. "Sexual Relations in Later Life: A National Study of Married Persons." *Journal of Gerontology* 46, 6 (November 1991): S338–S344.

Martelli, Leonard, et al. *When Someone You Know Has AIDS: A Book of Hope for Family and Friends.* New York: Crown, 1987.

Martin, Del. *Battered Wives.* San Francisco: New Glide, 1981.

Martin, Douglas. "Many Dads Struggle to Fit New Roles." *New York Times* (June 20, 1993): 11.

Martin, Peter, et al. "Family Stories: Events (Temporarily) Remembered." *Journal of Marriage and the Family* 50, 2 (May 1988): 533–541.

Martin, T. C., and L. L. Bumpass. "Recent Trends in Marital Disruption." *Demography* 26 (February 1989): 37–51.

Maruta, Toshihiko, and Mary Jane McHardy. "Sexual Problems in Patients with Chronic Pain." *Medical Aspects of Human Sexuality* 17, 2 (February 1983): 68J–68U ff.

Marwell, G., et al. "Legitimizing Factors in the Initiation of Heterosexual Relationships." Paper presented at the First International Conference on Personal Relationships, Madison, WI, July 1982.

Masheter, Carol. "Postdivorce Relationships Between Ex-Spouses: The Roles of Attachment and Interpersonal Conflict." *Journal of Marriage and the Family* 53 (1) April 1991: 103–110.

Mason, Karen, and Yu Hsia Lu. "Attitudes toward Women's Familial Roles: Changes in the United States, 1977–1985." *Gender and Society* 2, 1 (March 1988): 39–57.

Masse, Michelle, and Karen Rosenblum. "Male and Female Created They Them: The Depiction of Gender in the Advertising of Traditional Women's and Men's Magazines." *Women's Studies International Forum* 11, 2 (1988): 127–144.

Masters, John, et al. "The Role of the Family in Coping with Childhood Chronic Illness." In *Coping with Chronic Disease,* edited by T. Burish and L. Bradley. New York: Academic Press, 1983.

Masters, William, and Virginia Johnson. *Human Sexual Inadequacy.* Boston: Little, Brown, 1970.

Masters, William, et al. *Masters and Johnson on Sex and Human Loving.* Boston: Little, Brown, 1986.

Mathis, Richard D., and Zoe Tanner. "Cohesion, Adaptability, and Satisfaction of Family Systems in Later Life." *Family Therapy* 18, 1 (1991): 47–60.

Maticka-Tyndale, Eleanor. "Sexual Scripts and AIDS Prevention: Variations in Adherence to Safer-Sex Guidelines." *Journal of Sex Research* 28, 1 (February 1991): 145–166.

Matiella, Ana Consuelo. *Positively Different: Creating a Bias-Free Environment for Children.* Santa Cruz, CA: Network Publications, 1991.

Mattessich, Paul, and Reuben Hill. "Life Cycle and Family Development." In *Handbook of Marriage and the Family,* edited by M. Sussman and S. Steinmetz. New York: Plenum Press, 1987.

Matthews, Sarah, and Tana Rosner. "Shared Filial Responsibility: The Family as the Primary Caregiver." *Journal of Marriage and the Family* 50, 1 (February 1988): 185–195.

Mays, Vickie M., S. D. Cochran, G. Bellinger, and R. G. Smith. "The Language of Black Gay Men's Sexual Behavior: Implications for AIDS Risk Reduction." *Journal of Sex Research* 29, 3 (August 1992): 425–434.

Mazor, Miriam. "Barren Couples." *Psychology Today* (May 1979): 101–108, 112.

Mazor, Miriam, and Harriet Simons, eds. *Infertility: Medical, Emotional and Social Considerations.* New York: Human Sciences Press, 1984.

McAdoo, Harriette Pipes. "Changes in the Formation and Structure of Black Families: The Impact on Black Women." Working paper no. 182, Center for Research on Women, Wellesley College, Wellesley, MA, 1988.

———. "Ethnic Family Strengths That Are Found in Diversity." In *Family Ethnicity: Strength in Diversity,* edited by H. Pipes McAdoo. Newbury Park, CA: Sage, 1993.

———, ed. *Black Families,* 2nd ed. Beverly Hills, CA: Sage, 1988.

———. *Family Ethnicity: Strength in Diversity.* Newbury Park, CA: Sage, 1993.

———. *Black Families,* 3rd ed. Thousand Oaks, CA: Sage, 1996.

McAdoo, Harriette Pipes, and John McAdoo, eds. *Black Children: Social, Educational, and Parental Environments.* Beverly Hills, CA: Sage, 1985.

McAdoo, John Lewis. "Decision Making and Marital Satisfaction in African American Families." In *Family Ethnicity: Strength in Diversity,* edited by H. Pipes McAdoo. Newbury Park, CA: Sage, 1993.

McCaghy, Charles, Timothy Capron, and J. D. Jamieson. *Deviant Behavior: Crime, Conflict, and Interest Groups,* 5th ed. Needham Heights, MA: Allyn and Bacon, 2000.

McClary, Susan. "Living to Tell: Madonna's Resurrection of the Flesh." *Genders* 7 (March 1990): 1–21.

McCormick, John. "Why Parents Kill." *Newsweek* (November 14, 1994): 31–35.

McCormack, M. J., et al. "Patient's Attitudes Following Chorionic Villus Sampling." *Prenatal Diagnosis* 10, 4 (April 1990): 253–255.

McCubbin, Hamilton I., Constance Joy, and Elizabeth Cauble. "Family Stress and Coping: A Decade Review." *Journal of Marriage and the Family* 42, 4 (November 1980): 855–871.

McCubbin, Hamilton I., and Marilyn A. McCubbin. "Typologies of Resilient Families: Emerging Roles of Social Class and Ethnicity." *Family Relations* 37, 3 (July 1988): 247–254.

McCubbin, Marilyn A. "Family Stress, Resources, and Family Types: Chronic Illness in Children." *Family Relations* 37, 2 (April 1988): 203–210.

McEwan, K. L., C. G. Costello, and P. J. Taylor. "Adjustment to Infertility." *Journal of Abnormal Psychology* 96, 2 (May 1987): 108–116.

McGill, Michael. *The McGill Report on Male Intimacy.* New York: Henry Holt, 1985.

McGoldrick, Monica. "Normal Families: An Ethnic Perspective." In *Normal Family Processes,* edited by F. Walsh. New York: Guilford Press, 1982.

———. "Ethnicity and the Family Life Cycle." In *The Changing Family Life Cycle,* 2nd ed., edited by B. Carter and M. McGoldrick. Boston: Allyn and Bacon, 1989.

McGoldrick, Monica, and Randy Gerson. *Genograms in Family Assessment.* New York: Norton, 1985.

McGoldrick, Monica, J. K. Pearce, and J. Giordano, eds. *Ethnicity and Family Therapy.* New York: Guilford Press, 1982.

McGraw, J. Melbourne, and Holly A. Smith. "Child Sexual Abuse Allegations amidst Divorce and Custody Proceedings: Refining the Validation Process." *Journal of Child Sexual Abuse* 1, 1 (1992): 49–62.

McHale, Susan, and Ted Huston. "The Effect of the Transition to Parenthood on the Marriage and Family Relationship: A Longitudinal Study." *Journal of Family Issues* 6, 4 (December 1985): 409–433.

McIntosh, Everton. "An Investigation of Romantic Jealousy among Black Undergraduates." *Social Behavior and Personality* 17, 2 (1989): 135–141.

McIntosh, Everton, and Douglas T. Tate. "Correlates of Jealous Behaviors." *Psychological Reports* 66, 2 (April 1990): 601–602.

McIntosh, Everton G., and Calvin O. Matthews. "Use of Direct Coping Resources in Dealing with Jealousy." *Psychological Reports* 70, 3 pt. 2 (1992): 1037–1038.

McKim, Margaret K. "Transition to What? New Parents' Problems in the First Year." *Family Relations* 36, 1 (January 1987): 22–25.

McKinney, Kathleen, and Susan Sprecher. *Sexuality in Close Relationships.* Hillsdale, NJ: Erlbaum, 1991.

———, eds. *Human Sexuality: The Societal and Interpersonal Context.* Norwood, NJ: Ablex, 1989.

McLanahan. S. S., and G. Sandefur. *Growing Up with a Single Parent: What Hurts, What Helps.* Cambridge, MA: Harvard University Press, 1994.

McLanahan, Sara, and Karen Booth. "Mother-Only Families: Problems, Prospects, and Politics." In *Contemporary Families: Looking Forward, Looking Back,* edited by A. Booth. Minneapolis: National Council on Family Relations, 1991.

McLanahan, Sara, and Karen Booth. "Mother-Only Families: Problems, Prospects, and Politics." *Journal of Marriage and the Family* (51) 1989: 557–580.

McLanahan, Sara, et al. "Network Structure, Social Support, and Psychological Well-Being." *Journal of Marriage and the Family* 43, 3 (August 1981): 601–612.

McLeer, S. V., et al. "Sexually Abused Children at High Risk for Post-Traumatic Stress Disorder." *Journal of the American Academy of Child and Adolescent Psychiatry* 31, 5 (September 1992): 875–879.

McLeod, B. "The Oriental Express." *Psychology Today* 20 (July 1986): 48–52.

McLoyd, Vonnie, Ana Marie Cauce, David Takeuchi, and Leon Wilson. "Marital Processes and Parental Socialization in Families of Color: A Decade Review of Research." *Journal of Marriage and the Family* 62, 4 (November 2000): 1070–1093.

McLoyd, Vonnie, and Julia Smith. "Physical Discipline and Behavior Problems in African American, European American, and Hispanic Children: Emotional Support as a Moderator" *Journal of Marriage and the Family* 64, 1, (February 2002): 40–53.

McMahon, Kathryn. "The Cosmopolitan Ideology and the Management of Desire." *Journal of Sex Research* 27, 3 (August 1990): 381–396.

McNeely, R. L., and John N. Colen, eds. *Aging in Minority Groups.* Beverly Hills, CA: Sage , 1983.

McNeely, R. L., and Barbe Fogarty. "Balancing Parenthood and Employment Factors Affecting Company Receptiveness to Family-Related Innovations in the Workplace." *Family Relations* 37, 2 (April 1988): 189–195.

McRoy, R. G., H. D. Grotevant, S. Ayers-Lopez. *Changing Patterns in Adoption.* Austin, TX: Hogg Foundation for Mental Health, 1994.

Meacham, R. E., and L. I. Lipshultz. "Assisted Reproductive Technologies for Male Factor Infertility." *Current Opinion in Obstetrics and Gynecology* 3, 5 (October 1991): 656–661.

Mead, Barbara J., and Arlene A. Ignicio. "Children's Gender-Typed Perceptions of Physical Activity: Consequences and Implications." *Perceptual and Motor Skills* 75, 3 (1992): 1035–1042.

Mead, Margaret. *Male and Female.* New York: Morrow, 1975.

Mederer, Helen J. "Division of Labor in Two-Earner Homes: Task Accomplishment versus Household Management as Critical Variables in Perceptions about Family Work." *Journal of Marriage and the Family* 55, 1 (February 1993): 133–145.

Melichor, Joseph, and David Chiriboga. "Significance of Time in Adjustment to Marital Separation." *American Journal of Orthopsychiatry* 58, 2 (April 1988): 221–227.

Melito, Richard. "Adaptation in Family Systems: A Developmental Perspective." *Family Processes* 24 (1985): 89–100.

Melson, Gail F., Susan Peets, and Cheryl Sparks. "Children's Attachment to Their Pets: Links to Socio-Emotional Development." *Children's Environments Quarterly* 8, 2 (1991): 55–65.

Menaghan, Elizabeth, and Toby Parcel. "Parental Employment and Family Life Research in the 1980s." In *Contemporary Families: Looking Forward, Looking Back,* edited by A. Booth. Minneapolis: National Council on Family Relations, 1991.

Menning, Barbara Eck. *Infertility: A Guide for Childless Couples,* 2nd ed. New York: Prentice Hall, 1988.

Merritt, Bishetta, and Carolyn A. Stroman. "Black Family Imagery and Interactions on Television." *Journal of Black Studies* 23, 4 (June 1993): 492–499.

Messerschmidt, James. *Masculinities and Crime.* Nanham, MD: Rowman and Littlefield, 1993.

Metts, Sandra, and William Cupach. "The Role of Communication in Human Sexuality." In *Human Sexuality: The Social and Interpersonal Context,* edited by K. McKinney and S. Sprecher. Norwood, NJ: Ablex, 1989.

Meyer, Daniel R., and Judi Bartfeld. "Compliance with Child Support Orders in Divorce Cases." *Journal of Marriage and the Family* 58, 1 (February 1996): 201–212.

Meyer, Daniel R., Elizabeth Phillips, and Nancy L. Maritato. "The Effects of Replacing Income Tax Deductions with Children's Allowance." *Journal of Family Issues* 12, 4 (December 1991): 467–491.

Meyers, Marcia K. "Child Care in JOBS Employment and Training Program: What Difference Does Quality Make?" *Journal of Marriage and the Family* 55, 3 (August 1993): 767–783.

Meyrowitz, J. *No Sense of Place: The Impact of Electronic Media on Social Behavior.* New York: Oxford University Press, 1985.

Miall, Cherlene E. "The Stigma of Adoptive Parent Status: Perceptions of Community Attitudes Toward Adoption and the Experience of Informal Self-Sanctioning." *Family Relations* 36, 1 (January 1987): 34–39.

Michael, Robert, John Gagnon, Edward Laumann, and Gina Kolata. *Sex in America: The Definitive Survey.* Boston: Little, Brown, 1994.

Michaels, D., and C. Levine. "Estimates of the Number of Motherless Youth Orphaned by AIDS in the United States." *JAMA* 268, 40 (December 23–30, 1992): 3456–3461.

Milan, Richard Jr., and Peter Kilmann. "Interpersonal Factors in Premarital Contraception." *Journal of Sex Research* 23, 3 (August 1987): 321–389.

Milardo, Robert. "Changes in Social Networks of Women and Men Following Divorce: A Review." *Journal of Family Issues* 8 (March 1987): 78–96.

Miller, Brent. *Family Research Methods.* Beverly Hills, CA: Sage, 1986.

Miller, Brent C., and G. L. Fox. "Theories of Adolescent Heterosexual Behavior." *Adolescent Research,* 2 (1987): 269–282.

Miller, Glenn. "The Psychological Best Interests of the Child." *Journal of Divorce and Remarriage* 19, 1–2 (1993): 21–36.

Miller, Jean B. "Psychological Recovery in Low-Income Single Parents." *American Journal of Orthopsychiatry* 52, 2 (April 1982): 346–352.

Miller, Randi, and Michael Gordon. "The Decline of Formal Dating: A Study in Six Connecticut High Schools." *Marriage and Family Relationships* 10, 1 (April 1986): 139–154.

Miller, Ron. "Black and White Television." *San Jose Mercury News* (June 15, 1992): D1, D5.

Miller, Susan. "Viewer Discretion." *Newsweek* (December 23, 1996): 60.

Mills, David. "A Model for Stepparent Development." *Family Relations* 33 (1984): 365–372.

Min, Pyong Gap. "The Korean American Family." In *Ethnic Families in America: Patterns and Variations,* edited by C. Mindel, R. Habenstein, and R. Wright. 3rd ed. New York: Elsevier-North Holland, 1988.

Mindel, Charles H., Robert W. Habenstein, and Roosevelt Wright Jr., eds. *Ethnic Families in America: Patterns and Variations,* 3rd. ed. New York: Elsevier North Holland, 1976, 1981, 1988.

Minkler, Meredith, and Kathleen M. Roe. *Grandmothers as Caregivers: Raising Children of the Crack Cocaine Epidemic.* Family Caregiver Applications Series 2. Newbury Park, CA: Sage, 1993.

Minkler, Meredith, et al. "Profile of Grandparents: Raising Grandchildren in the United States." *The Gerontologist* 37, 3 (June 1997): 400–411.

Mintz, Steven, and Susan Kellogg. *Domestic Revolutions: A Social History of American Family Life.* New York: Free Press, 1988.

Minuchin, Salvador. *Family Therapy Techniques.* Cambridge, MA: Harvard University Press, 1981.

Miracle, Tina, A. Miracle, and R. Baumeister, *Human Sexuality: Meeting Your Basic Needs.* Upper Saddle River, NJ: Prentice Hall, 2003.

Mirandé, Alfredo. *The Chicano Experience: An Alternative Perspective.* Notre Dame, IN: University of Notre Dame Press, 1985.

Moen, Phyllis, and Kay Forest. "Strengthening Families: Policy Issues for the Twenty-First Century." In *Handbook of Marriage and the Family,* 2nd ed., edited by M. Sussman, S. Steinmetz, and G. Peterson. New York: Plenum. 1999.

Moffatt, Betty Clare, et al. *AIDS: A Self-Care Manual.* Los Angeles: IBS Press, 1987.

Moffatt, Michael. *Coming of Age in New Jersey: College and American Culture.* New Brunswick, NJ: Rutgers University Press, 1989.

Moller, Lora C., S. Hymel, and K. H. Rubin. "Sex Typing in Play and Popularity in Middle Childhood." *Sex Roles* 26, 7/8 (1992): 331–335.

Monahan, Thomas. "Are Interracial Marriages Really Less Stable?" *Social Forces* 48 (1970): 461–473.

Money, John. *Love and Lovesickness.* Baltimore: Johns Hopkins University Press, 1980.

Montagu, Ashley. *Touching,* 3rd ed. New York: Columbia University Press, 1986.

Montgomery, M. J., E. R. Anderson, E. M. Hetherington, and W. G. Clingempeel. "Patterns of Courtship for Remarriage: Implications for Child Adjustment and Parent-Child Relationships." *Journal of Marriage and the Family,* 54 (August 1992): 686–698.

Montgomery, Marilyn J., and Gwendoly T. Sorell. "Differences in Love Attitudes Across Family Life Stages." *Family Relations* 46, 1 (January 1997): 55–61.

Moore, Dianne, and Pamela Erickson. "Age, Gender, and Ethnic Differences in Sexual and Contraceptive Knowledge, Attitudes, and Behaviors." *Family and Community Health* 8, 3 (November 1985): 38–51.

Moore, Lisa J. "Protecting Babies from Hepatitis-B." *U.S. News and World Report* (May 9, 1988): 85.

Moore, M. M. "Nonverbal Courtship Patterns in Women: Context and Consequences." *Ethology and Sociobiology* 6, 2 (1985): 237–247.

Morgan, Carolyn, and Alexis Walker. "Predicting Sex Role Attitudes." *Social Psychology Quarterly* 46 (1983): 148–153.

Morgan, Edmund. *The Puritan Family.* New York: Harper and Row, 1966. (1944).

Morgan, Leslie. "Outcome of Marital Separation: A Longitudinal Test of Predictors." *Journal of Marriage and the Family* 50, 2 (May 1988): 493–498.

Morgan, S. Philip, Diane Lye, and Gretchen Condran. "Sons, Daughters, and the Risk of Marital Disruption." *American Journal of Sociology* 94 (July 1988): 110–129.

Morrison, Donna, and Daniel Lichter. "Family Migration and Female Employment: The Problem of Underemployment among Migrant Women." *Journal of Marriage and the Family* 50, 1 (February 1988): 161–172.

Morrison, Donna R., and Andres J. Cherlin. "The Divorce Process and Young Children's Well-Being: A Prospective Analysis." *Journal of Marriage and the Family* 57, 3 (August 1995): 800–812.

Mosher, D. L., and S. S. Tomkins. "Scripting the Macho Man: Hypermasculine Socialization and Enculturation." *Journal of Sex Research* 25 (February 1988): 60–84.

Mosher, Donald. "Sex Guilt and Sex Myths in College Men and Women." *Journal of Sex Research* 15, 3 (August 1979): 224–234.

Moss, Peter, et al. "Marital Relations During the Transition to Parenthood." *Journal of Reproductive and Infant Psychology* 4, 1–2 (September 1986): 57–67.

Moynihan, Daniel Patrick. *The Negro Family: The Case for National Action.*Washington, DC: U.S. Government Printing Office, 1965.

Muehlenhard, Charlene. "Misinterpreted Dating Behaviors and the Risk of Date Rape." *Journal of Social and Clinical Psychology* 9, 1 (1988): 20–37.

Muehlenhard, Charlene L., and S. W. Cook. "Men's Self-Reports of Unwanted Sexual Activity." *Journal of Sex Research* 24 (1988): 58–72.

Muehlenhard, Charlene L., and L. C. Hollabaugh. "Do Women Sometimes Say No When They Mean Yes? The Prevalence and Correlates of Women's Token Resistance to Sex." *Journal of Personality and Social Psychology* 54 (May 1988), 872–879.

Muehlenhard, Charlene L., and M. Linton. "Date Rape and Sexual Aggression in Dating Situations." *Journal of Consulting Psychology* 34 (April 1987): 186–196.

Muehlenhard, Charlene L., and M. L. McCoy. "Double Standard/ Double Bind." *Psychology of Women Quarterly* 15 (1991): 447–461.

Muehlenhard, Charlene L., I. G. Ponch, J. L. Phelps, and L. M. Giusti. "Definitions of Rape: Scientific and Political Implications." *Journal of Social Issues* 48, 1 (Spring 1992): 23–44.

Muehlenhard, Charlene L., and J. Schrag. "Nonviolent Sexual Coercion." In *Acquaintance Rape: The Hidden Crime,* edited by A. Parrot and L. Bechhofer. New York: Wiley, 1991.

Mueller, B. A., et al. "Risk Factors for Tubal Infertility: Influence of History of Prior Pelvic Inflammatory Disease." *Sexually Transmitted Diseases* 19, 1 (January 1992): 28–34.

Mullan, Bob. *The Mating Trade.* Boston: Routledge and Kegan Paul, 1984.

Mullen, Paul E. "The Crime of Passion and the Changing Cultural Construction of Jealousy." *Criminal Behavior and Mental Health* 3, 1 (1993): 1–11.

Mulligan, Thomas, and C. Renee Moss. "Sexuality and Aging in Male Veterans: A Cross-Sectional Study of Interest, Ability, and Activity." *Archives of Sexual Behavior* 20, 1 (February 1991): 17–25.

Mulligan, Thomas, and Robert Palguta. "Sexual Interest, Activity, and Satisfaction among Male Nursing Home Residents." *Archives of Sexual Behavior* 20, 2 (April 1991): 199–204.

"Muppet Gender Gap." *Media Report to Women* 21, 1 (1993): 8.

Muram, David, K., Miller, and A. Cutler. "Sexual Assault of the Elderly Victim." *Journal of Interpersonal Violence* 7, 1 (March 1992): 70–76.

Murdock, George. "World Ethnographic Sample." *American Anthropologist* 59 (1957): 664–687.

———. *Social Structure.* New York: Free Press, 1967.

Murnen, S. K., A. Perot, and D. Byrne. "Coping with Unwanted Sexual Activity: Normative Responses, Situational Determinants, and Individual Differences." *Journal of Sex Research,* 26 (1989): 85–106.

Murray, Charles. *Losing Ground: American Social Policy, 1950–1980.* New York: Basic Books, 1984.

Murray, Maresa, and Kathleen Gilbert. "Images of African American Families on Prime-Time Situation Comedies: Are

Our 'Roots' There?" Poster session presented at the annual meeting of the National Council on Family Relations, Baltimore, November 11, 1993.

Murry, Velma McBride. "Socio-Historical Study of Black Female Sexuality: Transition to First Coitus." In *The Black Family,* 4th ed., edited by R. Staples. Belmont, CA: Wadsworth, 1991.

Murstein, Bernard. *Who Will Marry Whom: Theories and Research in Marital Choice.* New York: Springer Publishing, 1976.

———. *Paths to Marriage: Family Studies Text Series,* vol 5. Beverly Hills, CA: Sage, 1986.

———. "A Clarification and Extension of the SVR Theory of Dyadic Pairing." *Journal of Marriage and the Family* 49 (1987): 929–933.

Mydans, Seth. "Surrogate Loses Custody Bid in Case Defining Motherhood." *New York Times* (October 23, 1990).

Myers-Walls, Judith, and Fred Piercy. "Mass Media and Prevention: Guidelines for Family Life Professionals." *Journal of Primary Prevention* 5, 2 (December 1984): 124–136.

Nadelson, Carol, and Maria Sauzier. "Intervention Programs for Individual Victims and Their Families." In *Violence in the Home: Interdisciplinary Perspectives,* edited by M. Lystad. New York: Brunner/Mazel, 1986.

Nadler, Arie, and Iris Dotan. "Commitment and Rival Attractiveness: Their Effects on Male and Female Reactions to Jealousy-Arousing Situations." *Sex Roles* 26, 7–8 (1992): 293–310.

Nakonezny, Paul, Robert Shull, and Joseph Lee Rodgers. "The Effect of No-Fault Divorce Law on the Divorce Rate Across the 50 States and Its Relation to Income, Education and Religiosity." *Journal of Marriage and the Family* 57, 2 (May 1995): 477–488.

Nanda, Serena. *Neither Man Nor Woman: The Hijras of India.* Belmont. CA: Wadsworth, 1990.

Napier, Augustus, and Carl Whitaker. *The Family Crucible.* New York: Harper and Row, 1978.

"A Nation Out of Balance." *Health* (October 1994).

National Center for Health Statistics. "Advance Report of Final Natality Statistics, 1991." *Monthly Vital Statistics Report* 42, 3 (Supplement, September 9, 1993).

———. *Healthy People 2000 Review.* Hyattsville, Maryland: Public Health Service, 1994.

National Center for Health Statistics, 2000.

———. "Births, Marriages, Divorces and Deaths: Provisional Data for 2001" *National Vital Statistics Report* 50, 14 (September 2002a).

———. "Births: Final Data for 2001." *National Vital Statistics Report* 51, 2. (December 2002b).

Needle, Richard, et al. "Drug Abuse: Adolescent Addictions and the Family." In *Stress and the Family: Coping with Catastrophe,* vol. 2, edited by C. R. Figley and H. McCubbin. New York: Brunner/Mazel, 1983.

Nelsen, Jane. *Positive Discipline.* New York: Ballantine, 1987.

Nelson, Margaret, and Gordon Nelson. "Problems of Equity in the Reconstituted Family: a Social Exchange Analysis." *Family Relations* 31, 2 (April 1982): 223–231.

"Neonatal Herpes Is Preventable." *U.S.A. Today* (February 1984): 8–9.

Neuman, W. Lawrence. *Social Research Methods: Qualitative and Quantitative Approaches* 4th ed. Needham Heights, MA: Allyn and Bacon. 2000.

Nevid, Jeffrey. "Sex Differences in Factors of Romantic Attraction." *Sex Roles* 11, 5/6 (1984): 401–411.

"New APA Position Statement Urges Actions to Reduce High Rates of Nicotine Dependence." *Psychiatric Services* 46, 2 (February 1995).

Newcomb, H. ed. *Television: The Critical View.* 4th ed. New York: Oxford University Press, 1987.

Newcomb, Michael. "Cohabitation in America: An Assessment of Consequences." *Journal of Marriage and the Family* 41, 3 (August 1979): 597–603.

Newcomb, Nora. *Child Development: Change Over Time,* 8th ed. New York: HarperCollins, 1996.

Newcomer, Susan, and Richard Udry. "Oral Sex in an Adolescent Population." *Archives of Sexual Behavior* 14, 1 (February 1985): 41–46.

Newman, David. *Sociology of Families.* Thousand Oaks, CA: Pine Forge, 1999.

Newman, Katherine. *Falling from Grace.* New York: Free Press, 1988.

Newton, Niles. *Maternal Emotions.* New York: Basic Books, 1955.

Ney, Philip G. "Transgenerational Abuse." In *Intimate Violence: Interdisciplinary Perspectives,* edited by E. C. Viano. Washington, DC: Hemisphere, 1992.

NiCarthy, Ginny. *Getting Free: A Handbook for Women in Abusive Relationships.* Seattle, WA: Seal Press, 1986.

Nichols, William C., and Mary A. Pace-Nichols. "Developmental Perspectives and Family Therapy: The Marital Life Cycle." *Contemporary Family Therapy: An International Journal* 15, 4 (1993): 299–315.

Nielsen, Joyce McCarl, Russell K. Endo, and Barbara L. Ellington. "Social Isolation and Wife Abuse: A Research Report." In *Intimate Violence,* edited by E. Viano. Washington, DC: Hemisphere Publishing, 1992.

Noble, Elizabeth. *Having Your Baby by Donor Insemination.* Boston: Houghton Mifflin, 1987.

Nobles, Wade W. "African-American Family Life: An Instrument of Culture." In *Black Families,* 2nd ed., edited by H. Pipes McAdoo. Newbury Park, CA: Sage, 1988.

Nock, Steven. "The Symbolic Meaning of Childbearing." *Journal of Family Issues* 8, 4 (December 1987): 373–393.

Noller, Patricia. *Nonverbal Communication and Marital Interaction.* Oxford, England:Pergamon, 1984

Noller, Patricia, and Mary Anne Fitzpatrick. "Marital Communication." In *Contemporary Families: Looking Forward, Looking Back,* edited by A. Booth. Minneapolis: National Council on Family Relations, 1991.

———. eds. *Perspectives on Marital Interaction.* Philadelphia: Multilingual Matters, 1988.

Norton, Arthur, and Jeanne Moorman. "Current Trends in American Marriage and Divorce Among American Women." *Journal of Marriage and the Family* 49 (February 1987): 3–14.

Norton, Arthur J. "Family Life Cycle: 1980." *Journal of Marriage and the Family* 45 (1983): 267–275.

Notarius, Clifford, and Jennifer Johnson. "Emotional Expression in Husbands and Wives." *Journal of Marriage and the Family* 44, 2 (May 1982): 483–489.

Notman, Malkah T., and Eva P. Lester. "Pregnancy: Theoretical Considerations." *Psychoanalytic Inquiry* 8, 2 (1988): 139–159.

Nugent, R., and J. Gramick. "Homosexuality: Protestant, Catholic, and Jewish Issues: A Fishbone Tale." *Journal of Homosexuality* 18 (1989): 7–46.

Nuland, Sherwin B. *How We Die: Reflections on Life's Final Chapter.* New York: Knopf, 1994.

Nurmi, Jari Erik. "Age Differences in Adult Life Goals, Concerns, and Their Temporal Extension: A Life Course Approach to Future-Oriented Motivation." *International Journal of Behavioral Development* 15, 4 (1992): 487–508.

Nye, F. Ivan. "Fifty Years of Family Research." *Journal of Marriage and the Family* 50, 2 (May 1988): 305–316.

Nye, F. Ivan, and Felix Berardo, eds. *Emerging Conceptual Frameworks in Family Analysis.* 2 vols. New York: Praeger, 1981.

Nyquist, Linda, et al. "Household Responsibilities in Middle-Class Couples: The Contribution of Demographic and Personality Variables." *Sex Roles* 12, 1/2 (1985): 15–34.

Oakley, Ann. *The Sociology of Housework.* New York: Pantheon, 1974.

——— ed. *Sex, Gender, and Society.* Rev. ed. New York: Harper and Row, 1985.

O'Farrell, Timothy J., Keith A. Choquette, and Gary R. Birchler. "Sexual Satisfaction and Dissatisfaction in the Marital Relationships of Male Alcoholics Seeking Marital Therapy." *Journal of Studies on Alcohol* 52, 5 (September 1991): 441–447.

O'Flaherty, Kathleen, and Laura Eells. "Courtship Behavior of the Remarried." *Journal of Marriage and the Family* 50, 2 (May 1988): 499–506.

Oggins, Jean, Joseph Veroff, and Douglas Leber. "Perceptions of Marital Interaction among Black and White Newlyweds." *Journal of Personality and Social Psychology* 65, 3 (1993): 494–511.

Ohninger, S., and N. J. Alexander. "Male Infertility: The Focus Shifts to Sperm Manipulation." *Current Opinion in Obstetrics and Gynecology* 3, 2 (1991): 182–190.

Oldenberg, Don. "Watch TV with Kids, Experts Say." *San Jose Mercury* (April 12, 1992).

Olds, S. W. *The Eternal Garden: Seasons of Our Sexuality.* New York: Times Books, 1985.

O'Leary, K. Daniel. "Through a Psychological Lens: Personality Traits, Personality Disorders, and Levels of Violence." In *Current Controversies in Family Violence,* edited by R. Gelles and D. Loseke. Newbury Park, CA: Sage, 1993.

Oliver, Mary Beth, and Janet Shibley Hyde. "Gender Differences in Sexuality: A Meta-analysis." *Psychological Bulletin* 114, 1 (1993): 29–51.

Olshansky, E. F. "Redefining the Concepts of Success and Failure in Infertility Treatment." *Naacogs Clinical Issues in Perinatal and Women's Health Nursing* 3, 2 (1992): 343–346.

Olshansky, S. Jay, Bruce A. Carnes, and Christine Cassel. "In Search of Methuselah: Estimating the Upper Limits to Human Longetivity." *Science* 250, 4981 (November 2, 1990): 634–641.

Olson, David. "Insiders' and Outsiders' Views of Relationships: Research Studies." In *Close Relations,* edited by G. Levinger and H. Rausch. Amherst, MA: University of Massachusetts Press, 1977.

Olson, David H., and John DeFrain. *Marriage and the Family; Diversity and Strengths,* 2nd ed. Mountain View, CA: Mayfield, 1997.

Olson, David H., Hamilton I. McCubbin, Howard Barnes, Andrea Larson, Maria Muxen, and Marc Wilson. *Families: What Makes Them Work.* Beverly Hills, CA: Sage, 1983.

Olson, Myrna R., and Judith A. Haynes. "Successful Single Parents." *Families in Society* 74, 5 (1993): 259–267.

O'Neil, Robin, and Ellen Greenberger. "Patterns of Commitment to Work and Parenting: Implications for Role Strain." *Journal of Marriage and the Family* 52, 1 (February 1994): 101–115.

Opie, Anne. "Ideologies of Joint Custody." *Family and Conciliation Courts Review* 31, 3 (1993): 313–326.

O'Reilly, Jane. "Wife Beating: The Silent Crime." *Time* (September 5, 1983).

Orthner, D. K., and J. F. Pittman. "Family Contributions to Work Commitment." *Journal of Marriage and the Family* 48 (1986): 573–581.

Ortiz, Silvia, and Jesus Manuel Casas. "Birth Control and Low-Income Mexican-American Women: The Impact of Three Values." *Hispanic Journal of the Behavioral Sciences* 12, 1 (February 1990): 83–92.

Ortiz, Vilma, and Rosemary Santana Cooney. "Sex Role Attitudes and Labor Force Participation among Young Hispanic Females and Non-Hispanic White Females." *Social Science Quarterly* 65, 2 (June 1984): 392–400.

Oster, Sharon. "A Note on the Determinants of Alimony." *Journal of Marriage and the Family* 49, 1 (February 1987): 81–86.

O'Sullivan, Denis A., and Eleanor O'Leary. "Love: A Dimension of Life." *Counseling and Values* 37, 1 (1992): 32–38.

O'Sullivan, Lucia, and E. Sandra Byers. "College Students' Incorporation of Initiator and Restrictor Roles in Sexual Dating Interactions." *Journal of Sex Research* 29, 3 (August 1992): 435–446.

Oswald, Ramona Faith. "Resilience within the Family Networks of Lesbians and Gay Men: Intentionality and Redefinition." *Journal of Marriage and the Family* 64, 2 (May 2002): 374–383.

"Outlook." *U.S. News and World Report* (June 6, 1994): 12.

Padilla, E. R., and K. E. O'Grady. "Sexuality among Mexican Americans: A Case of Sexual Stereotyping." *Journal of Personality and Social Psychology* 52 (1987): 5–10.

Pais, Shoba. "Asian Indian Families in America." In *Families in Cultural Context,* edited by M. K. DeGenova. Mountain View, CA: Mayfield, 1997.

Palkovitz, Rob. "Reconstructing 'Involvement': Expanding Conceptualizations of Men's Caring in Contemporary Families." In *Generative Fathering: Beyond Deficit Perspectives,* Vol. 3, *Current Issues in the Family,* edited by A. Hawkins and D. Dollahite. Thousand Oaks, CA: Sage, 1997: 200–216.

Pallow-Fleury, Angie. "Your Hospital Birth: Questions to Ask." *Mothering* (December 1983): 83–85.

Paludi, M. A. "Sociopsychological and Structural Factors Related to Women's Vocational Development." *Annals of New York Academy of Sciences,* 602 (1990): 157–168.

"Panel Says Nation Is Lagging in Children's Health." *New York Times* (March 29, 1992): A16.

Panuthos, Claudia, and Catherine Romeo. *Ended Beginnings: Healing Childbearing Losses.* New York: Warner, 1984.

Papanek, Hannah. "Men, Women, and Work: Reflections on the Two-Person Career," *American Journal of Sociology,* 78 (January 1975): 853–872.

Papernow, Patricia L. *Becoming a Stepfamily.* San Francisco: Jossey-Bass, 1993.

Pareles, Jon. "Indians' Heritage Survives in Songs." *New York Times* (December 4, 1990): B1.

Parker, Philip. "Motivation of Surrogate Mothers—Initial Findings." *American Journal of Psychiatry* 140, 1 (1983): 117–118.

Parsons, Jacqueline, ed. *The Psychobiology of Sex Differences and Sex Roles.*Washington, DC: Hemisphere, 1980.

Parsons, Talcott. "Family Structure and the Socialization of the Child." In *Family Socialization and Interaction Process,* edited by T. Parsons and R. F. Bales. Glencoe, IL: Free Press, 1955.

Parsons, Talcott, and R. F. Bales. *Family Socialization and Interaction Processes.* Glencoe, IL: Free Press, 1955.

Pasley, Kay. "Family Boundary Ambiguity: Perceptions of Adult Stepfamily Family Members." In *Remarriage and Stepparenting: Current Research and Theory,* edited by K. Pasley and M. Ihinger-Tallman. New York: Guilford Press, 1987.

Pasley, Kay, and Marilyn Ihinger-Tallman, eds. *Remarriage and Stepparenting: Current Research and Theory.* New York: Guilford Press, 1987.

Patterson, Charlotte. "Family Relationships of Lesbians and Gay Men." *Journal of Marriage and the Family,* 62, 4 (November 2000): 1052–1069.

Patterson, Charlotte, and Raymond Chan. "Families Headed by Lesbian and Gay Parents." In *Parenting and Child Development in "Nontraditional" Families,* edited by M. Lamb. Mahwah, NJ: Lawrence Erlbaum. 1999.

Patterson, C. J. "Children of Lesbian and Gay Parents." *Child Development* 63 (October 1992): 1025–1042.

Patterson, G. *Families Applications of Social Learning to Family Life.* Champaign, IL: Research Press, 1971.

Patterson, Joan M. "Integrating Family Resilience and Family Stress Theory." *Journal of Marriage and the Family* 64, 2 (May 2002): 349–360.

Patterson, Joan, and Hamilton McCubbin. "Chronic Illness: Family Stress and Coping." In *Stress and the Family: Coping with Catastrophe,* edited by C. R. Figley and H. McCubbin. New York: Brunner/Mazel, 1983.

Patton, Michael. "Twentieth-Century Attitudes toward Masturbation." *Journal of Religion and Health* 25, 4 (December 1986): 291–302.

Paulson, David. "Hot Tubs and Reduced Sperm Counts." *Medical Aspects of Human Sexuality* 14 (September 1980): 121.

Paulson, Morris J., Robert H. Coombs, and John Landsverk. "Youth Who Physically Assault Their Parents." *Journal of Family Violence* 5, 2 (June 1990): 121–134.

Payton, Isabelle. "Single-Parent Households: An Alternative Approach." *Family Economics Review* (December 1982): 11–16.

Pear, Robert. "U.S. Reports Poverty Is Down but Inequality Is Up. "*New York Times* (September 27, 1990): A12.

———. "Rich Got Richer in 80's: Others Held Even." *New York Times* (January 11, 1991): A1, A11.

Pearl, D., L. Bouthilet, and J. Lazar, eds. *Television and Behavior: Ten Years of Scientific Progress and Implications for the Eighties,* vol 1 (DHHS Publication No. ADM 82–1196). Washington, DC: U.S. Government Printing Office, 1982.

Peck, M. Scott. *The Road Less Traveled: A New Psychology of Love, Traditional Values, and Spiritual Growth.* New York: Simon and Schuster, 1978.

Peek, Charles, et al. "Teenage Violence Toward Parents: A Neglected Dimension in Family Violence." *Journal of Marriage and the Family* 47, 4 (November 1985): 1051–1058.

Peirce, Kate. "Sex Role Stereotyping of Children: A Content Analysis of the Roles and Attributes of Child Characters." *Sociological Spectrum* 9, 3 (September 1989): 321–328.

Pennington, Saralie Bisnovich. "Children of Lesbian Mothers." In *Gay and Lesbian Parents,* edited by F. W. Bozett. New York: Praeger, 1987.

Peplau, Letitia. "What Homosexuals Want." *Psychology Today,* 15, 3 (March 1977): 28–38.

———. "Research on Homosexual Couples." In *Gay Relationships,* edited by J. DeCecco. New York: Haworth Press, 1988.

Peplau, Letitia, and Susan Cochran. "Value Orientations in the Intimate Relationships of Gay Men." *Gay Relationships,* edited by J. DeCecco. New York: Haworth Press, 1988.

Peplau, Letitia Anne, and Steven Gordon. "The Intimate Relationships of Lesbians and Gay Men." In *Gender Roles and Sexual Behavior,* edited by E. Allgeier and N. McCormick. Palo Alto, CA: Mayfield, 1982.

Peplau, Letitia Anne, Charles T. Hill, and Zick Rubin. "Sex Role Attitudes in Dating and Marriage: A 15-Year Follow-Up of the Boston Couples Study." *Journal of Social Issues* 49, 3 (1993): 31–53.

Peplau, Letitia Anne, et al. "Sexual Intimacy in Dating Relationships." *Journal of Social Issues* 33, 2 (March 1977): 86–109.

Perkins, Kathleen. "Psychosocial Implications of Women and Retirement." *Social Work* 37, 6 (1992): 526–532.

Perlman, Daniel, and Steve Duck, eds. *Intimate Relationships: Development, Dynamics, and Deterioration.* Beverly Hills, CA: Sage, 1987.

Perlman, David. "Brave New Babies." *San Francisco Chronicle* (March 5, 1990): B3.

Perry-Jenkins, Maureen, and Karen Folk. "Class, Couples, and Conflict: Effects of the Division of Labor on Assessments of Marriage in Dual-Earner Marriages." *Journal of Marriage and the Family* 56, 1 (February 1994): 165–180.

Perry-Jenkins, Maureen, Rena Repetti, and Ann Crouter. "Work and Family in the 1990's." *Journal of Marriage and the Family,* 62, 4, (November 2000): 981–998.

Petchesky, R. P. *Abortion and Woman's Choice: The State, Sexuality, and Reproductive Freedom.* Rev. ed. Boston: Northeastern University Press, 1990.

Peters, Stefanie, et al. "Prevalance." In *Sourcebook on Child Sexual Abuse,* edited by D. Finkelhor. Beverly Hills, CA: Sage, 1986.

Petersen, James R., et al. "Playboy's Readers' Sex Survey." *Playboy* (March 1983): 178–184.

Petersen, Larry. "Interfaith Marriage and Religious Commitment among Catholics." *Journal of Marriage and the Family* 48, 4 (November 1986): 725–735.

Petersen, Virginia, and Susan B. Steinman. "Helping Children Succeed after Divorce: A Court-Mandated Educational Program for Divorcing Parents." *Family and Conciliation Courts Review* 32, 1 (1994): 27–39.

Peterson, Gary W., and Boyd C. Rollins. "Parent-Child Socialization." In *Handbook of Marriage and the Family,* edited by M. B. Sussman and S. K. Steinmetz. New York: Plenum Press, 1987.

Peterson, Marie Ferguson. "Racial Socialization of Young Black Children." In *Black Children: Social, Educational, and Parental Environments,* edited by H. Pipes McAdoo and J. McAdoo. Beverly Hills, CA: Sage, 1985.

Peterson, Richard R. "A Reevaluation of the Economic Consequences of Divorce." *American Sociological Review* 61, 3 (June 1996).

Petit, Charles. "New Study to Ask Why So Many Infants Die." *San Francisco Chronicle* (May 9, 1990): A3.

Petit, Ellen, and Bernard Bloom. "Whose Decision Was It? The Effect of Initiator Status on Adjustment to Marital Disruption." *Journal of Marriage and the Family* 46, 3 (August 1984): 587–595.

Peyrot, Mark, et al. "Marital Adjustment to Adult Diabetes: Interpersonal Congruence and Spouse Satisfaction." *Journal of Marriage and the Family* 50, 2 (May 1988): 363–376.

Pfaus, James, Myronuk, and Jacobs. "Soundtrack Contents and Depicted Sexual Violence." *Archives of Sexual Behavior* 15, 3 (June 1986): 231–237.

Pies, Cheri. "Considering Parenthood: Psychological Issues for Gay Men and Lesbians Choosing Alternative Fertilization." In *Gay and Lesbian Parents,* edited by F. W. Bozett. New York: Praeger, 1987.

Pilisuk, Marc. "The Delivery of Social Support: The Social Innoculation." *American Journal of Orthopsychiatry* 52, 1 (January 1982): 20–31.

Pill, Cynthia. "Stepfamilies: Redefining the Family." *Family Relations* 39, 2 (April 1990): 186–193.

Pillemer, Karl. "The Abused Offspring Are Dependent: Abuse Is Caused by the Deviance and Dependence of Abusive Caregivers." In *Current Controversies in Family Violence,* edited by R. Gelles and D. Loseke. Newbury Park, CA: Sage, 1993.

Pillemer, Karl, and J. Jill Suitor. "Elder Abuse." In *Handbook of Family Violence,* edited by V. B. Van Hasselt et al. New York: Plenum Press, 1988.

Pink, Jo Ellen, and Karen Smith Wampler. "Problem Areas in Stepfamilies: Cohesion, Adaptability, and the Stepfather-Adolescent Relationship." *Family Relations* 34, 3 (July 1985): 327–335.

Piotrkowski, Chaya, Robert Rapoport, and Rhona Rapoport. "Families and Work" In *Handbook of Marriage and the Family,* edited by M. Sussman and S. Steinmetz. New York: Plenum, 1987: 251–283.

Pirog-Good, Maureen A., and Jane Stets, eds. *Violence in Dating Relationships: Emerging Social Issues.* New York: Praeger, 1989.

Pistole, M. Carole. "Attachment in Adult Romantic Relationships: Style of Conflict Resolution and Relationship Satisfaction." *Journal of Social Personal Relationships* 6, 4 (1989): 505–512.

Piven, Frances Fox, and Richard Cloward. *Regulating the Poor.* New York: Vintage Books, 1972.

Pleck, Joseph. *Working Wives/Working Husbands.* Beverly Hills, CA: Sage, 1985.

Plummer, William, and Margaret Nelson. "A Mother's Priceless Gift." *People Weekly* (August 26, 1991).

Pogrebin, Letty Cottin. *Growing Up Free: Raising Your Child in the 1980s.* New York: McGraw-Hill, 1980.

———. *Family Politics.* New York: McGraw-Hill, 1983.

"Poll on Families: Small Is Best." *New York Times* (May 25, 1986): L22.

Pollack, William. *Real Boys: Rescuing Our Sons from the Myths of Boyhood.* New York: Random House, 1998.

Popenoe, David. "American Family Decline, 1960–1990: A Review and Appraisal." *Journal of Marriage and the Family* 55, 1993: 527–555.

———. *Life Without Father: Compelling New Evidence That Fatherhood and Marriage Are Indispensable for the Good of Children and Society.* New York: Free Press, 1996.

Popovich, Paula M., et al. "Assessing the Incidence and Perceptions of Sexual Harassment Behaviors among American Undergraduates." *Journal of Psychology* 120 (1986): 387–396.

———. "Perceptions of Sexual Harassment as a Function of Sex of Rater and Incident Form and Consequence." *Sex Roles* 27 (December 1992): 609–625.

Porter, Nancy L., and F. Scott Christopher. "Infertility: Towards An Awareness of a Need Among Family Life Practitioners." *Family Relations* 33, 2 (April 1984): 309–315.

Portes, P. R., S. C. Howell, J. H. Brown, S. Eichenberger, and C. A. Mas. "Family Functions and Children's Postdivorce Adjustment. *Journal of Orthopsychiatry* 62 (October 1992): 613–617.

"Poverty Helps Break Up Families, Report Says." *New York Times* (January 15, 1993): A7.

Powell, Rachel. "It's One Party Even the Recession Can't Spoil." *New York Times* (June 23, 1991): 10.

Pratt, Clara, and Vicki Schmall. "College Students' Attitudes toward Elderly Sexual Behavior: Implications for Family Life Education." *Family Relations* 38, 2 (April 1989): 137–141.

Press, Robert. "Hunger in America." *Christian Science Monitor* (February 15, 1985): 18–19.

Presser, Harriet. "Shift Work Among American Women and Child Care." *Journal of Marriage and the Family* 48, 3 (August 1986): 551–663.

———. "Some Economic Complexities of Child Care Provided by Grandmothers." *Journal of Marriage and the Family* 51, 3 (August 1989): 581–591.

———. "Nonstandard Work Schedules and Marital Instability." *Journal of Marriage and the Family,* 62, 1 (February 2000): 93–110.

Price, J. H., and P. A. Miller. "Sexual Fantasies of Black and White College Students." *Psychological Reports* 54 (1984): 1007–1014.

Price, Jane. "Who Wants to Have Children? and Why?" In *Relationships: The Marriage and Family Reader,* edited by J. Rosenfeld. Glenview, IL: Scott, Foresman, 1982.

Price, John. "North American Indian Families." In *Ethnic Families in America: Patterns and Variations,* 2nd ed., edited by C. Mindel and R. Haberstein. New York: Elsevier-North Holland, 1981.

Price-Bonham, S., and J. O. Balswick. "The Noninstitutions: Divorce, Desertion, and Remarriage." *Journal of Marriage and the Family* 42 (1980): 959–972.

Priest, Ronnie. "Child Sexual Abuse Histories among African-American College Students: A Preliminary Study." *American Journal of Orthopsychiatry* 62, 3 (July 1992): 475.

Pruett, Kyle. *The Nurturing Father: Journey toward Complete Man.* New York: Warner, 1987.

Puglisi, J. T., and D.W. Jackson. "Sex Role Identity and Self-Esteem in Adulthood." *Journal of Aging and Human Development* 12 (1981): 129–138.

Radway, J. A. *Reading the Romance.* Chapel Hill, NC: University of North Carolina, 1984.

Ramsey, Patricia G. *Teaching and Learning in a Diverse World: Multicultural Education for Young Children.* New York: Teachers College Press, 1987.

Rando, Therese A. "Death and Dying Are Not and Should Not Be Taboo Topics." In *Principles of Thanatology,* edited by A. H. Kutscher et al. New York: Columbia University Press, 1987.

Rank, Mark R. and Li-Chen Cheng. "Welfare Use Across Generations: How Important Are the Ties That Bind?" *Journal of Marriage and the Family* 57, 3 (August 1995): 673–684.

Rao, Kavitha, et al. "Child Sexual Abuse of Asians Compared with Other Populations." *Journal of the American Academy of Child and Adolescent Psychiatry* 31, 5 (September 1992): 880 ff.

Rapp, Carol, and Sally Lloyd. "The Role of 'Home as Haven' Ideology in Child Care Use." *Family Relations* 38, 4 (October 1989): 427–430.

Raschke, Helen. "Divorce." In *Handbook of Marriage and the Family,* edited by M. Sussman and S. Steinmetz. New York: Plenum Press, 1987.

"Rates of Cesarean Delivery—United States, 1993." Centers for Disease Control and Prevention. *Morbidity and Mortality Weekly Report,* 44 (1995): 303–307.

Raval, H., et al. "The Impact of Infertility on Emotions and the Marital and Sexual Relationship." *Journal of Reproductive and Infant Psychology* 5, 4 (October 1987): 221–234.

Raven, Bertram, et al. "The Bases of Conjugal Power." In *Power in Families,* edited by R. Cromwell and D. Olson. New York: Halstead Press, 1975.

Ravinder, Shashi. "Androgyny: Is It Really the Product of Educated Middle-Class Western Societies?" *Journal of Cross-Cultural Psychology* 18, 2 (June 1987): 208–220.

Rawlings, Steve. "Studies in Marriage and the Family: Single Parents and Their Children." *Current Population Reports.* (Series, P- 23.) Washington, DC: U.S. Census Bureau, 1989.

———. "Households and Families." *Current Population Reports* (Series P-20, No. 483). Washington, DC: U.S. Census Bureau, 1994a.

———. "Household and Family Characteristics, March 1993." *Current Population Reports.* Population Characteristics. (Series P20, No. 477). Washington, DC: U.S. Census Bureau, 1994b.

Real, Michael R. *Exploring Media Culture.* Thousand Oaks, CA: Sage, 1996.

Real, Terrence. *I Don't Want to Talk About It: Overcoming the Secret Legacy of Male Depression.* New York: Scribner. 1998.

Reed, David, and Martin Weinberg. "Premarital Coitus: Developing and Establishing Sexual Scripts." *Social Psychology Quarterly* 47, 2 (June 1984): 129–138.

Reedy, M., et al. "Age and Sex Differences in Satisfying Love Relationships across the Adult Life Span." *Human Development* 24 (1981): 52–86.

Reeves, Terrance, and Claudette Pennett. "The Asian and Pacific Island Population in the United States: March 2002." *Current Populations Reports* Washington, DC: Government Printing Office. www.census/gov/population/www/socdemo/race/api/html.

Regan, Mary, and Helen Roland. "Rearranging Family and Career Priorities: Professional Women and Men of the Eighties." *Journal of Marriage and the Family* 47, 4 (November 1985): 985–992.

Regan, Pamela. *The Mating Game: A Primer on Love, Sex, and Marriage.* Thousand Oaks, CA: Sage, 2003.

Reid, Pamela, and L. Comas-Diaz. "Gender and Ethnicity: Perspectives on Dual Status." *Sex Roles* 22, 7 (April 1990): 397–408.

Reilly, Mary E., et al. "Tolerance for Sexual Harassment Related to Self-Reported Sexual Victimization." *Gender and Society* 6, 1 (March 1992): 122–138.

Reisman, C. *Divorce Talk: Women and Men Make Sense of Personal Relationships.* New Brunswick, NJ: Rutgers University Press, 1990.

Reisman, C. K., and N. Gerstel. "Marital Dissolution and Health: Do Males or Females Have Greater Risk?" *Social Science and Medicine* 20 (1985): 627–635.

Reiss, Ira. *Family Systems in America.* 3rd ed. New York: Holt, Rinehart, and Winston, 1980a.

———. "A Multivariate Model of the Determinants of Extramarital Sexual Permissiveness." *Journal of Marriage and the Family* 42, 2 (May 1980b): 395–411.

Renzetti, Claire. "Violence in Gay and Lesbian Relationships." In *Vision 2010: Families and Violence, Abuse and Neglect,* edited by R. J. Gelles. Minneapolis: National Council on Family Relations, 1995.

Renzetti, Claire, and Daniel Curran. *Living Sociology.* Needham Heights, MA: Allyn and Bacon, 1998.

———. *Women, Men, and Society,* 5th ed. Needham Heights, MA: Allyn and Bacon, 2003.

Resnick, Sandven. "Informal Adoption among Black Adolescent Mothers." *American Journal of Orthopsychiatry* 60 (April 1990): 210–224.

Retik, Alan B., and Stuart B. Bauer. "Infertility Related to DES Exposure in Utero: Reproductive Problems in the Male." In *Infertility: Medical, Emotional, and Social Considerations,* edited by M. Mazor and H. Simons. New York: Human Sciences Press, 1984.

Rettig, K. D., and M. M. Bubolz. "Interpersonal Resource Exchanges as Indicators of Quality of Marriage." *Journal of Marriage and the Family* 45 (1983): 497–510.

Rexroat, Cynthia. "Race and Marital Status Differences in the Labor Force Behavior of Female Family Heads: The Effect of Household Structure." *Journal of Marriage and the Family* 52, 3 (August 1990): 591–601.

Rexroat, Cynthia, and Constance Shehan. "The Family Life Cycle and Spouses' Time in Housework." *Journal of Marriage and the Family* 49, 4 (November 1987): 737–750.

Rice, Susan. "Sexuality and Intimacy for Aging Women: A Changing Perspective." *Journal of Women and Aging* 1, 1–3 (1989): 245–264.

Rich, Adrienne. *Of Woman Born.* New York: Norton, 1976.

Richards, Leslie N., and Cynthia J. Schmiege. "Problems and Strengths of Single-Parent Families: Implications for Practice and Policy." *Family Relations* 42, 3 (July 1993): 277–285.

Richardson, Diana. "The Dilemma of Essentiality in Homosexual Theory." *Journal of Homosexuality* 9, 2/3 (December 1983): 79–90.

Richmond-Abbott, M. *Masculine and Feminine: Gender Roles over the Life Cycle,* 2nd ed. New York: McGraw-Hill, 1992.

Ridley, Jane, and Michael Crowe. "The Behavioural-Systems Approach to the Treatment of Couples." *Sexual and Marital Therapy* 7, 2 (1992): 125–140.

Riegle, Donald. "The Psychological and Social Effects of Unemployment." *American Psychologist* 37, 10 (October 1982): 1113–1115.

Riffer, Roger, and Jeffrey Chin. "Dating Satisfaction among College Students." *International Journal of Sociology and Social Policy* 8, 5 (1988): 29–36.

Riggs, David S. "Relationship Problems and Dating Aggression: A Potential Treatment Target." *Journal of Interpersonal Violence* 8, 1 (1993): 18–35.

Riley, Alan J. "Sexuality and the Menopause." *Sexual and Marital Therapy* 6, 2 (1991): 135–146.

Riportella-Muller, Roberta. "Sexuality in the Elderly: A Review." In *Human Sexuality: The Societal and Interpersonal Context,* edited by K. McKinney and S. Sprecher. Norwood, NJ: Ablex, 1989.

Riseden, Andrea D., and Barbara E. Hort. "A Preliminary Investigation of the Sexual Component of the Male Stereotype." Unpublished paper, 1992.

Risman, Barbara. "Can Men Mother? Life as a Single Father." *Family Relations* (35) 1986: 95–102.

———. *Gender Vertigo: American Families in Transition.* New Haven, CT: Yale University Press, 1998.

Risman, Barbara, and Danette Johnson-Sumerford. "Doing It Fairly: A Study of Post Gender Marriages." *Journal of Marriage and the Family* 60 (February 1998): 23–40.

Rivlin, Leanne. "The Significance of Home and Homelessness." *Marriage and Family Review* 15, 1/2 (1990): 39–57.

Roberts, Linda, and Lowell Krokoff. "A Time Series Analysis of Withdrawal, Hostility, and Displeasure in Satisfied and Dissatisfied Marriages." *Journal of Marriage and the Family* 52, 1 (February 1990): 95–105.

Roberts, Michael C., Kristi Lekander, and Debra Fanurik. "Evaluation of Commercially Available Materials to Prevent Child Sexual Abuse and Abduction." *American Psychologist* 45, 6 (June 1990): 782–783.

Roberts, Sam. "The Hunger beneath the Statistics." *New York Times* (May 23, 1988): A13.

Robertson, Elizabeth, et al. "The Costs and Benefits of Social Support in Families." *Journal of Marriage and the Family* 53, 2 (May 1991): 403–416.

Robinson, Bryan. *Teenage Fathers.* Lexington, MA: Lexington Books, 1987.

Robinson, I., K. Ziss, B. Ganza, and S. Katz. "20 Years of the Sexual Revolution, 1965–1985—An Update." *Journal of Marriage and the Family* 53, 1 (February 1991): 216–220.

Robinson, Linda C., and Pricilla White Blanton. "Marital Strengths in Enduring Marriages." *Family Relations* 42 (January 1993): 38–45.

Robinson, Pauline. "The Sociological Perspective." In *Sexuality in the Later Years: Roles and Behavior,* edited by R. Weg. New York: Academic Press, 1983.

Rodgers, Roy, and L. Conrad. "Courtship for Remarriage: Influences on Family Reorganization after Divorce." *Journal of Marriage and the Family* 48 (1986): 767–775.

Rodman, Hyman, and Cynthia Cole. "When School-Age Children Care for Themselves: Issues for Family Life Educators and Parents." *Family Relations* 36, 1 (January 1987): 92–96.

Roen, Philip. *Male Sexual Health.* New York: 1974.

Roenrich, L., and B. N. Kinder. "Alcohol Expectancies and Male Sexuality: Review and Implications for Sex Therapy." *Journal of Sex and Marital Therapy* 17 (1991): 45–54.

Rogers, Stacy J. "Mothers' Work Hours and Marital Quality: Variations by Family Structure and Family Size." *Journal of Marriage and the Family* 58, 3 (August 1996): 606–617.

Rogers, Susan M., and Charles F. Turner. "Male-Male Sexual Contact in the U.S.A.: Findings from Five Sample Surveys, 1970–1990." *Journal of Sex Research* 28, 4 (November 1991): 491–519.

Rolland, John. *Families, Illness, and Disability.* New York: Basic Books, 1994.

Roopnarine, J. L., and N. S. Mounts. "Current Theoretical Issues in Sex Roles and Sex Typing." In *Current Conceptions of Sex Roles and Sex Typing: Theory and Research,* edited by D. B. Carter. New York: Praeger, 1987.

Roopnarine, Jaipaul, et al. "Mothers' Perceptions of Their Children's Supplemental Care Experience Correlation with Spousal Relationship." *American Journal of Orthopsychiatry* 56, 4 (October 1986): 581–588.

Roos, Patricia, and Lawrence Cohen. "Sex Roles and Social Support as Moderates of Life Stress Adjustment." *Journal of Personality and Social Psychology* 52, 3 (March 1987): 576–585.

Root, Maria P., ed. *Filipino Americans: Transforming Identity.* Thousand Oaks, CA: Sage, 1997.

Roscoe, Bruce, and Nancy Benaske. "Courtship Violence Experienced by Abused Wives: Similarities in Patterns of Abuse." *Family Relations* 34, 3 (July 1985): 419–424.

Rose, Suzanna, and Irene Hanson Frieze. "Young Singles' Contemporary Dating Scripts." *Sex Roles* 28, 9–10 (May 1993): 499–510.

Rosen, M. P., et al. "Cigarette Smoking: An Independent Risk Factor for Atherosclerosis in the Hypograstric-Cavernous Arterial Bed of Men with Arteriogenic Impotence." *Journal of Urology* 145, 4 (April 1991): 759–776.

Rosenberg, Joshua D. "In Defense of Mediation." *Family and Conciliation Courts Review* 30, 4 (1992): 422–467.

Rosengren, A., H. Wedel, and L. Wilhelmsen. "Marital Status and Mortality in Middle-Aged Swedish Men." *American Journal of Epidemiology* 129, 1 (January 1989): 54–64.

Rosenthal, Carolyn. "Kinkeeping in the Familial Division of Labor." *Journal of Marriage and the Family* 47, 4 (November 1985): 965–947.

Rosenthal, Elisabeth. "Cost of High-Tech Fertility: Too Many Tiny Babies." *New York Times* (May 26, 1992): B5, B7.

Rosenthal, Kristine, and Harry F. Keshet. "The Not Quite Stepmother." *Psychology Today* 12 (1979): 82–86.

Rosenweig, P. M. *Married and Alone: The Way Back.* New York: Plenum, 1992.

Ross, Catherine. "Reconceptualizing Marital Status as a Continuum of Social Attachment," *Journal of Marriage and the Family* 57 (1) February 1995: 129–140.

Ross, Catherine, and John Mirowsky. "Parental Divorce, Life-Course Disruption, and Adult Depression. *Journal of Marriage and the Family* 61, 4 (November 1999): 1034–1045.

Ross, Catherine, John Mirowsky, and Karen Goldsteen. "The Impact of the Family on Health." In *Contemporary Families: Looking Forward, Looking Back,* edited by A. Booth. Minneapolis: National Council on Family Relations, 1991.

Ross, L., et al. "Television Viewing and Adult Sex-Role Attitudes." *Sex Roles* 8 (1982): 589–592.

Rossi, Alice. "Transition to Parenthood." *Journal of Marriage and the Family* 30 (1) February 1968: 26–39.

Rothenberg, R. B. "Those Other STDs." *American Journal of Public Health* 81 (October 1991): 1250–1251.

Rothschild, B. S., P. J. Fagan, and C. Woodall. "Sexual Functioning of Female Eating-Disordered Patients." *International Journal of Eating Disorders* 10 (1991): 389–394.

Rotter, Julian, B. Liverant, and D. P. Crowne. "The Growth and Extinction of Expectancies in Chance Controlled and Skilled Tests." *Journal of Psychology* 52 (1961): 161–177.

"Routine AZT Use Cuts Babies' HIV Risk, Study Finds." *San Mateo County Times* (July 10, 1996): A-6.

Rowan, Edward. "Editorial: Masturbation According to the Boy Scout Handbook." *Journal of Sex Education and Therapy* 15, 2 (June 1989): 77–81.

Rowland, Robyn. "Technology and Motherhood: Reproductive Choice Reconsidered." *Signs: Journal of Women in Culture and Society* 12, 3 (1987): 512–528.

Rubenstein, Carin, and Carol Tavris. "Special Survey Results: 26,000 Women Reveal the Secrets of Intimacy." *Redbook* (September 1987): 147–149 ff.

Rubin, A. M., and J. R. Adams. "Outcomes of Sexually Open Marriages." *Journal of Sex Research* 22 (1986): 311–319.

Rubin, Lillian. *Worlds of Pain: Life in the Working Class Family.* New York: Basic Books, 1976.

———. *Intimate Strangers: Men and Women Together.* New York: Perennial, 1983.

———. *Just Friends: The Role of Friendship in Our Lives.* New York: Harper and Row, 1985.

———. *Erotic Wars.* New York: Farrar, Straus and Giroux, 1990.

———. *Families on the Faultline: America's Working Class Speaks about the Family, the Economy, Race, and Ethnicity.* New York: HarperCollins, 1994.

Rubin, Sylvia. "Women's Health Goes Mainstream." *San Francisco Chronicle* (March 21, 1994): E9.

Rubin, Zick. *Liking and Loving: An Invitation to Social Psychology.* New York: Holt, Rinehart, and Winston, 1973.

———. "Self Disclosure in Dating Relationships: Sex Roles and the Ethic of Openness." *Journal of Marriage and the Family* 42 (1980): 305–317.

———. "Loving and Leaving: Sex Differences in Romantic Attachments." *Sex Roles* 7 (1981): 821–835.

Rubin, Zick, F. Provenzano, and Z. Luria. "The Eye of the Beholder: Parents' View of Sex of Newborn." *American Journal of Orthopsychiatry* 44 (September 1974): 512–519.

Runyan, William. "In Defense of the Case Study Method." *American Journal of Orthopsychiatry* 52, 3 (July 1982): 440–446.

Rush, D., et al. "The National WIC Evaluation." *American Journal of Clinical Nutrition* 48, 2 (Supplement, August 1988): 439–483.

Russel, C. "Transition to Parenthood." *Journal of Marriage and the Family* 36, 2 (May 1974): 294–302.

Russell, Diana. *Sexual Exploitation: Rape, Child Sexual Abuse, and Workplace Harassment.* Beverly Hills, CA: Sage, 1984.

Russell, Diana E. H. *The Secret Trauma: Incest in the Lives of Girls and Women.* New York: Basic Books, 1986.

———. *Rape in Marriage.* Rev. ed. Bloomington, IN: Indiana University Press, 1990.

Russell, Graeme. "Problems in Role-Reversed Families." In *Reassessing Fatherhood,* edited by C. Lewis and M. O'Brien. London: Sage, 1987: 161–179.

Russo, Nancy Felipe, J. D. Horn, and R. Schwartz. "U.S. Abortions in Context: Selected Characteristics." *Journal of Social Issues* 48, 3 (1992): 183–202.

Rutter, V. "Lessons from Stepfamilies." *Psychology Today* (May 1994): 30–33, 60.

Ryan, Michael. "We Need to Teach Doctors to Care." *Parade* (July 3, 1994): 8, 10.

Sabatelli, Ronald, and Erin Cecil-Pigo. "Relational Interdependence and Commitment in Marriage." *Journal of Marriage and the Family* 47, 4 (November 1985): 931–938.

Sadker, Myra, and David Sadker. *Failing at Fairness: How American Schools Cheat Girls.* New York: Touchstone, 1995.

Safilios-Rotschild, Constantina. "The Study of the Family Power Structure." *Journal of Marriage and the Family* 32, 4 (November 1970): 539–543.

———. "Family Sociology or Wives' Sociology? A Cross-Cultural Examination of Decisionmaking." *Journal of Marriage and the Family* 38 (1976): 355–362.

Saitoti,Tepelit Ole. *The Worlds of a Masai Warrior: An Autobiography.* Berkeley: University of California Press, 1986.

Salgado de Snyder, V. Nelly, Richard Cervantes, and Amado Padilla. "Gender and Ethnic Differences in Psychosocial Stress and Generalized Distress among Hispanics." *Sex Roles* 22, 7 (April 1990): 441–453.

Salholz, E. "The Future of Gay America." *Newsweek* (March 12, 1990): 20–25.

Salholz, Eloise. "The Marriage Crunch: If You're a Single Woman, Here Are Your Chances of Getting Married." *Newsweek* (June 2, 1986): 54–58.

Salovey, P., and J. Rodin. "Provoking Jealousy and Envy: Domain Relevance and Self-Esteem Threat." *Journal of Social and Clinical Psychology* 10, 4 (December 1991): 395–413.

Saluter, Arlene. "Marital Status and Living Arrangements: March 1989." *Current Population Reports.* (Population Characteristics Series P-20, No. 468.) Washington, DC: U.S. Census Bureau, 1992.

———. "Marital Status and Living Arrangements: March 1992." *Current Population Reports.* (Population Characteristics Series P-20). Washington, DC: U.S. Census Bureau, 1993.

———. "Marital Status and Living Arrangements: March 1993." *Current Population Reports.* (Population Characteristics Series P-20, No. 478). Washington, DC: U.S. Census Bureau, 1994a.

———. "Marital Status and Living Arrangements." *Current Population Reports* (Series P20–483). Washington, DC: U.S. Census Bureau, 1994b.

Sampson, Ronald. *The Problem of Power.* New York: Pantheon, 1966.

Samuels, M., and N. Samuels. *The New Well Pregnancy Book.* New York: Simon and Schuster, 1996.

Samuels, Shirley. *Ideal Adoption: A Comprehensive Guide to Forming an Adoptive Family.* New York: Insight Books, 1990.

Sanchez, Laura. "Feminism and Families." *Journal of Marriage and the Family* 59, 4 (1997): 1031–1032.

Sanchez, Laura, Steven Nock, James D. Wright, and Constance Gager. "Setting the Clock Forward or Back?" *Journal of Family Issues,* 23, 1 (January 2002): 91–120.

Sánchez-Ayéndez, Melba."Puerto Rican Elderly Women: Shared Meanings and Informal Supportive Networks." In *All American Women: Lives That Divide, Ties That Bind,* edited by J. B. Cole. New York: Free Press, 1986.

———."The Puerto Rican American Family." In *Ethnic Families in America: Patterns and Variations,* 3rd ed., edited by C. Mindel et al. Elsevier-North Holland, 1988.

Sander, William. "Catholicism and Intermarriage in the United States." *Journal of Marriage and the Family* 55, 4 (November 1993): 1037–1041.

Sanders, Gregory F., and Debra W. Trygstad. "Strengths in the Grandparent-Grandchild Relationship." *Activities, Adaptation and Aging* 17, 4 (1993): 43–53.

Sanders, Stephanie A., June M. Reinisch, and D. P. McWhirter. "Homosexuality/Heterosexuality: An Overview." In *Homosexuality/ Heterosexuality: Concepts of Sexual Orientation,* edited by D. P. McWhirter, S. A. Sanders, and J. M. Reinisch. New York: Oxford University Press, 1990.

Sanderson, Bettie, and Lawrence A. Kurdek. "Race and Gender as Moderator Variables in Predicting Relationship Satisfaction and Relationship Commitment in a Sample of Dating Heterosexual Couples." *Family Relations* 42, 3 (July 1993): 263–267.

Santrock, John, and Karen Sitterle. "Parent-Child Relationships in Stepmother Families." In *Remarriage and Stepparenting: Current Research and Theory,* edited by K. Pasley and M. Ihinger-Tallman. New York: Guilford Press, 1987.

Satir, Virginia. *The New Peoplemaking.* Rev. ed. Mountain View, CA: Science and Behavior Books, 1988.

———. *Peoplemaking.* Palo Alto, CA: Science and Behavior Books, 1972

Saunders, Daniel. "When Battered Women Use Violence: Husband-Abuse or Self-Defense." *Victim and Violence* 1, 1 (1986): 47–60.

Savin-Williams, Ritch, and Richard G. Rodriguez. "A Developmental, Clinical Perspective on Lesbian, Gay Male, and Bisexual Youths." In *Adolescent Sexuality,* edited by T. P. Gullotta et al. Newbury Park, CA: Sage, 1993.

Scanzoni, John. *Sexual Bargaining.* 2nd ed. Englewood Cliffs, NJ: Prentice Hall, 1980.

———. "Reconsidering Family Policy: Status Quo or Force for Change?" *Journal of Family Issues* 3, 3 (September 1982): 277–300.

Schaap, Cas, Bram Buunk, and Ada Kerkstra. "Marital Conflict Resolutions." In *Perspectives on Marital Interaction,* edited by P. Noller and M. A. Fitzpatrick. Philadelphia: Multilingual Matters, 1988.

Schaap, Cas, et al. "Marital Conflict Resolution." In *Perspectives on Marital Interaction,* edited by P. Noller and M. A. Fitzpatrick. Philadelphia: Multilingual Matters, 1988.

Schechter, S., and L. Gary. "A Framework for Understanding and Empowering Battered Women." In *Abuse and Victimization Across the Life Span,* edited by M. B. Straus. Baltimore: Johns Hopkins University Press, 1988.

Schenden, Laurie. "Gay Couples Making Adoption Gains." *Los Angeles Times* (September 29, 1993).

Schiavi, Raul C., P. Schreiner-Engle, J. Mandeli, J. Schanzer, and E. Cohen. "Chronic Alcoholism and Male Sexual Dysfunction." *Journal of Sex and Marital Therapy* 16, 1 (March 1990): 23–33.

Schickedanz, Judith A., David I. Schickedanz, Karen Hansen, and Peggy O Forstyh. *Understanding Children.* Mountain View, CA: Mayfield, 1993.

Schmalz, J. "Homosexuals Wake to See a Referendum: It's on Them." *New York Times* (January 31, 1993).

Schmitt, Bernard, and Robert Millard. "Construct Validity of the Bem Sex Role Inventory: Does the BSRI Distinguish between Gender-Schematic and Gender Aschematic Individuals?" *Sex Roles* 19, 9–10 (November 1988): 581–588.

Schneider, John. *Stress, Loss, and Grief: Understanding Their Origins and Growth Potential.* Baltimore: University Park Press, 1984.

Schneider, Peter. "Lost Innocents: The Myth of Missing Children." *Harper's Magazine* (February 1987): 47–53.

Schooler, Carmi. "Psychological Effects of Complex Environments during the Life Span: A Review and Theory." In *Cognitive Functioning and Social Structure over the Life Course,* edited by C. Schooler and K. Warner Schaie. Norwood, NJ: Ablex, 1987.

Schooler, Carmi, and K. Warner Schaie, eds. *Cognitive Functioning and Social Structure over the Life Course.* Norwood, NJ: Ablex, 1987.

Schooler, Carmi, et al. "Work for the Household: Its Nature and Consequences for Husbands and Wives." *American Journal of Sociology* 90, 1 (July 1984): 97–124.

Schor, Juliet B. *The Overworked American: The Unexpected Decline of Leisure.* New York: Basic Books, 1991.

Schrager, Cynthia D. "Questioning the Promise of Self-Help: A Reading of 'Women Who Love Too Much.'" *Feminist Studies* 19, 1 (1993): 176–192.

Schudson, Charles B. "Antagonistic Parents in Family Courts: False Allegations or False Assumptions about True Allegations of Child Sexual Abuse?" *Journal of Child Sexual Abuse* 1, 2 (1992): 113–116.

Schumm, Walter, et al. "His and Her Marriage Revisited." *Journal of Family Issues* 6, 2 (June 1985): 211–227.

Schvaneveldt, Jay. "The Interactional Framework in the Study of the Family." In *Emerging Conceptual Frameworks in Family Analysis,* 2nd ed., edited by F. I. Nye and F. Berardo. New York: Praeger, 1981.

Schvaneveldt, Jay, Shelley Lindauer, and Margaret Young. "Children's Understanding of AIDS: A Developmental Viewpoint." *Family Relations* 39, 3 (July 1990): 330–335.

Schvaneveldt, Jay, and Margaret H. Young. "Strengthening Families: New Horizons in Family Life Education." *Family Relations* 41, 4 (October 1992): 385–389.

Schwartz, Lita L. "Enabling Children of Divorce to Win." *Family and Conciliation Courts Review* 32, 1 (1994): 72–83.

Schwartz, Pepper. Interview with Bryan Strong. December 20, 1992.

———. *Peer Marriage: How Love Between Equals Really Works.* New York: Free Press, 1994.

Schwebel, Andrew I. Ryan L. Dunn, Barry F. Moss, and Maureena A. Renner. "Factors Associated with Relationship Stability in Geographically Separated Couples." *Journal of College Student Development* 33, 3 (1992): 222–230.

Schwebel, Andrew, Mark Fine, and Maureena Renner. "A Study of Perceptions of the Stepparent Role." *Journal of Family Issues* 23, 1 (March 1991): 43–57.

Scott, Clarissa, Lydia Shifman, Lavenda Orr et al. "Hispanic and Black American Adolescents' Beliefs Relating to Sexuality and Contraception." *Adolescence* 23, 91 (September 1998): 667–688.

Scott, Jacqueline. "Conflicting Beliefs about Abortion: Legal Approval and Moral Doubts." *Social Psychology Quarterly* 52, 4 (December 1989): 319–328.

Scott, Janny. "Low Birth Weight's High Cost." *Los Angeles Times,* (December 24, 1990a): 1.

———. "Trying to Save the Babies." *Los Angeles Times* (December 21, 1990b): 1, 18ff.

Scott, Joan. "Gender: A Useful Category of Historical Analysis." *American Historical Review* 91 (1986): 1053–1075.

Scott, S. G., et al. "Therapeutic Donor Insemination with Frozen Semen." *Canadian Medical Association Journal* 143, 4 (August 15, 1990): 273–278.

Scott-Jones, Diane, and Sherry Turner. "Sex Education, Contraceptive and Reproductive Knowledge, and Contraceptive Use among Black Adolescent Females." *Journal of Adolescent Research* 3, 2 (June 1988): 171–187.

Sears, David. "College Sophomores in the Laboratory: Influences of Narrow Data Base on Social Psychology's View of Human Nature." *Journal of Personality and Social Psychology* 51 (August 1986): 515–530.

Seaward, Brian Luke. *Managing Stress: Principles and Strategies for Health and Well-Being,* 2nd ed. Boston: Jones and Barlett, 1997.

Seecombe, Karen. "Families in Poverty in the 1990's: Trends, Causes, Consequences, and Lessons Learned." *Journal of Marriage and the Family*, 62, 4 (November 2000): 1094–1113.

———. "'Beating the Odds' Versus 'Changing the Odds': Poverty, Resilience, and Family Policy." *Journal of Marriage and the Family* 64, 2 (May 2002): 384–394.

Seidman, Steven. "Constructing Sex as a Domain of Pleasure and Self-Expression: Sexual Ideology in the Sixties." *Theory, Culture, and Society* 6, 2 (May 1989): 293–315.

Select Committee on Children, Youth and Families. *Families and Child Care: Improving the Options*. Rept. 98–1180, 98th Cong., 2nd session, 1985.

Seltzer, Judith. "Families Formed Outside of Marriage." *Journal of Marriage and the Family*, 62, 4 (November 2000): 1247–1268.

Semler, Tracy Chutorian. *All About Eve: The Complete Guide to Women's Health and Mental Well-Being*. New York: HarperCollins, 1995.

Sennett, Richard. *Authority*. New York: Knopf, 1980.

Sennett, Richard, and Jonathan Cobb. *The Hidden Injuries of Class*. New York: Vintage, 1972.

Serdahely, William, and Georgia Ziemba. "Changing Homophobic Attitudes through College Sexuality Education." *Journal of Homosexuality* 10, 1 (September 1984): 148 ff.

Serovich, Julianne M., Sharon J. Price, and Steven F. Chapman. "Former In-Laws as a Source of Support." *Journal of Divorce and Remarriage* 17, 1–2 (1991): 17–25.

Settles, Barbara H. "A Perspective on Tomorrow's Families." In *Handbook of Marriage and the Family*, edited by M. Sussman and S. Steinmetz. New York: Plenum Press, 1987.

Sexton, Christine, and Daniel Perlman. "Couples' Career Orientation, Gender Role Orientation, and Perceived Equity as Determinants of Marital Power." *Journal of Marriage and the Family* 51, 4 (November 1989): 933–941.

Shanis, B. S., et al. "Transmission of Sexually Transmitted Diseases by Donor Semen." *Archives of Andrology* 23, 3 (1989): 249–257.

Shannon, Thomas. *Surrogate Motherhood*. New York: Crossroad Publishing, 1988.

Shapiro, Adam D. "Explaining Psychological Distress in a Sample of Remarried and Divorced Persons. *Journal of Family Issues* 17, 2 (March 1996): 186–203.

Shapiro, David. "No Other Hope for Having a Child." *Newsweek* (January 19, 1987): 50–51.

Shapiro, Jerrold Lee. *The Measure of a Man: Becoming the Father You Wish Your Father Had Been*. New York: Delacorte, 1993.

Shapiro, Joanna. "Family Reactions and Coping Strategies in Response to the Physically Ill or Handicapped Child: A Review." *Social Science and Medicine* 17 (1983): 913–931.

Shapiro, Johanna, and Ken Tittle. "Maternal Adaptation to Child Disability in a Hispanic Population." *Family Relations* 39, 2 (April 1990): 179–185.

Sharpsteen, Don J. "Romantic Jealousy as an Emotion Concept: A Prototype Analysis." *Journal of Social and Personal Relationships* 10, 1 (1993): 69–82.

Shaver, Phillip, and Cindy Hazan. "Being Lonely, Falling in Love: Perspectives from Attachment Theory." *Journal of Social Behavior and Personality* 2, 2 Pt. 2 (1987): 105–124.

———. "A Biased Overview of the Study of Love." *Journal of Social and Personal Relationships* 5, 4 (1988): 473–501.

Shaver, Phillip, Cindy Hazan, and D. Bradshaw. "Love as Attachment: The Integration of Three Behavioral Systems." In *The Psychology of Love*, edited by R. Sternberg and M. Barnes. New Haven, CT: Yale University Press, 1988.

Shaver, Phillip R., and Kelly A. Brennan. "Attachment Styles and the 'Big Five' Personality Traits: Their Connections with Each Other and with Romantic Relationship Outcomes." *Personality and Social Psychology Bulletin* 18, 5 (1992): 536–545.

Shaw, David. "Despite Advances Stereotypes Still Used by Media." *Los Angeles Times* (December 12, 1990): A31.

Shaw, G. M., et al. "Preconceptional Vitamin Use, Dietary Folate, and the Occurrence of Neural Tube Defects." *Epidemiology* 6, 3 (1995): 219–226.

Shelp, Earl. *Born to Die?* New York: Free Press, 1986.

Shelton, Beth A., and Daphne John. "Does Marital Status Make a Difference? Housework among Married and Cohabiting Men and Women." *Journal of Family Issues* 14, 3 (September 1993): 401–420.

Sherman, Barry, and Joseph Dominick. "Violence and Sex in Music Videos: TV and Rock 'n' Roll." *Journal of Communication* 36, 1 (December 1986): 79–93.

Sherman, Beth. "Spanking Experts Say No." *San Jose Mercury News* (June 24, 1985).

Shifren, Kim, Robert Bauserman, and D. Bruce Carter. "Gender Role Orientation and Physical Health: A Study among Young Adults." *Sex Roles: A Journal of Research* 29, 5–6 (1993): 421–432.

Shilts, Randy. *And the Band Played On: Politics, People, and the AIDS Epidemic*. New York: St. Martin's Press, 1987.

Shon, Steven, and Davis Ja. "Asian Families." In *Ethnicity and Family Therapy*, edited by M. McGoldrick et al. New York: Guilford Press, 1982.

Shostak, A., G. McLouth, and L. Seng. *Men and Abortion: Lessons, Losses, and Love*. New York: Praeger, 1984.

Shostak, Arthur B. "Tommorow's Family Reforms: Marriage Course, Marriage Test, Incorporated Families, and Sex Selection Mandate." *Journal of Marital and Family Therapy* 7, 4 (October 1981): 521–528.

———. "Singlehood." In *Handbook of Marriage and the Family*, edited by M. Sussman and S. Steinmetz. New York: Plenum Press, 1987.

Sidorowicz, Laura, and G. Sparks Lunney. "Baby X Revisited." *Sex Roles* 6, 1 (February 1980): 67–73.

SIDS Resource Center. Personal communication with author Christine DeVault, 1996.

Sigelman, C. K., et al. "Courtesy Stigma: The Social Implications of Associating with a Gay Person." *Journal of Social Psychology* 131 (1991): 45–56.

Signorielli, Nancy. "Children, Television, and Gender Roles—Messages and Impact." *Journal of Broadcasting and Electronic Media* 33, 3 (June 1989a): 325–331.

———. "Television and Conceptions about Sex Roles—Maintaining Conventionality and the Status Quo." *Sex Roles* 21, 5 (September 1989b): 341–350.

Signorielli, Nancy, and Margaret Lears. "Children, Television, and Conceptions about Chores: Attitudes and Behaviors." *Sex Roles: A Journal of Research* 27, 3–4 (1992): 157–170.

Signorielli, Nancy, and N. Morgan, eds. *Cultivation Analysis: New Directions in Media Research.* Newbury Park, CA: Sage, 1990.

Silver, Donald, and B. Kay Campbell. "Failure of Psychological Gestation." *Psychoanalytic Inquiry* 8, 2 (1988): 222–223.

Silverstein, Judith L. "The Problem with In-Laws." *Journal of Family Therapy* 14, 4 (1992): 399–412.

Silverstein, Meril, Xuan Chen, and Kenneth Heller. "Too Much of a Good Thing? Intergenerational Social Support and the Psychological Well-Being of Older Parents." *Journal of Marriage and the Family* 58, 4 (November 1996): 970–982.

Simmons, Tavia, and Jane Lawler Dye, "Grandparents Living with Grandchildren, 2000," *Census 2000 Brief* (October 2003).

Simon, William. "Letters to the Editor: Reply to Muir and Eichel." *Archives of Sexual Behavior* 21, 6 (December 1992): 595–597.

Simon, William, and John Gagnon. "Sexual Scripts." *Society* 221, 1 (November 1984): 53–60.

Simons, Ronald, Kuei-Hsiu Lin, Leslie Gordon, Rand Conger, and Frederick Lorenz. "Explaining the Higher Incidence of Adjustment Problems Among Children of Divorce Compared with Those in Two-Parent Families." *Journal of Marriage and the Family* 61, 4 (November 1999): 1020–1033.

Simons, Ronald, and Les Whitbeck. "Sexual Abuse as a Precursor to Prostitution and Victimization Among Adolescent and Adult Homeless Women." *Journal of Family Issues* 12 (3) September 1991: 361–379.

Simpson, George Eaton, and J. Milton Yinger. *Racial and Cultural Minorities: An Analysis of Prejudice and Discrimination.* New York: Plenum Press, 1985.

Simpson, Jeffrey. "The Dissolution of Romantic Relationships: Factors Involved in Relationship Stability and Emotional Distress." *Journal of Personality and Social Psychology* 53, 4 (1987): 683–694.

Simpson, Jeffrey, B. Campbell, and E. Berscheid. "The Association between Romantic Love and Marriage: Kephart (1967) Twice Revisited." *Personality and Social Psychology Bulletin* 12 (1986): 363–372.

Simpson, William S., and Joanne A. Ramberg. "Sexual Dysfunction in Married Female Patients with Anorexia and Bulimia Nervosa." *Journal of Sex and Marital Therapy* 18, 1 (March 1992): 44–54.

Singer, Dorothy, et al. *Use TV to Your Child's Advantage.* Washington, DC: Acropolis Books, 1990.

Singer, Peter, and Deane Wells. *Making Babies: The New Science and Ethics of Conception.* New York: Scribner's, 1985.

Sinnott, Jan. *Sex Roles and Aging: Theory and Research from a Systems Perspective.* New York: Karger, 1986.

Skitka, L. J., and C. Maslach. "Gender Roles and the Categorization of Gender-Relevant Information." *Sex Roles* 22 (1990): 3–4.

Slater, Alan, Jan A. Shaw, and Joseph Duquesnel. "Client Satisfaction Survey: A Consumer Evaluation of Mediation and Investigative Services: Executive Summary." *Family and Conciliation Courts Review* 30, 2 (1992): 252–259.

Slater, Suzanne, and Julie Mencher. "The Lesbian Family Life Cycle: A Contextual Approach." *American Journal of Orthopsychiatry* 61, 3 (1991): 372–382.

"Sleeping on Back Saves 1,500 Babies." *San Mateo County Times* (June 25, 1996): A-4.

Sluzki, Carlos. "The Latin Lover Revisited." In *Ethnicity and Family Therapy,* edited by M. McGoldrick et al. New York: Guilford Press, 1982.

Small, Stephen, and Dave Riley. "Toward a Multidimensional Assessment of Work Spillover into Family Life." *Journal of Marriage and the Family* 52, 1 (February 1990): 51–61.

Smeeding, T. M. O'Higgins, and L. Rainwater, eds. *Poverty, Inequality and Income Distribution in Comparative Perspective; The Luxembourg Income Study.* Washington, DC: Urban Institute Press, 1990.

Smelser, Neil, and Erik Erikson, eds. *Themes of Work and Love in Adulthood.* Cambridge, MA: Harvard University Press, 1980.

Smith, Corless. "Sex and Genre on Prime Time." *Journal of Homosexuality* 21, 1–2 (1991): 119–138.

Smith, E. J. "The Black Female Adolescent: A Review of the Educational, Career, and Psychological Literature." *Psychology of Women Quarterly* 6 (1982): 261–288.

Smith, Herbert L., and S. Philip Morgan. "Children's Closeness to Father as Reported by Mothers, Sons and Daughters." *Journal of Family Issues* 15, 1 (March 1994): 3–29.

Smith, Ken, and Cathleen Zick. "The Incidence of Poverty among the Recently Widowed: Mediating Factors in the Life Course." *Journal of Marriage and the Family* 48 (1986): 619–630.

Smith, T. W. "Changing Racial Labels: From Negro to Black to African American." *Public Opinion Quarterly* 56, 4 (1992): 496–514.

SmithBattle, Lee. "Intergenerational Ethics of Caring for Adolescent Mothers and Their Children." *Family Relations* 45, 1 (January 1996): 56–64.

Smits, Jeroen, Wout Ultee, and Jan Lammers. "Effects of Occupational Status Differences Between Spouses on the Wife's Labor Force Participation and Occupational Achievement: Findings from 12 European Countries." *Journal of Marriage and the Family* 58, 1 (February 1996): 101–115.

Smock, P. J. "The Economic Costs of Marital Disruption for Young Women over the Past Two Decades." *Demography 30, 3* (August 1993): 353–371.

Smock, Pamela J. "Cohabitation in the United States: An Appraisal of Research Themes, Findings, and Implications." *Annual Review of Sociology* 26, Summer 2000.

Snarey, John, et al. "The Role of Parenting in Men's Psychosocial Development." *Developmental Psychology* 23, 4 (July 1987): 593–603.

Snitow, Ann. "Mass Market Romance: Pornography for Women is Different." In *Powers of Desire: The Politics of Sexuality,* edited by A. Snitow et al. New York: Monthly Review Press, 1983.

Sobol, Thomas. "Understanding Diversity." *Educational Leadership* 48, 3 (November 1990): 27–31.

Solon, G. "Intergenerational Income Mobility in the United States." *American Economic Review* 82, 3 (June 1992): 393–408.

Sommers-Flanagan, Rita, John Sommers-Flanagan, and Britta Davis. "What's Happening on Music and Television? A Gender Role Content Analysis." *Sex Roles* 28, 11–12 (June 1993): 745–753.

Sonenstein, Freya L. "Rising Paternity: Sex and Contraception Among Adolescent Males." In *Adolescent Fatherhood,* edited by A. B. Elster and M. E. Lamb. Hillsdale, NJ: Lawrence Erlbaum, 1986.

Sonenstein, Freya L., Joseph H. Pleck, and L. C. Ku. "Sexual Activity, Condom Use, and AIDS Awareness among Adolescent Males." *Family Planning Perspectives* 21, 4 (July 1989): 152–158.

Sontag, Susan. "The Double Standard of Aging." *Saturday Review* 55 (September 1972): 29–38.

Sophie, Joan. "Internalized Homophobia and Lesbian Identity." *Journal of Homosexuality* 14, 1–2 (September 1987): 53–65.

South, Scott, and Richard Felson. "The Racial Patterning of Rape." *Social Forces* 69, 1 (September 1990): 71–93.

South, Scott, and Glenna Spitze. "Determinants of Divorce over the Marital Life Course." *American Sociological Review* 47 (1986): 583–590.

Spallone, P., and D. L. Steinberg. *Made to Order: The Myth of Reproductive and Genetic Progress.* New York: Pergamon Press, 1987.

Spanier, Graham B. "Measuring Dyadic Adjustment." *Journal of Marriage and the Family* 38, 1 (February 1976): 15–28.

———. "Improve, Refine, Recast, Expand, Clarify—But Don't Abandon." *Journal of Marriage and the Family* 47, 4 (November 1985): 1073–1074.

Spanier, Graham B., and Frank Furstenberg Jr. "Remarriage After Divorce: A Longitudinal Analysis of Well-being." *Journal of Marriage and the Family* 44, 3 (August 1982): 709–720.

———. "Remarriage and Reconstituted Families." In *Handbook of Marriage and the Family,* edited by M. Sussman and S. Steinmetz. New York: Plenum Press, 1987.

Spanier, Graham B., and R. L. Margolis. "Marital Separation and Extramarital Sexual Behavior." *Journal of Sex Research* 19 (1983): 23–48.

Spanier, Graham B., and Linda Thompson. "A Confirmatory Analysis of the Dyadic Adjustment Scale." *Journal of Marriage and the Family* 44, 3 (August 1982): 731–738.

———. "Relief and Distress after Marital Separation." *Journal of Divorce* 7 (1983): 31–49.

———, eds. *Parting: The Aftermath of Separation and Divorce.* Beverly Hills, CA: Sage Publications, 1984.

———. *Parting: The Aftermath of Separation and Divorce.* Rev. ed. Newbury Park, CA: Sage, 1987.

Spanier, Graham B., et al. "Marital Trajectories of American Women: Variations in the Life Course." *Journal of Marriage and the Family* 47, 4 (November 1985): 993–1003.

Spector, I. P., and M. P. Carey. "Incidence and Prevalence of the Sexual Dysfunctions—A Critical Review of the Empirical Literature." *Archives of Sexual Behavior* 19, 4 (August 1990): 389–408.

Spence, Janet T. "Gender-Related Traits and Gender Ideology: Evidence for a Multifactorial Theory (Personality Processes and Individual Differences)." *Journal of Personality and Social Psychology* 64, 4 (1993): 624–636.

Spence, Janet, and L. L. Sawin. "Images of Masculinity and Femininity." In *Sex, Gender, and Social Psychology,* edited by V. O'Leary et al. Hillsdale, NJ: Lawrence Erlbaum, 1985.

Spence, Janet, et al. "Ratings of Self and Peers on Sex Role Attributes and the Relation to Self-Esteem and Conceptions of Masculinity and Feminity." *Journal of Personality and Social Psychology* 32 (1975): 29–39.

———. "Sex Roles in Contemporary Society." In *Handbook of Social Psychology,* edited by G. Lindzey and E. Aronson. New York: Random House, 1985.

Spencer, Paul. *The Masai of Matapato.* Bloomington: Indiana University Press, 1988.

Spencer, S. L., and A. M. Zeiss. "Sex Roles and Sexual Dysfunction in College Students." *Journal of Sex Research* 23, (1987): 338–347.

Spiegel, Lynn. *Installing the Television Set: Television and the Family Ideal in Postwar America.* Chicago: University of Chicago Press, 1991.

Spitzberg, Brian H., and William R. Cupach. "The Inappropriateness of Relational Intrusion." *Inappropriate Relationships: The Unconventional, the Disapproved, and the Forbidden.* Mahway, NJ: Lawrence Erlbaum, 2002.

Spitzberg, Brian H., Linda Marshall, and William R. Cupach. "Obsessive Relational Intrusion, Coping, and Sexual Coercion Victimization." *Communication Reports* 14, 1 (2001): 19–30.

Spitze, Glenna. "The Division of Task Responsibility in U.S. Households: Longitudinal Adjustments to Change." *Social Forces* 64 (1986): 689–701.

Spitze, Glenna, and Scott South. "Women's Employment, Time Expenditure, and Divorce." *Journal of Family Issues* 6, 3 (September 1985): 307–329.

Spock, Benjamin, and Michael Rothenberg. *Dr. Spock's Baby and Child Care.* New York: Pocket Books, 1985.

Sponaugle, G. C. "Attitudes Toward Extramarital Relations." In *Human Sexuality: The Societal and Interpersonal Context,* edited by K. McKinney and S. Sprecher. Norwood, NJ: Ablex, 1989.

Sprecher, Susan. "Sex Difference in Bases of Power in Dating Relationships." *Sex Roles* 12, 34 (1985): 449–462.

———. "A Revision of the Reiss Premarital Sexual Permissiveness Scale." *Journal of Marriage and the Family* 50, 3 (August 1988): 821–828.

———. "Influences on Choice of a Partner and on Sexual Decision Making in the Relationship." In *Human Sexuality: The Social and Interpersonal Context,* edited by K. McKinney and S. Sprecher. Norwood, NJ: Ablex, 1989.

Sprecher, Susan, and Kathleen McKinney. *Sexuality.* Newbury Park, CA: Sage, 1993.

Sprecher, Susan, et al. "Sexual Relationships." In *Human Sexuality: The Societal and Interpersonal Context,* edited by K. McKinney and S. Sprecher. Norwood, NJ: Ablex, 1989.

Sprey, Jetse. "Current Theorizing on the Family: An Appraisal." *Journal of Marriage and the Family* 50, 4 (November 1988): 875–890.

Springer, D., and D. Brubaker. *Family Caregivers and Dependent Elderly: Minimizing Stress and Maximizing Independence.* Beverly Hills, CA: Sage, 1984.

Stacey, Judith. "Good Riddance to the Family: A Response to David Popenoe," *Journal of Marriage and the Family* 55 (3) (August 1993): 545-547.

Stacey, Judith, and Timothy Biblarz "(How) Does the Sexual Orientation of Parents Matter?" *American Sociological Review* 66, 2 (2001): 159–183.

Stack, Carol B. *All Our Kin: Strategies for Survival in a Black Community.* New York: Harper and Row, 1974.

Stack, C. B., and L. M. Burton. "Kinscripts; Reflections on Family, Generation, and Culture." In *Mothering: Ideology, Experience, and Agency,* edited by E. Glenn, G. Chang, and L. R. Forcey. New York: Routledge, 1992.

Stafford, Laura, and James R. Reske."Idealization and Communication in Long-Distance Premarital Relationships." *Family Relations: Journal of Applied Family and Child Studies* 39, 3 (1990): .274–279.

Stallones, Lorann, Martin B. Marx, Thomas F. Garrity, and Timothy P. Johnson. "Pet Ownership and Attachment in Relation to the Health of U.S. Adults, 21 to 64 Years of Age." *Anthrozoos* 4, 2 (1990): 100–112.

Stanton, M. Colleen. "The Fetus: A Growing Member of the Family." *Family Relations* 34, 3 (July 1985): 321–326.

Staples, Robert. "The Black American Family." In *Ethnic Families in America,* edited by C. Mindel and R. Habenstein. New York: Elsevier-North Holland, 1976.

———. "The Black American Family." In *Ethnic Families in America: Patterns and Variations,* 3rd ed., edited by C. Mindel et al. New York: Elsevier-North Holland, 1988a.

———. "The Emerging Majority: Resources for Nonwhite Families in the United States." *Family Relations* 37, 3 (July 1988b): 348–354.

———. "The Sexual Revolution and the Black Middle Class." In *The Black Family,* edited by R. Staples. 4th ed. Belmont, CA: Wadsworth, 1991.

———, ed. *The Black Family: Essays and Studies.* 3rd ed. Belmont, CA: Wadsworth, 1988c.

Staples, Robert, and Leanor Boulin Johnson. *Black Families at the Crossroads: Challenges and Prospects.* San Francisco: Jossey-Bass, 1993.

Staples, Robert, and Alfredo Mirandé. "Racial and Cultural Variations among American Families: A Decennial Review of the Literature." *Journal of Marriage and the Family* 42 (1980): 887–922.

Starrels, Marjorie E. "Gender Differences in Parent-Child Relations." *Journal of Family Issues* 15, 1 (March 1994): 148–165.

Starrels, Marjorie E., Sally Bould, and Leon J. Nicholas. "The Feminization of Poverty in the United States: Gender, Race, Ethnicity, and Family Factors." *Journal of Family Issues* 15, 4 (December 1994): 590–607.

Steck, L., D. Levitan, D. McLane, and H. H. Kelley. "Care, Need, and Conceptions of Love." *Journal of Personality and Social Psychology* 43 (1982): 481–491.

Steele, Brandt F. "Psychodynamic Factors in Child Abuse." In *The Battered Child,* edited by C. H. Kempe and R. Helfer. Chicago: University of Chicago Press, 1980.

Steelman, Lala Carr, and Brian Powell. "The Social and Academic Consequences of Birth Order: Real, Artificial, or Both?" *Journal of Marriage and the Family* 47 (1985): 117–124.

Stein, Peter, ed. *Single.* Englewood Cliffs, NJ: Prentice Hall, 1976.

Stein, Peter. "Men and Their Friendships." In *Men In Families,* edited by R. Lewis and R. Salt. Beverly Hills, CA: Sage, 1986: 261–270.

Steinberg, Laurence, and Susan Silverberg. "Marital Satisfaction in Middle Stages of Family Life Cycle." *Journal of Marriage and the Family* 49, 4 (November 1987): 751–760.

Steinglass, Peter. "Families and Substance Abuse." In *Vision 2010: Families and Health Care,* edited by S. Price and B. Elliott. Minneapolis: National Council of Family Relations, 1993.

Steinglass, Peter, L. A. Bennett, S. J. Wolin, and D. Reiss. *The Alcoholic Family.* New York: Basic Books, 1987.

Steinman, Susan. "The Experience of Children in a Joint-Custody Arrangement: A Report of a Study." *American Journal of Orthopsychiatry* 51, 3 (July 1981): 403–414.

Steinmetz, Suzanne. "Family Violence." In *Handbook of Marriage and the Family,* edited by M. Sussman and S. Steinmetz. New York: Plenum Press, 1987.

Steinmetz, Suzanne, Sylvia Clavan, and K. Stein. *Marriage and Family Realities.* New York: Harper and Row, 1990.

Steinmetz, Suzanne K. "The Abused Elderly Are Dependent: Abuse Is Caused by the Perception of Stress Associated with Providing Care." In *Current Controversies in Family Violence,* edited by R. Gelles and D. Loseke. Newbury Park, CA: Sage,, 1993.

Stephen, Timothy. "Fixed-Sequence and Circular-Causal Models of Relationship Development: Divergent Views on the Role of Communication in Intimacy." *Journal of Marriage and the Family* 47, 4 (November 1985): 955–963.

Sternberg, Robert, and Michael Barnes, eds. "Real and Ideal Others in Romantic Relationships: Is Four a Crowd?" *Journal of Personality and Social Psychology* 49 (1985): 1589–1596.

———. "A Triangular Theory of Love." *Psychological Review* 93 (1986): 119–135.

———. *The Psychology of Love.* New Haven, CT: Yale University Press, 1988.

Sternberg, Robert, and S. Grajek. "The Nature of Love." *Journal of Personality and Social Psychology* 47 (1984): 312–327.

Stets, Jan E. "Verbal and Physical Aggression in Marriage." *Journal of Marriage and the Family* 52, 2 (May 1990): 501–514.

Stets, Jan E., and Maureen A. Pirog-Good. "Violence in Dating Relationships." *Social Psychology Quarterly* 50, 3 (September 1987): 237–246.

———. "Patterns of Physical and Sexual Abuse for Men and Women in Dating Relationships: A Descriptive Analysis." *Journal of Family Violence* 4, 1 (March 1989): 63–76.

Stets, Jan E., and Murray A. Straus. "The Marriage License as a Hitting License: A Comparison of Assaults in Dating, Cohabitation and Married Couples." *Journal of Family Violence* 4, 2 (June 1989): 161–180.

Stevens, Gillian, and Robert Schoen. "Linguistic Intermarriage in the United States." *Journal of Marriage and the Family* 50, 1 (February 1988): 267–280.

Stevenson, M. "Tolerance for Homosexuality and Interest in Sexuality Education." *Journal of Sex Education and Therapy* 16 (1990): 194–197.

Stewart, Susan. "Nonresident Mothers' and Fathers' Social Contact with Children" *Journal of Marriage and the Family* 61, 4 (November 1999): 894-907.

Stinnett, Nick, and John DeFrain. *Secrets of Strong Families.* Boston: Little, Brown, 1985.

Stockard, Janice. *Daughters of the Canton Delta: Marriage Patterns and Economic Strategies in South China, 1860–1930.* Stanford, CA: Stanford University Press, 1989.

Stockdale, M. S. "The Role of Sexual Misperceptions of Women's Friendliness in an Emerging Theory of Sexual Harassment." *Journal of Vocational Behavior* 42, 1 (February 1993): 84–101.

Strachan, Catherine E. "The Role of Power and Gender in Anger Responses to Sexual Jealousy." *Journal of Applied Social Psychology* 22, 22 (1992): 1721–1740.

Strasser, Susan. *Never Done: A History of American Housework.* New York: Pantheon Books, 1982.

Straus, Murray. "Preventing Violence and Strengthening the Family." In *Family Strengths, 4,* edited by N. Stinnett et al. Lincoln: University of Nebraska Press, 1980.

Straus, Murray A. "Physical Assaults by Wives: A Major Social Problem." In *Current Controversies in Family Violence,* edited by R. Gelles and D. Loseke. Newbury Park, CA: Sage, 1993.

Straus, Murray, and Carolyn Field. "Psychological Aggression by American Parents: National Data on Prevalence, Chronicity, and Severity." *Journal of Marriage and the Family* 65, 4 (November 2003): 795–808.

Straus, Murray, Richard Gelles, and Suzanne Steinmetz. *Behind Closed Doors.* Garden City, NY: Anchor Books, 1980.

Straus, Murray, and Carrie Yodanis. "Corporal Punishment in Adolescence and Physical Assaults on Spouses Later in Life: What Accounts for the Link?" *Journal of Marriage and the Family* 58, 4 (November 1996): 825–841.

Strom, Robert, Pat Collinsworth, Shirley Strom, and D. Griswold. "Strengths and Needs of Black Grandparents." *International Journal of Aging and Human Development* 36, 4 (1992–1993): 255–268.

Stroman, C. "The Socialization Influence of Television on Black Children." *Journal of Black Studies* 15 (September 1989): 79–100.

Strong, Bryan, and Christine DeVault. *Human Sexuality: Diversity in Contemporary America,* 2nd ed. Mountain View, CA: Mayfield, 1997.

Strube, Michael J., and Linda Barbour. "Factors Related to the Decision to Leave an Abusive Relationship." *Journal of Marriage and the Family* 46, 4 (November 1984): 837–844.

"Study Blames MTV for Video Violence." *San Francisco Chronicle* (May 15, 1997): E-1.

"Study Reveals Deep Scars of Divorce." *San Francisco Chronicle* (June 3, 1997): Al ff.

"Study Says 20% of Kids Get Poor Preventive Care." *San Jose Mercury News* (May 30, 1992): 8A.

Suarez, Zulema. "Cuban American Families." In *Families in Cultural Context,* edited by M. K. DeGenova. Mountain View, CA: Mayfield, 1997.

Sugarman, David B., and Gerald T. Hotaling. "Dating Violence: Prevalence, Context, and Risk Markers." In *Violence in Dating Relationships: Emerging Social Issues,* edited by M. Pirog-Good and J. Stets. New York: Praeger, 1989.

Suitor, J. Jill. "Marital Quality and Satisfaction with Division of Household Labor." *Journal of Marriage and the Family* 53, 1 (February 1991): 221–230.

Sulloway, Frank J. *Born to Rebel; Birth Order, Family Dynamics, and Creative Lives.* New York: David McKay Company, 1996.

Sung, B. L. *Mountains of Gold.* New York: Macmillan, 1967.

Sunoff, Alvin. "A Conversation with Jerome Kagan." *U.S. News and World Report* (March 25, 1985): 63–64.

Surra, Catherine. "Research and Theory on Mate Selection and Premarital Relationships in the 1980s." In *Contemporary Families: Looking Forward, Looking Back,* edited by A. Booth. Minneapolis: National Council on Family Relations, 1991.

Surra, Catherine, P. Arizzi, and L. L. Asmussen. "The Association between Reasons for Commitment and the Development and Outcome of Marital Relationships." *Journal of Social and Personal Relationships* 5 (1988): 47–63.

"Survey of 80,000 Cases Calls Birth-defect Test Safe." *San Jose Mercury News* (August 4, 1992): 3F.

Sussman, Marvin B. S. Steinmetz, and G. Peterson eds. *Handbook of Marriage and the Family,* 2nd ed. New York: Plenum Press, 1999.

Swain, Scott. "Covert Intimacy: Closeness in Men's Friendships." In *Gender in Intimate Relationships: A Microstructural Approach,* edited by B. Risman and P. Schwartz . Belmont, CA.: Wadsworth, 1989.

Sweet, J. A., L. L. Bumpass, and V. R. A. Call. *The Design and Content of the National Survey of Families and Households.* Working Paper NSFH-1. Madison: University of Wisconsin, Center for Demography and Ecology, 1988.

Swensen, C. H., Jr. "The Behavior of Love." In *Love Today: A New Exploration,* edited by H. A. Otto. New York: Association Press, 1972.

Swenson, Clifford, and Geir Trahaug. "Commitment and the Long-Term Marriage Relationship." *Journal of Marriage and the Family* 47, 4 (November 1985): 939–945.

Swift, Carolyn. "Preventing Family Violence: Family-Focused Programs." In *Violence in the Home: Interdisciplinary Perspectives,* edited by M. Lystad. New York: Brunner/Mazel, 1986.

Symons, Donald. *The Evolution of Human Sexuality.* New York: Oxford University Press, 1979.

Szapocznik, Jose, and Roberto Hernandez. "The Cuban American Family." In *Ethnic Families in America: Patterns and Variations,* 3rd ed,, edited by C. Mindel et al. New York: Elsevier North Holland, 1988.

Szinovacz, Maximiliane. "Family Power." In *Handbook of Marriage and the Family,* edited by M. Sussman and S. Steinmetz. New York: Plenum Press, 1987.

Takagi, Diana Y. "Japanese American Families." In *Minority Families in the United States: A Multicultural Perspective,* edited by R. L. Taylor. Englewood Cliffs, NJ: Prentice Hall, 1994.

"Talking to Children about Prejudice." *PTA Today* (December 1989–1990): 7–8.

Tallmer, Margot. "Grief as a Normal Response to the Death of a Loved One." In *Principles of Thanatology,* edited by A. H. Kutscher et al. New York: Columbia University Press, 1987.

Tanfer, Koray. "Patterns of Premarital Cohabitation among Never-Married Women in the United States." *Journal of Marriage and the Family* 49, 3 (August 1987): 683–697.

Tanke, E. D. "Dimensions of the Physical Attractiveness Stereotype: A Factor/Analytic Study." *Journal of Psychology* 110 (1982): 63–74.

Tanouye, Elyse. "Safety of Tests for Birth Defect Backed in Study." *Wall Street Journal* (August 27, 1992): B1.

Tavris, Carol. *The Mismeasure of Woman.* New York: Norton, 1992.

Tavris, Carol, and Carol Wade, eds. *The Longest War: Sex Differences in Perspective.* 2nd ed. New York: Harcourt Brace Jovanovich, 1984.

Taylor, Ella. "From the Nelsons to the Huxtables: Genre and Family Imagery in American Network Television." *Qualitative Sociology* 12, 1 (Spring 1989a): 13–28.

———. *Prime-Time Families: Television Culture in Postwar America.* Berkeley, CA: University of California Press, 1989b.

Taylor, J. R., A. P. Lockwood, and A. J. Taylor. "The Prepuce: Specialized Mucosa of the Penis and Its Loss to Circumcision," *British Journal of Urology 77* (1996): 291–295.

Taylor, Patricia. "It's Time to Put Warnings on Alcohol." *New York Times* (March 20, 1988): B2.

Taylor, Robert J. "Receipt of Support from Family among Black Americans." *Journal of Marriage and the Family* 48, 1 (February 1986): 67.

———. "Need for Support and Family Involvement among Black Americans." *Journal of Marriage and the Family* 52, 3 (August 1990): 584–590.

———. "Black American Families." In *Minority Families in the United States: A Multicultural Perspective,* edited by R. L. Taylor. Englewood Cliffs, NJ: Prentice Hall, 1994a.

———. "Minority Families in America." In *Minority Families in the United States: A Multicultural Perspective,* edited by R. L. Taylor. Englewood Cliffs, NJ: Prentice Hall, 1994b.

Taylor, Robert J., Linda M. Chatters, and James S. Jackson. "A Profile of Familial Relations Among Three-Generation Black Families." *Family Relations* 42, 3 (1993): 332–341.

Taylor, Robert J., Linda M. Chatters, Belinda Tucker, and Edith Lewis. "Developments in Research on Black Families." In *Contemporary Families: Looking Forward, Looking Back,* edited by A. Booth. Minneapolis: National Council on Family Relations, 1991.

Taylor, Ronald L., ed. *Minority Families in the United States: A Multicultural Perspective.* Englewood Cliffs, NJ: Prentice Hall, 1994.

Teachman, Jay. "Receipt of Child Support in the United States." *Journal of Marriage and the Family* 53, 3 (August 1991): 759–772.

Teachman, Jay, and Alex Heckert. "The Impact of Age and Children on Remarriage." *Journal of Family Issues* 6, 2 (June 1985): 185–203.

Teachman, Jay, Lucky Tedrow, and Kyle Crowder. "The Changing Demography of America's Families." *Journal of Marriage and the Family* 62, 4 (November 2000): 1234–1246.

Teachman, Jay D., and Karen A. Polonko. "Timing of the Transition to Parenthood: A Multidimensional Birth-interval Approach." *Journal of Marriage and the Family* 47, 4 (November 1985): 867–879.

———. "Cohabitation and Marital Stability in the United States." *Social Forces* 69, 1 (September 1990): 207–220.

Teachman, Jay D., R. Vaughn, A. Call, and Karen P. Carver. "Marital Status and Duration of Joblessness Among White Men." *Journal of Marriage and the Family* 56, 2 (May 1994): 415–428.

Teeser, Abraham, and Richard Reardon. "Perceptual and Cognitive Mechanisms in Human Sexual Attraction." In *The Bases of Human Sexual Attraction,* edited by M. Cook. New York: Academic Press, 1981.

Tessina, Tina. *Gay Relationships: For Men and Women. How to Find Them, How to Improve Them, How to Make Them Last.* Los Angeles: Jeremy P. Tarcher, 1989.

Testa, Ronald J., Bill N. Kinder, and G. Ironson. "Heterosexual Bias in the Perception of Loving Relationships of Gay Males and Lesbians." *Journal of Sex Research* 23, 2 (May 1987): 163–172.

Thayer, Leo. *On Communication.* Norwood, NJ: Ablex, 1986.

Thoits, Peggy A. "Identity Structures and Psychological Well-Being: Gender and Marital Status Comparisons." *Social Psychology Quarterly* 55, 3 (1992): 236–256.

Thomma, Steven, and Angie Cannon. "Mood of Gloom and Doom Persists, But America Is Better Than It Was." *San Jose Mercury* (October 30, 1994): C 1, 6.

Thompson, Anthony. "Extramarital Sex: A Review of the Research Literature." *Journal of Sex Research* 19, 1 (February 1983): 1–22.

———. "Emotional and Sexual Components of Extramarital Relations." *Journal of Marriage and the Family* 46, 1 (February 1984): 35–42.

Thompson, Edward, Christopher Grisanti, and Joseph Pleck. "Attitudes toward the Male Role and Their Correlates." *Sex Roles* 13, 7 (October 1985): 413–427.

Thompson, Linda. "Family Work: Women's Sense of Fairness," *Journal of Family Issues* 12 (2) June 1991a: 181–196.

———. "Gender in Families: Women and Men in Marriage, Work, and Parenthood." In *Contemporary Families: Looking Forward, Looking Back,* edited by A. Booth. Minneapolis: National Council on Family Relations, 1991b.

———. "Conceptualizing Gender in Marriage: The Case of Marital Care." *Journal of Marriage and the Family* 55, 3 (August 1993): 557–569.

———. "The Place of Feminism in Family Studies." *Journal of Marriage and the Family* 57 (4) November 1995: 847–866.

Thompson, Linda, et al. "Developmental Stage and Perceptions of Intergenerational Continuity." *Journal of Marriage and the Family* 47, 4 (November 1985): 913–920.

Thompson, Ross, and Deborah Laible. "Noncustodial Parents." In *Parenting and Child Development in "Nontraditional" Families* edited by M. Lamb. Mahwah, NJ: Lawrence Erlbaum.

Thomson, Keith. "Research on Human Embryos: Where to Draw the Line." *American Scientist* (March 1985): 187–189.

Thorne, B., and M. Yalom, eds. *Rethinking the Family: Some Feminist Questions.* New York: Longman, 1982.

Thornton, Arland. "Changing Attitudes toward Family Issues in the United States." *Journal of Marriage and the Family* 51, 4 (November 1989): 873–893.

———. "The Courtship Process and Adolescent Sexuality." *Journal of Family Issues* 11, 3 (September 1990): 239–273.

Thornton, Arland, William G. Axinn, and Daniel H. Hill. "Reciprocal Effects of Religiosity, Cohabitation, and Marriage." *American Journal of Sociology* 98, 3 (1992): 628–651.

Thornton, Arland, et al. "Causes and Consequences of Sex-Role Attitudes and Attitude Change." *American Sociological Review* 48 (1983): 211–227.

Thorton-Dill, B. "Fictive Kin, Paper Sons, and *Compadrazgo:* Women of Color and the Struggle for Family Survival." In *Women of Color in U.S. Society,* edited by M. Baca-Zinn and B. Thorton-Dill. Philadelphia: Temple University Press, 1994.

Tienda, Marta, and Ronald Angel. "Headship and Household Composition among Blacks, Hispanics, and Other Whites." *Social Forces* 61 (1982): 508–531.

Tienda, Marta, and Jennifer Glass. "Household Structure and Labor Force Participation of Black, Hispanic, and White Mothers." *Demography* 22 (1985): 281–394.

Tietjen, Anne, and Christine F. Bradley. "Social Support and Maternal Psychosocial Adjustment during the Transition to Parenthood." *Canadian Journal of Behavioral Science* 17, 2 (April 1985): 109–121.

Ting-Toomey, Stella. "An Analysis of Verbal Communication Patterns in High and Low Marital Adjustment Groups." *Human Communications Research* 9, 4 (June 1983): 306–319.

TIME Almanac 2000, edited by B. Brunner. Boston: Information Please, 1999.

Toback, B. M. "Recent Advances in Female Infertility Care." *Naacogs Clinical Issues in Perinatal and Women's Health Nursing* 3, 2 (1992): 313–319.

Toback, James. "James Toback on 'The Hunger.'" [Special Issue: "The Sexual Revolution in Movie, Music and TV"] *US* (August 1992): 56–58.

Tolman, Richard M. "Treatment Program for Men Who Batter." In *Vision 2010: Families and Violence, Abuse and Neglect,* edited by R. J. Gelles. Minneapolis: National Council on Family Relations, 1995.

Toney, G., and J. Weaver. "Effects of Gender and Gender Role Self-Perceptions on Affective Reactions to Rock Music Videos." *Sex Roles* 30, 7–8 (April 1994): 567–583.

Tooth, Geoffrey. "Why Children's TV Turns Off So Many Parents." *U.S. News and World Report* (February 18, 1985): 65.

Torres, Aida, and Jacqueline D. Forrest. "Why Do Women Have Abortions?" *Family Planning Perspectives* 20, 4 (July-August 1988): 169–176.

Torres, José. "A Letter to a Child Like Me." *Parade* (February 26, 1991): 8–9.

Torrey, Barbara. "Aspects of the Aged: Clues and Issues." *Population and Development Review* 14, 3 (September 1988): 489–497.

Toufexis, Anastasia. "Older—But Coming on Strong." *Time* (February 22, 1988): 76–79.

Tracy, Kathleen. *The Secret Story of Polygamy.* Naperville, IL: Sourcebooks, 2002.

Trafford, Abigail. "Medical Science Discovers the Baby." *U.S . News and World Report* (November 10, 1980): 59–62.

Tran, Than Van. "The Vietnamese American Family." In *Ethnic Families in America: Patterns and Variations,* 3rd ed., edited by C. H. Mindel et al. New York: Elsevier, 1988.

———. "The Vietnamese-American Family." In *Ethnic Families in America: Patterns and Variations,* 4th ed., edited by C. Mindel, R. Habenstein, and R. Wright.Upper Saddle River, NJ: Prentice Hall, 1998: 254–283.

Treas, Judith, and Vern L. Bengtson. "The Demography of Mid- and Late-Life Transitions." *Annals of the American Academy* 464 (November 1982): 11–21.

———. "The Family in Later Years." In *Handbook of Marriage and the Family,* edited by M. Sussman and S. Steinmetz. New York: Plenum Press, 1987.

Treas, Judith, and Deirdre Giesen, "Sexual Infidelity Among Married and Cohabiting Americans." *Journal of Marriage and the Family* 62, 1 (2000): 48-60.

"Trends in Pregnancies and Pregnancy Rates by Outcome: Estimates for the United States," 1976–96. Vol. 21, No. 56.

Centers for Disease Control and Prevention, National Center for Health Statistics

Tribe, Laurence, *Abortion: Clash of Absolutes.* New York: Norton, 1990.

Troiden, Richard. *Gay and Lesbian Identity: A Sociological Analysis.* New York: General Hall, 1988.

Troiden, Richard, and Erich Goode. "Variables Related to the Acquisition of a Gay Identity." *Journal of Homosexuality* 5 (June 1980): 383–392.

Troll, Lillian. "The Contingencies of Grandparenting." In *Grandparenthood,* edited by V. Bengston and J. Robertson. Beverly Hills, CA: Sage 1985.

Trost, Jan. "Abandon Adjustment." *Journal of Marriage and the Family* 47,4 (November 1985): 1072–1073.

True, Reiko Homma. "Psychotherapeutic Issues with Asian American Women." *Sex Roles* 22, 7 (April 1990): 477–485.

Trzcinski, Eileen, and Matia Finn-Stevenson. "A Response to Arguments against Mandated Parental Leave: Findings from the Connecticut Survey of Parental Leave Policies." *Journal of Marriage and the Family* 53, 2 (May 1991): 445–460.

Tucker, M. B., and R. J. Taylor. "Demographic Correlates of Relationship Status among Black Americans." *Journal of Marriage and the Family* 51 (August 1989): 655–665.

Tucker, Raymond K., M. G. Marvin, and B. Vivian. "What Constitutes a Romantic Act." *Psychological Reports* 89, 2 (October 1991): 651–654.

Tucker, Sandra. "Adolescent Patterns of Communication about Sexually Related Topics." *Adolescence* 24, 94 (June 1989): 269–278.

Tuleja, Tad. *Curious Customs.* New York: Harmony Books, 1987.

Turner, P. H., et al. "Parenting in Gay and Lesbian Families." Paper presented at the first meeting of the Future of Parenting Symposium, Chicago, March 1985.

Turner, R. "Rising Prevalence of Cohabitation in the United States May Have Partially Offset Decline in Marriage Rates." *Family Planning Perspectives* 22, 2 (March 1990): 90–91.

Turner, R. Jay, and William R. Avison. "Assessing Risk Factors for Problem Parenting: The Significance of Social Support." *Journal of Marriage and the Family* 47, 4 (November 1985): 881–892.

Turner, Robert L., and M. E. Fakouri. "Androgyny and Differences in Fantasy Patterns." *Psychological Reports* 73, 3 (1993): 1164–1166.

Udry, J. Richard. *The Social Context of Marriage.* Philadelphia: Lippincott, 1974.

———. "Marriage Alternatives and Marital Disruption." *Journal of Marriage and the Family* 43 (November 1981): 889–897.

Umberson, Debra. "Parenting and Well-Being: The Importance of Context." *Journal of Family Issues* 10, 4 (December 1989): 427–439.

Umberson, Debra, and Walter R. Gove. "Parenthood and Psychological Well-Being: Theory, Measurement, and Stage in the Family Life Course." *Journal of Family Issues* 10, 4 (December 1989): 440–462.

Unger, Donald G., and Marvin B. Sussman. "Introduction: A Community Perspective on Families." *Marriage and Family Review* 15, (1990): 1–2.

Unger, R. K. "Toward a Redefinition of Sex and Gender." *American Psychologist* 34 (1979): 1085–1094.

University of Michigan, "Living Together: Facts, Myths, About 'Living in Sin' Studied." News and Information Services, News Release: February 4, 2000.

"Update: Trends in Fetal Alcohol Syndrome—United States, 1979–1993." Centers for Disease Control and Prevention. *Morbidity and Mortality Weekly Report* 44, 13 (1995): 249–251.

Urwin, Charlene. "AIDS in Children: A Family Concern." *Family Relations* 37, 2 (April 1988): 154–159.

U.S. Bureau of Justice Statistics, *Violence by Intimates: Analysis of Data on Crimes by Current or Former Spouses, Boyfriends, and Girlfriends.* Washington, DC: Department of Justice, 1998.

U.S. Census Bureau, *Current Population Report,* 1998. www.census.gov.

———. "The Hispanic Population in the United States: March 1990." In *Current Population Reports,* Series P-20. Washington, DC: U.S. Government Printing Office, 1991.

———. "Race and Hispanic or Latino." U.S. Census Bureau, Census 2000 Summary File 1, Matrices P3, P4, PCT4, PCT5, PCT8, and PCT11.

———. *Infant Mortality.* International Data Base, 1996.

———. *Statistical Abstract of the United States.* Washington, DC: U.S. Government Printing Office, 1994.

———. *Statistical Abstract of the United States.* Washington, DC: U.S. Government Printing Office, 1995.

———. *Statistical Abstract of the United States.* Washington, DC: U.S. Government Printing Office, 1996.

———. *Statistical Abstract of the United States.* Washington, DC: U.S. Government Printing Office, 1999.

———. *Statistical Abstract of the United States.* Washington, DC: U.S. Government Printing Office, 2000.

———. *Statistical Abstract of the United States.* Washington, DC: U.S. Government Printing Office, 2001.

———. *Statistical Abstract of the United States.* Washington, DC: U.S. Government Printing Office, 2002.

———. *Statistical Abstract of the United States.* Washington, DC: U.S. Government Printing Office, 2003.

U.S. Commission on Civil Rights. *Child Care and Equal Opportunity for Women.*Washington, DC: U.S. Government Printing Office, 1981.

U.S. Department of Justice. "Sexual Offenses and Offenders." Washington, DC: Bureau of Justice Statistics, 1997.

U.S. Department of Labor. "Labor Force Participation of Fathers and Mothers Varies with Children's Ages." *Monthly Labor Review,* 6/3/99. http://stats.bls.gov.

———. "Women Usually Victimized by Offenders They Know." Washington, DC: Bureau of Justice Statistics, 1995.

Vaillant, Caroline O., and George E. Vaillant. "Is the U-Curve of Marital Satisfaction an Illusion? A 40–Year Study of Marriage." *Journal of Marriage and the Family* 55, 1 (February 1993): 230–240.

Valentine, Deborah. "The Experience of Pregnancy: A Developmental Process." *Family Relations* 31, 2 (April 1982): 243–248.

Van Buskirk. "Soap Opera Sex: Tuning In, Tuning Out." *US* (August 1992): 64–67.

Vande Berg, Leah R., and Diane Streckfuss. "Prime-Time Television's Portrayal of Women and the World of Work: A Demographic Profile." *Journal of Broadcasting and Electronic Media* (March 1992): 195–207.

Vandell, Deborah. "After School Care: Choices and Outcome for Third Graders." Paper presented to the Association for the Advancement of Science, May 27, 1985.

Van Hasselt, Vincent B., et al. *Handbook of Family Violence.* New York: Plenum Press, 1988.

Van Horn, K. Roger, Angela Arnone, Kelly Nesbitt, Laura Desilets, Tanya Sears, Michelle Giffin, and Rebecca Brudi "Physical Distance and Interpersonal Characteristics in College Students' Romantic Relationships." *Personal Relationships* 4,1 (1997): 25–34.

Vaselle-Augenstein, Renata, and Annette Ehrlich. "Male Batterers: Evidence for Psychopathology." In *Intimate Violence: Interdisciplinary Perspectives,* edited by E. C. Viano. Washington, DC: Hemisphere, 1992.

Vasquez-Nuthall, E., et al. "Sex Roles and Perceptions of Femininity and Masculinity of Hispanic Women: A Review of the Literature." *Psychology of Women Quarterly* 11 (1987): 409–426.

Vasta, Ross. "Physical Child Abuse: A Dual-Component Analysis." *Developmental Review* 2, 2 (June 1992): 125–149.

Vaughan, Diane. *Uncoupling: Turning Points in Intimate Relationships.* New York: University Press, 1986.

Veevers, Jean. *Childless by Choice.* Toronto: Butterworth, 1980.

Vega, W. A. "The Study of Latino Families." In *Understanding Latino Families,* edited by R. Zambrana. Thousand Oaks, CA: Sage, 1995.

Vega, William. "Hispanic Families in the 1980s: A Decade of Research." *Journal of Marriage and the Family* 52 (4) November 1990: 1015–1024.

———. "Hispanic Families." In *Contemporary Families: Looking Forward, Looking Back,* edited by A. Booth. Minneapolis: National Council on Family Relations, 1991.

Velsor, Ellen, and Angela O'Rand. "Family Life Cycle, Work Career Patterns, and Women's Wages at Midlife." *Journal of Marriage and the Family* 46, 2 (May 1984): 365–373.

Vemer, Elizabeth, et al. "Marital Satisfaction in Remarriage: A Meta-analysis." *Journal of Marriage and the Family* 53, 3 (August 1989): 713–726.

Ventura, Jacqueline N. "The Stresses of Parenthood Reexamined." *Family Relations* 36, 1 (January 1987): 26–29.

Ventura, S., T. Hamilton. "Teenage Births in the United States: Trends, 1991–2000, an Update." *National Vital Statistics Reports;* 50, 9 National Center for Health Statistics (2002).

Vera, Hernan, et al. "Age Heterogamy in Marriage." *Journal of Marriage and the Family* 47, 3 (August 1985): 553–566.

Verbugge, Lois. "Marital Status and Health." *Journal of Marriage and the Family* 41, 2 (May 1979): 267–285.

———. "From Sneezes to Adieu: Stages of Health for American Men and Women." *Social Science and Medicine* 22 (1986): 1195–1212.

Viano, Emilio C., ed. *Intimate Violence: Interdisciplinary Perspectives.*Washington, DC: Hemisphere, 1992.

Vincent, Richard, et al. "Sexism on MTV: The Portrayal of Women in Rock Video." *Journalism Quarterly* 64, 4 (December 1987): 750–755, 941.

"Violence Kills More U.S. Kids." *San Francisco Chronicle* (February 7, 1997): A-1.

Visher, Emily B., and John S. Visher. *Stepfamilies: A Guide to Working with Stepparents and Stepchildren.* New York: Brunner/Mazel, 1979.

———.*How to Win as a Stepfamily.* New York: Brunner/Mazel, 1991.

Vobejda, Barbara. "Census Bureau Says Rapid Changes in Family Size, Style Are Slowing." *Washington Post* (June 24, 1993): A21.

Voeller, Bruce. "Society and the Gay Movement." In *Homosexual Behavior,* edited by J. Marmor. New York: Basic Books, 1980.

Vogel, D. A., M. A. Lake, and S. Evans. "Children's and Adults' Sex-Stereotyped Perceptions of Infants." *Sex Roles* 24 (1991): 605–616.

Voydanoff, Patricia.———. *Work and Family Life.* Newbury Park, CA: Sage, 1987.

———. "Work Role Characteristics, Family Structure Demands, and Work/Family Conflict." *Journal of Marriage and the Family* 50, 3 (August 1988): 749–761.

———."Economic Distress and Family Relations: A Review of the Eighties." In *Contemporary Families: Looking Forward, Looking Back,* edited by A. Booth. Minneapolis: National Council on Family Relations, 1991.

Voydanoff, Patricia, and Brenda Donnelly. "Work and Family Roles and Psychological Distress." *Journal of Marriage and the Family* 51, 4 (November 1989): 933–941.

———. *Adolescent Sexuality and Pregnancy.* Newbury Park, CA: Sage, 1990.

Voydanoff, Patricia, and Linda Majka, eds. *Families and Economic Distress: Coping Strategies and Social Policy.* Beverly Hills, CA: Sage, 1988.

Vredevelt, P. *Empty Arms: Emotional Support for Those Who Have Suffered Miscarriage or Stillbirth.* Sisters, OR: Questar, 1994.

Wade, R. C. *For Men about Abortion.* Boulder, CO: R. C. Wade, 1978.

Wagner, Roland M. "Psychosocial Adjustments during the First Year of Single Parenthood: A Comparison of Mexican-American and Anglo Women." *Journal of Divorce and Remarriage* 19, 1–2 (1993): 121–142.

Waite, Linda. "Does Marriage Matter?" *Demography* 32, 4 (November 1995): 483–507.

Waldman, Steven, and Lucy Shackelford. "Welfare Booby Traps." *Newsweek* (December 12, 1994): 34–35.

Waldman, Steven, and Lincoln Caplan. "The Politics of Adoption." *Newsweek* (March 21, 1994): 64–65.

Walker, Alexis J. "Reconceptualizing Family Stress." *Journal of Marriage and the Family* 47, 4 (November 1985): 827–837.

———. "Teaching About Race, Gender, and Class Diversity in United States Families." *Family Relations* 42, 3 (1993): 342–350.

———. "Gender and Family Relationships." *The Handbook of Marriage and the Family,* 2nd ed., edited by M. Sussman, S. Steinmetz and G. Peterson. New York: Plenum, 1999.

Walker, Alexis J., et al. "Feminist Programs for Families." *Family Relations* 37, 1 (January 1988): 17–22.

———. "Perceptions of Relationship Change and Caregiver Satisfaction." *Family Relations* 39, 2 (April 1990): 147–152.

Walker, E. A., et al. "Medical and Psychiatric Symptoms in Women with Childhood Sexual Abuse." *Psychosomatic Medicine* 54, 6 (November 1992): 658–664.

Walker, Janet. "Co-operative Parenting Post-Divorce: Possibility or Pipedream?" *Journal of Family Therapy* 15, 3 (1993): 273–292.

Walker, Karen. "Men, Women, and Friendship: What They Say, What They Do," *Gender and Society* 8, 2 (June 1994).

Walker, Lenore. *The Battered Woman Syndrome.* New York: Harper Colophon, 1979.

———. *The Battered Woman.* New York: Springer Publishing, 1984.

———. "Psychological Causes of Family Violence." In *Violence in the Home: Interdisciplinary Perspectives,* edited by M. Lystad. New York: Brunner/Mazel, 1986.

———. "The Battered Woman Syndrome Is a Psychological Consequence of Abuse." In *Current Controversies in Family Violence,* edited by R. Gelles and D. Loseke. Newbury Park, CA: Sage Publications, 1993.

Wall, Jack C. "Maintaining the Connection: Parenting as a Noncustodial Father." *Child and Adolescent Social Work Journal* 9, 5 (1992): 441–456.

Waller, Willard, and Reuben Hill. *The Family: A Dynamic Interpretation.* New York: Dryden Press, 1951.

Wallerstein, Judith. "Children of Divorce: Report of a Ten-Year Follow-up of Early Latency-Age Children." *American Journal of Orthopsychiatry* 57, 2 (April 1987): 199–211.

Wallerstein, Judith, and Sandra Blakeslee. *Second Chances: Men, Women, and Children a Decade after Divorce.* New York: Ticknor & Fields, 1989.

Wallerstein, Judith, and Joan Kelly. "Effects of Divorce on the Visiting Father-Child Relationship." *American Journal of Psychiatry* 137, 12 (December 1980a): 1534–1539.

———. *Surviving the Breakup: How Children and Parents Cope with Divorce.* New York: Basic Books, 1980b.

Wallerstein, Judith, Julia Lewis, and Sandra Blakeslee. *The Unexpected Legacy of Divorce: A Twenty-Five Year Landmark Study.* New York: Hyperion Press, 2000.

Walling, Mary, et al. "Hormonal Replacement Therapy for Postmenopausal Women: A Review of Sexual Outcomes and Related Gynecologic Effects." *Archives of Sexual Behavior* 19, 2 (1990): 119–127.

Wallis, Claudia. "Children Having Children." *Time* (December 9, 1985): 78–79.

Walsh, Froma, ed. *Normal Family Processes.* New York: Guilford Press, 1982.

Walster, Elaine, and G. William Walster. *A New Look at Love.* Reading, MA: Addison-Wesley, 1978.

Walzer, Susan. *Thinking About the Baby: Gender and Transitions into Parenthood.* Philadelphia: Temple University, 1998.

Wardle, Francis. "Helping Children Respect Differences." *PTA Today* (December 1989): 5–6.

Warren, Jennifer A. and Phyllis J. Johnson. "The Impact of Workplace Support on Work-Family Role Strain." *Family Relations* 44, 2 (April 1995): 163–169.

Watkins, Susan Cotts, and Jane Menken. "Demographic Foundations of Family Change." *American Sociological Review* 52 (1987): 346–358.

Watkins, William G., and Arnon Bentovim. "The Sexual Abuse of Male Children and Adolescents: A Review of Current Research." *Journal of Child Psychology and Psychiatry and Allied Disciplines* 33, 1 (January 1992): 197–248.

Wedemeyer, Nancy, and Harold Grotevant. "Mapping the Family System: A Technique for Teaching Family Systems Theory Concepts." *Family Relations* 31, 2 (April 1982): 185–193.

Weeks, Jeffrey. *Sexuality and Its Discontents.* London: Routledge, 1985.

Weeks, M. O'Neal, and Bruce Gage. "A Comparison of the Marriage-Role Expectations of College Women Enrolled in a Functional Marriage Course in 1961, 1972, and 1978." *Sex Roles* 11, 5/6 (1984): 377–388.

Weg, Ruth. "The Physiological Perspective." In *Sexuality in the Later Years: Roles and Behavior,* edited by R. Weg. New York: Academic Press, 1983.

———. *Sexuality in the Later Years: Roles and Behavior.* New York: Academic Press, 1983a.

Wegscheider, Sharon. *Another Chance: Hope and Health for the Alcoholic Family,* 2nd ed. Palo Alto, CA: Science and Behavior Books, 1989.

Weinberg, Martin, and Colin Wilson. "Black Sexuality: A Test of Two Theories." *Journal of Sex Research* 25, 2 (May 1988): 197–218.

Weinberg, Martin S., C. J. Williams, and Douglas W. Pryor. *Dual Attraction: Understanding Bisexuality.* New York: Oxford University Press, 1994.

Weiner, L., B. A. More, and P. Garrido. "FAS/FAE: Focusing Prevention on Women at Risk." *International Journal of the Addictions* 24, 5 (May 1989): 385–395.

Weingarten, Helen. "Remarriage and Well-Being." *Journal of Family Issues* 1, 4 (December 1980): 533–559.

———. "Marital Status and Well-Being: A National Study Comparing First-Married, Currently Divorced and Remarried Adults." *Journal of Marriage and the Family* 47, 3 (August 1985): 653–662.

Weir, John. "Gay-Bashing, Villainy and the Oscars." *New York Times* (March 29, 1992): H17.
Weishaus, Sylvia, and Dorothy Field. "A Half Century of Marriage: Continuity or Change?" *Journal of Marriage and the Family* 50, 3 (August 1988): 763–774.
Weiss, David. "Open Marriage and Multilateral Relationships: The Emergence of Nonexclusive Models of the Marital Relationship." In *Contemporary Families and Alternative Lifestyles,* edited by E. Macklin and R. Rubin. Beverly Hills, CA: Sage, 1983.
Weiss, David, and Joan Jurich. "Size of Community as a Predictor of Attitudes toward Extramarital Sexual Relations." *Journal of Marriage and the Family* 47, 1 (February 1985): 173–178.
Weiss, Robert. "The Fund of Sociability." *Transactions* (July 1969): 36–43.
———. *Marital Separation.* New York: Basic Books, 1975.
———. "Men and the Family." *Family Processes* 24 (1985): 49–58.
Weitzman, Lenore. ———. *The Marriage Contract : Spouses, Lovers, and the Law.* New York: Macmillan, 1981.
———. *The Divorce Revolution: The Unexpected Social and Economic Consequences for Women and Children in America.* New York: Free Press, 1985.
Weitzman, Lenore, and Ruth Dixon. "The Transformation of Legal Marriage through No-Fault Divorce." In *Family in Transition,* edited by A. Skolnick and J. Skolnick. Boston: Little, Brown, 1980.
Weitzman, Susan. *Not to People Like Us: Hidden Abuse in Upscale Marriages.* New York: Basic Books, 2000.
Weizman, R., and J. Hart. "Sexual Behavior in Healthy Married Elderly Men." *Archives of Sexual Behavior* 16, 1 (February 1987): 39–44.
Werner, Carol M., Barbara B. Brown, Irwin Altman, and Brenda Staples. "Close Relationships in Their Physical and Social Contexts: A Transactional Perspective." *Journal of Social and Personal Relationships* 9, 3 (1992): 411–431.
Wertz, Richard, and Dorothy Wertz *Lying-In: A History of Childbirth in America.* New York: Harper and Row, 1978.
West, Candace. and Don Zimmerman. "Doing Gender." *Gender and Society* 1 (1987): 125–151.
Whelan, Elizabeth. *Boy or Girl?* New York: Pocket Books, 1986.
"When Parenthood Extends to Raising Grandchildren." *San Francsico Chronicle* (June 3, 1997): A-1.
Whitbourne, Susan, and Joyce Ebmeyer. *Identity and Intimacy in Marriage: A Study of Couples.* New York: Springer-Verlag, 1990.
Whitchurch, Gail, and Fran Dickson. "Family Communication." In *Handbook of Marriage and the Family,* 2nd ed., edited by M. Sussman, S. Steinmetz, and G. Peterson. New York: Plenum, 1999.
White, Charles. "Sexual Interest, Attitudes, Knowledge, and Sexual History in Relation to Sexual Behavior of the Institutionalized Aged." *Archives of Sexual Behavior* 11 (February 1982): 11–21.
White, Gregory. "Inducing Jealousy: A Power Perspective." *Personality and Social Psychology Bulletin* 6, 2 (June 1980a): 222–227.
———. "Physical Attractiveness and Courtship Progress." *Journal of Personality and Social Psychology* 39, 4 (October 1980b): 660–668.
———. "Jealousy and Partner's Perceived Motives for Attraction to a Rival." *Social Psychology Quarterly* 44, 1 (March 1981): 24–30.
White, Jacquelyn W. "Feminist Contributions to Social Psychology." *Contemporary Social Psychology* 17, 3 (September 1993): 74–78.
White, Jacquelyn W., and R. Farmer. "Research Methods: How They Shape Views of Sexual Violence." *Journal of Social Issues* 48 (Spring 1992): 45–59.
White, James. "Premarital Cohabitation and Marital Stability in Canada." *Journal of Marriage and the Family* 49 (August 1987): 641–647.
White, James, and D. Klein. *Family Theories,* 2nd ed. Thousand Oaks, CA:Sage, 2002.
White, Joseph, and Thomas Parham. *The Psychology of Blacks: An African-American Perspective,* 2nd ed. Englewood Cliffs, NJ: Prentice Hall, 1990.
White, Lynn. "Determinants of Divorce." In *Contemporary Families: Looking Forward, Looking Back,* edited by A. Booth. Minneapolis: National Council on Family Relations, 1991.
White, Lynn, and Alan Booth. "The Transition to Parenthood and Marital Quality." *Journal of Family Issues* 6 (1985): 435–449.
———. "Divorce over the Life Course: The Role of Marital Happiness." *Journal of Family Issues* 12, 1 (March 1991): 5–22.
White, Lynn, and Bruce Keith. "The Effect of Shift Work on the Quality and Stability of Marital Relations." *Journal of Marriage and the Family* 52, 2 (May 1990): 453–462.
Whitehead, Barbara Dafoe. "How to Rebuild a 'Family Friendly' Society." *Des Moines Sunday Register* (October 7, 1990): C3.
———. "Dan Quayle Was Right," *Atlantic Monthly.* April 1993: 47–84.
———. *The Divorce Culture.* New York: Knopf, 1997.
Whiteside, Mary F. "Family Rituals as a Key to Kinship Connections in Remarried Families." *Family Relations* 38, 1 (January 1989): 34–39.
Whitman, David. "The Coming of the 'Couch People.'" *U.S. News and World Report* (August 3, 1987): 19–21.
"Who's Minding the Kids? Primary Child Care Arrangements of Children Under 5." Accessed on Grolier Web site at http://grolier.go.com.
Wiehe, Vernon. *Sibling Abuse: The Hidden Physical, Emotional, and Sexual Trauma.* Lexington, MA: Lexington Books, 1990.
Wiehe, Vernon R. "Religious Influence on Parental Attitudes Toward the Use of Corporal Punishment." *Journal of Family Violence* 5, 2 (June 1990): 173 ff.
Wilcox, Allen, et al. "Incidents of Early Loss of Pregnancy." *New England Journal of Medicine* 319, 4 (July 28, 1988): 189–194.

Wilkerson, Isabel. "Infant Mortality: Frightful Odds in Inner City." *New York Times* (June 26, 1987): 1.

Wilkie, Colleen F., and Elinor W. Ames. "The Relationship of Infant Crying to Parental Stress in the Transition to Parenthood." *Journal of Marriage and the Family* 48, 3 (August 1986): 545–550.

Wilkinson, Doris. "Family Ethnicity in America." In *Family Ethnicity: Strength in Diversity,* edited by H. Pipes McAdoo. Newbury Park, CA: Sage, 1993.

Wilkinson, Doris Y. "American Families of African Descent." In *Families in Cultural Context,* edited by M. K. DeGenova. Mountain View, CA: Mayfield, 1997.

Wilkinson, Doris, Maxine Baca Zinn, and E. N. L. Chow. "Race, Class, and Gender: Introduction." *Gender and Society* 6, 3 (September 1992): 341–345.

Wilkinson, Doris, et al., eds. "Transforming Social Knowledge: The Interlocking of Race, Class, and Gender." *Gender and Society* [Special issue] (September 1992).

Willemsen, Tineke M. "On the Bipolarity of Androgyny: A Critical Comment on Kottke (1988)." *Psychological Reports* 72, 1 (February 1993): 327–332.

Willetts, Marion. "An Exploratory Investigation of Heterosexual Licensed Domestic Partners." *Journal of Marriage and the Family* 65, 4 (November 2003): 939–952.

Williams, John, and Arthur Jacoby. "The Effects of Premarital Heterosexual and Homosexual Experience on Dating and Marriage Desirability." *Journal of Marriage and the Family* 51 (May 1989): 489–497.

Williams, Juanita. "Middle Age and Aging." In *Women: Behavior in a Biosocial Context,* edited by J. Williams. New York: Norton, 1977.

Wilson, Glenn. *The Secrets of Sexual Fantasy.* London: J. M. Dent, 1978.

Wilson, Melvin N., et al. "Flexibility and Sharing of Childcare Duties in Black Families." *Sex Roles* 22, 7–8 (April 1990): 409–425.

Wilson, Pamela. "Black Culture and Sexuality." *Journal of Social Work and Human Sexuality* 4, 3 (March 1986): 29–46.

Wilson, S. M., and N. P. Medora. "Gender Comparisons of College Students' Attitudes toward Sexual Behavior." *Adolescence* 25, 99 (September 1990): 615–627.

Wilson, Yumi. "This Time Domestic Violence Stayed in the Spotlight." *San Francisco Chronicle* (February 6, 1997): A-10.

Winch, Robert. *Mate Selection: A Study of Complementary Needs.* New York: Harper and Row, 1958.

Wineberg, H. "Marital Reconciliation in the United States: Which Couples Are Successful?" *Journal of Marriage and the Family* 56, 1 (February 1994): 80–88.

Winn, Rhoda, and Niles Newton. "Sexuality and Aging: A Study of 106 Cultures." *Archives of Sexual Behavior* 11, 4 (August 1982): 283–298.

Winton, Chester. *Frameworks for Studying Families.* Guilford, CT: Dushkin, 1995.

———. *Children as Caregivers: Parental and Parentified Children.* Boston: Pearson Allyn and Bacon, 2002.

Wise, P. H., and A. Meyer. "Poverty and Child Health." *Pediatric Clinics of North America* 35, 6 (December 1988): 1169–1186.

Wisensale, Steven. "The Family in the Think Tank." *Family Relations* 40, 2 (April 1990a): 199–207.

———. "Approaches to Family Policy in State Government: A Report on Five States." *Family Relations* 39, 2 (April 1990b): 136–140.

Wisensale, Steven, and Michael Allison. "Family Leave Legislation State and Federal Initiatives." *Family Relations* 38, 2 (April 1989): 182–189.

Wishy, Bernard. *The Child and the Republic.* Philadelphia: Lippincott, 1968.

Woititz, Janet. *Adult Children of Alcoholics.* Deerfield, FL: Health Communications, 1983.

Wolf, Michelle, and Alfred Kielwasser. "Introduction: The Body Electric: Human Sexuality and the Mass Media." *Journal of Homosexuality* 21, 1/2 (1991): 7–18.

Wolf, Rosalie S. "Abuse and Neglect of the Elderly." In *Vision 2010: Families and Violence, Abuse and Neglect,* edited by R. J. Gelles. Minneapolis: National Council on Family Relations, 1995.

Wolf, Rosalie S., and Karl A. Pillemer. "Intervention, Outcome, and Elder Abuse." In *Coping with Family Violence,* edited by G. T. Hotaling et al. Newbury Park, CA: Sage, 1988.

Wolkind, Stephen and Eva Zajicek, eds. *Pregnancy: A Psychological and Social Study.* New York: Grune and Stratton, 1981.

Wong, Morrison G. "The Chinese American Family." In *Ethnic Families in America: Patterns and Variations,* 3rd ed., edited by C. H. Mindel et al. New York: Elsevier, 1988.

Wood, J. T. *Gendered Lives: Communication, Gender, and Culture.* Belmont, CA: Wadsworth, 1994.

Woods, Stephen C., and James Guy Mansfield. "Ethanol and Disinhibition: Physiological and Behavioral Links," edited by Robin Room. *Proceedings of Alcoholism and Drug Abuse Conference, Berkeley/Oakland (February 11, 1981).* Washington, DC: U.S. Department of Health and Human Services, 1981.

Woodward, K. L., V. Quade, and B. Kantrowitz. "Q: When Is a Marriage Not a Marriage?" *Newsweek* (March 13, 1995): 58–59.

Woodward, Kenneth. "New Rules for Making Love and Babies." *Newsweek* (March 23, 1987): 42–43.

Wright, J., C. Duchesne, S. Sabourin, F. Bissonette, J. Benoit, and Y. Girard. "Psychosocial Distress and Infertility: Men and Women Respond Differently." *Fertility and Sterility* 55, 1 (January 1991): 100–108.

Wright, Julia. "Getting Engaged: A Case Study and a Model of the Engagement Period as a Process of Conflict-Resolution." *Counselling Psychology Quarterly* 3, 4 (1990): 399–408.

Wu, Zheng, and T. R. Balakrishnan. "Attitudes towards Cohabitation and Marriage in Canada." *Journal of Comparative Family Studies* 23, 1 (1992): 1–12.

Wyatt, Gail E. "The Sociocultural Context of African American and White American Women's Rape." *Journal of Social Issues,* 48, 1 (March 1992): 77–91.

Wyatt, Gail Elizabeth. "The Aftermath of Child Sexual Abuse of African American and White Women: The Victim's Experience." *Journal of Family Violence* 5, 1 (March 1990): 61–81.

Wyatt, Gail, and Sandra Lyons-Rowe. "African American Women's Sexual Satisfaction as a Dimension of Their Sex Roles," *Sex Roles* 22, 7–8 (April 1990): 509–524.

Wyatt, Gail, Stephanie Peters, and Donald Guthrie. "Kinsey Revisited I: Comparisons of the Sexual Socialization and Sexual Socialization and Sexual Behavior of Black Women over 33 Years." *Archives of Sexual Behavior* 17, 3 (June 1988a): 201–239.

———. "Kinsey Revisited II: Comparison of the Sexual Socialization and Sexual Socialization and Sexual Behavior of Black Women over 33 Years." *Archives of Sexual Behavior* 17, 4 (August 1988b): 289–332.

Wyatt, Gail Elizabeth, et al. "Differential Effects of Women's Child Sexual Abuse and Subsequent Sexual Revictimization." *Journal of Consulting and Clinical Psychology* 60, 2 (April 1992): 167.

Wyche, Karen F. "Psychology and African-American Women: Findings from Applied Research." *Applied and Preventive Psychology* 2, 3 (June 1993): 115–121.

Yawn, Barbara P., and Roy A. Yawn. "Adolescent Pregnancy: A Preventable Consequence?" *The Prevention Researcher.* Eugene, OR: Integrated Research Services, 1997.

Ybarra, Lea. "When Wives Work: The Impact on the Chicano Family." *Journal of Marriage and the Family* 44, 1 (February 1982): 169–178.

Yellowbird, Michael, and C. Matthew Snipp. "American Indian Families." In *Minority Families in the United States: A Multicultural Perspective,* edited by R. L. Taylor. Englewood Cliffs, NJ: PrenticeHall, 1994.

Yerby, J., N. Buerkel-Rothfuss, and A. P. Bochner. *Understanding Family Communication.* Scottsdale, AZ: Gorsuch Scarisbrick, 1990.

Ÿllo, Kersti. "Through a Feminist Lens: Gender, Power, and Violence." In *Current Controversies in Family Violence,* edited by R. Gelles and D. Loseke. Newbury Park, CA: Sage , 1993.

———. "Marital Rape." In *Vision 2010: Families and Violence, Abuse and Neglect,* edited by R. J. Gelles. Minneapolis: National Council on Family Relations, 1995.

Yogev, Sara, and Jane Brett. "Perceptions of the Division of Housework and Child Care and Marital Satisfaction." *Journal of Marriage and the Family* 47, 3 (August 1985): 609–618.

Young, L. "Sexual Abuse and the Problem of Embodiment." *Child Abuse and Neglect* 16, 1 (1992): 89–100.

Yulsman, Tom. "A Little Help for Creation." *Good Health Magazine (New York Times)* (October 7, 1990): 22ff.

Yura, Michael. "Family Subsystem Functions and Disabled Children: Some Conceptual Issues." *Marriage and Family Review* 11, 1 (1987): 135–151.

Zaslow, Martha, et al. "Depressed Mood in New Fathers." Unpublished paper, Society for Research in Child Development, April 1981.

Zavella, Patricia. *Women's Work and Chicano Families.* Ithaca, NY: Cornell University Press, 1987.

Zayas, Luis H., and Josephine Palleja. "Puerto Rican Familism: Considerations for Family Therapy." *Family Relations* 37, 3 (1988): 260–264.

———. "Puerto Rican Familism: Considerations for Family Therapy." In *Minority Families in the United States: A Multicultural Perspective,* edited by R. L. Taylor. Englewood Cliffs, NJ: Prentice Hall, 1994.

Zelnik, Melvin. *Sex and Pregnancy in Adolescence.* Beverly Hills, CA: Sage, 1981.

Zilbergeld, Bernie. *The New Male Sexuality.* New York: Bantam Books, 1992.

Zill, Nicholas, Donna R. Morrison, and Mary J. Coiro. "Long-Term Effects of Parental Divorce on Parent-Child Relationships, Adjustment, and Achievement in Young Adulthood." *Journal of Family Psychology* 7, 1 (1993): 91–103.

Zimmerman, Shirley L. "State Level Public Policy Choices as Predictors of State Teen Birth Rates." *Family Relations* 37 (July 1988a): 315–321.

———. *Understanding Family Policy: Theoretical Approaches.* Beverly Hills, CA: Sage, 1988b.

———. "The Welfare State and Family Breakup: The Mythical Connection." *Family Relations* 40, 2 (April 1991): 139–147.

Zimmerman, Shirley L., and Phyllis Owens. "Comparing the Family Policies of Three States: A Content Analysis." *Family Relations* 38, 2 (April 1989): 190–195.

Zinn, Maxine Baca. "Family, Feminism, and Race." *Gender and Society* 4 (1990): 68–82.

———. "Adaptation and Continuity in Mexican-Origin Families." In *Minority Families in the United States: A Multicultural Perspective,* edited by R. L. Taylor. Englewood Cliffs, NJ: Prentice Hall, 1994.

Zvonkovic, Anisa N., Kathleen M. Greaves, Cynthia J. Schmiege, and Leslie D. Hall. "The Marital Construction of Gender through Work and Family Decisions: A Qualitative Analysis." *Journal of Marriage and the Family* 58, 1 (February 1996): 91–100.

Zvonkovic, Anisa, et al. "Making the Most of Job Loss: Individual and Marital Features of Job Loss." *Family Relations* 37, 1 (January 1988): 56–61.

PHOTO CREDITS

Chapter 1

2 The Wedding: Lovers Quilt #1, 1986 © Faith Ringgold, 1986
6 My Big Fat Greek Wedding (2001), Directed by Joel Zwick Shown: Bridal Party with Nia Vardalos, Lainie Kazan CREDIT: Photofest
11 © Phill Snell/Getty Images
13 © Corbis. All Rights Reserved.
14 © Christine Mendes/Buena Vista Photography
16 © David Young Wolff/PhotoEdit
18 © Michael Newman/PhotoEdit

Chapter 2

26 Norma Rannie
29L Copyright © Everett Collection/Everett Collection
29R 1996- ©2003 CBS Worldwide Inc. All Rights Reserved. Courtesy: Everett Collection
33 © Laurie DeVault Photography
38 © Laurie DeVault Photography
40 © Laura Dwight/PhotoEdit
44 © Michael Newman/PhotoEdit
47 © David Young-Wolff/PhotoEdit
56L © Jose Luis Pelaez, Inc. CORBIS
56R © Jonathan Nourok/PhotoEdit

Chapter 3

62 © Bob Daemmrich/Stock Boston, LLC.
68 © Hulton Archive/Getty Images
70 National Park Service: Statue of Liberty National Monument
72L © CORBIS
72R © Lambert/Hulton Archive/Getty Images
82L. © Bill Bachmann/PhotoEdit
82R © A. Ramey/Stock Boston, LLC.
89 © Tony Freeman/PhotoEdit

Chapter 4

100 © Mark Burnett/Stock Boston, LLC.
103 © Ellis Hedwig/Stock Boston, LLC.
108 © David Young-Wolff/PhotoEdit
111L © Bob Daemmrich/Stock Boston, LLC.
111R © Bob Daemmrich/Stock Boston, LLC.
118 © Myrleen Ferguson Cate/PhotoEdit
124L © Paul Conklin/PhotoEdit
124R © M. Greenlar/The Image Works

Chapter 5

130 © Comstock
134 © Esbin Anderson/The Image Works
153 © Comstock
158 © Jean Higgings/Envision
160 © David Young-Wolff/PhotoEdit

Chapter 6

164 © Christopher Thomas Getty Images/Stone
170 © Mary Kate Denny/PhotoEdit
173 © Scott Barrow
175 © Tony Freeman/PhotoEdit
179 © Laurie DeVault Photography
188 © Laurie DeVault Photography

Chapter 7

198 © Scott Barrow
202 © Claudia Kunin/Getty Images/Stone
212 © Michael Newman/PhotoEdit
216 © Skjold/The Image Works
226 © David Young-Wolff/PhotoEdit

Chapter 8

232 © Ron Chapple Photography
242 © Miriam Grosman
244L © BANG! Inc./Getty Images
244R © Evan Agostini/Getty Images
248 Keri Pickett
252 © Laurie DeVault Photography
254 © Jerrican/Photo Researchers, Inc.

Chapter 9

278 © Rafael Macia/Photo *Researchers, Inc.*
281 © David Hanover/Getty Images/Stone
284 © Laurie DeVault Photography
290 © Esbin Anderson/The Image Works
292 © Mark Richards/PhotoEdit
312 © Ken Fisher/ Getty Images/ Stone

Chapter 10

316 © Laurie DeVault Photography
320 © Laurie DeVault Photography
322 © Scott Barrow
328 © Dennis MacDonald/PhotoEdit
329 How the Husband Assists in the Birth of a Child by Guadalupe, widow of Ramón Medina Silva, after a yarn painting by Ramón. 23 1/2 × 23 1/2" (59.7 × 59.7 cm.). Yarn on plywood, beeswax. #74.21.12. Fine Arts Museums of San Francisco. Gift of Peter F. Young.
334 © Laurie DeVault Photography
337 © Myrleen Ferguson Cate/PhotoEdit

Chapter 11

342 © Jose Luis Pelaez, Inc./CORBIS
345 © Laurie DeVault Photography
349 © Christine Mendes/Buena Vista Photography
354 © Paul Conklin/PhotoEdit
369 © Paul Conklin/PhotoEdit
371 © Spencer Grant/PhotoEdit
374 © Michael Newman/ PhotoEdit

Chapter 12

378 © Myrleen Ferguson Cate/PhotoEdit
381 © David Young-Wolff/PhotoEdit
385 © Laurie DeVault Photography
389 © Cindy Charles/PhotoEdit
400 © Okoniwieski/The Image Works
401 © Tom McCarthy/PhotoEdit
403 © Laurie DeVault Photography
408 © Tony Freeman/PhotoEdit

Chapter 13

420 © Christina Mendes/Buena Vista Photography
423 © Scott Barrow, Inc./Superstock
427 © Gary Connor/PhotoEdit
436 © Mark Peterson/Saba Press Photos
440 © Michael Newman/PhotoEdit
441 Barbara Sayed
451 © Lawrence Migdale/Stock Boston, LLC.

Chapter 14

458 © Scott Barrow
461 © Ghislain & Marie David de Lossy/Getty Images/The Image Bank
472 © Walter Hodges/Getty Images/Stone
481 © Kindra Clineff/Index Stock
492 © Pictor Images/ImageState

Chapter 15

500 © Laurie DeVault Photography
503 © Laurie DeVault Photography
511 © Christine Mendes/Buena Vista Photography

Chapter 16

528 © Comstock
534 © Michael Newman/PhotoEdit
538L © Esbin Anderson/The Image Works
538R © Bill Bachmann/Stock Boston, LLC.
540 © Laurie DeVault Photography
541 © Tom McCarthy/PhotoEdit
549 Courtesy of Keegan Family/ © Christine Mendes/Buena Vista Photography

Appendix B

565 © Lennart Nilsson/Albert Bonniers Forlag AB

NAME INDEX

C

D

E

F

G

H

I

J

K

L

M

N

O

P

Q

R

S

T

U

V

W

Y

Z

SUBJECT INDEX

Boldface numbers indicate definitions of terms.

A

B

C

D

E

F

G

L

M

N

O

P

Q

R

S

T

U

V

W

Y